MATHEMATICS IN INDUSTRY 25

THE EUROPEAN CONSORTIUM
FOR MATHEMATICS IN INDUSTRY

SUBSERIES

More information about this series at http://www.springer.com/series/4650

Matthias Ehrhardt • Michael Günther •
E. Jan W. ter Maten
Editors

Novel Methods in Computational Finance

Springer

ECMI
EUROPEAN CONSORTIUM FOR MATHEMATICS IN INDUSTRY

Editors
Matthias Ehrhardt
Lehrstuhl für Angewandte
Mathematik/Numerische Analysis
Bergische Universität Wuppertal
Wuppertal, Germany

Michael Günther
Lehrstuhl für Angewandte
Mathematik/Numerische Analysis
Bergische Universität Wuppertal
Wuppertal, Germany

E. Jan W. ter Maten
Lehrstuhl für Angewandte
Mathematik/Numerische Analysis
Bergische Universität Wuppertal
Wuppertal, Germany

ISSN 1612-3956 ISSN 2198-3283 (electronic)
Mathematics in Industry
The European Consortium for Mathematics in Industry
ISBN 978-3-319-61281-2 ISBN 978-3-319-61282-9 (eBook)
DOI 10.1007/978-3-319-61282-9

Library of Congress Control Number: 2017947686

Mathematics Subject Classification (2010): 65NXX, 91BXX

Printed on acid-free paper

This Springer imprint is published by Springer Nature
The registered company is Springer International Publishing AG
The registered company address is: Gewerbestrasse 11, 6330 Cham, Switzerland

Foreword

The traditional view of computational finance is based on making rather simplistic assumptions about markets and then, using hedging arguments, deriving risk-neutral pricing equations. From a computational point of view, this leads to the numerical solution of linear partial integro-differential equations (PIDEs) for low-dimensional problems, and Monte Carlo techniques for high-dimensional applications.

However, once realistic market effects are modelled, the standard paradigm is no longer applicable. In fact, perhaps the most general way to think about finance problems is as a form of optimal stochastic control. In the context of option pricing, the objective is to design a hedging strategy (the control) which results in the smallest profit and loss (P&L) risk. Right from the start, we are assuming that the hedge may not be perfect, which is one of the essential complete market assumptions which form the basis of the Black-Scholes analysis. Given the framework of optimal control, it is straightforward to consider liquidity effects and valuation adjustments, such as debt value adjustment (DVA) and credit value adjustment (CVA). This modelling approach naturally gives rise to nonlinear PIDEs, often of the Hamilton-Jacobi-Bellman (HJB) type. In the low-dimensional case, these nonlinear PDEs can be attacked directly. For higher-dimensional problems, methods based on backward stochastic differential equations (BSDEs) look promising.

Of course, these ideas also have relevance to problems in wealth management. With the trend towards replacing defined benefit pension plans with defined contribution (DC) plans, there is a need for automated asset allocation techniques. Since most DC plan participants have strict constraints on allowable strategies (e.g. no leverage), it goes without saying that any asset allocation algorithm must be able to handle realistic constraints. These optimal asset allocation problems also give rise to nonlinear HJB PIDEs.

Many problems in the insurance industry result in massive computational issues. For example, determining the distribution of the P&L for a large book of variable annuities involves simulation of the asset prices in the objective measure while at the same time assuming an imperfect hedge. Determining the hedging parameters is usually done using a Monte Carlo (MC) technique. The standard approach for solving this problem involves a nested MC simulation, which may require weeks of computational time using current algorithms and hardware. In addition, since retail clients rarely exercise their optionality in optimal fashion, machine/statistical learning techniques are the method of choice for analysing customer behaviour.

Consequently, the future of computational finance is bright. We face challenges associated with the development of numerical algorithms for solution of nonlinear PIDEs, possibly using BSDE approaches. However, in order to produce timely results, it will be necessary to exploit the latest hardware advances (e.g. GPUs) and programming environments. This will all have to be coupled with data science methods.

This book focuses on these new techniques, with specific chapters devoted to many of the issues listed above. Students, academics and practitioners will find much of interest here, with a glimpse of the future of computational finance.

Waterloo, ON, Canada
January 2017

Peter A. Forsyth

Preface

This book originated from the Marie Curie European research network STRIKE *Novel Methods in Computational Finance*, which was coordinated by the University of Wuppertal and continued from 2013 to 2016. This initial network was based on the beneficiary partners (11 universities) and the associated partners (3 universities, 6 SMEs and 1 bank) from 11 European countries.

In recent years, the computational complexity of mathematical models employed in financial mathematics has witnessed tremendous growth. Advanced numerical techniques are indispensable to most present-day applications in the financial industry.

The motivation for this European training network STRIKE was the need for a network of highly educated European scientists in the field of financial mathematics and computational science, so as to exchange and discuss current insights and ideas and to lay the groundwork for future long-term collaborations. Here, the challenge lays in the necessity of combining transferable techniques and skills such as mathematical analysis, sophisticated numerical methods and stochastic simulation methods with a deep qualitative and quantitative understanding of mathematical models arising from financial markets.

In STRIKE, the aim was to achieve a better understanding of complex (mostly nonlinear) financial models and to develop effective and robust numerical schemes for solving linear and nonlinear problems arising from the mathematical theory of pricing financial derivatives and related financial products. This aim was accomplished by means of financial modelling, mathematical analysis and numerical simulations, optimal control techniques and validation of models.

Special attention was devoted to a uniform methodology for both testing the latest achievements and simultaneously educating young researchers: Most of the mathematical codes emerging from the STRIKE projects are linked into a new computational finance toolbox (CFT, cf. Chap. 30), which is provided in MATLAB and PYTHON in an open access licence.

Generalized nonlinear option pricing models are capable of capturing several important phenomena like transaction costs, investor's risk from unprotected portfolios, investor's expected utility maximization, illiquid markets, large traders' feedback influence, etc. Such generalizations can be mathematically stated in the form of a nonlinear generalization of the Black-Scholes (BS) equation, in which the local volatility is a function of the option price and its derivatives. In STRIKE, we investigated a wide range of financial derivatives described by a nonlinear BS equation (see the survey Chap. 1). In this regard, an important aspect was the development of high-order compact finite difference methods (FDMs) and transformation techniques for numerically solving nonlinear BS equations.

One of our principal goals in STRIKE was to provide training in advanced methods and techniques in scientific computing with applications to complex financial models. As such, STRIKE provided intensive courses in large-scale financial computing in the many-core graphical processing units (GPUs) cluster used for Monte Carlo simulations, finite difference and fast Fourier numerical methods.

In the context of our research training programme, we covered the complete development cycle for novel financial derivative products starting from setting up stochastic differential equations (SDEs) for underlying assets, and modelling new product prices, to the calibration of model parameters with respect to financial market data by means of a hedge test, scenario studies, inverse problem techniques, etc. This typical development cycle is reflected in the structure of this book. Accordingly, we grouped the chapters into the 8 topics: modelling, analysis, transformation methods and special discretizations, numerical methods in finance, compact FDMs and splitting schemes, scientific computing, high-performance computing and software.

With an eye to the future, we established an ECMI Special Interest Group on Computational Finance, providing a network and basis for further fruitful collaborations. Additionally, we founded a conference series, the International Conference on Computational Finance (ICCF); see www.iccf.eu. The first conference took place in December 2015 in Greenwich, London, and will be followed by a second instalment in Lisbon in September 2017 and a third conference in A Coruña in July 2019.

The editors wish to thank their colleagues from the STRIKE network for their past cooperation and their valued contributions in the chapters of this handbook: Daniel Sevčovič of Bratislava, Lucas Jódar Sánchez of Valencia, Lyuben Vulkov of Ruse, Maria do Rosario Lourenço Grossinho of Lisbon, Ljudmila Bordag of Zittau, Ansgar Jüngel of Vienna, Kees Oosterlee of Delft, Choi-Hong Lai of Greenwich, Alfio Borzì of Würzburg, Karel in 't Hout of Antwerp, Olivier Pironneau of Paris, Bertram Düring of Sussex, Carlos Vázquez Cendón of A Coruña, Jörg Kienitz of Quaternion and Jacques Du Toit of NAG.

Let us close the preface with a citation from a STRIKE PhD fellow:

> The highlights of the STRIKE network are several. STRIKE is a serious boost to 17 young people – 12 ESRs and 5 ERs – in their professional development in computational finance. These researchers work extensively together with their supervisors and professionals from the associated partners in order to get results of high importance to industry (novel models, novel numerical algorithms and theoretical results).

Further, last but not least important, STRIKE strengthens seriously the interaction and transfer of knowledge of methods and algorithms between all parties involved in the consortium. This is going to produce very good results in the long term as collaboration and team work is a recipe for success.
STRIKE is a big boost in their professional career of ESRs and ERs who are in the focus of the project.
STRIKE may provide financial assistance in the several remaining months through mobilization of all resources in order to support the researchers in completing their tasks, in completing some courses (or industry experience) which are essential for a professional in computational finance and further in finding a position in the academy or industry.

Acknowledgement The authors were partially supported by the European Union in the FP7-PEOPLE-2012-ITN Program under Grant Agreement Number 304617 (FP7 Marie Curie Action, Project Multi-ITN STRIKE Novel Methods in Computational Finance).

http://www.itn-strike.eu/

Wuppertal, Germany
January 2017

Matthias Ehrhardt
Michael Günther
E. Jan W. ter Maten

Contents

Contributors

Saeed Amen Cuemacro Limited, Level39, London, UK

Mario Annunziato Dipartimento di Matematica, Universitá degli Studi di Salerno, Fisciano, Italy

Mark Beinker d-fine GmbH, Frankfurt, Germany

Paul Bilokon The Thalesians, London, UK

Ljudmila A. Bordag Faculty of Natural and Environmental Sciences, University of Applied Sciences Zittau/Görlitz, Zittau, Germany

Alfio Borzì Lehrstuhl für Mathematik IX (Chair Scientific Computing), University of Würzburg, Würzburg, Germany

Zuzana Bučková (Zíková) Lehrstuhl für Angewandte Mathematik/Numerische Analysis, Bergische Universität Wuppertal, Wuppertal, Germany

Nicola Cantarutti Instituto Superior de Economia e Gestão (ISEG) Lisbon, Lisbon, Portugal

Carlos Vázquez Cendón Faculty of Informatics, Department of Mathematics, University of A Coruña, Coruña, Spain

Rafael Company Instituto de Matemática Multidisciplinar, Universitat Politècnica de València, Valencia, Spain

Bertram Düring Department of Mathematics at University of Sussex, Brighton, UK

Vera N. Egorova Instituto de Matemática Multidisciplinar, Universitat Politècnica de València, Valencia, Spain

Matthias Ehrhardt Lehrstuhl für Angewandte Mathematik/Numerische Analysis, Bergische Universität Wuppertal, Wuppertal, Germany

Mohamed El-Fakharany Faculty of Science, Mathematics Department, Tanta University, Tanta, Egypt

Yaser Faghan Instituto Superior de Economia e Gestão (ISEG) Lisbon, Lisbon, Portugal

Peter A. Forsyth David R. Cheriton School of Computer Science, University of Waterloo, Waterloo, ON, Canada

Beatrice Gaviraghi Lehrstuhl für Mathematik IX (Chair Scientific Computing), University of Würzburg, Würzburg, Germany

Maria do Rosario Lourenço Grossinho Instituto Superior de Economia e Gestão (ISEG) Lisbon, Lisbon, Portugal

Lech A. Grzelak ING, Quantitative Analytics, Amsterdam and TU Delft, Delft Institute of Applied Mathematics, Delft, The Netherlands

João Guerra Instituto Superior de Economia e Gestão (ISEG) Lisbon, Lisbon, Portugal

Manuel Guerra Instituto Superior de Economia e Gestão (ISEG) Lisbon, Lisbon, Portugal

Michael Günther Lehrstuhl für Angewandte Mathematik/Numerische Analysis, Bergische Universität Wuppertal, Wuppertal, Germany

James Gwinnutt HSBC Bank plc, London, UK

Christian Hendricks Lehrstuhl für Angewandte Mathematik/Numerische Analysis, Bergische Universität Wuppertal, Wuppertal, Germany

Christof Heuer Lehrstuhl für Angewandte Mathematik/Numerische Analysis, Bergische Universität Wuppertal, Wuppertal, Germany

Karel in 't Hout Department of Mathematics and Computer Science, University of Antwerp, Antwerp, Belgium

Daniel Jones University of Oxford, Oxford, UK

Ansgar Jüngel Institute for Analysis and Scientific Computing, TU Wien, Vienna, Austria

Jörg Kienitz Quaternion Risk Management, Krefeld, Germany

University of Cape Town, Cape Town, South Africa

University of Wuppertal, Wuppertal, Germany

Natkunam Kokulan Department of Mathematical Sciences, University of Greenwich, Greenwich, London, UK

Miglena Koleva Department of Numerical Analysis and Statistics, Ruse University Angel Kanchev, Ruse, Bulgaria

Choi-Hong Lai School of Computing and Mathematical Sciences, University of Greenwich, Greenwich, London, UK

José Germán López-Salas Faculty of Informatics, Department of Mathematics, University of A Coruña, Coruña, Spain

E. Jan W. ter Maten Lehrstuhl für Angewandte Mathematik/Numerische Analysis, Bergische Universität Wuppertal, Wuppertal, Germany

Thomas McWalter University of Cape Town and University of Johannesburg, Cape Town, South Africa

James Miles Department of Mathematics, University of Sussex, Brighton, UK

Walter Mudzimbabwe Department of Numerical Analysis and Statistics, Ruse University Angel Kanchev, Ruse, Bulgaria

Kees Oosterlee Delft Institute of Applied Mathematics (DIAM), TU Delft, Delft, The Netherlands

Pedro Pólvora Department of Applied Mathematics and Statistics, Comenius University Bratislava, Bratislava, Slovakia

André M.S. Ribeiro School of Computing and Mathematical Sciences, University of Greenwich, Greenwich, London, UK

Álvaro Leitao Rodríguez Delft Institute of Applied Mathematics (DIAM), TU Delft, Delft, The Netherlands

Lucas Jódar Sánchez Instituto de Matemática Multidisciplinar, Universitat Politècnica de València, Valencia, Spain

Sebastian Schlenkrich d-fine GmbH, Frankfurt, Germany

Daniel Sevčovič Department of Applied Mathematics and Statistics, Comenius University Bratislava, Bratislava, Slovakia

Roelof Sheppard Standard Bank, Johannesburg, South Africa

José Pedro Campos Moreira da Silva Lehrstuhl für Angewandte Mathematik/Numerische Analysis, Bergische Universität Wuppertal, Wuppertal, Germany

Fazlollah Soleymani Instituto de Matemática Multidisciplinar, Universitat Politècnica de València, Valencia, Spain

Shih-Hau Tan School of Computing and Mathematical Sciences, University of Greenwich, Greenwich, London, UK

Long Teng Lehrstuhl für Angewandte Mathematik/Numerische Analysis, Bergische Universität Wuppertal, Wuppertal, Germany

Lara Trussardi Centre de Recherche Inria de Paris, REO Project-Team, Paris, France

Radoslav L. Valkov Department of Mathematics and Computer Science, University of Antwerp, Antwerp, Belgium

Lyuben Vulkov Department of Numerical Analysis and Statistics, Ruse University Angel Kanchev, Ruse, Bulgaria

Ivan P. Yamshchikov Max Plank Institute for Mathematics in Sciences, Leipzig, Germany

Part I
Modelling

Chapter 1
Nonlinear Parabolic Equations Arising in Mathematical Finance

Daniel Ševčovič

Abstract This survey chapter is focused on qualitative and numerical analyses of fully nonlinear partial differential equations of parabolic type arising in financial mathematics. The main purpose is to review various non-linear extensions of the classical Black-Scholes theory for pricing financial instruments, as well as models of stochastic dynamic portfolio optimization leading to the Hamilton-Jacobi-Bellman (HJB) equation. After suitable transformations, both problems can be represented by solutions to nonlinear parabolic equations. Qualitative analysis will be focused on issues concerning the existence and uniqueness of solutions. In the numerical part we discuss a stable finite-volume and finite difference schemes for solving fully nonlinear parabolic equations.

1.1 Nonlinear Generalization of the Black-Scholes Equation for Pricing Financial Instruments

According to the classical theory developed by Black, Scholes and Merton the value $V(S,t)$ of an option in the idealized financial market can be computed from a solution to the well-known Black-Scholes (BS) linear parabolic equation:

$$\partial_t V + \frac{1}{2}\sigma^2 S^2 \partial_S^2 V + (r-q)S\partial_S V - rV = 0, \qquad t \in [0,T), \quad S > 0, \tag{1.1}$$

derived by Black and Scholes and, independently by Merton (cf. [29, 38]). Here $\sigma > 0$ is the volatility of the underlying asset driven by the geometric Brownian motion, $r > 0$ is the risk-free interest rate of zero-coupon bond and $q \geq 0$ is the dividend rate. Similarly, as in the case of the HJB equation the solution is subject to the terminal condition $V(S,T) = \bar{V}(S)$ at $t = T$.

The linear Black-Scholes equation with constant volatility σ has been derived under several restrictive assumptions like e.g., frictionless, liquid and complete

D. Ševčovič (✉)
Department of Applied Mathematics and Statistics, Comenius University, 842 48 Bratislava, Slovakia
e-mail: sevcovic@fmph.uniba.sk

M. Ehrhardt et al. (eds.), *Novel Methods in Computational Finance*, Mathematics in Industry 25, DOI 10.1007/978-3-319-61282-9_1

markets, etc. We also recall that the linear Black-Scholes equation provides a solution corresponding to a perfectly replicated hedging portfolio which need not be a desirable property. In the last decades some of these assumptions have been relaxed in order to model, for instance, the presence of transaction costs (see e.g. Leland [18, 29] and Avellaneda and Paras [5]), feedback and illiquid market effects due to large traders choosing given stock-trading strategies (Schönbucher and Willmott [40], Frey and Patie [16], Frey and Stremme [15], imperfect replication and investor's preferences (Barles and Soner [8]), risk from the unprotected portfolio (Jandačka and Ševčovič [22]). Another nonlinear model in which transaction costs are described by a decreasing function of the number of shares has been derived by Amster et al. [2]. In all aforementioned generalizations of the linear BS equation (1.1) the constant volatility σ is replaced by a nonlinear function:

$$\sigma = \sigma(S\partial_S^2 V) \tag{1.2}$$

depending on the second derivative $\partial_S^2 V$ of the option price itself.

One of the first nonlinear models taking into account transaction costs is the Leland model for pricing the call and put options. This model was further extended by Hoggard et al. [18] for general type of derivatives. In this model the variance σ^2 is given by

$$\sigma(S\partial_S^2 V)^2 = \sigma_0^2\left(1 - \mathrm{Le}\,\mathrm{sgn}\left(S\partial_S^2 V\right)\right) = \begin{cases} \sigma^2(1-\mathrm{Le}), & \text{if } \partial_S^2 V > 0, \\ \sigma^2(1+\mathrm{Le}), & \text{if } \partial_S^2 V < 0, \end{cases} \tag{1.3}$$

where $\mathrm{Le} = \sqrt{\frac{2}{\pi}}\frac{C_0}{\sigma\sqrt{\Delta t}}$ is the so-called Leland number, σ_0 is a constant historical volatility, $C_0 > 0$ is a constant transaction costs per unit dollar of transaction in the underlying asset market and Δt is the time-lag between consecutive portfolio adjustments. The nonlinear model with the volatility function given as in (1.3) can be also viewed as a jumping volatility model investigated by Avellaneda and Paras [5].

The important contribution in this direction has been presented in the work by Amster et al. [2], where the transaction costs are assumed to be a non-increasing linear function of the form $C(\xi) = C_0 - \kappa\xi$, $(C_0, \kappa > 0)$, depending on the volume of trading stock $\xi \geq 0$ needed to hedge the replicating portfolio. A disadvantage of such a transaction costs function is the fact that it may attain negative values when the amount of transactions exceeds the critical value $\xi = C_0/\kappa$. In the model studied by Amster et al. [2] (see also Averbuj [4], Mariani et al. [33]) the volatility function has the following form:

$$\sigma(S\partial_S^2 V)^2 = \sigma_0^2\left(1 - \mathrm{Le}\,\mathrm{sgn}\left(S\partial_S^2 V\right) + \kappa S\partial_S^2 V\right). \tag{1.4}$$

In the recent paper [39] Ševčovič and Žitnanská investigated a model for pricing option under variable transaction costs.

$$\sigma(S\partial_S^2 V)^2 = \sigma_0^2 \left(1 - \sqrt{\frac{2}{\pi}}\tilde{C}(\sigma S|\partial_S^2 V|\sqrt{\Delta t})\frac{\operatorname{sgn}(S\partial_S^2 V)}{\sigma\sqrt{\Delta t}}\right) \tag{1.5}$$

where $\tilde{C}$ is the mean value modification of the transaction cost function $C = C(\xi)$ defined as follows: $\tilde{C}(\xi) = \int_0^\infty C(\xi x)x\, e^{-x^2/2}dx$. As an example one can consider the piecewise linear transaction cost function of the form:

$$C(\xi) = \begin{cases} C_0, & \text{if} \quad 0 \le \xi \le \xi_-, \\ C_0 - \kappa(\xi - \xi_-), & \text{if} \quad \xi_- \le \xi \le \xi_+, \\ \underline{C}_0, & \text{if} \quad \xi \ge \xi_+. \end{cases} \tag{1.6}$$

Bakstein and Howison [7] investigated a parametrized model for liquidity effects arising from the asset trading. In their model σ is a quadratic function of the term $H = S\partial_S^2 V$:

$$\begin{aligned} \sigma(S\partial_S^2 V)^2 =& \sigma_0^2 \Bigg(1 + \bar{\gamma}^2(1-\alpha)^2 + 2\lambda S\partial_S^2 V + \lambda^2(1-\alpha)^2 \left(S\partial_S^2 V\right)^2 \\ & + 2\sqrt{\frac{2}{\pi}}\bar{\gamma}\operatorname{sgn}\left(S\partial_S^2 V\right) + 2\sqrt{\frac{2}{\pi}}\lambda(1-\alpha)^2\bar{\gamma}\left|S\partial_S^2 V\right|\Bigg). \end{aligned} \tag{1.7}$$

The parameter λ corresponds to a market depth measure, i.e. it scales the slope of the average transaction price. Next, the parameter $\bar{\gamma}$ models the relative bid-ask spreads and it is related to the Leland number through relation $2\bar{\gamma}\sqrt{2/\pi} = \text{Le}$. Finally, α transforms the average transaction price into the next quoted price, $0 \le \alpha \le 1$.

The risk adjusted pricing methodology (RAPM) model takes into account the risk from the unprotected portfolio was proposed by Kratka [28]. It was generalized and analyzed by Jandačka and Ševčovič [22]. In this model the volatility function has the form:

$$\sigma(S\partial_S^2 V)^2 = \sigma_0^2 \left(1 + \mu \left(S\partial_S^2 V\right)^{\frac{1}{3}}\right), \tag{1.8}$$

where $\sigma_0 > 0$ is the constant historical volatility of the asset price return and $\mu = 3(C_0^2 R/2\pi)^{\frac{1}{3}}$, where C_0, $R \ge 0$ are non-negative constants representing the transaction cost measure and the risk premium measure, respectively.

If transaction costs are taken into account perfect replication of the contingent claim is no longer possible and further restrictions are needed in the model. By assuming that investor's preferences are characterized by an exponential utility function Barles and Soner (cf. [8]) derived a nonlinear Black-Scholes equation with the volatility σ given by

$$\sigma(S\partial_S^2 V, S, t)^2 = \sigma_0^2 \left(1 + \Psi(a^2 e^{r(T-t)} S^2 \partial_S^2 V)\right) \tag{1.9}$$

where Ψ is a solution to the ODE:

$$\Psi'(x) = \frac{\Psi(x) + 1}{2\sqrt{x\Psi(x)} - x}, \qquad \Psi(0) = 0,$$

and $a > 0$ is a given constant representing risk aversion. Notice that $\Psi(x) = O(x^{\frac{1}{3}})$ for $x \to 0$ and $\Psi(x) = O(x)$ for $x \to \infty$.

All the nonlinear volatility models mentioned in this section can be written in the form of a solution to the fully nonlinear parabolic equation:

$$\partial_t V + \frac{1}{2}\sigma(\partial_S^2 V)^2 S^2 \partial_S^2 V + (r-q)S\partial_S V - rV = 0, \quad t \in [0, T),\ S > 0. \tag{1.10}$$

Jandačka and Ševčovič [22] proposed the method of transformation of equation (1.10) into a quasi-linear parabolic equation for the second derivative $\partial_S^2 V$ (the so-called Gamma of an option) of a solution. Indeed, if we introduce the new variables $H(x,\tau) = S\partial_S^2 V(S,t)$, $x = \ln S$ and $\tau = T - t$ then Eq. (1.10) can be transformed into the so-called Gamma equation:

$$\partial_\tau H = \partial_x^2 \beta(H) + \partial_x \beta(H) + (r-q)\partial_x H - qH, \quad x \in R, \tau \in (0, T), \tag{1.11}$$

where

$$\beta(x, H) = \frac{1}{2}\sigma(H)^2 H$$

(cf. [10, 22]). Recall that the Gamma equation can be obtained by twice differentiation with respect to x of the Black-Scholes equation (1.18) with the volatility of the general type (1.2). A solution $H(x,\tau)$ to the Cauchy problem for (1.11) is subject to the initial condition $H(x, 0) = H_0(x)$.

1.2 Nonlinear Hamilton-Jacobi-Bellman Equation and Optimal Allocation Problems

Optimal allocation and optimal investment problems with state constraints attracted a lot of attention from both theoretical as well as application point of view. The main purpose is to maximize the total expected discounted utility of consumption for the optimal portfolio investment consisting of several stochastic assets, over infinite or finite time horizons. It is known that the value function of the underlying stochastic control problem is the unique smooth solution to the corresponding Hamilton-Jacobi-Bellman (HJB) equation and the optimal consumption and portfolio are presented in feedback form (Zariphopoulou [44]).

Let us consider the stylized financial market in which the aim of a portfolio manager is to maximize the expected value of the terminal wealth of a portfolio, measured by a prescribed utility function U. In particular, if n is the number of assets entering the portfolio, T the investment horizon, the goal is to find an optimal trading strategy $\{\theta\} = \{\theta_t \in \mathbb{R}^n \mid t \in [0, T]\}$ belonging to a set $A = A_{0,T}$ of strategies $A_{t,T} = \{\{\theta\}|\theta_s \in S^n, s \in [t, T]\}$, where $S^n = \{\theta_t \in \mathbb{R}^n|\theta_t \in [0, 1]^n, 1^T\theta_t = 1\}$ is a convex compact simplex such that $\{\theta\}$ maximizes the expected terminal utility from the portfolio:

$$\max_{\{\theta\}\in A} \mathbb{E}\left[U(X_T^\theta)|X_0^\theta = x_0\right]. \tag{1.12}$$

Here $X_t = \ln Y_t$ represents a stochastic process governed by the following stochastic differential equation

$$\mathrm{d}X_t^\theta = \left(\mu(\theta) - \frac{1}{2}\sigma(\theta)^2\right)\mathrm{d}t + \sigma(\theta)\mathrm{d}W_t$$

for a logarithmic portfolio value, where x_0 is its initial value at the time $t = 0$. Here $\mu(\theta)$ and $\sigma(\theta)$ are the expected return and volatility of the portfolio. As a typical example, one can consider functions $\mu(\theta) = \mu^T\theta$ and $\sigma^2(\theta) = \theta^T\Sigma\theta$, where μ is a vector of mean returns and Σ is a covariance matrix. It is known from the theory of stochastic dynamic programming that the so-called value function

$$V(x, t) := \sup_{\{\theta\}\in A_{t,T}} \mathbb{E}\left[U(X_T^\theta)|X_t^\theta = x\right] \tag{1.13}$$

subject to the terminal condition $V(x, T) := U(x)$ can be used for solving the stochastic dynamic optimization problem (1.12) (cf. Bertsekas [9], Fleming and Soner [14]). Moreover, it is also known, that the value function $V = V(x, t)$ satisfies the following Hamilton-Jacobi-Bellman equation:

$$\partial_t V + \max_{\theta\in S^n}\left\{\left(\mu(\theta) - \frac{1}{2}\sigma(\theta)^2\right)\partial_x V + \frac{1}{2}\sigma(\theta)^2\partial_x^2 V\right\} = 0, \tag{1.14}$$

for all $x \in \mathbb{R}, t \in [0, T)$ and it satisfies the terminal condition $V(., T) := U(.)$ (see e.g. [20, 32]).

In general, explicit solutions to HJB equations are not available and this is why various numerical approaches have to be adopted. Regarding numerical approaches for solving HJB equations associated with portfolio optimization, we can mention and refer to finite difference methods for approximating its viscosity solution developed and analyzed by Tourin and Zariphopoulou [42], Crandall et al. [12], Nayak and Papanicolaou [36]. Other approach based on Markov chain approximation techniques was investigated by Song [41] and Fleming and Soner [14]. Classical methods for solving HJB equations are discussed by Benton [44]. In [34], Musiela and Zariphopoulou applied the power-like transformation in order to linearize the non-linear PDE for the value function in the case of an exponential utility function. Muthamaran and Sunil [35] solved a multi-dimensional portfolio optimization problem with transaction costs. They used finite element method and iterative procedure that converts a free-boundary problem into a sequence of fixed boundary problems. Peyrl et al. [37] applied a successive approximation algorithm for solving the corresponding HJB equation. The fixed point-policy iteration scheme for solving discretized HJB equations is discussed in Huang et al. [19]. Witte and Reisinger [43] presented a penalty approach for the numerical solution of discrete continuously controlled HJB equations.

In the recent paper [23] Kilianová and Ševčovič transformed the fully nonlinear HJB equation (1.14) into the Cauchy problem for the quasi-linear parabolic equation:

$$\partial_t \varphi + \partial_x^2 \beta(\varphi) + \partial_x[(1-\varphi)\beta(\varphi)] = 0, \quad x \in R, t \in [0,T), \tag{1.15}$$

$$\varphi(x,T) = 1 - \frac{U''(x)}{U'(x)}, \quad x \in R. \tag{1.16}$$

To this aim we introduced the following transformation:

$$\varphi(x,t) = 1 - \frac{\partial_x^2 V(x,t)}{\partial_x V(x,t)}.$$

It is referred to as the Riccati transformation and it has been proposed and studied in [1, 32] and further analyzed by Ishimura and Ševčovič in [20]. The resulting equation was solved numerically by an iterative method based on the finite volume approximation. Furthermore, it follows from the analysis [23] by Kilianová and Ševčovič that the diffusion function $\beta(\varphi)$ is the value function of the following parametric optimization problem:

$$\beta(\varphi) = \min_{\theta \in S^n} \{-\mu(\theta) + \frac{\varphi}{2}\sigma(\theta)^2\}. \tag{1.17}$$

The dispersion function $\theta \mapsto \sigma(\theta)^2$ is assumed to be strictly convex and $\theta \mapsto \mu(\theta)$ is a linear function. Therefore problem (1.17) belongs to a class of parametric convex optimization problems (cf. Bank et al. [6], Hamala and Trnovská [17]). Useful generalization of the HJB equation (1.14) in case the covariance matrix Σ belongs to some set P of (e.g. ellipsoidal sets) of covariance matrices was studied by Kilianová and Trnovská in [24] with regard to application to the so-called "worst case variance" portfolio model in which the diffusion function (1.17) has the form:

$$\beta(\varphi) = \min_{\theta \in S^n} \max_{\Sigma \in P} -\mu^T \theta + \frac{\varphi}{2}\theta^T \Sigma \theta .$$

They showed this problem can be analyzed by the methods of semidefinite programming. The value function $\beta(\varphi)$ need not be sufficiently smooth and its second derivative can have jumps.

In fact, the Riccati transformation is the logarithmic derivative of the derivative of the value function. In the context of a class of HJB equations with range constraints, the Riccati transformation has been analyzed recently by Ishimura and Ševčovič in [20] where a traveling wave solution to the HJB equation was constructed. Concerning numerical methods for solving the quasi-linear parabolic PDE obtained from the HJB equation by means of the Riccati transformation we mention recent papers by Ishimura et al. [21, 25]. In [25], Koleva considered a similar nonlinear parabolic equation, obtained by means of a Riccati-like transformation of the HJB equation, arising in pension saving management. In contrary to our model problem, she considered a problem without constraints on the optimal decision. She applied two iterative numerical methods, namely the fully implicit Picard method and the mixed Picard-Newton method and discussed their accuracy and effectiveness.

In summary, the nonlinear volatility generalization of the Black-Scholes equation as well as the Hamilton-Jacobi-Bellman equation can be transformed into the quasilinear parabolic equation for the unknown function $H = H(x, \tau)$ representing either the Gamma of the portfolio $H = S\partial_S^2 V$ (nonlinear volatility Black-Scholes models) or the relative risk aversion function $H = 1 - \partial_x^2 V/\partial_x V$ (Hamilton-Jacobi-Bellman equation). The resulting quasilinear parabolic equation has the form:

$$\partial_\tau H = \partial_x^2 \beta(H) + f(x, H, \partial_x H), \quad x \in R, \tau \in (0, T), \tag{1.18}$$

where β is a suitable nonlinear function.

1.3 Existence of Classical Solutions, Comparison Principle

In this section we recall results on existence of classical smooth solutions to the Cauchy problem for the quasilinear parabolic equation (1.18). Following the methodology based on the so-called Schauder's type of estimates (cf. Ladyzhenskaya et al. [30]), we shall proceed with a definition of function spaces we will work with. Let $\Omega = (x_L, x_R) \subset \mathbb{R}$ be a bounded interval. We denote $Q_T = \Omega \times (0, T)$ the space-time cylinder. Let $0 < \lambda < 1$. By $\mathscr{H}^\lambda(\Omega)$ we denote the Banach space consisting of all continuous functions H defined on $\bar{\Omega}$ which are λ-Hölder continuous. It means that their Hölder semi-norm $\langle H \rangle^{(\lambda)} = \sup_{x,y\in\Omega, x\neq y} |H(x) - H(y)|/|x - y|^\lambda$ is finite. The norm in the space $\mathscr{H}^\lambda(\Omega)$ is then the sum of the maximum norm of H and the semi-norm $\langle H \rangle^{(\lambda)}$. The space $\mathscr{H}^{2+\lambda}(\Omega)$ consists of all twice continuously differentiable functions H in $\bar{\Omega}$ whose second derivative $\partial_x^2 H$ belongs to $\mathscr{H}^\lambda(\Omega)$. The space $\mathscr{H}^{2+\lambda}(\mathbb{R})$ consists of all functions $H : \mathbb{R} \to \mathbb{R}$ such that $H \in \mathscr{H}^{2+\lambda}(\Omega)$ for any bounded domain $\Omega \subset \mathbb{R}$.

The parabolic Hölder space $\mathscr{H}^{\lambda,\lambda/2}(Q_T)$ of functions defined on a bounded cylinder Q_T consists of all continuous functions $H(x,\tau)$ in $\bar{Q}_T$ such that H is λ-Hölder continuous in the x-variable and $\lambda/2$-Hölder continuous in the t-variable. The norm is defined as the sum of the maximum norm and corresponding Hölder semi-norms. The space $\mathscr{H}^{2+\lambda,1+\lambda/2}(Q_T)$ consists of all continuous functions on $\bar{Q}_T$ such that $\partial_\tau H, \partial_x^2 H \in \mathscr{H}^{\lambda,\lambda/2}(Q_T)$. Finally, the space $\mathscr{H}^{2+\lambda,1+\lambda/2}(\mathbb{R}\times[0,T])$ consists of all functions $H:\mathbb{R}\times[0,T]\to\mathbb{R}$ such that $H\in\mathscr{H}^{2+\lambda,1+\lambda/2}(Q_T)$ for any bounded cylinder Q_T (cf. [30, Chap. I]).

In the nonlinear models discussed in the previous sections one can derive useful lower and upper bounds of a solution H to the Cauchy problem (1.18). The idea of proving upper and lower estimates for $H(x,\tau)$ is based on construction of suitable sub- and super-solutions to the parabolic equation (1.18) (cf. [30]).

$$\lambda_- \le \beta'(H) \le \lambda_+$$

for any $H \ge 0$ where $\lambda_\pm > 0$ are constants. This implies strong parabolicity of the governing nonlinear parabolic equation.

Theorem 1.1 ([39, Theorem 3.1]) *Suppose that the initial condition $H(.,0) \ge 0$ belongs to the Hölder space $\mathscr{H}^{2+\lambda}(\mathbb{R})$ for some $0<\lambda<\min(1/2,\varepsilon)$ and $\overline{H} = \sup_{x\in\mathbb{R}} H(x,0) < \infty$. Assume that $\beta, f \in C^{1,\varepsilon}$ and β satisfies $\lambda_- \le \beta'(H) \le \lambda_+$ for any $0 \le H \le \overline{H}$ where $\lambda_\pm > 0$ are constants.*

Then there exists a unique classical solution $H(x,\tau)$ to the quasilinear parabolic equation (1.18) satisfying the initial condition $H(x,0)$. The function $\tau \mapsto \partial_\tau H(x,\tau)$ is $\lambda/2$-Hölder continuous for all $x\in\mathbb{R}$ whereas $x\mapsto \partial_x H(x,\tau)$ is Lipschitz continuous for all $\tau\in[0,T]$. Moreover, $\beta(H(.,.)) \in \mathscr{H}^{2+\lambda,1+\lambda/2}(\mathbb{R}\times[0,T])$ and $0 < H(x,\tau) \le \overline{H}$ for all $(x,\tau)\in\mathbb{R}\times[0,T)$.

The proof is based on the so-called Schauder's theory on existence and uniqueness of classical Hölder smooth solutions to a quasi-linear parabolic equation of the form (1.18). It follows the same ideas as the proof of [23, Theorem 5.3] where Kilianová and Ševčovič investigated a similar quasilinear parabolic equation obtained from a nonlinear HJB equation in which a stronger assumption $\beta \in C^{1,1}$ is assumed.

1.4 Numerical Full Space-Time Discretization Scheme for Solving the Gamma Equation

In this section we present an efficient numerical scheme for solving the Gamma equation. The construction of numerical approximation of a solution H to (1.18) is based on a derivation of a system of difference equations corresponding to (1.18) to be solved at every discrete time step. We make use of the numerical scheme adopted from the paper by Jandačka and Ševčovič [22] in order to solve the Gamma equation (1.18) for a general function $\beta = \beta(H)$ including, in particular, the case of the model with variable transaction costs. The efficient numerical discretization is based on the finite volume approximation of the partial derivatives entering (1.18). The resulting scheme is semi-implicit in a finite-time difference approximation scheme.

Other finite difference numerical approximation schemes are based on discretization of the original fully nonlinear Black-Scholes equation in non-divergence form. We refer the reader to recent publications by Ankudinova and Ehrhardt [3], Company et al. [11], Düring et al. [13], Liao and Khaliq [31], Zhou et al. [45]. Recently, a quasilinearization technique for solving the fully nonlinear parabolic equation was proposed and analyzed by Koleva and Vulkov [26]. Our approach is based on a solution to the quasilinear Gamma equation written in the divergence form, so we can use existing finite volume based numerical scheme to solve the problem efficiently (cf. Jandačka and Ševčovič [22], Kútik and Mikula [27]).

For numerical reasons we restrict the spatial interval to $x \in (-L, L)$ where $L > 0$ is sufficiently large. Since $S = Ee^x \in (Ee^{-L}, Ee^L)$ it is sufficient to take $L \approx 2$ in order to include the important range of values of S. For the purpose of construction of a numerical scheme, the time interval $[0, T]$ is uniformly divided with a time step $k = T/m$ into discrete points $\tau_j = jk$, where $j = 0, 1, \ldots, m$. We consider the spatial interval $[-L, L]$ with uniform division with a step $h = L/n$, into discrete points $x_i = ih$, where $i = -n, \ldots, n$.

The proposed numerical scheme is semi-implicit in time. Notice that the term $\partial_x^2 \beta$, can be expressed in the form $\partial_x^2 \beta = \partial_x (\beta'(H)\partial_x H)$, where β' is the derivative of $\beta(H)$ with respect to H. In the discretization scheme, the nonlinear terms $\beta'(H)$ are evaluated from the previous time step τ_{j-1} whereas linear terms are solved at the current time level.

Such a discretization scheme leads to a solution of a tridiagonal system of linear equations at every discrete time level. First, we replace the time derivative by the time difference, approximate H in nodal points by the average value of neighboring segments, then we collect all linear terms at the new time level τ_j and by taking all the remaining terms from the previous time level τ_{j-1}. We obtain a tridiagonal system for the solution vector $H^j = (H^j_{-n+1}, \ldots, H^j_{n-1})^T \in \mathbb{R}^{2n-1}$:

$$a_i^j H_{i-1}^j + b_i^j H_i^j + c_i^j H_{i+1}^j = d_i^j, \quad H_{-n}^j = 0, \ H_n^j = 0\,, \tag{1.19}$$

where $i = -n+1, \ldots, n-1$ and $j = 1, \ldots, m$. The coefficients of the tridiagonal matrix are given by

$$a_i^j = -\frac{k}{h^2}\beta'_H(H_{i-1}^{j-1}) + \frac{k}{2h}r \quad c_i^j = -\frac{k}{h^2}\beta'_H(H_i^{j-1}) - \frac{k}{2h}r, \quad b_i^j = 1 - (a_i^j + c_i^j)\,,$$
$$d_i^j = H_i^{j-1} + \frac{k}{h}\left(\beta(H_i^{j-1}) - \beta(H_{i-1}^{j-1})\right).$$

It means that the vector H^j at the time level τ_j is a solution to the system of linear equations $\mathbf{A}^{(j)} H^j = d^j$, where the $(2n-1)\times(2n-1)$ matrix $\mathbf{A}^{(j)} = \text{tridiag}(a^j, b^j, c^j)$.

In order to solve the tridiagonal system in every time step in a fast and effective way, we can use the efficient Thomas algorithm.

In [39] the authors showed that the option price $V(S, T-\tau_j)$ can be constructed from the discrete solution H_i^j by means of a simple integration scheme:

$$\text{(call option)} \qquad V(S, T-\tau_j) = h\sum_{i=-n}^{n}(S - Ee^{x_i})^+ H_i^j, \quad j = 1, \dots, m,$$

$$\text{(put option)} \qquad V(S, T-\tau_j) = h\sum_{i=-n}^{n}(Ee^{x_i} - S)^+ H_i^j, \quad j = 1, \dots, m.$$

1.5 Numerical Results for the Nonlinear Model with Variable Transaction Costs

In this section we present the numerical results for computation of the option price for the nonlinear volatility Black-Scholes model with variable transaction costs derived and analyzed by Ševčovič and Žitnanská in the recent paper [39]. As an example for numerical approximation of a solution we consider variable transaction costs described by the piecewise linear non-increasing function, depicted in Fig. 1.1. The function $\beta(H)$ corresponding to the variable transaction costs function $C(\xi)$ has the form

$$\beta(H) = \frac{\sigma_0^2}{2}\left(1 - \sqrt{\frac{2}{\pi}}\tilde{C}(\sigma|H|\sqrt{\Delta t})\frac{\operatorname{sgn}(H)}{\sigma\sqrt{\Delta t}}\right)H,$$

where $\tilde{C}$ is the modified transaction costs function.

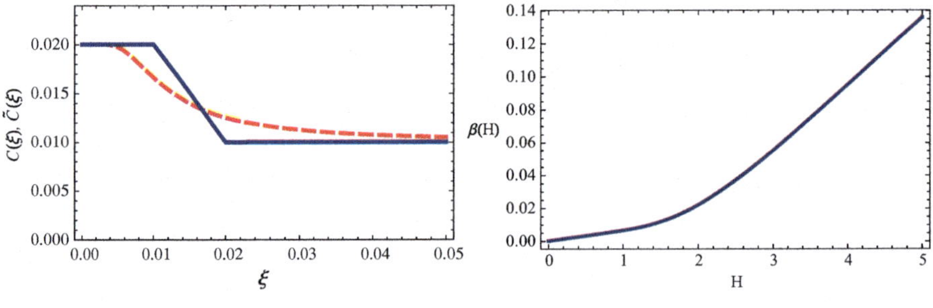

Fig. 1.1 *Left*: The piecewise linear transaction costs function C (*solid line*), its mean value modification $\tilde{C}$ (*dashed line*). *Right*: the graph of the corresponding function $\beta(H)$. Source [39]

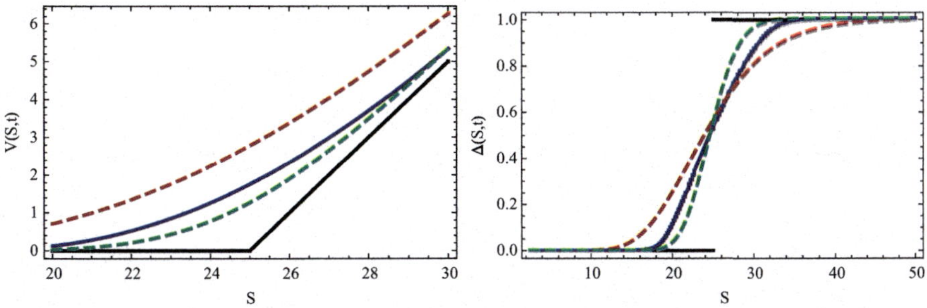

Fig. 1.2 The call option price $V(S,t)$ as a function of S for $t = 0$ (*left*) and its delta $\Delta(S,t) = \partial_S V(S,t)$. Source [39]

In our computations we chose the following model parameters describing the piecewise transaction costs function: $C_0 = 0.02$, $\kappa = 0.3$, $\xi_- = 0.05$, $\xi_+ = 0.1$. The length of the time interval between two consecutive portfolio rearrangements: $\Delta t = 1/261$. The maturity time $T = 1$, historical volatility $\sigma = 0.3$ and the risk-free interest rate $r = 0.011$. As for the numerical parameters we chose $L = 2.5$, $n = 250$, $m = 200$. The parameters C_0, σ, κ, $\xi_\pm$ and Δt correspond to the Leland numbers $\mathrm{Le} = 0.85935$ and $\underline{\mathrm{Le}} = 0.21484$. In Fig. 1.2 we plot the solution $V_{vtc}(S,t)$ and the option price delta factor $\Delta(S,t) = \partial_S V(S,t)$, for $t = 0$. The upper dashed line corresponds to the solution of the linear Black-Scholes equation with the higher volatility

$$\hat{\sigma}^2_{max} = \sigma^2 \left(1 - \underline{C}_0 \sqrt{\frac{2}{\pi}} \frac{1}{\sigma\sqrt{\Delta t}}\right),$$

where $\underline{C}_0 = C_0 - \kappa(\xi_+ - \xi_-) > 0$, whereas the lower dashed line corresponds to the solution with a lower volatility

$$\hat{\sigma}^2_{min} = \sigma^2 \left(1 - C_0 \sqrt{\frac{2}{\pi}} \frac{1}{\sigma\sqrt{\Delta t}}\right).$$

Acknowledgements This research was supported by the European Union in the FP7-PEOPLE-2012-ITN project STRIKE—Novel Methods in Computational Finance (304617).

References

1. Abe, R., Ishimura, N.: Existence of solutions for the nonlinear partial differential equation arising in the optimal investment problem. Proc. Jpn. Acad. Ser. A. **84**, 11–14 (2008)
2. Amster, P., Averbuj, C.G., Mariani, M.C., Rial, D.: A Black-Scholes option pricing model with transaction costs. J. Math. Anal. Appl. **303**, 688–695 (2005)

3. Ankudinova, J., Ehrhardt, M.: On the numerical solution of nonlinear Black-Scholes equations. Comput. Math. Appl. **56**, 799–812 (2008)
4. Averbuj, C.G.: Nonlinear integral-differential evolution equation arising in option pricing when including transaction costs: a viscosity solution approach. Rev. Bras. Econ. Empresas **12**, 81–90 (2012)
5. Avellaneda, M., Levy, A., Paras, A.: Pricing and hedging derivative securities in markets with uncertain volatilities. Appl. Math. Fin. **2**, 73–88 (1995)
6. Bank, B., Guddat, J., Klatte, D., Kummer, B., Tammer, K.: Nonlinear Parametric Optimization. Birkhäuser Verlag, Basel/Boston, MA (1983)
7. Bakstein, D., Howison, S.: A non-arbitrage liquidity model with observable parameters. Working paper, http://eprints.maths.ox.ac.uk/53/
8. Barles, G., Soner, H.M.: Option Pricing with transaction costs and a nonlinear Black-Scholes equation. Finance Stochast. **2**, 369–397 (1998)
9. Bertsekas, D.P.: Dynamic Programming and Stochastic Control. Academic, New York (1976)
10. Bordag, L.A., Frey, R.: Study of the risk-adjusted pricing methodology model with methods of Geometrical Analysis. Stoch. Int. J. Probab. Stoch. Process **83**, 333–345 (2011)
11. Company, R., Navarro, E., Pintos, J.R., Ponsoda, E.: Numerical solution of linear and nonlinear Black-Scholes option pricing equations. Comput. Math. Appl. **56**, 813–821 (2008)
12. Crandall, M.C., Ishii, H., Lions, P.L.: User's guide to viscosity solutions of second order partial differential equations. Bull. Am. Math. Soc. **1**, 1–67 (1992)
13. Düring, B., Fournié, M., Jüngel, A.: High order compact finite difference schemes for a nonlinear Black-Scholes equation. Int. J. Theor. Appl. Finance **7**, 767–789 (2003)
14. Fleming, W.H., Soner, H.M.: Controlled Markov Processes and Viscosity Solutions, 2nd edn. Springer, New York (2005)
15. Frey, R., Patie, P.: Risk management for derivatives in illiquid markets: a simulation study. In: Advances in Finance and Stochastics, pp. 137–159. Springer, Berlin (2002)
16. Frey, R., Stremme, A.: Market volatility and feedback effects from dynamic hedging. Math. Fin. **4**, 351–374 (1997)
17. Hamala, M., Trnovská, M.: Nonlinear Programming, Theory and Algorithms. Epos, Bratislava (2013)
18. Hoggard, T., Whalley, A.E., Wilmott, P.: Hedging option portfolios in the presence of transaction costs. Adv. Futures Opt. Res. **7**, 21–35 (1994)
19. Huang, Y., Forsyth, P.A., Labahn, G.: Combined fixed point and policy iteration for HJB equations in finance. SIAM J. Numer. Anal. **50**, 1861–1882 (2012)
20. Ishimura, N., Ševčovič, D.: On traveling wave solutions to a Hamilton-Jacobi-Bellman equation with inequality constraints. Jpn. J. Industr. Appl. Math. **30**, 51–67 (2013)
21. Ishimura, N., Koleva, M.N., Vulkov, L.G.: Numerical solution via transformation methods of nonlinear models in option pricing. AIP Conf. Proc. **1301**, 387–394 (2010)
22. Jandačka, M., Ševčovič, D.: On the risk adjusted pricing methodology based valuation of vanilla options and explanation of the volatility smile. J. Appl. Math. **2005**, 235–258 (2005)
23. Kilianová, S., Ševčovič, D.: A method of solving Hamilton-Jacobi-Bellman equation for constrained optimal investment problem via Riccati transformation. Anziam J. **55**, 14–38 (2013)
24. Kilianová, S., Trnovská, M.: Robust portfolio optimization via solution to the Hamilton-Jacobi-Bellman equation. Int. J. Comput. Math. **93**, 725–734 (2016)
25. Koleva, M.: Iterative methods for solving nonlinear parabolic problem in pension saving management. AIP Conf. Proc. **1404**, 457–463 (2011)
26. Koleva, M.N., Vulkov, L.G.: Quasilinearization numerical scheme for fully nonlinear parabolic problems with applications in models of mathematical finance. Math. Comput. Model. **57**, 2564–2575 (2013)
27. Kútik, P., Mikula, K.: Finite volume schemes for solving nonlinear partial differential equations in financial mathematics. In: Finite Volumes for Complex Applications VI Problems and Perspectives. Springer Proceedings in Mathematics, vol. 4, pp. 643–651. Springer, Berlin (2011)

28. Kratka, M.: No mystery behind the smile. Risk **9**, 67–71 (1998)
29. Kwok, Y.K.: Mathematical Models of Financial Derivatives. Springer, New York (1998)
30. Ladyženskaja, O.A., Solonnikov, V.A., Ural'ceva, N.N.: Linear and Quasilinear Equations of Parabolic Type (Translated from the Russian by S. Smith. Translations of Mathematical Monographs), vol. 23. American Mathematical Society, Providence, RI (1968)
31. Liao, W., Khaliq, A.Q.M.: High-order compact scheme for solving nonlinear Black-Scholes equation with transaction costs. Int. J. Comput. Math. **86**, 1009–1023 (2009)
32. Macová, Z., Ševčovič, D.: Weakly nonlinear analysis of the Hamilton-Jacobi-Bellman equation arising from pension saving management. Int. J. Numer. Anal. Model. **4**, 619–638 (2010)
33. Mariani, M.C., Ncheuguim, E., Sengupta, I.: Solution to a nonlinear Black-Scholes equation. Electron. J. Differ. Equ. **158**, 1–10 (2011)
34. Musiela, M., Zariphopoulou, T.: An example of indifference prices under exponential preferences. Finance Stochast. **8**, 229–239 (2004)
35. Muthuraman, K., Kumar, S.: Multi-dimensional portfolio optimization with proportional transaction costs. Math Fin. **16**, 301–335 (2006)
36. Nayak, S., Papanicolaou, G.: Market influence of portfolio optimizers. Appl. Math. Fin. **15**, 21–40 (2008)
37. Peyrl, H., Herzog, F., Geering, H.P.: Numerical solution of the Hamilton-Jacobi-Bellman equation for stochastic optimal control problems. In: WSEAS International Conference on Dynamical Systems and Control, Venice, Italy, 2–4 Nov 2005, pp. 489–497
38. Ševčovič, D., Stehlíková, B., Mikula, K.: Analytical and Numerical Methods for Pricing Financial Derivatives. Nova Science Publishers, Inc., Hauppauge (2011)
39. Ševčovič, D., Žitnanská, M.: Analysis of the nonlinear option pricing model under variable transaction costs. Asia-Pacific Finan. Markets **23**, 153–174 (2016)
40. Schönbucher, P., Wilmott, P.: The feedback-effect of hedging in illiquid markets. SIAM J. Appl. Math. **61**, 232–272 (2000)
41. Song, Q.S.: Convergence of Markov chain approximation on generalized HJB equation and its applications. Automatica **44**, 761–766 (2008)
42. Tourin, A., Zariphopoulou, T.: Numerical schemes for investment models with singular transactions. Comput. Econ. **4**, 287–307 (1994)
43. Witte, J.H., Reisinger, Ch.: Penalty methods for the solution of discrete HJB equations – continuous control and obstacle problems. SIAM J. Numer. Anal. **50**, 595–625 (2012)
44. Zariphopoulou, T.: Consumption-investment models with constraints. SIAM J. Control. Optim. **1**, 59–85 (1994)
45. Zhou, S., Han, L., Li, W., Zhang, Y., Han, M.: A positivity-preserving numerical scheme for option pricing model with transaction costs under jump-diffusion process. Comput. Appl. Math. **34**, 881–900 (2015)

Chapter 2
Modeling of Herding and Wealth Distribution in Large Markets

Ansgar Jüngel and Lara Trussardi

Abstract The dynamics of the number of participants in a large market is described by nonlinear partial differential equations of kinetic and diffusive type. The results on the modeling, analysis, and numerical simulation of three market models are briefly reviewed. The interplay of the agents with external sources, herding phenomena, and irrationality of the individuals as well as the exchange of knowledge and wealth is explored mathematically. The focus lies on the mathematical understanding of the differential equations rather than on the modeling of real economic situations, aiming at identifying models which are able to produce the desired effects.

2.1 Introduction

The modeling of markets with a large number of agents became very vital in recent years with the aim to understand inefficient markets or irrational behavior of agents, for instance. The dynamics of such markets may be described by agent-based models, kinetic equations, or diffusive systems. Agent-based models specify the behavior of individuals by using elements of game theory and Monte-Carlo simulation techniques [31]. In kinetic modeling, the analogy with statistical mechanics is exploited: Interactions between market agents are interpreted as collisions between gas particles, and conservation laws for income and/or wealth may hold [26, 28]. Diffusive systems are often derived from kinetic equations in the so-called grazing collision limit , and they illustrate the behavior on a macroscopic level [33]. In this section, we summarize the results of Boudin et al. [7], Düring et al. [15], Jüngel et al. [25] on kinetic and diffusive equations modeling socio-economic scenarios.

A. Jüngel (✉)
Institute for Analysis and Scientific Computing, Vienna University of Technology, Wiedner Hauptstr. 8-10, 1040 Wien, Austria
e-mail: juengel@tuwien.ac.at

L. Trussardi
Centre de Recherche Inria de Paris, REO project-team, 2 rue Simone IFF, 75012 Paris, France
e-mail: lara.trussardi@inria.fr

M. Ehrhardt et al. (eds.), *Novel Methods in Computational Finance*,
Mathematics in Industry 25, DOI 10.1007/978-3-319-61282-9_2

The first scenario is the herding in financial markets. Herding is characterized by a homogenization of the actions of the market participants, which behave at a certain time in the same way. Herding may lead to strong trends with low volatility of asset prices, but eventually also to abrupt corrections, so it promotes the occurrence of bubbles and crashes. Numerous socio-economic papers [4, 8, 30] and research in biological sciences [1, 19] show that herding interactions play a crucial role in social scenarios. Herding behavior is often irrational because people are not basing their decision on objective criteria.

A full understanding of herding behavior needs the ability to understand two levels: the microscopic one, which considers each individual of the crowd separately; and the macroscopic level, which deals with the group of individuals, i.e. the herd. The first level usually represents the individual as a particle, and a microscopic particle-type or mesoscopic kinetic description may be useful. The latter one may be represented by a density function depending (continuously) on space and time, leading to diffusive equations. We consider a diffusive herding model in Sect. 2.2 and a kinetic herding equation in Sect. 2.3.

The second scenario addressed in this review is the distribution of wealth. Most of the models in the literature are agent-based models [9], mean-field games [18], or kinetic equations [27]. In kinetic modeling, binary collisions are replaced by trades between agents by defining rules which specify how wealth is exchanged in trades. The output of the model are the statistics of the wealth distribution in the market. It turns out that in many models, the stationary profile has an overpopulated tail (called fat tail or Pareto tail), which is interpreted as the existence of an upper class of very wealthy people [27]. Pareto tails appear under various assumptions, assuming wealth conservation in the mean or pointwise wealth conservation [14].

Binary wealth exchange models go back to the work [3]. Later, the relation to statistical mechanics was highlighted [24], and strictly conservative exchange models were developed [10]. The strict conservation was relaxed in [12] to conservation in the mean. Our contribution is to combine wealth and knowledge of agents in a society and to examine the interaction of these qualities; see Sect. 2.4.

We stress the fact that the models that we are proposing and analyzing are quite simple. Certainly, the socio-economic behavior of real market agents is extremely complex and includes psychological and social phenomena. Still, we believe that a large number of agents may be described to some extent in an averaged sense—at least in simplified situations. Our aim is to understand the mathematical phenomena arising from the new terms in the models rather than devising models that include as many features as possible. Our analysis shows which terms produce the desired effects and henceforth can be included in more realistic models. The hope is that this analysis helps to identify irregularities in (financial) markets or in societies and to lead to improved market regulations and counter-actions to avoid financial crashes.

2.2 A Cross-Diffusion Herding Model

A very simple model for herding behavior is given by the cross-diffusion system [25]

$$\partial_t u = \operatorname{div}(\nabla u - g(u)\nabla v), \quad \partial_t v = \operatorname{div}(\delta\nabla u + \kappa\nabla v) + f(u) - \alpha v, \tag{2.1}$$

where $u(x,t)$ represents the normalized density of individuals with information variable $x \in \Omega$ at time $t \geq 0$ ($\Omega \subset \mathbb{R}^d$ being a bounded domain), and $v(x,t)$ is an influence function which modifies the information state of the individuals. The influence function acts through the cross-diffusion term $g(u)\nabla v$ in the first equation in (2.1). We assume that the influence becomes weak if the number of individuals at a fixed state x is very low or close to the maximal value $u = 1$. Thus, we suppose that $g(0) = g(1) = 0$. The influence function is modified by diffusive effects also due to the random behavior of the agents with parameter $\delta > 0$, by the nonnegative source term $f(u)$, time relaxation with rate $\alpha > 0$, and diffusion with coefficient $\kappa > 0$. Our aim is to understand whether the above model exhibits herding phenomena, i.e. regions in which the density of the agents is very low or close to the maximal value.

The equations are supplemented by no-flux boundary and initial conditions:

$$\begin{aligned}
&(\nabla u - g(u)\nabla v)\cdot \nu = 0, \quad (\delta\nabla u + \kappa\nabla v)\cdot\nu = 0 \quad \text{on } \partial\Omega, \\
&u(\cdot,0) = u_0, \quad v(\cdot,0) = v_0 \quad \text{in } \Omega,\ t > 0,
\end{aligned} \tag{2.2}$$

where ν denotes the exterior unit normal vector to $\partial\Omega$.

If $\delta = 0$, system (2.1) correspond to a nonlinear chemotaxis Keller-Segel model, where u represents the cell density and v the concentration of the chemoattractant [20]. While the original Keller-Segel model exhibits finite-time blow-up of the solutions, the nonlinear mobility $g(u) = u(1-u)$ prevents blow up [34]. Equations (2.1) with $\delta > 0$ can be derived from stochastic partial differential equations describing interacting particles, at least for constant mobility functions $g(u)$ [17]. The case $\delta > 0$ and $g(u) = u$ was analyzed in [23]. A typical example in the present situation is $g(u) = u(1-u)$ since this function satisfies $g(0) = g(1) = 0$.

In the work [25], the following results have been obtained.

2.2.1 *Existence of Solutions*

If f and g are smooth, bounded, nonnegative functions such that there exists $m \in (0,1)$ satisfying

$$g(0) = g(1) = 0, \quad \int_0^m \frac{ds}{g(s)} = \int_m^1 \frac{ds}{g(s)} = \infty, \tag{2.3}$$

and $u_0, v_0 \in L^\infty(\Omega)$, then there exists a global weak solution (u, v) to (2.1)–(2.2) satisfying $0 \le u \le 1$ in Ω, $t > 0$, as long as $\delta > -\kappa/\gamma$, where $\gamma = \max_{s\in[0,1]} g(s)$. The function $g(u) = u(1-u)$ satisfies (2.3).

The restriction on δ ensures that the real parts of the eigenvalues of the diffusion matrix from (2.1) are positive, such that the system is parabolic in the sense of Petrovskii and local existence of solutions can be expected [2]. The challenge is to prove the existence of global (weak) solutions. A key element of the proof is the observation that Eq. (2.1) admit a Lyapunov functional (called an entropy),

$$H(u, v) = \int_\Omega \left(h(u) + \frac{v^2}{2\delta_0} \right), \quad \text{where } h(u) = \int_m^s \int_m^\sigma \frac{dt}{g(t)} d\sigma,$$

and $\delta_0 = \delta$ if $\delta > 0$, $\delta_0 = \kappa/\gamma$ if $\delta < 0$. A computation shows that for $\delta > -\kappa/\gamma$, there exists $c_\delta > 0$ such that

$$\frac{dH}{dt} + c_\delta \int_\Omega \left(\frac{|\nabla u|^2}{g(u)} + \frac{|\nabla v|^2}{\delta_0^2} \right) dx \le c,$$

where $c_\delta > 0$ also depends on Ω, f, and g. The gradient estimate is needed to prove the compactness of the fixed-point operator needed to apply the Leray-Schauder fixed-point theorem [25]. The exponential decay of the solutions (in terms of $H(u, v)$) to a constant steady state holds for sufficiently large values of $\delta > 0$. We wish to understand what happens if δ becomes small positive or negative large. This is done via a bifurcation analysis.

2.2.2 *Bifurcation Analysis*

Choosing δ as a bifurcation parameter, we can apply bifurcation theory to show that the stationary solutions bifurcate from the constant steady state (u^*, v^*) for $\delta \neq \delta_d := -\kappa/g(u^*)$. For this result, we employed in [25] the local bifurcation theory of Crandell and Rabinowitz and the global bifurcation theory for nonlinear Fredholm mappings from Shi and Wang [32]. The difficulty here is that (u^*, v^*) is not an isolated bifurcation branch as a function of δ, since fixing any initial mass, there is a family of homogeneous steady states with $u^* = \int_\Omega u(x)dx/\text{meas}(\Omega)$. For the numerical bifurcation analysis, this degeneracy is resolved by introducing a small relaxation term $\rho(u - u^*)$ in the first equation of (2.1) with very small $\rho > 0$ and by applying a homotopy continuation step to achieve solutions for $\rho = 0$.

Numerically, there exist local bifurcation points on the branch of homogeneous steady states if $\delta < \delta_d$ for sufficiently large α and if $\delta > \delta_d$ for sufficiently small α. The results have been obtained by using the software AUTO; for details we refer to [25]. Here, we only depict one stationary density u in Fig. 2.1, showing that there is indeed a region in which the number of individuals with a certain information state is very small, which indicates some herding phenomenum.

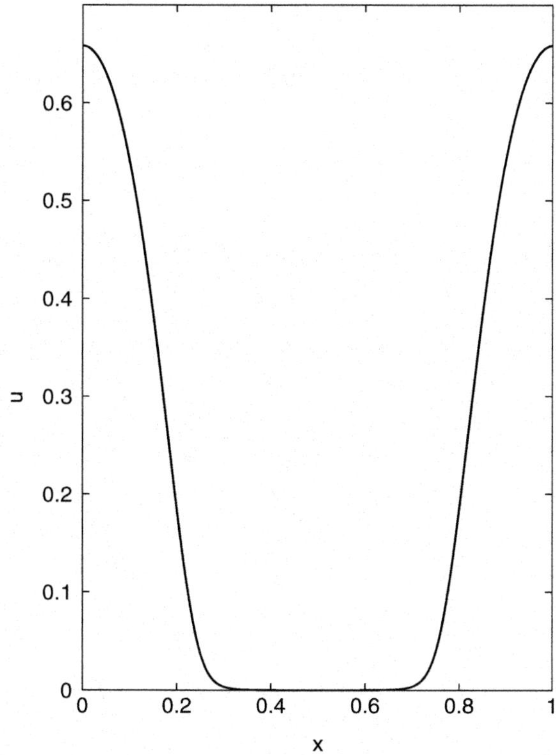

Fig. 2.1 Stationary density of individuals for the model (2.1) with parameters $\alpha = 0.001$, $\kappa = 1$, $\delta = 9$, and $\Omega = (0, 50)$

2.3 A Kinetic Model with Irrationality and Herding

A second approach to model herding consists in using kinetic equations. We describe the evolution of the distribution function $f(x, w, t)$ of the agents depending on the rationality $x \in \mathbb{R}$ and the estimated asset value $w \in \mathbb{R}_+ := [0, \infty)$, assigned to the asset by an individual. The agent behaves rational when $x > 0$ and irrational when $x < 0$. The time evolution is given by the inhomogeneous kinetic equation

$$\partial_t f + (\Phi(x, w)f)_x = Q_I(f) + Q_H(f, f), \quad (x, w) \in \mathbb{R} \times \mathbb{R}_+, \ t > 0, \tag{2.4}$$

with the boundary and initial conditions

$$f(x, 0, t) = 0, \ f(x, w, 0) = f_0(x, w) \quad \text{for } (x, w) \in \mathbb{R} \times \mathbb{R}_+, \ t > 0. \tag{2.5}$$

The second term in (2.4) models the irrationality of the agents. When the asset price w lies within a certain range $|w - W| < R$ around a "fair" prize $W > 0$ which is determined by fundamentals, the agents are supposed to behave more irrational because of psychological biases like overconfidence or limited attention [22]. This is modeled by a negative drift field $\Phi(x, w)$. When the asset value is outside of the "fair" prize region, it is believed to be driven by speculation. The agents will recognize this fact at a certain point and are becoming more rational. Thus, the drift field is positive. An example for such a function is

$$\Phi(x, w) = \begin{cases} -\delta\kappa & \text{if } |w - W| < R, \\ \kappa & \text{if } |w - W| \geq R, \end{cases} \tag{2.6}$$

where δ and κ are some positive numbers.

The first term on the right-hand side of (2.4) describes an interaction that is solely based on economic fundamentals, and the second term describes binary interactions of the agents modeling the exchange of information and possibly leading to herding. The precise modeling is as follows.

2.3.1 Public Information and Herding

Let w be the estimated asset value of an agent before the interaction and w^* the asset value after exchanging information with a public source. Similarly as in [11], the interaction is given by

$$w^* = w - \alpha P(|w - W|)(w - W) + \eta d(w), \tag{2.7}$$

where $P \in [0, 1]$ measures the compromise propensity, $\alpha > 0$ measures the strength of this effect, η is a random variable with distribution μ with variance σ_I^2 and zero mean taking values in $\mathbb{R}$, and $d(w) \in [0, 1]$ models the modification of the asset prize due to diffusion. For instance, we may choose $P(|w - W|)$ as the characteristic function $\mathbf{1}_{\{|w-W|<r\}}$ on $\{|w - W| < r\}$ for some $r > 0$. The above interaction rule means that if a market agent trusts an information source, she/he will update her/his estimated value to bring it closer to the one suggested by the information source. A rational investor is supposed to follow such a strategy.

The second interaction rule models herding effects by taking into account the interaction between an agent and other investors. We choose, similarly as in [33],

$$\begin{aligned} w^* &= w - \beta\gamma(v, w)(w - v) + \eta_1 d(w), \\ v^* &= v - \beta\gamma(v, w)(v - w) + \eta_2 d(v). \end{aligned} \tag{2.8}$$

Here, (w, v) and (w^*, v^*) denote the asset values of two arbitrary agents before and after the interaction, respectively. The constant $\beta \in [0, 1/2]$ measures the attitude of the market participants to change their mind because of herding mechanisms, η_1, η_2 are random variables with the same distribution with variance σ_H^2 and zero mean, and the function d is as above. The function $\gamma \in [0, 1]$ describes a socio-economic scenario where individuals are highly confident in the asset. In [13], the example $\gamma(v, w) = \mathbf{1}_{\{w<v\}} v f(w)$ is suggested, where f is nonincreasing, $f(0) = 1$,

and $\lim_{w\to\infty} f(w) = 0$. The meaning of this choice is as follows: If an agent has an asset value w smaller than v, the function γ will push the agent to assume a higher value w^* than that one before the interaction. This means that the agent trusts other agents that assign a higher value. If w is larger than v, the agent hesitates to lower her/his asset value and nothing changes. For a discussion of the nonnegativity of w^* and v^*, we refer to [15, Sect. 2].

With the above interaction rules, we can define the interaction operators Q_I and Q_H in the weak form. Let $\phi(w) := \phi(x, w)$ be a regular test function and set $\Omega = \mathbb{R} \times \mathbb{R}_+$, $z = (x, w)$. Then

$$\int_\Omega Q_I(f)\phi(w)dz = \frac{1}{\tau_H}\left\langle \int_{\mathbb{R}_+}\int_\Omega \big(\phi(w^*) - \phi(w)\big)M(W)f(x, w, t)dzdW \right\rangle,$$

$$\int_\Omega Q_H(f,f)\phi(w)dz = \frac{1}{\tau_I}\left\langle \int_{\mathbb{R}_+}\int_\Omega \big(\phi(w^*) - \phi(w)\big)f(x, w, t)f(x, v, t)dzdW \right\rangle,$$

where $\langle\cdot\rangle$ is the expectation value with respect to the random variable η and $M(W) \geq 0$ is a fixed background satisfying $\int_{\mathbb{R}_+} M(W)dW = 1$.

We have obtained in [15] the following results.

2.3.2 *Grazing Collision Limit*

The analysis of the Boltzmann equation (2.4) is quite involved, and we expect that its main features are contained in the limiting equation derived in the diffusion limit $(\alpha, \beta, \sigma_H^2, \sigma_I^2) \to 0$. More precisely, we scale the variables according to $t \mapsto \alpha t$ and $x \mapsto \alpha x$. Performing a Taylor expansion in the collision integrals and passing to the limit $(\alpha, \beta, \sigma_H^2, \sigma_I^2) \to 0$ such that $\lambda_I = \sigma_I^2/\alpha$ and $\lambda_H = \sigma_H^2/\alpha$ are fixed, the limiting equation for the function $g(x, w, t)$ reads as

$$\partial_t g + (\Phi(x, w)g)_x = (K[g]g + H(w))_w + (D(w)g)_{ww}, \tag{2.9}$$

where $(x, w) \in \mathbb{R} \times \mathbb{R}_+$, $t > 0$, $D(w) = \frac{1}{2}(\lambda_I/\tau_I + \lambda_H\rho/\tau_H)d(w)^2$, $\rho = \int_\Omega f dz$,

$$\begin{aligned} K[g] &= \int_0^\infty \Gamma(v, w)g(v)dv, \quad \Gamma(v, w) = \frac{k}{\tau_H}\gamma(v, w)(v - w), \\ H(w) &= \frac{1}{\tau_I}\int_{\mathbb{R}_+} P(|w - W|)(w - W)M(W)dW. \end{aligned} \tag{2.10}$$

The equation is supplemented by the boundary and initial conditions

$$g(x, 0, t) = 0, \;\; g(x, w, 0) = g_0(x, w) \quad \text{for } (x, w) \in \mathbb{R} \times \mathbb{R}_+,\; t > 0. \tag{2.11}$$

2.3.3 Existence of Weak Solutions

Equation (2.9) is nonlinear, nonlocal, degenerate in w, and anisotropic in x (incomplete diffusion) and hence, its analysis is challenging. Partial diffusion may lead to singularity formation [21], and often solutions have very low regularity [16]. As the transport in x is linear in (2.9), our situation is better but still delicate. In particular, we need the hypothesis that $D(w)$ is strictly positive to get rid of the degeneracy in w. Assuming additionally that the functions in (2.10) are smooth, $\Gamma \geq 0$ and $\partial\Gamma/\partial w \leq 0$, and the initial datum g_0 is nonnegative and bounded, there exists a weak solution g to (2.9)–(2.11) such that $g \in L^2(0,T;H^1(\Omega))$, $\partial_t g \in L^2(0,T;H^1(\Omega)')$ and $0 \leq g(x,w,t) \leq \|g_0\|_{L^\infty}e^{\lambda t}$ for $(x,w) \in \Omega$, $t > 0$, for some $\lambda > 0$ and for all $T > 0$.

The idea of the proof is to regularize Eq. (2.9) by adding a second-order derivative with respect to x, to truncate the nonlinearity, and to solve the equation in the finite interval $w \in (0,R)$. Then we pass to the deregularization limit. The key step of the proof is the derivation of H^1 estimates uniform in the approximation parameters, which allow for the compactness argument. These estimates are derived by analyzing the differential equation satisfied by g_x and by making crucial use of the boundary conditions. For details, we refer to [15].

2.3.4 Numerical Simulations

We illustrate the behavior of the solution to the kinetic model (2.4) numerically by using an operator splitting ansatz, i.e., we split (2.4) into a drift part and the collisional parts $\partial_t f = Q_I(f)/\tau_I$ and $\partial_t f = Q_H(f,f)/\tau_H$. The collisional parts are solved by using the interaction rules (2.7), (2.8), respectively, and a slightly modified Bird scheme [5]. The transport part $\partial_t f = (\Phi(x,w)f)_x$ is numerically solved by a flux-limited Lax-Wendroff/upwind scheme. The parameters and functions are chosen as follows: $\tau_H = \tau_I = 1$ and

$$P(|w-W|) = 1, \quad d(w) = 4w(1-w), \quad \gamma(v,w) = \mathbf{1}_{\{w<v\}}v(1-w),$$

Φ is given by (2.6), and we choose the time-dependent background

$$W(t) = (\sin(t/200) + 0.5 * \exp(t/500))/30.$$

The time evolution of the first moment

$$m(f(t)) = \int_\Omega f(x,w,t)d(x,w)$$

is shown in Fig. 2.2. The mean asset value stays within the range $[W(t)-R, W(t)+R]$ if $W(t)$ is increasing but it has the tendency to become larger than $W(t)+R$ if $W(t)$

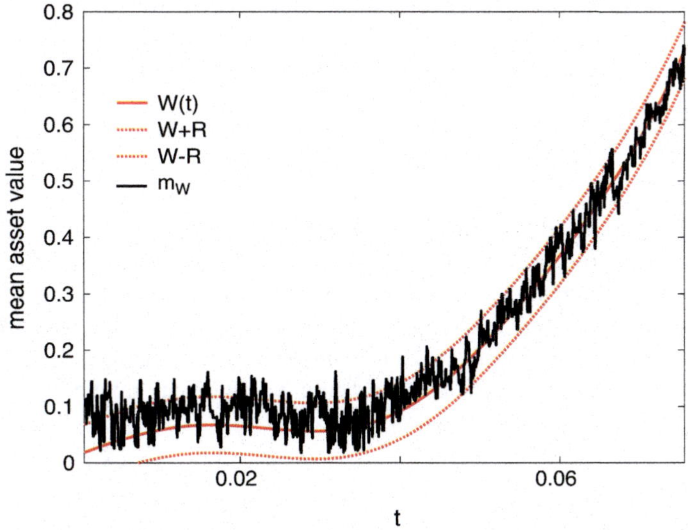

Fig. 2.2 Mean asset value $m(f(t))$ versus time for $\kappa = 1$, $\alpha = 0.3$, $\beta = 0.2$, $\delta = 1.5$, $R = 0.05$, and $\eta = \pm 0.061$

is not varying much. Furthermore, if α is "small", $m(f(t))$ usually does not leave the interval $[W(t)-R, W(t)+R]$ (see [15]). Large values of α mean that the compromise propensity is larger and thus, herding may be become more likely.

2.4 A Kinetic Model with Wealth and Knowledge Exchanges

The effect of wealth and knowledge exchange in a closed society may be described by kinetic equations. Let $f(x, w, t)$ be the distribution function depending on the knowledge variable $x \in \mathbb{R}_+$, the wealth $v \in \mathbb{R}_+$, and time $t > 0$. We assume that the evolution of f is given by the homogeneous Boltzmann-type equation

$$\partial_t f = Q_K(f,f) + Q_W(f,f), \quad (x, w) \in \mathbb{R}_+ \times \mathbb{R}_+, \ t > 0, \tag{2.12}$$

where the operators Q_K and Q_W model the interaction of the agents with respect to knowledge and wealth, respectively. The exchange rules, defining these operators, are as follows.

Let (x, v) and (y, w) denote the knowledges and wealths of two agents, respectively. The knowledges x^* and y^* after the interaction are, similarly as in (2.8), given by

$$x^* = x + \kappa(w)(y - x), \quad y^* = y + \kappa(v)(x - y),$$

where κ is a nondecreasing function of the wealth variable, modeling the confidence, i.e., agent (y, w) trusts agent (x, v) more if the latter agent is wealthier than the former one. The wealth values v^* and w^* after the interaction are defined by

$$v^* = (1 - \gamma\Psi(x))v + \gamma\Psi(y)w, \quad w^* = \gamma\Psi(x)v + (1 - \gamma\Psi(y))w,$$

where $\gamma \in (0, 1)$ is fixed and $\Psi : \mathbb{R}_+ \to (0, 1]$ is a nonincreasing continuous function of the knowledge variable. This rule is exactly that one used in [29] without the random risk parameter. The quantity $\gamma\Psi(x)$ can be understood as the saving/risk-taking propensity of agent (x, v). The monotonicity of Ψ means that the higher is the knowledge of an agent, the less risky is the wealth exchange for her/him. We observe that the microscopic total wealth is conserved during the exchange, $v^* + w^* = v + w$. With the above exchange rules, the interaction operators are defined in weak form, for some smooth test function ϕ, as

$$\int_{(\mathbb{R}_+)^2} Q_K(f,f)\phi dz_1 = \nu_K \int_{(\mathbb{R}_+)^4} \big(\phi(x^*, v) - \phi(x, v)\big) f(x, v, t) f(y, w, t) dz_1 dz_2,$$

$$\int_{(\mathbb{R}_+)^2} Q_W(f,f)\phi dz_2 = \nu_W \int_{(\mathbb{R}_+)^4} \big(\phi(x, v^*) - \phi(x, v)\big) f(x, v, t) f(y, w, t) dz_1 dz_2,$$

where ν_K, ν_W are some rate parameters and $dz_1 = dxdv$, $dz_2 = dydw$.

2.4.1 Existence of Solutions

If Ψ is lower bounded by a positive constant and the initial datum $f_0 \in L^1(\mathbb{R}_+^2)$ is nonnegative, there exists a nonnegative solution $f \in L^\infty(0, T; L^1(\mathbb{R}_+^2))$ to (2.12), $f(x, v, 0) = f_0(x, v)$ for $(x, v) \in \mathbb{R}_+^2$. This result is shown similarly as in [6]. The idea is to solve (2.12) iteratively, thus defining a sequence (f_n) which is bounded and satisfies $f_{n+1} \geq f_n$. The monotone convergence theorem then ensures the existence of a limit function which solves (2.12) in a distributional sense in time and in a weak sense in $L^1(\mathbb{R}_+^2)$.

2.4.2 Numerical Simulations

Equation (2.12) is numerically solved by a particle method [5], approximating the distribution function by a sum of Dirac masses,

$$f(x, w, t) \approx \sum_{p=1}^{N} \delta_{(x_p(t), w_p(t))}(x, v),$$

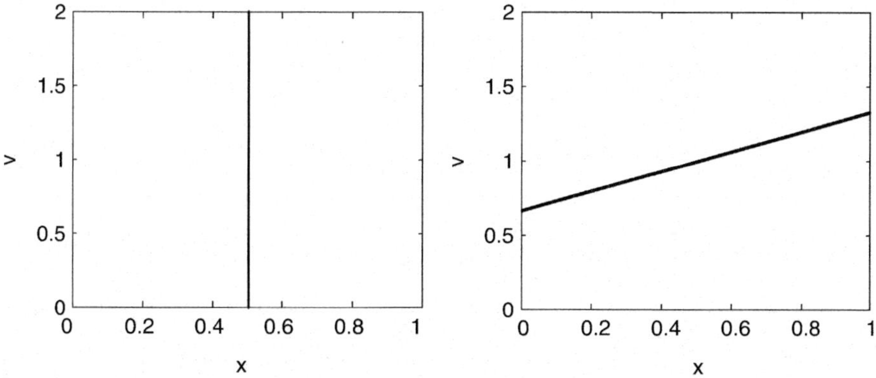

Fig. 2.3 Level sets for the stationary distribution function at $f = 1$ if only knowledge (*left*) or wealth (*right*) is exchanged

where $N \in \mathbb{N}$ is the number of agents and $x_p(t)$, $w_p(t)$ are the knowledge and wealth of the pth agent at time $t > 0$, respectively. The simulations are repeated 30 times for $N = 2000$ agents with a fixed set of parameters, and the results are averaged. The simulations are performed until we have approximately reached the stationary state. The functions are defined by $\kappa(v) = 0.15$ and $\Psi(x) = (1+x)^{-\beta}$, and we have taken the parameters $\beta = 1$, $\gamma = 0.9$, and $\nu_k = \nu_w = 1$. Figure 2.3 shows the level set $f = 1$ of the stationary distribution function if only interactions for the knowledge Q_K (left figure) or for the wealth Q_W (right figure) are present, i.e., we have considered only one type of collisions in each simulation series. The collision rule for the knowledge induces a concentration of the agents at the average knowledge, which equals 0.5, having no effect on the wealth distribution. If only the wealth collision rule is applied, the agents aggregate again on a line, but they do not have the same wealth. The less informed agents are poorer, while the more informed are more wealthy. Choosing other values for β will not give a line but a curve, which allows for more flexibility in the modeling. For instance, for $\beta > 1$, the wealth increases superlinearly with the knowledge, i.e., even a small improvement of the knowledge leads to a significant increase of the wealth. Thus, Fig. 2.3 (right) presents a situation which seems to be not unrealistic, giving rise to the hope that the model may be applicable to more complex socio-economic scenarios.

References

1. Altshuler, E., Ramos, O., Núñez, Y., Fernández, J., Batista-Leyva, A., Noda, C.: Symmetry breaking in escaping ants. Am. Nat. **166**, 643–649 (2005)
2. Amann, H.: Dynamic theory of quasilinear parabolic systems. III. Global existence. Math. Z. **202**, 219–250 (1989)
3. Angle, J.: The surplus theory of social stratification and the size distribution of personal wealth. Soc. Forces **65**, 293–326 (1986)

4. Banerjee, A.: A simple model of herd behavior. Q. J. Econ. **107**, 797–817 (1992)
5. Bird, G.: Molecular Gas Dynamics and the Direct Simulation of Gas Flows. Oxford University Press, Oxford (1995)
6. Boudin, L., Salvarani, F.: A kinetic approach to the study of opinion formation. Math. Model. Numer. Anal. **43**, 507–522 (2009)
7. Boudin, L., Jüngel, A., Trussardi, L.: A kinetic model with wealth and knowledge exchange. Draft, 15 pp. (2016)
8. Brunnermeier, M.: Asset Pricing Under Asymmetric Information: Bubbles, Crashes, Technical Analysis, and Herding. Oxford University Press, Oxford (2001)
9. Chakraborti, A., Muni Toke, I., Patriarca, M., Abergel, F.: Econophysics review: II. Agent-based models. Q. Finance **11**, 1013–1041 (2011)
10. Chatterjee, A., Chabkrabarti, B.: Statistical mechanics of money: how saving propensity affects its distribution. Eur. Phys. J. **17**, 167–170 (2000)
11. Comincioli, V., Della Croce, L., Toscani, G.: A Boltzmann-like equation for choice formation. Kinet. Relat. Models **2**, 135–149 (2009)
12. Cordier, S., Pareschi, L., Toscani, G.: On a kinetic model for a simple market economy. J. Stat. Phys. **120**, 253–277 (2005)
13. Delitala, M., Lorenzi, T.: A mathematical model for value estimation with public information and herding. Kinet. Relat. Models **7**, 29–44 (2014)
14. Düring, B., Matthes, D., Toscani, G.: Kinetic equations modelling wealth redistribution: a comparison of approaches. Phys. Rev. E **78**, 056103, 12 pp. (2008)
15. Düring, B., Jüngel, A., Trussardi, L.: A kinetic equation for economic value estimation with irrationality and herding. Kinet. Relat. Models **10**, 239–261 (2017)
16. Escobedo, M., Vázquez, J.-L., Zuazua, E.: Entropy solutions for diffusion-convection equations with partial diffusivity. Trans. Am. Math. Soc. **343**, 829–842 (1994)
17. Galiano, G., Selgas, V.: On a cross-diffusion segregation problem arising from a model of interacting particles. Nonlinear Anal. Real World Appl. **18**, 34–49 (2014)
18. Gomes, D., Nurbekyan, L., Pimentel, E.: Economic models and mean-field games theory. In: 30th Brazilian Mathematics Colloquium. Instituto Nacional de Matemática Pura e Aplicada (IMPA), Rio de Janeiro, 135 pp. (2015)
19. Hamilton, W.: Geometry for the selfish herd. J. Theor. Biol. **31**, 295–311 (1971)
20. Hillen, T., Painter, K.: Volume filling and quorum sensing in models for chemosensitive movement. Can. Appl. Math. Q. **10**, 501–543 (2002)
21. Hillen, T., Painter, K., Winkler, M.: Anisotropic diffusion in oriented environments can lead to singularity formation. Eur. J. Appl. Math. **24**, 371–413 (2013)
22. Hirshleifer, D.: Investor psychology and asset pricing. J. Finance **56**, 1533–1597 (2001)
23. Hittmeir, S., Jüngel, A.: Cross diffusion preventing blow up in the two-dimensional Keller-Segel model. SIAM J. Math. Anal. **43**, 997–1022 (2011)
24. Ispolatov, S., Krapivsky, P., Redner, S.: Wealth distribution in asset exchange models. Eur. Phys. J. B **2**, 267–276 (1998)
25. Jüngel, A., Kuehn, C., Trussardi, L.: A meeting point of entropy and bifurcation in cross-diffusion herding. Eur. J. Appl. Math. **28**, 317–356 (2017)
26. Mandelbrot, B.: The Pareto-Lévy law and the distribution of income. Int. Econ. Rev. **1**, 79–106 (1960)
27. Pareschi, L., Toscani, G.: Self-similarity and power-like tails in nonconservative kinetic models. J. Stat. Phys. **124**, 747–779 (2006)
28. Pareschi, L., Toscani, G.: Interacting Multiagent Systems. Oxford University Press, Oxford (2013)
29. Pareschi, L., Toscani, G.: Wealth distribution and collective knowledge: a Boltzmann approach. Philos. Trans. Roy. Soc. A **372**, 20130396, 15 pp. (2014)
30. Rook, L.: An economic psychological approach to herd behavior. J. Econ. **40**, 75–95 (2006)
31. Samanidou, E., Zschischang, E., Stauffer, D., Lux, T.: Agent-based models of financial markets. Rep. Prog. Phys. **70**, 409–450 (2007)

32. Shi, J., Wang, X.: On the global bifurcation for quasilinear elliptic systems on bounded domains. J. Differ. Equ. **246**, 2788–2812 (2009)
33. Toscani, G.: Kinetic models of opinion formation. Commun. Math. Sci. **4**, 481–496 (2006)
34. Wrzosek, D.: Global attractor for a chemotaxis model with prevention of overcrowding. Nonlinear Anal. **59**, 1293–1310 (2004)

Chapter 3
Indifference Pricing in a Market with Transaction Costs and Jumps

Nicola Cantarutti, João Guerra, Manuel Guerra, and Maria do Rosário Grossinho

Abstract We present an approach for pricing a European call option in presence of proportional transaction costs, considering the dynamics of the stock price following a general exponential Lévy process. The model is a generalization of the celebrated work of Davis, Panas and Zariphopoulou, where the value of the option is defined as the utility indifference price. This approach requires the solution of two stochastic singular control problems in finite time, satisfying the same Hamilton-Jacobi-Bellman equation with different terminal conditions. Numerical results are obtained by Markov chain approximation methods. Option prices are computed for both writer and buyer, when the returns follow a Brownian motion and a Variance Gamma process.

3.1 Introduction

Option pricing in complete markets uses the concept of replication, whereby a portfolio in stocks and cash replicates the terminal payoff of the option. The unique price of the option is given by the initial wealth necessary to create the *replicating portfolio*. The most famous example of a complete market model is the Black-Scholes (BS) model [2], where all the market frictions are ignored, the risk free interest rate is constant and the asset returns are modelled by a Brownian motion.

However, the real markets are incomplete, and it is not possible to perfectly hedge an option. The presence of any kind of transactions costs (proportional costs, fixed commissions, bid/ask spread) as well as portfolio constraints and price jumps, are the main sources of incompleteness that a trader can face in the markets. In this situation, the notion of pricing by replication falls apart. Continuous time trading is infinitely costly due to the transaction costs. Moreover, there are sources of risk, such as the jumps in the underlying, that are impossible to hedge with the common

N. Cantarutti (✉) • J. Guerra • M. Guerra • M. do Rosário Grossinho
CEMAPRE - Centre for Applied Mathematics and Economics, ISEG - University of Lisbon, Lisbon, Portugal
e-mail: nicolacantarutti@gmail.com; jguerra@iseg.ulisboa.pt; mguerra@iseg.ulisboa.pt; mrg@iseg.ulisboa.pt

M. Ehrhardt et al. (eds.), *Novel Methods in Computational Finance*, Mathematics in Industry 25, DOI 10.1007/978-3-319-61282-9_3

delta-hedging approach. A completely different approach to option pricing is to introduce preferences. The investor's risk profile is described by a *utility function*. The hedging problem is formulated as a finite time portfolio optimization problem, where the investor aims to maximize the expectation of the utility of his portfolio net value at the terminal time, corresponding to the expiration of the option. The *indifference price*, also known as the reservation price, is the price at which an agent would have the same expected utility of his final wealth by selling (buying) the option or by not doing so, and then trading in the optimal way. Thus the indifference price is not unique. This definition produces two prices, for the writer and for the buyer, which is a more realistic property. A general overview of the indifference pricing concept applied to several incomplete models can be found in [4]. Applications for processes with jumps are presented in [6].

Hodges and Neuberger [13] were the first to compute indifference prices in a market with proportional transaction costs. They assumed Gaussian distributed returns. They used an exponential utility function, which has the property that the risk aversion coefficient is constant and does not depend on the total wealth. This choice simplifies the problem reducing by one the number of state variables. The model was further developed by Davis et al. [11]. They formulated the utility maximization problem rigorously as a singular stochastic optimal control problem. They proved also that the value functions can be interpreted as the viscosity solutions of the associated Hamilton-Jacobi-Bellman (HJB) equation, and that the numerical solution, based on the *Markov chain approximation*, converges to the viscosity solution. Other results within the framework of this model are obtained in [5, 9, 10, 17]. Barles and Soner [1] developed an asymptotic analysis of this model for small levels of transaction cost, reducing the complicated HJB equation, which is a variational inequality, to a simpler non-linear PDE.

The model has been generalized in [3] for an underlying asset following a general exponential Lévy process. Considering processes with jumps allows the possibility of bankruptcy for the portfolio, which is an important innovation in the model. The drawback is an additional complexity in the general equation, that cannot be simplified by the choice of the exponential utility function. In order to be able to do numerical computations, the authors considered the simplified case of a huge firm such that they could ignore the possibility of default and thus use the exponential utility for the variable reduction. Numerical results are presented for the case of a Merton jump-diffusion process.

In Sect. 3.2, we briefly present the general theory following [3]. In Sect. 3.3, we explain the algorithm to solve the optimization problem. In Sect. 3.4 we show some numerical results for the well known case of diffusion dynamics, and for a Variance Gamma (VG) process.

3.2 The Model

In this section we present the utility maximization problem formulated as a singular stochastic control problem and derive the associated HJB equation. In this framework, we model the portfolio dynamics considering the proportional transaction costs. Then we explain the indifference pricing method and, introducing the assumption of an always solvent investor, derive a simpler HJB equation. We obtain explicit expressions for the writer and buyer option prices. More details can be found in [3].

3.2.1 Portfolio Dynamics

Consider a portfolio with one risk-free asset B in the cash account, paying an interest rate $r > 0$, and a stock S. We call Y the number of shares of the stock. The investor creates the portfolio at time t_0, when the option is sold (or purchased), and liquidates it at the time of option expiration T. For $t \in [t_0, T]$, the state of the portfolio is indicated as (B_t^π, Y_t^π, S_t), where the superscript π indicates the presence of a control. The portfolio dynamics is:

$$\begin{cases} dB_t^\pi &= rB_t dt - (1+\theta_b)S_t dL_t + (1-\theta_s)S_t dM_t \\ dY_t^\pi &= dL_t - dM_t \\ dS_t &= S_t\left(\mu dt + \sigma dW_t + \int_{\mathbb{R}}(e^z - 1)\tilde{N}(dt, dz)\right). \end{cases} \tag{3.1}$$

The parameters θ_b, $\theta_s \geq 0$ are the proportional transaction costs when buying and selling, respectively. The process $\pi(t) = (L(t), M(t))$ is the trading strategy and represents the cumulative number of shares bought ($L(t)$) and sold ($M(t)$) in $[t_0, T]$. These processes are right-continuous, nondecreasing, and progressively measurable. By convention $L(t_0^-) = M(t_0^-) = 0$ and we allow a possible initial transaction at t_0. The price S_t follows an exponential Lévy process with finite mean and variance:

$$S_t = S_0 e^{X_t}. \tag{3.2}$$

The Lévy process X_t has the characteristic Lévy triplet (b, σ, ν), where $b \in \mathbb{R}$, $\sigma \geq 0$ and ν is the *Lévy measure*. The drift parameter μ and the Lévy triplet are related by

$$\mu = b + \frac{1}{2}\sigma^2 + \int_{\mathbb{R}} \left(e^z - 1 - z\mathbb{1}_{\{|z|<1\}}\right)\nu(dz). \tag{3.3}$$

The last line in (3.1) is the differential form of S_t. It can be obtained applying the Itô lemma on (3.2), considering the *Lévy-Itô decomposition* for X_t.[1] The term $\tilde{N}(dt, dz)$ is the compensated Poisson martingale measure, defined by

$$\tilde{N}(dt, dz) = N(dt, dz) - dt\, \nu(dz), \tag{3.4}$$

where $N(dt, dz)$ is the Poisson random measure with intensity $dt\, \nu(dz)$.

3.2.2 Utility Maximization

Before introducing the cost functional that we want to maximize, we have to define some other important concepts. Define the *cash value* as the value in cash when long positions are sold and short positions are covered:

$$c(y, s) = \begin{cases} (1 + \theta_b) ys, & \text{if } y \leq 0 \\ (1 - \theta_s) ys, & \text{if } y > 0. \end{cases} \tag{3.5}$$

We consider three different portfolios:

- *No option*: A portfolio with just cash and shares, with initial value (B_0, Y_0, S_0).
- *Writer*: A portfolio with cash, shares and short in a European vanilla option with strike K and expiration T, with initial value $(B_0 + p^w, Y_0, S_0)$. The value p^w is the price at which the option is sold.
- *Buyer*: A portfolio with cash, shares and long in a European vanilla option with strike K and expiration T, with initial value $(B_0 - p^b, Y_0, S_0)$. The value p^b is the price at which the option is purchased.

For $t \in [t_0, T]$, we define the *total wealth* process (in cash) for the "no option" portfolio:

$$W_t^0 = B_t + c(Y_t, S_t). \tag{3.6}$$

The wealth process for the writer and buyer portfolios considers the additional option payoff:

$$W_t^w = B_t + c(Y_t, S_t) \mathbb{1}_{\{t<T, c(1,S_T) \leq K\}} + \Big(c\big(Y_t - 1, S_t\big) + K \Big) \mathbb{1}_{\{t=T, c(1,S_T)>K\}}, \tag{3.7}$$

$$W_t^b = B_t + c(Y_t, S_t) \mathbb{1}_{\{t<T, c(1,S_T) \leq K\}} + \Big(c\big(Y_t + 1, S_t\big) - K \Big) \mathbb{1}_{\{t=T, c(1,S_T)>K\}}. \tag{3.8}$$

[1] For a review of these concepts, see [6].

In case of proportional transaction costs, the option is not exercised when $S_T > K$, but when $c(1, S_T) = S_T(1-\theta_s) > K$. In this model we require the wealth to be always greater than a fixed constant $-C$, with $C \geq 0$ for all $t \in [t_0, T]$, as a condition for the solvency of the investor. The *solvency region* is defined for the three cases:

$$\mathscr{S}^j = \Big\{(B_t, Y_t, S_t) \in \mathbb{R} \times \mathbb{R} \times \mathbb{R}^+ : W_t^j \geq -C\Big\}, \tag{3.9}$$

with $j = 0, w, b$. Define the first exit time from the solvency region as

$$\tau^j = \inf\{t \in [t_0, T] : W_t^j \notin \mathscr{S}^j\}. \tag{3.10}$$

Define the set of *admissible trading strategies* $\Pi^j(B_t, Y_t, S_t)$, as the set of all right-continuous, nondecreasing, measurable processes $L(t)$ and $M(t)$, such that the triplet $(B^\pi(t), Y^\pi(t), S(t))$ is a solution of (3.1) for $t \in [t_0, \tau^j \wedge T]$, and with initial values $(\tilde{B}_0^j, Y_0, S_0)$. The auxiliary variable $\tilde{B}_0^j$ can assume the values B_0, $B_0 + p^w$ or $B_0 - p^b$ depending on the portfolio we consider, respectively for $j = 0, w, b$.

This is a dynamic set, so at every time $t \in [t_0, \tau^j \wedge T]$ it depends on the current state. Later on we assume that the investor has such a big size that we can ignore the possibility of insolvency. In this case, the set of portfolios at terminal time $(B^\pi(T), Y^\pi(T), S(T))$ is completely determined by the starting value $(\tilde{B}_0^j, Y_0, S_0)$, and we can substitute the dynamic set $\Pi^j(B_t, Y_t, S_t)$ with the static set $\Pi^j(\tilde{B}_0^j, Y_0, S_0)$. We will use this assumption later in order to simplify the problem.

The investor wishes to maximize the expected utility of the wealth of his portfolio at $\tau^j \wedge T$ over all the admissible strategies. This expectation is conditioned on the initial value of cash, number of shares and value of the stock. The value function of the maximization problem is:

$$V^j(t_0, \tilde{B}_0^j, Y_0, S_0) = \sup_{\pi \in \Pi(B_t, Y_t, S_t)} \mathbb{E}_{\tilde{B}_0^j, Y_0, S_0}\left[\mathscr{U}(W_T^j)\mathbb{1}_{\{\tau^j > T\}} + \mathscr{U}(-C)e^{r(T-\tau^j)}\mathbb{1}_{\{\tau^j \leq T\}}\right] \tag{3.11}$$

for $j = 0, w, b$. The function $\mathscr{U} : \mathbb{R} \to \mathbb{R}$ is a concave and increasing utility function, such that $\mathscr{U}(0) = 0$. The associated HJB equation is a variational inequality:

$$\begin{aligned} \max \Bigg\{ & \frac{\partial V^j}{\partial t} + rb\frac{\partial V^j}{\partial b} + \mu s\frac{\partial V^j}{\partial s} + \frac{1}{2}\sigma^2 s^2 \frac{\partial^2 V^j}{\partial s^2} \\ & + \int_{\mathbb{R}} \left[V^j(t, b, y, se^z) - V^j(t, b, y, s) - s(e^z - 1)\frac{\partial V^j}{\partial s}\right]\nu(dz)\,, \\ & \frac{\partial V^j}{\partial y} - (1+\theta_b)s\frac{\partial V^j}{\partial b}\,, -\left(\frac{\partial V^j}{\partial y} - (1-\theta_s)s\frac{\partial V^j}{\partial b}\right)\Bigg\} = 0, \end{aligned} \tag{3.12}$$

for $(t, b, y, s) \in [t_0, T] \times \mathscr{S}^j$ and $j = 0, w, b$.

A rigorous derivation of the previous equation can be found in [12]. The terminal and lateral conditions are given by (3.11). This HJB equation is a partial integro-differential equation (PIDE). The presence of the integral operator implies that the lateral conditions have to be defined not only on the boundary of the solvency region, but also beyond. This condition reads:

$$V^j(t,b,y,s) = e^{r(T-t)}\mathscr{U}(-C) \quad \text{for} \quad t \in [t_0,T], \; (b,y,s) \notin \mathscr{S}^j, \; j = 0,w,b. \tag{3.13}$$

3.2.3 Indifference Price

With this model we can compute two option prices: the price for the writer p^w and the price for the buyer p^b. The indifference price is defined as the initial amount of money required to have the same final expected utility of wealth trading in the optimal way in the portfolio with the option, as if trading in the optimal way in the portfolio without the option. The two prices can be obtained implicitly by the conditions

$$V^0(t_0,B_0,Y_0,S_0) = V^w(t_0,B_0+p^w,Y_0,S_0), \tag{3.14}$$

$$V^0(t_0,B_0,Y_0,S_0) = V^b(t_0,B_0-p^b,Y_0,S_0). \tag{3.15}$$

3.2.4 Variable Reduction

We now assume the simple case of a big investor that cannot default. The dynamic set $\Pi^j(B_t,Y_t,S_t)$ can be replaced by the static set $\Pi^j(\tilde{B}_0^j,Y_0,S_0)$. As long as the portfolio never goes into bankruptcy ($\tau^j > T$), we can ignore the lateral boundary conditions (3.13). Moreover we can use the properties of the exponential utility function to reduce the number of variables of the problem. The exponential utility is defined by

$$\mathscr{U}(w) = 1 - e^{-\gamma w}. \tag{3.16}$$

The exponential utility has the property that the coefficient of risk aversion

$$\gamma = -\mathscr{U}''(x)/\mathscr{U}'(x)$$

is constant, and does not depend on the wealth w. This means that the amount invested in the risky asset at time T, is independent of the total wealth at time T. As long as the amount in the risky asset is independent of the total wealth, the amount in the cash account is irrelevant to the trading strategy. We can thus remove B_t from the state dynamics. The integral representation of the evolution of B_t in (3.1) is:

$$B^{\pi}(T) = \frac{B_0}{\delta(t_0, T)} - \int_{t_0}^{T} (1+\theta_b)\frac{S(t)}{\delta(t,T)} dL(t) + \int_{t_0}^{T} (1-\theta_s)\frac{S(t)}{\delta(t,T)} dM(t), \quad (3.17)$$

where $\delta(t,T) = e^{-r(T-t)}$. Putting these expressions into (3.11), we get

$$V^j(t_0, \tilde{B}_0^j, Y_0, S_0) = 1 - e^{-\gamma \frac{\tilde{B}_0^j}{\delta(t_0,T)}} Q^j(t_0, Y_0, S_0), \quad (3.18)$$

with

$$Q^j(t_0, Y_0, S_0) = \inf_{\pi \in \Pi(\tilde{B}_0^j, Y_0, S_0)} \mathbb{E}_{Y_0,S_0}\left[e^{-\gamma\left[-\int_{t_0}^T (1+\theta_b)\frac{S_t}{\delta(t,T)} dL_t + \int_{t_0}^T (1-\theta_s)\frac{S_t}{\delta(t,T)} dM_t\right]} \right. \quad (3.19)$$

$$\left. \times H^j(Y^{\pi}(T), S(T)) \right].$$

The first factor can be interpreted as a discount factor, while the function $H^j(y,s) = Q^j(T,y,s)$ is the terminal cost.

- No option:

$$H^0(y,s) = e^{-\gamma\, c(y,s)}. \quad (3.20)$$

- Writer:

$$H^w(y,s) = e^{-\gamma\left[c(y,s)\mathbb{1}_{\{c(1,s)\le K\}} + \left(c(y-1,s)+K\right)\mathbb{1}_{\{c(1,s)>K\}}\right]}. \quad (3.21)$$

- Buyer:

$$H^b(y,s) = e^{-\gamma\left[c(y,s)\mathbb{1}_{\{c(1,s)\le K\}} + \left(c(y+1,s)-K\right)\mathbb{1}_{\{c(1,s)>K\}}\right]}. \quad (3.22)$$

Using conditions (3.18) together with (3.14), (3.15), we obtain the price of the option as:

$$p^w(t_0,y,s) = \frac{\delta(t_0,T)}{\gamma} \log\left(\frac{Q^w(t_0,y,s)}{Q^0(t_0,y,s)}\right), \quad (3.23)$$

$$p^b(t_0,y,s) = \frac{\delta(t_0,T)}{\gamma} \log\left(\frac{Q^0(t_0,y,s)}{Q^b(t_0,y,s)}\right). \quad (3.24)$$

Passing to a log-variable $x = \log(s)$, the derivative operators change to

$$s\frac{\partial}{\partial s} = \frac{\partial}{\partial x}, \qquad s^2\frac{\partial^2}{\partial s^2} = \frac{\partial^2}{\partial x^2} - \frac{\partial}{\partial x}. \tag{3.25}$$

The simplified HJB equation is

$$\begin{aligned} \min \Bigg\{ & \frac{\partial Q^j}{\partial t} + (\mu - \frac{1}{2}\sigma^2)\frac{\partial Q^j}{\partial x} + \frac{1}{2}\sigma^2\frac{\partial^2 Q^j}{\partial x^2} \\ & + \int_{\mathbb{R}} \left[Q^j(t,y,x+z) - Q^j(t,y,x) - (e^z - 1)\frac{\partial Q^j}{\partial x} \right] \nu(dz) \,, \\ & \frac{\partial Q^j}{\partial y} + (1+\theta_b)e^x \frac{\gamma}{\delta(t,T)} Q^j \,, \; -\left(\frac{\partial Q^j}{\partial y} + (1-\theta_s)e^x \frac{\gamma}{\delta(t,T)} Q^j \right) \Bigg\} = 0, \end{aligned} \tag{3.26}$$

with $j = 0, w, b$. It is convenient to consider the integral representation

$$\begin{aligned} Q^j(t,y,x) = \min \Bigg\{ & \mathbb{E}_{y,x}\left[Q^j\big(t + \Delta t, y, x + \Delta X\big) \right], \\ & \exp\left(\frac{\gamma}{\delta(t,T)}(1+\theta_b)e^x \Delta L_t^* \right) Q^j\big(t, y + \Delta L_t^*, x\big), \\ & \exp\left(- \frac{\gamma}{\delta(t,T)}(1-\theta_s)e^x \Delta M_t^* \right) Q^j\big(t, y - \Delta M_t^*, x\big) \Bigg\}, \end{aligned} \tag{3.27}$$

where each term inside the "min" is the solution of the corresponding term in the differential equation taken equal to zero. The values ΔL_t^* and ΔM_t^* are the optimal number of shares bought or sold at time t.

3.3 The Algorithm

In this section we describe the method to solve the minimization problems (3.19). The solution can be found by discretizing the dynamic programming equation (3.27). To this purpose, we use the Markov chain approximation method for singular control problems developed by Kushner and Dupuis in [14], where the portfolio dynamics is approximated by a discrete state controlled Markov chain in discrete time. The method consists in creating a backward recursive dynamic programming algorithm, in order to compute the value function at time t, given its value at time $t + \Delta t$. With the variable reduction introduced in the previous section (B_t is removed from the state variables) and the change to the log-variable (3.25), the portfolio stochastic differential equation (SDE) (3.1) has the simpler form

$$\begin{cases} dY_t^{\pi} &= dL_t - dM_t, \\ dX_t &= \left(\mu - \frac{1}{2}\sigma^2 - \int_{\mathbb{R}} \left(e^z - 1 - z\right)\nu(dz)\right) dt + \sigma dW_t + \int_{\mathbb{R}} z\tilde{N}(dt, dz). \end{cases} \tag{3.28}$$

We have to discretize the time and space to create a Markov chain approximation for this portfolio dynamics. The theory developed by Kushner and Dupuis considers only Lévy processes with finite activity of jumps, which means that $\int_{\mathbb{R}} \nu(z)dz < \infty$. For Lévy processes with infinite activity, it is possible to approximate the small jumps with a Brownian motion, as explained in [6]. This helps to remove the singularity of the Lévy measure near the origin, and therefore allows us to use the Markov chain framework of Kushner and Dupuis.

For $n = 0, 1, \ldots N \in \mathbb{N}$, define the discrete time step $\Delta t = \frac{T-t_0}{N}$ such that $t_n = t_0 + n\Delta t$. Define the set $\Sigma_x = \{-K_1 h_x, \cdots - h_x, 0, h_x, \cdots + K_2 h_x\}$, where we consider the discrete log-price step $h_x > 0$ and $K_1, K_2 \in \mathbb{N}$. The values of K_1 and K_2 can be different to capture the possible asymmetry of the jump sizes. Define also the set $\Sigma_y = \{-K_3 h_y, \cdots - h_y, 0, h_y, \cdots + K_4 h\}$, where $h_y > 0$ is a discrete step and $K_3, K_4 \in \mathbb{N}$. The number of shares Y_n takes values in Σ_y. The discretized version of the SDE (3.28) is

$$\begin{cases} \Delta Y_n = \Delta L_n - \Delta M_n \\ \Delta X_n = \hat{\mu}\Delta t + \hat{\sigma}\Delta W_n + \Delta \tilde{J}_n \end{cases} \tag{3.29}$$

where we consider the increment $\Delta X_n = X(t_n + \Delta t) - X(t_n)$, and the new drift $\hat{\mu}$ and volatility $\hat{\sigma}$ parameters. The term $\Delta W_n = W(t_n + \Delta t) - W(t_n) \in \Sigma_x$ assumes only the three possible values $\{-h_x, 0, h_x\}$, and $\Delta \tilde{J}_n$ is the compensated Poisson jump term with finite activity λ, that assumes all the values in Σ_x.

The two increments ΔL_n, ΔM_n which describe the change in the number of shares bought or sold are positive multiples of h_y. The action of the control is supposed to happen instantaneously: $\Delta L_n = L(t_n) - L(t_n^-)$ and $\Delta M_n = M(t_n) - M(t_n^-)$ happen at the same time t_n. We indicate by $L(t_n^-)$ and $M(t_n^-)$ the number of shares just before the transaction.

The Markov chain approximation has to satisfy two conditions:

1. The transition probabilities p^X are represented as:

$$p^X(x, z) = (1 - \lambda \Delta t)\, p^W(x, z) + (\lambda \Delta t)\, p^J(x, z), \tag{3.30}$$

   where $\lambda > 0$ is the jumps activity, p^W and p^J are the transition probabilities of the Brownian and jump components respectively.
2. The transition probabilities have to be *locally consistent* with the SDE (3.28). This means that, at each time step, the first two moments of the Markov chain

have to be equal to the first two moments of the continuous process:

$$\mathbb{E}_n\big[\Delta X_n\big] = \hat{\mu}\,\Delta t, \tag{3.31}$$

$$\mathbb{E}_n\Big[\big[\Delta X_n - \mathbb{E}[\Delta X_n]\big]^2\Big] = \Big(\hat{\sigma}^2 + \int_{\mathbb{R}} z^2 \nu(dz)\Big)\,\Delta t. \tag{3.32}$$

In the construction of the chain X_n, the transition probabilities and the sizes of time and space steps have to be chosen such that the Markov chain satisfies the two properties. A possible technique to obtain those parameters is to discretize the infinitesimal generator of the process by an explicit finite difference method. In the numerical examples of the next section we model the discrete dynamics of the diffusion process with a *binomial tree*. For the discretization of the infinite activity Variance Gamma process, we first have to approximate it with a jump-diffusion process in order to use the Kushner framework, and then we obtain the transition probabilities by discretizing the infinitesimal generator, following the procedure described in [3]. The discrete version of the dynamic programming equation (3.27) is:

$$Q(t_n, Y_n, X_n) = \min \Bigg\{ \mathbb{E}_n\Big[Q\big(t_{n+1}, Y_n, X_n + \Delta X_n\big)\Big], \tag{3.33}$$

$$\min_{\Delta L_n} \exp\Big(\frac{\gamma}{\delta(t_n,T)}(1+\theta_b)e^{X_n}\Delta L_n\Big)\mathbb{E}_n\Big[Q\big(t_{n+1}, Y_n + \Delta L_n, X_n + \Delta X_n\big)\Big],$$

$$\min_{\Delta M_n} \exp\Big(-\frac{\gamma}{\delta(t_n,T)}(1-\theta_s)e^{X_n}\Delta M_n\Big)\mathbb{E}_n\Big[Q\big(t_{n+1}, Y_n - \Delta M_n, X_n + \Delta X_n\big)\Big]\Bigg\}.$$

Algorithm 1

Input: r, (b, θ, ν), S_0, K, T, θ_b, θ_s, γ, N, $\bar{L}$, $\bar{M}$,
Output: $Q(t_0, y, s)$ for $j = 0, w, b$

1: Compute the transition probabilities and the steps Δt and $h_x = \text{Sd}[\Delta X_t]$.
2: Create the log-price tree for the Markov chain (3.29).
3: Create the $\big(N(\bar{L}-1)+1\big) \times \bar{M}$ grid with terminal conditions (3.20), (3.21) or (3.22).
4: **for** n = N-1 to 0 **do**
5: $\quad W_{j,i+l} = \sum_{k=-K_1}^{K_2} p_k\, Q^{n+1}_{j,i+k}$
6: $\quad Q^n_{j,i} = \min\{W_{j,i-m},\ \min_l F(x_i, l, n) W_{j,i-m},\ \min_m G(x_i, m, n) W_{j,i-m}\}$
$\quad$ At time n the grid has size $\big(n(\bar{L}-1)+1\big) \times \bar{M}$.
7: **end for**
8: Once obtained $Q^0(t_0, y, s)$, $Q^w(t_0, y, s)$, $Q^b(t_0, y, s)$ use (3.23) and (3.24) for p^w and p^b.

We can solve it with a backward induction algorithm, where we used the short notation: $Q(t_n, y_j, x_i) = Q^n_{j,i}$, the parameters $\bar{L} = K_1 + K_2 + 1$ and $\bar{M} = K_3 + K_4 + 1$ and defined the functions $F(x, l, n) = e^{\gamma\, l\,(1+\theta_b)\, e^x/\delta(n,N)}$ and $G(x, m, n) = e^{-\gamma\, m\,(1-\theta_s)\, e^x/\delta(n,N)}$.

3.4 Numerical Results

In this section we implement the algorithm for two different Lévy processes: the Brownian motion and the Variance Gamma.

3.4.1 *Brownian Motion*

The diffusion case has been extensively studied theoretically and numerically in [9, 11, 13, 17] and related works. The Brownian motion with drift is a Lévy process with triplet $(b, \sigma, 0)$. The Lévy measure is identically zero, so the path of the process has no jumps. The parameter b is related with the drift $\mu = b + \frac{1}{2}\sigma^2$ by formula (3.3). A simple discretization of the Brownian motion can be obtained by the "binomial tree", see [8]. The binomial method with zero transaction costs converge to the Black-Scholes price. For the numerical computations, we use the following set of input parameters:

K	T	r	μ	σ	N	$\bar{M}$	γ
15	1	0.1	0.1	0.25	1000	1000	0.01

Considering the Black-Scholes price as the reference price, we implement Algorithm 1 for different levels of transaction costs and assuming $\sigma_b = \sigma_s$.

In Figs. 3.1 and 3.2, we obtain values respectively for the writer and for the buyer of the option. We can see that when the level of transaction costs is zero, the algorithm reproduces the Black-Scholes prices.[2] The option price is an increasing function of the level of transaction costs for the writer, while for the buyer is a decreasing function. Therefore the spread is bigger when the market has bigger transaction costs. In our computation we chose $\mu = r$. In [3], the authors show that the parameter μ is not relevant for the value of the option.

[2]The BS curve in this case has been computed solving the BS PDE using a finite difference implicit method.

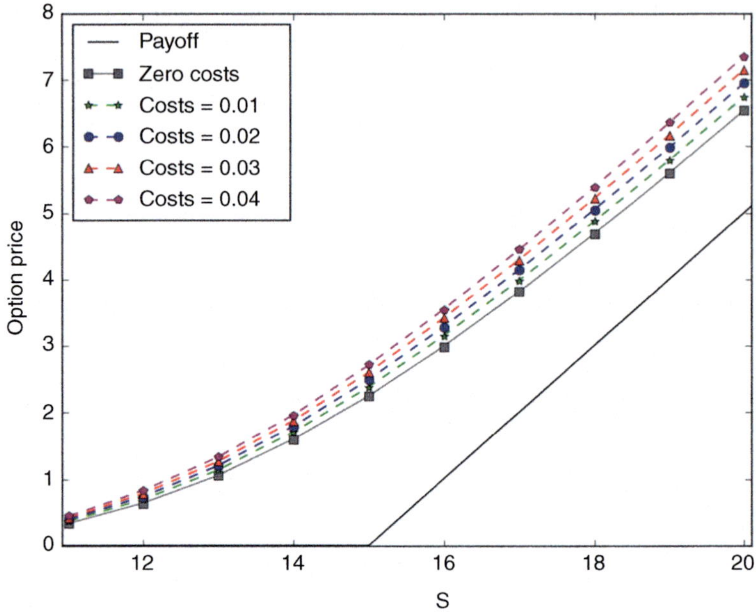

Fig. 3.1 Writer prices for different transaction costs. The *continuous line* is the solution of the Black-Scholes PDE

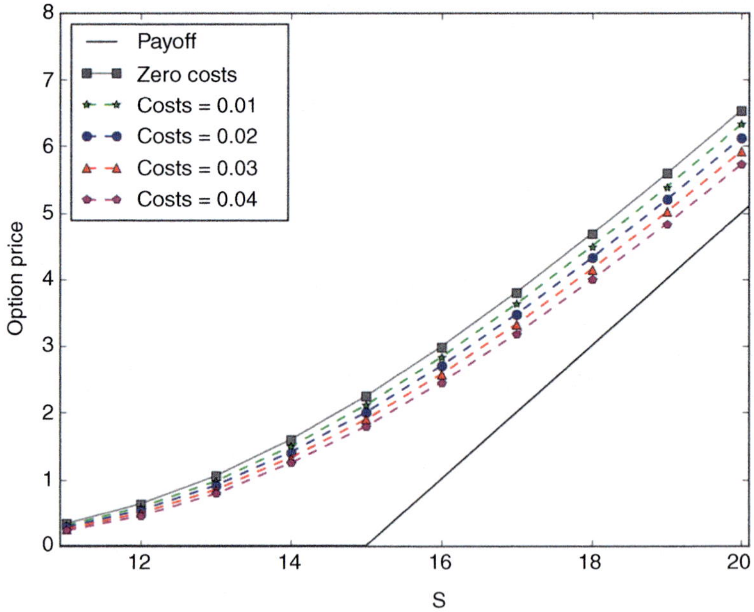

Fig. 3.2 Buyer prices for different transaction costs. The *continuous line* is the solution of the Black-Scholes PDE

3.4.2 Variance Gamma

The Variance Gamma process is a pure jump Lévy process with infinite activity. Applications of the VG process to financial modelling can be found for example in [6, 15, 16]. This process can be obtained by time subordination of Brownian motion or as the difference of two Gamma processes. These two methods are equivalent (see [16]). We choose the first representation. Consider a Brownian motion with drift $X_t = \theta t + \sigma W_t$ and substitute the time variable with the gamma subordinator $T_t \sim \Gamma(t, \kappa)$. We obtain the *Variance Gamma* process:

$$X_{T_t} = \theta T_t + \sigma W_{T_t}. \tag{3.34}$$

It depends on three parameters:

- θ, the drift of the Brownian motion,
- σ, the volatility of the Brownian motion,
- κ, the variance of the Gamma process.

The VG Lévy measure is:

$$\nu(dz) = \frac{e^{\frac{\theta z}{\sigma^2}}}{\kappa|z|} \exp\left(-\frac{\sqrt{\frac{2}{\kappa} + \frac{\theta^2}{\sigma^2}}}{\sigma} |z| \right) dz, \tag{3.35}$$

and completely describe the process. Even if the process has been created by Brownian subordination, it has no diffusion components. The Lévy triplet is $(\int_{|z|<1} z\nu(dz), 0, \nu)$

We now approximate the small jumps with a Brownian motion. Fixing a truncation parameter $\epsilon > 0$, the infinitesimal generator of the process [the first term in Eq. (3.26)] becomes:

$$\mathscr{L}Q = \frac{\partial Q}{\partial t} + \mu \frac{\partial Q}{\partial x} + \int_{|z|<\epsilon} [Q(t,y,x+z) - Q(t,y,x) - (e^z - 1)\frac{\partial Q}{\partial x}]\nu(dz)$$
$$+ \int_{|z|\geq\epsilon} [Q(t,y,x+z) - Q(t,y,x) - (e^z - 1)\frac{\partial Q}{\partial x}]\nu(dz),$$

where $\sigma = 0$ and μ can be computed using (3.3). In the integral term on the domain $|z| < \epsilon$, use the Taylor approximation:

- $Q(t,y,x+z) = Q(t,y,x) + \frac{\partial Q}{\partial x} z + \frac{1}{2}\frac{\partial^2 Q}{\partial x^2} z^2 + \mathscr{O}(z^3)$.
- $e^z - 1 = z + \frac{z^2}{2} + \mathscr{O}(z^3)$.

Considering only the terms up to the second order, the integral for $|z| < \epsilon$ is

$$\int_{|z|<\epsilon} \frac{z^2}{2}\left[\frac{\partial^2 Q}{\partial x^2} - \frac{\partial Q}{\partial x}\right]\nu(dz) = \frac{\sigma_\epsilon^2}{2}\left[\frac{\partial^2 Q}{\partial x^2} - \frac{\partial Q}{\partial x}\right],$$

where we called the second moment of the Lévy measure

$$\sigma_\epsilon^2 = \int_{|z|<\epsilon} z^2 \nu(dz). \tag{3.36}$$

We can define also the parameters:

$$w_\epsilon = \int_{|z|\geq\epsilon} (e^z - 1)\nu(dz)$$

and

$$\lambda_\epsilon = \int_{|z|\geq\epsilon} \nu(dz).$$

We obtain an infinitesimal generator of a jump-diffusion process:

$$\begin{aligned}\mathscr{L}Q &= \frac{\partial Q}{\partial t} + (\mu - \frac{1}{2}\sigma_\epsilon^2 - w_\epsilon)\frac{\partial Q}{\partial x} + \frac{1}{2}\sigma_\epsilon^2 \frac{\partial^2 Q}{\partial x^2} \\ &+ \int_{|z|\geq\epsilon} Q(t,y,x+z)\nu(dz) - \lambda_\epsilon Q(t,y,x).\end{aligned} \tag{3.37}$$

In the numerical computations, we use the following parameters:

K	T	r	γ	θ	σ	κ	N	$\bar{L}$	$\bar{M}$
15	1	0.05	0.02	−0.1	0.2	0.1	138	33	146

The transition probabilities of the Markov chain are obtained by a finite difference discretization of (3.37). For more details, see [3].

Figures 3.3 and 3.4 show the prices with the presence of transaction costs. The continuous line is the solution of the VG PIDE for the price of the option. We solved the PIDE using an implicit/explicit scheme and the Brownian approximation as proposed in [7]. In order to estimate the numbers of time steps N and branches for the multinomial tree $\bar{L}$, we consider the space step $h_X = \sigma_X\sqrt{\Delta t}$, where $\sigma_X = \sqrt{\sigma^2 + \theta^2\kappa}$ and $\Delta t = T/N$. We demand that $h_X \cdot L > 3\sigma_J$, where $\sigma_J^2 = \int_{|z|\geq\epsilon} z^2\nu(dz)$ is the variance of the jump process. Of course $\sigma_X^2 = \sigma_\epsilon^2 + \sigma_J^2$. Putting all this together, we obtain $\bar{L} \geq \frac{3\sigma_J}{\sigma_X}\sqrt{N}$. Using our values for the parameters, this corresponds to $\bar{L} \geq 2.8\sqrt{N}$.

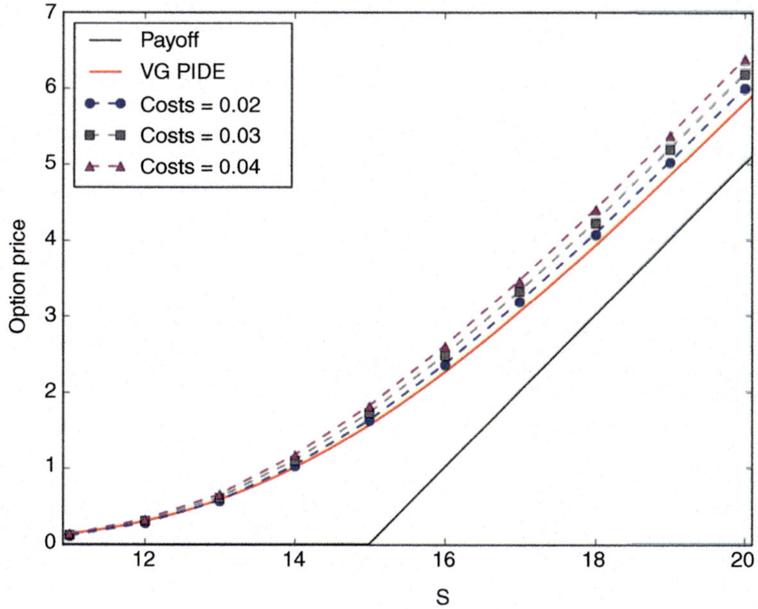

Fig. 3.3 Writer prices with different transaction costs. The *continuous line* is the solution of the VG PIDE

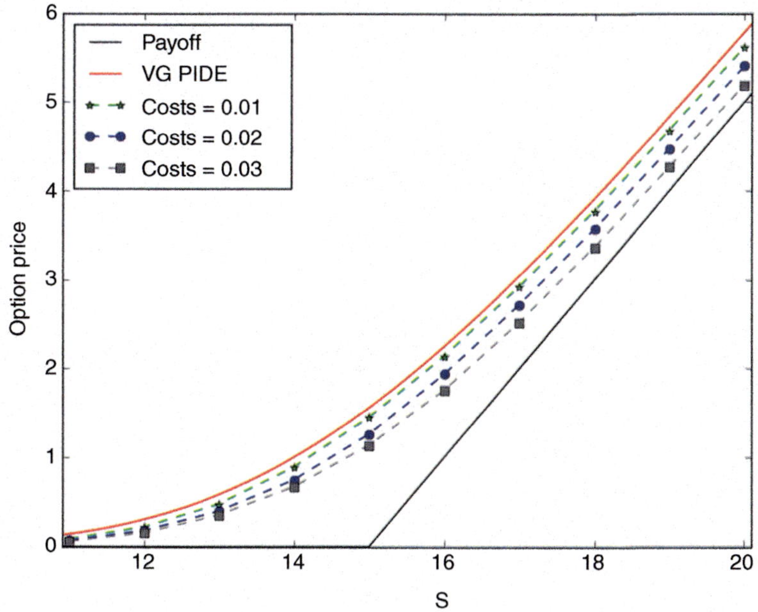

Fig. 3.4 Buyer prices with different transaction costs. The *continuous line* is the solution of the VG PIDE

References

1. Barles, G., Soner, H.M.: Option pricing with transaction costs and a nonlinear Black-Scholes equation. Finance Stochast. **2**(4), 369–397 (1998)
2. Black, F., Scholes, M.: The pricing of options and corporate liabilities. J. Polit. Econ. **81**(3), 637–654 (1973)
3. Cantarutti, N., Guerra, J., Guerra, M., Grossinho, M.R.: Option pricing in exponential Lévy models with transaction costs (2016). arXiv:1611.00389 [q-fin.MF]
4. Carmona, R.: Indifference Pricing. Princeton University Press, Princeton (2009)
5. Clewlow, L., Hodges, S.D.: Optimal delta hedging under transaction costs. J. Econ. Dyn. Control. **21**, 1353–1376 (1997)
6. Cont, R., Tankov, P.: Financial Modelling with Jump Processes. Chapman and Hall/CRC, Boca Raton (2003)
7. Cont, R., Voltchkova, E.: A finite difference scheme for option pricing in jump diffusion and exponential Lévy models. SIAM J. Numer. Anal. **43**(4), 1596–1626 (2005)
8. Cox, J.C., Ross, S.A., Rubinstein, M.: Option pricing: a simplified approach. J. Finance Econ. **7**, 229–263 (1979)
9. Damgaard, A.: Optimal portfolio choice and utility based option pricing in markets with transaction costs. Ph.D. thesis, School of Business and Economics, Odense University (1998)
10. Davis, M.H.A., Panas, V.G.: The writing price of a European contingent claim under proportional transaction costs. Comput. Appl. Math. **13**(2), 0101–8205 (1994)
11. Davis, M.H.A., Panas, V.G., Zariphopoulou, T.: European option pricing with transaction costs. SIAM J. Control Optim. **31**(2), 470–493 (1993)
12. Fleming, W.H., Soner, M.H.: Controlled Markov Processes and Viscosity Solutions, 2nd edn. Springer, Berlin (2005)
13. Hodges, S.D., Neuberger, A.: Optimal replication of contingent claims under transaction costs. Rev. Future Mark. **8**(2), 223–236 (1989)
14. Kushner, H., Dupuis, P.G.: Numerical Methods for Stochastic Control Problems in Continuous Time. Springer, Berlin (2001)
15. Madan, D.B., Seneta, E.: The Variance Gamma (V.G.) model for share market returns. J. Bus. **63**(4), 511–524 (1990)
16. Madan, D.B., Carr, P., Chang, E.C.: The variance Gamma process and option pricing. Eur. Finance Rev. **2**, 79–105 (1998)
17. Monoyios, M.: Efficient option pricing with transaction costs. J. Comput. Finance **7**, 107–128 (2003)

Chapter 4
Negative Rates: New Market Practice

Jörg Kienitz

Abstract Considering the current interest rate environment it has become necessary to extend option pricing models for strikes in $(-\infty, 0]$. We consider the new market environment and show the changes in the modelling landscape. Our scope is on models that can accommodate negative rates. Especially, we focus on a recently proposed model which extends the classical SABR model of Hagan et al. (Wilmott Mag 1:84–108, 2002). This model was introduced in (Antonov et al., Risk, 2015) and is called the Free Boundary SABR model.

Since for practitioners it is necessary to frequently calibrate a model to market data fast approximation methods together with benchmark methods for their performance and testing their accuracy are essential. In this chapter we consider two approximation formulae for the Bachelier volatility, also known as Gaussian or Normal volatility, produced by this model. The latter numbers can be used as input to the Bachelier pricing formula. Together with the current forward value and the time to maturity this leads to prices of European Call and Put options.

We have to stress the fact that the approximation formulae can serve for calibration purposes where fast calculation of prices is essential. However, the inapplicability to certain parameter ranges have to be taken into account. The numerical approach proposed by Antonov et al. (Risk, 2015) does not lead to implied volatilities. The implied volatilities have to be inferred by numerical methods from option prices where this method also suffers from the fact that not all values of the parameters may be covered.

4.1 Introduction

If we consider recent market data we find that certain interest rates have negative values. This was assumed to be impossible in the past at least if one considers the major currencies, e.g. EUR or CHF. Figure 4.1 shows the short end of the OIS

J. Kienitz (✉)
Department of Actuarial Science, University of Cape Town, Cape Town, South Africa

Applied Mathematics and Numerical Analysis, University of Wuppertal, Wuppertal, Germany
e-mail: joerg.kienitz@gmx.de

M. Ehrhardt et al. (eds.), *Novel Methods in Computational Finance*,
Mathematics in Industry 25, DOI 10.1007/978-3-319-61282-9_4

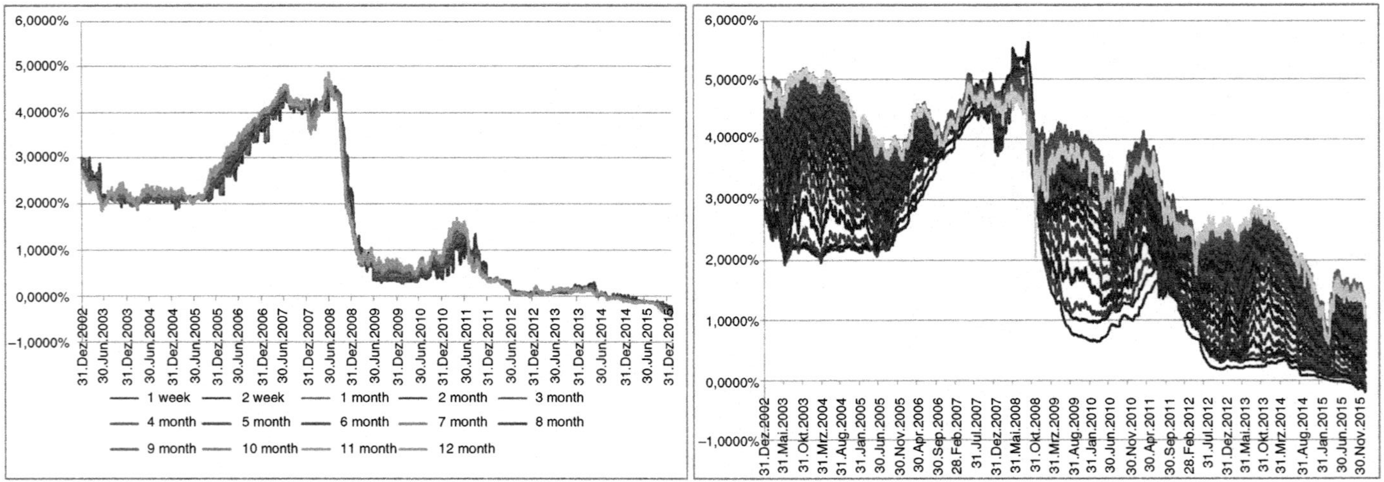

Fig. 4.1 Realisation of negative interest rates. On the *left* the EONIA Swap Rates for short times are plotted over time and on the *right* the Swap Rates against the 6M Euribor are displayed

curve and the 6M curves for EUR. Significant parts of the curve exhibit negative values. For the OIS curve the rates up to 6 years are negative while for the 6M curve the rates are negative up to 4 years. Negative rates do have further consequences. One consequence is that the quoting mechanism using implied logarithmic normal volatility is not possible anymore. This is due to the fact that this model does not allow negative values to be realized by the underlying rate and, thus, for instance Floors having a strike equal or less than 0 would simply be worth 0. The current market quotes in fact assign non-zero values to such options.

While logarithmic normal models cannot be used any longer the market has started to apply the Bachelier model, respectively the Displaced Diffusion/Shifted Lognormal model to quote option prices in terms of volatility for interest rate derivatives. Before the market adopted models other than the standard logarithmic normal model it was observed that the corresponding implied logarithmic normal volatilities increased, even exploded and finally disappeared. This is due to the fact that values for options are bounded in this model, European Call options can only be as large as the forward subtracted by the strike, $F(T) - K$. This means that it does not matter how big we choose the volatility the bound is determined by the current forward asset price and the strike. But actually the market quoted prices that were larger. We come back to this issue in Sect. 4.1.2. We introduce the Bachelier model in Sect. 4.1.1 and the Displaced Diffusion model in Sect. 4.1.2. Furthermore, standard approaches to model the skew/smile such as the SABR model, [5], cannot be used either and extensions to such models have to be explored. We consider a version of the SABR model in Sect. 4.2.

The following sections highlighting the pricing methods are based on [7]. For numerical implementations see [9] and the accompanying software.

4.1.1 Bachelier Model

For the Bachelier model which is also sometimes also called Gaussian or Normal model the dynamic of an asset S is given by

$$dS(t) = \sigma_{\mathrm{N}} dW(t), \quad S(0) = s_0 \tag{4.1}$$

For a European Call, respectively Put option price denoted by V_C, respectively V_P with strike K, maturity T and dynamic given by (4.1) we have

$$C_N(S(0), K, T, \sigma) = (S(0) - K\exp(-rT))\mathcal{N}(d(\sigma_{\mathrm{N}})) + \sigma\sqrt{T}n(d(\sigma_{\mathrm{N}})) \tag{4.2}$$

$$P_N(S(0), K, T, \sigma) = (K - S(0))\mathcal{N}(-d(\sigma_{\mathrm{N}})) + \sigma\sqrt{T}n(d(\sigma_{\mathrm{N}})) \tag{4.3}$$

with $d(\sigma) = \frac{S(0)-K}{\sigma_{\mathrm{N}}\sqrt{T}}$. Since $S(T)$ is distributed with respect to a Gaussian distribution realizations of $S(T)$ can become negative. The volatilities σ_{impl} matching a given price, that is $V_C = C(S(0), K, T, \sigma_{\mathrm{impl}})$ are called Bachelier implied volatilities or bp (basis point) volatilities and are currently used to quote prices for interest rate options. The fact that $S(T)$ can become negative is heavily important here since using this model it is possible to quote values for strikes $K \leq 0$.

Example 4.1 Let us take $S(0) = 0.03$, $K = 0.032$, $T = 2$, $r = 0$ and $\sigma = 0.2$. We obtain $d = -0.01767767$ which leads to a Call option price of 0.110355547.

For the Bachelier model to every price p of a European Call/Put option it is possible to find a value σ_N such that put into the corresponding valuation formula, (4.2) or (4.3), leads to the price p. The Bachelier model leads to unbounded option prices.

4.1.2 Displaced Diffusion Model: Shifted Lognormal Model

Another approach, see for instance [11], to take negative rates into account is to consider the following dynamics:

$$dS(t) = (S(t) + b)\sigma_{dd} dW(t), \quad S(0) = s_0. \tag{4.4}$$

Together with the volatility σ_{dd} there is a *displacement* parameter b. The pricing equations for European Call and Put options for the model given by (4.4) are:

$$C_{DD}(S(0), K, T, \sigma) = (S(0) + b)\mathcal{N}(d_1) - (K + b)\mathcal{N}(d_2) \tag{4.5}$$

$$P_{DD}(S(0), K, T, \sigma) = (K + b)\mathcal{N}(-d_2) - (S(0) + b)\mathcal{N}(-d_1) \tag{4.6}$$

with $d_1 = \frac{\log\left(\frac{S+b}{K+b}\right) + \frac{\Sigma_{\mathrm{dd}}^2}{2}}{\Sigma_{\mathrm{dd}}}$ and $d_2 = d_1 - \Sigma_{\mathrm{dd}}$, $\Sigma_{\mathrm{dd}} = \sigma_{dd}\sqrt{T}$. Data providers have started to quote σ together with the displacement parameter b. Then, the option prices can be obtained by (4.5), respectively (4.6). We observe that the pricing within the displaced diffusion model is very close to the Black76 model. It is just to replace the input data and adjust the data by the displacement coefficient. For a given displacement parameter there is only one input parameter—the volatility—which can uniquely be used to determine prices for European Call and Put options. Using the same methodology as for the Black and the Bachelier models we define the Displaced Diffusion/Shifted Lognormal implied volatility.

Example 4.2 Let us take $S(0) = 0.03$, $K = 0.032$, $T = 2$, $r = 0$, $\sigma = 0.2$ and $b = 0.005$. We obtain $d_1 = -0.33068341$ and $d_2 = -0.613526122$ which leads to a Call option price of 0.001671505.

Let us consider an ATM option with a given implied Bachelier volatility σ_N. Then, the price can be expressed in terms of a shifted/displaced log-normal implied volatility if and only if

$$F + d > \sqrt{\frac{T}{2\pi}}\sigma_N \tag{4.7}$$

This means: Only if (4.7) holds there is a number σ_{DD} which lead to the same price as calculate with the formula (4.2). In this case there is a Displaced Diffusion volatility σ_{DD} and

$$\sigma_{DD} = \frac{2\sqrt{2}}{\sqrt{T}} \cdot \mathrm{erf}^{-1}\left(\frac{\sigma_N}{F+d}\sqrt{\frac{T}{2\pi}}\right)$$

This implicitly leads to a lower bound on the displacement. This is given by

$$d > \frac{\sqrt{\frac{T}{2\pi}}\sigma_N}{F}$$

If we would chose a smaller displacement it would not be possible to recover the option price. Thus, the Displaced Diffusion/Shifted Black model suffers from the same shortcoming as the standard lognormal model. With fixed displacement it might not be possible to recover the prices for European Call/Put options. Methods for converting between different volatilities are considered in [3, 4].

Finally, the Displaced Diffusion as well as the Bachelier model are limited in terms of modelling the skew/smile. To handle this problem it is possible to consider a Displaced Diffusion version of the SABR or any other smile model. The next section introduces a variant of the SABR model which takes a different approach.

4.2 The Free Boundary SABR Model

It is possible to consider the standard SABR model from [5] extended by a displacement. But let us consider the model introduced in [2]. It is given by the system of Stochastic Differential Equations:

$$dF(t) = \alpha(t)|F(t)|^{\beta}dW_1(t) \tag{4.8}$$

$$d\alpha(t) = \nu\alpha(t)dW_2(t) \tag{4.9}$$

$$\langle dW_1(t), dW_2(t)\rangle = \rho dt \tag{4.10}$$

$$F(0) = f_0$$

$$\alpha(0) = \alpha_0$$

It is called the Free Boundary SABR model abbreviated fSABR. The dynamic for the forward rate $F(t)$, (4.8), is a CEV type process. The special feature here is that not only forward value but its absolute value is chosen to govern the dynamic. The volatility, (4.9), evolves as in the original SABR model and is modelled as Geometric Brownian Motion. The Brownian motions driving the processes are correlated, (4.10). The initial values are given by f_0 and α_0.

In [2] the authors propose to use the analytic solution for the case $\rho = 0$. This is done by deducing a formula for the time value $\mathcal{O}_F^{fSABR}(T, K)$ of a European Call option:

$$\mathcal{O}_F^{SABR}(T, K) = \frac{1}{\pi}\sqrt{|Kf_0|}\{1_{K\geq 0}A_1 + \sin(\nu\pi)A_2\} \tag{4.11}$$

with $\gamma = \left|-\frac{1}{2}(1-\beta)^{-1}\right|$ and

$$A_1 = \int_0^{\pi} \frac{\sin(\phi)\sin(\gamma\phi)}{b - \cos(\phi)} \frac{G(T\nu^2, s(\phi))}{\cosh(s(\phi))} d\phi$$

$$A_2 = \int_0^{\infty} \frac{\sinh(\psi)\left(1_{K\geq 0}\cosh(\gamma\psi) + 1_{K<0}\sinh(\gamma\psi)\right)}{b + \cosh(\psi)} \frac{G(T\nu^2, s(\psi))}{\cosh(s(\psi))} d\psi$$

and

$$\sinh(s(\phi)) = \frac{\nu}{\alpha}\sqrt{2\frac{|f_0K|^{1-\beta}}{(1-\beta)^2}(b - \cos(\phi))}$$

$$\sinh(s(\psi)) = \frac{\nu}{\alpha}\sqrt{2\frac{|f_0K|^{1-\beta}}{(1-\beta)^2}(b - \cosh(\psi))}$$

$$G(t, s) = 2\sqrt{2}\frac{e^{-t/8}}{t\sqrt{2\pi t}} \int_s^{\infty} ue^{-u^2/(2t)} \sqrt{\cosh(u) - \cosh(s)} du$$

$$b = \frac{|f_0|^{2(1-\beta)} + |K|^{2(1-\beta)}}{2|f_0K|^{1-\beta}}$$

Since (4.11) does only hold for $\rho = 0$ it is not applicable for the general case $\rho \neq 0$. To apply the model for the remaining cases the authors show that (Markovian) projection technique (see [10]) can be utilized to find a Free Boundary SABR model with zero correlation which is in a sense close to the model with $\rho \neq 0$. This technique allows to use a model, call it the p-model, where valuation techniques are well known. The model under consideration is then projected onto the p-model. This projection is essentially expressing the parameters of the p-model as functions of the model's parameters. The resulting distributions at maturity are very close.

However, they remark that the approach not necessarily leads to a model which is free of arbitrage and, furthermore, it can be shown that the technique is not stable or not even applicable for all possible combinations of parameters. Especially for large values of $|\rho|$ that is close to 1. This is not addressed further but we consider a method in Chap. 15 that can be applied for the full range of parameter values including $-1 \leq \rho \leq 1$.

We introduce two approximation formulae for calculating the Bachelier volatility in this model. These formulae are in the spirit as [1, 5, 6]. The final section gives numerical results and we also provide an application for the calibration to market data for Caplets.

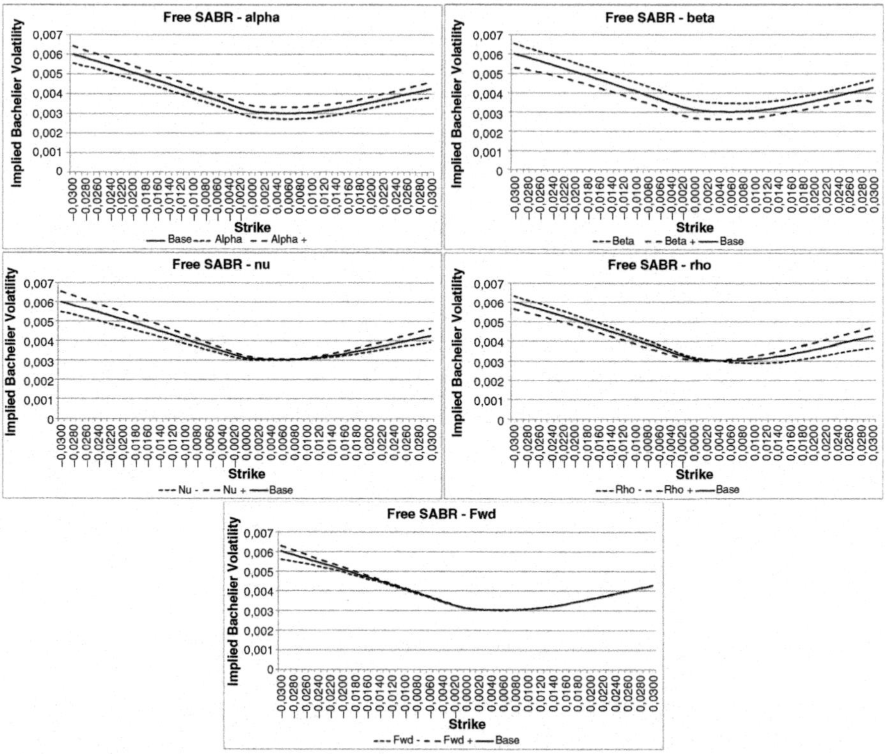

Fig. 4.2 Intuition for the effect of the parameters on the implied Bachelier volatility for the Free SABR model α (*upper left*), β (*upper right*), ν (*mid left*), ρ (*mid right*) and f_0 (*lower left*)

4.2.1 The Parameters

We wish to gain intuition what the role of the each parameter plays in this model. To this end we consider a base scenario and change the parameters. As a base scenario we take $\alpha = 0.005095939$, $\beta = 0.1$, $\nu = 0.3$, $\rho = -0.3$, $f_0 = 0.005$ and $T = 5$. The results are shown in Fig. 4.2. This is taken from [8]

The results displayed in Fig. 4.2 illustrate the effects a change in a given parameter has on the implied volatility for a fixed time to maturity.

4.2.2 Applicability

The integration limits need to be adjusted with respect to the parameters and we observe that small values of the CEV parameter lead to erroneous results as well as small values for the forward rates also leverage this phenomenon.

Now, we wish to consider the model with $\rho \neq 0$. In [2] the general recipe for the projection is to choose a mimicking zero correlation model such that

$$\begin{aligned}
\tilde{\beta} &= \beta \\
\tilde{\nu} &= \sqrt{\nu^2 - \frac{3}{2}\left(\rho^2\nu^2 + \alpha\nu\rho(1-\beta)F^{\beta-1}\right)} \\
\tilde{\alpha}^{(0)} &= \frac{2\theta\Delta\tilde{\nu}}{\theta^2 - 1} \\
\left.\frac{\tilde{\nu}^{(1)}}{\tilde{\nu}^{(0)}}\right|_{K=F} &= \frac{1}{12}\left(1 - \frac{\tilde{\nu}^2}{\nu^2} - \frac{3}{2}\rho^2\right)\nu^2 + \frac{1}{4}\beta\rho\alpha\nu F^{\beta-1}
\end{aligned} \tag{4.12}$$

with

$$\Delta = \frac{K^{1-\beta} - F^{1-\beta}}{1-\beta}, \qquad \theta = \left(\frac{\alpha_{\min} + \rho\nu + \nu\Delta}{(1+\rho)\alpha}\right)^{\frac{\tilde{\nu}}{\nu}}$$

$$\alpha_{\min} = \sqrt{\nu^2\Delta^2 + 2\rho\nu\Delta\alpha + \alpha^2}$$

Figure 4.3 shows $\tilde{\nu}^2$ from Formula (4.12) and it is clear that this expression can become negative, thus, it is not possible to get a real square root.

We have observed that the standard integration formula depends on the choice of the integration upper bound. Figure 4.4 illustrates this. We plotted the calculated implied Bachelier volatilities for the prices calculated with different upper bounds for the integration.

Again we based this subsection on [8].

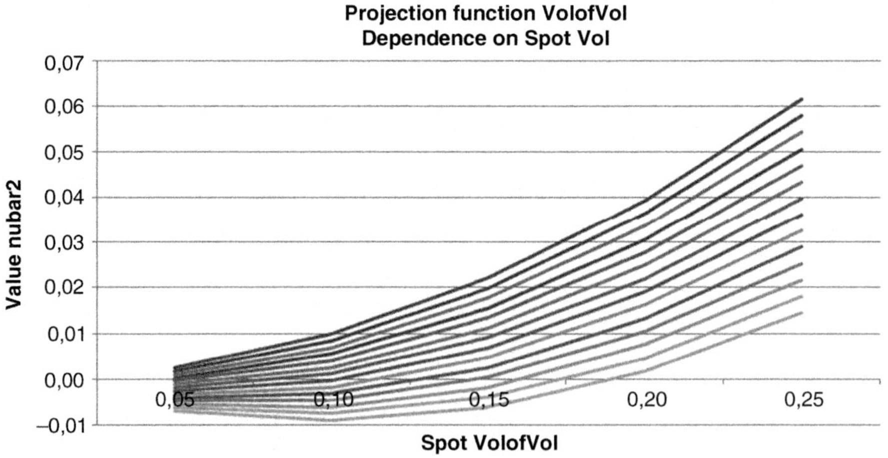

Fig. 4.3 Instabilities using Numerical Integration of the valuation equation (4.11). We show the implied Bachelier volatility calculated from prices using different upper bounds for integration

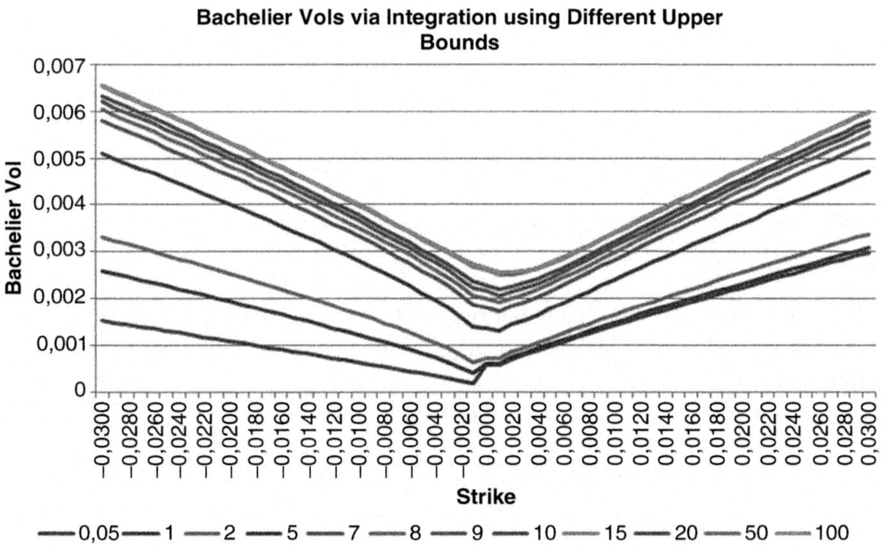

Fig. 4.4 Instabilities using Numerical Integration of the valuation equation (4.11). We show the implied Bachelier volatility calculated from prices using different upper bounds for integration

4.3 Approximation Formulae

In this section we consider approximation formulae for the implied Bachelier volatility. These formulae are straightforward generalizations of results obtained in [5] and [1]. The approximation formulae suffer from the well known shortcomings. This includes inaccuracy, resp. inapplicability for large values of volatility, long time to maturity and may lead to negative densities around 0 for certain parameter settings.

4.3.1 Approximation 1

First, we consider the approximation formula for Bachelier volatility by following the same lines of thought as in [5] and [6]. We obtain the formula

$$\sigma_B^H(T,K) \approx \frac{\alpha_0(f_0-K)}{I}\frac{\xi}{x(\xi)}\left(1+\left(g\alpha_0-\frac{\rho\nu\alpha_0}{4}\Gamma(K)+\frac{2-3\rho^2}{24}\nu^2\right)T\right), \tag{4.13}$$

with

$$\begin{aligned} g &= \log\left((f_0 - K)^{-1} I |f_0|^\beta |K|^\beta\right)/I^2 \\ \xi &= \nu\alpha_0^{-1} I \\ x(\xi) &= (1-\rho)^{-1}(\sqrt{1-2\rho\xi+\xi^2} - \rho + \xi) \\ I &:= \int_K^{f_0} |x|^{-\beta}\, dx. \end{aligned}$$

Substituting (4.13) into the Bachelier pricing formula gives the price of a European Call option. This is a computationally cheap approach that can be used for calibration purposes.

To be able to use (4.13) we take $C(f) = |f|^{-\beta}$ and find that

$$I = \begin{cases} \frac{(-f)^{1-\beta} - (-K)^{1-\beta}}{1-\beta} & K < 0, f < 0 \\ \frac{f^{1-\beta} + (-K)^{1-\beta}}{1-\beta} & K < 0, f > 0 \\ \frac{f^{1-\beta} - K^{1-\beta}}{1-\beta} & K > 0, f > 0 \end{cases} \tag{4.14}$$

The analytic solution (4.14) is used to calculate (4.13) and putting it into the Bachelier pricing formula gives the price of a European Call option.

4.3.2 Approximation 2

We wish to consider another approximation formula within the framework of the Free Boundary SABR model. To this end we follow the same lines of thought as in [1]. Here with T being the maturity and K the strike an expression for the Bachelier volatility is obtained. We have

$$\sigma_B^{AB}(T, K) \approx \frac{\Sigma_0(K) u^{1/2}(T) + \Sigma_1(K) u^{3/2}(T)}{\sqrt{T}} \tag{4.15}$$

For the free boundary SABR model we determine $\Sigma_0(K)$, $\Sigma_1(K)$ and u to get an expression for the approximate volatility which can be put into the Bachelier pricing formula. We have:

$$\sigma_B^{AB}(T, K) = \begin{cases} \frac{|f_0| + |K|}{\xi_K}\left(1 + (g_K + \frac{1}{4}\rho\nu\alpha_0\Gamma_K)T\right) & K < 0, f_0 \geq 0 \\ \frac{|f_0| - |K|}{\xi_{KK}}(1 + (g_K + \frac{1}{4}\rho\nu\alpha_0\Gamma_K)T) & \text{else} \end{cases} \tag{4.16}$$

with the corresponding function

$$\xi_{KK} = \frac{\log(\sqrt{1 - 2\rho\xi_K + \xi_K^2} - \rho + \xi_K)}{\nu(1-\rho)}$$

and $Df = \alpha|f_0|^\beta$ and $Dk = \sqrt{\alpha_0^2 + 2\alpha_0\rho\nu y_K + \nu^2 y_K^2}|K|^\beta$ with the expressions for g_K and y_K depending on the sign of the strike K, namely for the case $K < 0$:

$$g_K = -\log\left(\frac{|f_0| + |K|}{\xi_{KK}\sqrt{Df\,DK}}\right)/\xi_{KK}^2, \quad \Gamma_K = \frac{-|K|^\beta - |f_0|^\beta}{1-\beta}$$

$$y_K = \frac{-|K|^{1-\beta} - |f_0|^{1-\beta}}{1-\beta}, \quad \xi_K = \frac{\nu}{\alpha(1-\beta)}(|f_0|^{1-\beta} + |K|^{1-\beta})$$

and $K > 0$:

$$g_K = -\log\left(\frac{|f_0| - |K|}{\xi_{KK}\sqrt{Df\,Dk}}\right)/\xi_{KK}^2, \quad \Gamma_K = \frac{|K|^\beta - |f_0|^\beta}{1-\beta}$$

$$y_K = \frac{|K|^{1-\beta} - |f_0|^{1-\beta}}{1-\beta}, \quad \xi_K = \frac{\nu}{\alpha(1-\beta)}(|f_0|^{1-\beta} - |K|^{1-\beta})$$

Again this leads to a computational effective way that can be used for calibration purposes.

4.4 Numerical Results

Now, we consider the Free Boundary SABR model and test the proposed numerical methods discussed in this chapter, Sects. 4.3.1 and 4.3.2. To this end we consider the model parameters $\alpha = 0.005095939$, $\beta = 0.1$, $\nu = 0.3$, $\rho = -0.3$ with forward $f = 0.005$ and maturity $T = 3$.

Table 4.1 displays the results for pricing a European Call option for different strike values obtained using (4.11). We displayed the true prices obtained by a benchmark method which we describe later in this book. This method is based on a partial differential equation efficiently solved numerically that is considered in Chap. 15.

To consider the performance of the approximations we calculate the Bachelier volatilities using formulae (4.13) and (4.16) as well as the corresponding prices. Furthermore, we also calculate the absolute values of the differences with respect to the prices obtained by integration and by applying the PDE solution. We summarize our findings in Table 4.2. We took many strike values for a given parameter set. All methods lead to nearly the same option prices and work very well on the used parameter set.

4.4.1 Approximations vs Integration

First, we consider the differences of all methods compared to the PDE approach. Figure 4.5 shows the option prices as well as the differences for all strike values.

Table 4.1 Applying the PDE and the numerical integration method

Strike	Benchmark	Integration	Rel Diff
−0.03	0.03500118	0.03500238	0.00003432
−0.029	0.03400137	0.03400269	0.00003882
−0.028	0.03300159	0.03300304	0.00004411
−0.027	0.03200185	0.03200345	0.0000501
−0.026	0.03100216	0.03100393	0.00005694
−0.025	0.03000254	0.03000448	0.00006475
−0.024	0.02900298	0.02900513	0.00007406
−0.023	0.02800351	0.02800588	0.0000846
−0.022	0.02700415	0.02700676	0.00009673
−0.021	0.02600492	0.0260078	0.00011073
−0.02	0.02500585	0.02500902	0.00012688
−0.019	0.02400698	0.02401048	0.00014555
−0.018	0.02300834	0.0230122	0.00016777
−0.017	0.02201002	0.02201427	0.00019306
−0.016	0.02101208	0.02101676	0.00022235
−0.015	0.02001464	0.02001975	0.00025538
−0.014	0.01901778	0.01902339	0.00029483
−0.013	0.01802169	0.01802783	0.00034062
−0.012	0.01702656	0.01703327	0.00039361
−0.011	0.01603269	0.01603998	0.00045475
−0.01	0.01504043	0.01504833	0.00052516
−0.009	0.01405025	0.01405877	0.00060611
−0.008	0.01306495	0.01307194	0.00053405
−0.007	0.01207884	0.01208867	0.00081257
−0.006	0.01109958	0.01111012	0.00094888
−0.005	0.01012658	0.0101379	0.0011169
−0.004	0.00916197	0.00917424	0.00133784
−0.003	0.00820878	0.00822238	0.00165386
−0.002	0.00727138	0.00728704	0.00214869
−0.001	0.00635677	0.00637561	0.00295456
0	0.00547749	0.00550237	0.0045226
0.001	0.00466845	0.00469153	0.00492124
0.002	0.00391954	0.00393808	0.00470724
0.003	0.00323394	0.00324754	0.00418695
0.004	0.00261781	0.00262651	0.00330935
0.005	0.00207666	0.00217132	0.04359626
0.006	0.00161351	0.00161347	−0.00002349
0.007	0.00122821	0.0012248	−0.00278862
0.008	0.00091696	0.000911	−0.00653648
0.009	0.00067271	0.00066511	−0.01142512
0.01	0.0004862	0.00047782	−0.01754904
0.011	0.0003473	0.00033878	−0.02514202
0.012	0.00024589	0.00023782	−0.03393193

The final column show the absolute differences

Table 4.2 Numerical results, prices, Bachelier volatilities and differences to the integration as well as the PDE approach

K	Approx 1 Vol	Approx 1 Price	Approx 1 –Integral	Approx 2 Vol	Approx 2 Price	Approx 2 –Integral
−0.03	0.0062	0.035	−0.0000009	0.0061	0.035	−0.0000012
−0.029	0.0061	0.034	−0.000001	0.006	0.034	−0.0000013
−0.028	0.006	0.033	−0.0000011	0.0059	0.033	−0.0000014
−0.027	0.0059	0.032	−0.0000012	0.0058	0.032	−0.0000016
−0.026	0.0058	0.031	−0.0000013	0.0057	0.031	−0.0000017
−0.025	0.0057	0.03	−0.0000014	0.0056	0.03	−0.0000019
−0.024	0.0056	0.029	−0.0000016	0.0055	0.029	−0.000002
−0.023	0.0055	0.028	−0.0000017	0.0054	0.028	−0.0000022
−0.022	0.0054	0.027	−0.0000019	0.0053	0.027	−0.0000024
−0.021	0.0053	0.026	−0.000002	0.0052	0.026	−0.0000026
−0.02	0.0052	0.025	−0.0000022	0.0051	0.025	−0.0000028
−0.019	0.0051	0.024	−0.0000024	0.005	0.024	−0.0000031
−0.018	0.0049	0.023	−0.0000027	0.0049	0.023	−0.0000033
−0.017	0.0048	0.022	−0.0000029	0.0048	0.022	−0.0000035
−0.016	0.0047	0.021	−0.0000032	0.0047	0.021	−0.0000038
−0.015	0.0046	0.02	−0.0000035	0.0046	0.02	−0.000004
−0.014	0.0045	0.019	−0.0000038	0.0045	0.019	−0.0000042
−0.013	0.0044	0.018	−0.0000042	0.0044	0.018	−0.0000043
−0.012	0.0043	0.017	−0.0000046	0.0043	0.017	−0.0000043
−0.011	0.0042	0.016	−0.0000051	0.0042	0.016	−0.0000042
−0.01	0.0041	0.015	−0.0000057	0.0041	0.015	−0.0000039
−0.009	0.004	0.0141	−0.0000065	0.004	0.014	−0.0000033
−0.008	0.0038	0.0131	−0.0000074	0.0039	0.013	−0.0000023
−0.007	0.0037	0.0121	−0.0000086	0.0038	0.012	−0.0000008
−0.006	0.0036	0.0111	−0.0000103	0.0037	0.011	0.0000015
−0.005	0.0035	0.0101	−0.0000126	0.003	0.010	−0.0000804
−0.004	0.0034	0.0092	−0.0000158	0.0035	0.009	0.0000092
−0.003	0.0033	0.0082	−0.0000202	0.0034	0.008	0.0000147
−0.002	0.0032	0.0073	−0.000026	0.0033	0.007	0.0000195
−0.001	0.0031	0.0063	−0.000032	0.0032	0.006	0.0000124
0	0.003	0.0054	−0.0000551	0.0031	0.006	−0.0000052
0.001	0.0031	0.0047	0.0000192	0.003	0.005	−0.000031
0.002	0.0031	0.004	0.0000277	0.003	0.004	−0.0000041
0.003	0.0031	0.0033	0.0000288	0.003	0.003	0.0000069
0.004	0.0031	0.0027	0.000027	0.003	0.003	0.0000122
0.005	0.003	0.0021	−0.0000669	0.003	0.002	−0.000076
0.006	0.003	0.0016	0.0000213	0.003	0.002	0.0000163
0.007	0.003	0.0012	0.0000188	0.003	0.001	0.000017
0.008	0.003	0.0009	0.0000167	0.003	0.001	0.000017
0.009	0.003	0.0007	0.000015	0.003	0.001	0.0000164
0.01	0.0031	0.0005	0.0000135	0.0031	0.001	0.0000154
0.011	0.0031	0.0004	0.0000121	0.0031	0.000	0.0000139
0.012	0.0031	0.0002	0.0000107	0.0031	0.000	0.0000122

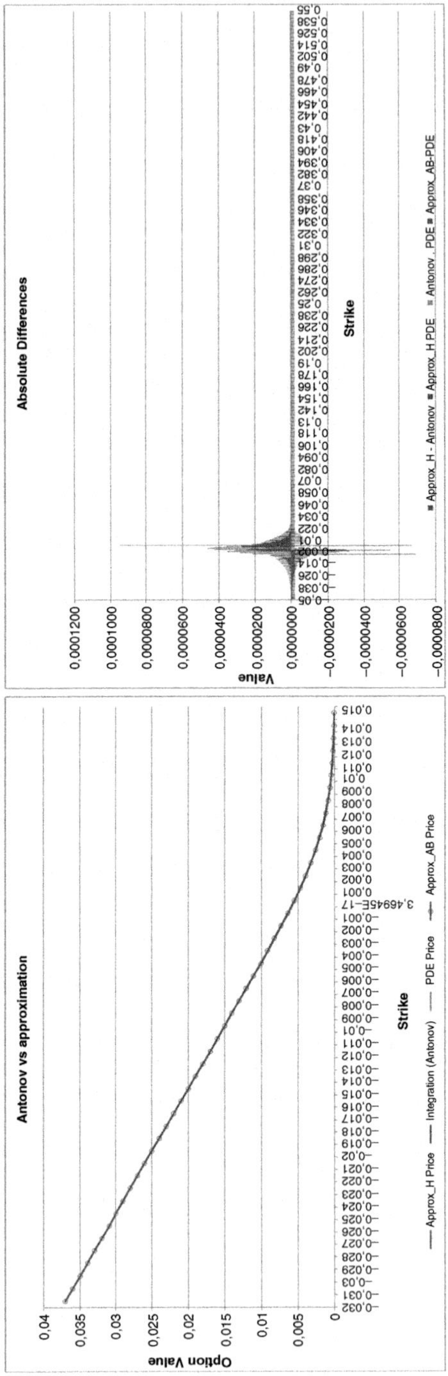

Fig. 4.5 European Call option prices (*left*) and differences (*right*) using the PDE and the integration approach for $\alpha = 0.005095939$, $\beta = 0.1$, $\nu = 0.3$, $\rho = -0.3$ with forward 0.005 and $T = 3$

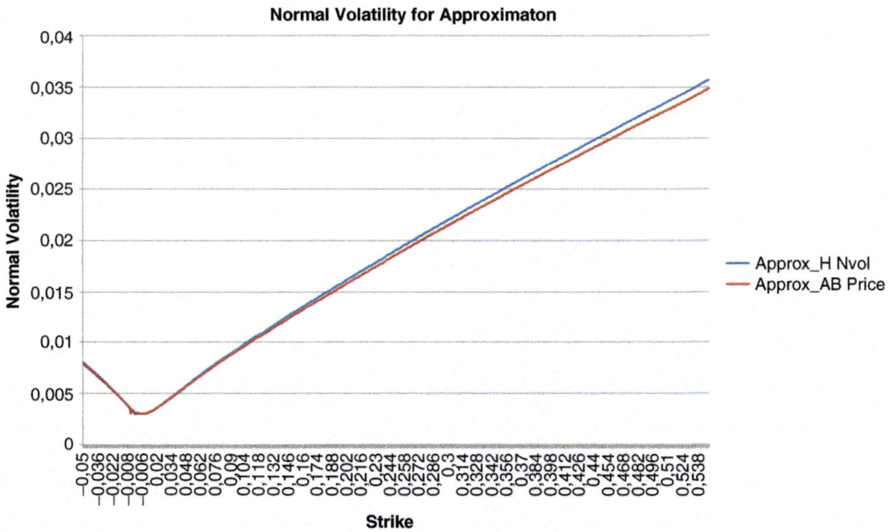

Fig. 4.6 Differences of the approximation formulas (4.13) and (4.15) with respect to the PDE and the integration approach for $\alpha = 0.005095939$, $\beta = 0.1$, $\nu = 0.3$, $\rho = -0.3$ with forward 0.005 and $T = 3$

Now, we consider the approximation formulae with respect to the results they produce on European Call option prices for our example. Figure 4.6 shows the results for the absolute differences. The difference observed on the right wing gets worse for large values of T and/or ν.

4.4.2 Calibration

We performed a calibration using market data of the EUR from Cap/Floor markets and applied the integration as well as the approximation formulae. Figure 4.7 shows the result. We have showed that at least for a restricted parameter set the calibration can be achieved using the approximation formulae presented in this chapter.

4.5 Conclusions

We have considered the standard ways of quoting option prices after the crisis. This includes negative interest rates being observed even for the major currencies. The market standard methods are using either Bachelier implied volatility or Displaced Diffusion/Shifted Log-Normal volatility. For the second method the quotation needs two parameters, namely the volatility and the displacement. Without fixing the

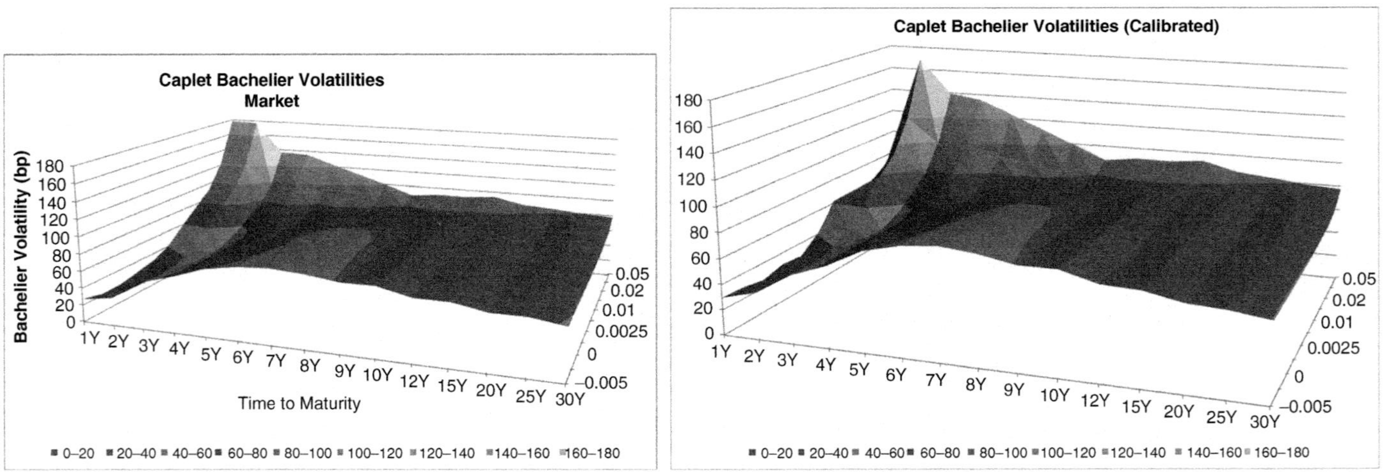

Fig. 4.7 Market data for EUR Cap/Floor (*left*) in terms of implied Bachelier volatility and the calibrated surface (*right*) using the approximation formulae

displacement the volatility is not unique. Then, we reviewed an extension of the SABR model which is a generalization to negative rates. We review the proposed valuation technique and show that straight forward approximation formulae exist. However, these formulae suffer from the same problems as the approximation techniques for the standard SABR model. Finally, we included some numerical results.

References

1. Andersen, L.B.G., Brotherton-Ratcliffe, R.: Extended Libor market models with stochastic volatility. J. Comput. Finance **9**(1), 29–50 (2005)
2. Antonov, A., Konikov, M., Spector, M.: Free boundary SABR. Risk (2015)
3. Dimitroff, G., Fries, C., Lichtner, M., Rodi, N.: Lognormal vs normal volatilities and sensitivities in practice. SSRN preprint (2016)
4. Hagan, P., Woodward, D.: Equivalent black volatilities. Appl. Math. Finance **6**(3), 147–157 (1999)
5. Hagan, P.S., Kumar, D., Lesniewski, A.S., Woodward, D.E.: Managing smile risk. Wilmott Mag. **1**, 84–108 (2002)
6. Hagan, P., Kumar, D., Lesniewski, A.S., Woodward, D.E.: Arbitrage free SABR. Wilmott (2014)
7. Kienitz, J.: Interest Rates Explained Volume 1: Products and Markets. Palgrave MacMillan, Basingstoke (2014)
8. Kienitz, J., Caspers, P.: Interest Rates Explained Volume 2: Term Structure Models. Palgrave MacMillan, Basingstoke (2017)
9. Kienitz, J., Wetterau, D.: Financial Modeling – Theory, Implementation and Practice – (With Matlab Source). Wiley, New York (2012)
10. Piterbarg, V.: Markovian projection method for volatility calibration. Available at SSRN (2006)
11. Rubinstein, M.: Displaced diffusion option pricing. J. Finance **38**(1), 213–217 (1983)

Chapter 5
Accurate Vega Calculation for Bermudan Swaptions

Mark Beinker and Sebastian Schlenkrich

Abstract Short rate models are widely used for pricing of Bermudan Swaptions. In addition to prices, traders and risk managers need sensitivities for hedging and risk management. Vega is the sensitivity of the price with respect to changes in market volatilities (i.e. implied Black'76 or Bachelier volatilities). This sensitivity is of particular importance for Bermudan Swaptions.

It is common practice to evaluate Vega by shifting market data and re-evaluating model prices. Even though this procedure is often used, in practice it is inefficient here since the model calibration process flattens out the shift of single volatility surface grid points. Thus this procedure may underestimate sensitivities. In this chapter, we demonstrate how Adjoint Algorithmic Differentiation can be used to calculate accurate and stable Vegas without loss of performance.

5.1 Financial Models and Algorithmic Differentiation

Hedging and risk management require knowledge of the sensitivities of financial instrument prices with respect to their key risk factors (i.e., Greeks). Mathematically, these sensitivities can be identified with partial derivatives with respect to input parameters of the pricing model. From that perspective it does not come as a surprise that methods of Algorithmic Differentiation (AD) attracted some attention in financial engineering recently. Some studies in that field are, for example, [2, 3, 5, 8, 12].

5.1.1 Financial Models and Sensitivities

Prices of exotic financial derivatives are typically determined by means of a specific financial model rather than by balancing offer and demand for this product. Such a model could be described by the dynamics of certain risk factors and parameters

M. Beinker (✉) • S. Schlenkrich
d-fine GmbH, An der Hauptwache 7, 60313 Frankfurt, Germany
e-mail: mark.beinker@d-fine.de; sebastian.schlenkrich@d-fine.de

M. Ehrhardt et al. (eds.), *Novel Methods in Computational Finance*,
Mathematics in Industry 25, DOI 10.1007/978-3-319-61282-9_5

involved. Common risk factors are stock prices, foreign exchange rates, or interest rates. Exotic derivatives will also frequently depend on volatilities implied by liquidly traded Plain Vanilla options (*implied volatilities*). Often, implied volatilities of plain vanilla options are directly quoted instead of prices. Thus it might not be necessary to imply these volatilities, because they are directly observable. Still, throughout this chapter, we use the term implied volatilities synonymously for market volatilities.

The implementation of the model should be able to price plain vanilla options as well as exotic financial derivatives in order to verify the calibration quality of the model or a given set of calibrated model parameters. Here, we focus on pricing Bermudan swaptions. Then, the particular market observable parameters are the yield curve and implied volatilities

$$\sigma_{\mathrm{Mkt}} = \left(\sigma_{\mathrm{Mkt},1}, \ldots, \sigma_{\mathrm{Mkt},N}\right)^{\top}$$

of plain vanilla European swaptions observed at the valuation date t. These implied volatilities correspond to either log-normal volatilities based on Black's formula or normal volatilities based on Bachelier's formula.[1] We require our pricing model to match a set of market prices of plain vanilla European swaptions (called *benchmark swaptions*) given as a function of the implied volatility Market(σ_{Mkt}):

$$\mathrm{Market}\left(\sigma_{\mathrm{Mkt}}\right) = \begin{pmatrix} \mathrm{Swaption}_1\left(\sigma_{\mathrm{Mkt}}\right), \\ \vdots \\ \mathrm{Swaption}_N\left(\sigma_{\mathrm{Mkt}}\right) \end{pmatrix}.$$

Here, Swaption$_i$(.) represents either Black's or Bachelier's formula with appropriate parameters for the ith swaption. Since we are only interested in the volatility dependence, all other parameters are suppressed.

In general, a specific pricing model depends on a set of M model parameters. These parameters are calibrated such that the model prices for a given set of benchmark products matches the observed market prices as good as possible. This yields to a minimization problem $\min_{x_{\mathrm{Mdl}}} \chi\left(x_{\mathrm{Mdl}}, x_{\mathrm{Mkt}}\right)$. Here χ is a differentiable cost functional depending on market observables x_{Mkt} and model parameters x_{Mdl}. Then, the first order optimality condition yields the calibration function $\Psi\left(x_{\mathrm{Mdl}}, x_{\mathrm{Mkt}}\right) = \nabla_{x_{\mathrm{Mdl}}} \chi\left(x_{\mathrm{Mdl}}, x_{\mathrm{Mkt}}\right) = 0$. A typical choice for the cost function is

$$\chi\left(x_{\mathrm{Mdl}}, x_{\mathrm{Mkt}}\right) = \frac{1}{2}\left(F(x_{\mathrm{Mdl}}) - y(x_{\mathrm{Mkt}})\right)^{\top}\left(F(x_{\mathrm{Mdl}}) - y(x_{\mathrm{Mkt}})\right)$$

[1] Volatilities based on the shifted log-normal model are also common nowadays. Since the shifted log-normal model is related to the log-normal model by a simple variable transformation $x \mapsto x + a$ with some constant offset a, we will not consider shifted log-normal models here since this wouldn't add much value, but would complicate notation unnecessarily.

with $N \geq M$. In that setting $F(x_{\mathrm{Mdl}})$ and $y(x_{\mathrm{Mkt}})$ are model- and market-implied reference prices, respectively. This yields a calibration function of the form

$$\Psi\left(x_{\mathrm{Mdl}}, x_{\mathrm{Mkt}}\right) = F'(x_{\mathrm{Mdl}})^{\top}\left(F(x_{\mathrm{Mdl}}) - y(x_{\mathrm{Mkt}})\right) = 0 \,.$$

In the special case that we have the same number of model parameters as reference prices we may simplify the calibration function $\Psi(\cdot)$. Suppose $M = N$ and $F'(x_{\mathrm{Mdl}})$ is regular then we may write the calibration function equivalently as

$$\Psi\left(x_{\mathrm{Mdl}}, x_{\mathrm{Mkt}}\right) = F(x_{\mathrm{Mdl}}) - y(x_{\mathrm{Mkt}}) = 0.$$

In particular for our application we use this reduced setting.

The Hull White interest rate model (see Sect. 5.2), which we focus on, specifies the dynamics of the instantaneous short rate. Its model parameters are the short rate volatility function and the mean reversion speed. In our setting the short rate volatility is chosen piecewise constant in time. In particular we have the same number of short rate volatility pieces as market implied volatilities. We denote the Hull White short rate model volatilities by

$$\sigma_{\mathrm{Mdl}} = \left(\sigma_1, \ldots, \sigma_N\right)^{\top}.$$

The mean reversion speed is chosen constant and predefined by the user.[2]

The Hull White model allows the pricing of European coupon bond options (CBO) that are equivalent to our plain vanilla European swaptions. Of course, the model prices depend on the model parameters used. Hence, we get a function

$$\mathrm{Model}(\sigma_{\mathrm{Mdl}}) = \begin{pmatrix} \mathrm{CBO}_1(\sigma_{\mathrm{Mdl}}) \\ \vdots \\ \mathrm{CBO}_N(\sigma_{\mathrm{Mdl}}) \end{pmatrix}$$

summarizing the model prices of our reference plain vanilla instruments.

In this setting, the short rate model volatilities σ_{Mdl} are chosen such that the coupon bond option prices equal the equivalent reference European swaption prices. That is we consider a calibration function Ψ with

$$\Psi\left(\sigma_{\mathrm{Mdl}}, \sigma_{\mathrm{Mkt}}\right) = \mathrm{Model}(\sigma_{\mathrm{Mdl}}) - \mathrm{Market}\left(\sigma_{\mathrm{Mkt}}\right).$$

[2]Since we have N benchmark prices and $N + 1$ model parameters, we have to fix one parameter, otherwise the optimization problem would not be well defined. Experience shows that adding an additional benchmark swaption does not lead to a stable calibration process, i.e. calibrating the mean reversion parameter to swaption prices is generally not a good choice. However, it is possible to calibrate the parameter to Bermudan swaption prices, if available from an external source.

The model is calibrated to the market observable implied volatility parameters if

$$\Psi\left(\sigma_{\mathrm{Mdl}}, \sigma_{\mathrm{Mkt}}\right) = 0 .$$

After the model is calibrated to the market, we may use the determined model parameters to price exotic derivatives. The mathematical formulation of the exotic option price is often more complex than the formulation of the plain vanilla option prices. Typically, a numerical scheme implements an approximate solution. We determine the Bermudan swaption price

$$V = \mathrm{Exotic}\left(\sigma_{\mathrm{Mdl}}\right)$$

by successive numerical integration.

5.1.1.1 Evaluating Sensitivities

For hedging and risk management purposes the sensitivities of exotic option prices with respect to market parameters are of particular interest. Evaluation of these sensitivities requires the differentiation of the Exotic option pricing procedure as well as the calibration procedure of the model applied. In this study we consider the sensitivity of the Bermudan price with respect to changes in the market implied volatilities of our reference European swaptions. Hence we want to evaluate

$$\frac{dV}{d\sigma_{\mathrm{Mkt}}} = \mathrm{Exotic}'\left(\sigma_{\mathrm{Mdl}}\right) \cdot \frac{d\sigma_{\mathrm{Mdl}}}{d\sigma_{\mathrm{Mkt}}}.$$

The term $d\sigma_{\mathrm{Mdl}}/d\sigma_{\mathrm{Mkt}}$ may be considered the differentiation of the calibration procedure. We assume Ψ is differentiable and $\Psi_{\sigma_{\mathrm{Mdl}}}$ is non-singular at the solution. Then the implicit function theorem yields

$$\begin{aligned}\frac{d\sigma_{\mathrm{Mdl}}}{d\sigma_{\mathrm{Mkt}}} &= -\left[\frac{\partial}{\partial\sigma_{\mathrm{Mdl}}}\Psi\left(\sigma_{\mathrm{Mdl}}, \sigma_{\mathrm{Mkt}}\right)\right]^{-1} \cdot \frac{\partial}{\partial\sigma_{\mathrm{Mkt}}}\Psi\left(\sigma_{\mathrm{Mdl}}, \sigma_{\mathrm{Mkt}}\right) \\ &= \mathrm{Model}'\left(\sigma_{\mathrm{Mdl}}\right)^{-1} \cdot \mathrm{Market}'\left(\sigma_{\mathrm{Mkt}}\right).\end{aligned}$$

Consequently, we get

$$\frac{dV}{d\sigma_{\mathrm{Mkt}}} = \mathrm{Exotic}'\left(\sigma_{\mathrm{Mdl}}\right) \cdot \mathrm{Model}'\left(\sigma_{\mathrm{Mdl}}\right)^{-1} \cdot \mathrm{Market}'\left(\sigma_{\mathrm{Mkt}}\right).$$

Thus, the evaluation of Bermudan swaption Vegas requires derivatives of the Market(), Model(), and Exotic() functions, as well as a solution of a linear system. In practice these functions are implemented as computer programs. Although we probably do not have a closed form representation of these functions, we can apply the chain rule of differentiation to the individual operations of the computer programs. This approach is facilitated by the methods of Algorithmic Differentiation.

5.1.2 Algorithmic Differentiation at a Glance

Algorithmic Differentiation (AD) provides principles and techniques to augment computer models with additional code that allows to compute sensitivities of output variables with respect to inputs of the model. Sensitivities computed by AD-enhanced models (AD models) are computed as numerical values (not symbolic) for the evaluation point given by the values of the model inputs. They are exact up to machine precision. In contrast, widely used approximations of derivatives computed by finite difference schemes often suffer from rounding and cancellation errors (especially for higher order derivatives).

The key idea of Algorithmic Differentiation is applying the chain rule of differentiation to the individual or atomic functions of a computer program. Two basic modes are distinguished in this context: the forward mode and the reverse (or adjoint) mode. In the forward mode, the chain-rule is applied in the same order as the function evaluation itself, starting at the independent variables and evaluating the dependent variables. On the contrary, the chain-rule can also be applied in the reverse order of the function evaluation. This means, the sensitivities of the intermediate variables with respect to the dependent variables are computed successively and in reverse order. However, this procedure requires a preceding function evaluation where all overwritten program variables are stored for restoration in the reverse sweep.

In the reverse mode the additional computational cost for gradient derivative evaluation is independent of the number of input variables and roughly bounded by five times the cost of the function evaluation itself. This property makes Algorithmic differentiation particularly appealing for finance applications. A typical use case is that sensitivities of many inputs (like interest rates and volatilities) with respect to a single output (present value) are desired.

From a technical point of view, two fundamental concepts of AD tools have to be distinguished: Source transformation tools are directly applied to the model code.[3] A source transformation AD tool generates the AD model as new source code. AD tools based on overloading of operators and intrinsic functions provide a new (often called *active*) data type. This active data type is embedded into a sensitivity aware arithmetic. This means, all relevant operators and intrinsic functions are defined for the active data type and the intrinsic data types. The AD model is derived by activating a copy of the original model. In practice, the data types of relevant floating point variables are changed into the active data type. This can be realized, for example, by hand, global definitions, or some scripting. Another approach is

[3]Often the original code has to be adapted slightly to meet the requirements of the AD tool applied.

based on template metaprogramming. This allows binding classes and functions to active data types at compile time.

An introduction to Algorithmic Differentiation can be found in [4, 9]. For this study we apply the operator overloading AD tool dco/C++ [10] developed by NAG and RWTH Aachen.

5.2 Pricing Bermudan Swaptions with a Hull White Model

In this section we present the details of our Hull White model implementation. We aim at pricing Bermudan swaptions. A Bermudan swaption gives the option holder the right to enter an interest rate swap at predefined dates. The underlying swap exchanges a fixed rate cash flows based on simple compounding rate R against a variable cash flow based on a float rate index L_i. Typical floating rate indices are interbank offered rates (Ibor), such as Euribor or Libor. Forthcoming we use the general term Ibor rate. The subscript i indicates the dependency of the floating rate on fixing date t_i at which the Ibor rate is determined. Each accrual period has its own rate. The Ibor rate is typically fixed between 0 and 3 business days prior to accrual start date.

Fixed leg coupon payment dates are denoted by $S_1, \ldots, S_M$ and S_0 is the start date of the first coupon period. Year fractions associated with the fixed leg coupon periods are $\tau_1, \ldots, \tau_M$. Here, we assume that fixed leg payment dates are a subset of the floating rate payment dates. In the Euro market fixed payments are usually annual and the fixed leg day count convention is 30/360.

Floating leg payment dates are given by $\hat{S}_1, \ldots, \hat{S}_{\hat{M}}$ and $\hat{S}_0 = S_0$ is again the start date of the first coupon period. Corresponding year fractions are $\eta_1, \ldots, \eta_{\hat{M}}$. Fixed and float leg have identical maturity dates, i.e. $S_M = \hat{S}_{\hat{M}}$.

The price of a risk free zero coupon bond at observation time $t = 0$ with maturity T $(t \leq T)$ is given by $P(t, T)$. The mapping $T \mapsto P(t, T)$ represents the yield curve at observation time t. A yield curve may be inferred from deposit, forward rate, and swap rates quoted in the market. We assume it to be given as an interpolated set of zero coupon bond prices (i.e., discount factors).

To simplify notations we work in a single yield curve setting. Individual curves for forward rate projection and discounting can easily be incorporated into the pricing of Bermudans. This feature has effects on European swaption pricing as well as the specification of the Hull White model. Details on tenor and funding spread modeling are discussed, for example, in [13].

The price of the underlying swap at time $t \leq S_0$ is determined by discounting the fixed and forward floating leg cash flows. For a (fixed) receiver swap it becomes

$$\text{Swap}(t) = \underbrace{R \sum_{i=1}^{M} \tau_i P(t, S_i)}_{\text{FixedLeg}(t)} - \underbrace{\sum_{j=1}^{\hat{M}} L_j(t) \eta_j P(t, \hat{S}_j)}_{\text{FloatLeg}(t)} \; .$$

In a single interest rate curve setting the forward Ibor rate $L_j(t)$ is given by

$$L_j(t) = \left[\frac{P\left(t, \hat{S}_j\right)}{P\left(t, \hat{S}_{j-1}\right)} - 1 \right] \frac{1}{\eta_j} .$$

Consequently, the floating leg of the swap can be simplified to

$$\text{FloatLeg}(t) = \sum_{j=1}^{\hat{M}} L_j(t)\eta_j P(t, \hat{S}_j) = P(t, S_0) - P(t, S_M) .$$

Thus, we obtain the swap pricing formula

$$\begin{aligned} \text{Swap}(t) &= R \sum_{i=1}^{M} \tau_i P(t, S_i) \;-\; [P(t, S_0) - P(t, S_M)] \\ &= R\, a_0(t) \;-\; [P(t, S_0) - P(t, S_M)] \end{aligned}$$

which depends on the fixed rate R, the fixed leg schedule $S_0, \ldots, S_M$, and the yield curve $T \mapsto P(t, T)$. Here, we have introduced the annuity $a_i(t) = \sum_{i+1}^{M} \tau_i P(t, S_i)$. Rearranging terms yields that the swap can also be interpreted as the time t price of a risk free (forward) bond contract with unit bond price paid at S_0, fixed coupons $R\tau_i$ paid at S_i for $i = 1, \ldots, M$ and unit notional payment at S_M. That is

$$\text{Bond}(t) = \underbrace{-P(t, S_0)}_{\text{bond price}} + \underbrace{R\, a_0(0, t) + P(t, S_M)}_{\text{coupons and notional}}$$

and

$$\text{Swap}(t) = \text{Bond}(t) . \tag{5.1}$$

A swaption gives its holder the right to enter a swap at a given strike rate R. The swaption is considered to be of European style if the right may be exercised at a single predefined date. A Bermudan swaption gives its holder the right to enter into a fixed maturity swap at one of several predefined exercise dates. Therefore, the holder has to decide at each exercise date, whether it is more favourable to enter into the swap right now instead of keep holding the option to enter the swap at some later exercise date.[4] At an exercise date T with $S_{i_0} = \min_{i=1,\ldots,M} \{S_i | S_i \geq T\}$ the exercise

[4]Here we assume physical settlement rather than cash settlement, where the option holder receives a cash compensation only in case of exercise.

value of the Bermudan swaption equals the value of a European swaption given by

$$\begin{aligned}\text{Swaption}(T) &= [\omega\ \text{Swap}(T)]^+ \\ &= [\omega\ \{R\, a_{i_0}(T) - [P(T, S_{i_0}) - P(T, S_M)]\}]^+ \ .\end{aligned}$$

Here $\omega \in \{-1, +1\}$ distinguishes between a payer (-1) and receiver $(+1)$ swaption and $[\,\cdot\,]^+$ abbreviates $\max\{\,\cdot\,, 0\}$.

Exploiting the equivalence between swaps and bonds yields that the swaption may be interpreted as a coupon bond option with payoff

$$\text{CBO}(T) = \left[\omega \left\{ \underbrace{R\, a_{i_0}(T)}_{\text{coupons}} + \underbrace{P(T, S_M)}_{\text{notional}} - \underbrace{P(T, S_{i_0})}_{\text{strike}} \right\}\right]^+ .$$

The underlying bond coupons equal the fixed leg payments of the swap. At maturity the unit notional is paid and the option strike equals the unit notional. The exercise date is assumed to coincide with the accrual start date of the first period, i.e. $T = S_{i_0}$.

5.2.1 Market Formulas for European Swaptions

The payoff of the swaption is rewritten as

$$\text{Swaption}(T) = a_{i_0}(T) \left[\omega \left(R - \frac{P(T, S_{i_0}) - P(T, S_M)}{a_{i_0}(T)}\right)\right]^+ .$$

In this representation the (forward) par swap rate is denoted by

$$Y(T) = \frac{P(T, S_{i_0}) - P(T, S_M)}{a_{i_0}(T)} ,$$

and the swaption payoff becomes

$$\text{Swaption}(T) = a_{i_0}(T)\ [-\omega\ \ (Y(T) - R)]^+ .$$

Thus a European receiver (payer) swaption is equivalent to a European put (call) on the forward par swap rate $Y(T)$ with strike R.

The present value of the swaptions is derived by discounting the expectation of the payoff as

$$\text{Swaption}(t) = a_{i_0}(t)\ \mathbb{E}^A \left[-\omega\ \ (Y(T) - R)^+\right] .$$

The expectation is taken in the so-called annuity measure which uses $a_{i_0}(t)$ as numéraire. Since the annuity $a_{i_0}(t)$ can be interpreted as the weighted sum of zero bond prices, it is indeed a valid choice as numéraire. It is market practice to evaluate the expectation $\mathbb{E}^A\left[-\omega\ (Y(T)-R)^+\right]$ by means of Black's or Bachelier's formula and corresponding implied volatilities σ^B and σ^N respectively. Black's formula is given by

$$\text{Black}(Y,R,\sigma^B,T,\omega) = \omega\cdot(Y\cdot N(\omega\cdot d_1) - R\cdot N(\omega\cdot d_2)),$$

$$d_{1,2} = \frac{\ln(Y/R)}{\sigma^B\cdot\sqrt{T}} \pm \frac{\sigma^B\cdot\sqrt{T}}{2},$$

and Bachelier's formula is given by

$$\text{Bachelier}(Y,R,\sigma^N,T,\omega) = \omega\cdot(Y-R)\cdot N(\omega\cdot d) + N'(\omega\cdot d)\cdot\sigma^N\cdot\sqrt{T},$$

$$h = \frac{Y-R}{\sigma^N\cdot\sqrt{T}}.$$

5.2.2 *Analytical Pricing Formulas for the Hull White Model*

The Hull White model [6] specifies a stochastic process for the short rate $r(t)$. The model is given by

$$dr(t) = [\theta(t) - ar(t)]\,dt + \sigma(t)dW(t).$$

Here $\theta(t)$ denotes the risk neutral drift and is fully determined by the current interest rate curve, a the constant mean reversion parameter, and $\sigma(t)$ the volatility of the short rate. The stochasticity is given by the Brownian motion increment $dW(t)$ under the risk neutral probability measure. It is common to assume that the volatility is piecewise constant between two exercise dates of a Bermudan swaption. Let $T_0 = t$ and denote the Bermudan exercise dates with $T_1,\ldots,T_N$ then we have that

$$\sigma(t) = \sigma_j \quad \text{for} \quad t\in(T_{j-1},T_j],\ j=1,\ldots,N.$$

With a given set of model parameters (i.e. short rate volatilities and mean reversion speed) as of time t, the price of a (future) zero coupon bond at the future date $T_j > t$, with maturity S, and realized short rate r at time t is

$$\text{ZCB}(t;T_j,S,r) = A(t;T_j,S)\,e^{-B(T_j,S)\,r}$$

with

$$\begin{aligned}
A(t;T_j,S) &= \frac{P(t,S)}{P(t,T_j)} \exp\left\{ B(T_j,S)f(t,T_j) - \frac{B(T_j,S)^2}{2} e^{-2a\,\delta(t,T_j)} C(t;T_j) \right\}, \\
B(T_j,S) &= \frac{1}{a}\left(1 - e^{-a\,\delta(T_j,S)}\right), \\
C(t;T_j) &= \sum_{k=1}^{j} \frac{\sigma_k^2}{2a}\left(e^{2a\,\delta(t,T_k)} - e^{2a\,\delta(t,T_{k-1})}\right), \\
f(t,T_j) &= -\frac{\partial \log(P(t,T_j))}{\partial T_j}.
\end{aligned} \tag{5.2}$$

A coupon bond with cash flows c_i at coupon payment dates S_i is determined by the sum of the scaled zero coupon bond prices, i.e.

$$\mathrm{CB}(t;T_j,S_1,\ldots,S_M,r) = \sum_{S_i \ge T_j} c_i\,\mathrm{ZCB}(t;T_j,S_i,r).$$

The time t price of an option on a zero coupon bond with exercise date T_j, bond maturity S, and strike price K paid at T_j is given by

$$\mathrm{ZCO}(t;T_j,S,K,\omega) = P(t,T_j)\,\mathrm{Black}(P(t,S)/P(t,T_j),K,\sigma_P,1,\omega) \tag{5.3}$$

with

$$\sigma_P = \frac{1}{a}\left(e^{-a\,\delta(t,T_j)} - e^{-a\,\delta(t,S)}\right)\sqrt{C(t;T_j)}.$$

Note that the notation $\tau = 1$ in the Black formula implies no loss of generality. The temporal scaling of the volatility σ_P is already incorporated in the terms $\left(e^{-a\,\delta(t,T_j)} - e^{-a\,\delta(t,S)}\right)$ and $C(t;T_j)$.

An option on a coupon bond with cash flows c_i at coupon payment dates S_i, exercise date T_j, and strike price K may be valued using Jamshidian's decomposition [7]. This approach requires to solve the equation

$$\mathrm{CB}(t;T_j,S_1,\ldots,S_M,r^\star) = K$$

for the short rate $r^\star$. Using the resulting short rate $r^\star$ we can evaluate corresponding individual strikes K_i by

$$K_i = \mathrm{ZCB}(t;T_j,S_i,r^\star).$$

With these individual strikes the coupon bond option can be priced as a sum of zero coupon bond options, i.e.

$$\mathrm{CBO}(t, T_j, S_1, \ldots, S_M, \omega) = \sum_{S_i \geq T_j} c_i \, \mathrm{ZCO}(t, T_j, S_i, K_i, \omega).$$

For the calibration procedure it is important to note that the price of a coupon bond option with exercise date T_j depends only on short rate volatilities σ_1 to σ_j. The price is independent on volatilities corresponding to times larger than T_j.

5.2.3 *Pricing Bermudan Swaptions*

Once the Hull White model is calibrated to European swaptions, it may be applied to price Bermudan swaptions. In Sect. 5.2 we demonstrated the equivalence between European swaptions and European bond options. Analogously, we find that a Bermudan swaption can be represented as a Bermudan option to buy or sell a coupon bond at predefined exercise dates with given strike prices.

Bermudan swaptions are frequently traded as hedges against call rights of multi-callable bonds. They are probably the most frequently traded interest rate derivatives which require a term structure model for pricing. The term structure model is necessary to determine the exercise boundary which separates the two regions where exercise is optimal and where exercise should be postponed. This decision depends on the interest rate level. At each exercise date, the value of the Bermudan option equals the maximum of hold value and exercise value.

The exercise value can easily be determined, since it is just the value of the underlying swap at the exercise date. The hold value, i.e. the value of the Bermudan option if exercise is postponed, is more difficult to determine, since it is essentially the value of a Bermudan option with one exercise date less. But there is one exception: at the last exercise date T_{M-1}, no further exercise dates are left, therefore the hold value must be zero. It follows that at the last exercise date, the value of the Bermudan swaptions equals the maximum of the exercise value and zero. For numerical methods running backwards in time, this Bermudan value can be rolled back to the previous exercise date T_{M-2}. Here, the new Bermudan swaption value equals the maximum of the rolled-back value (which is the hold value) and the exercise value at T_{M-2}, and so forth to the valuation date t. Numerical methods which apply a forward simulation approach (e.g., Monte Carlo methods) would require instead some additional method to estimate the exercise boundary in order to price Bermudan options.

In this subsection we sketch our approach for the pricing of Bermudan bond options in the Hull White model. The method is based on the reformulation of the Hull White model in the time-T neutral measure. For references, see for example [1] or [11, Appendix C]. The fundamental theorem of asset pricing yields for the price $V(t, r(t))$ of a security depending on the time t and a (also time-dependent) risk factor $r(t)$ that

$$V(t, r(t)) = \text{ZCB}(0; t, T, r(t)) \cdot \mathbb{E}^T [V(T, r(T))] \quad \text{for} \quad t < T \,.$$

In this representation ZCB$(0; t, T, r(t))$ is the time-t price of a zero coupon bond maturing at time T. The zero coupon bond price depends on the model calibration at time-0 and the time-t state of the risk factor $r(t)$. For the Hull White model the risk factor is the short rate. The analytical formula for ZCB() was elaborated in Sect. 5.2.2.

The expectation $\mathbb{E}^T$, conditional on the information at time t, is evaluated in the time-T neutral measure. That is, the numéraire applied is the zero coupon bond maturing at time T. In our setting the price of the numéraire is given by the ZCB() formula. For the pricing of the option in this setting we require the dynamics of the short rate $r(t)$ in the time-T neutral measure.

Provided we can evaluate $\mathbb{E}^T [V(T, r(T))]$ for a given option price or payoff at time T then we can also price Bermudan bond options. We discretize the short rate by a grid $r_0, \ldots, r_n$. Analogously to a PDE approach, we start at the last exercise date T_N and work backwards in time. We evaluate the auxiliary option price

$$\tilde{V}(T_{N-1}, r_j) \;=\; \text{ZCB}(0; T_{N-1}, T_N, r_j) \cdot \mathbb{E}^{T_N} [p_N(r(T_N))]$$

for $j = 0, \ldots, n$. Here $p_N(r)$ is the Nth payoff function of the Bermudan option. We consider a sequence of coupon bond options. Therefore

$$p_i(r) = [\omega\, (\text{CB}(0; T_i, S_1, \ldots, S_M, r) - K_i)]^+ \quad \text{for} \quad i = 1, \ldots, N$$

with $\omega \in \{-1, +1\}$ distinguishing put and call options, K_i determining the (dirty) strike price of the bond, and CB$(\ldots)$ defined in Sect. 5.2.2. The option price at T_{N-1} then becomes

$$V(T_{N-1}, r_j) \;=\; \max\left\{\tilde{V}(T_{N-1}, r_j),\; p_{N-1}(r_j)\right\} \quad \text{for} \quad j = 0, \ldots, n.$$

The resulting discrete points $V(T_{N-1}, r_0), \ldots, V(T_{N-1}, r_n)$ are interpolated to model the option price function $V(T_{N-1}, r)$ at time T_{N-1} and intermediate short rate points r. We proceed by evaluating

$$\tilde{V}(T_{N-2}, r_j) \;=\; \text{ZCB}(0; T_{N-2}, T_{N-1}, r_j) \cdot \mathbb{E}^{T_{N-1}} [V(T_{N-1}, r(T_{N-1}))]\,,$$
$$V(T_{N-2}, r_j) \;=\; \max\left\{\tilde{V}(T_{N-2}, r_j),\; p_{N-2}(r_j)\right\}$$

for $j = 0, \ldots, n$. These steps are repeated until $V(T_1, r)$ is available. The desired price of the Bermudan option is finally determined as

$$V(0, r(0)) = P(0, T_1) \cdot \mathbb{E}^{T_1}\left[V(T_1, r(T_1))\right].$$

The computationally crucial step is the evaluation of the expectation

$$\mathbb{E}^{T_{N-1}}\left[V(T_{N-1}, r(T_{N-1}))\right]$$

depending on the stochastic short rate $r(T_{N-1})$ given the information at time T_{N-2}. From the dynamics of the Hull White model we know that the short rate is normally distributed. Moreover, we may derive corresponding mean and variance parameters

$$\mu = \mathbb{E}^{T_{N-1}}\left[r(T_{N-1})\right]$$

and

$$\sigma = Var\left[r(T_{N-1})\right] = \mathbb{E}^{T_{N-1}}\left[\left(r(T_{N-1}) - \mathbb{E}^{T_{N-1}}\left[r(T_{N-1})\right]\right)^2\right].$$

For a derivation of these quantities, see e.g. [11, Appendix C]. Given the distribution parameters the expectation of the option payoff is evaluated as

$$\mathbb{E}^{T_{N-1}}\left[V(T_{N-1}, r(T_{N-1}))\right] = \frac{1}{2\pi\sigma^2} \cdot \int_{-\infty}^{\infty} V(T_{N-1}, r) \cdot e^{-\frac{(r-\mu)^2}{2\sigma^2}} dr.$$

In the implementation the integral is solved by interpolating $V(T_{N-1}, \cdot)$ and numerical quadrature.

5.3 Pricing and Vega Calculation Example

In this section we illustrate an example for price and Vega calculation. As example, we choose a 5 years into 10 years Bermudan swaption with annual exercise dates. Market data are selected for EUR as of September 2016 month end.

As mentioned in Sect. 5.2, we work in a single interest rate curve setting. We model the 6m swap curve which is derived from par quotes of forward rate agreements and Vanilla swaps. Figure 5.1 shows the calibrated yield curve in terms of continuous compounded zero rates. Note that rates are negative up to about 7y maturity; a phenomenon typical for current low interest rate markets.

In addition to interest rates, we need to incorporate volatility information. Current low interest rates are not compatible with log-normal implied volatilities and corresponding models. The market provides quotes for shifted log-normal as well as normal implied volatilities. For this analysis we choose normal volatilities to avoid the additional dependency on the volatility shift parameter. Nevertheless, the concepts can be adapted easily to shifted log-normal volatilities as well.

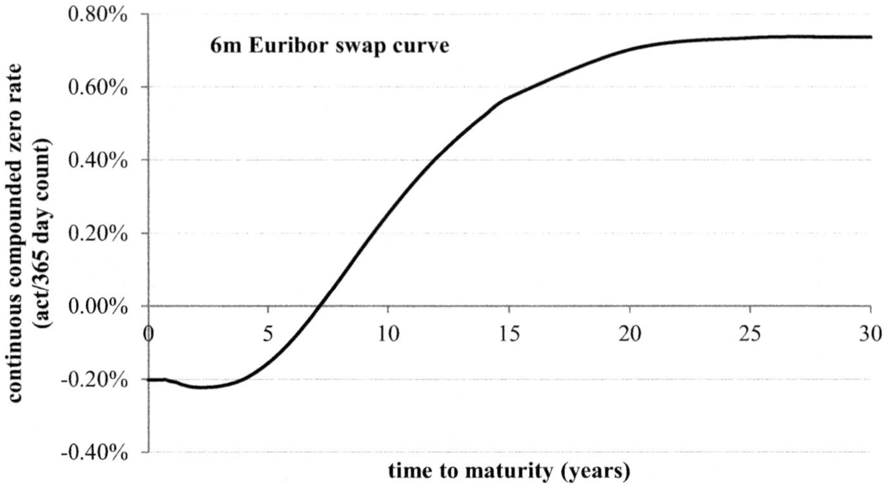

Fig. 5.1 EUR 6 months swap curve

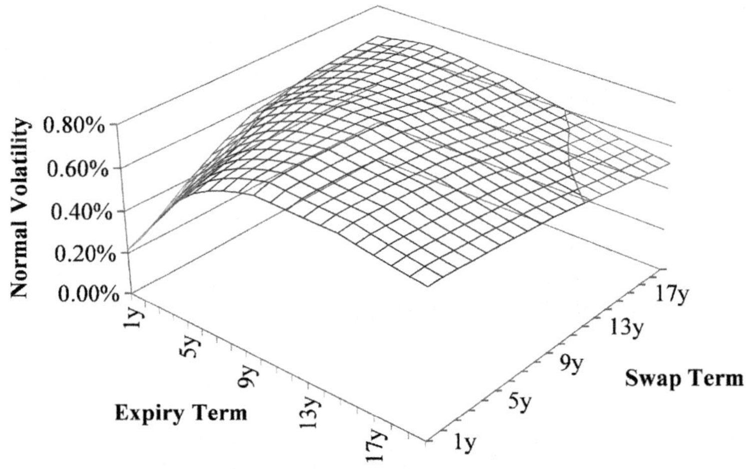

Fig. 5.2 EUR at-the-money normal swaption volatilities

Figure 5.2 illustrates the at-the-money swaption volatility surface. For brevity, we omit the volatility smile, i.e. the market implied dependency of the volatilities on the option strike.

The Hull White model is set up with 1% mean reversion speed and piecewise constant (backward flat) short rate volatilities. We calibrate the model to co-terminal European swaptions corresponding to the exercises of the Bermudan.

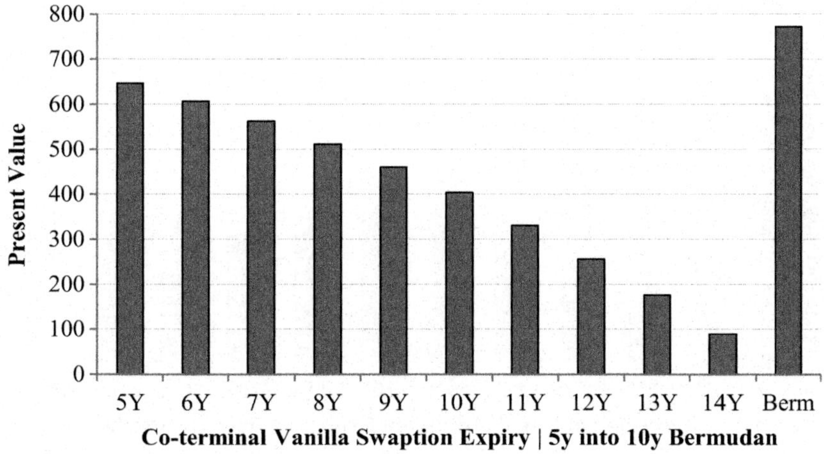

Fig. 5.3 European and Bermudan swaption present value

Given that we have a 5 years into 10 years Bermudan with annual exercises we get exercise dates 5 years, 6 years, ..., 14 years. Market volatilities for the reference European swaption expiries are derived by interpolating the at-the-money implied normal volatility surface. The resulting market volatilities $\sigma_{\text{Mkt,i}}$ are fed into Bachelier's formula to get the reference European swaption prices corresponding to the $\text{Swaption}_i(\cdot)$ market prices in Sect. 5.1.1. We set the short rate volatility grid equal to the expiry dates of the reference European swaptions and calibrate the short rate volatility parameters to match the market prices.

Before looking at Vega calculation we analyze the Bermudan swaption pricing. Figure 5.3 illustrates the European swaption present values $\text{Swaption}_i(\cdot)$ as well as the Bermudan value denoted by $V = \text{Exotic}(\cdot)$ in Sect. 5.1.1. The result nicely illustrates that the Bermudan value exceeds the maximum European value by the switch option value, i.e. the value of the option to postpone the exercise decision.

Figure 5.4 compares the Vegas of the European swaptions with the flat Vega of the Bermudan swaption. European Vega is evaluated by differentiating Bachelier's formula. Bermudan Vega is derived along the lines elaborated in Sect. 5.1.1 utilizing Algorithmic Differentiation to derive the gradient $\text{Exotic}'(\cdot)$ and the Jacobian $\text{Model}'(\cdot)$. This procedure naturally yields sensitivities w.r.t. all the individual input volatilities. A flat Bermudan Vega is derived by summing up the individual Vega contributions.

The results in Fig. 5.4 show a similar pattern as for the present value: The Bermudan Vega exceeds the Vega of the maximum European swaption Vega.

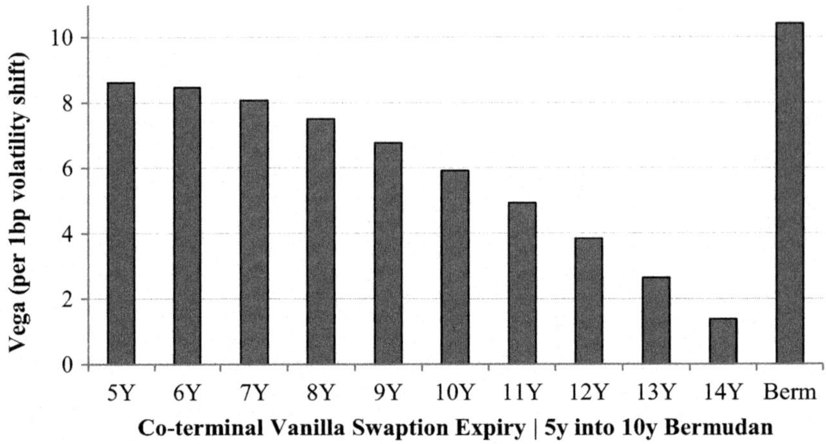

Fig. 5.4 European and (flat) Bermudan swaption Vega (per 1bp normal vol shift)

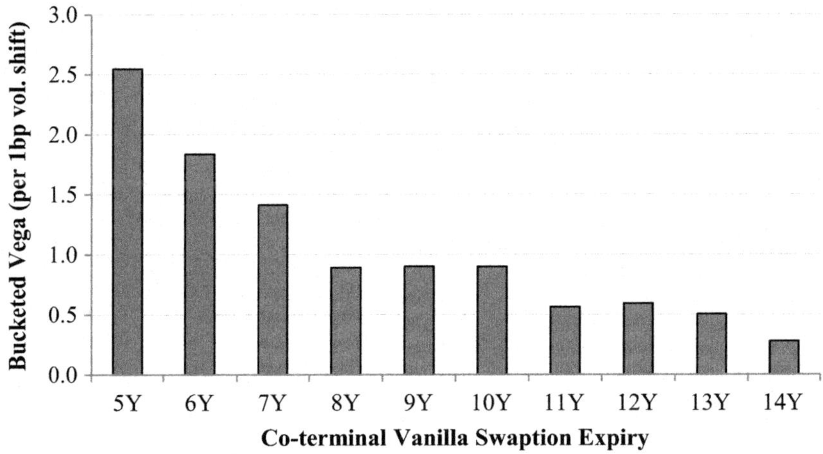

Fig. 5.5 Bucketed Bermudan Vega (per 1bp normal vol shift)

For risk management and hedging it is of particular interest to understand the risk contribution of the individual buckets. This information is naturally obtained by the sensitivity calculation procedure applied. The corresponding bucketed Bermudan Vega $dV/d\sigma_{\text{Mkt}}$ is shown in Fig. 5.5.

Note that the Bucketed normal Vega w.r.t. market volatilities differs structurally from the Bermudan sensitivity w.r.t. the short rate model volatilities. The latter is denoted by $dV/d\sigma_{\text{Mdl}}$ and given in Fig. 5.6. The short rate volatility of the initial 5y no-call period dominates the overall impact on the Bermudan value.

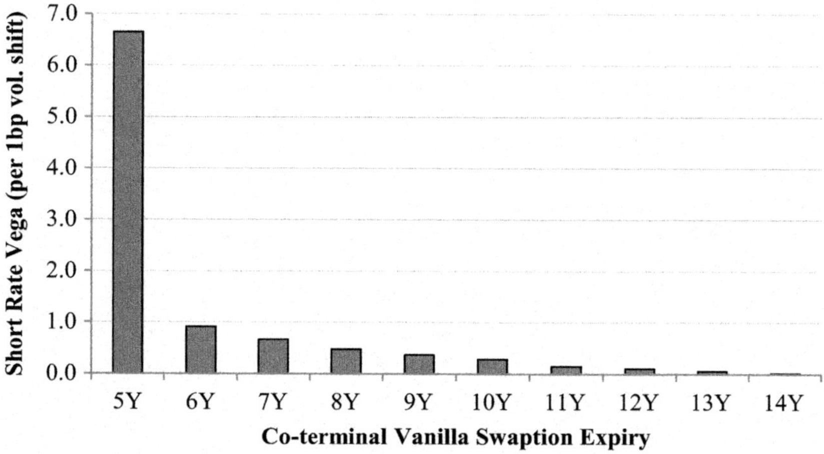

Fig. 5.6 Bucketed Bermudan short rate Vega (per 1bp normal vol shift)

5.3.1 Implementation and Computational Costs

The methodology described is implemented in C++ and exported to Excel. The main computational effort lies in the evaluation of the Bermudan value. Model calibration and matrix factorization (or inversion) are usually not an issue for this kind of application. AD theory predicts an increase of the AD-enabled code compared to the original code by a factor of 4–5. In our application we observe a factor of about 10.

References

1. Brigo, D., Mercurio, F.: Interest Rate Models – Theory and Practice. Springer, Berlin (2007)
2. Capriotti, L.: Fast greeks by algorithmic differentiation. J. Comput. Finance **14**, 3–35 (2011)
3. Giles, M.B., Glasserman, P.: Smoking adjoints: fast Monte Carlo greeks. Risk **19**, 88–92 (2006)s
4. Griewank, A., Walther, A.: Evaluating Derivatives: Principles and Techniques of Algorithmic Differentiation, 2nd edn. SIAM, Philadelphia (2008)
5. Henrard, M.: Adjoint algorithmic differentiation: calibration and implicit function theorem. Risk **17**, 37–47 (2014)
6. Hull, J.C., White, A.: Pricing interest-rate-derivative securities. Rev. Finance Stud. **3**, 573–592 (1990)
7. Jamshidian, F.: An exact bond option pricing formula. J. Finance **44**, 205–209 (1989)
8. Leclerc, M., Liang, Q., Schneider, I.: Fast Monte Carlo Bermudan greeks. Risk **22**, 84–88 (2009)
9. Naumann, U.: The Art of Differentiating Computer Programs: An Introduction to Algorithmic Differentiation. SIAM, Philadelphia (2012)
10. Naumann, U., Leppkes, K., Lotz, J.: dco/C++ user guide. Technical Report, RWTH Aachen (2014). Technical Report AIB-2014-03

11. Schlenkrich, S.: Evaluating sensitivities of Bermudan swaptions. Master's thesis, University of Oxford (2011)
12. Schlenkrich, S.: Efficient calibration of the Hull White model. Optim. Control Appl. Methods **33**, 249–374 (2012)
13. Schlenkrich, S., Miemiec, A.: Choosing the right spread. Wilmott Mag. 60–67 (2015)

Chapter 6
Modelling and Calibration of Stochastic Correlation in Finance

Long Teng, Matthias Ehrhardt, and Michael Günther

Abstract This chapter deals with the modelling and calibration of stochastic correlations. Correlation plays essential role in pricing derivatives on multi-assets. Market observations give evidence that the correlation is hardly a deterministic quantity, however, a constant or deterministic correlation has been widely used, although it may lead to correlation risk. It has been recently proposed to model correlation by a stochastic process, similar to stochastic volatility process. In this chapter, we review the concept of stochastic correlation process including calibration via the transition density function and its application for pricing the European-style Quanto option. As an illustrating example, we compare the Quanto option prices between using constant and stochastic correlation and analyze the effect of considering stochastic correlations on pricing the Quanto option.

6.1 Introduction

For two random variables X_1 and X_2 with finite variances, the correlation of them is defined as

$$\rho_{1,2} = \mathrm{Corr}(X_1, X_2) = \frac{\mathrm{Cov}(X_1, X_2)}{\sigma_1 \sigma_2}\,, \tag{6.1}$$

with covariance

$$\mathrm{Cov}(X_1, X_2) = \mathbb{E}\big[(X_1 - \mu_1)(X_2 - \mu_2)\big], \tag{6.2}$$

L. Teng (✉) • M. Ehrhardt • M. Günther
Lehrstuhl für Angewandte Mathematik/Numerische Analysis, Bergische Universität Wuppertal, Wuppertal, Germany
e-mail: teng@math.uni-wuppertal.de; ehrhardt@math.uni-wuppertal.de; guenther@math.uni-wuppertal.de

M. Ehrhardt et al. (eds.), *Novel Methods in Computational Finance*, Mathematics in Industry 25, DOI 10.1007/978-3-319-61282-9_6

where μ_i and σ_i are the expectation and standard deviation of X_i, $i = 1, 2$. Here $\rho_{1,2}$ denotes a coefficient number in the interval $[-1, 1]$. The boundaries -1 and 1 will be reached if and only if X_1 and X_2 are indeed perfectly negatively and positively related. The greater the absolute value of $\rho_{1,2}$ the stronger the dependence between X_i, $i = 1, 2$ is. Hence, the concept can be used to quantify the relationship between financial quantifies driven by random variables.

In finance, the relationship between financial quantities plays usually an essential role, see, e.g., wrong way and right way risk in portfolio credit models [2, 15]. The correlation concept (6.1) known in statistics as the (linear) Pearson correlation coefficient [10], had been widely applied for the relationship in financial applications. However, by doing that, there are several disadvantages or fallacies, we state only some of them:

- If the random variables X_1 and X_2 are independent, then it follows $\rho_{1,2} = 0$. However, the converse implication does not hold, since in (6.1) only the two first moments are included. For example, we compute that $\rho_{1,2} = 0$ for $X_2 = X_1^2$. Indeed, X_1 and X_2 depend even almost perfectly on each other. This simple example illustrates that the correlation coefficient only recognizes linear dependences between random variables.
- Correlation is invariant under strictly increasing linear transformations, but, in contrast to Copula methods, not invariant under nonlinear strictly increasing transformations. For example, in general the correlation of the random variables X_1 and X_2 does not equal the correlation of the random variables $\ln X_1$ and $\ln X_2$, i.e. after a transformation of the financial data the correlation may change.
- Usually, the given marginal distributions and pairwise correlations of a random vector cannot determine its joint distribution.
- Finally, as stated above, the variances of the two random variables X_1 and X_2 has to be finite. This assumptions is not fulfilled for every standard distribution, e.g. the Students's t-distribution with $v \leq 2$ degrees of freedom possess an infinite variance.

For more detailed information about the disadvantages or fallacies we refer to [9].

Using the correlated Brownian motions (BMs) is a standard way to insert the correlation concept for relationship in financial models. We illustrate this by a simple example: the coupled stochastic processes for the European-style Quantity (Quanto) adjusting option in the Black-Scholes framework

$$\begin{cases} dS_t = \mu_S S_t \, dt + \sigma_S S_t \, dW_t^S \\ dR_t = \mu_R R_t \, dt + \sigma_R R_t \, dW_t^R, \end{cases} \tag{6.3}$$

with positive constants μ_S, μ_R, σ_S and σ_R. The first stochastic differential equation (SDE) describes the price of the traded asset in a currency A. The second SDE is used to model the exchange rate between currency A and another currency B. The dependence between the series is given by the correlated BMs W^S and W^R, with the symbolic notion

$$dW_t^S \, dW_t^R = \rho_{S,R} \, dt. \tag{6.4}$$

We see that the BMs are assumed to be correlated by a constant correlation $\rho_{S,R} \in [-1, 1]$ which is a measure of co-movements between S_t and R_t.

In financial markets, the first problem of using a correlation concept is the "observability". Unlike other quantities as price, exchange rate and so on, the correlation can not be obtained directly from the market and can only be measured in the context of a model. The easiest estimator of the correlation is the sample correlation coefficient. Given a series of N measurements of X_1 and X_2, which are observable quantities in the market, and denoting the measurements by $x_{1,j}$ and $x_{2,j}$, $j = 1, 2, \ldots, N$, the sample coefficient correlation reads

$$\hat{\rho}_{1,2} = \frac{\sum_{j=1}^{N}(x_{1,j} - \bar{\mu}_1)(x_{2,j} - \bar{\mu}_2)}{\sqrt{\sum_{j=1}^{N}(x_{1,j} - \bar{\mu}_1)^2 \sum_{j=1}^{N}(x_{2,j} - \bar{\mu}_2)^2}}, \tag{6.5}$$

where $\bar{\mu}_1$ and $\bar{\mu}_2$ are the sample means of X_1 and X_2.

As we explained above, the constant correlation coefficient defined by (6.1) only captures linear relationships between X_1 and X_2. Therefore, in the model (6.3) a linear dependence between S_t and R_t is assumed. However, from the market we realize that there is often a non-linear dependence between returns. Specifically, a constant correlation means that the two return processes are jointly stationary which is generally not true in the real world. Thus, the correlation is hardly a fixed constant, i.e. the constant correlation may not be an appropriate measure of co-dependence. There exist already some works which show that the correlation should not be constant and even changes over a small time interval as the volatility, see e.g. [14]. Several approaches generalize the constant correlation to a time-varying and stochastic concept, like conditional correlation in [3, 5].

How to estimate a time-varying correlation from the market data using the estimator (6.5)? At time t, using the n_T times most recent daily returns, the correlation at time t is given by the following estimator

$$\hat{\rho}_t = \frac{\sum_{j=1}^{n_T}(\hat{s}_{t-j} - \frac{1}{n_T}\sum_{j=1}^{n_T}\hat{s}_{t-j})(\hat{r}_{t-j} - \frac{1}{n_T}\sum_{j=1}^{n_T}\hat{r}_{t-j})}{\sqrt{\sum_{j=1}^{n_T}(\hat{s}_{t-j} - \frac{1}{n_T}\sum_{j=1}^{n_T}\hat{s}_{t-j})^2 \sum_{j=1}^{n_T}(\hat{r}_{t-j} - \frac{1}{n_T}\sum_{j=1}^{n_T}\hat{r}_{t-j})^2}}. \tag{6.6}$$

We then just need to roll it to the time $t + 1$, and so on to obtain a series of correlations through the time, which is known as historical correlation. And the following question is how long should the time window N_T be? To address this question we firstly give an example of historical correlations between *S&P 500 index* and *Euro/US-Dollar exchange rate* on a daily basis. We use $\bar{s}$ and $\bar{r}$ to denote the daily return series of S&P 500 and Euro/US-Dollar exchange rate and fix a size time window n_T, e.g. $n_T = 60$ for 60-day historical correlation. In Fig. 6.1, the 15-day, 30-day and 60-day historical correlations are displayed.

Firstly, we see that the longer a time window (the value of n_T) the less volatile a historical correlation is. In Fig. 6.1, the 15-day historical correlation is more variable than the 30-day historical correlation which is again more variable than the 60-day correlation. With a longer averaging period a long-term correlation is calculated. If we choose $n_T = 10$ or 15 days, the estimated correlation for each time t using (6.6), could be seen as a short-term correlation of the current market phenomena whose immediate past returns are used for the estimation. It is worthwhile noting that the events, especially, some extreme events in a time window will affect the correlation which would be estimated in the following time windows, even has a delayed effect on the long-term correlation. If one assumes that the phenomena in the past could be a reflection of the future, one would like to use the historical correlation as a forecast for the future.

The behaviour of the historical correlations in Fig. 6.1 gives evidence that one should describes the correlation using a mean-reverting stochastic process. Thereby, not only the variation of the short-term correlation can be reflected, also the attributes of long-term correlation is determined by the long-term parameter values, like long-term mean value and mean reversion speed. Besides, implied correlation in the context of a model also shows us that the correlation should be time-varying and behaves like a stochastic process.

To see more properties, which a mean-reverting stochastic process should have to be a stochastic correlation process (SCP), we plot its empirical density functions

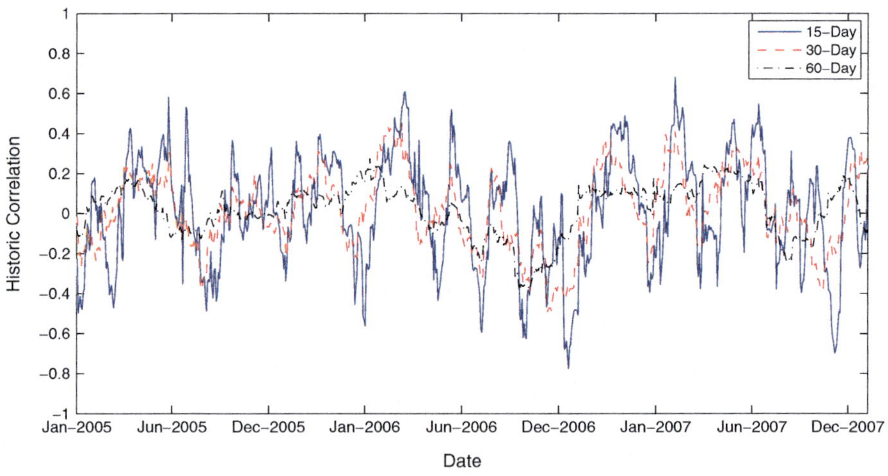

Fig. 6.1 Historical correlation between S&P 500 and Euro/US-Dollar exchange rate. (Source of data: www.yahoo.com)

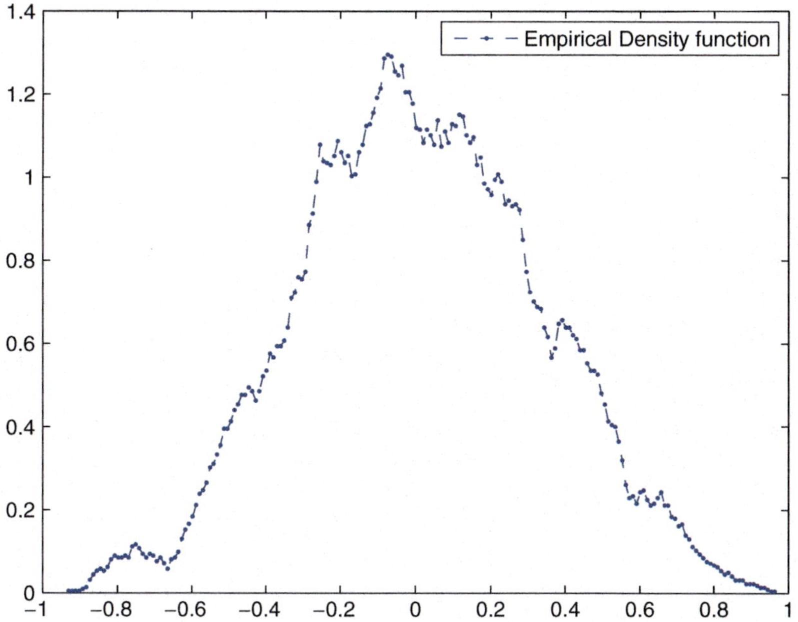

Fig. 6.2 Empirical density function of the historical correlation between S&P 500 and Euro/US-Dollar exchange rate with bandwidth $1/40$

in Figs. 6.2 and 6.3, using different bandwidths. We refer to [1] for the detailed information about the estimation of density function from historical data. From studying the empirical density functions we request that the SCP should satisfy the following properties:

1. only takes values in the interval $(-1, 1)$,
2. varies around a mean value,
3. the probability mass tends to zero at the boundaries -1, $+1$.

6.2 Stochastic Correlation Models

For modelling correlation as a stochastic quantity, we firstly refer to the dynamic conditional correlation model by Engle [5] and the Wishart autoregressive process proposed by Gourieroux [7]. Moreover, as the previous section indicates that one could model correlation as a proper stochastic process. In [8, 21], the authors suggested to use a modified Jacobi process to model stochastic correlations. A restriction on the parameter range has been found to ensure that the boundaries -1 and 1 of the correlation process are not attractive and unattainable. A more general stochastic correlation process was proposed by Teng et al. [16], which relies on

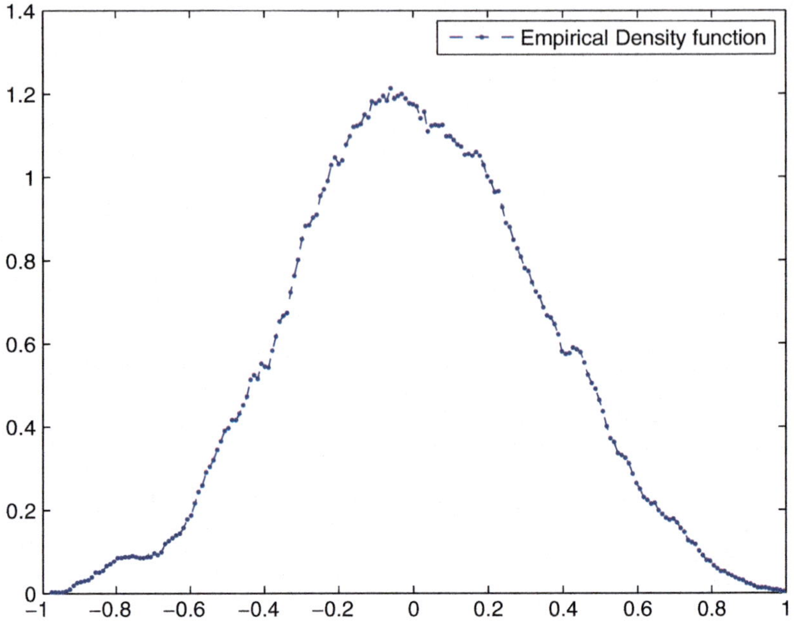

Fig. 6.3 Empirical density function of the historical correlation between S&P 500 and Euro/US-Dollar exchange rate with bandwidth $1/30$

the hyperbolic transformation with the function tangens hyperbolicus of any mean-reverting process with positive and negative value, the properties (1)–(3) above can be thus directly satisfied without facing any additional parameter restrictions. It is more general because it had been proved that the type of modified Jacobi process in [8, 21] turns out to be a special case of the hyperbolic transformation of a stochastic process, cf. [17].

How to incorporate a SCP into financial models? The most intuitive way is to generalise the concept of the correlated BMs (6.4). Based on two independent Brownian motions $W_{2,t}$ and $W_{3,t}$ we define

$$W_{1,t} = \int_0^t \rho_s \, dW_{2,s} + \int_0^t \sqrt{1-\rho_s^2} \, dW_{3,s}, \tag{6.7}$$

where ρ_t is a SCP.

Lemma 6.1 *$W_{1,t}$ satisfies*

(1) $W_{1,0} = 0$,
(2) $\mathbb{E}\left[(W_{1,t})^2\right] = t$,
(3) $\mathbb{E}[W_{1,t}|\mathscr{F}_s] = W_{1,s}$, *for* $s \leq t$.

Proof

(1) is obvious.
We calculate the two expected values as follows:

(2)

$$\mathbb{E}\left[(W_{1,t})^2\right] = \mathbb{E}\left[\left(\int_0^t \rho_s\, dW_{2,s}\right)^2 + \left(\int_0^t \sqrt{1-\rho_s^2}\, dW_{3,s}\right)^2 + 2\int_0^t \rho_s\, dW_{2,s} \int_0^t \sqrt{1-\rho_s^2}\, dW_{3,s}\right]$$

$$= \mathbb{E}\left[\int_0^t \rho_s^2\, ds + \int_0^t (1-\rho_s^2)\, ds\right] + \underbrace{\mathbb{E}\left[2\int_0^t \rho_s\, dW_{2,s} \int_0^t \sqrt{1-\rho_s^2}\, dW_{3,s}\right]}_{=0,\ \text{since}\ W_2 \perp W_3.}$$

$$= \int_0^t 1\, ds = t.$$

(3)

$$\mathbb{E}[W_{1,t}|\mathcal{F}_s] = W_{1,s} + \underbrace{\mathbb{E}\left[\int_s^t \rho_{s_1} dW_{2,s_1} + \int_s^t \sqrt{1-\rho_{s_1}^2} dW_{3,s_1} |\mathcal{F}_s\right]}_{:=0}.$$

Regarding the two independent BMs $W_{2,t}$ and $W_{3,t}$, the new BM $W_{1,t}$ has been defined which satisfies

$$\mathbb{E}\left[W_{1,t} \cdot W_{2,t}\right] = \mathbb{E}\left[\int_0^t \rho_s\, ds\right]. \tag{6.8}$$

One can straightly see that (6.8) agrees for

$$\mathbb{E}\left[W_{1,t} \cdot W_{2,t}\right] = \rho_{1,2}\, t, \tag{6.9}$$

where $W_{1,t}$ and $W_{2,t}$ are correlated by the constant $\rho_{1,2}$. Thus we use the symbolic notion

$$dW_{1,t}\, dW_{2,t} = \rho_t\, dt. \tag{6.10}$$

for the definition of that the Brownian motions $W_{1,t}$ and $W_{2,t}$ are correlated by the SCP ρ_t. Through this stochastically correlated BMs we can directly include exogenous stochastic correlations into financial models.

6.3 A General Stochastic Correlation Process

In this section, we study the hyperbolic transformation proposed in [16, 17] of a mean-reverting process to be a SCP. We fix a probability space $(\Omega, \mathscr{F}, \mathbb{P})$ and an information filtration $(\mathscr{F}_t)_{t \in \mathbb{R}+}$ satisfying the standard conditions, see e.g., [11].

6.3.1 *The Transformed Mean-Reverting Process*

For the motivations and the properties (1)–(3) in Sect. 6.1, Teng et al. [16] proposed to use the tangens hyperbolicus function of a mean-reverting stochastic process X_t with positive and negative values)

$$dX_t = a(t, X_t)dt + b(t, X_t)\, dW_t, \quad t \geq 0,\ X_0 = x_0, \tag{6.11}$$

to model the correlation as

$$\rho_t = \tanh(X_t), \quad \rho_0 = \tanh(x_0) \in (-1, 1). \tag{6.12}$$

Obviously, the properties (1)–(3) are fulfilled due to the range of values of the tangens hyperbolicus and mean reversion of the process. Besides, the function tanh is symmetrical and measurable. Although the function tanh can not really attain -1 and 1 which presents perfect negative and perfect positive dependence, respectively. It should make no difference to use this function for modelling correlations, because the correlation equal to -1 or 1 is indeed an extreme event which happens very rarely in the real market, see e.g., Fig. 6.1. Besides, the function *tanh* tends to the boundaries -1 and 1 at infinity.

Applying Itô's Lemma with (6.12)

$$d\rho_t = d\tanh(X_t) = \frac{\partial \tanh(X_t)}{\partial t}dt + \frac{\partial \tanh(X_t)}{\partial x}dX_t + \frac{1}{2}\frac{\partial^2 \tanh(X_t)}{\partial x^2}(dX_t)^2, \tag{6.13}$$

we obtain the SCP

$$d\rho_t = (1 - \rho_t^2)\left((\tilde{a} - \rho_t \tilde{b}^2)dt + \tilde{b}dW_t\right), \quad t \geq 0, \tag{6.14}$$

where $\rho_0 \in (-1, 1)$, $\tilde{a} = a(t, \text{artanh}(\rho_t))$ and $\tilde{b} = b(t, \text{artanh}(\rho_t))$. From (6.14) we see that there is a suitable number of free parameters to calibrate the model to market data. Besides, it is obvious, in this approach any mean-reverting process (with positive and negative values) can be considered without facing any additional parameter restrictions. The free parameters are hidden in the functions a and b, see the example (6.16) in Sect. 6.3.2 and (6.26) in Sect. 6.3.3. Why is the function

$\tanh(x)$ chosen for modelling correlation, we refer to [16, 17]. In the next sections we present two examples of SCPs.

6.3.2 The van Emmerich's Correlation Model

As the first example, we show that van Emmerich's correlation model can be obtained by transforming a special mean-reverting process (6.15), i.e. the van Emmerich's correlation process is just a special case of the general transformation [16]. To do so, we define the following mean-reverting process

$$dX_t = \frac{\kappa\big(\mu - \tanh(X_t)\big)}{1 - \tanh^2(X_t)} dt + \frac{\sigma}{\sqrt{1 - \tanh^2(X_t)}} dW_t, \quad t \geq 0, \ X_0 = x_0, \tag{6.15}$$

where κ and σ are positive, $\mu \in (-1, 1)$. Next, we transform (6.15) with $\rho_t = \tanh(X_t)$. Again, applying Itô's Lemma we obtain

$$d\rho_t = \big[(\kappa(\mu - \rho_t)) - \sigma^2 \rho_t\big] dt + \sigma \sqrt{1 - \rho_t^2}\, dW_t, \tag{6.16}$$

the calculation is straight but a little tedious. If we define

$$\kappa^* = \kappa + \sigma^2, \tag{6.17}$$

$$\mu^* = \frac{\kappa\mu}{\kappa + \sigma^2}, \tag{6.18}$$

$$\sigma^* = \sigma, \tag{6.19}$$

the correlation process (6.16) can be rewritten as

$$d\rho_t = \kappa^*(\mu^* - \rho_t)dt + \sigma^* \sqrt{1 - \rho_t^2}\, dW_t, \tag{6.20}$$

which is exactly the van Emmerich's correlation process in [21]. Due to the transformation with the function *tanh*, the correlations provided by (6.20), whose coefficients are found in (6.17)–(6.19), are obviously located in the interval $(-1, 1)$. van Emmerich [21] derived the analytic condition

$$\kappa^* \geq \frac{\sigma^*}{1 \pm \mu*} \tag{6.21}$$

of that the boundaries -1 and 1 can not be unattainable. We see that the correlation process (6.20) must have already satisfied the condition (6.21): substituting

(6.17)–(6.19) in (6.21) we get

$$\frac{\sigma^2}{\kappa(1 \pm \mu) + \sigma^2} \leq 1, \tag{6.22}$$

which always holds whilst κ is positive and $\mu \in (-1, 1)$.

6.3.3 *The Transformed Modified Ornstein-Uhlenbeck Process*

As the other example, we consider the SCP obtained by transforming the modified Ornstein-Uhlenbeck (OU) process The OU process [20] is defined by the stochastic differential equation

$$dX_t = \kappa(\mu - X_t)\,dt + \sigma\,dW_t, \tag{6.23}$$

where $\kappa, \sigma > 0$ and $X_0, \mu \in \mathbb{R}$. If we want to restrict the mean value μ to be only in $(-1, 1)$, it is reasonable to modify the Ornstein-Uhlenbeck process (6.23) as

$$dX_t = \kappa\big(\mu - \tanh(X_t)\big)\,dt + \sigma\,dW_t, \tag{6.24}$$

where $\kappa, \sigma > 0$ and $X_0, \mu \in (-1, 1)$.

Lemma 6.2 *Applying Itô's Lemma with* $\rho_t = \tanh(X_t)$,

$$d\rho_t = \frac{\partial \tanh(X_t)}{\partial x} dX_t + \frac{1}{2}\frac{\partial^2 \tanh(X_t)}{\partial x^2}\sigma^2\,dt \tag{6.25}$$

gives the stochastic correlation process as

$$d\rho_t = (1 - \rho_t^2)\big(\kappa(\mu - \rho_t) - \sigma^2\rho_t\big)\,dt + (1 - \rho_t^2)\sigma\,dW_t, \tag{6.26}$$

where $t \geq 0$, $\rho_0 \in (-1, 1)$, $\kappa, \sigma > 0$ *and* $\mu \in (-1, 1)$.

Proof

$$\begin{aligned}
(6.25) &= \operatorname{sech}^2(X_t)\kappa\big(\mu - \tanh(X_t)\big)dt - \operatorname{sech}^3(X_t)\sinh(X_t)\sigma^2 dt + \operatorname{sech}^2(X_t)\sigma^2 dW_t \\
&= \operatorname{sech}^2(X_t)\kappa\big(\mu - \tanh(X_t)\big)dt - \operatorname{sech}^2(X_t)\frac{\sinh(X_t)}{\cosh(X_t)}\sigma^2 dt + \operatorname{sech}^2(X_t)\sigma^2 dW_t \\
&= (1 - \rho_t^2)\kappa(\mu - \rho_t)dt - (1 - \rho_t^2)\rho_t\sigma^2 dt + (1 - \rho_t^2)\sigma^2 dW_t \\
&= (6.26).
\end{aligned}$$

Again, we define

$$\kappa^* = \kappa + \sigma^2, \tag{6.27}$$

$$\mu^* = \frac{\kappa\mu}{\kappa + \sigma^2}, \tag{6.28}$$

$$\sigma^* = \sigma, \tag{6.29}$$

and rewrite (6.26) as

$$\frac{d\rho_t}{1-\rho_t^2} = \kappa^*(1-\mu^*)dt + \sigma^*\, dW_t, \tag{6.30}$$

where $t \geq 0$, $\rho_0 \in (-1, 1)$, $\kappa^*, \sigma^* > 0$ and $\mu^* \in (-1, 1)$.

6.4 Calibration Via Density Function

In this section, we show how to fit a SCP to historical market data via its density function. As an example, we consider the calibration of the SCP (6.30).

6.4.1 Transition Density Function

Let us assume that the stochastic correlation process (6.30) possesses a transition density $f(t, \tilde{\rho}|\rho_0)$ which satisfies the following Fokker-Planck equation

$$\frac{\partial}{\partial t}f(t,\tilde{\rho}) + \frac{\partial}{\partial \tilde{\rho}}\big(\hat{a}(t,\tilde{\rho})f(t,\tilde{\rho})\big) - \frac{1}{2}\frac{\partial^2}{\partial \tilde{\rho}^2}\big(\hat{b}(t,\tilde{\rho})^2 f(t,\tilde{\rho})\big) = 0, \tag{6.31}$$

with

$$\hat{a}(t,\tilde{\rho}) = \kappa^*(1-\mu^*)(1-\tilde{\rho}^2), \tag{6.32}$$

$$\hat{b}(t,\tilde{\rho}) = (1-\tilde{\rho}^2)\sigma^*. \tag{6.33}$$

For the calibration purpose we consider the stationary density (for $t \to \infty$)

$$f(\tilde{\rho}) := \lim_{t\to\infty} f(t,\tilde{\rho}|\rho_0). \tag{6.34}$$

With the above construction the SCP (6.30) is also a mean-reverting process. Thus one can show that every two solutions of (6.31) are the same for $t \to \infty$, i.e., a unique stationary solution $f(\tilde{\rho})$ exists, cf. [13]. In [17], $f(\tilde{\rho})$ has been given in a closed form

$$f(\tilde{\rho}) = \frac{(1+\tilde{\rho})^{a+b}(1-\tilde{\rho})^{a-b}}{M}, \tag{6.35}$$

where

$$a = \frac{\kappa^* - 2\sigma^{*2}}{\sigma^{*2}}, \quad b = \frac{\kappa^* \mu^*}{\sigma^{*2}}, \tag{6.36}$$

$$\begin{aligned} M :=& \frac{\Gamma(1+a-b)F(1,-a-b,2+a-b,-1)}{\Gamma(2+a-b)} \\ &+ \frac{\Gamma(1+a+b)F(1,-a+b,2+a+b,-1)}{\Gamma(2+a+b)}, \end{aligned} \tag{6.37}$$

with the hypergeometric function $F(a,b,c,x) = \sum_{k=0}^{\infty} \frac{x^k}{k!} \frac{(a)_k (b)_k}{(c)_k}$, $|x| < 1$, and the Gamma function Γ. $(\cdot)_k$ denotes the Pochhammer symbol $(a)_k = a(a+1)(a+2)\cdots(a+k-1)$, $(a)_0 = 1$. The condition of existence of the density function (35) is

$$a \pm b > 1 \tag{6.38}$$

which is equivalent to $\mu \in (-1, 1)$.

This result can be transmitted to the SCP in the form of

$$\frac{d\rho_t}{1-\rho_t^2} = \kappa(1-\mu)dt + \sigma\, dW_t. \tag{6.39}$$

with the arbitrary parameter coefficients $\kappa > 0$, $\mu \in (-1, 1)$ and $\sigma > 0$ instead of the defined κ^*, μ^*, σ^*. We check the existence condition of the density function: for this case we have for a and b, like defined in (6.36), as

$$a = \frac{\kappa - 2\sigma^2}{\sigma^2}, \quad b = \frac{\kappa\mu}{\sigma^2}. \tag{6.40}$$

We perform a similar calculation for checking the condition (6.38):

$$\begin{aligned} a+b > -1 &\Leftarrow \tfrac{\kappa-2\sigma^2}{\sigma^2} + \tfrac{\kappa\mu}{\sigma^2} > -1 \Leftarrow \kappa(1+\mu) > \sigma^2 \Leftarrow \kappa > \tfrac{\sigma^2}{1+\mu}, \\ a-b > -1 &\Leftarrow \tfrac{\kappa-2\sigma^2}{\sigma^2} - \tfrac{\kappa\mu}{\sigma^2} > -1 \Leftarrow \kappa(1-\mu) > \sigma^2 \Leftarrow \kappa > \tfrac{\sigma^2}{1-\mu}. \end{aligned}$$

Thus, the process (6.39) could be employed for the stochastic correlation if the condition

$$\kappa > \frac{\sigma^2}{1 \pm \mu} \tag{6.41}$$

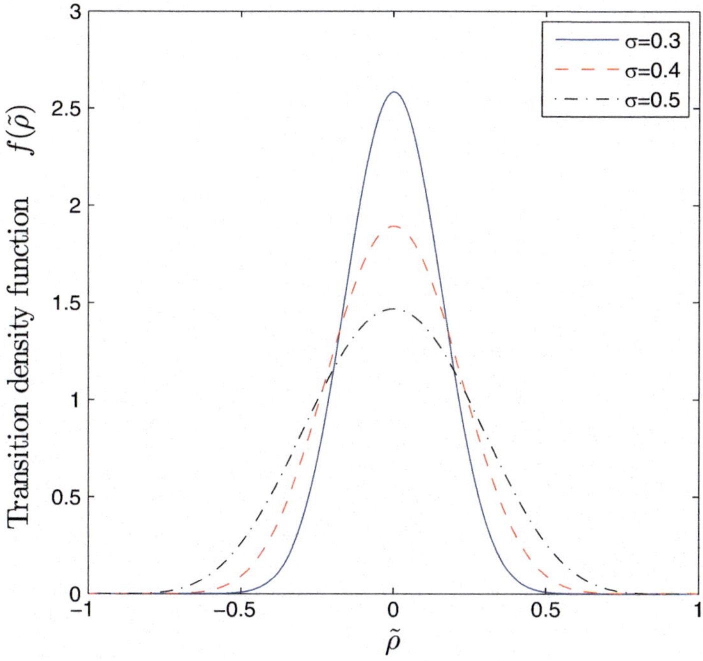

Fig. 6.4 Comparison of $f(\tilde{\rho})$ for different values of σ ($\kappa = 2$ and $\mu = 0$)

is fulfilled. We find that this condition dovetails nicely with that condition in [21], which ensures that the boundaries -1 and 1 are unattainable.

As an example, we let $\kappa = 2$ and $\mu = 0$ and display $f(\tilde{\rho})$ with different values of σ, which is equal to 0.3, 0.4 and 0.5, respectively. The behaviour of $f(\tilde{\rho})$ is displayed in Fig. 6.4. Obviously, σ shows the magnitude of variation from the mean value $\mu = 0$.

6.4.2 *Calibration*

We assume that the correlation is itself observable. Under this assumption the transition density can be used for calibration purposes. One uses usually *maximum-likelihood estimation (MLE)* when the density function is known. Considering the density function (6.35), it will be tedious to get its likelihood-function. An alternative approach to estimate the parameters is to fit the empirically observed density to the stationary density (6.35). As an example we fit the historical data from Fig. 6.2 to (6.35). This fitting works very well, see Fig. 6.5.

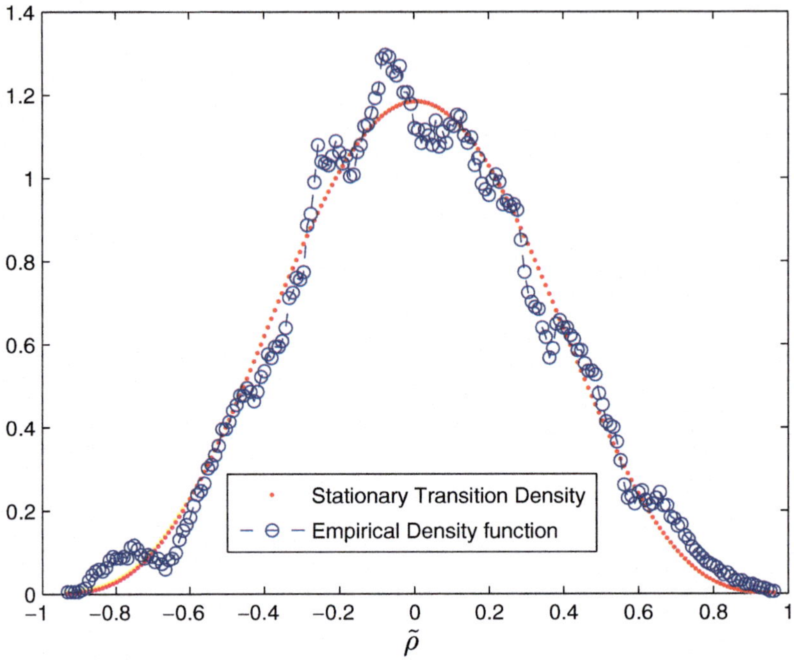

Fig. 6.5 Correlation between S&P 500 and Euro/US-Dollar exchange rate, empirical density compared to density (6.35) computed with $\kappa = 7.937$, $\mu = 0.003$ and $\sigma = 1.186$ (Mean Squared error: 2.46e-06)

6.5 Pricing Quantos with Stochastic Correlation

To illustrate the impact of using stochastic correlation on pricing, we use European-style Quantos as an example. These options hedge the exchange rate risk when investing in financial products not valued in the domestic currency. We extend the model (6.3) by including stochastic correlations driven by, e.g., the SCP (6.20) as

$$\begin{cases} dS_t = \mu_s S_t\, dt + \sigma_s S_t\, dW_t^s, \\ dR_t = \mu_r R_t\, dt + \sigma_r R_t\, dW_t^r, \\ d\rho_t = \kappa_\rho(\mu_\rho - \rho_t)dt + \sigma_\rho\sqrt{1-\rho_t^2}dW_t^\rho, \qquad \rho_0 \in [-1, 1], \end{cases} \tag{6.42}$$

where

$$dW_t^s dW_t^r = \rho_t\, dt, \quad dW_t^s dW_t^\rho = \rho_{s\rho}\, dt, \quad dW_t^r dW_t^\rho = \rho_{r\rho}\, dt. \tag{6.43}$$

And the parameters of $d\rho_t$ are assumed to satisfy the condition (6.41). In this model setting, the underlying asset process and the exchange rate process are assumed to be correlated stochastically, driven by the correlation process ρ_t which is by itself correlated with the underlying asset process by $\rho_{s\rho}$ and with the exchange rate by $\rho_{r\rho}$, respectively.

For this model (6.42), Teng et al. [19] have found the well approximated pricing formula in a closed form. In the sequel we present their results. The model can be straightforwardly specified under the risk-neutral measure as

$$\begin{cases} dS_t = (r_f - \sigma_s\sigma_r\rho_t)S_t\,dt + \sigma_s S_t\,d\tilde{W}_t^s, \\ dR_t = (r_h - r_f)R_t\,dt + \sigma_r R_t\,d\tilde{W}_t^r, \\ d\rho_t = (\kappa_\rho(\mu_\rho - \rho_t) - \lambda(S_t, R_t, \rho_t, t))dt + \sigma_\rho\sqrt{1-\rho_t^2}d\tilde{W}_t^\rho, \end{cases} \tag{6.44}$$

with

$$d\tilde{W}_t^s d\tilde{W}_t^r = \rho_t\,dt, \quad d\tilde{W}_t^s d\tilde{W}_t^\rho = \rho_{s\rho}\,dt, \quad d\tilde{W}_t^r d\tilde{W}_t^\rho = \rho_{r\rho}\,dt, \tag{6.45}$$

where $\lambda(S_t, R_t, \rho_t, t)$ represents the price of the correlation risk and could be assumed to be $\lambda\rho_t$, with a constant λ. For notational simplicity we let $\lambda = 0$. With the transform $x_t = \ln(S_t)$ and $y_t = \ln(R_t)$ the model (6.44) can be represented by

$$\begin{cases} dx_t = (r_f - \frac{1}{2}\sigma_s^2 - \sigma_s\sigma_r\rho_t)\,dt + \sigma_s\,d\tilde{W}_t^s, \\ dy_t = (r_h - r_f - \frac{1}{2}\sigma_r^2)\,dt + \sigma_r\,d\tilde{W}_t^r, \\ d\rho_t = \kappa_\rho(\mu_\rho - \rho_t)dt + \sigma_\rho\sqrt{1-\rho_t^2}d\tilde{W}_t^\rho. \end{cases} \tag{6.46}$$

We know that the underlying asset S is denominated in the foreign currency (denoted by F). Let the exchange rate R be the number of units of the domestic or home currency (denoted by H) per unit of F, namely $R = H/F$. Let $U(\ln(S_t), \ln(R_t), \rho_t, t)$ denote the value of any contract with the underlying asset in F but paid in H, obviously, based on (6.46), U must satisfy the partial differential equation (PDE)

$$\begin{aligned} &\frac{\partial U}{\partial t} + (r_f - \frac{\sigma_s^2}{2} - \sigma_s\sigma_r\rho_t)\frac{\partial U}{\partial x} + (r_h - r_f - \frac{\sigma_r^2}{2})\frac{\partial U}{\partial r} + \kappa_\rho(\mu_\rho - \rho_t)(t, \rho_t)\frac{\partial U}{\partial \rho} \\ &+ \frac{\sigma_s^2}{2}\frac{\partial^2 U}{\partial x^2} + \frac{\sigma_r^2}{2}\frac{\partial^2 U}{\partial r^2}\frac{\sigma_\rho^2(1-\rho_t^2)(t, \rho_t)}{2}\frac{\partial^2 U}{\partial \rho^2} + \sigma_s\sigma_r\rho_t\frac{\partial^2 U}{\partial x \partial r} \\ &+ \sigma_s\sigma_\rho\sqrt{1-\rho_t^2}\rho_{s\rho}\frac{\partial^2 U}{\partial x \partial \rho} + \sigma_r\sigma_\rho\sqrt{1-\rho_t^2}\rho_{r\rho}\frac{\partial^2 U}{\partial r \partial \rho} - r_h U = 0. \end{aligned} \tag{6.47}$$

We denote the value of a standard Quanto option by $V(S_t, R_t, \rho_t, t)$ which yields

$$V(S_t, R_t, \rho_t, t) = R_0 \cdot \mathrm{e}^{-r_h(T-t)} \cdot \mathbb{E}^H[\alpha(S_T - K)^+] \tag{6.48}$$

with the terminal condition $R_0 \cdot (\alpha(S_T - K)^+)$, where R_0 is the fixed exchange rate for the payment, e.g., one can take the today's rate, $\mathbb{E}^H[\cdot]$ is the expectation under domestic risk-neutral probability measure and $\alpha = 1$ for Quanto calls and $\alpha = -1$

for Quanto puts. Obviously, as a contract with the underlying asset in a foreign currency but paid in a domestic currency, the value of Quanto option (6.48) must satisfy the pricing PDE (6.47). As an example, we consider Quanto calls and without loss of generality we assume $R_0 = 1$, we thus have

$$C(S_t, R_t, \rho_t, t) = \mathrm{e}^{-r_h(T-t)} \cdot \mathbb{E}^H[(S_T - K)^+], \tag{6.49}$$

the price of Quanto puts can be determined straightforwardly from the put-call parity.

It is assumed that a solution of (6.49) has the form

$$\begin{aligned} C(S_t, R_t, \rho_t, t) =& \mathrm{e}^{x_t+(r_f-r_h)\tau}\mathbb{E}[\mathrm{e}^{-\sigma_s\sigma_r\int_t^T \rho_s ds}]P_1(x_T \geq \ln(K)) \\ & - \mathrm{e}^{-r_h\tau}KP_2(x_T \geq \ln(K)) \end{aligned} \tag{6.50}$$

with the time to maturity $\tau = T - t$. Due to the embedded stochastic correlation process, the probabilities P_1 and P_2 are not immediately available in a closed form. We know that not only P_j, but also their corresponding characteristic functions $\phi_j(x, r, \rho, t; u) = E[\mathrm{e}^{iux_T}|\mathscr{F}_t]$ satisfy the PDE (6.47) subject to the terminal condition

$$\phi_j(x, r, \rho, T; u) = \mathrm{e}^{iux_T}, \quad j = 1, 2. \tag{6.51}$$

Thus, one can obtain

$$\begin{aligned} &\frac{\partial \phi_1}{\partial t} + (r_f + \frac{\sigma_s^2}{2} - \sigma_s\sigma_r\rho_t)\frac{\partial \phi_1}{\partial x} + (r_h - r_f - \frac{\sigma_r^2}{2} + \sigma_s\sigma_r\rho_t)\frac{\partial \phi_1}{\partial r} \\ &+ (\kappa_\rho\mu_\rho - \kappa_\rho\rho_t + \sigma_s\sigma_\rho\rho_{s\rho})\frac{\partial \phi_1}{\partial \rho} + \frac{\sigma_s^2}{2}\frac{\partial^2 \phi_1}{\partial x^2} + \frac{\sigma_r^2}{2}\underline{(1-\rho_t^2)}\frac{\partial^2 \phi_1}{\partial r^2} + \frac{\sigma_\rho^2}{2}\frac{\partial^2 \phi_1}{\partial \rho^2} \\ &+ \sigma_s\sigma_r\rho_t\frac{\partial^2 \phi_1}{\partial x \partial r} + \sigma_s\sigma_\rho\underline{\sqrt{1-\rho_t^2}\rho_{s\rho}}\frac{\partial^2 \phi_1}{\partial x \partial \rho} + \sigma_r\sigma_\rho\underline{\sqrt{1-\rho_t^2}\rho_{r\rho}}\frac{\partial^2 \phi_1}{\partial r \partial \rho} = 0 \end{aligned} \tag{6.52}$$

and

$$\begin{aligned} &\frac{\partial \phi_2}{\partial t} + (r_f - \frac{\sigma_s^2}{2} - \sigma_s\sigma_r\rho_t)\frac{\partial \phi_2}{\partial x} + (r_h - r_f - \frac{\sigma_r^2}{2})\frac{\partial \phi_2}{\partial r} + \kappa_\rho(\mu_\rho - \rho_t)\frac{\partial \phi_2}{\partial \rho} \\ &+ \frac{\sigma_s^2}{2}\frac{\partial^2 \phi_2}{\partial x^2} + \frac{\sigma_r^2}{2}\frac{\partial^2 \phi_2}{\partial r^2} + \frac{\sigma_\rho^2}{2}\underline{(1-\rho_t^2)}\frac{\partial^2 \phi_2}{\partial \rho^2} + \sigma_s\sigma_r\rho_t\frac{\partial^2 \phi_2}{\partial x \partial r} \\ &+ \sigma_s\sigma_\rho\underline{\sqrt{1-\rho_t^2}\rho_{s\rho}}\frac{\partial^2 \phi_2}{\partial x \partial \rho} + \sigma_r\sigma_\rho\underline{\sqrt{1-\rho_t^2}\rho_{r\rho}}\frac{\partial^2 \phi_2}{\partial r \partial \rho} = 0. \end{aligned} \tag{6.53}$$

Obviously, due to the nonlinear coefficients ρ_t^2 and $\sqrt{1-\rho_t^2}$, the corresponding characteristic functions $\phi_j(x, r, \rho_t, t; u)$ can not be derived in a closed form. However, as indicated by Teng et al. in [18], such nonlinear coefficient could be linearized by its expectation which further can be well approximated by a linear combination of exponential functions.

Proposition 6.1 *Denote the original solution of* $\mathbb{E}[\rho_t^2]$ *in* (6.52) *and* (6.53) *by* $f_\rho(t)$ *which can be approximated by*

$$e^{-mt} + be^{-nt} + a, \tag{6.54}$$

where

$$a = \frac{(\sigma_\rho^2 + \kappa_\rho)(\sigma_\rho^2 + 2\kappa_\rho\mu_\rho^2)}{\sigma_\rho^4 + 3\kappa_\rho\sigma_\rho^2 + 2\kappa_\rho^2}, \quad b = \rho_0^2 - a - 1, \tag{6.55}$$

$$m = -2\log\left(\gamma_1 - be^{-\frac{n}{2}}\right), \quad n = -2\log\left(\frac{b\gamma_1 - \sqrt{b^2\gamma_1^2 - \gamma_2\gamma_3}}{\gamma_2}\right), \tag{6.56}$$

with

$$\gamma_1 := f_\rho(0.5) - a, \quad \gamma_2 := b + b^2, \quad \gamma_3 := \gamma_1^2 + a - f_\rho(1). \tag{6.57}$$

Proposition 6.2 *Denote the original solution of* $\mathbb{E}[\sqrt{1-\rho_t^2}]$ *in* (6.52) *and* (6.53) *by* $g_\rho(t)$ *which can be approximated by*

$$e^{-\hat{m}t} + \hat{b}e^{-\hat{n}t} + \hat{a}, \tag{6.58}$$

where

$$\begin{aligned} \hat{a} &= \sqrt{1 - \frac{(\sigma_\rho^2 + \kappa_\rho)(\sigma_\rho^2 + 2\kappa_\rho\mu_\rho^2) - \mu_\rho^4(\sigma_\rho^4 + 3\kappa_\rho\sigma_\rho^2 + 2\kappa_\rho^2)}{(1-\mu_\rho^2)(\sigma_\rho^4 + 3\kappa_\rho\sigma_\rho^2 + 2\kappa_\rho^2)}}, \\ \hat{b} &= \sqrt{1-\rho_0^2} - \hat{a} - 1, \end{aligned} \tag{6.59}$$

$$\hat{m} = -2\log\left(\zeta_1 - \hat{b}e^{-\frac{\hat{n}}{2}}\right), \quad \hat{n} = -2\log\left(\frac{\hat{b}\zeta_1 - \sqrt{\hat{b}^2\zeta_1^2 - \zeta_2\zeta_3}}{\zeta_2}\right), \tag{6.60}$$

with

$$\zeta_1 := g_\rho(0.5) - \hat{a}, \quad \zeta_2 := \hat{b} + \hat{b}^2, \quad \zeta_3 := \zeta_1^2 + \hat{a} - g_\rho(1). \tag{6.61}$$

For the proof and measure quality of the approximation we refer to [18]. Based on the Propositions 6.1 and 6.2, the corresponding approximated characteristic functions $\phi_j(x, r, \rho_t, t; u)$ of P_j, $j = 1, 2$ can thus be obtained by solving PDEs.

Lemma 6.3 *The characteristic function ϕ_1 reads*

$$\phi_1(x, r, \rho, t; u) = e^{D_1(\tau,u)+C_1(\tau,u)\rho_t+iux_t}, \tag{6.62}$$

with

$$C_1(u, \tau) = \underbrace{\frac{iu\sigma_s\sigma_r}{\kappa_\rho}}_{:=c_1}(e^{-\kappa_\rho\tau} - 1) \tag{6.63}$$

and

$$\begin{aligned} D_1(u, \tau) = &\frac{d_1(u)(1 - e^{-\kappa_\rho\tau})}{\kappa_\rho} + \frac{\sigma_\rho^2 c_1^2(1 - a)(1 - e^{-2\kappa_\rho\tau})}{4\kappa_\rho} + \frac{\sigma_\rho^2 c_1^2 e^{-mT+(m-\kappa_\rho)\tau}}{m - \kappa_\rho} \\ &+ \frac{\sigma_\rho^2 c_1^2 e^{-mT+(m-2\kappa_\rho)\tau}}{2(m - 2\kappa_\rho)} + \frac{b\sigma_\rho^2 c_1^2 e^{-nT+(n-\kappa_\rho)\tau}}{n - \kappa_\rho} + \frac{b\sigma_\rho^2 c_1^2 e^{-nT+(n-2\kappa_\rho)\tau}}{2(n - 2\kappa_\rho)} \\ &+ \frac{\sigma_s\sigma_\rho\rho_{s\rho}c_1(1 + iu)e^{-\hat{m}T+(\hat{m}-\kappa_\rho)\tau}}{\hat{m} - \kappa_\rho} + \frac{\sigma_s\sigma_\rho\rho_{s\rho}c_1\hat{b}(1 + iu)e^{-\hat{n}T+(\hat{n}-\kappa_\rho)\tau}}{\hat{n} - \kappa_\rho} \\ &- \frac{\sigma_\rho^2 c_1^2 e^{-(T-\tau)m}}{2m} - \frac{b\sigma_\rho^2 c_1^2 e^{-(T-\tau)n}}{2n} - \frac{\sigma_s\sigma_\rho\rho_{s\rho}c_1(1 + iu)e^{-(T-\tau)\hat{m}}}{\hat{m}} \\ &- \frac{\hat{b}\sigma_s\sigma_\rho\rho_{s\rho}c_1(1 + iu)e^{-(T-\tau)\hat{n}}}{\hat{n}} + d_2(u)\tau + d_3(u), \end{aligned} \tag{6.64}$$

where

$$d_1(u) = \kappa_\rho\mu_\rho c_1 + \sigma_s\sigma_\rho\rho_{s\rho}\hat{a}c_1(1 + iu) + \sigma_\rho^2 c_1^2(a - 1), \tag{6.65}$$

$$d_2(u) = r_f iu - \kappa_\rho\mu_\rho c_1 + (\frac{\sigma_s^2 iu}{2} - \sigma_s\sigma_\rho\rho_{s\rho}\hat{a}c_1)(1 + iu) + \sigma_\rho^2 c_1^2(a - 1) \tag{6.66}$$

$$\begin{aligned} d_3(u) = &-\frac{\sigma_\rho^2 c_1^2 e^{-mT}}{m - \kappa_\rho} + \frac{\sigma_\rho^2 c_1^2 e^{-mT}}{2(m - 2\kappa_\rho)} - \frac{b\sigma_\rho^2 c_1^2 e^{-nT}}{n - \kappa_\rho} + \frac{\sigma_\rho^2 c_1^2 e^{-nT}}{2(n - 2\kappa_\rho)} \\ &- \frac{\sigma_s\sigma_\rho\rho_{s\rho}c_1(1 + iu)e^{-\hat{m}T}}{\hat{m} - \kappa_\rho} - \frac{\sigma_s\sigma_\rho\rho_{s\rho}c_1\hat{b}(1 + iu)e^{-\hat{n}T}}{\hat{n} - \kappa_\rho} + \frac{\sigma_\rho^2 c_1^2 e^{-mT}}{2m} \\ &+ \frac{b\sigma_\rho^2 c_1^2 e^{-nT}}{2n} + \frac{\sigma_s\sigma_\rho\rho_{s\rho}c_1(1 + iu)e^{-\hat{m}T}}{\hat{m}} + \frac{\hat{b}\sigma_s\sigma_\rho\rho_{s\rho}c_1(1 + iu)e^{-\hat{n}T}}{\hat{n}}. \end{aligned} \tag{6.67}$$

Analogously,

Lemma 6.4 *The characteristic function ϕ_2 reads*

$$\phi_2(x, r, \rho, t; u) = e^{D_2(\tau,u)+C_2(\tau,u)\rho_t+iux_t} \tag{6.68}$$

with $C_2(u, \tau) = C_1(u, \tau)$ given in (6.63) *and*

$$\begin{aligned}
D_2(u, \tau) &= \frac{(d_1(u) - \rho_{\rho x}\sigma_s\sigma_r\hat{a}c_1)(1 - e^{-\kappa_\rho\tau})}{\kappa_\rho} + \frac{\sigma_\rho^2 c_1^2(1-a)(1 - e^{-2\kappa_\rho\tau})}{4\kappa_\rho} \\
&+ \frac{\sigma_\rho^2 c_1^2 e^{-mT+(m-\kappa_\rho)\tau}}{m-\kappa_\rho} + \frac{\sigma_\rho^2 c_1^2 e^{-mT+(m-2\kappa_\rho)\tau}}{2(m-2\kappa_\rho)} + \frac{b\sigma_\rho^2 c_1^2 e^{-nT+(n-\kappa_\rho)\tau}}{n-\kappa_\rho} \\
&+ \frac{b\sigma_\rho^2 c_1^2 e^{-nT+(n-2\kappa_\rho)\tau}}{2(n-2\kappa_\rho)} + \frac{\sigma_s\sigma_\rho\rho_{s\rho}c_1 iue^{-\hat{m}T+(\hat{m}-\kappa_\rho)\tau}}{\hat{m}-\kappa_\rho} \\
&+ \frac{\sigma_s\sigma_\rho\rho_{s\rho}c_1\hat{b}iue^{-\hat{n}T+(\hat{n}-\kappa_\rho)\tau}}{\hat{n}-\kappa_\rho} - \frac{\sigma_\rho^2 c_1^2 e^{-(T-\tau)m}}{2m} - \frac{b\sigma_\rho^2 c_1^2 e^{-(T-\tau)n}}{2n} \\
&- \frac{\sigma_s\sigma_\rho\rho_{s\rho}c_1 iue^{-(T-\tau)\hat{m}}}{\hat{m}} - \frac{\hat{b}\sigma_s\sigma_\rho\rho_{s\rho}c_1 iue^{-(T-\tau)\hat{n}}}{\hat{n}} \\
&+ (d_2(u) + \rho_{\rho x}\sigma_s\sigma_r\hat{a}c_1 - \sigma_s^2 iu)\tau + d_3(u) + \frac{\sigma_s\sigma_\rho\rho_{s\rho}c_1 e^{-\hat{m}T}}{\hat{m}-\kappa_\rho} \\
&+ \frac{\sigma_s\sigma_\rho\rho_{s\rho}c_1\hat{b}1e^{-\hat{n}T}}{\hat{n}-\kappa_\rho} - \frac{\sigma_s\sigma_\rho\rho_{s\rho}c_1 e^{-\hat{m}T}}{\hat{m}} - \frac{\hat{b}\sigma_s\sigma_\rho\rho_{s\rho}c_1 e^{-\hat{n}T}}{\hat{n}},
\end{aligned} \tag{6.69}$$

where $d_1(u), d_2(u)$ and $d_3(u)$ are defined in the last lemma.
We refer to [19] for the proof. Now, the both probabilities in (6.50) can be computed by

$$P_j(x_T \geq \ln K) = \frac{1}{2} + \frac{1}{\pi}\int_0^\infty \Re\left[\frac{\mathrm{e}^{-iu\ln K}\tilde{\phi}_j(x, r, \rho, t; u)}{iu}\right] du, \quad j = 1, 2 \tag{6.70}$$

using e.g., Fourier techniques [4, 6]. For computing the Quanto price in (6.50), the task remaining is to compute $\mathbb{E}[\mathrm{e}^{-\sigma_s\sigma_r\int_0^t \rho_s ds}]$ which is addressed in the following lemma.

Lemma 6.5 *Let $\mathscr{R}_t := \int_0^t \rho_s ds$ be a integrated SCP in the form of* (6.20)*. We have*

$$\mathbb{E}[e^{-\sigma_s\sigma_r\mathscr{R}_t}] = e^{-\psi(t)-\rho_0\eta(t)} \tag{6.71}$$

with

$$\eta(t) = \frac{\sigma_s\sigma_r}{\kappa_\rho}(1 - e^{-\kappa_\rho t}), \tag{6.72}$$

$$\begin{aligned}\psi(t) = \frac{\sigma_s^2\sigma_r^2\sigma_\rho^2}{2\kappa_\rho^2}\Bigg(&\frac{2(e^{-(\kappa_\rho+m)t} - 1)}{\kappa_\rho + m} + \frac{2b(e^{-(\kappa_\rho+n)t} - 1)}{\kappa_\rho + n} - \frac{e^{-(2\kappa_\rho+m)t} - 1}{2\kappa_\rho + m}\\ &- \frac{b(e^{-(2\kappa_\rho+n)t} - 1)}{2\kappa_\rho + n} + \frac{2(a-1)(e^{-\kappa_\rho t} - 1)}{\kappa_\rho} - \frac{(a-1)(e^{-2\kappa_\rho t} - 1)}{2\kappa_\rho}\\ &- \frac{e^{-mt} - 1}{m} - \frac{b(e^{-nt} - 1)}{n} + (a-1)t\Bigg) + \frac{\sigma_s\sigma_r t}{\kappa_\rho} + \frac{\sigma_s\sigma_r(e^{-\kappa_\rho t} - 1)}{\kappa_\rho^2},\end{aligned} \tag{6.73}$$

where a, b, m and n have been defined in Proposition 6.1.
The proof is available in [19].

6.6 Numerical Results

To recognize the effect of stochastic correlations we compare the Quanto prices between using constant and stochastic correlation. The pricing formula of the Quanto based on model (6.3) can be given by the extended Black-Scholes formula with a special dividend yield of

$$r_h - r_f + \sigma_S\sigma_R\,\rho_{S,R}, \tag{6.74}$$

in a closed form, cf. [12, 22]. For the constant correlation, we apply the sample coefficient correlation (6.5) to estimate a constant correlation using the whole historical data (Jan 2003–Mar 2013) of S&P 500 and Euro/US-Dollar exchange rate, which is 0.025. For initializing the SCP we take the estimates in Fig. 6.5 and let the SCP starting from the first correlation in the historical correlations.

In Fig. 6.6, we compare the both prices for several maturities between using constant and stochastic correlation and present the relative price differences in Fig. 6.7. We can observe, whilst the maturity T is shorter than 2 years, the price with constant correlation is higher than the price with stochastic correlation. Then, from nearly $T = 1.8$, the price calculated with constant correlation becomes lower than the corresponding price calculated with stochastic correlation. The reason for this, before the time $T = 1.8$, the SCP provides the correlations which are closed to the initial correlation $\rho_0 = 0.3$ which is larger than the constant correlation $\rho = 0.025$. That's why is the price with stochastic correlation lower than the price with constant correlation before $T = 1.8$ due to the fact that the price of quanto Put-option decreases direct proportional with that correlation. As the time increases, the generated

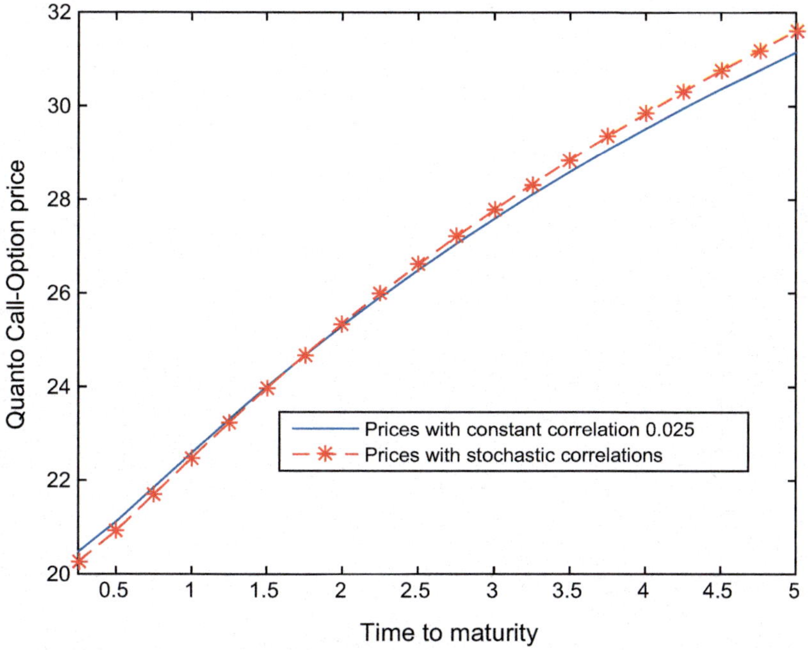

Fig. 6.6 Black-Scholes parameters: $K = 80$, $S_0 = 100$, $R_0 = 1$, $r_d = 0.05$, $r_e = 0.03$, $\sigma_S = 0.2$, $\sigma_R = 0.4$, Correlation process parameters: $\kappa = 7.937$, $\mu = 0.003$, $\sigma = 1.186$ and $\rho_0 = 0.3$

correlations tend to the mean value μ. Obviously, the parameter κ determines the speed of approach as the correlation nears the mean value. The price differences could be seen as the correlation risk by misusing the constant correlation.

6.7 Conclusions

In this chapter we considered modelling stochastic correlations as a stochastic process. In particular, we studied the general SCP established by applying hyperbolic transformation of a mean-reverting process, including the fitting to the historical market data via the density function.

As an application example, we showed how to price European-style Quanto including stochastic correlations. For the numerical results we compared the both prices of Quanto calls between using constant and stochastic correlation. The results show that the correlation risk caused by using a wrong (constant) correlation model cannot be neglected.

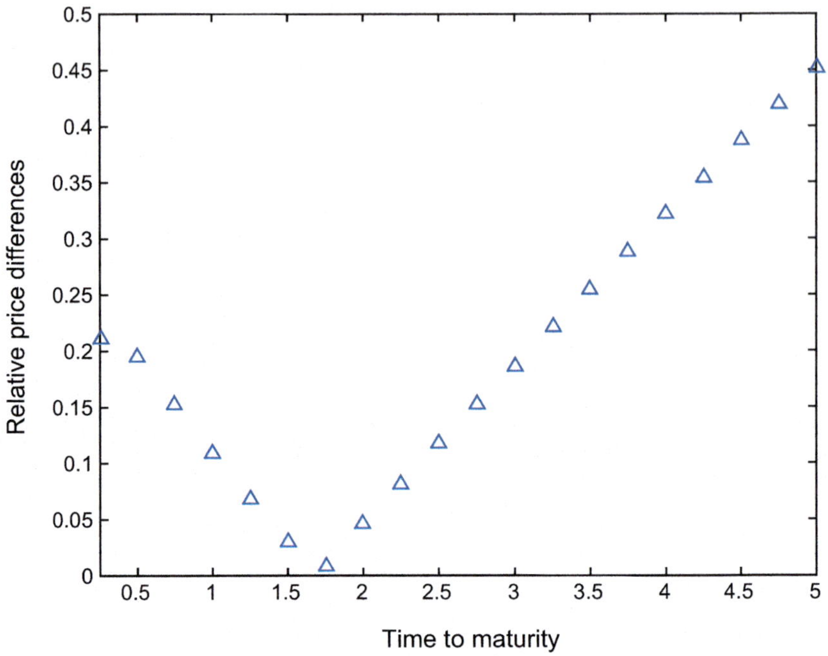

Fig. 6.7 The relative price difference between using constant and stochastic correlation in Fig. 6.6

Acknowledgements The authors were partially supported by the European Union in the FP7-PEOPLE-2012-ITN Program under Grant Agreement Number 304617 (FP7 Marie Curie Action, Project Multi-ITN STRIKE—Novel Methods in Computational Finance).

References

1. Bowman, A.W., Azzalini, A.: Applied Smoothing Techniques for Data Analysis. Oxford University Press, New York (1997)
2. Brigo, D., Chourdakis, K.: Counterparty risk for credit default swaps: impact of spread volatility and default correlation. Int. J. Theor. Appl. Finance **12**, 1007–1026 (2009)
3. Campbell, R.A.J., Forbes, C.S., Koedijk, K.G., Kofman, P.: Increasing correlations or just fat tails? J. Empir. Fin. **15**, 287–309 (2008)
4. Carr, P., Madan, D.: Option valuation using the fast Fourier transform. J. Comput. Finance **2**(4), 61–73 (1999)
5. Engle, R.F.: Dynamic conditional correlation: a simple class of multivariate GARCH. J. Bus. Econ. Stat. **17**, 425–446 (2002)
6. Fang, F., Oosterlee, C.W.: A novel pricing method for European options based on Fourier-Cosine series expansions. SIAM J. Sci. Comput. **31**, 826–848 (2008)
7. Gourieroux, C., Jasiak, J., Sufana, R.: The Wishart autoregressive process of multivariate stochastic volatility. J. Econ. **150**, 167–181 (2009)

8. Ma, J.: Pricing foreign equity options with stochastic correlation and volatility. Ann. Econ. Finance **10**, 303–327 (2009)
9. McNeil, A.J., Frey, R., Embrechts, P.: Quantitative Risk Management: Concepts, Techniques, and Tools. Princeton University Press, Princeton (2005)
10. Nelsen, R.: An Introduction to Copulas, 2nd edn. Springer, Berlin (2006)
11. Øksendal, B.: Stochastic Differential Equations. Springer, Berlin (2000)
12. Reiner, E.: Quanto mechanics. Risk **5**(3), 59–63 (1992)
13. Risken, H.: The Fokker-Planck Equation. Springer, Berlin (1989)
14. Schöbel, R., Zhu, J.: Stochastic volatility with an Ornstein Uhlenbeck process: an extension. Eur. Finance Rev. **3**, 23–46 (1999)
15. Teng, L., Ehrhardt, M., Günther, M.: Bilateral counterparty risk valuation of CDS contracts with simultaneous defaults. Int. J. Theor. Appl. Finance **16**(7), 1350040 (2013)
16. Teng, L., van Emmerich, C., Ehrhardt, B., Günther, M.: A general approach for stochastic correlation using hyperbolic functions. Int. J. Comput. Math. **93**(3), 524–539 (2014)
17. Teng, L., Ehrhardt, M., Günther, M.: Modelling stochastic correlation. J. Math. Ind. **6**(1), 1–18 (2016)
18. Teng, L., Ehrhardt, M., Günther, M.: On the Heston model with stochastic correlation. Int. J. Theor. Appl. Finance **19**(6), 1650033 (2016)
19. Teng, L., Ehrhardt, M., Günther, M.: Quanto pricing in stochastic correlation models. Preprint 16/15, University of Wuppertal (2016)
20. Uhlenbeck, G.E., Ornstein, L.S.: On the theory of Brownian motion. Phys. Rev. **36**, 823–841 (1930)
21. van Emmerich, C.: Modelling correlation as a stochastic process. Preprint 06/03, University of Wuppertal (2006)
22. Wilmott, P.: Paul Wilmott on Quantitative Finance, 2nd edn. Wiley, New York (2006)

Part II
Analysis

Chapter 7
Lie Group Analysis of Nonlinear Black-Scholes Models

Ljudmila A. Bordag and Ivan P. Yamshchikov

Abstract In this chapter we discuss the problem of financial illiquidity and give an overview of different modeling approaches to this problem. We focus on one of the approaches to an optimization problem of a portfolio with an illiquid asset sold in an exogenous random moment of time. We formulate the problem in a mathematically rigorous way and apply it to the problems of liquidity in such context for the first time to our knowledge. We provide a compact summary of achieved results. We have demonstrated the uniqueness and existence of the viscosity solution for this problem under certain conditions. The formulation of such problem gives rise to a number of three dimensional nonlinear partial differential equations (PDEs) of Black-Scholes type. Such equations rather challenging for further studies with analytical or numerical methods. One of the standard techniques to reduce the complexity of the problem is to find an inner symmetry of the equation with a help of Lie group analysis. We carried out a complete Lie group analysis of PDEs describing value function as well as investment and consumption strategies for a portfolio with an illiquid asset that is sold in an exogenous random moment of time with a prescribed liquidation time distribution. The admitted Lie algebra of the studied PDEs and the optimal system of subalgebras of this algebra provides a complete set of different invariant reductions of three dimensional PDEs to lower dimensional ones. We provide two examples of such reductions for the case of a logarithmic utility function.

L.A. Bordag (✉)
Faculty of Natural and Environmental Sciences, University of Applied Sciences Zittau/Görlitz, Theodor-Körner-Allee 16, 02763 Zittau, Germany
e-mail: LBordag@hszg.de

I.P. Yamshchikov
Max Plank Institute for Mathematics in Sciences, Inselstrasse 22, 04109 Leipzig, Germany
e-mail: ivan@yamshchikov.info

M. Ehrhardt et al. (eds.), *Novel Methods in Computational Finance*,
Mathematics in Industry 25, DOI 10.1007/978-3-319-61282-9_7

7.1 Economical Setting of the Optimization Problem for a Portfolio with an Illiquid Asset with a Given Liquidation Time Distribution

One of the challenging problems of modern mathematical finance is a management of a portfolio that includes an illiquid asset. The practical importance of such problems became especially obvious during the global financial crisis of 2008–2009 that has demonstrated a significant deficit of solid mathematical models addressing this problem. Despite the fact that financial institutes deal with illiquid assets on a daily basis there are no widely accepted approaches to the models of such assets especially if they provide stochastic income or down payments.

Indeed it is rather difficult to incorporate the illiquidity in a mathematically rigorous way. One can easily form an intuitive understanding of this phenomenon, yet there is still no widely accepted way of defining illiquidity of an asset as a measurable empiric parameter. This might be explained with the fact that illiquidity is connected with different sale mechanisms and some of the deals tend to have an essential time-lag. Such a variety of different assets is extremely tempting for the empirical research. One can find more and more works about certain type of illiquid assets and ad-hoc descriptions of their behavior, see, for example, [9] or [14]. However, there seems to be no integrated approach to such problems.

In this particular work we focus on one of the approaches to the problem of financial illiquidity, namely, an optimization of a portfolio with an illiquid asset sold in an exogenous random moment of time. We believe this approach could be a good unified method to work with the problems of illiquid assets. This is an approach that is well developed in academic literature yet seems to be industrially applicable and can be used as unifying methodology for different types of the assets.

The model that we formulate further represents an interesting class of optimization problems that go in line with the so-called adapted resource allocation problem developed by Pickenhain et al. in [21]. Indeed the idea to work with an infinite horizon problem with certain weight function, see [25] and [26] for the details, seems very promising and fruitful. The problem of a portfolio optimization with an asset that has an exogenous random liquidation time that we describe further, can be regarded as an infinite horizon problem with a special weight-function.

We manage to formulate a problem of portfolio optimization in a framework mentioned above and show the existence and uniqueness of the solution for such problem under certain assumptions of the liquidation time distribution.

The formulation of such problem gives rise to a number of three dimensional nonlinear partial differential equations (PDEs) of Black-Scholes types. We carried out a complete Lie group analysis of PDEs describing value function and investment and consumption strategies for a portfolio with an illiquid asset that is sold in an exogenous random moment of time with a prescribed liquidation time distribution in [4]. Such three dimensional nonlinear Hamilton-Jacobi-Bellman (HJB) equations are not only tedious for analytical methods but are also quite challenging from a numeric point of view. One usually uses certain substitutions to reduce the three-dimensional problem to a two-dimensional one or even to an ordinary differential equation (ODE), however the methods used to find such substitutions are rarely discussed by the authors.

We found in [4] the admitted Lie algebra for a broad class of liquidation time distributions in cases of HARA- and log-utility functions and formulate corresponding theorems for all these cases. We use found Lie algebras to obtain reductions of the studied equations. Several of similar substitutions were used in other papers before whereas others are new to our knowledge. It is important to note that this method gives us the possibility to provide a complete set of non-equivalent substitutions and reduced equations.

Let us also note that we list and discus main results here and address the reader to the publications [4–6] for detailed proofs and more detailed reasoning.

7.1.1 Theoretical Approaches to Liquidity

The understanding of the liquidity or the illiquidity of a given asset is still a matter of a debate among practitioner as well as among academics. One of the first authors to define liquidity was Keynes [18] in 1930, who said that an asset is more liquid if it is 'more certainly realizable at short notice without loss'. This intuitive definition stayed unaltered for the next 50 years, indeed in 'Wall Street Words: An A to Z Guide to Investment Terms for Today's Investor' by Scott [28] written 50 years after one can read: *Illiquid asset is an asset that is difficult to buy or sell in a short period of time without its price being affected.* This formulation is also far from mathematical rigor yet it points out two important aspects of illiquidity: temporal and monetary. The first attempts to define liquidity in a mathematically correct way were separately taken in these two directions.

In 1986 Lippman and McCall [20] defined the environment characterized by four different objects: c_i, T_i, X_i and β. All of them were described in the discrete time framework. c_i is a cost of owning or operating the asset during the period number i. It can also be considered as the cost of the attempt to sell the asset. The offers come at every moment that is in the set $\{S_i : i = 1, 2, \dots\}$ of arrival times. These random variables S_i satisfy

$$S_i = \Sigma^i{}_{j=1} T_j,$$

where the integer valued random variables $T_i \geqslant 0$ neither need to be independent nor identically distributed.

X_i are positive independent identically distributed random variables that correspond to the price offered in the ith moment. All the expenditures are discounted at the rate β so that a present value of a dollar received in period i is β^i. The discounted net receipts $R(\tau)$ associated with a stopping time τ is given by

$$R(\tau) = \beta^{\tau} Y_{N(\tau)} - \Sigma^{\tau}{}_{j=1}\beta^i c_i, \tag{7.1}$$

$$Y_i = \begin{cases} X_i, & \text{if recall is not allowed,} \\ max(X_1, \ldots, X_i), & \text{if recall is allowed,} \end{cases}$$

where $N(\tau) = max\{n : S_n \leqslant \tau\}$ is the random number of offers that the seller observes when employing the decision rule τ and the random variable $Y_{N(\tau)}$ is the size of the accepted offer. Consequently, the seller chooses a stopping rule τ^* in the set T_i of all stopping rules such that

$$E[R(\tau^*)] = max\{E[R(\tau)] : \tau \in T_i\}.$$

Obviously, the time it takes to estimate the asset's value and to convert the asset into cash is defined by the random variable τ^*. Lippman and McCall [20] proposed to regard the expectation of this variable, $E[\tau^*]$ as the measure of an assets' illiquidity. According to this definition as $E[\tau^*]$ increases (i.e. one need to wait longer until the asset is sold) a liquidity of a corresponding asset is to decrease.

In 1994 Hooker and Kohn [16] addressed a monetary aspect of illiquidity. The authors introduce an index of liquidity, so-called $\lambda(I_t)$, as

$$\lambda(I_t) = \frac{V(I_t) - L(I_t)}{V(I_t)},$$

where $V(I_t)$ is the value of the asset under optimal sale, as a function of the information set I_t and $L(I_t)$ is a loss from immediate sale of the asset. Since λ depends on the information set I_t they call this index the *conditional liquidity* of the asset. The authors also introduce the *expected liquidity* of an asset, Λ, which, naturally, is

$$\Lambda = E[\lambda(I_t)].$$

These two approaches to illiquidity that give rise to a number of others that quite often base on one or another approach. As we have mentioned before, illiquidity is still broadly discussed and there seems to be no agreement among the academics and practitioners. For example, since the beginning of 2000s the idea to estimate the liquidity of the assets through the bid-ask spread became rather popular, see Bangia et al. [1] or Coppejans et al. [8]. However, the definitions addressing either one or another aspect of illiquidity have certain mishaps. In particular, different assets can demonstrate different temporal and monetary behavior under same market circumstances and a definition that addresses only one of the aspects could not be used as an unified approach. However, a portfolio optimization problem for a portfolio with an illiquid asset can incorporate both aspects and this is the approach that we want to focus on.

In 1974 Miller in [23] formulated a problem of optimal consumption with a stochastic income. He showed that an upper bound on consumption is lower than the value of optimal consumption in the case where the random labour income is replaced by its mean. To our knowledge, this was a first work that formulated the problem of a portfolio optimization in presence of a *stochastic income*.

In 1987 Grossman and Baroque [15] analyzed a model of optimal consumption and portfolio selection in which consumption services are generated by holding a durable good. This problem was very close to the problem of a *stochastic income* that we discuss later in detail, but it did not take into consideration any random effects associated with illiquidity. These effects could be associated with liquidation time or price discount, yet were out of the scope of this moment.

Two years later Zeldes provided the first numerical solution for the problem of optimal consumption with stochastic income and constant relative risk aversion in [33].

Finally, in 1993 Duffie and Zariphopoulou [10] propose the framework of the optimal consumption with undiversifiable income risk (also called a *stochastic income* model) as an extension for the continuous time model, developed by Merton [22]. Under an assumption of an infinite time horizon the authors showed the existence and uniqueness of the viscosity solution of the associated HJB equation for the class of concave utility functions $U(c)$ satisfying the following conditions: U in c is strictly concave; $U(c) \in C^2(0, +\infty)$, $U(c) \leq M(1+c)^\gamma$, with $0 < \gamma < 1, M > 0$; $U(0) \geq 0$, $\lim\limits_{c\to 0} U'(c) = +\infty$, $\lim\limits_{c\to\infty} U'(c) = 0$.

In 1997, in [11] the authors extended the problem of hedging in incomplete markets with hyperbolic absolute risk aversion (so called HARA) utility function. The stochastic income in this case cannot be replicated by trading available securities. An investor receives stochastic income in moment t at a rate Y_t, where

$$dY_t = \mu Y_t dt + \eta Y_t dW_t^1, \quad t \geq 0, Y_0 = y, \quad y \geq 0$$

and $\mu, \eta > 0 - const$, here W_t^1 is a standard Brownian motion. The riskless bank account has a constant continuously compound interest rate r. A traded security has a price S given by

$$dS_t = \alpha S_t dt + \sigma S_t (\rho dW_t^1 + \sqrt{1-\rho^2} dW_t^2),$$

α, $\sigma > 0 - const$ and W_t^2 is an independent standard Brownian motion, $\rho \in (-1, 1)$ is a correlation between price processes S_t and Y_t. The investor utility function for consumption process c_t is given by

$$\mathscr{U}(c(t)) = E\left[\int_0^\infty e^{-\kappa t} U(c(t)) dt\right], \quad U(c(t)) = c(t)^\gamma,$$

where $\gamma \in (0, 1)$ and κ is a discount factor $\kappa > r$.

Remark 7.1 The notation of the strategy (π, c) is standard for the problems of such kind. We also further denote the amount of the investment in a liquid risky asset as π and investor's consumption as c. Both controls do depend on time, so to emphasize it to the reader we might also use $(\pi(t), c(t))$ or even (π_t, c_t) from time to time.

The investors wealth process L evolves

$$dL_t = [rL_t + (\alpha + \delta - r)\pi_t - c_t + Y_t]\, dt \\ + \sigma\pi_t \left(\rho dW_t^1 + \sqrt{1-\rho^2}\, dW_t^2\right), \qquad t \geq 0, \quad L_0 = l,$$

where δ could be regarded as the dividends paid constantly from an illiquid asset or as the possession costs, l is an initial wealth endowment and π_t represents an investment in the risky asset S, with the remaining wealth held in riskless borrowing or lending. The goal is to characterize an investor value function $V(l, y) = \sup_{(\pi,c)\in\mathscr{A}(l,y)} \mathscr{U}(c)$. The set $\mathscr{A}(l, y)$ is a set of admissible controls (π, c) such that $L_t \geq 0$.

The authors in [11] proved the smoothness of the viscosity solution of the associated Hamilton-Jacobi-Bellman (HJB) equation in the case of the HARA utility function and the infinite time horizon. This proof heavily relies on a reduction of the initial HJB equation to an ODE. After this reduction the main result follows from the uniform convergence of the classical solution of a uniformly elliptic equation to the viscosity solution, which is unique. Here one should mention that the authors use the discount factor $e^{-\kappa t}$ in [11] as a technical factor which is not related to stochastic income. The economical setting of this problem does not imply any liquidation of an illiquid asset which provides stochastic income Y_t. Further we demonstrate in [5] that such a discounting could be interpreted as an exponential liquidation time distribution and is a unique situation that allows to reduce the three-dimensional PDE to an ODE (see also [4]).

In 2007 Schwarz and Tebaldi in [27] broadened a model of random income proposed before and connected it to the problems of illiquidity. They assumed that the non-traded illiquid asset generates a flow of random income in the form of dividends, until it is sold at a fixed moment of time. This idea allowed to build models for a portfolios with illiquid assets, using the results obtained for the problems with random income. One of a huge challenges connected with optimizations problems in presence of illiquidity is the question of pricing of illiquid assets that is a serious mathematical problem in itself. Assuming that the asset generates a certain dividends, connected with its fair price authors could elegantly incorporate illiquid asset in the model. Further, the authors define illiquid asset as an asset that can not be sold neither piece by piece nor at once before the investment's horizon, denoted as T, which is a fixed deterministic value at which the asset generates a random cash-flow equal to its' paper-value at this moment T (the cash-flow is denoted as H_T). With this economical reasoning behind it this model of illiquidity looks rather promising yet needs a more exact qualitative and quantitative description. In [31] we have broadened this framework for the case of logarithmic utility and finite deterministic liquidation time.

Later in 2008 Schwarz et al. [7] applied the approach very close to the one formulated in [27] to the problem of housing choice for a household. The constrained of a deterministic time was abandoned as the idea of the model was to compare two 'realities': one, where housing was purely illiquid and another 'thought-experiment' reality, where the household could sell the real-estate partially. It was demonstrated that optimal strategies for two models differ significantly.

In this work we focus on the case when a time-horizon is an exogenous random variable. We would like to note that the set-up with exogenous time is actually economically motivated. For example, standard inheritance procedures in several EU countries assume that the illiquid assets are sold and the cash is then divided between the heirs. Naturally the sale occurs in a random moment of time and the inheritance manager splits the cash between the heirs immediately after the sale. Another example of an exogenous liquidation time that justifies our model are shares-for-the-loan auctions. This phenomenon is typical for the emerging markets where governmentally owned businesses are at some point privatized fully or partially. For example, it was very typical for a post-soviet markets in their transition period and is still relevant for a number of states in the Eastern Europe [30].

7.1.2 *Portfolio Optimization in the Case of an Illiquid Asset with a Given Liquidation Time Distribution: Problem Setting*

Here we describe the proposed model. We assume that the investor's portfolio consists of a riskless bond, a risky asset and a non-traded asset that generates stochastic income, i.e. dividends. However, in contrast with the previous works we replace the liquidation time that was deterministic before with a stochastically distributed time T. A risk-free bank account B_t with the interest rate r and a stock price S_t describe the classical Black-Scholes market [2]

$$dB_t = rB_t dt, \;\; dS_t = S_t(\alpha dt + \sigma dW_t^1), \quad t \leq T, \tag{7.2}$$

where the interest rate r, the continuously compounded rate of return $\alpha > r$ and the standard deviation σ are assumed to be constant; r, α, σ – *const*. An illiquid asset H_t that can not be traded up to the time T and which paper value is correlated with the stock price and follows

$$\frac{dH_t}{H_t} = (\mu - \delta)dt + \eta(\rho dW_t^1 + \sqrt{1-\rho^2} dW_t^2), \quad t \leq T, \tag{7.3}$$

where μ is the expected rate of return of the risky illiquid asset, (W_t^1, W_t^2) are two independent standard Brownian motions, δ is the rate of dividend paid by the illiquid asset, η is the continuous standard deviation of the rate of return, and $\rho \in (-1; 1)$ is the correlation coefficient between the stock index and the illiquid risky asset. The parameters μ, δ, η, ρ are all assumed to be constant. The liquidation time T is an exogenous random-distributed continuous variable which does not depend on the Brownian motions (W_t^1, W_t^2). The probability density function of liquidation time distribution T is denoted by $\phi(t)$ whereas $\Phi(t)$ denotes the cumulative distribution function, and $\overline{\Phi}(t)$ the survival function also known as *a reliability function* $\overline{\Phi}(t) = 1 - \Phi(t)$. We omit here the explicit notion of the possible parameters of distribution in order to make the formulae shorter.

Since the filtration $\{\mathscr{F}_t\}$ generated by the Brownian motion $W = (W_t^1, W_t^2)$ we assume that the consumption process is an element of the space $\mathscr{L}_+$ of non-negative $\{\mathscr{F}_t\}$-progressively measurable processes c_t such that

$$E\left(\int_0^s c(t)dt\right) < \infty, \quad s \in [0, T], \tag{7.4}$$

where E denotes a mathematical expectation with respect to filtration $\{\mathscr{F}_t\}$. The investor wants to maximize the average utility consumed up to the time of liquidation, given by

$$\mathscr{U}(c) := \mathscr{E}\left[\int_0^T U(c(t))dt\right]. \tag{7.5}$$

Here we used $\mathscr{E}$ to indicate that we are averaging over all random variables including T. The wealth process L_t is the sum of cash holdings in bonds, stocks and *random* dividends from the non-traded asset minus the consumption stream. Thus, we can write

$$dL_t = (rL_t + \delta H_t + \pi_t(\alpha - r) - c_t)dt + \pi_t \sigma dW_t^1. \tag{7.6}$$

The set of admissible policies is standard and consists of investment strategies (π_t, c_t) such that

1. c_t belongs to $\mathscr{L}_+$,
2. π_t is $\{\mathscr{F}_t\}$-progressively measurable and $\int_t^s (\pi_\tau)^2 d\tau < \infty$ a.s. for any $t \le \tau \le T$,
3. L_t, defined by the stochastic differential equation (7.6) and initial conditions $L_t = l > 0$, $H_t = h > 0$ a.e. $(t \le T)$.

We proved in [5] that one can explicitly average (7.5) over T and with the certain conditions (which are formulated later in Proposition 7.1) posed on $\overline{\Phi}$ and $U(c)$ the problem (7.5) is equivalent to the maximization of

$$\mathscr{U}(c) := E\left[\int_0^\infty \overline{\Phi}(t) U(c(t))dt\right], \tag{7.7}$$

where E is an expectation over space coordinates excluding T.

Remark 7.2 If T is exponentially distributed we get *precisely* the problem of optimal consumption with random income that was studied in [11] and already discussed in the introduction.

Remark 7.3 There several works dealing with the problems of random time horizon, for example, Ekeland in [12, 13] has shown the possibility to work with different discounting factors. Skiba and Tobacman [29] also mentioned a non-exponential discounting in the loan context, yet the authors do not provide any mathematically strict way to model these effects. To our knowledge the idea of a discounting different from an exponential one in a framework of illiquidity was never proposed before.

Proposition 7.1 ([5]) *The problems (7.5) and (7.7) are equivalent provided*

$$\lim_{t\to\infty} \overline{\Phi}(t)E\left[U(c(t)\right] = 0. \tag{7.8}$$

Remark 7.4 Consumption $c(t)$ is usually bounded as time goes to infinity. For all these models condition (7.8) is satisfied automatically. Yet if one regards absolute values of consumption and it grows as time goes to infinity this constraint is needed.

In this work we regard the problem (7.5) with random liquidation time T that has a distribution satisfying the condition (7.8) in Proposition 7.1 and, therefore, corresponds to the *value function* $V(t, l, h)$ which is defined as

$$V(t,l,h) = \max_{(\pi,c)} E\left[\int_t^{\infty} \overline{\Phi}(\tau)U(c(\tau))d\tau | L(t) = l, H(t) = h\right]. \tag{7.9}$$

For the value function $V(t, l, h)$ we can derive a HJB equation on which we focus in this work

$$\begin{aligned} V_t(t,l,h) &+ \frac{1}{2}\eta^2 h^2 V_{hh}(t,l,h) + (rl+h)V_l(t,l,h) + (\mu-\delta)hV_h(t,l,h) \\ &+ \max_{\pi} G[\pi] + \max_{c\geq 0} H[c] = 0, \end{aligned} \tag{7.10}$$

$$\begin{aligned} G[\pi] &= \frac{1}{2}V_{ll}(t,l,h)\pi^2\sigma^2 + V_{lh}(t,l,h)\eta\rho\pi\sigma h \\ &+ \pi(\alpha - r)V_l(t,l,h), \end{aligned} \tag{7.11}$$

$$H[c] = -cV_l(t,l,h) + \overline{\Phi}(t)U(c), \tag{7.12}$$

with the boundary condition

$$V(t,l,h) \to 0, \text{ as } t \to \infty.$$

The value function for a problem of such kind as (7.10) is a viscosity solution if the control and state variables are uniformly bounded. However, this is not the case for the optimal consumption problem and thus a more sophisticated proof is needed.

Similar problems were previously studied in [10, 11, 32]. The main difficulties in our case come from the non-exponential time discounting we are using in the utility functional (7.9). Now we list the main results of [5] were we proved the existence and uniqueness of viscosity solutions in details.

Theorem 7.1 *There exists a unique viscosity solution of the corresponding HJB equation (7.10) if*

1. *$U(c)$ is strictly increasing, concave and twice differentiable in c,*
2. $\lim_{t\to\infty}\overline{\Phi}(t)E[U(c(t))] = 0$, $\overline{\Phi}(t) \sim e^{-\kappa t}$ *or faster as $t \to \infty$,*
3. *$U(c) \leq M(1+c)^{\gamma}$ with $0 < \gamma < 1$ and $M > 0$,*
4. $\lim_{c\to 0} U'(c) = +\infty$, $lim_{c\to+\infty} U'(c) = 0$.

The proof of this statement is to be done in three steps. At first we need to establish certain properties of the value function $V(t,l,h)$ that corresponds to our problem. These properties are formulated and proved in Lemma 7.1 that follows. Then we show that the value function with such properties is a viscosity solution of the problem, this is done in Lemma 7.2. The uniqueness of this solution follows from the *comparison principle* that is actually a very useful tool by itself and is formulated and proved in Theorem 7.2.

Lemma 7.1 *Under the conditions* (1)–(4) *from Theorem 7.1 the value function $V(t,l,h)$ (7.9) has the following properties:*

1. *$V(t,l,h)$ is concave and non-decreasing in l and in h,*
2. *$V(t,l,h)$ is strictly increasing in l,*
3. *$V(t,l,h)$ is strictly decreasing in t starting from some point,*
4. *$0 \leq V(t,l,h) \leq O(|l|^{\gamma} + |h|^{\gamma})$ uniformly in t.*

Now the existence of the viscosity solution of the problem (7.10) can be proved.

Lemma 7.2 *Under the conditions of Lemma 7.1 the function $V(t,l,h)$ is a viscosity solution of (7.10) on the domain $D = (0,\infty)\times(0,\infty)\times(0,\infty)$.*

The third result that is needed to finalize the proof of Theorem 7.1 is a *comparison principle* formulated below as Theorem 7.2. Results of this type are well-known in general for bounded controls, but due to the unbounded controls, classical proofs require adaptations for our case.

Theorem 7.2 (Comparison Principle) *Let $u(t,l,h)$ be an upper-semicontinuous concave viscosity subsolution of (7.10) on D and $V(t,l,h)$ is a supersolution of (7.10) on D which is bounded from below, uniformly continuous on D, and locally Lipschitz in D, such that $u(t,l,h) \to 0$, $V(t,l,h) \to 0$ as $t \to \infty$ and $|u(t,l,h)| + |V(t,l,h)| \leq O(|l|^{\gamma} + |h|^{\gamma})$ for large l,h, where $0 < \gamma < 1$, uniformly in t. Then $u \leq v$ on $\overline{D}$.*

For the detailed proofs of these results and for applications of these principles to the specific cases we address the reader to [5].

7.2 Lie Group Analysis of the HJB Equations with HARA- and Log-Utility Functions

In a number of the results listed above the dimension reduction of the original HJB problem plays a crucial role. A majority of authors dealing with three dimensional HJB equations come up with variable substitutions yet they generally do not have any remark on how to get similar substitution in other cases or why they use this or that substitution. Let us note here that the smooth point transformations with continuous parameter admitted by linear or nonlinear PDEs can be found algorithmically using Lie group analysis. This is a well-known fact demonstrated by Lie in [19]. Such procedure that find a symmetry group admitted by a PDE is described in many textbooks, see, for example, [17, 24] or [3]. However practical application of these procedures is connected with tedious and voluminous calculations which can be only slightly facilitated with the help of modern computer packages. Preparing our work [4] we have, for example, used the program `IntroToSymmetry` to obtain the determining system of equations. Nevertheless solving these determining systems of partial differential equations is usually hard and can rarely be done algorithmically, but the possibility to find the system of determining equations facilitates the work of a researcher since the systems are quite voluminous. For example, in the studied cases the systems had more then a hundred equations. Now we provide a short summary of results proved in [4].

7.2.1 *The Case of HARA Utility Function*

A utility function $U(c)$ where the risk tolerance $R(c)$ is defined as $R(c) = -\frac{U'(c)}{U''(c)}$ and is a linear function of c, is called a HARA (hyperbolic absolute risk aversion) utility function. We use two types of utility functions: a HARA utility function $U^{HARA}(c)$ and the log-utility function $U^{LOG}(c) = \log(c)$. Let us note here that the log-utility function is often regarded as a limit case of HARA utility function. One can indeed choose HARA utility in such a way that allows a formal transition from HARA utility to log-utility as parameter γ of HARA utility goes to zero, but in general this transition does not hold for any form of HARA utility. We demonstrated this transition on different levels in [4] and because of that further in this chapter we work with HARA utility in the form

$$U^{HARA}(c) = \frac{1-\gamma}{\gamma}\left(\left(\frac{c}{1-\gamma}\right)^{\gamma} - 1\right), \tag{7.13}$$

with the risk tolerance $R(c) = \frac{c}{1-\gamma}$, $0 < \gamma < 1$. One can easily see that as $\gamma \to 0$ HARA-utility function written as (7.13) tends the to log-utility

$$U^{HARA}(c) \underset{\gamma \to 0}{\longrightarrow} U^{LOG}(c), \tag{7.14}$$

and by $\gamma \to 1$ we obtain $U^{HARA}(c) \to c$. The HJB equation (7.10) where we insert the HARA utility in the form (7.13) after formal maximization procedure (see also [5]) will take the form

$$V_t(t,l,h) + \frac{1}{2}\eta^2 h^2 V_{hh}(t,l,h) + (rl + \delta h)V_l(t,l,h) + (\mu - \delta)hV_h(t,l,h)$$
$$- \frac{(\alpha - r)^2 V_l^2(t,l,h) + 2(\alpha - r)\eta\rho h V_l(t,l,h)V_{lh}(t,l,h) + \eta^2\rho^2\sigma^2 h^2 {V_{lh}}^2(t,l,h)}{2\sigma^2 V_{ll}(t,l,h)}$$
$$+ \frac{(1-\gamma)^2}{\gamma}\overline{\Phi}(t)^{\frac{1}{1-\gamma}} V_l(t,l,h)^{-\frac{\gamma}{1-\gamma}} - \frac{1-\gamma}{\gamma}\overline{\Phi}(t) = 0, \quad V \underset{t\to\infty}{\longrightarrow} 0. \tag{7.15}$$

Here the investment $\pi(t,l,h)$ and consumption $c(t,l,h)$ strategies look as follows in terms of the value function $V(t,l,h)$

$$\pi(t,l,h) = -\frac{\eta\rho\sigma h V_{lh}(t,l,h) + (\alpha - r)V_l(t,l,h)}{\sigma^2 V_{ll}(t,l,h)}, \tag{7.16}$$

$$c(t,l,h) = (1-\gamma)V_l(t,l,h)^{-\frac{1}{1-\gamma}}\overline{\Phi}(t)^{\frac{1}{1-\gamma}}. \tag{7.17}$$

Equation (7.15) is a nonlinear three dimensional PDE with three independent variables t, l, h. To reduce the dimension of the Eq. (7.15) we use Lie group analysis, that allows us to find the generators of the corresponding symmetry algebra admitted by this equation. In detail one can find the description of this method applied to similar PDEs in [3]. Here we formulate the main theorem of Lie group analysis for the optimization problem with HARA type utility function which we proved in [4].

Theorem 7.3 *The Eq. (7.15) admits the three dimensional Lie algebra L_3^{HARA} spanned by generators $\mathbf{U}_1, \mathbf{U}_2, \mathbf{U}_3$, i.e. $L_3^{HARA} =< \mathbf{U}_1, \mathbf{U}_2, \mathbf{U}_3 >$, where*

$$\mathbf{U}_1 = \frac{\partial}{\partial V}, \quad \mathbf{U}_2 = e^{rt}\frac{\partial}{\partial l}, \quad \mathbf{U}_3 = l\frac{\partial}{\partial l} + h\frac{\partial}{\partial h} + \left(\gamma V - (1-\gamma)\int \overline{\Phi}(t)dt\right)\frac{\partial}{\partial V}, \tag{7.18}$$

for any liquidation time distribution. Moreover, if and only if the liquidation time distribution has the exponential form, i.e. $\overline{\Phi}(t) = de^{-\kappa t}$, where d, κ are constants the studied equation admits a four dimensional Lie algebra L_4^{HARA} with an additional generator

$$\mathbf{U}_4 = \frac{\partial}{\partial t} - \kappa V\frac{\partial}{\partial V}, \tag{7.19}$$

i.e. $L_4^{HARA} =< \mathbf{U}_1, \mathbf{U}_2, \mathbf{U}_3, \mathbf{U}_4 >$.

Except finite dimensional Lie algebras (7.18) and (7.19) correspondingly Eq. (7.15) admits also an infinite dimensional algebra $L_\infty = < \psi(h,t)\frac{\partial}{\partial V} >$ *where the function* $\psi(h,t)$ *is any solution of the linear PDE*

$$\psi_t(h,t) + \frac{1}{2}\eta^2 h^2 \psi_{hh}(h,t) + (\mu - \delta)h\psi_h(h,t) = 0. \tag{7.20}$$

The Lie algebra L_3^{HARA} *has the following non-zero commutator relations*

$$[\mathbf{U}_1, \mathbf{U}_3] = \gamma \mathbf{U}_1, \quad [\mathbf{U}_2, \mathbf{U}_3] = \mathbf{U}_2. \tag{7.21}$$

The Lie algebra L_4^{HARA} *has the following non-zero commutator relations*

$$[\mathbf{U}_1, \mathbf{U}_3] = \gamma \mathbf{U}_1, \ [\mathbf{U}_1, \mathbf{U}_4] = -\kappa \mathbf{U}_1, \ [\mathbf{U}_2, \mathbf{U}_3] = \mathbf{U}_2, \ [\mathbf{U}_2, \mathbf{U}_4] = -r\mathbf{U}_2. \tag{7.22}$$

7.2.2 The Case of the Log-Utility Function

A logarithmic utility could be regarded as a limit case of HARA-utility (7.14). Yet logarithm has particular properties than make it rather popular utility function in financial mathematics therefore we analyze it separately.

The proposed approach is similar to the method described above therefore we omit some details here. In the case of the log-utility function the HJB equation after the formal maximization procedure will take the following form

$$\begin{aligned} &V_t(t,l,h) + \frac{1}{2}\eta^2 h^2 V_{hh}(t,l,h) + (rl + \delta h)V_l(t,l,h) + (\mu - \delta)hV_h(t,l,h) \\ &- \frac{(\alpha - r)^2 V_l^2(t,l,h) + 2(\alpha - r)\eta\rho h V_l(t,l,h)V_{lh}(t,l,h) + \eta^2\rho^2\sigma^2 h^2 {V_{lh}}^2(t,l,h)}{2\sigma^2 V_{ll}(t,l,h)} \\ &- \overline{\Phi}(t)\left(\log V_l - \log \overline{\Phi}(t) + 1\right) = 0, \quad V \underset{t\to\infty}{\longrightarrow} 0. \end{aligned} \tag{7.23}$$

Here the investment $\pi(t,l,h)$ and consumption $c(t,l,h)$ look as follows in terms of the value function V(t,l,h)

$$\pi(t,l,h) = -\frac{\eta\rho\sigma h V_{lh}(t,l,h) + (\alpha - r)V_l(t,l,h)}{\sigma^2 V_{ll}(t,l,h)}, \quad c(t,l,h) = \frac{\overline{\Phi}(t)}{V_l(t,l,h)}. \tag{7.24}$$

Remark 7.5 HARA-utility is chosen in such a way that (7.14) holds and the maximization procedure that transforms HJB equation to PDE preserves this property as well. If we formally take a limit of (7.15) as $\gamma \to 0$ we obtain (7.23).

As we proved in [4] a log-utility can be regarded as a limit case of HARA utility and this correspondence holds for all analytical and algebraic structures connected to these problems.
Analogously to the previous paragraph one can formulate the main theorem of Lie group analysis for this PDE which was proved in [4].

Theorem 7.4 *The Eq. (7.23) admits the three dimensional Lie algebra L_3^{LOG} spanned by generators $\mathbf{U}_1, \mathbf{U}_2, \mathbf{U}_3$, i.e. $L_3^{LOG} =< \mathbf{U}_1, \mathbf{U}_2, \mathbf{U}_3 >$, where*

$$\mathbf{U}_1 = \frac{\partial}{\partial V}, \quad \mathbf{U}_2 = e^{rt}\frac{\partial}{\partial l}, \quad \mathbf{U}_3 = l\frac{\partial}{\partial l} + h\frac{\partial}{\partial h} - \int \overline{\Phi}(t)dt\frac{\partial}{\partial V}, \tag{7.25}$$

for any liquidation time distribution. Moreover, if and only if the liquidation time distribution has the exponential form, i.e. $\overline{\Phi}(t) = de^{-\kappa t}$, where d, κ are constants, the studied equation admits a four dimensional Lie algebra L_4^{LOG} with an additional generator

$$\mathbf{U}_4 = \frac{\partial}{\partial t} - \kappa V\frac{\partial}{\partial V}, \tag{7.26}$$

i.e. $L_4^{LOG} =< \mathbf{U}_1, \mathbf{U}_2, \mathbf{U}_3, \mathbf{U}_4 >$.

Except finite dimensional Lie algebras L_3^{LOG} and L_4^{LOG} correspondingly the Eq. (7.23) admits also an infinite dimensional algebra $L_\infty =< \psi(h,t)\frac{\partial}{\partial V} >$ where the function $\psi(h,t)$ is any solution of the linear PDE

$$\psi_t(h,t) + \frac{1}{2}\eta^2h^2\psi_{hh}(h,t) + (\mu - \delta)h\psi_h(h,t) = 0. \tag{7.27}$$

The Lie algebra L_3^{LOG} has one non-zero commutator relation $[\mathbf{U}_2, \mathbf{U}_3] = \mathbf{U}_2$.

The Lie algebra L_4^{LOG} has the following non-zero commutator relations

$$[\mathbf{U}_1, \mathbf{U}_4] = -\kappa\mathbf{U}_1, \quad [\mathbf{U}_2, \mathbf{U}_3] = \mathbf{U}_2, \quad [\mathbf{U}_2, \mathbf{U}_4] = -r\mathbf{U}_2.$$

Remark 7.6 If we compare the form of Lie algebras generators in the cases of HARA- and log- utilities, i.e. formulas (7.18) and (7.25) as well as (7.19) and (7.26), we can see that the formal limit procedure holds for them as well and the generators for HARA-utility transfer to generators for log-utility under a formal limit $\gamma \to 0$.

When the Lie algebra admitted by the studied PDE is found we can find all non equivalent variable substitutions which reduce the dimension of the given PDE, if there are any. Using the corresponding exponential map of the adjoint representation of the admitted Lie algebra we can find the symmetry group or subgroups of the equation as well. It is rather fortunate that we do not have to look for an explicit form of the admitted symmetry group to find possible reductions or simplification of the studied PDEs and invariant solutions of the equations. It is enough to know and to use the properties of the admitted Lie algebra which corresponds to the symmetry group. The optimal system of subalgebras of this algebra gives rise to a complete set of non equivalent substitutions and as a result a set of different reductions of the studied PDE. Reductions of the studied three dimensional PDEs (7.15) and (7.23) for different liquidation time distributions we discuss in the next section.

7.3 Symmetry Reductions for Different Liquidation Time Distributions

We obtain reductions of studied three dimensional PDEs if we replace original variables with new independent and dependent variables which are invariant under the action of one of the subgroups of the symmetry group admitted by the equation. The solutions of reduced PDEs are called invariant solutions because they are invariant under the action of a given subgroup. We find the admitted Lie algebras for a certain class of liquidation time distributions in cases of HARA- and log-utility functions and formulate corresponding results in Theorems 7.3 and 7.4 in previous section and in details in [4]. In [4] we provided the optimal system of subalgebras for a general case of a liquidation time distribution in both cases of HARA and logarithmic utility functions. We separately regarded there a case of an exponential distribution of a liquidation time where the corresponding PDE admits an extended Lie algebra. It leads to certain distinguishing properties that give rise to non trivial reductions of three dimensional PDEs to two dimensional equations and even to ODEs in some cases.

As it was shown in Theorem 7.3 the admitted Lie algebra for the case of HARA-utility function and with a general form of liquidation time distribution is three dimensional. As it was proved in [4] the optimal system of subalgebras of this algebra has four non equivalent one dimensional and three non equivalent two dimensional subalgebras. Just one of these subalgebras allowed a meaningful reduction of the three dimensional PDE (7.15) to a two dimensional one (see [4]). Any further reductions of this PDE in the framework of Lie group analysis are not possible.

Theorem 7.3 demonstrates that the admitted Lie algebra is four dimensional if and only if we use an exponential liquidation time distribution. In this case the optimal system of subalgebras is richer and we can reduce (7.15) to an ODE. All the reductions and corresponding conditions are listed in [4].

In fact, the case of log-utility function is very similar to the case with HARA-utility function. As it was proved in Theorem 7.4 for the general form of the liquidation time distribution we obtain just one meaningful reduction of (7.23) to a two dimensional PDE.

Now let us look at the case when the liquidation time T is a random Weibull-distributed variable independent of the Brownian motions (W_t^1, W_t^2).

The probability density function of the Weibull distribution is defined as follows

$$\phi(t,\lambda,k) = \begin{cases} \frac{k}{\lambda}\left(\frac{t}{\lambda}\right)^{k-1} e^{-(t/\lambda)^k}, & \text{if } t \geq 0,\\ 0, & \text{if } t < 0, \end{cases}$$

where $\lambda > 0$ and $k, \lambda - const$ and the survival function $\overline{\Phi}(t) = 1 - \Phi(t)$ has a form $\overline{\Phi}(t) = e^{-(t/\lambda)^k}$. We will often omit the constant parameters λ and k in notations for shortness.

Indeed, when $k = 1$ the Weibull-distribution turns into exponential one, and for $k > 1$ its probability density has a local maximum. This situation corresponds to our economical motivation.

The Eq. (7.10) is the same as before but the term that corresponds to $\overline{\Phi}$ is naturally replaced by Weibull survival function $\overline{\Phi}(t) = e^{-(t/\lambda)^k}$

$$V_t(t,l,h) + \frac{1}{2}\eta^2 h^2 V_{hh}(t,l,h) + (rl+h)V_l(t,l,h) + (\mu-\delta)hV_h(t,l,h)$$
$$+ \max_{\pi} G[\pi] + \max_{c\geq 0} H[c] = 0, \tag{7.28}$$

$$G[\pi] = \frac{1}{2}V_{ll}(t,l,h)\pi^2\sigma^2 + V_{lh}(t,l,h)\eta\rho\pi\sigma h$$
$$+ \pi(\alpha - r)V_l(t,l,h), \tag{7.29}$$

$$H[c] = -cV_l(t,l,h) + e^{-(t/\lambda)^k}U(c), \tag{7.30}$$

Proposition 7.2 ([5]) *Theorem 7.1 holds for the case of the Weibull distribution with $k > 1$ and, therefore, there exists a unique solution for the problem (7.28).*

We can use one dimensional subalgebra of the admitted Lie algebra (see Theorem 7.4) and introduce the invariant variables

$$z = \frac{l}{h}, \qquad W(t,z) = V(t,l,h) - \frac{\lambda}{k}\log h\Gamma\left(\frac{1}{k}, \left(\frac{t}{\lambda}\right)^k\right).$$

So we obtain the two dimensional PDE on $W(t,z)$

$$W_t + \frac{1}{2}\eta^2\left(2zW_z + z^2W_{zz}\right) + (rz+\delta)W_z - (\mu-\delta)zW_z$$
$$- \frac{(\alpha-r)^2W_z^2 - 2(\alpha-r)\eta\rho W_z(W_z + zW_{zz}) + \eta^2\rho^2\sigma^2(W_z - zW_{zz})^2}{2\sigma^2 W_{zz}}$$
$$- e^{-(t/\lambda)^k}\log W_z + \frac{\lambda}{k}\left(\frac{\eta^2}{2} - \mu + \delta\right)\Gamma\left(\frac{1}{k}, \left(\frac{t}{\lambda}\right)^k\right) - e^{-(t/\lambda)^k)}\left(1 + \left(\frac{t}{\lambda}\right)^k\right) = 0,$$

where $\Gamma\left(\frac{1}{k}, \left(\frac{t}{\lambda}\right)^k\right)$ is an incomplete gamma function. A similar substitution we have used in [5]. However one can not reduce the original PDE to an ordinary differential equation as one could do with exponential liquidation time distribution. But also a two dimensional PDE is better for further analytical or numerical studies as a three dimensional one.

When there is an exponential liquidation time distribution and just in this case one can reduce (7.23) to an ODE. Let us now briefly show the reduction in the case of an exponential liquidation time distribution and log-utility function.

The value function (7.9) in this case is defined as follows

$$V(t,l,h) = \max_{(\pi,c)} E\left[\int_t^{\infty} e^{-\kappa t}\log(c)dt|L(t) = l, H(t) = h\right], \quad \kappa > 0. \tag{7.31}$$

The HJB equation and the resulting PDE (7.23) is now homogeneous in time. We introduce $\tilde{V}(l,h)$

$$\tilde{V}(l,h) = \max_{(\pi,c)} E\left[\int_t^{\infty} e^{-\kappa(s-t)}\log(c)ds\right] = \max_{(\pi,c)} E\left[\int_0^{\infty} e^{-\kappa v}\log(c)dv\right],$$

which is independent on time and substituting $V(t,l,h) = e^{-\kappa t}\tilde{V}(l,h)$ into the HJB equation (7.10) we arrive at a time-independent PDE on $\tilde{V}(l,h)$. With a slight abuse of notation, hereafter we will use the same letter V for $\tilde{V}$. Now using substitution for $V(l,h)$ in the form

$$V(l,h) = v(z) + \frac{\log h}{\kappa} + \frac{1}{\kappa^2}\left(\mu - \delta - \frac{\eta^2}{2}\right), \quad z = l/h, \tag{7.32}$$

we obtain that $v(z)$ satisfies the ordinary differential equation of second order

$$\frac{\eta^2}{2}z^2 v'' + \max_{\pi}\left[\frac{1}{2}\pi^2\sigma^2 v' - \pi\left((v' + zv'')\eta\rho\sigma + (\alpha - r)v'\right)\right] \\ + \max_{c\geq -\delta}\left[-cv_z + \log(c+\delta)\right] = \kappa v, \tag{7.33}$$

where $v' = v_z$ and the dimension of the problem is reduced to one. It is important to note that such reduction was possible due to existence of the corresponding two dimensional Lie subalgebra of the four dimensional admitted Lie algebra provided in Theorem 7.4 (see [4] for the proofs). Performing a formal maximization of the quadratic part (7.33) and coming back to the original independent variables we obtain the optimal policies in the form

$$c_\star(l,h) = \frac{h}{v'(l/h)}, \quad \pi_\star(l,h) = -\frac{\eta\rho}{\sigma}l - h\frac{d_1}{\sigma}\frac{v'(l/h)}{v''(l/h)}. \tag{7.34}$$

Summing up, we formulate the following theorem

Theorem 7.5 ([5]) *Suppose $r - (\mu - \delta) > 0$ and $d_1 = \frac{\alpha - r - \eta\rho\sigma}{\sigma^2} \neq 0$.*

- *There is the unique $C^2(0, +\infty)$ solution $v(z)$ of (7.33) in a class of concave functions.*
- *For $l, h > 0$ the value function is given by (7.32). For $h = 0, l > 0$ the value function $V(l, 0)$ coincides with the classical Merton solution*

$$V(l,0) = \frac{1}{\kappa^2}\left[r + \frac{1}{2}\frac{(\alpha - r)^2}{\sigma^2} - \kappa\right] + \frac{\log(\kappa l)}{\kappa} = M + \frac{\log(\kappa l)}{\kappa}. \tag{7.35}$$

- *If the ratio between the stochastic income and the total wealth tends to zero, the policies $(\pi^\star, c^\star)$ given by (7.34) tend to the classical Merton's policies*

$$c_\star(l,0) \sim \kappa l, \ \pi_\star(l,0) \sim -\frac{(\alpha - r) l}{\sigma^2}\frac{V_l^2}{V_{ll}}. \tag{7.36}$$

- *Policies (7.34) are optimal.*

We do not provide the proofs here and for all further details we address the reader to [5].

7.4 Conclusions

In this chapter we have formulated an optimization problem for a portfolio with an illiquid asset sold at an exogenous random moment of time. One can prove that under certain conditions there exists a unique viscosity solution for the problem of such type. We have also closely regarded a family of three dimensional nonlinear PDEs that such problem gives rise to in [4, 5] and discussed here the main results. The study of the Lie algebraic structure of such PDEs gives a possibility to find invariant variables and reduce three dimensional PDEs to two dimensional or in the case of an exponential liquidation time distribution to ODEs. We provided two examples of such reductions.

References

1. Bangia, A., Diebold, F.X., Schuermann, T., Stroughair, J.D.: Modeling liquidity risk, with implications for traditional market risk measurement and management. Technical Report, Stern School of Business (1998). http://www.ssc.upenn.edu/~fdiebold/papers/paper25/bds.pdf
2. Black, F., Scholes, M.: The pricing of options and corporate liabilities. J. Polit. Econ. **81**, 637–654 (1973)
3. Bordag, L.A.: Geometrical properties of differential equations. Applications of Lie group analysis in Financial Mathematics. World Scientific, Singapore (2015)

4. Bordag, L.A., Yamshchikov, I.P.: Optimization problem for a portfolio with an illiquid asset: Lie group analysis. J. Math. Anal. Appl. **453**(2), 668–699 (2017). http://dx.doi.org/10.1016/j.jmaa.2017.04.014
5. Bordag, L.A., Yamshchikov, I.P., Zhelezov, D.: Portfolio optimization in the case of an asset with a given liquidation time distribution. Int. J. Eng. Math. Modell. **2**(2), 31–50 (2015)
6. Bordag, L.A., Yamshchikov, I.P., Zhelezov, D.: Optimal allocation-consumption problem for a portfolio with an illiquid asset. Int. J. Comput. Math. **93**(5), 749–760 (2016)
7. Cauley, S.D., Pavlov, A.D., Schwartz, E.S.: Homeownership as a constraint on asset allocation. J. Real Estate Financ. Econ. **34**(3), 283–311 (2007)
8. Coppejans, M., Domowitz, I., Madhavan, A.: Liquidity in an automated auction. Technical Report, Pennsylvania State University (2000). http://ssrn.com/abstract=295765
9. Dick-Nielsen, J., Feldhütter, P., Lando, D.: Corporate bond liquidity before and after the onset of the subprime crisis. J. Financ. Econ. **103**(3), 471–492 (2012)
10. Duffie, D., Zariphopoulou, T.: Optimal investment with undiversifiable income risk. Math. Financ. **3**, 135–148 (1993)
11. Duffie, D., Fleming, W., Soner, H.M., Zariphopoulou, T.: Hedging in incomplete markets with HARA utility. J. Econ. Dyn. Control. **21**, 753–782 (1997)
12. Ekeland, I., Pirvu, T.A.: On a non-standard stochastic control problem. Technical Report, Department of Mathematics, The University of British Columbia (2008). http://arxiv.org/abs/0806.4026v1
13. Ekeland, I., Pirvu, T.A.: Investment and consumption without commitment. Math. Finan. Econ. **2**(1), 57–86 (2008)
14. Goyenko, R., Sarkissian, S.: Treasury bond illiquidity and global equity returns. J. Financ. Quant. Anal. **49**, 1227–1253 (2014)
15. Grossman, S.J., Laroque, G.: Asset pricing and optimal portfolio choice in the presence of illiquid durable consumption goods, National Bureau of Economic Research; w2369
16. Hooker, M.A., Kohn, M.: An empirical measure of asset liquidity. Technical Report, Dartmouth College Department of Economics (1994). http://ssrn.com/abstract=5544
17. Ibragimov, N.H.: Lie Group Analysis of Differential Equations. CRC Press, Boca Raton (1994)
18. Keynes, J.: A Treatise on Money. Macmillan, London (1930)
19. Lie, S.: Vorlesungen über Differentialgleichungen mit bekannten infinitesimalen Transformationen, Teubner, Stuttgart (1891)
20. Lippman, S.A., McCall, J.J.: An operational measure of liquidity. Am. Econ. Rev. **76**(1), 43–55 (1986)
21. Lykina, V., Pickenhain, S., Wagner, M.: On a resource allocation model with infinite horizon. Appl. Math. Comput. **204**(2), 595–601 (2008)
22. Merton, R.: Optimum consumption and portfolio rules in a continuous time model. J. Econ. Theory **3**, 373–413 (1971)
23. Miller, B.L.: Optimal consumption with a stochastic income stream. Econometrica **42**, 253–266 (1974)
24. Olver, P.J.: Applications of Lie Groups to Differential Equations. Springer, New York (2000)
25. Pickenhain, S.: Sufficiency conditions for weak local minima in multidimensional optimal control problems with mixed control-state restrictions. Z. Anal. Anwend. **11**(4), 559–568 (1992)
26. Pickenhain, S.: Infinite horizon optimal control problems in the light of convex analysis in Hilbert spaces. Set Valued Var. Anal. **23**(1), 169–189 (2015)
27. Schwartz, E.S., Tebaldi, C.: Illiquid assets and optimal portfolio choice. Technical Report, NBER Working Paper Series (2006). http://www.nber.org/papers/w12633
28. Scott, D.: Wall Street Words: An A to Z Guide to Investment Terms for Today's Investor. Houghton Mifflin Corporation, Boston (2003)
29. Skiba, P.M., Tobacman, J.: Payday loans, uncertainty and discounting: explaining patterns of borrowing, repayment, and default. Vanderbilt Law and Economics Research Paper No. 08-33 (2008)

30. Spicer, A., McDermott, G.A., Kogut, B.: Entrepreneurship and privatization in central Europe: The tenuous balance between destruction and creation. Acad. Manage. Rev. **25**(3), 630–649 (2000)
31. Yamshchikov, I.P., Zhelezov, D.: Liquidity and optimal consumption with random income. Project Report IDE1149, Technical Report, Department of Mathematics, Physics and Electrical Engineering, IDE, Halmstad University, Sweden (2011). http://urn.kb.se/resolve?urn=urn:nbn:se:hh:diva-16108
32. Zariphopoulou, T.: Optimal investment and consumption models with non-linear stock dynamics. Math. Meth. Oper. Res. **50**, 271–296 (1999)
33. Zeldes, S.P.: Optimal consumption with stochastic income: deviations from certainty equivalence. Q. J. Econ. **104**, 275–298 (1989)

Chapter 8
Analytical and Numerical Results for American Style of Perpetual Put Options Through Transformation into Nonlinear Stationary Black-Scholes Equations

Maria do Rosário Grossinho, Yaser Faghan, and Daniel Ševčovič

Abstract We analyze and calculate the early exercise boundary for a class of stationary generalized Black-Scholes equations in which the volatility function depends on the second derivative of the option price itself. A motivation for studying the nonlinear Black Scholes equation with a nonlinear volatility arises from option pricing models including, e.g., non-zero transaction costs, investors preferences, feedback and illiquid markets effects and risk from unprotected portfolio. We present a method how to transform the problem of American style of perpetual put options into a solution of an ordinary differential equation and implicit equation for the free boundary position. We finally present results of numerical approximation of the early exercise boundary, option price and their dependence on model parameters.

8.1 Introduction

In this chapter we are concerned with a financial option with no fixed maturity and no exercise limit, called the perpetual option. This type of an option, which can be exercised at any time, can be considered as the American style of an option. However, in this case, the time to maturity has no impact on the price of the option. From the mathematical point of view, this leads to a solution of the stationary Black-Scholes problem. More precisely, the valuation problem is transformed into the free

M. do Rosário Grossinho (✉) • Y. Faghan
Instituto Superior de Economia e Gestão and CEMAPRE, Universidade de Lisboa, Lisbon, Portugal
e-mail: mrg@iseg.ulisboa.pt; yaser.kord@yahoo.com

D. Ševčovič
Department of Applied Mathematics & Statistics, Comenius University, 842 48 Bratislava, Slovakia
e-mail: sevcovic@fmph.uniba.sk

M. Ehrhardt et al. (eds.), *Novel Methods in Computational Finance*,
Mathematics in Industry 25, DOI 10.1007/978-3-319-61282-9_8

boundary problem that consists of the construction of the function $V(S)$ together with the early exercise boundary point ϱ satisfying the following conditions:

$$\frac{1}{2}\sigma^2 S^2 \partial_S^2 V + rS\partial_S V - rV = 0, \qquad S > \varrho,$$

and

$$V(\varrho) = E - \varrho, \quad \partial_S V(\varrho) = -1, \quad V(+\infty) = 0$$

(cf. [5, 11, 18]). The function V is defined in the domain $S > \varrho$, where ϱ is the free boundary position. If the diffusion coefficient $\sigma > 0$ is constant then we are, in fact, considering stationary solutions of the classical linear Black-Scholes parabolic equation. However, we suppose that σ depends on the asset price S and the product of the asset price S and the second derivative (Gamma) of the option price $H = S\partial_S^2 V$, i.e.

$$\sigma = \sigma(S, H) = \sigma(S, S\partial_S^2 V). \tag{8.1}$$

Let us mention our motivation for studying a nonlinear volatility of the form (8.1). As it is known, the classical linear Black-Scholes model (cf. [12, 14]) was derived under several restrictive assumptions that did not reflect the real market. In fact, no transaction costs were considered, the volatility was supposed to be constant, only liquid and complete markets were considered. Since then, several results have appeared in the literature relaxing these assumptions in order to overcome some drawbacks they created in practice. Regarding the volatility, it has been justified in practice that it is not constant and it may depend on the asset price itself. With this volatility function (8.1), the classical model is generalized in such a way that it allows to consider non-zero transaction costs, market feedback and illiquid market effects due to large trading volumes, risk from investors preferences, etc.. Mathematically, the problem will lose its linear feature, since the equation becomes a nonlinear partial differential equation (see e.g. [18]).

One of the first nonlinear models taking into account non-trivial transaction costs was proposed by Leland [16] for put or call options, later extended for more general types of option by Hoggard, Whalley and Wilmott [10]. Avellaneda and Paras [2] proposed the jumping volatility model in which the volatility changes with respect to the sign of the Gamma of the option. Frey and Patie [6], Frey and Stremme [7] developed models dealing with feedback and illiquid market impact due to large trading (see also [17]). We also mention the so-called the risk adjusted pricing model (RAPM) derived by Kratka [13] and Jandačka and Ševčovič [12] in which both the transaction costs as well as the risk from unprotected portfolio are taken into account. In the RAPM model the volatility function depends on $H = S\partial_S^2 V$ only, and it has the form:

$$\sigma(H)^2 = \sigma_0^2(1 + \lambda H^{\frac{1}{3}}) = \sigma_0^2(1 + \lambda(S\partial_S^2 V)^{\frac{1}{3}}), \tag{8.2}$$

where $\sigma_0 > 0$ is the constant historical volatility of the underlying asset and λ is a model parameter depending on the transaction cost rate and the unprotected portfolio risk exposure. Recently, explicit solutions to European style of options described by the nonlinear Black-Scholes equation with varying volatility have been derived by Bordag et al. [4] for the Frey and Patie as well as the RAPM models.

Barles and Soner [3] proposed a model assuming that investor's preferences are shown by an exponential utility function. In this model, the volatility function depends on $H = S\partial_S^2 V$ as well as S, and it has the following form:

$$\sigma(S,H)^2 = \sigma_0^2 \left(1 + \Psi(a^2 SH)\right) = \sigma_0^2 \left(1 + \Psi(a^2 S^2 \partial_S^2 V)\right), \tag{8.3}$$

where the function Ψ is the unique solution to the ODE:

$$\Psi'(x) = (\Psi(x) + 1)/(2\sqrt{x\Psi(x)} - x), \qquad \Psi(0) = 0$$

and $a \geq 0$ is a constant depending transaction costs and investor's risk aversion parameter (see [3] for details). Note that $\Psi(x) \geq 0$ for all $x \geq 0$ and it has the following asymptotic: $\Psi(x) = O(x^{\frac{1}{3}})$ for $x \to 0$ and $\Psi(x) = O(x)$ for $x \to \infty$.

Finally, we also mention the nonlinear volatility model developed by Amster et al. [1], where the transaction costs depend on the volume of trading assets in a linear decreasing way. Recently, it was generalized for arbitrary transaction cost functions by Ševčovič and Žitňanská [19].

This chapter is organized as follows. In the next section, we recall the mathematical formulation of the perpetual American put option pricing model. Furthermore, we prove the existence and uniqueness of a solution to the free boundary problem. We derive a formula for the option price and a single implicit equation for the free boundary position ϱ. In Sect. 8.3 we construct suitable sub- and upper-solutions based on Merton's explicit solutions with constant volatility. Finally, in Sect. 8.4, we present computational results of the free boundary position ϱ, the option price $V(S)$ and their dependence on model parameters.

8.2 Perpetual American Put Option

In this section we analyze the problem of the American style of perpetual put options. As referred previously, perpetual options are financial options with no fixed maturity and no exercise limit. As they can be exercised at any time, they have infinite maturity $T = +\infty$.

Consider the American style of a put option with the volatility σ of the form (8.1). Suppose that there exist a limit of the solution V and an early exercise boundary position S_f for the maturity $T \to \infty$. The pair consisting of the limiting price $V = V(S) = \lim_{T-t\to\infty} V(S,t)$ and the limiting early exercise boundary position $\varrho = \lim_{T-t\to\infty} S_f(t)$ of the perpetual put option is a solution to the stationary nonlinear Black-Scholes problem (cf. [8]):

$$\frac{1}{2}\sigma(S, S\partial_S^2 V)S^2\partial_S^2 V + rS\partial_S V - rV = 0, \qquad S > \varrho, \tag{8.4}$$

and

$$V(\varrho) = E - \varrho, \quad \partial_S V(\varrho) = -1, \quad V(+\infty) = 0 \tag{8.5}$$

(cf. [14, 15, 18]). We shall prove that under certain assumptions made on the volatility function the perpetual American put option problem (8.4)–(8.5) has the unique solution $(V(.), \varrho)$. We will present its explicit formula for the case when $\sigma = \sigma(H)$, i.e. the volatility depends on the term $H = S\partial_S^2 V$ only. Furthermore, we will also present comparison results with explicit Merton's solutions recently obtained by the authors in [9].

Throughout the chapter we will assume that the volatility function $\sigma = \sigma(S, H)$ fulfills the following assumption:

Assumption 8.1 *The volatility function $\sigma = \sigma(S, H)$ in* (8.4) *is assumed to be a C^1 smooth nondecreasing function in the $H > 0$ variable and $\sigma(S, H) \geq \sigma_0 > 0$ for any $S > 0$ and $H \geq 0$ where σ_0 is a positive constant.*

If we extend the volatility function $\sigma(S, H)$ by $\sigma(S, 0)$ for negative values of H, i.e. $\sigma(S, H) = \sigma(S, 0)$ for $H \leq 0$ then the function

$$\mathbb{R} \ni H \longmapsto \frac{1}{2}\sigma(S, H)^2 H \in \mathbb{R}$$

is strictly increasing and therefore there exists the unique inverse function $\beta : \mathbb{R} \to \mathbb{R}$ such that

$$\frac{1}{2}\sigma(S, H)^2 H = w \qquad \text{if and only if} \qquad H = \beta(x, w), \quad \text{where } S = e^x. \tag{8.6}$$

Note that the function β is a continuous and increasing function such that $\beta(0) = 0$.

Concerning the inverse function we have the following useful lemma:

Lemma 8.1 *Assume the volatility function $\sigma(S, H)$ satisfies Assumption 8.1. Then the inverse function β has the following properties:*

1. *$\beta(x, 0) = 0$ and $\frac{\beta(x,w)}{w} \leq \frac{2}{\sigma_0^2}$ for all $x, w \in \mathbb{R}$;*
2. *$\beta'_w(x, w) \leq \frac{2}{\sigma_0^2}$ for all $x \in \mathbb{R}$ and $w > 0$.*

Proof Clearly, $\beta(x, 0) = 0$. For $w > 0$ we have $\beta(x, w) > 0$ and $w = \frac{1}{2}\sigma(e^x, \beta(x, w))^2\, \beta(x, w) \geq \frac{\sigma_0^2}{2}\beta(x, w)$ and so $\frac{\beta(x,w)}{w} \leq \frac{2}{\sigma_0^2}$. If $w < 0$ then $\beta(x, w) < 0$ and we can proceed similarly as before.

Differentiating the equality $w = \frac{1}{2}\sigma(e^x, \beta(x,w))^2\beta(x,w) \geq \frac{\sigma_0^2}{2}\beta(x,w)$ with respect to $w > 0$ yields:

$$1 = \frac{1}{2}\sigma^2(e^x, \beta(x,w))\beta'_w(x,w) + \partial_H\left(\frac{1}{2}\sigma(e^x, H)^2\right)H \geq \frac{1}{2}\sigma_0^2\beta'_w(x,w)$$

for $H = \beta(x,w) > 0$ and the proof of the second statement of Lemma follows.

The key step how to solve the perpetual American put option problem (8.4)–(8.5) consists in introduction of the following variable:

$$W(x) = \frac{r}{S}\left(V(S) - S\partial_S V(S)\right) \quad \text{where } S = e^x. \tag{8.7}$$

Lemma 8.2 *Let $x^0 \in \mathbb{R}$ be given. The function $V(S)$ is a solution to Eq. (8.4) for $S > \varrho = e^{x_0}$ satisfying the boundary condition:*

$$V(S) - S\partial_S V(S) = E, \quad \text{at } S = \varrho,$$

iff and only if the transformed function $W(x)$ is a solution to the initial value problem for the ODE:

$$\begin{aligned} \partial_x W(x) &= -W(x) - r\beta(x, W(x)), \quad x > x_0, \\ W(x_0) &= rEe^{-x_0}. \end{aligned} \tag{8.8}$$

Proof As $\partial_x = S\partial_S$ we obtain

$$\begin{aligned} \partial_x W(x) &= rS\partial_S(S^{-1}V(S) - \partial_S V(S)) = rSS^{-1}\partial_S V(S) - rS^{-1}V(S) - rS\partial_S^2 V(S) \\ &= -W(x) - rS\partial_S^2 V(S) = -W(x) - r\beta(x, W(x)), \end{aligned}$$

because $\beta(x, W(x)) = H \equiv S\partial_S^2 V(S)$ if and only if $\frac{1}{2}\sigma(S,H)^2 H = W(x)$ and V solves (8.4), i.e.

$$\frac{1}{2}\sigma(S,H)^2 H + \frac{r}{S}\left(S\partial_S V(S) - V(S)\right) = 0.$$

Finally, $W(x_0) = \frac{r}{S}\left(V(S) - S\partial_S V(S)\right) = rEe^{-x_0}$ where $S = \varrho = e^{x_0}$, as claimed.

Notice the equivalence of conditions:

$$V(S) - S\partial_S V(S) = E \text{ and } V(S) = E - S \iff \partial_S V(S) = -1 \text{ and } V(S) = E - S. \tag{8.9}$$

Concerning the solution W of the ODE (8.8) we have the following auxiliary result:

Lemma 8.3 *Assume $x^0 \in \mathbb{R}$. Let $W = W_{x_0}(x)$ be the unique solution to the ODE (8.8) for $x \in \mathbb{R}$ satisfying the boundary condition $W(x_0) = rEe^{-x_0}$ at the initial point x_0. Then*

1. *$W_{x_0}(x) > 0$ for any $x \in \mathbb{R}$,*
2. *the function $x_0 \mapsto W_{x_0}(x)$ is increasing in the x_0 variable for any $x \in \mathbb{R}$,*
3. *if the volatility function depends on $H = S\partial_S^2 V$ only, i.e. $\sigma = \sigma(H)$, then*

$$W_{x_0}(x)=F^{-1}(x_0-x) \text{ where } F(W)=\int_{W_0}^{W} \frac{1}{w+r\beta(w)}dw,\ \ W_0 = W(x_0) = rEe^{-x_0}.$$

Proof According to Lemma 8.1 we have $\beta(x,w)/w \le 2/\sigma_0^2$ for any $x \in \mathbb{R}$ and $w \neq 0$. Hence

$$\partial_x|\ln(W(x)| = -\left(1 + r\frac{\beta(x, W(x))}{W(x)}\right) \ge -(1+\gamma)$$

where $\gamma = 2r/\sigma_0^2$. Therefore

$$|W(x)| \ge |W(x_0)|e^{-(1+\gamma)(x-x_0)} > 0,$$

and this is why the function $W(x)$ does not change the sign. As $W(x_0) = rEe^{-x_0} > 0$ we have $W_{x_0}(x) > 0$ as well.

The solution $W_{x_0}(x)$ to the ODE (8.8) can be expressed in the form

$$\begin{aligned} W_{x_0}(x) &= W_{x_0}(x_0) - \int_{x_0}^{x} (W_{x_0}(\xi) + r\beta(\xi, W_{x_0}(\xi)))d\xi \\ &= rEe^{-x_0} - \int_{x_0}^{x} (W_{x_0}(\xi) + r\beta(\xi, W_{x_0}(\xi)))\, d\xi. \end{aligned}$$

Let us introduce the auxiliary function

$$y(x) = \partial_{x_0} W_{x_0}(x).$$

Then

$$\begin{aligned} y(x) &= -rEe^{-x_0} + W_{x_0}(x_0) + r\beta(x_0, W_{x_0}(x_0)) - \int_{x_0}^{x} \left(1 + r\beta_w'(\xi, W_{x_0}(\xi))\right) y(\xi)d\xi \\ &= r\beta(x_0, W_{x_0}(x_0)) - \int_{x_0}^{x} \left(1 + r\beta_w'(\xi, W_{x_0}(\xi))\right) y(\xi)d\xi. \end{aligned}$$

Hence y is a solution to the ODE:

$$\partial_x y(x) = -\left(1 + r\beta'_w(x, W_{x_0}(x))\right) y(x), \quad x \in \mathbb{R}, \tag{8.10}$$
$$y(x_0) = r\beta(x_0, rEe^{-x_0}) > 0.$$

With regard to Lemma 8.1 we have $\beta'_w(x, W_{x_0}(x)) \leq 2/\sigma_0^2$. Therefore the function y is a solution to the differential inequality:

$$\partial_x y(x) \geq -(1+\gamma)y(x), \quad x \in \mathbb{R},$$

where $\gamma = 2r/\sigma_0^2$. As a consequence we obtain

$$|y(x)| \geq |y(x_0)|e^{-(1+\gamma)(x-x_0)} > 0 \tag{8.11}$$

and this is why the function $y(x)$ does not change the sign. Therefore $\partial_{x_0} W_{x_0}(x) = y(x) > 0$ and the proof of the statement (2) follows.

Finally, if $\sigma = \sigma(H)$ we have $\beta = \beta(w)$ and so

$$\partial_x F(W(x)) = \frac{1}{W(x) + r\beta(W(x))}\partial_x W(x) = -1.$$

Hence $F(W(x)) = F(W(x_0)) - (x - x_0) = x_0 - x$ and the statement (3) follows.

Lemma 8.4 *Under Assumption 8.1, there exists the unique root* $x_0 \in \mathbb{R}$ *of the implicit equation*

$$\int_{x_0}^{\infty} \beta(x, W_{x_0}(x))dx = 1. \tag{8.12}$$

Proof Denote $\phi(x_0) = \int_{x_0}^{\infty} \beta(x, W_{x_0}(x))dx$. Then $\phi(\infty) = 0$ and

$$\phi'(x_0) = -\beta(x_0, W_{x_0}(x_0)) + \int_{x_0}^{\infty} \beta'_w(x, W_{x_0}(x))y(x)dx$$

where $y(x) = \partial_{x_0} W_{x_0}(x)$ is the solution to (8.10). That is

$$\partial_x y(x) = -\left(1 + r\beta'_w(x, W_{x_0}(x))\right) y(x)$$

and $y(x_0) = r\beta(x_0, W_{x_0}(x_0)) = r\beta(x_0, rEe^{-x_0})$. Therefore

$$\begin{aligned}\phi'(x_0) &= -\beta(x_0, W_{x_0}(x_0)) - \frac{1}{r}\int_{x_0}^{\infty} \partial_x y(x) + y(x)dx \\ &= -\frac{1}{r}y(\infty) - \frac{1}{r}\int_{x_0}^{\infty} y(x)dx \leq -\frac{1}{r}\int_{x_0}^{\infty} y(x)dx.\end{aligned}$$

As $y(x) = \partial_{x_0} W_{x_0}(x) \geq y(x_0)e^{-(1+\gamma)(x-x_0)}$ we have

$$\phi'(x_0) \leq -\frac{1}{r}\frac{y(x_0)}{1+\gamma} = -\frac{\beta(x_0, W_{x_0}(x_0))}{1+\gamma}.$$

It means that the function ϕ is strictly decreasing. Since

$$\frac{1}{2}\sigma(e^{x_0}, \beta(x_0, W_{x_0}(x_0)))^2\beta(x_0, W_{x_0}(x_0)) = W_{x_0}(x_0) = rEe^{-x_0} \to +\infty \quad \text{as } x_0 \to -\infty,$$

we have $\lim_{x_0\to-\infty} \beta(x_0, W_{x_0}(x_0)) = \infty$ and therefore $\lim_{x_0\to-\infty} \phi'(x_0) = -\infty$. Therefore $\phi(-\infty) = \infty$. In summary, there exists the unique root x_0 of the equation $\phi(x_0) = 1$, as claimed.

Now we are in a position to state our main result on unique solvability of the perpetual American put option problem (8.4)–(8.5).

Theorem 8.1 *Assume the volatility function σ satisfies Assumption 8.1. Then there exists the unique solution $(V(.), \varrho)$ to the perpetual American put option problem (8.4)–(8.5). The function $V(S)$ is given by*

$$V(S) = \frac{S}{r}\int_{\ln S}^{\infty} W_{x_0}(x)dx, \quad \textit{for } S \geq \varrho = e^{x_0},$$

where $W_{x_0}(x)$ is the solution to the ODE (8.8) *and x_0 is the unique root of the implicit equation* (8.12).

Proof Differentiating the above expression for $V(S)$ we obtain

$$\partial_S V(S) = \frac{1}{r}\int_{\ln S}^{\infty} W_{x_0}(x)dx - \frac{1}{r}W_{x_0}(\ln S)$$
$$S\partial_S^2 V(S) = -\frac{1}{r}\left(W_{x_0}(x) + \partial_x W_{x_0}(x)\right) = \beta(x, W_{x_0}(x)),$$

where $x = \ln S$. Hence

$$\frac{1}{2}\sigma(S, S\partial_S^2 V)^2 S^2\partial_S^2 V + rS\partial_S V - rV$$
$$= S\left(\frac{1}{2}\sigma(e^x, \beta(x, W_{x_0}(x)))^2\beta(x, W_{x_0}(x)) - W_{x_0}(x)\right) = 0,$$

i.e. $V(S)$ is the solution to (8.4) for $S > \varrho = e^{x_0}$.

Furthermore,

$$[V(S) - S\partial_S V(S)]_{S=\varrho} = V(\varrho) - \frac{\varrho}{r}\int_{\ln \varrho}^{\infty} W_{x_0}(x)dx + \frac{\varrho}{r}W_{x_0}(\ln \varrho) = E\varrho e^{-\ln \varrho} = E,$$

and,

$$V(\varrho) = \frac{\varrho}{r}\int_{\ln\varrho}^{\infty} W_{x_0}(x)dx = \frac{\varrho}{r}\int_{\ln\varrho}^{\infty} -\partial_x W_{x_0}(x) - r\beta(x, W_{x_0}(x))dx$$
$$= \frac{\varrho}{r}W_{x_0}(\ln\varrho) - \varrho\int_{\ln\varrho}^{\infty}\beta(x, W_{x_0}(x))dx = E - \varrho$$

because x_0 is the unique solution to (8.12). With regard to the equivalence (8.9) we have $\partial_S V(S) = -1$ at $S = \varrho$. In summary, $(V(.), \varrho)$ is the unique solution to the perpetual American put option problem (8.4)–(8.5).

Remark 8.1 In the case the volatility function depends on $H = S\partial_S^2 V$ only, i.e. $\sigma = \sigma(H)$, then Eq. (8.12) can be simplified by introducing the change of variables $w = W_{x_0}(x)$. Indeed, $\beta = \beta(w)$ and $dw = \partial_x W_{x_0}(x)dx = -(W_{x_0}(x) + r\beta(W_{x_0}(x)))dx$. Therefore

$$\int_{x_0}^{\infty}\beta(W_{x_0}(x))dx = -\int_{W_{x_0}(x_0)}^{0}\frac{\beta(w)}{w + r\beta(w)}dw = \int_0^{\frac{rE}{\varrho}}\frac{\beta(w)}{w + r\beta(w)}dw.$$

Equation (8.12) can be rewritten in the following form

$$\int_0^{\frac{rE}{\varrho}}\frac{\beta(w)}{w + r\beta(w)}dw = 1. \tag{8.13}$$

This is the condition for the free boundary position ϱ recently derived by the authors in [9].

8.3 The Merton Explicit Solution, Sub- and Super-Solutions

In this section we recall recent results due to the authors [9] dealing with comparison of the solution $(V(.), \varrho)$ to the perpetual American put option problem (8.4)–(8.5) for the case when the volatility function depends on $H = S\partial_S^2 V$ only, i.e. $\sigma = \sigma(H)$.

Suppose that the volatility $\sigma \equiv \sigma_0$ is constant, then for the function $V(S)$ and the limiting early exercise boundary position ϱ the free boundary value problem (8.4)–(8.5) has the explicit solution presented by Merton (cf. [14, 18]), which has the closed form:

$$V_\gamma(S) = \begin{cases} E - S, & 0 < S \le \varrho_\gamma, \\ \frac{E}{1+\gamma}\left(\frac{S}{\varrho_\gamma}\right)^{-\gamma}, & S > \varrho_\gamma, \end{cases} \tag{8.14}$$

where

$$\varrho_\gamma = E\frac{\gamma}{1+\gamma}, \qquad \gamma = \frac{2r}{\sigma_0^2}. \tag{8.15}$$

Our next goal is to establish sub- and super-solutions to the perpetual American put option pricing problem. Let $\gamma > 0$ is a positive constant and denote by V_γ the explicit Merton solution defined before. It is clear that the pair $(V_\gamma(\cdot), \varrho_\gamma)$ is the explicit Merton solution with constant volatility $\sigma_0^2 = 2r/\gamma$.

Then, for the transformed function $W_\gamma(x)$ we have

$$W_\gamma(x) = rE\varrho_\gamma^\gamma e^{-(1+\gamma)x}, \quad \text{for } x = \ln S > x_{0\gamma} = \ln \varrho_\gamma.$$

Furthermore W_γ is a solution to the ODE:

$$\partial_x W_\gamma + W_\gamma + \gamma W_\gamma = 0. \tag{8.16}$$

Applying the Eq. (8.16) we can construct a super-solution W_{γ^+} and a sub-solution W_{γ^-} to the solution W of the equation:

$$\partial_x W = -W - r\beta(W)$$

using the Merton solution W_γ. Here γ^+ is the unique root of the equation

$$\gamma^+\sigma(1+\gamma^+)^2 = 2r$$

and γ^- satisfies

$$\gamma^-\sigma(0)^2 = 2r.$$

As a consequence, the following inequalities hold. For more details, we refer to [9].

$$\begin{cases} \partial_x W_{\gamma^+}(x) \geq -W_{\gamma^+}(x) - r\beta(W_{\gamma^+}(x)), \text{ for } x > x_{0\gamma^+} = \ln \varrho_{\gamma^+}, \\ \partial_x W_{\gamma^-}(x) \leq -W_{\gamma^-}(x) - r\beta(W_{\gamma^-}(x)), \text{ for } x > x_{0\gamma^-} = \ln \varrho_{\gamma^-}. \end{cases} \tag{8.17}$$

Moreover, it can be proved that

$$\varrho_{\gamma^+} \leq \varrho \leq \varrho_{\gamma^-}.$$

Since, for initial conditions we have $W_{\gamma^\pm}(x_{0\gamma^\pm}) = \frac{rE}{\varrho_{\gamma^\pm}}$ and $W(x_0) = \frac{rE}{\varrho}$ and so

$$W_{\gamma^-}(x_{0\gamma^-}) \leq W(x_0) \leq W_{\gamma^+}(x_{0\gamma^+}).$$

Using the comparison principle for solutions of ordinary differential inequalities in (8.17) we conclude

$$W_{\gamma^-}(x) \le W(x) \le W_{\gamma^+}(x).$$

Then taking into account the explicit solution of the function $V(S)$ from Theorem 8.1 we present the following result:

Theorem 8.2 ([9, Theorem 3]) *Let* $(V(\cdot), \varrho)$ *be the solution to the perpetual American pricing problem* (8.4)–(8.5). *Then for any* $S \ge 0$ *we have*

$$V_{\gamma^-}(S) \le V(S) \le V_{\gamma^+}(S)$$

and

$$\varrho_{\gamma^+} \le \varrho \le \varrho_{\gamma^-}$$

where $(V_{\gamma^\pm}(.), \varrho_{\gamma^\pm})$ *are explicit Merton's solutions with constant volatilities.*

8.4 Numerical Approximation Scheme and Results

In the last section, our aim is to present an efficient numerical scheme for constructing a solution to the perpetual American put option problem (8.4)–(8.5) for the case when the volatility function has the form: $\sigma = \sigma(H)$ where $H = S\partial_S^2 V$. The numerical results were obtained by the authors in [9].

Our scheme is based on transformation $H = \beta(w)$, i.e. $w = \frac{1}{2}\sigma(H)^2 H$ and $dw = \frac{1}{2}\partial_H(\sigma(H)^2 H)dH$ by using this we can rewrite the Eq. (8.13) for the free boundary position ϱ as follows:

$$\int_0^{\beta(rE/\varrho)} \frac{H}{\frac{1}{2}\sigma(H)^2 H + rH} \frac{1}{2}\partial_H(\sigma(H)^2 H)dH = 1. \tag{8.18}$$

Similarly, the expression (see Theorem 8.1) for the price of the option can be rewritten in terms of the H variable as follows:

$$V(S) = \frac{S}{r}\int_0^{\beta(F^{-1}(\ln(\varrho/S)))} \frac{\frac{1}{2}\sigma(H)^2 H}{\frac{1}{2}\sigma(H)^2 H + rH} \frac{1}{2}\partial_H(\sigma(H)^2 H)dH. \tag{8.19}$$

When the inverse function $\beta(w)$ is not given by a closed form formula by applying this transformation we can avoid computational complexity.

In what follows we recall numerical results of computation of the solution to the perpetual American put option problem (8.4)–(8.5) for the RAPM model with the nonlinear volatility function of the form:

$$\sigma(H)^2 = \sigma_0^2 \left(1 + \lambda H^{\frac{1}{3}}\right), \tag{8.20}$$

We propose the results of numerical calculation for the risk adjusted pricing methodology model (RAPM). We would like to show the position of the free boundary ϱ and the value of the perpetual option V evaluated at exercise price $S = E$. The option values are computed for various values of the model $\lambda \in [0, 2]$ for the RAPM model. The rest of the model parameters were chosen as: $r = 0.1$, $E = 100$ and $\sigma_0 = 0.3$. In computations shown in Table 8.1 we present results of the free boundary position and the perpetual American put option price $V(E)$ for the RAPM model.

Finally, in Fig. 8.1 we show the option price $V(S)$ for the Risk adjusted pricing methodology model with closed form explicit Merton's solutions with constant volatility.

Table 8.1 The perpetual put option free boundary position ϱ and the option price $V(S)$ evaluated at $S = E$ for various values of the model parameter $\lambda \geq 0$ for the RAPM model (Source [9])

λ	0.00	0.20	0.40	0.60	1.20	1.60	2.00
ϱ	68.9655	64.7181	61.2252	58.2647	51.1474	47.2975	44.5433
$V(E)$	13.5909	15.4853	17.1580	18.6669	22.5461	24.7444	26.6804

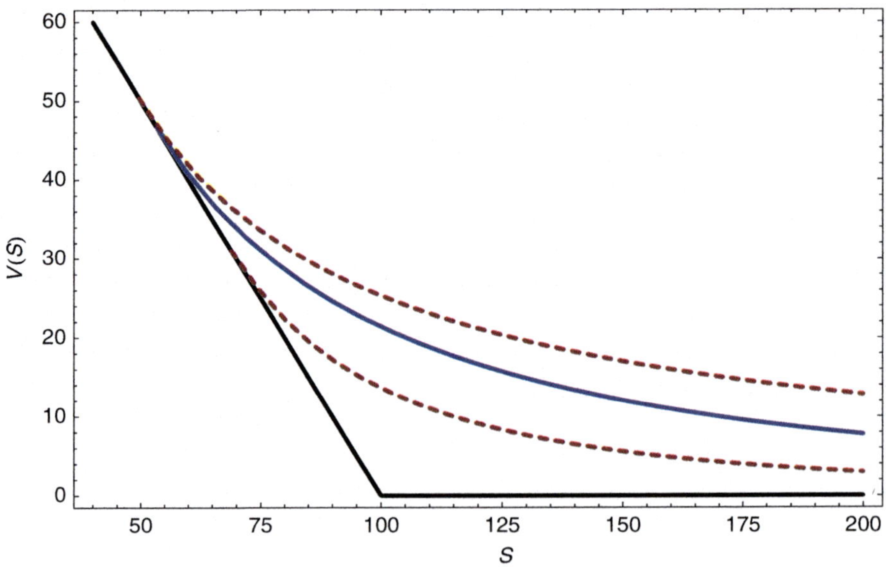

Fig. 8.1 *Solid curve* represents a graph of a perpetual American put option $V(S)$ for the RAPM model with $\lambda = 1$. Sub- and super- solutions V_{γ^-} and V_{γ^+} are depicted by *dashed curves*. The model parameters: $r = 0.1$, $E = 100$ and $\sigma_0 = 0.3$ (Source [9])

8.5 Conclusions

In this chapter we analyzed the problem of American style perpetual options when the nonlinear volatility is a function of the second derivative. We studied the free boundary problem that models this type of options, by transforming it into a single implicit equation for the free boundary position and explicit integral expression for the option price.

Acknowledgements This research was supported by the European Union in the FP7-PEOPLE-2012-ITN project STRIKE—Novel Methods in Computational Finance (304617), the project CEMAPRE MULTI/00491 financed by FCT/MEC through national funds and the Slovak research Agency Project VEGA 1/0780/15.

References

1. Amster, P., Averbuj, C.G., Mariani, M.C., Rial, D.: A Black-Scholes option pricing model with transaction costs. J. Math. Anal. Appl. **303**, 688–695 (2005)
2. Avellaneda, M., Paras, A.: Dynamic hedging portfolios for derivative securities in the presence of large transaction costs. Appl. Math. Finance **1**, 165–193 (1994)
3. Barles, G., Soner, H.M.: Option pricing with transaction costs and a nonlinear Black-Scholes equation. Finance Stochast. **2**, 369–397 (1998)
4. Bordag, L.A., Chmakova, A.Y.: Explicit solutions for a nonlinear model of financial derivatives. Int. J. Theor. Appl. Finance **10**(1), 1–21 (2007)
5. Dewynne, J.N., Howison, S.D., Rupf, J., Wilmott, P.: Some mathematical results in the pricing of American options. Eur. J. Appl. Math. **4**, 381–398 (1993)
6. Frey, R., Patie, P.: Risk management for derivatives in illiquid markets: a simulation study. Advances in Finance and Stochastics. Springer, Berlin (2002)
7. Frey, R., Stremme, A.: Market volatility and feedback effects from dynamic hedging. Math. Financ. **4**, 351–374 (1997)
8. Grossinho, M.R., Morais, E.: A note on a stationary problem for a Black-Scholes equation with transaction costs. Int. J. Pure Appl. Math. **51**, 557–565 (2009)
9. Grossinho, M., Kord Faghan, Y., Ševčovič, D.: Pricing perpetual put options by the Black-Scholes equation with a nonlinear volatility function. (submitted, 2016). arXiv.org:1611.00885
10. Hoggard, T., Whalley, A.E., Wilmott, P.: Hedging option portfolios in the presence of transaction costs. Adv. Futures Options Res. **7**, 21–35 (1994)
11. Hull, J.: Options, Futures and Other Derivative Securities. Prentice Hall, New York (1989)
12. Jandačka, M., Ševčovič, D.: On the risk adjusted pricing methodology based valuation of vanilla options and explanation of the volatility smile. J. Appl. Math. **3**, 235–258 (2005)
13. Kratka, M.: No mystery behind the smile. Risk **9**, 67–71 (1998)
14. Kwok, Y.K.: Mathematical Models of Financial Derivatives. Springer, Heidelberg (1998)
15. Lauko, M., Ševčovič., D.: Comparison of numerical and analytical approximations of the early exercise boundary of American put options. ANZIAM J. **51**, 430–448 (2011)
16. Leland, H.E.: Option pricing and replication with transaction costs. J. Financ. **40**, 1283–1301 (1985)

17. Schönbucher, P., Wilmott, P.: The feedback-effect of hedging in illiquid markets. SIAM J. Appl. Math. **61**, 232–272 (2000)
18. Ševčovič, D., Stehlíková, B., Mikula, K.: Analytical and numerical methods for pricing financial derivatives. Nova Science, Hauppauge (2011)
19. Ševčovič, D., Žitňanská, M.: Analysis of the nonlinear option pricing model under variable transaction costs. Asia-Pacific Finan. Markets **23**(2), 153–174 (2016)

Chapter 9
Stochastic Dynamic Programming and Control of Markov Processes

Manuel Guerra

Abstract This chapter contains a brief discussion of the basic mathematical ideas behind dynamic programming methods for optimal control of Markov processes. It is based on lectures given by the author at the Summer School on Computational Finance held at Smolenice Castle, Slovakia, in September 2014.

The key theoretical ideas behind the approach are outlined in a somewhat abstract setting. We hope that this will help the reader to understand the key points of the dynamic programming principle and Hamilton-Jacobi-Bellman equations and their potential as tools to solve a large array of optimization problems, without paying too much attention to the technical difficulties that often arise in concrete applications.

9.1 Introduction

The most common type of stochastic optimal control problem concerns the control of diffusions. For this type of control problems, a rather complete theory is available (see [6, 9, 10], among others). The main approach to these problems is the so-called dynamic programming approach which relies on the dynamic programming principle and yields the solution of the optimal control problem in the form of a solution of a nonlinear second order PDE, known as the Hamilton-Jacobi-Bellman equation.

The same approach may, in principle, be applied to many other types of optimal control problems, leading to other types of Hamilton-Jacobi-Bellman equations. There is a large number of recent publications dealing with different types of extensions and applications of the basic method. A survey of this literature is out of the scope of this text, but we mention a very general if short discussion in [5]. About more specific results, we note that generalizations to optimal control of Lévy processes as draw particular attention, as in [2, 3, 7], among others.

This paper presents the dynamic programming approach in a very general setting, from a point of view that has similarities to [5], but uses different mathematical

M. Guerra (✉)
ISEG and CEMAPRE—Universidade de Lisboa, Rua do Quelhas 6, 1200-781 Lisboa, Portugal
e-mail: mguerra@iseg.ulisboa.pt

M. Ehrhardt et al. (eds.), *Novel Methods in Computational Finance*,
Mathematics in Industry 25, DOI 10.1007/978-3-319-61282-9_9

tools. It differs from the classical approach in the fact that our emphasis is on control of the law of the process, rather than control of the sample paths.

We discuss dynamic programming in the class of Feller processes. This is a broad class that includes Lévy processes as particular cases. By contrast, the regularity assumptions on the optimization criterion are quite restrictive. This option is motivated by the belief that this setting is quite convenient to present the basic ideas without having to deal with too much technical detail. Also, the class of optimization criteria that can be dealt with is sensitive to the type of process being considered. Hence, it is difficult to present a discussion that is simultaneously general with respect to the processes and with respect to the optimization criteria.

The paper is organized as follows. Section 9.2 contains a brief outline of the dynamic programming principle and Hamilton-Jacobi-Bellman equation for diffusions, in the form it is usually presented in the literature [6, 9, 10]. This serves as a background of concepts and results that will be discussed and generalized in the following sections. Section 9.3 deals with Feller processes, their operator semigroups and infinitesimal generators. These are the basic concepts from which control systems will be build in Sect. 9.4. In that section, a suitable class of optimal control problems is introduced. The dynamic programming principle and the relation between solutions of the optimal control problem and solutions of the appropriate Hamilton-Jacobi-Bellman equation are proved. A final section contains some global remarks and references about the contents of this text.

9.2 Optimal Control of Diffusions

In this section, we sketch the optimal stochastic control problem and the main results concerning its solution by dynamic programming, in the case of diffusions. The Theorems 9.1, 9.2 and 9.3 below constitute the main topic of discussion for the subsequent sections of this paper. Readers who are not familiar with these results and want a deeper understanding of the material in this section may see [6] or [9].

9.2.1 The Optimal Control Problem, the Dynamic Programming Principle, and the Hamilton-Jacobi-Bellman Equation

Consider a filtered probability space $(\Omega, \mathscr{F}, \{\mathscr{F}_t\}_{t\geq 0}, P)$, supporting an m-dimensional $\{\mathscr{F}_t\}$-adapted Brownian motion W. Let $A \subset \mathbb{R}^n$, $U \subset \mathbb{R}^k$ be nonempty sets, with A open, and fix functions $b : [0,T] \times A \times U \mapsto \mathbb{R}^n$, $\sigma : [0,T] \times A \times U \mapsto \mathbb{R}^{n\times m}$. The process to be controlled is the solution of the stochastic differential equation

$$dX_s = b(s, X_s, u_s)ds + \sigma(s, X_s, u_s)dW_s, \tag{9.1}$$

where u is the "control", i.e., a process to be chosen among a class $\mathscr{U}$ of $\{\mathscr{F}_t\}$-progressively measurable U-valued processes.

The functions b, σ are assumed to satisfy some regularity and growth conditions ensuring that for each control process $u \in \mathscr{U}$, each initial time $t \in [0, T[$ and state $x \in A \subset \mathbb{R}^n$, there is one unique A-valued process $X_s^{t,x,u}$ solving Eq. (9.1) in the time interval $[t, T]$ with initial condition $X_t = x$.

Take functions $f : [0, T] \times A \times U \mapsto \mathbb{R}$, $g : A \mapsto \mathbb{R}$, $\rho : [0, T] \times A \times U \mapsto \mathbb{R}$, and for each $u \in \mathscr{U}$ and $0 \leq t \leq s \leq T$, let

$$\beta^u_{t,s} = e^{-\int_t^s \rho(\theta, X_\theta^{t,x,u}, u_\theta)d\theta}.$$

For each initial time and state $(t, x) \in [0, T[\times A$, the optimal control problem consists of finding the control $\hat{u} \in \mathscr{U}$ maximizing the functional

$$J(t, x, u) = \mathbb{E}\left[\int_t^T \beta^u_{t,s} f(s, X_s^{t,x,u}, u_s)ds + \beta^u_{t,T} g(X_T^{t,x,u})\right]. \tag{9.2}$$

Thus, $\beta^u_{t,s}$ is a (stochastic) discount factor corresponding to the stochastic interest rate $\rho(s, X_s^{t,x,u}, u_s)$. This comes naturally in problems arising from finance or economics. Again, the functions f, g, ρ are assumed to satisfy regularity and growth conditions ensuring that the functional (9.2) is well defined for every $(t, x) \in [0, T] \times A$ and every $u \in \mathscr{U}$.

To solve the problem (9.1)–(9.2), we introduce the *value function*

$$V(t, x) = \sup_{u \in \mathscr{U}} J(t, x, u),$$

and the differential operators $\{\varphi \mapsto \mathscr{L}^u \varphi\}_{u \in U}$, defined as

$$\mathscr{L}^u \varphi(t, x) = -\rho(t, x, u)\varphi(t, x) + \frac{\partial}{\partial x}\varphi(t, x)b(t, x, u) + \frac{1}{2} Tr\left(\sigma\sigma^*(t, x, u)\frac{\partial^2}{\partial x^2}\varphi(t, x)\right)$$

The *dynamic programming principle* is a theorem of the following type:

Theorem 9.1 *(Under Suitable Assumptions)*
For every $(t, x) \in [0, T[\times A$ and every $\{\mathscr{F}_t\}$-stopping time such that $t \leq \tau \leq T$:

$$V(t, x) = \sup_{u \in \mathscr{U}} \mathbb{E}\left[\int_t^\tau \beta^u_{t,s} f(s, X_s^{t,x,u}, u_s)ds + \beta^u_{t,T} V(\tau, X_\tau^{t,x,u})\right]. \tag{9.3}$$

Thus, the dynamic programming principle allows for the decomposition of problem (9.1)–(9.2) into a pair of problems in any non-overlapping stochastic intervals $[t, \tau]$, $]\tau, T]$.

Using the dynamic programming principle, it is possible to prove the following theorems, which characterize the solution of the problem (9.1)–(9.2).

Theorem 9.2 *(Under Suitable Assumptions)*
If the value function is $C^{1,2}$, then it solves the Hamilton-Jacobi-Bellman (HJB) equation:

$$\frac{\partial}{\partial t}V(t,x) + \sup_{u\in U}\big(f(t,x,u) + \mathscr{L}^u V(t,x)\big) = 0, \tag{9.4}$$

$$V(T,x) = g(x). \tag{9.5}$$

Theorem 9.3 (Verification Theorem) *(Under Suitable Assumptions)*
If a function V solves the HJB equation (9.4)–(9.5), then it is the value function. Further, if there is a measurable function $\hat{u} : [0,T[\times\mathbb{R}^n \mapsto U$ such that

$$f\big(s,x,\hat{u}(s,x)\big) + \mathscr{L}^{\hat{u}(s,x)}V(s,x) = \max_{u\in U}\big(f(s,x,u) + \mathscr{L}^u V(s,x)\big), \quad \forall (s,x)$$

then the process $\hat{u}_s = \hat{u}\left(s, X_s^{t,x,\hat{u}}\right)$ is an optimal control.

Usually, the Theorems 9.2 and 9.3 can be proved only under somewhat different assumptions. Therefore, they are not the exact converse of each other.

9.2.2 Time-Homogeneous Problems

Before proceeding to the discussion and generalization of the theorems above, we introduce some simplifications of problem (9.1)–(9.2) to obtain a formulation more suitable for the abstractions we have in mind.

Consider the autonomous version of problem (9.1)–(9.2). That is, the controlled stochastic differential equation

$$dX_s = b(X_s,u_s)ds + \sigma(X_s,u_s)dW_s. \tag{9.6}$$

As before, let $X_s^{t,x,u}$ denote the unique solution of (9.6) for the particular control $u \in \mathscr{U}$ and initial condition $X_t = x$. Now, we seek to maximize the simplified functional

$$J(t,x,u) = \mathbb{E}\left[g\left(X_T^{t,x,u}\right)\right]. \tag{9.7}$$

To see that there is no loss of generality in substituting the problem (9.6)–(9.7) for the problem (9.1)–(9.2), introduce the additional processes

$$\theta_s = s, \qquad \beta_s = e^{-\int_t^s \rho(r,X_r,u_r)dr}, \qquad Y_s = \int_t^s \beta_r f(r,X_r,u_r)dr.$$

The vector $(\theta_s, \beta_s, Y_s, X_s)$ solves the SDE

$$d\begin{pmatrix}\theta_s\\ \beta_s\\ Y_s\\ X_s\end{pmatrix} = \begin{pmatrix}1\\ -\rho(\theta_s, X_s, u_s)\beta_s\\ f(\theta_s, X_s, u_s)\beta_s\\ b(\theta_s, X_s, u_s)\end{pmatrix} ds + \begin{pmatrix}0\\ 0\\ 0\\ \sigma(\theta_s, X_s, u_s)\end{pmatrix} dW_s =$$
$$=\tilde{b}(\theta_s, \beta_s, Y_s, X_s)ds + \tilde{\sigma}(\theta_s, \beta_s, Y_s, X_s)dW_s,$$

and the objective function (9.2) can be represented in the form

$$J(t,x,u) = \mathbb{E}\left[Y_T^{t,x,u} + \beta_T^{t,x,u} g(X_T^{t,x,u})\right] = \mathbb{E}\left[\tilde{g}(\beta_T^{t,x,u}, Y_T^{t,x,u}, X_T^{t,x,u})\right].$$

This reduces the problem (9.1)–(9.2) to a problem of type (9.6)–(9.7). For this type of problem, the dynamic programming principle (9.3) takes the form

$$V(t,x) = \sup_{u\in\mathscr{U}} \mathbb{E}\left[V(\tau, X_\tau^{t,x,u})\right] \tag{9.8}$$

for every $\{\mathscr{F}_t\}$-stopping time $t \le \tau \le T$. The Hamilton-Jacobi-Bellman equation (9.4) reduces to

$$\frac{\partial}{\partial t} V(t,x) + \sup_{u\in U} \mathscr{L}^u V(t,x) = 0, \tag{9.9}$$

with

$$\mathscr{L}^u \varphi(t,x) = \frac{\partial}{\partial x}\varphi(t,x) b(x,u) + \frac{1}{2} Tr\left(\sigma\sigma^*(x,u)\frac{\partial^2}{\partial x^2}\varphi(t,x)\right). \tag{9.10}$$

9.3 Feller Processes, Feller Semigroups, and Infinitesimal Generators

Feller processes are a class of stochastic processes that are suitable to be "controlled", as shown in Sect. 9.4 below. Other classes may be considered to outline similar arguments (see Sect. 9.5), but Feller processes fit particularly well in the general point of view adopted in this paper.

This section contains only the briefest outline of the topic. For a complete account, see [1, 4, 8], among others.

9.3.1 Feller Processes and Feller Semigroups

Let $(\Omega, \mathscr{F}, \{\mathscr{F}_t\}_{t\geq 0}, \mu)$ be a filtered probability space satisfying the following conditions

(A1) The filtration $\{\mathscr{F}_t\}$ is complete, i.e., $\mathscr{F}_0$ contains all the μ-null sets;
(A2) The filtration $\{\mathscr{F}_t\}$ is right continuous, i.e., $\mathscr{F}_{t+} = \mathscr{F}_t$ for every $t \geq 0$.

Let $C_0(\mathbb{R}^n)$ be the space of all continuous functions $f : \mathbb{R}^n \mapsto \mathbb{R}$ vanishing at infinity, provided with the topology of uniform convergence. That is, $C_0(\mathbb{R}^n)$ is provided the norm $\|f\| = \max_{x\in\mathbb{R}^n} |f(x)|$.

Any $\mathbb{R}^n$-valued random variable $X : \Omega \mapsto \mathbb{R}^n$ induces a probability measure in $\mathbb{R}^n$, defined as

$$\mu^X(A) = \mu\{\omega \in \Omega : X(\omega) \in A\},$$

for every A, a Borel subset of $\mathbb{R}^n$. Thus, for any $\mathbb{R}^n$-valued stochastic process X, any $f \in C_0(\mathbb{R}^n)$, and any $s, t \geq 0$, the function $x \mapsto \mathbb{E}[f(X_t)| X_s = x]$ is well defined μ^{X_s}-almost everywhere in $\mathbb{R}^n$. As usual, we say that such a function is bounded (continuous, differentiable, etc.) if it coincides μ^{X_s}-almost everywhere with a bounded (continuous, differentiable, etc.) function.

Let $X : [0, +\infty[\times\Omega \mapsto \mathbb{R}^n$ be an $\{\mathscr{F}_t\}$-adapted process such that:

(A3) X is a càdlàg process, i.e., for μ-almost every $\omega \in \Omega$, the trajectory $t \mapsto X_t(\omega)$ is right-continuous and has finite left limit at every point $t \in [0, +\infty[$;
(A4) X is a Markov process, i.e., $\Pr\{X_t \in A| \mathscr{F}_s\} = \Pr\{X_t \in A| X_s\}$ for every $0 \leq s \leq t < +\infty$ and every A, a Borel subset of $\mathbb{R}^n$.
(A5) X has the *Feller property*, i.e., for every $f \in C_0(\mathbb{R}^n)$ and every $0 \leq s \leq t < +\infty$, the function $x \mapsto \mathbb{E}[f(X_t)| X_s = x]$ is continuous and vanishes at infinity.

We denote the function $x \mapsto \mathbb{E}[f(X_t)| X_s = x]$ by $P_{s,t}f$. From the above it follows that X induces a family of continuous linear operators

$$\{P_{s,t} : C_0(\mathbb{R}^n) \mapsto C_0(\mathbb{R}^n)\}_{0\leq s\leq t<+\infty}.$$

We consider two further conditions on the process X:

(A6) For every $f \in C_0(\mathbb{R}^n)$, $0 \leq s \leq t < +\infty$, we have $\lim_{u\to t} \|P_{s,u}f - P_{s,t}f\| = 0$ if $t > s$, or $\lim_{u\to t+} \|P_{s,u}f - P_{s,t}f\| = 0$ if $t = s$;
(A7) X is *time-homogeneous*, i.e., for every $f \in C_0(\mathbb{R}^n)$ and every $0 \leq s \leq t < +\infty$, $P_{s,t}f = P_{0,(t-s)}f$.

A process satisfying conditions **(A1)**–**(A6)** is called a *Feller process*. If it satisfies also **(A7)**, it is called a *time-homogeneous Feller process*. Notice that the continuity condition **(A6)** does not require X to have continuous sample paths. Many important

Markov processes are Feller. For example, every Lévy process (including Brownian motion but also many processes with jumps) is a Feller process.

The condition **(A7)** is not restrictive: for any $\mathbb{R}^n$-valued Feller process X, (t, X_t) is a time-homogeneous $\mathbb{R}^{n+1}$-valued Feller process. For time homogeneous Feller processes, we write P_t instead of $P_{0,t}$, and the Markov property **(A4)** implies that $\{P_t\}_{t\geq 0}$ is a semigroup with respect to composition of operators:

$$P_0 f = f, \qquad P_s P_t f = P_{s+t} f$$

for every $f \in C_0(\mathbb{R}^n)$, $s, t \geq 0$. Notice that all the operators P_t are positive contractions, i.e.,

$$0 \leq P_t f \leq \|f\| \qquad \forall t \in [0, +\infty[,\ f \in C_0(\mathbb{R}^n), f \geq 0. \tag{9.11}$$

In the time-homogeneous case, the continuity condition **(A6)** reduces to the apparently simpler condition

$$\lim_{t\to 0+} \|P_t f - f\| = 0 \qquad \forall f \in C_0(\mathbb{R}^n). \tag{9.12}$$

A semigroup of linear operators $\{P_t : C_0(\mathbb{R}^n) \mapsto C_0(\mathbb{R}^n)\}_{t\geq 0}$ satisfying (9.11)–(9.12) is called a *Feller semigroup*. It can be proved that given an initial distribution μ^{X_0}, every Feller semigroup is the semigroup of a unique (in law) Feller process. Further, for an appropriate modification of the Feller process, the minimal filtration satisfying conditions **(A1)**–**(A2)** is the natural filtration generated by the process augmented with all the null sets. Thus, every statement about Feller semigroups is a statement about Feller processes an vice-versa.

9.3.2 *Infinitesimal Generators*

It turns out that Feller semigroups are characterized by their infinitesimal generators, a concept that we now introduce.

Theorem 9.4 *Let $\{P_t : C_0(\mathbb{R}^n) \mapsto C_0(\mathbb{R}^n)\}_{t\geq 0}$ be a Feller semigroup. There is a dense linear subspace $\mathscr{D} \subset C_0(\mathbb{R}^n)$ such that for each $f \in \mathscr{D}$ there is $g \in C_0(\mathbb{R}^n)$ such that $\lim_{t\to 0+} \left\| \frac{1}{t}(P_t f - f) - g \right\| = 0$.*

The transformation

$$f \mapsto \mathscr{L} f = \lim_{t\to 0+} \frac{1}{t}(P_t f - f)$$

is a (usually unbounded) linear operator from $\mathscr{D}$ into $C_0(\mathbb{R}^n)$. It is called the *infinitesimal generator* of the semigroup $\{P_t\}_{t\geq 0}$ (or, of the corresponding Feller process X).

Example 9.1 (Compound Poisson Process) Consider the process

$$X_t = X_0 + \sum_{i=1}^{N_t} Y_i,$$

where N_t is a Poisson process with intensity $\lambda > 0$, and Y_i, $i = 1, 2, \ldots$ are i.i.d. real-valued random variables. This is a càdlàg time-homogeneous Markov process. A simple computation shows that for any $f \in C_0(\mathbb{R}^n)$:

$$\mathbb{E}\left[f(X_t)|X_0 = x\right] = f(x)e^{-\lambda t} + \sum_{n=1}^{+\infty} \mathbb{E}\left[f\left(x + \sum_{i=1}^{n} Y_i\right)\right] \frac{\lambda^n t^n}{n!} e^{-\lambda t}.$$

Thus, it is a Feller process with infinitesimal generator

$$\mathscr{L}f(x) = \lambda\left(\mathbb{E}\left[f(x+Y)\right] - f(x)\right),$$

where Y is any random variable with the same law as Y_i, $i = 1, 2, \ldots$. This is a continuous operator and its domain is the whole space $C_0(\mathbb{R}^n)$.

Example 9.2 (Diffusion) Consider the stochastic differential equation

$$dX_t = b(X_t)dt + \sigma(X_t)dW_t. \tag{9.13}$$

If there is some constant $C < \infty$ such that

$$|b(x) - b(y)| + |\sigma(x) - \sigma(y)| < C|x - y| \qquad \forall x, y \in \mathbb{R}^n,$$

then Eq. (9.13) admits one unique solution. This solution is a time-homogeneous Markov process.

Let $C_0^2(\mathbb{R}^n)$ be the space of all functions twice continuously differentiable, vanishing at infinity and with partial derivatives up to second order vanishing at infinity. This is a dense linear subspace of $C_0(\mathbb{R}^n)$. Due to Itô's formula, for every $f \in C_0^2(\mathbb{R}^n)$ and $t > 0$:

$$f(X_t) = f(X_0) + \int_0^t \left(Dfb + \frac{1}{2}\mathrm{Tr}\left(\sigma\sigma^* D^2 f\right)\right)(X_s)ds + \int_0^t (Df\sigma)(X_s)dW_s.$$

The stochastic integral in the right-hand side is a martingale and therefore

$$\mathbb{E}\left[f(X_t)|\, X_0 = x\right] = f(x) + \mathbb{E}\left[\int_0^t \left(Dfb + \frac{1}{2}\mathrm{Tr}\left(\sigma\sigma^* D^2 f\right)\right)(X_s)ds \,\middle|\, X_0 = x\right].$$

From this, a simple computation shows that the infinitesimal generator of X is

$$\mathscr{L}f(x) = Df(x)b(x) + \frac{1}{2}Tr\left(\sigma(x)\sigma(x)^* D^2 f(x)\right).$$

Its domain contains at least all the functions in $C_0^2(\mathbb{R}^n)$. Notice that in this case, the infinitesimal generator is not a bounded operator: uniform convergence does not imply convergence of partial derivatives.

The similitude between the expression above and Eq. (9.10) is not a coincidence.

As mentioned above and illustrated in Example 9.2, the infinitesimal generator is in general, an unbounded operator. However, it has some regularity. Recall that an operator $\Phi : D \subset X \mapsto Y$ is *closed* if for every sequence $\{x_k \in D\}_{k\in\mathbb{N}}$, if $\{x_k\}$ converges towards some $x \in X$ and $\{\Phi x_k\}$ converges towards some $y \in Y$, then $x \in D$ and $\Phi x = y$. An operator is said to be *closable* if it admits some extension that is closed. The *closure* of an operator is its minimal closed extension. It turns out that infinitesimal generators of Feller semigroups are always closable. Indeed, if we set $\mathscr{D}$ to be the maximal subspace with the property described in Theorem 9.4, then the infinitesimal generator is closed. This is the most usual definition of infinitesimal generators, but since two closable operators coinciding in a dense subset must have the same closure, we do not insist on this point.

The relation between infinitesimal generators and Feller processes goes both ways: the next theorem shows that the infinitesimal generator characterizes the Feller semigroup and therefore it characterizes the law of the Feller process.

Theorem 9.5 *Let $\{P_t : C_0(\mathbb{R}^n) \mapsto C_0(\mathbb{R}^n)\}_{t\geq 0}$ be a Feller semigroup with closed infinitesimal generator $\mathscr{L} : \mathscr{D} \mapsto C_0(\mathbb{R}^n)$. Then:*

(a) For every $f \in C_0(\mathbb{R}^n)$ and $t \geq 0$, we have $\int_0^t P_s f ds \in \mathscr{D}$ and $P_t f = f + \mathscr{L}\int_0^t P_s f ds$;
(b) For every $f \in \mathscr{D}$ and $t \in [0, +\infty[$, we have $P_t f \in \mathscr{D}$, $\mathscr{L}P_t f = P_t\mathscr{L}f$, and $\{P_t\}_{t\geq 0}$ is the unique family of operators satisfying

$$P_t f = f + \int_0^t \mathscr{L}P_s f ds = f + \int_0^t P_s\mathscr{L}f ds. \tag{9.14}$$

The integral $\int_0^t P_s f ds$ is understood in the weak sense: for every $x \in \mathbb{R}^n$, the mapping $s \mapsto (P_s f)(x)$ is a continuous function from $[0, +\infty[$ into $\mathbb{R}$. Thus, the integral $\int_0^t (P_s f)(x) ds$ is well defined. $\int_0^t P_s f ds$ denotes the function $x \mapsto \int_0^t (P_s f)(x) ds$. The definition of the integral $\int_0^t \mathscr{L}P_s f ds$ is analogous.

The Eq. (9.14) is called the (integral form of the) *Kolmogorov equation*. It implies that two Feller semigroups with the same infinitesimal generator must be identical. The following theorem gives a characterization of infinitesimal generators

Theorem 9.6 (Hille-Yosida) *A linear operator* $\mathscr{L} : \mathscr{D} \subset C_0(\mathbb{R}^n) \mapsto C_0(\mathbb{R}^n)$ *is the infinitesimal generator of some Feller semigroup if and only if it satisfies all the following conditions:*

(a) $\mathscr{D}$ *is dense in* $C_0(\mathscr{R}^n)$*;*
(b) There is some $\lambda > 0$ *such that the image* $(\lambda Id - \mathscr{L})(\mathscr{D})$ *is dense in* $C_0(\mathscr{R}^n)$*;*
(c) For every $f \in \mathscr{D}$ *and every* $\lambda > 0$, $\|(\lambda Id - \mathscr{L})f\| \geq \lambda \|f\|$.

An operator satisfying condition *(c)* of Theorem 9.6 is called a *dissipative operator*. Unfortunately, this condition is not easy to check in practice. Thus, it is useful to have the following sufficient condition.

Proposition 9.1 *Consider a linear operator* $\mathscr{L} : \mathscr{D} \subset C_0(\mathbb{R}^n) \mapsto C_0(\mathbb{R}^n)$.

If $\mathscr{L}f(x) \leq 0$ *for every* $f \in \mathscr{D}$ *with a positive maximum and every x a maximizer of* f*, then* $\mathscr{L}$ *is dissipative.*

Notice that the infinitesimal generators in Examples 9.1 and 9.2 satisfy the condition in Proposition 9.1.

9.3.3 Properties of Feller Processes

Feller processes have very good properties. In particular, the Propositions 9.2, 9.3, and 9.4 below play a central role in the concept of stochastic control system presented in the next section.

Proposition 9.2 *Let X be a time-homogeneous Feller process, and let* $\{\mathscr{F}_t^X\}_{t\geq 0}$ *be its natural filtration. X has the* strong Markov property*:*

$$\Pr\{X_{\tau+t} \in A | \mathscr{F}_\tau^X\} = \Pr\{X_{\tau+t} \in A | X_\tau\}$$

for every finite $\{\mathscr{F}_t^X\}$*-stopping time* τ*, every* $t \geq 0$*, and every Borel set A. If* P_t *is the Feller semigroup generated by X, then*

$$\mathbb{E}\left[f(X_{\tau+t}) | \mathscr{F}_\tau^X\right] = P_t f(X_\tau) \tag{9.15}$$

for every $f \in C_0(\mathbb{R}^n)$.

Proposition 9.3 *Let X be a time-homogeneous Feller process with infinitesimal generator* $\mathscr{L} : \mathscr{D} \mapsto C_0(\mathbb{R}^n)$*. For every* $f \in \mathscr{D}$*, the process*

$$M_t = f(X_t) - \int_0^t (\mathscr{L}f)(X_s)ds$$

is a martingale.

Proof The proposition follows easily from the definitions of semigroup of a Feller process and infinitesimal generator.

Let $\{P_t\}_{t\geq 0}$ be the semigroup of X. Then,

$$\begin{aligned}
&\mathbb{E}\left[M_{s+t}\middle|\,\mathscr{F}_s\right] = \\
=&\mathbb{E}\left[f(X_{s+t})\middle|\,\mathscr{F}_s\right] - \int_0^s (\mathscr{L}f)(X_u)du - \int_s^{s+t} \mathbb{E}\left[(\mathscr{L}f)(X_u)\middle|\,\mathscr{F}_s\right] du = \\
=&(P_t f)(X_s) - \int_0^s (\mathscr{L}f)(X_u)du - \int_s^{s+t} \left(P_{u-s}(\mathscr{L}f)\right)(X_s)du = \\
=&f(X_s) + \int_0^t \left(P_u(\mathscr{L}f)\right)(X_s)du - \int_0^s (\mathscr{L}f)(X_u)du - \\
&- \int_0^t \left(P_u(\mathscr{L}f)\right)(X_s)du = M_s.
\end{aligned}$$

Since X is càdlàg, it follows that M is a càdlàg martingale and therefore, Doob's optional sampling theorem states that

$$\mathbb{E}\left[f(X_\tau) - \int_0^\tau (\mathscr{L}f)(X_s)ds \,\middle|\, \mathscr{F}_\theta \right] = f(X_\theta) - \int_0^\theta (\mathscr{L}f)(X_s)ds \tag{9.16}$$

for every bounded stopping times $0 \leq \theta \leq \tau$.

Rearranging equality (9.16), we obtain the following.

Proposition 9.4 *Let X be a time-homogeneous Feller process with infinitesimal generator $\mathscr{L} : \mathscr{D} \mapsto C_0(\mathbb{R}^n)$. For every $f \in \mathscr{D}$ and every bounded stopping times $0 \leq \theta \leq \tau$:*

$$\mathbb{E}\left[f(X_\tau)\middle|\,\mathscr{F}_\theta\right] = \mathbb{E}\left[f(X_\tau)\middle|\,X_\theta\right] = f(X_\theta) + \mathbb{E}\left[\int_\theta^\tau (\mathscr{L}f)(X_s)ds \,\middle|\, X_\theta \right]. \tag{9.17}$$

9.3.4 The Feynman-Kac Equation

Before introducing controlled stochastic processes, it is useful to consider a simpler problem.

Let X be a Feller process with semigroup $\{P_t\}_{t\geq 0}$ and infinitesimal generator $\mathscr{L} : \mathscr{D} \mapsto C_0(\mathbb{R}^n)$. Fix $f \in \mathscr{D}$, $T \in]0, +\infty[$, and consider the function $V : [0, T] \times \mathbb{R}^n \mapsto \mathbb{R}$ defined as

$$V(t, x) = \mathbb{E}\left[f(X_T)\middle|\,X_t = x\right]. \tag{9.18}$$

Clearly, V is a well-defined function. The problem is to find a suitable characterization of this function. This is given by the following theorem.

Theorem 9.7 *The function V defined by* (9.18) *is the unique solution of the terminal value problem*

$$\frac{\partial V}{\partial t}(t,x) + \mathscr{L}V(t,x) = 0 \qquad \forall (t,x) \in [0,T[\times\mathbb{R}^n, \tag{9.19}$$

$$V(T,x) = f(x) \qquad \forall x \in \mathbb{R}^n. \tag{9.20}$$

Proof By definition, $V(t,x) = P_{T-t}f(x)$. Thus, equality (9.20) must hold. By the Theorem 9.5, the function $t \mapsto \mathscr{L}P_t f(x)$ is continuous for every $x \in \mathbb{R}^n$. Using Theorem 9.5, we see that every $t, s < T$:

$$\begin{aligned} V(t,x) - V(s,x) =& P_{T-t}f(x) - P_{T-s}f(x) = \int_{T-s}^{T-t} \mathscr{L}P_u f(x)du = \\ =& -(t-s)\mathscr{L}P_{T-t}f(x) + o(t-s) \end{aligned}$$

when $s \to t$. This is equality (9.19).

Uniqueness of the solution is a consequence of uniqueness of the solution of the Kolmogorov equation (9.14).

The Eq. (9.19) is known as the *Feynman-Kac* equation.

Example 9.3 (Compound Poisson Process) For the process described in Example 9.1, the Feynman-Kac equation is an integro-differential equation

$$\frac{\partial V}{\partial t}(t,x) + \lambda\left(\mathbb{E}[V(t,x+Y)] - V(t,x)\right) = 0.$$

Example 9.4 (Diffusion) For the diffusion described in Example 9.2, the Feynman-Kac equation is the parabolic PDE:

$$\frac{\partial V}{\partial t}(t,x) + D_x V(t,x)b(x) + \frac{1}{2}Tr\left(\sigma(x)\sigma(x)^* D_x^2 V(t,x)\right) = 0.$$

9.4 Optimal Control of Feller Processes

Now, we are ready to introduce a general definition of controlled stochastic process and obtain the corresponding analogous to Theorems 9.1, 9.2, and 9.3.

9.4.1 Controlled Processes

Let U be a nonempty set, and consider a family of Feller semigroups indexed by U, $\{P_t^u\}_{u\in U}$. All semigroups P_t^u are understood to operate on the same space $C_0(\mathbb{R}^n)$.

We use $\mathscr{L}^u$ to denote the infinitesimal generator of the semigroup P_t^u and X^u to denote any Feller process with semigroup P_t^u. For each $u \in U$, let $\mathscr{D}^u$ be the maximal domain of the infinitesimal generator $\mathscr{L}^u$ (thus, $\mathscr{L}^u : \mathscr{D}^u \mapsto C_0(\mathbb{R}^n)$ is a closed operator for every $u \in U$).

We assume that the collection $\{\mathscr{D}^u\}_{u\in U}$ satisfies the following condition:

(A8) The linear subspace $\mathscr{D} = \bigcap_{u\in U} \mathscr{D}^u$ is dense in $C_0(\mathbb{R}^n)$.

From these elements, we construct new processes in the following way.

Pick an element $u \in U$ and consider the corresponding process X^u and its natural filtration $\{\mathscr{F}_t^{X^u}\}_{t\geq 0}$. Pick τ, a finite $\{\mathscr{F}_t^{X^u}\}$-stopping time, and a new element $v \in U$. Then (provided the underlying probability space is sufficiently rich), there is a càdlàg process Y (unique in law) such that

$$\begin{aligned} &Y_t = X_t^u \qquad \text{for } t \leq \tau, \\ &\mathbb{E}\left[f(Y_{\tau+t}) | \mathscr{F}_\tau^{X^u} \right] = P_t^v f(X_\tau^u) \qquad \text{for every } t \geq 0,\ f \in C_0(\mathbb{R}^n). \end{aligned}$$

The idea behind this construction is simple: create a new process which coincides with the process X^u until the stopping time τ. After the stopping time, Y follows a new process that behaves like X^v, starting from the point where X^u left at time τ.

The process Y is not time homogeneous, but the expression

$$P_{s,t}^Y f(x) = \mathbb{E}\left[f(Y_t) | Y_s = x \right]$$

defines a two-parameter family of linear operators

$$\{P_{s,t}^Y : C_0(\mathbb{R}^n) \mapsto C_0(\mathbb{R}^n)\}_{0\leq s\leq t<+\infty}$$

such that

$$\begin{aligned} &P_{s,s}^Y = Id \qquad \forall s \in [0, +\infty[, \\ &P_{s,u}^Y \circ P_{u,t}^Y = P_{s,t}^Y \qquad \forall 0 \leq s \leq u \leq t < +\infty. \end{aligned}$$

Further, the process Y satisfies a time-dependent version of equality (9.17): for any bounded $\{\mathscr{F}_t^Y\}$-stopping times $\theta_1 \leq \theta_2$,

$$\begin{aligned} &\mathbb{E}\left[f(Y_{\theta_2}) | \mathscr{F}_{\theta_1}^Y \right] = \mathbb{E}\left[f(Y_{\theta_2}) | Y_{\theta_1} \right] = \\ &= f(Y_{\theta_1}) + \mathbb{E}\left[\int_{\theta_1}^{\theta_2} (\mathscr{L}^u f)(Y_s)\chi_{s\leq\tau} + (\mathscr{L}^v f)(Y_s)\chi_{s>\tau} ds \,\middle|\, Y_{\theta_1} \right]. \end{aligned} \tag{9.21}$$

The procedure above can be iterated any finite number of times: take two processes Y^1, Y^2 of the type above, pick θ, a finite $\left\{\mathscr{F}_t^{Y^1}\right\}$-stopping time, and a constant $s \geq 0$. From these, construct a new càdlàg process Y^3 such that

$$Y_t^3 = Y_t^1 \qquad \text{for } t \leq \theta,$$

$$\mathbb{E}\left[f(Y^3_{\theta+t}) \middle| \mathscr{F}_\theta^{Y^1} \right] = P^{Y^2}_{s,s+t} f(Y^1_\theta) \qquad \text{for every } t \geq 0,\ f \in C_0(\mathbb{R}^n).$$

Thus, we obtain a collection of stochastic processes (u, X^u) such that

(a) u is a U-valued process of type $u_t = u_0\chi_{[0,\tau_0]}(t) + \sum_{i=1}^{k} u_i \chi_{]\tau_{i-1},\tau_i]}(t)$, where all the u_i are fixed elements of U and $0 \leq \tau_0 \leq \tau_1 \leq \cdots \leq \tau_k < +\infty$ are finite $\{\mathscr{F}_t^{X^u}\}$-stopping times.
(b) (u, X^u) satisfies the trivial generalization of equality (9.21):

$$\begin{aligned}\mathbb{E}\left[f(X^u_{\theta_2}) \middle| \mathscr{F}^{X^u}_{\theta_1} \right] =& \mathbb{E}\left[f(X^u_{\theta_2}) \middle| X^u_{\theta_1} \right] = \\ =& f(X^u_{\theta_1}) + \mathbb{E}\left[\int_{\theta_1}^{\theta_2} (\mathscr{L}^{u_s} f)(X^u_s) ds \middle| X^u_{\theta_1} \right],\end{aligned} \tag{9.22}$$

for any bounded $\{\mathscr{F}_t^{X^u}\}$-stopping times $\theta_1 \leq \theta_2$.

The process u above is called a *simple control process*, while the corresponding process X^u is the *controlled process*. The set of all such pairs is denoted by $\mathscr{X}_S$. The set of admissible pairs of control and controlled processes can be greatly expanded by considering limits in some appropriate sense (e.g., convergence in probability, convergence in mean-squares, etc.) over sequences of pairs (u, X^u) with number of stopping times going to infinity. In that case, the set of all admissible pairs is denoted by $\mathscr{X}$.

Now, we can give a general formulation of the stochastic control problem:

Problem 9.1 For each $(t, x) \in [0, T[\times\mathbb{R}^n$, find the pair(s) $(\hat{u}, \hat{X}) \in \mathscr{X}$ maximizing the functional

$$J(t, x, u, X) = \mathbb{E}\left[g(X_T) \middle| X_t = x \right].$$

Here $g : \mathbb{R}^n \mapsto \mathbb{R}$ is a given function. To avoid technical difficulties, we assume that $g \in C_0(\mathbb{R}^n)$.

We will not go into the details of the construction of the space $\mathscr{X}$. Instead, we will take the following assumption:

(A9) For every $(t, x) \in [0, T[\times\mathbb{R}^n$ and every $(u, X) \in \mathscr{X}$, there is a sequence $\{(u_k, X_k) \in \mathscr{X}_S\}_{k\in\mathbb{N}}$ such that $\lim_{k\to\infty} J(t, x, u_k, X_k) \geq J(t, x, u, X)$.

Assumption **(A9)** allows us to deal with the main arguments in this text using only simple controls.

Remark 9.1 The fact that simple controls are left continuous while controlled processes are right continuous is significant.

To see this, consider a controlled version of the compound Poisson process from Example 9.1:

$$X_t^u = X_0 + \sum_{i=1}^{N_t} (u_{T_i} + Y_i),$$

where N_t is a Poisson process with intensity $\lambda > 0$, T_i is the time of the ith jump of N_t, Y_i, $i = 1, 2, \ldots$ are i.i.d. real-valued random variables, and u_t is a real-valued control process. Thus, the effect of the control is to change the expected value of the jumps $(X_t - X_{t-})$.

Left continuity of the control means that this shifting is decided *before* the jump. Clearly, if the choice of the shift could be delayed until the jump occurs, then it could be made to fully compensate for the value of the "default" jump Y_i, completely changing the nature of the problem.

In the setting above, the filtration being considered is always the natural filtration $\{\mathscr{F}_t^X\}$ of the process $(u, X) \in \mathscr{X}$ being considered. Thus, a change in control can lead to a change in filtration. For this reason, we need to consider a type of stopping times that can be applied to every filtration generated by elements of $\mathscr{X}$.

For each $(u, X) \in \mathscr{X}$, let $\mathscr{T}_X$ be the set of all $\{\mathscr{F}_t^X\}$-stopping times. A *universal stopping time* is a mapping $\tau : \mathscr{X} \mapsto \bigcup_{(u,X)\in\mathscr{X}} \mathscr{T}_X$ such that:

(a) $\tau_{(u,X)} \in \mathscr{T}_X$ for every $(u, X) \in \mathscr{X}$,
(b) If (u, X) and (u', X') are identical in law, then $\left(u, X, \tau_{(u,X)}\right)$ and $\left(u', X', \tau_{(u',X')}\right)$ are identical in law.

For example, for any open set $B \subset \mathbb{R}^n$, the first exit time of B, $\tau_X = \inf\{s : X_s \notin B\}$ is a universal stopping time. Also, deterministic times are universal stopping times and the class of universal stopping times is closed under lattice operations: for any universal stopping times τ, θ, $\tau \wedge \theta$ and $\tau \vee \theta$ are universal stopping times.

Consider a finite number of universal stopping times $0 = \tau^0 \leq \tau^1 \leq \tau^2 \leq \ldots \leq \tau^k$ and $u^1, u^2, \ldots, u^k \in U$. For each $(u, X) \in \mathscr{X}_S$, these elements define a unique U-valued process

$$w_t = \sum_{i=1}^{k} u^i \chi_{]\tau^{i-1}_{(u,X)}, \tau^i_{(u,X)}]}(t). \tag{9.23}$$

That is, (9.23) defines a collection of U-valued processes parameterized by $\mathscr{X}_S$. Such a collection is called a (simple) *feedback control*, and we use the short notation $w_t = \sum_{i=1}^{k} u^i \chi_{]\tau^{i-1},\tau^i]}(t)$. It can be checked that there is one unique process (in law) X such that $(w, X) \in \mathscr{X}_S$.

Notice that the notion of controlled process introduced in this section is in fact a way of controlling families of linear operators

$$\{P_{s,t} : C_0(\mathbb{R}^n) \mapsto C_0(\mathbb{R}^n)\}_{0\leq s\leq t<+\infty} .$$

In other words, we are concerned with control of laws of processes rather than control of the sample paths of a given process.

9.4.2 *The Dynamic Programming Principle*

Now, we present and prove the theorem analogous to Theorem 9.1 for the stochastic optimal control problem (Problem 9.1).

First, we introduce some notation: for any function $\varphi : [0, T] \times \mathbb{R}^n \mapsto \mathbb{R}$, let

$$\underline{\varphi}(t,x) = \liminf_{(s,y)\to(t,x)} \varphi(s,y), \qquad \overline{\varphi}(t,x) = \limsup_{(s,y)\to(t,x)} \varphi(s,y),$$

be, respectively, the lower semi-continuous and the upper semi-continuous envelopes of φ. Notice that every semi-continuous function is Borel-measurable.

We introduce the *value function* $V : [0, T[\times\mathbb{R}^n \mapsto \overline{\mathbb{R}}$, defined as

$$V(t,x) = \sup_{(u,X)\in\mathscr{X}} J(t,x,u,X) = \sup_{(u,X)\in\mathscr{X}} \mathbb{E}\left[g(X_T)\middle| X_t = x\right]. \tag{9.24}$$

Notice that, due to assumption **(A9)**, the supremum in (9.24) can be taken over the smaller set $\mathscr{X}_S$. The value function extends trivially to $t = T$ by $V(T,x) = g(x)$.

The dynamic programming principle takes the following form:

Theorem 9.8 (Dynamic Programming Principle) *For every* $(t,x) \in [0,T[\times\mathbb{R}$ *and every universal stopping time* τ *such that* $t \leq \tau \leq T$*:*

$$\sup_{(u,X)\in\mathscr{X}_S} \mathbb{E}\left[\underline{V}(\tau,X_\tau)|X_t = x\right] \leq V(t,x) \leq \sup_{(u,X)\in\mathscr{X}_S} \mathbb{E}\left[\overline{V}(\tau,X_\tau)|X_t = x\right]$$

Proof The right-hand side inequality is easy to prove.

Fix a universal stopping time τ with $t \leq \tau \leq T$. For every $(u,X) \in \mathscr{X}_S$,

$$\mathbb{E}\left[g(X_T)\middle| X_t = x\right] = \mathbb{E}\left[\mathbb{E}\left[g(X_T)\middle| \mathscr{F}_\tau^X\right]\middle| X_t = x\right] =$$
$$= \mathbb{E}\left[\mathbb{E}\left[g(X_T)\middle| X_\tau\right]\middle| X_t = x\right] \leq \mathbb{E}\left[\overline{V}(\tau,X_\tau)\middle| X_t = x\right].$$

The inequality follows by taking the supremum over $(u,X) \in \mathscr{X}_S$.

The proof of the left-hand side inequality is more labourious.

Due to Urysohn's lemma, every lower semicontinuous function $f : \mathbb{R}^n \mapsto \mathbb{R}$, bounded from below is the pointwise limit of a monotonically increasing sequence of continuous functions.

Notice that $|V(t,x)| \leq \|g\|$ and therefore, the same holds for $\underline{V}$. Thus, we can pick a bounded continuous function $\varphi : [0,T] \times \mathbb{R}^n \mapsto \mathbb{R}$ such that $\varphi \leq \underline{V}$. Fix a constant $\varepsilon > 0$. For each $(t,x) \in [0,T] \times \mathbb{R}^n$, pick $(u^{t,x}, X^{t,x}) \in \mathscr{X}_S$ such that $\mathbb{E}\left[g(X_T^{t,x}) \middle| X_t^{t,x} = x\right] > V(t,x) - \varepsilon$.

For every fixed $(u,X) \in \mathscr{X}_S$, the function $(t,x,s) \mapsto \mathbb{E}\left[g(X_s) | X_t\right]$ is continuous. Thus, for each $(t,x) \in [0,T] \times \mathbb{R}^n$, there is a constant $r^{t,x} > 0$ such that

$$\mathbb{E}\left[g(X_{T'}) | X_{t'} = x'\right] > \mathbb{E}\left[g(X_T) | X_t = x\right] - \varepsilon,$$
$$\varphi(t',x') < \varphi(t,x) + \varepsilon,$$

whenever $|t'-t| + |x'-x| + |T'-T| < r^{t,x}$.

The collection of balls $\left\{B_{\frac{1}{2}r^{(t,x)}}(t,x)\right\}_{(t,x)\in[0,T]\times\mathbb{R}^n}$ is an open cover of $[0,T] \times \mathbb{R}^n$. Hence it contains a countable subcover $\left\{B_{\frac{1}{2}r^{(t_i,x_i)}}(t_i,x_i)\right\}_{i\in\mathbb{N}}$. We use the short notation $B_i = B_{\frac{1}{2}r^{(t_i,x_i)}}(t_i,x_i)$, $r_i = r^{(t_i,x_i)}$, $(u^i, X^i) = (u^{t_i,x_i}, X^{t_i,x_i})$. The sequence $\left\{A_i = B_i \setminus \bigcup_{j<i} B_j\right\}_{i\in\mathbb{N}}$ is a countable partition of $[0,T] \times \mathbb{R}^n$ into Borel subsets.

For every $(t,x) \in B_i$, we have:

$$\begin{aligned}\mathbb{E}\left[g(X^i_{T+t_i-t}) \middle| X^i_t = x\right] &> \mathbb{E}\left[g(X^i_T) \middle| X^i_{t_i} = x_i\right] - \varepsilon > V(t_i,x_i) - 2\varepsilon \\ &\geq \varphi(t_i,x_i) - 2\varepsilon > \varphi(t,x) - 3\varepsilon. \end{aligned} \tag{9.25}$$

Fix a universal stopping time τ such that $t \leq \tau \leq T$, and $(u,X) \in \mathscr{X}_S$. Notice that for every $i \in \mathbb{N}$, the event $\{(\tau_X, X_{\tau_X}) \in A_i\}$ is $\mathscr{F}^X_{\tau_X}$-measurable. Therefore, the random variables τ_i, defined as

$$\tau_i = \begin{cases} T, & \text{if } (\tau_X, X_{\tau_X}) \notin A_i, \\ \tau_X, & \text{if } (\tau_X, X_{\tau_X}) \in A_i, \end{cases} \qquad i \in \mathbb{N}$$

are $\left\{F^X_t\right\}$-stopping times. Using these stopping times and the sequence of processes $\{(u^i, X^i)\}$ chosen above, we obtain a new sequence $\{(v^i, Y^i) \in \mathscr{X}_S\}_{i\in\mathbb{N}}$ such that

$$\begin{aligned}
&Y^1_t = X_t, && \text{for } t \leq \tau_1, \\
&\mathbb{E}\left[f(Y^1_{\tau_1+t}) \middle| \mathscr{F}^X_{\tau_1}\right] = P^{X^1}_{t_1,t_1+t} f(X_{\tau_1}), && \text{for } t \geq 0,\ f \in C_0(\mathbb{R}^n); \\
&Y^{i+1}_t = Y^i_t, && \text{for } t \leq \tau_{i+1}, \\
&\mathbb{E}\left[f(Y^{i+1}_{\tau_{i+1}+t}) \middle| \mathscr{F}^{Y^i}_{\tau_{i+1}}\right] = P^{X^{i+1}}_{t_{i+1},t_{i+1}+t} f(Y^i_{\tau_{i+1}}), && \text{for } t \geq 0,\ f \in C_0(\mathbb{R}^n),\ i \in \mathbb{N}.
\end{aligned}$$

Notice that every τ_X is a $\left\{\mathscr{F}_t^{Y^i}\right\}$-stopping time for every $i \in \mathbb{N}$. Hence, the same holds for every τ_i.

Also,

$$\mathbb{E}\left[g(Y_T^k)\middle|\,\mathscr{F}_{\tau_X}^{Y^k}\right] = \sum_{i=1}^{k}\left(P_{t_i,T+t_i-\tau_X}^{X^i} g(X_{\tau_X})\right)\chi_{A_i}(\tau_X, X_{\tau_X})$$
$$+\left(P_{\tau_X,T-\tau_X}^{X} g(X_{\tau_X})\right)\chi_{\bigcup_{i=k+1}^{\infty} A_i}(\tau_X, X_{\tau_X}).$$

Hence, inequality (9.25) shows that

$$\mathbb{E}\left[g(Y_T^k)\middle|\,\mathscr{F}_{\tau_X}^{Y^k}\right] > \varphi(\tau_X, X_{\tau_X})\chi_{\bigcup_{i=1}^{k} A_i}(\tau_X, X_{\tau_X}) - \|g\|\,\chi_{\bigcup_{i=k+1}^{\infty} A_i}(\tau, X_\tau^v) - 3\varepsilon$$

almost certainly. Therefore,

$$\begin{aligned} V(t,x) \geq & \mathbb{E}\left[g(Y_T^k)|Y_t^k = x\right] = \mathbb{E}\left[\mathbb{E}\left[g(Y_T^k)\middle|\,\mathscr{F}_{\tau_X}^{Y^k}\right]\middle|\, Y_t^k = x\right] \geq \\ \geq & \mathbb{E}\left[\varphi(\tau_X, X_{\tau_X})|\, X_t = x\right] - \\ & -C\Pr\left\{(\tau_X, X_{\tau_X}) \in \bigcup_{i=k+1}^{\infty} A_i \middle|\, X_t = x\right\} - 3\varepsilon, \end{aligned}$$

where C is a finite constant, independent of k. Since $\{A_i\}_{i\in\mathbb{N}}$ is a partition of $[0,T]\times\mathbb{R}^n$, $\lim_{k\to\infty}\Pr\left\{(\tau_X, X_{\tau_X}) \in \bigcup_{i=k+1}^{\infty} A_i \middle|\, X_t = x\right\} = 0$. Therefore,

$$V(t,x) \geq \mathbb{E}\left[\varphi(\tau_X, X_{\tau_X})|\, X_t = x\right] - 3\varepsilon.$$

Since ε is arbitrary, we conclude that $V(t,x) \geq \mathbb{E}\left[\varphi(\tau_X, X_{\tau_X})|\, X_t = x\right]$.

Finally, take a monotonically increasing sequence of bounded continuous functions $\{\varphi_k\}_{k\in\mathbb{N}}$, converging pointwise to $\underline{V}$. Due to Lebesgue's monotone convergence theorem, we obtain

$$V(t,x) \geq \lim_{k\to\infty}\mathbb{E}\left[\varphi_k(\tau_X, X_{\tau_X})|\, X_t = x\right] = \mathbb{E}\left[\underline{V}(\tau_X, X_{\tau_X})|\, X_t = x\right],$$

Since this holds for every $(u,X) \in \mathscr{X}_S$, the proof is complete.

Notice that the statement in Theorem 9.8 reduces to

$$V(t,x) = \sup_{(u,X)\in\mathscr{X}_S}\mathbb{E}\left[V(\tau, X_\tau)|\, X_t = x\right],$$

when V is a continuous function. Obviously, this is a general form of equality (9.8).

9.4.3 The Hamilton-Jacobi-Bellman Equation

Now we will present our versions of Theorems 9.2 and 9.3. In the setting of Problem 9.1, the Theorem 9.2 takes the following form.

Theorem 9.9 *Let $V : [0,T] \times \mathbb{R}^n \mapsto \mathbb{R}$ be the value function* (9.24)*, and suppose that:*

(a) for every $x \in \mathbb{R}^n$, the map $t \mapsto V(t,x)$ is absolutely continuous;
(b) for every $t \in [0,T[$, the map $x \mapsto V(t,x)$ is an element of $\mathscr{D}$.

Then, the value function satisfies the Hamilton-Jacobi-Bellman equation

$$\frac{\partial V}{\partial t}(t,x) + \sup_{u\in U} \mathscr{L}^u V(t,x) = 0, \tag{9.26}$$

for every (t,x), a continuity point of the functions $(t,x) \mapsto \frac{\partial V}{\partial t}$, $(t,x) \mapsto \sup_{u\in U} \mathscr{L}^u V(t,x)$.

Proof Fix $(t,x) \in [0,T[\times\mathbb{R}^n$, a continuity point of the function $(t,x) \mapsto \frac{\partial V}{\partial t}$, and fix small $h, \varepsilon > 0$. Let

$$\tau = \min\left(t+h, \inf\{s \geq t : |X_s - x| \geq \varepsilon\}\right).$$

This is a universal stopping time, and therefore the dynamic programming principle (Theorem 9.8) states that

$$V(t,x) \geq \mathbb{E}\left[V(\tau, X_\tau^u) | X_t = x\right]$$

for every $u \in U$. This implies

$$V(t,x) \geq \mathbb{E}\left[V(\tau, X_\tau^u) - V(t, X_\tau^u) | X_t = x\right] + \mathbb{E}\left[V(t, X_\tau^u) | X_t = x\right].$$

Due to the assumptions above and equality (9.22), this is

$$\begin{aligned} V(t,x) \geq & \mathbb{E}\left[\int_t^\tau \frac{\partial V}{\partial t}(s, X_\tau^u) ds \middle| X_t = x\right] + \\ & + V(t,x) + \mathbb{E}\left[\int_t^\tau (\mathscr{L}^u V)(t, X_s^u) ds \middle| X_t = x\right] = \\ = & V(t,x) + h\left(\frac{\partial V}{\partial t}(t,x) + \mathscr{L}^u V(t,x)\right) + o(h), \end{aligned}$$

when h goes to zero. Since u is arbitrary, this shows that

$$\frac{\partial V}{\partial t}(t,x) + \sup_{u\in U} \mathscr{L}^u V(t,x) \leq 0.$$

Now, suppose there is some $(t,x) \in [0,T[\times\mathbb{R}^n$, a continuity point of the functions $(t,x) \mapsto \frac{\partial V}{\partial t}$, $(t,x) \mapsto \sup_{u\in U} \mathscr{L}^u V(t,x)$, such that $\frac{\partial V}{\partial t}(t,x) + \sup_{u\in U} \mathscr{L}^u V(t,x) < 0$. Then, there is some $\varepsilon > 0$ such that

$$\frac{\partial V}{\partial t}(t,x) + \sup_{u\in U} \mathscr{L}^u V(t,x) < -\varepsilon.$$

By continuity, there is some $\eta > 0$ such that

$$\frac{\partial V}{\partial t}(s,y) + \sup_{u\in U} \mathscr{L}^u V(s',y') < -\frac{\varepsilon}{2}$$

whenever $|s-t| + |y-x| \leq \eta$ and $|s'-t| + |y'-x| \leq \eta$.

For every $k \in \mathbb{N}$, there is a function $\varphi_k \in C_0\left([0,T[\times\mathbb{R}^n\right)$ such that

$$\varphi(t,x) = 0, \qquad \varphi(s,y) = 1 \;\text{ whenever } \eta \leq |s-t| + |y-x| \leq k.$$

Since $\mathscr{D}$ is dense, there is a function $\phi_k : [0,T[\times\mathbb{R}^n \mapsto \mathbb{R}$ such that

(i) for every $s \in [0,T[$ the map $y \mapsto \phi_k(s,y)$ is an element of $\mathscr{D}$;
(ii) for every $y \in \mathbb{R}^n$ the map $s \mapsto \phi_k(s,y)$ is continuously differentiable;
(iii) $\|\phi_k - \varphi_k\| < \frac{1}{4}$.

Consider the universal stopping time

$$\tau = \inf\{s \geq t : s - t + |X_s - x| \geq \eta\},$$

and fix $(u,X) \in \mathscr{X}_S$. Then, for every $\delta > 0$:

$$\begin{aligned}
&V(t,x) \geq V(t,x) + \delta\left(\phi_k(t,k) - \frac{1}{4}\right) = \\
=&\mathbb{E}\left[V(\tau,X_\tau)|\,X_t = x\right] + \delta\mathbb{E}\left[\varphi_k(\tau,X_\tau)|\,X_t = x\right] - \\
&- \left(\mathbb{E}\left[(V+\delta\phi_k)(\tau,X_\tau)|\,X_t = x\right] - \mathbb{E}\left[(V+\delta\phi_k)(t,X_\tau)|\,X_t = x\right]\right) - \\
&- \left(\mathbb{E}\left[(V+\delta\phi_k)(t,X_\tau)|\,X_t = x\right] - (V+\delta\phi_k)(t,x)\right) - \frac{\delta}{4} = \\
=&\mathbb{E}\left[V(\tau,X_\tau)|\,X_t = x\right] + \delta\mathbb{E}\left[\varphi_k(\tau,X_\tau)|\,X_t = x\right] - \\
&- \mathbb{E}\left[\int_t^\tau \frac{\partial}{\partial t}(V+\delta\phi_k)(s,X_\tau)ds\,\middle|\, X_t = x\right] - \\
&- \mathbb{E}\left[\int_t^\tau \mathscr{L}^u(V+\delta\phi_k)(t,X_s)ds\,\middle|\, X_t = x\right] - \frac{\delta}{4}.
\end{aligned}$$

Notice that $\liminf_{k\to\infty} \mathbb{E}\left[\varphi_k(\tau, X_\tau)|\, X_t = x\right] \geq \frac{3}{4}$. Hence, for k sufficiently large

$$\begin{aligned} V(t,x) \geq & \mathbb{E}\left[V(\tau, X_\tau)|\, X_t = x\right] + \frac{\delta}{4} - \\ & - \mathbb{E}\left[\left.\int_t^\tau \frac{\partial}{\partial t}(V + \delta\phi_k)(s, X_\tau)ds \right| X_t = x\right] - \\ & - \mathbb{E}\left[\left.\int_t^\tau \mathscr{L}^u(V + \delta\phi_k)(t, X_s)ds \right| X_t = x\right] \geq \\ \geq & \mathbb{E}\left[V(\tau, X_\tau)|\, X_t = x\right] + \frac{\delta}{4} + \mathbb{E}\left[\tau - t|\, X_t = x\right]\left(\frac{\varepsilon}{2} - \delta \left\|\mathscr{L}^u\phi_k\right\|\right). \end{aligned}$$

Thus, for sufficiently small $\delta > 0$,

$$V(t,x) \geq \mathbb{E}\left[V(\tau, X_\tau)|\, X_t = x\right] + \frac{\delta}{4}.$$

This contradicts the dynamic programming principle and therefore

$$\frac{\partial V}{\partial t}(t,x) + \sup_{u\in U} \mathscr{L}^u V(t,x) \geq 0$$

must hold for every $(t,x) \in [0, T[\times\mathbb{R}^n$, a continuity point of $\frac{\partial V}{\partial t}$ and $\sup_{u\in U} \mathscr{L}^u V$.

Example 9.5 Consider the controlled compound Poisson process of Remark 9.1. For each $u \in \mathbb{R}$, the generator $\mathscr{L}^u$ is

$$\mathscr{L}^u f(x) = \lambda\left(\mathbb{E}\left[f(x + u + Y)\right] - f(x)\right).$$

Thus, if the control process is allowed to take values in some subset $U \subset \mathbb{R}$, the Hamilton-Jacobi-Bellman equation becomes the integro-differential equation

$$\frac{\partial V}{\partial t}(t,x) + \lambda\left(\sup_{u\in U} \mathbb{E}\left[f(x + u + Y)\right] - f(x)\right) = 0.$$

Example 9.6 Consider the problem (9.6)–(9.7). For each fixed $u \in U$, the infinitesimal generator $\mathscr{L}^u$ is given by (9.10). Thus, (9.9) is the Hamilton-Jacobi-Bellman equation for this problem.

The verification theorem can be stated as follows.

Theorem 9.10 *Let $v : [0, T] \times \mathbb{R}^n \mapsto \mathbb{R}$ be a continuous function such that:*

(a) for every $t \in [0, T[$, the function $x \mapsto v(t,x)$ is an element of $\mathscr{D}$;
(b) for every $x \in \mathbb{R}^n$, the function $t \mapsto v(t,x)$ is differentiable and $\frac{\partial v}{\partial t}$ is continuous and bounded in $[0, T[\times\mathbb{R}^n$;
(c) the function $(t,x) \mapsto \sup_{u\in U} \mathscr{L}^u v(t,x)$ is continuous in $[0, T[\times\mathbb{R}^n$.

If v solves the Hamilton-Jacobi-Bellman equation (9.26) *with terminal condition*

$$v(T,x) = g(x) \qquad \forall x \in \mathbb{R}^n, \tag{9.27}$$

then v is the value function.

Further, if there is a function $\hat{u} : [0,T[\times\mathbb{R}^n \mapsto U$ such that

$$\mathscr{L}^{\hat{u}(t,x)} v(t,x) = \sup_{u\in U} \mathscr{L}^u v(t,x) \qquad \forall (t,x) \in [0,T[\times\mathbb{R}^n,$$

and there is a process $\hat{X}$ such that $\left(\hat{u}(t,\hat{X}_t),\hat{X}_t\right) \in \mathscr{X}$, then the process $\left(\hat{u}(t,\hat{X}_t),\hat{X}_t\right)$ is a solution for Problem 9.1.

Proof Let $V : [0,T] \times \mathbb{R}^n \mapsto \mathbb{R}$ be the value function (9.24), and fix $(u,X) \in \mathscr{X}_S$.

Under the assumptions *(a)–(c)*, v satisfies the analogous to equality (9.22):

$$\mathbb{E}\left[v(\theta_2, X_{\theta_2}) | X_{\theta_1}\right] = v(\theta_1, X_{\theta_1}) + \mathbb{E}\left[\int_{\theta_1}^{\theta_2} \frac{\partial v}{\partial t}(s,X_s) + \mathscr{L}^{u_s} v(s,X_s) ds \,\middle|\, X_{\theta_1}\right],$$

for every universal stopping times $0 \le \theta_1 \le \theta_2 \le T$. Therefore,

$$\begin{aligned}
&\mathbb{E}\left[g(X_T) | X_t = x\right] = \mathbb{E}\left[v(T,X_T) | X_t = x\right] = \\
=&v(t,x) + \mathbb{E}\left[\int_t^T \frac{\partial v}{dt}(s,X_s) + (\mathscr{L}^{u_s} v)(s,X_s) ds \,\middle|\, X_t = x\right] \le \\
\le& v(t,x) + \mathbb{E}\left[\int_t^T \frac{\partial v}{dt}(s,X_s) + \sup_{u\in U}(\mathscr{L}^u v)(s,X_s) ds \,\middle|\, X_t = x\right] = v(t,x).
\end{aligned}$$

Since (u,X) is arbitrary, this proves that $V(t,x) \le v(t,x)$. Further, if the function $\hat{u}$ and the process $\hat{X}$ described in the theorem exist, the inequality above becomes an equality in the particular case $(u_t,X_t) = \left(\hat{u}(t,\hat{X}_t),\hat{X}_t\right)$. Hence, the result follows.

To prove the inequality $v(t,x) \le V(t,x)$ in the general case, we build a sequence $\left\{(w^k,X^k) \in \mathscr{X}_S\right\}_{k\in\mathbb{N}}$ as follows.

Suppose that v solves (9.26)–(9.27), and fix $\varepsilon > 0$. For each $(t,x) \in [0,T[\times\mathbb{R}^n$, there is some $r_{t,x} > 0$ and $u^{t,x} \in U$ such that

$$\sup_{u\in U} \mathscr{L}^u v(s,y) < \mathscr{L}^{u^{t,x}} v(s,y) + \varepsilon \qquad \forall (s,y) \in B_{r^{t,x}}(t,x).$$

The collection $\left\{B_{\frac{1}{2}r^{t,x}}(t,x)\right\}_{(t,x)\in[0,T[\times\mathbb{R}^n}$ is an open cover of $[0,T[\times\mathbb{R}^n$. Hence, it admits a countable subcover $\left\{B_{\frac{1}{2}r^{t_i,x_i}}(t_i,x_i)\right\}_{i\in\mathbb{N}}$. We use the short notation $r_i = r^{t_i,x_i}$, $B_i = B_{r^{t_i,x_i}}(t_i,x_i)$, $A_i = B_{\frac{1}{2}r_i}(t_i,x_i)$, $u^i = u^{t_i,x_i}$.

Fix $(t,x) \in [0,T[\times\mathbb{R}^n$. Without loss of generality, we set $(t_1,x_1) = (t,x)$, and construct the sequence of feedback controls $\left\{w^k\right\}_{k\in\mathbb{N}}$ as follows.

We set $w^1 = u^1$, a constant control, and define the universal stopping time $\tau_1 = \min(T, \inf\{s \geq t : (s, X_s) \notin B_1\})$. Let w^2 be the feedback

$$w_t^2 = \begin{cases} w_t^1, & \text{for } t \leq \tau_1, \\ u^2, & \text{for } t > \tau_1 \text{ if } (\tau_1, X_{\tau_1}) \in A_2, \\ w_t^1, & \text{for } t > \tau_1 \text{ if } (\tau_1, X_{\tau_1}) \notin A_2, \end{cases}$$

For $k \geq 2$, and $i = 1, 2, \ldots, k$, set

$$\tau_{k,i} = \begin{cases} \tau_{k-1} & \text{if } (\tau_{k-1}, X_{\tau_{k-1}}) \notin A_i, \\ \inf\{s > \tau_{k-1} : (s, X_s) \notin B_i\} & \text{if } (\tau_{k-1}, X_{\tau_{k-1}}) \in A_i. \end{cases}$$

Let

$$\tau_k = \min(T, \max(\tau_{k,1}, \tau_{k,2}, \ldots, \tau_{k,k})),$$

$$w_t^{k+1} = \begin{cases} w_t^k, \text{ for } t \leq \tau_k, \\ u^i, \text{ for } t > \tau_k \text{ if } (\tau_k, X_{\tau_k}) \in A_i \setminus \bigcup_{j<i} A_j, \quad i = 1, 2, \ldots, k+1, \\ w_t^k, \text{ for } t > \tau_k \text{ if } (\tau_k, X_{\tau_k}) \notin \bigcup_{i=1}^{k+1} A_i. \end{cases}$$

Let $\tilde{w}^k = \begin{cases} w_t^k \text{ for } t \leq \tau_k, \\ u^1 \text{ for } t > \tau_k, \end{cases}$ and let X^k be the unique (in law) process such that $(\tilde{w}^k, X^k) \in \mathscr{X}_S$. Then, $\{\tau_k\}$ is a monotonically increasing sequence converging to T and for every $s \in [0, \tau_k]$:

$$\mathscr{L}^{\tilde{w}_s^k} v(s, X_s^k) > \sup_{u \in U} \mathscr{L}^u v(s, X_s^k) - \varepsilon.$$

Therefore,

$$\begin{aligned} V(t,x) \geq & \mathbb{E}\left[g(X_T^k) \middle| X_t^k = x\right] = \mathbb{E}\left[v(T, X_T^k) \middle| X_t^k = x\right] = \\ = & \mathbb{E}\left[v(\tau_k, X_{\tau_k}^k) \middle| X_t^k = x\right] + \mathbb{E}\left[v(T, X_T^k) - v(\tau_k, X_{\tau_k}^k) \middle| X_t^k = x\right] = \\ = & v(t,x) + \mathbb{E}\left[\int_t^{\tau_k} \frac{\partial v}{\partial t}(s, X_s^k) + \mathscr{L}^{\tilde{w}_s^k} v(s, X_s^k) ds \middle| X_t^k = x\right] + \\ & + \mathbb{E}\left[v(T, X_T^k) - v(\tau_k, X_{\tau_k}^k) \middle| X_t^k = x\right] \geq \\ \geq & v(t,x) + \mathbb{E}\left[\int_t^{\tau_k} \frac{\partial v}{\partial t}(s, X_s^k) + \sup_{u \in U} \mathscr{L}^u v(s, X_s^k) - \varepsilon ds \middle| X_t^k = x\right] + \\ & + \mathbb{E}\left[v(T, X_T^k) - v(\tau_k, X_{\tau_k}^k) \middle| X_t^k = x\right] \geq \\ \geq & v(t,x) - (T-t)\varepsilon + \mathbb{E}\left[v(T, X_T^k) - v(\tau_k, X_{\tau_k}^k) \middle| X_t^k = x\right]. \end{aligned}$$

Now,

$$\begin{aligned}&\mathbb{E}\left[v(T,X_T^k)-v(\tau_k,X_{\tau_k}^k)\middle|X_t^k=x\right]=\\=&\mathbb{E}\left[\mathbb{E}\left[v(T,X_T^k)\middle|X_{\tau_k}\right]-v(\tau_k,X_{\tau_k}^k)\middle|X_t^k=x\right],\end{aligned}$$

the sequence $\mathbb{E}\left[v(T,X_T^k)\middle|X_{\tau_k}\right]-v(\tau_k,X_{\tau_k}^k)$ is uniformly bounded and converges pointwise to 0 when k goes to infinity. Therefore,

$$\lim_{k\to\infty}\mathbb{E}\left[v(T,X_T^k)-v(\tau_k,X_{\tau_k}^k)\middle|X_t^k=x\right]=0$$

and, since ε is arbitrary, we conclude that $V(t,x)\geq v(t,x)$.

Notice, due to Theorem 9.10, any solution of the Hamilton-Jacobi-Bellman equation with the regularity stated in the theorem is unique, since it must coincide with the value function. Also, the first part of the theorem does not require the Problem 9.1 to have a solution. Indeed, the proof gives a method to construct ε-optimal solutions, given the solution of the Hamilton-Jacobi-Bellman equation.

9.5 Conclusions

Though we presented the dynamic programming principle and the Hamilton-Jacobi-Bellman equation in a setting that does not depend on the particular type of process being controlled, we did not try to achieve maximal generality. Many variants of the results above can be obtained along the same lines but in different settings.

We introduced Feller processes as $\mathbb{R}^n$-valued processes. Therefore, our controlled processes are also $\mathbb{R}^n$-valued processes. This is just for convenience of writing. All the arguments used in the text are entirely valid if $\mathbb{R}^n$ is replaced by any separable locally compact metric space.

More importantly, the approach outlined above can be extended to Markov processes that are not Feller processes. Control systems analogous to those described in Sect. 9.4 can be constructed for any class of càdlàg processes, provided that class corresponds to a class of semigroups acting on a suitable Banach space of functions such that equality (9.15) holds.

The class of optimization criteria treated in this text is quite restricted, since we assume that the function g is an element of $C_0(\mathbb{R}^n)$. This assumption fails in many cases of practical interest, namely in economics and financial applications where optimization criteria are frequently unbounded and/or have discontinuities. Also very restrictive are the assumptions concerning the regularity of the value function required for Theorem 9.9 and 9.10. Naturally, these two issues are related. They can be partially overcome on a case-by-case approach. Often, it is possible to extend the results taking sequences of problems with regular optimization criteria and regular value functions and taking limits in some appropriate sense.

Issues related to the regularity of the value function are very present in all dynamic programming approaches to optimal control problems, including control of deterministic systems and control of diffusions. In the deterministic and diffusion cases, the Hamilton-Jacobi-Bellman equation is a PDE. It can be proved (under suitable assumptions) that appropriate versions of Theorems 9.9 or 9.10 hold if solutions of the Hamilton-Jacobi-Bellman equation in the viscosity sense are considered, instead of classical solutions (see [6, 9], among others). Viscosity solutions and their application to optimal control problems are the topic of a vast and growing volume of published literature, including generalizations to some types of Lévy processes.

Readers interested in the control of diffusions may see the monograph by Fleming and Soner [6] or the more recent book by Touzi [9]. These books contain a rigorous and comprehensive study of the topic, including solutions of the Hamilton-Jacobi-Bellman equation in the viscosity sense.

A study of Markov processes from the point of view outlined in Sect. 9.3 can be found in the lecture notes [1]. The monograph by Ethier and Kurtz [4] contains a very complete account of the same topic.

The study of general Markov processes from the point of view outlined in Sect. 9.4 above, has not been, to our best knowledge, the topic of any comprehensive publication. This material is can be rigorously constructed from the material contained in the references provided for Sect. 9.3.

References

1. Bovier A.: Markov processes. Lecture Notes Summer 2012, Bonn, Institute for Applied Mathematics, University of Bonn (2012). http://wt.iam.uni-bonn.de/bovier/lecture-notes/
2. Dumitrescu, R., Quenez, M.-C., Sulem, A.: A weak dynamic programming principle for combined optimal stopping/stochastic control with $\mathcal{E}^f$-expectations. SIAM J. Control Optim. **54**(4), 2090–2115 (2015)
3. de Vallière, D., Kabanov, Y., Lépinette, E.: Consumption-investment problem with transaction costs for Lévy-driven price process. Finance Stochast. **20**, 705–740 (2016)
4. Ethier, S.N., Kurtz, T.G.: Markov Processes. Characterization and Convergence. Wiley-Interscience, New York (2005)
5. El Karoui, N., Tan, X.: Capacities, measurable selection and dynamic programming. Part I: abstract framework. arXiv:1310.3363v1 (2013)
6. Fleming, W.H., Soner, H.M.: Controlled Markov Processes and Viscosity Solutions, 2nd edn. Springer, New York (2006)
7. Goldys, B., Wu, W.: Dynamic programming principle for stochastic control problems driven by general Lévy noise. Stoch. Anal. Appl. **34**(6), 1083–1093 (2016)
8. Rogers, L.C.G., Williams, D.: Diffusions, Markov Processes, and Martingales. Volume 1: Foundations, 2nd edn. Wiley, New York (1994)
9. Touzi, N.: Optimal Stochastic Control, Stochastic Target Problems, and Backward SDE. Springer, New York (2013)
10. Yong, J., Zhou, X.Y.: Stochastic Controls: Hamiltonian Systems and HJB Equations. Springer, New York (1999)

Part III
Transformation Methods and Special Discretizations

Chapter 10
Numerical Analysis of Novel Finite Difference Methods

Rafael Company, Vera N. Egorova, Mohamed El Fakharany, Lucas Jódar, and Fazlollah Soleymani

Abstract The core target of this chapter is numerical analysis and computing of novel finite difference methods related to several different option pricing models, including jump-diffusion, regime switching and multi-asset options. A special attention is paid to positivity, consistency and stability of the proposed methods. The consideration of jump processes leads to partial integro-differential equation (PIDE) for the European option pricing problem. The problem is solved by using quadrature formulas for the approximation of the integrals and matching the discretization of the integral and differential part of the PIDE problem. More complicated model under assumption that the volatility is a stochastic process derives to a PIDE problem where the volatility is also an independent variable. Such a problem is solved by introducing appropriate change of variables. Moreover, American options are considered proposing various front-fixing transformations to treat a free boundary. This free boundary challenge can be treated also by a recent rationality parameter approach that takes into account the irrational behavior of the market. Dealing with multidimensional problems the core difficulty is the appearance of the cross derivative terms. Appropriate transformations allow eliminating the cross derivative terms and reduce of the computational cost and the numerical instabilities. After using a semidiscretization approach, the time exponential integration method and appropriate quadrature integration formulas, the stability of the proposed method is studied independent to the problem dimension.

10.1 Introduction

This chapter deals with numerical analysis and computing of novel finite difference methods related to several option pricing models that correct the lack of adaptability of the classic Black-Scholes (BS) model to the reality of the market. As the best

R. Company (✉) • V.N. Egorova • M. El Fakharany • L. Jódar • F. Soleymani
Universitat Politècnica de València, Camino de Vera, s/n, 46022 Valencia, Spain
e-mail: rcompany@imm.upv.es; egorova.vn@gmail.com; fakharany@aucegypt.edu; ljodar@mat.upv.es; fazl_soley_bsb@yahoo.com

M. Ehrhardt et al. (eds.), *Novel Methods in Computational Finance*, Mathematics in Industry 25, DOI 10.1007/978-3-319-61282-9_10

model may be wasted with a disregarded analysis, we will pay attention to important issues such as consistency and stability of the proposed methods.

Dealing with prices, the guarantee of positivity of the numerical solution is a necessity that will be always considered here. After the 2008 financial crisis, the multidimensional option pricing problems became more relevant for both market industries and academia claiming for comfortable methods that be quick and reliable at the same time.

In Sect. 10.2, we consider finite difference methods for solving partial integro-differential equations (PIDEs) related to a wide class of Lévy processes introducing jump processes in the changes of the underlying assets. The consideration of jump processes motivates the appearance of the integral part of the PIDE. In Sect. 10.2.1, we solve the problem by introducing quadrature formulas for the approximation of the integrals and matching the discretization of the integral and differential part of the PIDE problem. Sect. 10.2.2 assumes that the volatility is a stochastic process deriving to a PIDE problem where the volatility is also an independent variable.

In Sect. 10.3, for dealing with American option pricing problems we follow the front-fixing approach initiated by [40] adding the numerical analysis in the numerical treatment of the problem and another transformation of the original PDE problem. To our knowledge we are the first users of the front-fixing method for regime-switching models fitting better the changing reality of the market.

Section 10.4 incorporates the rationality parameter approach recently proposed by [30] having the relevant issue that American option pricing problems can be approximated by solving a PDE instead of partial differential inequalities. This approach takes into account the irrational behavior of the market.

Section 10.5 addresses the challenge of the dimensionality. Firstly, in Sect. 10.5.1 the elimination of the cross derivative terms of the multidimensional PDE by using appropriate transformations allows the reduction of the computational cost and the numerical instabilities. After using a semidiscretization approach, the time exponential integration method and appropriate quadrature integration formulas, the stability of the proposed method is studied independent to the problem dimension.

10.2 Solving PIDE Option Pricing Using Finite Difference Schemes

The financial markets show that the underlying assets do not behave like a Brownian motion with a drift and a constant volatility. This fact motivates the emergence of alternative models to the pioneering Black-Scholes model [3]. Alternative models are stochastic volatility [35], deterministic volatility [17], jump diffusion [46, 64] and infinite activity Lévy models. Jump diffusion and Lévy models are characterized by a partial integro-differential equation (PIDE). This PIDE involves two major parts, namely, the differential part as in the Black-Scholes model and the non-local integral part due to the assumption of having assets with jumps. The option pricing under jump diffusion has been studied using the double discretization [7] and the integral term is approximated using the trapezoidal rule.

In this section, we propose positive stable and consistent methods to solve a wide class of infinite activity Lévy models using Gauss-Laguerre quadrature for approximating the integral part. Furthermore, the Bates model that incorporates both stochastic volatility and Jump diffusion is studied.

10.2.1 Solving PIDE for a Wide Class of Infinite Activity Lévy Processes

One of the most relevant and versatile Lévy models is the one proposed by Carr et al. the so called CGMY [6], that belongs to the family of KoBoL models [4]. Apart from these models, other Lévy processes such as Meixner [44, 57], Hyperbolic and Generalized Hyperbolic (GH) are used to obtain better estimation for the stock returns [56]. The Meixner process was introduced in 1998, it is used when the environment is changing stochastically over the time showing a reliable valuation for some indices such as Nikkei 225 [57].

The option pricing partial integro-differential equation (PIDE) unified model for several Lévy measures $\nu(y)$, given by [14, Chap. 12]

$$\frac{\partial \mathscr{C}}{\partial \tau}(S,\tau) = \frac{\sigma^2}{2}S^2\frac{\partial^2 \mathscr{C}}{\partial S^2}(S,\tau) + (r-q)S\frac{\partial \mathscr{C}}{\partial S}(S,\tau) - r\mathscr{C}(S,\tau)$$
$$+\int_{-\infty}^{+\infty}\nu(y)\left[\mathscr{C}(Se^y,\tau) - \mathscr{C}(S,\tau) - S(e^y-1)\frac{\partial \mathscr{C}}{\partial S}(S,\tau)\right]dy,\ S\in(0,\infty),\ \tau\in(0,T], \tag{10.1}$$

$$\mathscr{C}(S,0) = f(S) = (S-E)^+, \quad S\in(0,\infty), \tag{10.2}$$

$$\mathscr{C}(0,\tau) = 0;\ \lim_{S\to\infty}\mathscr{C}(S,\tau) = Se^{-q\tau} - Ee^{-r\tau}, \tag{10.3}$$

where $\mathscr{C}$ is the value of a contingent claim, S is the underlying asset and $\tau = T-t$ is the time to the maturity. The Lévy measures $\nu(y)$ are given in Table 10.1.

Table 10.1 The forms of $\nu(y)$

Model	The corresponding Lévy measure
KoBoL	$\nu(y) = \frac{C_- e^{-\mathscr{G}\lvert y\rvert}}{\lvert y\rvert^{1+Y}}\mathbf{1}_{y<0} + \frac{C_+ e^{-\mathscr{M}\lvert y\rvert}}{\lvert y\rvert^{1+Y}}\mathbf{1}_{y>0}$
Meixner	$\nu(y) = \frac{Ae^{-ay}}{y\sinh(by)}$
GH process	$\nu(y) = \frac{e^{\beta y}}{\lvert y\rvert}\left(\int_0^\infty \frac{e^{-\sqrt{2\zeta+\alpha^2}\lvert y\rvert}}{\pi^2\zeta\left(J^2_{\lvert\lambda\rvert}(\delta\sqrt{2\zeta}) + Y^2_{\lvert\lambda\rvert}(\delta\sqrt{2\zeta})\right)}d\zeta + \max(0,\lambda)e^{-\alpha\lvert y\rvert}\right)$

Note that the Hyperbolic process is obtained from the GH process when $\beta = 0$ and $\lambda = -1$.

The KoBoL model and in particular the CGMY, see Table 10.1 with parameter $C_- = C_+$, has been widely studied because its versatile and includes the finite and infinite activity cases as well as the finite and infinite variation, obtained by changing the value of Yor parameter $Y < 2$. A fairly complete revision of the methods used to solve the CGMY model can be found in [9, 15, 53, 65].

In this study we focus on the numerical analysis of the unified model (10.1)–(10.3) for the European case, by proposing a consistent, explicit and conditionally positive and stable finite difference scheme while the integral part is approximated using Gauss-Laguerre quadrature formula. We also include the computation of the linear complementarity problem (LCP) for the American option case using both the projected successive over relaxation method (PSOR) and the multigrid method (MG). The discretization for the differential operator is done using the three-level approximation, while the integral part is discretized as the same as in the European case. So, the integral part of the PIDE operator for the American and European cases is discretized using the Gauss-Laguerre quadrature. Although the three-level method is widely used and it is argued that the approximation error is of order two, however such method has two unsuitable properties, in fact as the method needs the first time step that must be obtained using another method (usually by implicit Euler method), in practice the accuracy is reduced.

Let us begin by transforming the PIDE (10.1) into a simpler one. Since the kernel of the integral in (10.1) presents a singularity at $y = 0$, a useful technique is to split the real line, for an arbitrary small parameter $\varepsilon > 0$, into two regions $\Omega_1 = [-\varepsilon, \varepsilon]$ and $\Omega_2 = \mathbb{R}\backslash\Omega_1$, the complementary set of Ω_1 in the real line. The integral on Ω_1 is replaced by a suitable coefficient in the diffusion term of the differential part of (10.1) obtained by Taylor expansion of $V(Se^y, \tau)$ about S, see [9, 15, 53, 65]. This coefficient depending on ε is a convergent integral and takes the form

$$\breve{\sigma}^2(\varepsilon) = \int_{-\varepsilon}^{\varepsilon} \nu(y)(e^y - 1)^2 dy = \varepsilon \int_{-1}^{1} \nu(\varepsilon\phi)(e^{\varepsilon\phi} - 1)^2 d\phi. \tag{10.4}$$

The resulting approximating PIDE is given by

$$\begin{aligned} \frac{\partial \mathscr{C}}{\partial \tau} = \frac{\hat{\sigma}^2}{2} S^2 \frac{\partial^2 \mathscr{C}}{\partial S^2} + (r - q - \gamma(\varepsilon)) S \frac{\partial \mathscr{C}}{\partial S} - (r + \lambda(\varepsilon)) \mathscr{C} \\ + \int_{\Omega_2} \nu(y) \mathscr{C}(Se^y, \tau) dy, \end{aligned} \tag{10.5}$$

where

$$\hat{\sigma}^2 = \sigma^2 + \breve{\sigma}^2(\varepsilon), \ \gamma(\varepsilon) = \int_{\Omega_2} \nu(y)(e^y - 1) dy, \ \lambda(\varepsilon) = \int_{\Omega_2} \nu(y) dy. \tag{10.6}$$

The convergent integrals (10.4) and (10.6) are evaluated using Gauss quadrature approximation. In order to obtain an approximation for $\breve{\sigma}^2(\varepsilon)$, the Gauss-Legendre quadrature approximation is used, so the weighting function $w(\phi) = 1$ such that

$$\breve{\sigma}^2(\varepsilon) \approx \varepsilon \sum_{m=1}^{M} \omega_m \nu(\varepsilon\phi_m)(e^{\varepsilon\phi_m} - 1)^2, \tag{10.7}$$

where ϕ_m are the roots of the Legendre polynomial $P_M(\phi)$ of degree M and ω_m is calculated based on [1, Eq. (25.4.29), p. 887]. Here M is chosen to be an even number so that zero is not a root of P_M. The improper integrals $\lambda(\varepsilon)$ and $\gamma(\varepsilon)$ are approximated using the shifted Gauss-Laguerre quadrature [19, p. 226]. Note that under change of variables $\eta = -y - \varepsilon$ for $y < 0$ and $\eta = y - \varepsilon$ for $y > 0$ then $\lambda(\varepsilon)$ and $\gamma(\varepsilon)$ have the following forms

$$\lambda(\varepsilon) = \int_0^\infty (\nu(-\eta - \varepsilon) + \nu(\eta + \varepsilon))\, d\eta \tag{10.8}$$

and

$$\gamma(\varepsilon) = \int_0^\infty \left[\nu(-\eta - \varepsilon)(e^{-(\eta+\varepsilon)} - 1) + \nu(\eta + \varepsilon)(e^{\eta+\varepsilon} - 1)\right] d\eta. \tag{10.9}$$

From (10.8), (10.9) and since the weighting function is $w(\eta) = e^{-\eta}$, then we have

$$\lambda(\varepsilon) \approx \sum_{m=1}^{M} \varpi_m F(\eta_m, \varepsilon), \;\; \gamma(\varepsilon) \approx \sum_{m=1}^{M} \varpi_m \mathscr{F}(\eta_m, \varepsilon), \tag{10.10}$$

where

$$\begin{aligned} F(\eta, \varepsilon) &= e^\eta(\nu(-\eta - \varepsilon) + \nu(\eta + \varepsilon)) \\ \mathscr{F}(\eta, \varepsilon) &= e^\eta \left(\nu(-\eta - \varepsilon)(e^{-(\eta+\varepsilon)} - 1) + \nu(\eta + \varepsilon)(e^{\eta+\varepsilon} - 1)\right). \end{aligned}$$

Here η_m are the roots of the Laguerre polynomial $L_M(\eta)$ of degree M and the weighting function ϖ_m is given in [1, Eq. (25.4.45), p. 890].

Coming back to (10.5) in order to eliminate the convection and reaction terms, using the transformation defined by

$$x = \exp[(r - q - \gamma(\varepsilon))\tau]S, \;\; V(x, \tau) = \exp[(r + \lambda(\varepsilon))\tau]\mathscr{C}(S, \tau), \tag{10.11}$$

one gets

$$\frac{\partial V}{\partial \tau} = \frac{\hat{\sigma}^2}{2} x^2 \frac{\partial^2 V}{\partial x^2} + \int_{\Omega_2} \nu(y) V(xe^y, \tau) dy, \;\; x \in (0, \infty), \;\; \tau \in (0, T], \tag{10.12}$$

with the initial and boundary conditions

$$V(x,0) = f(x) = (x-E)^{+} \tag{10.13}$$

$$V(0,\tau) = 0; \quad \lim_{x \to \infty} V(x,\tau) = e^{\lambda(\varepsilon)\tau}(xe^{\gamma(\varepsilon)\tau} - E). \tag{10.14}$$

Next, for the sake of convenience in the numerical treatment we rewrite the integral part of (10.12) as follows

$$\int_{\Omega_2} \nu(y)V(xe^y,\tau)dy = \int_{-\infty}^{\infty} \hat{\nu}(y)V(xe^y,\tau)dy, \tag{10.15}$$

where

$$\hat{\nu}(y) = \begin{cases} \nu(y), & y \in \Omega_2 \\ 0, & y \in \Omega_1 \end{cases}. \tag{10.16}$$

After that, in order to match the interval of the integration with the spatial domain of the problem, we use the following substitution $\phi = xe^y$ into (10.15), obtaining

$$\int_{\Omega_2} \nu(y)V(xe^y,\tau)dy = \int_0^{\infty} \hat{\nu}(\ln(\frac{\phi}{x}))V(\phi,\tau)\frac{d\phi}{\phi}. \tag{10.17}$$

Hence the PIDE for the European option under Lévy model, takes the following form

$$\frac{\partial V}{\partial \tau} = \frac{\hat{\sigma}^2}{2}x^2\frac{\partial^2 V}{\partial x^2} + \int_0^{\infty} \hat{\nu}(\ln(\frac{\phi}{x}))V(\phi,\tau)\frac{d\phi}{\phi}. \tag{10.18}$$

Now, we are in a good situation to construct an efficient explicit numerical scheme for the transformed problem (10.18) after choosing our numerical domain $[0, x_{\max}] \times [0, T]$ for large enough value of $x_{\max}$. For the time discretization, we take $\tau^n = nk$, $n = 0, 1, \ldots, N_\tau$ where $k = \frac{T}{N_\tau}$ and the spatial variable x is discretized by $x_j = jh$, $j = 0, 1, 2, \ldots, N_x$, $h = \frac{x_{\max}}{N_x}$.

Since the Laguerre-Gauss quadrature will be used for approximating the integral part of (10.18), then we have the sequence of roots $\{\phi_m\}_{m=1}^{M}$ of the Laguerre polynomial $L_M(\phi)$. The suitable value for M is selected such that $E < \phi_M < x_{\max}$.

By using explicit forward approximation for the time derivative of V and the central difference approximation for second spatial derivative, one gets

$$\frac{\partial V}{\partial \tau}(x_j,\tau^n) \approx \frac{V_j^{n+1} - V_j^n}{k}, \quad \frac{\partial^2 V}{\partial x^2}(x_j,\tau^n) \approx \frac{V_{j+1}^n - 2V_j^n + V_{j-1}^n}{h^2}. \tag{10.19}$$

In order to approximate the integral part of (10.18) matching the discretization of the integral and differential parts, taking into account that zeroes of Laguerre polynomial do not need to be nodes of the mesh, we use linear Lagrange interpolation polynomial. For any m, $1 \leq m \leq M$, let us denote by ℓ_m the last integer such that the mesh point $x_{\ell_m} < \phi_m$. The approximating value $V^n(\phi_m)$ is given by

$$V^n(\phi_m) = \tilde{a}_{\ell_m} V^n_{\ell_m} + \hat{a}_{\ell_m} V^n_{\ell_m+1}, \tag{10.20}$$

where the interpolation coefficients are

$$\tilde{a}_{\ell_m} = \frac{(x_{\ell_m+1} - \phi_m)}{h}; \ \hat{a}_{\ell_m} = \frac{(\phi_m - x_{\ell_m})}{h}. \tag{10.21}$$

Note that the linear interpolation approximation (10.20) has an error of order $\mathcal{O}(h^2)$ that coincide with the associated error of the central approximation of the spatial derivative (10.19). Hence the discretization for the integral part is given by

$$I^n_j = \sum_{m=1}^{M} \hat{\upsilon}(\ln \frac{\phi_m}{x_j}) \frac{e^{\phi_m}}{\phi_m} \varpi_m \left(\tilde{a}_{\ell_m} V^n_{\ell_m} + \hat{a}_{\ell_m} V^n_{\ell_m+1}\right). \tag{10.22}$$

Summarizing, from (10.19) to (10.22), the discretization of (10.18) with (10.13) and (10.14) takes the form

$$V^{n+1}_j = \alpha_j (V^n_{j+1} + V^n_{j-1}) + \beta_j V^n_j + k \sum_{m=1}^{M} \hat{\upsilon}(\ln \frac{\phi_m}{x_j}) \frac{e^{\phi_m}}{\phi_m} \varpi_m \left(\tilde{a}_{\ell_m} V^n_{\ell_m} + \hat{a}_{\ell_m} V^n_{\ell_m+1}\right), \tag{10.23}$$

$1 \leq j \leq N_x - 1, \ 0 \leq n \leq N_\tau - 1$, where

$$\alpha_j = \frac{k}{2h^2} \hat{\sigma}^2 x_j^2, \quad \beta_j = 1 - 2\alpha_j, \tag{10.24}$$

satisfying

$$V^0_j = (x_j - E)^+, \ V^n_0 = 0, \quad V^n_{N_x} = e^{\lambda(\varepsilon)\tau^n} (x_{\max} e^{\gamma(\varepsilon)\tau^n} - E). \tag{10.25}$$

In what follows, we state that the solution is conditionally positive and stable. The proof of this statement and consistency of the scheme can be found in [28].

Theorem 10.1 *The numerical solution* $\{V^n_j\}$ *of the scheme* (10.23)–(10.25) *is nonnegative under the condition.*

$$\frac{k}{h^2} \leq \frac{1}{\hat{\sigma}^2 x^2_{\max}}. \tag{10.26}$$

Based on Von Neumann approach, the stability of the numerical scheme (10.23) has been studied and summarized in the following theorem.

Theorem 10.2 *Under the positivity condition* (10.26)*, the numerical scheme* (10.23) *for* (10.18) *is conditionally stable see [29].*

The objective of the first example is to exhibit the importance of the positivity condition (10.26) for the three studied Lévy models.

Example 10.1 Here, we have an European option with $E = 30$, $T = 0.5$, $r = 0.08$, $q = 0$, $\sigma = 0.2$, $x_{\text{min}} = 0$, $x_{\text{max}} = 90$, $M = 15$, $\varepsilon = 0.5$ and $N_x = 128$. The parameters for Lévy models are given in Table 10.2.

Figures 10.1, 10.2, and 10.3 display the behavior of the option price $\mathscr{C}$ evaluated by the proposed explicit scheme when the positivity condition (10.26) holds for $N_\tau = 25e3$ and when it is broken for $N_\tau = 1e3$ represented by the solid and dot curves respectively under several Lévy processes.

The aim of the next example is to show the variation of the error for the Variance Gamma VG model as the stepsizes h and k change. The VG is obtained from the CGMY model when $Y = 0$, the reference option values for $S = \{20, 30, 40, 50\}$ are obtained using the closed form solution given in [45].

Table 10.2 The parameters for Lévy models used in Example 10.1

Model	Parameters
CGMY	$C = 0.5$, $\mathscr{G} = 15$, $\mathscr{M} = 25$ and $Y = 1.2945$
Meixner	$A = 0.5$, $a = -2.5$ and $b = 8$
GH	$\alpha = 4$, $\beta = -3.2$, $\delta = 0.4775$ and $\lambda = 2$

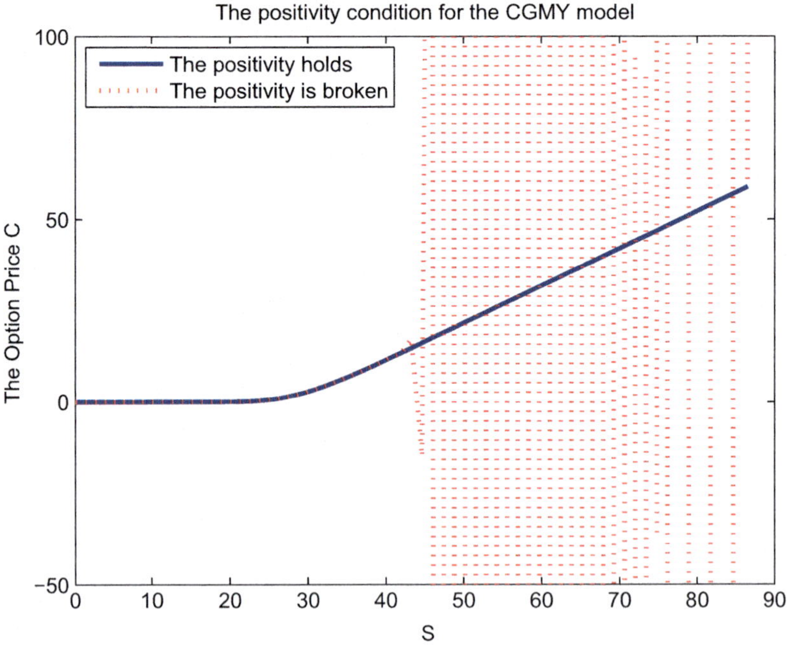

Fig. 10.1 About positivity condition of the explicit scheme under CGMY process

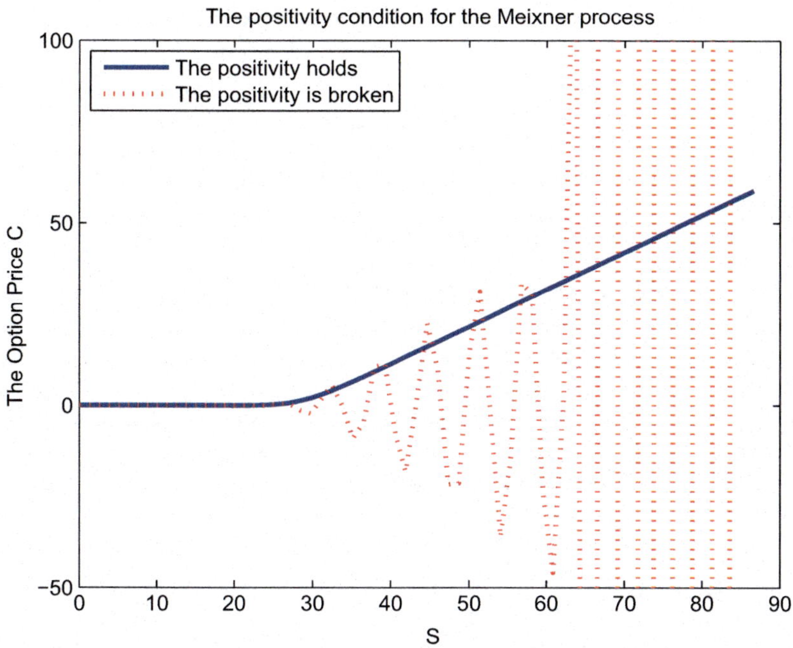

Fig. 10.2 The positivity condition of the explicit scheme under Meixner process

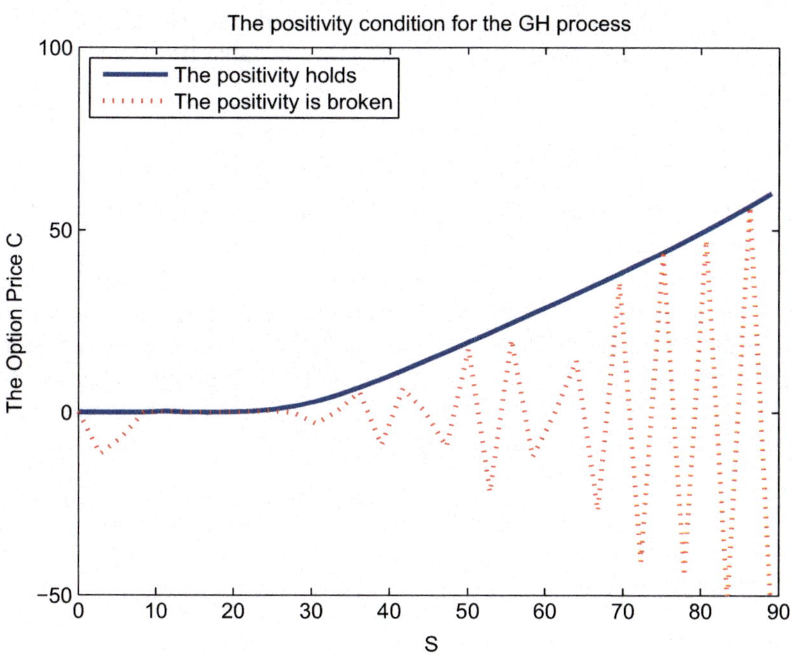

Fig. 10.3 The effect of positivity condition on the option price under GH process

Table 10.3 Errors and convergence rates for the VG model for several values of N_x

S	20		30		40		5		CPU
N_x	AE	α	AE	α	AE	α	AE	α	in s
32	$8.909e-4$	–	$1.926e-3$	–	$3.742e-3$	–	$4.386e-3$	–	1.84
64	$2.409e-4$	1.89	$5.335e-4$	1.85	$1.022e-3$	1.87	$1.181e-3$	1.89	4.63
128	$6.363e-5$	1.92	$1.413e-4$	1.92	$2.710e-4$	1.91	$3.053e-4$	1.95	10.85
256	$1.552e-5$	2.04	$3.698e-5$	1.93	$6.952e-5$	1.96	$7.603e-5$	2.01	18.99

Table 10.4 Errors and convergence rates for the VG model for various values of N_τ

S	20		30		40		50		CPU
N_τ	AE	β	AE	β	AE	β	AE	β	in s
$1.2e3$	$2.161e-4$	–	$4.790e-4$	–	$9.243e-4$	–	$1.151e-3$	–	4.06
$2.4e3$	$1.154e-4$	0.91	$2.552e-4$	0.89	$4.883e-4$	0.92	$6.049e-4$	0.93	7.28
$4.8e3$	$5.883e-5$	0.97	$1.304e-4$	0.94	$2.519e-4$	0.95	$3.072e-4$	0.98	12.65
$9.6e3$	$2.916e-5$	1.02	$6.462e-5$	0.96	$1.288e-4$	0.97	$1.489e-5$	1.04	20.37

Example 10.2 Consider an European option under the VG process with parameters $E = 30, T = 0.5, r = 0.1, q = 0, \sigma = 0.25, C_- = C_+ = 11.718, \mathscr{G} = 15$ and $\mathscr{M} = 25, x_{\min} = 0, x_{\max} = 90, M = 15, \varepsilon = 0.35$.

Table 10.3 reveals the variation of the absolute error (AE) as h changes as well as the spatial numerical convergence rate α and the CPU time while $N_\tau = 4.5e3$ for the explicit scheme (10.23). The change of the error due to the variation of N_τ, its convergence rate β and the elapsed time are shown in Table 10.4 while $N_x = 128$.

10.2.2 *Positive Finite Difference Schemes for Partial Integro-Differential Option Pricing Bates Model*

The Bates model is considered one of the effective mathematical models that has ability to describe the behavior of real markets of options usually of complex types for instance, currency options. In the Bates model, the Heston stochastic volatility model [35] and the Merton jump-diffusion model [46] are combined to describe the behavior of the underlying asset S and its variance ν [2]. The PIDE for the unknown option price $U(S, \nu, \tau)$ under Bates model is given by

$$\begin{aligned}\frac{\partial U}{\partial \tau} =& \frac{1}{2}\nu S^2 \frac{\partial^2 U}{\partial S^2} + \rho\sigma\nu S \frac{\partial^2 U}{\partial S \partial \nu} + \frac{1}{2}\sigma^2 \nu \frac{\partial^2 U}{\partial \nu^2} + (r - q - \lambda\xi) S \frac{\partial U}{\partial S} \\ &+ \kappa(\theta - \nu)\frac{\partial U}{\partial \nu} - (r + \lambda)U + \lambda \int_0^\infty U(S\eta, \nu, \tau) f(\eta) d\eta,\end{aligned} \tag{10.27}$$

and the density function $f(\eta)$ is given by

$$f(\eta) = \frac{1}{\sqrt{2\pi}\hat{\sigma}\eta}\exp[-\frac{(\ln\eta - \mu)^2}{2\hat{\sigma}^2}], \tag{10.28}$$

where μ is the mean of the jump and $\hat{\sigma}$ is the standard deviation. For the European call option we consider the initial condition

$$U(S, \nu, 0) = g_1(S, \nu) = \max\{S - E, 0\}, \tag{10.29}$$

where E is the strike price. We assume the boundary conditions applied to the Heston model, see [20], but modified for $\nu = 0$ due to the additional integral term appearing in Bates model. For the boundaries $S = 0$ and $S \to \infty$ one gets

$$U(0, \nu, \tau) = 0, \quad \lim_{S\to\infty}\frac{\partial U}{\partial S}(S, \nu, \tau) = 1. \tag{10.30}$$

Note that this last condition means a linear behavior of the option price for large values of S with slope 1 when no dividend payments are considered, $q = 0$. Based on that fact, we replace it by the following condition, see [66, Chap. 3, p. 54]

$$U(S, \nu, \tau) = e^{-q\tau}S. \tag{10.31}$$

For $\nu \to \infty$ and $\nu = 0$, the corresponding boundary conditions are imposed as follows

$$\lim_{\nu\to\infty} U(S, \nu, \tau) = S, \tag{10.32}$$

$$\frac{\partial U}{\partial \tau}(S, 0, \tau) = (r - q - \lambda\xi)S\frac{\partial U}{\partial S}(S, 0, \tau) - (r + \lambda)U(S, 0, \tau) + \kappa\theta\frac{\partial U}{\partial \nu}(S, 0, \tau)$$
$$+\frac{\lambda}{\sqrt{2\pi}\hat{\sigma}}\int_0^\infty U(\varphi, 0, \tau)\exp[-\frac{(\ln\varphi - \ln S - \hat{\mu})^2}{2\hat{\sigma}^2}]\frac{d\varphi}{\varphi}, \tag{10.33}$$

where $\varphi = S\eta$.

The model (10.28)–(10.33) has two challenges from the numerical analysis point of view. Firstly, the presence of a mixed spatial derivative term involves the existence of negative coefficient terms into the numerical scheme deteriorating the quality of the numerical solution such as spurious oscillation and slow convergence, see the introduction of [70]. Secondly, the discretization of the improper integral part should be adequate with the bounded numerical domain and the incorporation of the initial and boundary conditions.

Dealing with prices, guaranty of the positivity of the solution is essential. In this chapter we construct an explicit difference scheme that guarantees positive solutions. We transform the PIDE (10.27) into a new PIDE without mixed spatial derivative before the discretization, following the idea of [10], and avoiding the above quoted drawbacks. Furthermore, this strategy has additional computational advantage of the reduction of the stencil scheme points, from nine [22] or seven [54] to just five.

We begin this section by eliminating the mixed spatial derivative term of (10.28), inspired by the reduction of second order linear partial differential equation in two independent variables to canonical form, see [31, Chap. 3] and [10] for details. Let us consider the following transformation

$$x = \tilde{\rho}\sigma \ln S;\ y = \rho\sigma \ln S - \nu;\ w(x, y, \tau) = e^{(r+\lambda)\tau} U(S, \nu, \tau), \tag{10.34}$$

where $\tilde{\rho} = \sqrt{1-\rho^2}$, $0 < |\rho| < 1$, obtaining the following transformed equation

$$\frac{\partial w}{\partial \tau} = \frac{\tilde{\rho}^2 \nu\sigma^2}{2}\left(\frac{\partial^2 w}{\partial x^2} + \frac{\partial^2 w}{\partial y^2}\right) + \hat{\delta}\frac{\partial w}{\partial x} + \tilde{\delta}\frac{\partial w}{\partial y} + I(w), \tag{10.35}$$

with

$$I(w) = \lambda \int_0^\infty w(x + \sigma\tilde{\rho}\ln\eta,\ y + \rho\sigma \ln\eta, \tau) f(\eta) d\eta, \tag{10.36}$$

where

$$\hat{\delta} = \sigma\tilde{\rho}(\hat{\xi} - \frac{\nu}{2}),\ \tilde{\delta} = \sigma\rho(\hat{\xi} - \frac{\nu}{2}) - \kappa(\theta - \nu) \text{ and } \hat{\xi} = r - q - \lambda\xi. \tag{10.37}$$

For the sake of convenience in the matching of the further discretization of the differential and integral parts of (10.35), we consider now the substitution

$$\phi = x + \sigma\tilde{\rho}\ln\eta. \tag{10.38}$$

Hence from (10.29) and (10.36) one gets

$$I(w) = \frac{\lambda}{\sqrt{2\pi}\hat{\sigma}\tilde{\rho}\sigma} \int_{-\infty}^{\infty} w(\phi, y + m(\phi - x), \tau) \exp\left[\frac{-1}{\hat{\sigma}^2}\left(\frac{\phi - x}{\sigma\tilde{\rho}} - \mu\right)^2\right] d\phi, \tag{10.39}$$

where $m = \frac{\rho}{\tilde{\rho}}$. Note that from (10.34), we have $y = mx - \nu$.

The initial and boundary conditions (10.29)–(10.33) are transformed into the corresponding conditions using (10.34) and (10.38).

$$w(x, y, 0) = \max\{e^{\frac{x}{\sigma\tilde{\rho}}} - E, 0\},\quad \lim_{x\to-\infty} w(x, y, \tau) = 0, \tag{10.40}$$

$$w(x, y, \tau) \approx \exp\left[\frac{x}{\sigma\tilde{\rho}} + (r - q + \lambda)\tau\right],\ x \to \infty, \tag{10.41}$$

$$w(x, y, \tau) \approx \exp\left[\frac{x}{\sigma\tilde{\rho}} + (r + \lambda)\tau\right],\ mx - y \to \infty, \tag{10.42}$$

$$\frac{\partial w}{\partial \tau} \approx \sigma\tilde{\rho}\hat{\xi}\frac{\partial w}{\partial x} + (\sigma\rho\hat{\xi} - \kappa\theta)\frac{\partial w}{\partial y}$$
$$+\frac{\lambda}{\sqrt{2\pi}\hat{\sigma}\tilde{\rho}\sigma}\int_{-\infty}^{\infty} w(\phi, m\phi - \nu, \tau)\exp\left[\frac{-1}{\hat{\sigma}^2}\left(\frac{\phi - x}{\sigma\tilde{\rho}} - \mu\right)^2\right] d\phi,\ \nu \to 0. \tag{10.43}$$

From [27] a suitable bound for the underlying asset variable S is available and generally accepted. In an analogous way, considering an admissible range of the variance ν, we can identify a convenient-bounded numerical domain $\mathscr{R} = [S_1, S_2] \times [\nu_1, \nu_2]$ in the $S - \nu$ plane. Under the transformation (10.34) as it is shown in [10] the rectangle $\mathscr{R}$ is transformed into the rhomboid $ABCD$ see [28]. In light of the transformation (10.34) we use a discretization of the numerical domain where the space step sizes $h = \Delta x$ and $h_y = \Delta y = |m|h$ are related by the slope $m = \frac{\rho}{\tilde{\rho}}$. Here we subdivide space-time axes into uniform spaced points using

$$\begin{aligned} x_i &= a + ih,\ 0 \le i \le N_x, & y_j &= y_0 + j|m|h,\ i \le j \le N_y + i, \\ \nu_{i,j} &= mx_i - y_j, & \tau^n &= nk,\ 0 \le n \le N_\tau, \end{aligned} \tag{10.44}$$

where $h = \frac{b-a}{N_x}$, $y_0 = ma - \nu_2$, $N_y = \frac{\nu_2 - \nu_1}{|m|h}$ and $k = \frac{T}{N_\tau}$. Note that any mesh point in the computational spatial domain has the form

$$(x_i, y_j) = (a + ih, mx_i - \nu_2 + (j - i)|m|h).$$

By denoting the approximate value of w at a representative mesh point $P(x_i, y_j, \tau^n)$ by $W_{i,j}^n$, we implement the center difference approximation for spatial partial derivatives. On the other hand the improper integral $I(w)$ (10.39) is truncated into $[a, b]$, then the composite four points integration formula of open type has been implemented using the same step size for the variable x as in the differential part. Hence the corresponding finite difference equation for (10.35) is given by

$$W_{i,j}^{n+1} = \beta_{i,j}W_{i,j}^n + \hat{\alpha}_{i,j}W_{i+1,j}^n + \breve{\alpha}_{i,j}W_{i-1,j}^n + \alpha_{i,j}W_{i,j-1}^n + \gamma_{i,j}W_{i,j+1}^n + \hat{\lambda}J_{i,j}^n, \tag{10.45}$$

$$1 \le i \le N_x - 1,\ i + 1 \le j \le N_y + i - 1,\ 0 \le n \le N_\tau - 1,$$

where

$$\begin{aligned} \beta_{i,j} &= 1 - \tfrac{k\sigma^2}{h^2m^2}\nu_{i,j} = (1 - \tfrac{k}{h^2}\tilde{a}_{ij}), \\ \hat{\alpha}_{i,j} &= \tfrac{k\sigma\tilde{\rho}}{2h}\left[\tfrac{(2\tilde{\rho}\sigma - h)}{2h}\nu_{i,j} + \hat{\xi}\right] = \tfrac{k}{h}(\tfrac{\rho^2}{2h}\tilde{a}_{ij} + \tilde{b}_{ij}) \\ \breve{\alpha}_{i,j} &= \tfrac{k\sigma\tilde{\rho}}{2h}\left[\tfrac{(2\tilde{\rho}\sigma + h)}{2h}\nu_{i,j} - \hat{\xi}\right] = \tfrac{k}{h}(\tfrac{\rho^2}{2h}\tilde{a}_{ij} - \tilde{b}_{ij}), \\ \alpha_{i,j} &= \tfrac{k}{2|m|h}\left[\left(\tfrac{\sigma^2\tilde{\rho}^2}{|m|h} + \tfrac{\sigma\rho}{2} - \kappa\right)\nu_{i,j} - \sigma\rho\hat{\xi} + \kappa\theta\right] = \tfrac{k}{h}(\tfrac{\tilde{\rho}^2}{2h}\tilde{a}_{ij} - \tfrac{m}{|m|}\tilde{b}_{ij} + \tilde{c}_{ij}), \\ \gamma_{i,j} &= \tfrac{k}{2|m|h}\left[\left(\tfrac{\sigma^2\tilde{\rho}^2}{|m|h} - \tfrac{\sigma\rho}{2} + \kappa\right)\nu_{i,j} + \sigma\rho\hat{\xi} - \kappa\theta\right] = \tfrac{k}{h}(\tfrac{\tilde{\rho}^2}{2h}\tilde{a}_{ij} + \tfrac{m}{|m|}\tilde{b}_{ij} - \tilde{c}_{ij}), \end{aligned} \tag{10.46}$$

$$\hat{\lambda} = \frac{5kh\lambda}{24\sqrt{2\pi}\hat{\sigma}\tilde{\rho}\sigma}, \tag{10.47}$$

and the integral part is given by

$$J_{i,j}^n = \sum_{\ell=0}^{N_x/5-1} \Big(11 g_{i,5\ell+1} W_{5\ell+1,5\ell+1+j-i}^n + g_{i,5\ell+2} W_{5\ell+2,5\ell+2+j-i}^n + g_{i,5\ell+3} W_{5\ell+3,5\ell+3+j-i}^n + 11 g_{i,5\ell+4} W_{5\ell+4,5\ell+4+j-i}^n \Big), \tag{10.48}$$

assuming that N_x has been previously chosen as a multiple of 5. The weight function $g_{i,\ell}$ is given by

$$g_{i,\ell} \equiv g(x_i, \phi_\ell) = \exp\left[\frac{-1}{2\hat{\sigma}^2} \left(\frac{\phi_\ell - x_i}{\sigma\tilde{\rho}} - \mu \right)^2 \right], \; 0 \le \ell \le N_x. \tag{10.49}$$

The following theorem is established in order to guarantee nonnegative numerical solutions such that

Theorem 10.3 *If stepsizes h and k satisfy*

C1. $h \le \min\left\{ \frac{2\sigma\tilde{\rho}\nu_i}{|2\hat{\xi} - \nu_i|}, \frac{\sigma^2\tilde{\rho}^2\nu_i}{2m^2|\alpha\nu_i + \beta|}, \; i = 1, 2 \right\}$

C2. $k \le \min\left\{ \frac{m^2h^2}{\sigma^2\nu_2}, \frac{2h}{3\sigma\tilde{\rho}|\hat{\xi}|}, \frac{|m|h}{3\kappa\theta} \right\}$,

then the numerical solution $\{W_{i,j}^n\}$ *of the scheme* (10.45) *is nonnegative.*

The numerical scheme (10.45) is written in a matrix form in order to study its stability, see [28]. It has been shown that under the positivity conditions, the infinite norm of the vector solution is bounded such that

$$\frac{\|\mathbf{W}^n\|_\infty}{\|\mathbf{W}^0\|_\infty} \le \exp\left((r + \lambda + \lambda_1)T\right).$$

Establishing a conditional strong uniform stable scheme.

Example 10.3 The parameters are selected as follows $T = 0.5$, $E = 100$, $r = 0.05$, $q = 0$, $\theta = 0.05$, $\kappa = 2.0$, $\sigma = 0.3$, $\hat{\sigma} = 0.35$, $\mu = -0.5$, $\lambda = 0.2$ and $\rho = -0.5$ with a tolerance error $\varepsilon = 10^{-4}$.

The boundary points a and b of the spatial computational domain are obtained from [28], while $\nu_1 = 0.1$ and $\nu_2 = 1$. Table 10.5 shows the variation of the RMSRE for several values of the time step sizes, for fixed $N_x = 70$ and $N_y = 35$, with respect to reference values computed at $(N_x, N_y, N_\tau) = (500, 146, 7000)$.

The variation of error due to the change of the spatial step sizes, while $N_\tau = 500$ has been presented in Table 10.6.

Table 10.5 The RMSRE for several values of N_τ

N_τ	RMSRE	Ratio	CPU (s)
500	2.485×10^{-3}	–	6.66
1000	1.322×10^{-3}	1.88	6.94
2000	6.429×10^{-4}	2.06	7.28
4000	3.296×10^{-4}	1.95	7.69
8000	1.569×10^{-4}	2.10	7.91

Table 10.6 The associated RMSRE for different values of (N_x, N_y)

(N_x, N_y)	RMSRE	Ratio	CPU (s)
(40, 20)	1.526×10^{-2}	–	0.32
(60, 30)	3.459×10^{-3}	4.412	1.83
(80, 40)	9.271×10^{-4}	3.371	6.95
(100, 50)	3.589×10^{-4}	2.583	19.64
(120, 60)	8.473×10^{-5}	4.236	46.72

10.3 Front-Fixing Methods for American Option Pricing Problems

American option pricing problem leads to a free boundary value problem, that is challenging because one has to find the solution of a PDE that satisfies auxiliary initial conditions and boundary conditions on a fixed boundary as well as on an unknown free boundary. This complexity is reduced by transforming the problem into a new nonlinear PDE where the free boundary appears as a new variable of the PDE problem.

This technique which originated in physics problems is the so called front-fixing method based on the Landau transform [41] to fix the optimal exercise boundary on a vertical axis. The front-fixing method has been applied successfully to a wide range of problems arising in physics (see Crank [18]) and finance (see [11, 59, 68], etc.) In this section the front-fixing method combined with the use of an explicit finite difference scheme avoid the drawbacks of alternative algebraic approaches since it avoids the use of iterative methods and underlying difficulties such as how to initiate the algorithm, when to stop it and which is the error after the stopping.

10.3.1 Front-Fixing Methods for American Vanilla Options

First of all, classical Black-Scholes model for American call option (10.50)–(10.53) is considered. The option price $C(S, \tau)$, where $\tau = T - t$ is the time to maturity, with constant dividend yield q is the solution of linear PDE of the second order

$$\frac{\partial c}{\partial \tau} = \frac{1}{2}\sigma^2 S^2 \frac{\partial^2 C}{\partial S^2} + (r-q)S\frac{\partial C}{\partial S} - rC, \qquad S < S_f(\tau), \quad 0 < \tau \leq T, \qquad (10.50)$$

supplied with the following initial conditions

$$C(S,0) = \max(S - E, 0), \quad S_f(0) = E \max\left(\frac{r}{q}, 1\right), \tag{10.51}$$

and the boundary conditions

$$C(S_f(\tau), \tau) = S_f(\tau) - E, \quad \lim_{S\to\infty} P(S,\tau) = 0. \tag{10.52}$$

Since an additional unknown function $S_f(\tau)$ is included in the free boundary formulation, one extra condition is necessary. This condition is called *smooth pasting condition* and requires that the slope of the option price curve at the free boundary coincides with the slope of payoff function. Thus, for put option it is presented as follows

$$\frac{\partial C}{\partial S}\left(S_f(\tau), \tau\right) = 1. \tag{10.53}$$

A dimensionless Landau transformation [41] is proposed as follows

$$x = \ln\frac{S_f(\tau)}{S}, \quad c(x,\tau) = \frac{C(S,\tau)}{E}, \quad s_f(\tau) = \frac{S_f(\tau)}{E}. \tag{10.54}$$

The spatial variable x transfers the free boundary domain $S < S_f(\tau)$ to the fixed, but unbounded domain $(0;\infty)$. In new coordinates (x,τ) the problem (10.50)–(10.53) is rewritten in the following normalized form

$$\frac{\partial c}{\partial \tau} = \frac{1}{2}\sigma^2\frac{\partial^2 c}{\partial x^2} + \left(r - q - \frac{\sigma^2}{2}\right)\frac{\partial c}{\partial x} - rc + \frac{s_f'}{s_f}\frac{\partial c}{\partial x}, \qquad x > 0, \quad 0 < \tau \le T, \tag{10.55}$$

where s_f' denotes the derivative of s_f with respect to τ. The new transformed equation (10.55) is a nonlinear PDE on the domain $(0,\infty)\times(0,T]$ since s_f and its derivative are involved. The problem (10.55) is solved by explicit FDM.

Further, let us consider American call option problem with another dimensionless transformation that allows to fix the computational domain as in [68] and to simplify the boundary conditions like [66, p. 122],

$$x = \ln\frac{S_f(\tau)}{S}, \qquad c(x,\tau) = \frac{C(S,\tau) - S + E}{E}, \quad s_f(\tau) = \frac{S_f(\tau)}{E}. \tag{10.56}$$

Using transformation (10.56) the problem (for call option) can be rewritten in normalized form

$$\frac{\partial c}{\partial \tau} = \frac{\sigma^2}{2}\frac{\partial^2 c}{\partial x^2} - \left(r - q - \frac{\sigma^2}{2} + \frac{s_f'}{s_f}\right)\frac{\partial c}{\partial x} - rc - qs_f e^{-x} + r, \quad x > 0,\ 0 < \tau \le T, \tag{10.57}$$

with new initial conditions

$$s_f(0) = \max\left(\frac{r}{q}, 1\right), \quad c(x,0) = \begin{cases} 1 - e^{-x}, & r \le q, \\ g(x), & r > q, \end{cases} \quad x \ge 0, \tag{10.58}$$

$$g(x) = \max\left(1 - \frac{r}{q}e^{-x}, 0\right). \tag{10.59}$$

Analytical or closed form solution of the transformed problems (10.55) or (10.57) does not exist. Therefore explicit and fully implicit FDMs are employed for constructing effective and stable numerical solution.

The problem (10.57)–(10.59) can be numerically studied on the fixed domain $[0, x_{max}] \times [0, \tau]$. The value x_{max} is chosen big enough to guarantee the boundary condition. The computational grid of $M+1$ spatial points and $N+1$ time levels is chosen to be uniform with respective step sizes h and k:

$$h = \frac{x_{max}}{M}, \quad k = \frac{T}{N}, \tag{10.60}$$

$$x_j = hj, \quad j = 0, \dots, M, \quad \tau^n = kn, \quad n = 0, \dots, N. \tag{10.61}$$

The approximate value of option price at the point x_j and time τ^n is denoted by $c_j^n \approx c(x_j, \tau^n)$ and the approximate value of the free boundary is denoted by $S_f^n \approx S_f(\tau^n)$. Then a forward two-time level and centred in a space explicit scheme is constructed for internal spatial nodes as follows

$$c_j^{n+1} = a_1^n c_{j-1}^n + b c_j^n + a_2^n c_{j+1}^n + k\left(r - qS_f^n e^{-x_j}\right), \quad 1 \le j \le M-1, \tag{10.62}$$

where

$$a_1^n = \frac{k}{2h^2}\left(\sigma^2 + \left(r - q - \frac{\sigma^2}{2}\right)h\right) + \frac{S_f^{n+1} - S_f^n}{2hS_f^n} = a_1 + \frac{S_f^{n+1} - S_f^n}{2hS_f^n},$$

$$b = 1 - \sigma^2\frac{k}{h^2} - rk, \tag{10.63}$$

$$a_2^n = \frac{k}{2h^2}\left(\sigma^2 - \left(r - q - \frac{\sigma^2}{2}\right)h\right) - \frac{S_f^{n+1} - S_f^n}{2hS_f^n} = a_2 - \frac{S_f^{n+1} - S_f^n}{2hS_f^n}.$$

Special attention is paid to study positivity and monotonicity of the numerical solution as well as stability and consistency of the proposed schemes. Note, that using expressions (10.63) it is easy to obtain that the constants of the scheme a_1, b and a_2 are positive for both cases: $r \le q$ and $r > q$ under following conditions

$$h < \frac{\sigma^2}{\left|r - q - \frac{\sigma^2}{2}\right|}, \quad r \ne q + \frac{\sigma^2}{2}, \quad k < \frac{h^2}{\sigma^2 + rh^2}, \tag{10.64}$$

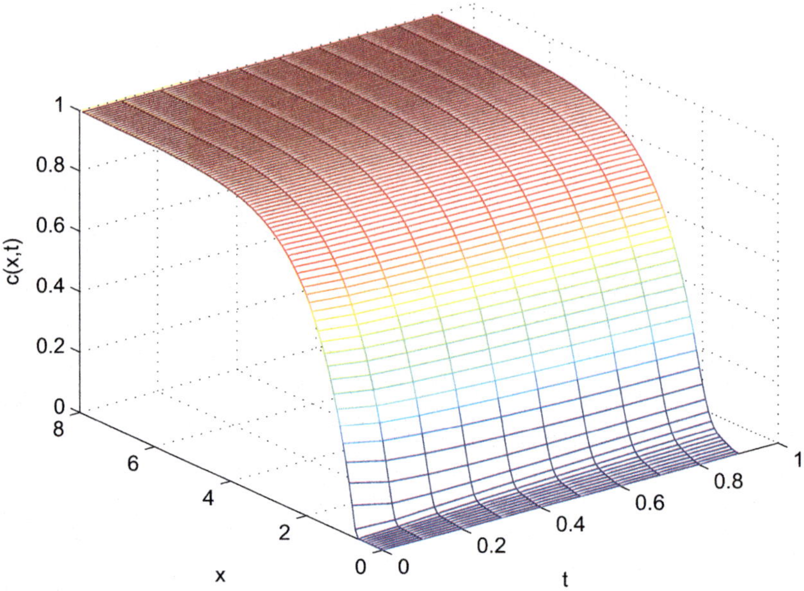

Fig. 10.4 The function $c(x, \tau)$ calculated by the proposed fully implicit method

If $r = q + \frac{\sigma^2}{2}$, then under the condition (10.64), coefficients a_1, b and a_2 are positive. Note, that these conditions are sufficient also for stability of the proposed explicit scheme. The details of the stability and consistency analysis can be found in [12].

Stability conditions on step sizes for explicit methods have been found. The implicit method is unconditionally stable, that allows to reduce computational time. But, there exist additional calculations of the inverse Jacobian matrix on each iteration. It has been shown that for the same step sizes the explicit method is ten times faster than the implicit one. The solution of (10.57) calculated by the proposed fully implicit method is shown in Fig. 10.4.

10.3.2 Moving Boundary Transformation for Nonlinear Models

For the case of American options with constant volatility various front-fixing transformations have been studied in [12, 40, 47, 58]. In this section an efficient front-fixing method for a nonlinear Black-Scholes equation is proposed. Under the transformation the free boundary is replaced by a time-dependent known boundary. In the resulting equation there is no reaction term and the convection term is simplified in a such way that the operator splitting technique is not required. This ensured a single numerical scheme is suitable for the entire equation. The connection between the transformed boundary conditions with the transformed option price and the free boundary does not require additional information.

The proposed formulation of the nonlinear problem allows the use of a versatile numerical treatment. In this chapter an explicit Euler and alternating direction explicit (ADE) method [21, 49] together with implicit methods are studied.

With the previous notation, nonlinear American call option pricing models may be formulated as the free boundary PDE problem

$$\frac{\partial C}{\partial \tau} = \frac{\tilde{\sigma}^2}{2} S^2 \frac{\partial^2 C}{\partial S^2} + (r-q) S \frac{\partial C}{\partial S} - rC, \quad 0 \leq S < S_f(\tau), \quad 0 < \tau \leq T, \tag{10.65}$$

where the adjusted volatility function is given by $\tilde{\sigma}^2 = \tilde{\sigma}^2\left(\tau, S, C_{SS}\right)$. Two nonlinear models with different adjusted volatility functions are considered:

- RAPM model: $\tilde{\sigma}^2 = \sigma_0^2 \left(1 + \mu \left(S \frac{\partial^2 C}{\partial S^2}\right)^{\frac{1}{3}}\right)$,
- Barles and Soner model: $\tilde{\sigma}^2 = \sigma_0^2 \left(1 + \Psi\left(e^{r\tau} a^2 S^2 C_{SS}\right)\right)$,
  where $a = \mu\sqrt{\gamma N}$, γ is the risk aversion factor and N denotes the number of options to be sold. The function Ψ is the solution of the nonlinear singular initial value problem

$$\Psi'(A) = \frac{\Psi(A) + 1}{2\sqrt{A\Psi(A)} - A}, \quad A \neq 0, \quad \Psi(0) = 0. \tag{10.66}$$

Taking advantages of the Landau transformation [41] with modifications in the exponential factors like those described in [10], it is possible to remove the reaction term and partially the convection term by using the transformation given below.

$$x = e^{(r-q)\tau} \frac{S}{S_f(\tau)}, \quad V(x,\tau) = \frac{e^{r\tau}}{E} C(S,\tau), \quad s_f(\tau) = \frac{S_f(\tau)}{E}. \tag{10.67}$$

Using transformation (10.67) the equation (10.65) takes the form

$$V_\tau = \frac{\sigma^2}{2} x^2 V_{xx} + \frac{s_f'}{s_f} x V_x, \quad 0 \leq x < e^{(r-q)\tau}, \quad 0 < \tau \leq T, \tag{10.68}$$

where

$$\sigma^2 = \sigma^2\left(\tau, x, V_{xx}\right) = \tilde{\sigma}^2(\tau, S, C_{SS}).$$

Note that the transformation described in (10.67) transforms the original free boundary value problem to a known moving boundary problem. In the case $r > q$ the computational domain increases with respect to time, otherwise it decreases. The numerical solution of the transformed problem can be found by explicit, ADE and implicit methods.

In Table 10.7 the results and comparisons are presented. Since the domain is changing in time and is covered by an equidistant grid, the spatial step size h_n is

Table 10.7 Root Mean Square Error (RMSE) with respect to CPU-time for different step sizes h_0 and fixed $k = 0.0001$, published in [26]

h_0	0.08	0.04	0.02	0.01
Explicit method				
RMSE	0.04984	0.02629	0.01232	0.00464
CPU-time, s	15.810	27.566	51.476	99.434
ADE method				
RMSE	0.16816	0.08172	0.02099	0.00620
CPU-time, s	15.129	27.776	53.865	104.247
Implicit method				
RMSE	0.04984	0.02355	0.00958	0.00445
CPU-time, s	34.099	60.030	112.728	257.880
Newton-like method				
RMSE	0.11376	0.06026	0.01389	0.00471
CPU-time, s	33.869	58.141	107.561	315.505

We acknowledge the permission of Taylor & Francis Ltd (http://www.tandfonline.com)

varying in time. The time step is fixed at $k = 0.0001$ to guarantee stability of all numerical solutions. For the implicit method the tolerance was chosen as $\epsilon = 10^{-4}$.

Note that the main part of the computational time is pertained for the calculation of $\Psi(A)$. For the implicit methods it has to be calculated on each iteration of Newton's method. Thus, their computational costs may be noticeably reduced by choosing another model. The details of the proposed methods can be found in [26].

10.3.3 Front-Fixing Method for Regime-Switching Model

An American put option on the asset $S_t = S$ with strike price E and maturity $T < \infty$ is considered under regime-switching model. Let $V_i(S, \tau)$ denote the option price functions, where $\tau = T - t$ denotes the time to maturity, the asset price S and the regime $\alpha_t = i$. Then, $V_i(S.\tau)$, $1 \le i \le I$, satisfy the following free boundary problem:

$$\frac{\partial V_i}{\partial \tau} = \frac{\sigma_i^2}{2} S^2 \frac{\partial^2 V_i}{\partial S^2} + r_i S \frac{\partial V_i}{\partial S} - r_i V_i + \sum_{l \neq i} q_{il}(V_l - V_i), \quad S > S_i^*(\tau), 0 < \tau \le T, \tag{10.69}$$

where $S_i^*(\tau)$ denote optimal stopping boundaries of the option. Initial conditions are

$$V_i(S, 0) = \max(E - S, 0), \quad S_i^*(0) = E, \quad i = 1, \ldots, I. \tag{10.70}$$

In spite of the apparent complexity of the transformed problem due to the appearance of new spatial variables, one for each equation, the explicit numerical scheme constructed becomes easy to implement, computationally cheap and accurate when one compares with the more relevant existing methods. Implicit weighted schemes have been developed for the sake of performance comparison.

Based on the transformation used by the authors in [11, 68] for the case of just one equation, the following multi-variable transformation is considered

$$x^i = \ln \frac{S}{S_i^*(\tau)}, \quad 1 \le i \le I. \tag{10.71}$$

Note that the new variables x^i lie in the fixed positive real line. The price V_i of i-th regime involved in i-th equation of the system and i-th free boundary are related by the dimensionless transformation

$$P_i(x^i, \tau) = \frac{V_i(S, \tau)}{E}, \quad X_i(\tau) = \frac{S_i^*(\tau)}{E}, \quad 1 \le i \le I. \tag{10.72}$$

Then the value of option l-th regime appearing in i-th coupled equation, $l \neq i$, becomes $P_{l,i}(x^i, \tau) = \frac{V_l(S,\tau)}{E}$.

Since from (10.72), $\frac{V_l(S,\tau)}{E} = P_l(x^l, \tau)$ and taking into account transformation (10.71) for indexes i and l one gets that

$$P_{l,i}(x^i, \tau) = P_l(x^l, \tau), \tag{10.73}$$

and it occurs when the variables are related by the equation

$$x^l = x^i + \ln \frac{X_i(\tau)}{X_l(\tau)}, \quad 1 \le i, l \le I. \tag{10.74}$$

From (10.71) to (10.73) the problem (10.69) for $1 \le i \le I$ takes a new form

$$\begin{aligned} \frac{\partial P_i}{\partial \tau}(x^i, \tau) &= \frac{\sigma_i^2}{2} \frac{\partial^2 P_i}{\partial (x^i)^2}(x^i, \tau) + \left(r_i - \frac{\sigma_i^2}{2} + \frac{X_i'(\tau)}{X_i(\tau)} \right) \frac{\partial P_i}{\partial x^i}(x^i, \tau) - r_i P_i(x^i, \tau) \\ &+ \sum_{l \neq i} q_{il}(P_{l,i}(x^i, \tau) - P_i(x^i, \tau)) = 0, \quad x^i > 0, \quad 0 < \tau \le T. \end{aligned} \tag{10.75}$$

PDE problem (10.75) is solved by the explicit FDM. Let us denote $u_{i,j}^n \approx P_i(x_j, \tau^n)$ the approximation of P_i in i-th equation at mesh point $(x^i = x_j, \tau = \tau^n)$ and $\tilde{u}_{l_i,j}^n \approx P_{l,i}(x_j, \tau^n)$ be the approximation of P_l in i-th equation evaluated at the point $(x^i = x_j, \tau = \tau^n)$. The discretization of the transformed optimal stopping boundary is denoted by $X_i^n \approx X_i(\tau^n)$. Then an explicit finite difference scheme can be written in the form

$$\begin{aligned} \frac{u_{i,j}^{n+1} - u_{i,j}^n}{k} &= \frac{\sigma_i^2}{2} \frac{u_{i,j+1}^n - 2u_{i,j}^n + u_{i,j-1}^n}{h^2} + \left(r_i - \frac{\sigma_i^2}{2} + \frac{X_i^{n+1} - X_i^n}{kX_i^n} \right) \frac{u_{i,j+1}^n - u_{i,j-1}^n}{2h} \\ &- r_i u_{i,j}^n + \sum_{l \neq i} q_{il}(\tilde{u}_{l_i,j}^n - u_{i,j}^n), \end{aligned} \tag{10.76}$$

where

$$\tilde{u}^n_{l_i,j} \approx P_{l,i}(x_j, \tau^n) = P_l\left(x_j + \ln \frac{X^n_i}{X^n_l}, \tau^n\right),$$

are obtained by linear interpolation of values $u^n_{l,j}$ at the point $x_j + \ln \frac{X^n_i}{X^n_l}$ known from the previous time level n.

We have studied the stability of the proposed explicit scheme following the von Neumann analysis approach originally applied to schemes with constant coefficients. However, such approach can be used also for the variable coefficients case by freezing at each level (see [61, p. 59], [24, 34]).

In order to avoid notational misunderstanding among the imaginary unit with the regime index i used in previous section, here we denote the regime index by R. An initial error vector for every regime g^0_R, $R = 1, \ldots, I$, is expressed as a finite complex Fourier series, so that at x_j the solution $u^n_{i,j}$ can be rewritten as follows

$$u^n_{R,j} = g^n_R e^{ij\theta}, \quad j = 1, \ldots, M-1, \; R = 1, \ldots, I, \tag{10.77}$$

where $i = (-1)^{1/2}$ is the imaginary unit and θ is phase angle. Then the scheme is stable if for every regime $R = 1, \ldots, I$ the amplification factor $G_R = \frac{g^{n+1}_R}{g^n_R}$ satisfies the relation

$$|G_R| \leq 1 + Kk = 1 + O(k), \tag{10.78}$$

where the positive number K is independent of h, k and θ, see [60, p. 68], [61, p. 50].

After some manipulations, one gets

$$|G| \left|1 - \frac{i \sin\theta}{h}\right| \leq |A(k, h, \theta)| + C(n)k,$$

where $C(n) = \left|\frac{g^n_{l_0(n)}}{g^n}\right| |q_{R,R}|$ is independent of θ, h and k and depends only on the index n.

$$|A(k, h, \theta)|^2 = \left(1 - 2\frac{\sigma^2 k \sin^2 \frac{\theta}{2}}{h^2} - (r-q)k\right)^2 + \frac{\sin^2\theta}{h^2}\left(\left(r - \frac{\sigma^2}{2}\right)^2 k^2 - 2k\left(r - \frac{\sigma^2}{2}\right) + 1\right).$$

Thus, in agreement with (10.78) the scheme is stable, if

$$\begin{cases} \sigma^2 k\left((r-q)-\frac{\sigma^2}{h^2}\right)-\sigma^2 \leq 0, \\ \left(\left(r-\frac{\sigma^2}{2}\right)^2+(r-q)\sigma^2\right)k-2r \leq 0. \end{cases} \tag{10.79}$$

Summarizing the following result can be established:

Theorem 10.4 *With previous notation the scheme* (10.76) *is conditionally stable under the constraint*

$$k \leq \min_{1 \leq R \leq I}\left(\frac{h^2}{\sigma_R^2+(r_R-q_{R,R})h^2}, \frac{2r_R}{\left(r_R-\frac{\sigma_R^2}{2}\right)^2+(r_R-q_{R,R})\sigma_R^2}\right). \tag{10.80}$$

Stability conditions on step sizes are found and proven by numerical experiments (see Figs. 10.5 and 10.6). Consistency of the proposed scheme is studied in [25].

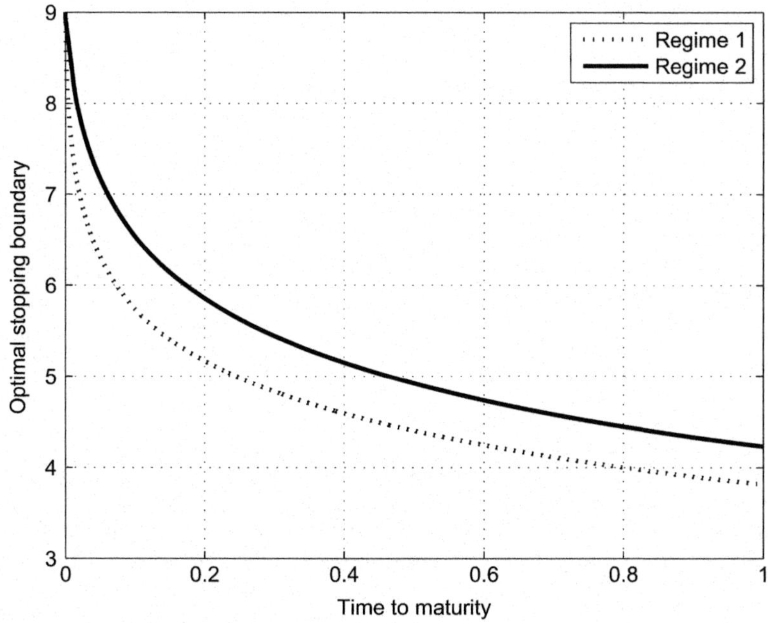

Fig. 10.5 Optimal stopping boundary for regime 1 and regime 2 (stability condition is fulfilled)

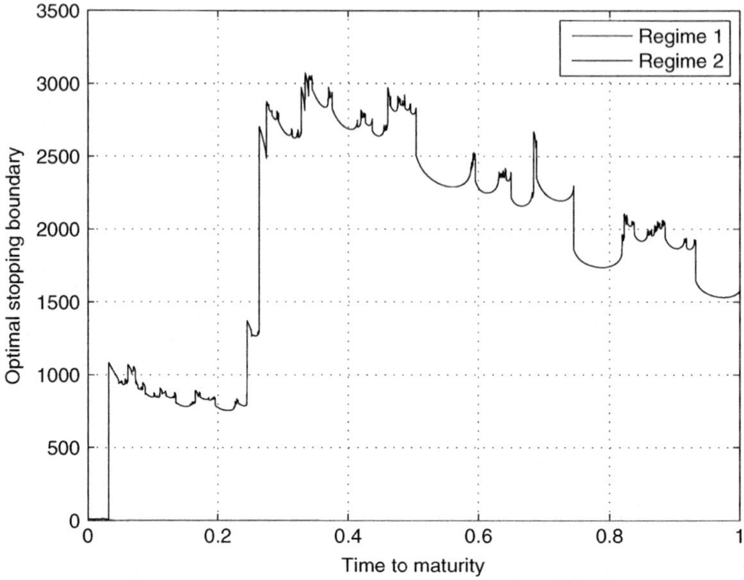

Fig. 10.6 Optimal stopping boundary for regime 1 and regime 2 (stability condition is violated)

10.4 Rationality Parameter Approach

Recently, in [30] a new nonlinear BS model that takes into account irrational exercise behaviour is proposed. We confirm numerically that the solution of the irrational problem proposed in [30] for large values of rationality parameter tends to the solution of the rational American option problem. This technique has been successfully applied to a regime switching model described in previous subsection.

With the previous notation, let $(f^\lambda)_{\lambda>0}$ be a family of positive deterministic intensity functions. For each $\lambda > 0$, let the stochastic intensity process be given by

$$\mu_t^\lambda = f^\lambda\left((E - S_t)^+ - P^\lambda(t, S_t)\right),$$

where $P^\lambda(t, S_t) = P^\lambda(t, S_t; \tau^*(\lambda)$ and $\tau^*(\lambda)$ is the exercise strategy of the American put given as the first jump time of a point process with intensity μ^λ. Let $\nu_\lambda(x) = 1_{(x<0)} \sup_{y\le x} f^\lambda(y) + 1_{(x\ge 0)} \sup_{y\ge x} f^\lambda(y)$ and assume that

- $\nu_\lambda(0+) \to \infty$ as $\lambda \to \infty$.
- There exists a function $\epsilon : (0, \infty) \to (0, \infty)$ such that $\nu_\lambda(-\epsilon(\lambda))) \to 0$ and $\epsilon(\lambda)\nu_\lambda(0-) \to 0$ as $\lambda \to \infty$.

Then λ is a rationality parameter in the sense that for every $t \in [0, T]$ we have that $P^\lambda(t, S_t)$ tends to $P^A(t, S_t)$ when $\lambda \to \infty$. Moreover, if f^λ is increasing then $f^\lambda = \nu_\lambda$.

We first consider the two cases proposed in [30] and we additionally propose two alternative expressions:

$$f_1^\lambda(x) = \begin{cases} \lambda, & \text{for } x \geq 0, \\ 0, & \text{for } x < 0; \end{cases} \qquad f_2^\lambda(x) = \lambda e^{\lambda^2 x}. \tag{10.81}$$

We have proposed two intensity functions that are the smooth analogue of the stepwise function (10.81):

$$f_3^\lambda(x) = \frac{2\lambda}{1+e^{-\lambda^2 x}}, \quad f_4^\lambda(x) = \lambda\left(1+\frac{2}{\pi}\arctan\lambda^2 x\right). \tag{10.82}$$

This irrational behaviour model of the American put option is studied and solved numerically in the following subsection. Then we apply this approach to model of the American option under regime-switching.

10.4.1 Irrational Behaviour Model of American Put Option

The following nonlinear Black-Scholes equation in the unbounded domain $\Omega = (0, +\infty) \times (0, T)$ is considered (sub-index and super-index of f are skipped):

$$\frac{\partial P}{\partial \tau} = \frac{\sigma^2}{2}S^2\frac{\partial^2 P}{\partial S^2} + (r-q)S\frac{\partial P}{\partial S} - rP + f\left((E-S)^+ - P\right)\left((E-S)^+ - P\right). \tag{10.83}$$

when $S \to 0$, the standard condition for American options, $P(0,\tau) = E$, is no longer valid in the irrational case, as prices bellow exercise price may occur due to irrational exercise, which is more evident when the rationality parameter tends to zero. The typical boundary condition for European options $P(0,\tau) = Ee^{-r\tau}$ is not consistent with the equation for $\lambda \to -\infty$, as the solution converges to the one of the rational case of American options. Since Eq. (10.83) is nonlinear and describes option pricing with rationality parameter, a new boundary condition has to be established. Therefore, we propose to pass to the limit in Eq. (10.83) when $S \to 0$:

$$\frac{\partial P}{\partial \tau}(0,\tau) = -rP(0,\tau) + f\left(E - P(0,\tau)\right)\left(E - P(0,\tau)\right).$$

The previous equation allows to adapt the option price when $S = 0$ according to rationality of the holder.

We introduce the new variable

$$x = \ln\frac{S}{E}, \quad u(x,\tau) = \frac{P(S,\tau)}{E}.$$

Then the original problem is transformed to the following problem for $x \in \mathbb{R}$:

$$\frac{\partial u}{\partial \tau} = \frac{\sigma^2}{2}\frac{\partial^2 u}{\partial x^2} + \left(r - q - \frac{\sigma^2}{2}\right)\frac{\partial u}{\partial x} - ru + f\left(E(1-e^x)^+ - Eu\right)\left((1-e^x)^+ - u\right). \tag{10.84}$$

For the transformed problem the numerical solution is constructed by the explicit FDM.

In the previous notation, let us denote $u_j^n \approx u(x_j, \tau^n)$, then the explicit finite difference scheme can be written in the form

$$u_j^{n+1} = b_1 u_{j-1}^n + b_2 u_j^n + b_3 u_{j+1}^n + kf_j^n, \quad j = 1, \ldots, N_x - 1, \tag{10.85}$$

where

$$\begin{aligned} b_1 &= \frac{\sigma^2}{2}\frac{k}{h^2} - \left(r - q - \frac{\sigma^2}{2}\right)\frac{k}{2h}, \\ b_2 &= 1 - \left(\sigma^2 \frac{k}{h^2} + rk\right), \\ b_3 &= \frac{\sigma^2}{2}\frac{k}{h^2} + \left(r - q - \frac{\sigma^2}{2}\right)\frac{k}{2h}. \end{aligned} \tag{10.86}$$

Note that under conditions

$$h < \frac{\sigma^2}{\left|r - q - \frac{\sigma^2}{2}\right|}, \quad k < \frac{h^2}{\sigma^2 + rh^2}, \quad k \le k_i,\ i = 1, \ldots 4, \tag{10.87}$$

where depending on the chosen rationality function

$$k_1 = \frac{1}{r + \lambda}, \quad k_2 = \frac{1}{\left(r + \lambda e^{\lambda^2 r k_2}\right)}, \quad k_3 = \frac{1}{r + 2\lambda}, \quad k_4 = \frac{1}{r + \lambda + \frac{2}{\pi E \lambda}}, \tag{10.88}$$

the coefficients b_1, b_2 and b_3 defined in (10.86) are positive and rationality term $k||f^n||$ is bounded.

In order to study the stability of the scheme we first choose the minimum index m, such that $u_m^{n+1} = ||u^{n+1}||$. Note that if $m = 0$ or $m = N_x$, then the scheme is stable by the definition.

Suppose for the index $1 \le m \le N_x - 1$, then taking into account that all coefficients are positive, one gets

$$|u_m^{n+1}| = |b_1 u_{m-1}^n + b_2 u_m^n + b_3 u_{m+1}^n + kf_m^n| \le (1 - rk)||u^n|| + k|f_m^n|,$$

The connection between $(n+1)$-th and n-th level is obtained:

$$||u^{n+1}|| = |u_m^{n+1}| \le ||u^n|| + k||f^n||. \tag{10.89}$$

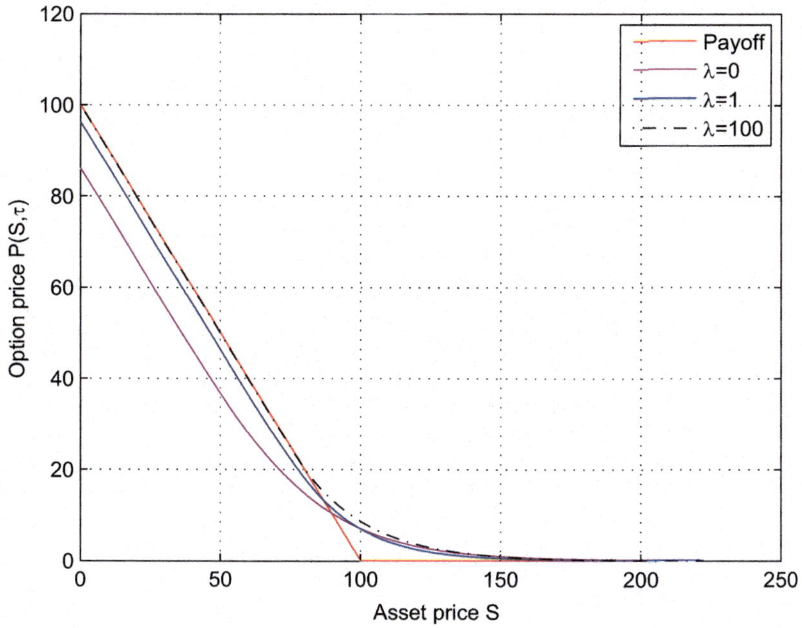

Fig. 10.7 Numerical solution for the intensity function belonging to family f_2 (10.81) for various values of λ

Therefore, under conditions (10.87), the scheme (10.85) is stable.

Assuming that $u(x, \tau)$ is continuously differentiable four times with respect to x and twice with respect to τ and following the procedure of consistency study, one finds that the truncation error behaves

$$T_j^n(\tilde{u}) = O(h^2) + O(k).$$

The aim of this part of work is to study numerically the rationality parameter approach and to prove the convergence of the solution to American option price with growing rationality parameter λ, that is presented in Fig. 10.7.

10.4.2 Rationality Parameter Approach for Regime-Switching Model

For an intensity function $f : [-E, E] \to [0, \infty)$ in the regime switching setting we assume that the relation between the profitability and the stochastic exercise intensity is $f((E-S)^+ - V_i(S, \tau))$ for each regime. After incorporating this term to the system of PDEs satisfied by the option price in describing the regime switching model (10.69) one gets for $i = 1, \ldots, I$,

$$\frac{\partial V_i}{\partial \tau} = \frac{\sigma_i^2}{2} S^2 \frac{\partial^2 V_i}{\partial S^2} + r_i S \frac{\partial V_i}{\partial S} - r_i V_i + f\left((E-S)^+ - V_i\right)\left((E-S)^+ - V_i\right)$$
$$+ \sum_{l \neq i} q_{i,l}(V_l - V_i), \quad S > 0,\ 0 < \tau \leq T. \tag{10.90}$$

In order to construct an effective FDM with constant coefficients in the differential part, let us introduce the following normalized transformation

$$x = \ln \frac{S}{E}, \quad u_i = \frac{V_i(S,\tau)}{E}, \quad i = 1, \ldots, I.$$

Then, problem (10.90) takes the following equivalent form:

$$\frac{\partial u_i}{\partial \tau} = \frac{\sigma_i^2}{2} \frac{\partial^2 u}{\partial x^2} + \left(r_i - \frac{\sigma_i^2}{2}\right) \frac{\partial u_i}{\partial x} - r_i u_i + \sum_{l \neq i} q_{i,l}(u_l - u_i) \tag{10.91}$$
$$+ f\left(E(1-e^x)^+ - E u_i\right)\left((1-e^x)^+ - u_i\right), \quad i = 1, \ldots, I.$$

The resulting nonlinear system of PDEs is solved by a weighted FDM, also known as θ-method. In order to avoid the need of an iterative method for the nonlinear system, the term with rationality parameter and the coupling term are treated explicitly. Next, the resulting linear system is solved by the Thomas algorithm. Stability conditions for the numerical scheme are studied by using the von Neumann approach.

Consistency of the θ-scheme for the PDE system is established and the truncation error takes the following form

$$T_j^n(\tilde{u}_i) = (1-2\theta)k \frac{\partial^2 u_i}{\partial \tau^2}(x_j, \tau^{n+\theta}) + O(k^2) + O(h^2) \quad \forall i = 1, \ldots, I.$$

Numerical experiments illustrate the efficiency and accuracy of the proposed method. In order to find computational convergence rate, a series of numerical results has been provided with fixed time step and various spatial steps h. The convergence rate γ_h has been calculated by formula

$$\gamma_h = \log_2 \frac{\|U_{h/2} - U_h\|}{\|U_{h/4} - U_{h/2}\|}, \tag{10.92}$$

for the proposed scheme with $\theta = 0,\ 0.5,\ 1$. The results are collected in Table 10.8 showing the expected orders for the approximation with $\lambda = 10^3$ and various intensity function families.

Table 10.8 Convergence rate in space of the proposed θ-scheme for $\lambda = 10^3$

	Regime 1			Regime 2		
θ	0	0.5	1	0	0.5	1
f_1	2.0084	2.0003	2.0007	2.0143	2.0004	2.0015
f_2	2.0083	2.0003	2.0005	2.0142	2.0007	2.0013
f_3	2.0079	2.0002	2.0001	2.0156	2.0005	2.0004

Table 10.9 Convergence rate in time of the proposed θ-scheme for $\lambda = 10^3$

	Regime 1			Regime 2		
θ	0	0.5	1	0	0.5	1
f_1	1.0013	1.7795	1.0007	1.0013	1.8889	1.0010
f_2	1.0009	1.7802	1.0007	1.0009	1.9017	1.0007
f_3	1.0010	1.8543	1.0001	1.0010	1.8943	1.0000

An analogous formula can be used in order to estimate the convergence rate in time, γ_k, for a fixed space step h:

$$\gamma_k = \log_2 \frac{\|U_{k/2} - U_k\|}{\|U_{k/4} - U_{k/2}\|}. \tag{10.93}$$

The convergence rates γ_k of the proposed method for various intensity function families (10.81) and (10.82) are presented in Table 10.9. The numerical convergence rate are in agreement with the theoretical study of consistency.

10.5 A Semi-Discretization Technique for Multi-Asset Option Pricing Problems

10.5.1 Removing Transformation Techniques for Multi-Asset Option Pricing

This section mainly covers removing the cross derivative terms in the formulation of an option pricing problem where the exercise value depends on more than one risky asset.

Basically the techniques for transformations aim at constructing the corresponding PDE with constant coefficients and also at removing the mixed derivative terms from the structure. Each of these transformations have some pros and cons.

The merit of transformations for removing the cross derivative terms is that the re-constructed PDE is easy to handle numerically since it has fewer number of terms which obviously ends in fewer mesh nodes in stencil in contrast to its non-transformed version. Furthermore, it may avoid the oscillation and spurious behaviors [39, 50] of the numerical solutions in the presence of mixed derivatives.

A transformation of spatial variables based on obtaining the canonical form of the second order PDE [31] can be used for the two correlated assets problems.

Technically speaking, it should be noticed that enforcing any types of transformations would change the initial and boundary conditions for the Black-Scholes multi-dimensional PDE problem.

In this section, we handle the new boundary conditions in order to obtain accurate and stable numerical solutions.

Considering a system of stochastic ordinary differential equations for an option pricing problem with two state variables, the authors in [38] used a transformation (with the Itô lemma and standard arbitrage arguments) that makes the instantaneous standard deviation of each to be constant. To be more precise, they suggest that a transformation be carried out to diagonalize a correlation matrix (tensor) in order to remove the cross derivative terms. This corresponds to a stretching and rotation of the coordinate system.

In the well-known stochastic volatility model (Heston model) [35], two space variables are existed in the presence of a cross derivative term. Such models are basically in the form of a the partial integro-differential equations (PIDEs) while they do not endure the pitfall of not capturing features like heavy tails and asymmetries observed in market-data log-returns densities unlike the normality of the log returns considered originally by Black-Scholes.

In [10], the authors applied two transformations in order to remove the reaction term and the cross derivative term from the Heston model and construct an elliptic form of it which is defined on rhomboid domain but with fewer terms which yielded to the construction of a stable and accurate numerical scheme.

One of the state-of-the-art techniques to remove the cross derivative terms is the use of eigenvalue decomposition [43, 52] which is also an algebraic transformation. In this technique, the eigenvalue decomposition of the diffusion matrix is constructed and used for deriving the multi-asset option pricing PDE without mixed derivative terms. We recall that the diffusion matrix in a multidimensional second order PDE is a symmetric matrix containing the coefficients of the second order derivatives in the PDE.

The multi-asset Black-Scholes PDE is expressed as follows [23, 62]:

$$\frac{\partial V}{\partial \tau} = \frac{1}{2} \sum_{i,j=1}^{M} \rho_{ij}\sigma_i\sigma_j S_i S_j \frac{\partial^2 V}{\partial S_i \, \partial S_j} + \sum_{i=1}^{M} (r - q_i) S_i \frac{\partial V}{\partial S_i} - rV, \tag{10.94}$$

where T, V, S_i, q_i, r, σ_i ρ are the maturity, the value of the option price, the i-th asset, the constant dividend yield of i-th asset, risk-free rate, the i-th volatility, the correlation parameter, respectively, while $\tau = T - t$ and $\rho_{ii} = 1$, $\rho_{ij} = \rho_{ji}$, $i \neq j$, and

$$|\rho_{ij}| \leq 1. \tag{10.95}$$

The mixed derivative terms appearing in (10.94) show the correlation among the prices of the assets S_i.

The logarithmic transformation [8] could transform the multi-dimensional PDE (10.94) into a PDE with constant coefficients as follows:

$$x_i = \frac{\log S_i}{\sigma_i}, \qquad 1 \le i \le M, \tag{10.96}$$

with $V(S, \tau) = W(X, \tau)$, where $X = (x_1, x_2, \ldots, x_M)^\top$, and we may achieve

$$\frac{\partial W}{\partial \tau} = \frac{1}{2}\sum_{i,j=1}^{M} \rho_{ij} \frac{\partial^2 W}{\partial x_i\, \partial x_j} + \sum_{i=1}^{M}\left(r - q_i - \frac{\sigma_i^2}{2}\right)\frac{1}{\sigma_i}\frac{\partial W}{\partial x_i} - rW, \tag{10.97}$$

Note that for a M-dimensional Black-Scholes PDE, the number of cross derivative terms is

$$\frac{1}{2}(M-1)M. \tag{10.98}$$

This evidently shows that by increasing the number underlying assets, the number of mixed derivative terms gets bigger which could cause several certain issues in the process of solving (10.94).

Apart from the appearance of instability and inaccuracy due to the presence of the cross derivative terms mentioned above, the number of stencil nodes or matrices that must be filled and computed in the development of the numerical schemes would be higher and subsequently relinquish further computational burden [67].

Here the main objective is to remove the mixed derivatives so as to reduce unsuitable instability drawbacks for the (10.94). Essentially, this may be pursued by applying transformations. This new transformation is different from the eigenvalue transformation and it is based on $LDL^\top$ factorization.

Toward this goal, let us consider the symmetric positive semi-definite correlation matrix [55]:

$$R = (\rho_{ij})_{1 \le i,j \le M}, \tag{10.99}$$

as the diffusion matrix corresponding to the PDE (10.97). Accordingly, in this section we present a general way by means of an easy to implement transformation based on Gaussian elimination and pivoting strategies [37] to remove the cross derivative terms.

Let us first recall the definition of the $LDL^\top$ factorization in what follows. If R be a symmetric positive semidefinite matrix in $\mathbb{R}^{M\times M}$. Then, there exists a unit lower triangular matrix L and a diagonal matrix $D = (d_{ij})$ in $\mathbb{R}^{M\times M}$ with $d_{ii} \ge 0$, $1 \le i \le M$, such that [33]:

$$R = LDL^\top. \tag{10.100}$$

Generally speaking, if the matrix R is only positive semidefinite, then (10.100) is not valid, but when R is positive definite then it is unique.

Here in order to have stable computation of this factorization [36], basically a permuted version of (10.100) on the matrix R should be performed, viz, this permuted factorization could be written as comes next:

$$PRP^{\top} = LDL^{\top}, \tag{10.101}$$

where P is a permutation matrix, $|l_{ij}| \leq 1$ and

$$d_{11} \geq d_{22} \geq \cdots \geq d_{nn} \geq 0. \tag{10.102}$$

Now in order to remove the cross derivative terms in the parabolic second-order constant-coefficient PDE (10.96), we take into account a linear transformation as follows:

$$Y = CX, \qquad C = (c_{ij})_{1 \leq i,j \leq M}, \tag{10.103}$$

where C is the matrix that should be computed such that the mixed derivative terms get vanished.

Now by applying (10.103), the PDE (10.97) reads

$$\frac{\partial U}{\partial \tau} = \frac{1}{2} \sum_{i,j=1}^{M} (\mathbf{c}_i R \mathbf{c}_j^{\top}) \frac{\partial^2 U}{\partial y_i \, \partial y_j} + \sum_{i,j=1}^{M} \left(\frac{r - q_j - \sigma_j^2/2}{\sigma_j} \right) c_{ij} \frac{\partial U}{\partial y_i} - rU, \tag{10.104}$$

where $U(Y, \tau) = W(X, \tau)$ and $\mathbf{c}_i = (c_{i1}, c_{i2}, \cdots, c_{iM})$ is the ith row vector of matrix C. Here, $\mathbf{c}_i$ denotes the ith row of matrix $L^{-1}P$:

$$\mathbf{c}_i = (L^{-1}P)_i. \tag{10.105}$$

Using

$$(L^{-1}P)R(L^{-1}P)^{\top} = D, \tag{10.106}$$

we obtain

$$\mathbf{c}_i R \mathbf{c}_j^{\top} = \begin{cases} 0, & i \neq j, \\ d_{ii}, & i = j. \end{cases} \tag{10.107}$$

Hence, Eq. (10.104) becomes:

$$\frac{\partial U}{\partial \tau} = \frac{1}{2} \sum_{i=1}^{M} (d_{ii}) \frac{\partial^2 U}{\partial y_i^2} + \sum_{i=1}^{M} \left(\sum_{j=1}^{M} \frac{(r - q_j - \sigma_j^2/2) c_{ij}}{\sigma_j} \right) \frac{\partial U}{\partial y_i} - rU. \tag{10.108}$$

Here we remark that the discussed transformation based on the permuted Cholesky factorization has several upsides from the computational cost and stability points of view, but it is not the only way to eliminate mixed derivative terms. In fact, if one uses the standard diagonalization transform of

$$R = FDF^{-1}, \tag{10.109}$$

even when $F^{-1} = F^T$ is available, the transformation

$$C = F^{-1}, \tag{10.110}$$

also transforms Eq. (10.94) into a PDE without cross derivative terms.

Example 10.4 In this experiment, we consider the general multi-asset option pricing problem (10.94) with $M = 7$ underlying assets, where the correlation matrix R is given by

$$R = \begin{pmatrix} 1.00 & -0.65 & 0.25 & 0.2 & 0.25 & -0.05 & 0.05 \\ -0.65 & 1.00 & 0.5 & 0.1 & 0.25 & 0.11 & -0.016 \\ 0.25 & 0.5 & 1.00 & 0.37 & 0.25 & 0.21 & 0.076 \\ 0.2 & 0.1 & 0.37 & 1.00 & 0.25 & 0.27 & 0.13 \\ 0.25 & 0.25 & 0.25 & 0.25 & 1.00 & 0.14 & -0.04 \\ -0.05 & 0.11 & 0.21 & 0.27 & 0.14 & 1.00 & 0.19 \\ 0.05 & -0.016 & 0.076 & 0.13 & -0.04 & 0.19 & 1.00 \end{pmatrix}, \tag{10.111}$$

with the parameters

$$\sigma = (\sigma_1, \ldots, \sigma_7) = (0.25, 0.35, 0.20, 0.25, 0.20, 0.21, 0.27), \tag{10.112}$$

$r = 0.045$, $T = 1$ year, and

$$q = (q_1, \ldots, q_7) = (0.05, 0.07, 0.04, 0.06, 0.04, 0.03, 0.02). \tag{10.113}$$

Applying the factorization (10.101), (10.96) and (10.103), one gets

$$D = diag(1.000, 0.998, 0.960, 0.907, 0.861, 0.787, 0.00786). \tag{10.114}$$

Subsequently, the transformation matrix which is a lower triangular matrix can be expressed as:

$$C = L^{-1}P = \begin{pmatrix} 1.000 & 0 & 0 & 0 & 0 & 0 & 0 \\ 0.050 & 0 & 0 & 0 & 0 & 1.000 & 0 \\ -0.060 & 0 & 0 & 0 & 0 & -0.190 & 1.000 \\ -0.260 & 0 & 0 & 0 & 1.000 & -0.170 & 0.085 \\ -0.210 & 0 & 1.000 & 0 & -0.170 & -0.190 & -0.036 \\ -0.110 & 0 & -0.270 & 1.000 & -0.130 & -0.190 & -0.074 \\ 0.900 & 1.000 & -0.680 & 0.021 & -0.330 & 0.120 & -0.017 \end{pmatrix}. \tag{10.115}$$

Now, the corresponding problem (10.94) is transformed into the following compact notation form

$$\frac{\partial U}{\partial \tau} = \frac{1}{2}(D\nabla) \cdot \nabla U + (CQ) \cdot \nabla U - rU, \tag{10.116}$$

where $\nabla = \left(\frac{\partial}{\partial y_1}, \frac{\partial}{\partial y_2}, \ldots, \frac{\partial}{\partial y_M}\right)^\top$, $U(Y, \tau) = V(S, \tau)$ and $Q = (Q_1, Q_2, \ldots, Q_M)^\top$. The interesting point is that the original multi-asset PDE with 37 terms has now been re-constructed into one with only 16 terms.

In the rest of this section, we discuss the cases when the diffusion matrix is symmetric possibly indefinite. This would be practical in the general case of solving PDEs with cross derivative terms [42]. Let us consider the equation

$$\sum_{i,j=1}^{M} a_{ij} \frac{\partial^2 v}{\partial x_i \partial x_j} + \sum_{i=1}^{M} b_i \frac{\partial v}{\partial x_i} + cv = 0, \tag{10.117}$$

where $A = (a_{ij})_{1 \le i,j \le M}$ is a real symmetric matrix, $\mathbf{b} = (b_1, \ldots, b_M)^\top \in \mathbb{R}^M$ and $c \in \mathbb{R}$.

In this case, the matrix A could be indefinite. So, the factorization (10.100) breaks [37] but we may use an alternative as discussed below which is called as Bunch-Kaufman factorization [5]. This approach does not always provide a diagonal factorization of A, but only a block-diagonal matrix B with 1×1 or 2×2 diagonal blocks such that

$$PAP^\top = LBL^\top, \tag{10.118}$$

where the permutation matrix P provides a partial pivoting strategy. Thus, one gets a more efficient method than other diagonal pivoting strategies as complete pivoting. In this way, only a part of mixed derivative terms are removed. However, with the use of eigenvalues decomposition on the final 2×2 block, we may remove all the mixed derivative terms and obtain a corresponding PDE without such terms.

The next example is related to multi-asset cross currency option pricing [38, Chap. 29] with indefinite sample correlation matrix.

Example 10.5 ([67]) Consider Eq. (10.94) for $M = 3$, with indefinite sample correlation matrix

$$R = \begin{pmatrix} 1 & \frac{3}{10} & \frac{9}{10} \\ \frac{3}{10} & 1 & \frac{9}{10} \\ \frac{9}{10} & \frac{9}{10} & 1 \end{pmatrix}. \tag{10.119}$$

Using Bunch-Kaufman strategy, one gets the transformation matrix C and the resulting matrix B,

$$C = L^{-1}P = \begin{pmatrix} 1 & 0 & 0 \\ -\frac{3}{10} & 1 & 0 \\ -\frac{9}{13} & -\frac{9}{13} & 1 \end{pmatrix}, \qquad B = diag\,(1, 91/100, -16/65) = D. \tag{10.120}$$

Hence, the original partial differential equation is transformed into a new one without cross derivative terms.

10.5.2 Stability and Numerical Example

Options with multi assets are based upon more than one underlying asset, unlike the well-known standard vanilla options. In this situation, due to the curse of dimensionality which is of exponential growth, the complexity of the problem grows when the dimensionality increases. That is to say, the number of unknowns for solving the corresponding PDE simultaneously grows exponentially [63].

One of the main restrictions here in the process of solving a multi-asset option pricing problem is that the mixed derivative of the solution has to be bounded and its presence could cause instability and further computational burdensome.

Efficient pricing of American and European options that are dependent on more than one asset is discussed in this section. The holder of a multi-asset contract has the right to buy a set of assets if the conditions are profitable which is known as a basket of assets.

To formulate this problem, we may choose $\mathbf{S} = (S_1, \ldots, S_M)$ to be the vector consisting the asset prices, where M is the number of assets in a portfolio while $P(\mathbf{S}, \tau)$ is the value of the option pricing.

This class of basket options (for put) can be described by a general equation for the contract function [48]

$$P(\mathbf{S}, 0) = \left(E - \sum_{i=1}^{M} \alpha_i S_i\right)^{+}, \tag{10.121}$$

where E is the exercise price of the complete basket and α_i are the percentages in the set of assets. The option price $P(\mathbf{S}, \tau)$ is the solution of the following PDE problem

$$\frac{\partial P}{\partial \tau} = \frac{1}{2} \sum_{i=1,j=1}^{M} \rho_{ij}\sigma_i\sigma_j S_i S_j \frac{\partial^2 P}{\partial S_i \partial S_j} + \sum_{i=1}^{M}(r - q_i)S_i \frac{\partial P}{\partial S_i} - rP + F(P), \tag{10.122}$$

$$S_i > 0, \quad i = 1, \ldots, M, \quad 0 < \tau \leq T,$$

where σ_i is the volatility of S_i, $\rho_{i,j}$ is the correlation between S_i and S_j, r is the risk free rate, q_i is the constant dividend yield of i-th asset and $F(P)$ is the rationality parameter term.

In the formulation (10.122), we applied the penalty approach [50] in order to handle the American options by transforming the free boundary value problem into a nonlinear PDE. In fact, due to opportunity to exercise at any time to maturity, American option pricing problems introduce a free exercise boundary which is more difficult than European options.

In this work, we consider the rationality term as follows [30]:

$$F(P) = \lambda\left(P(\mathbf{S}, 0) - P(\mathbf{S}, \tau)\right)^{+}, \tag{10.123}$$

which is a simpler version of the following general form

$$F(P) = f^{\lambda}\left(\left(E - \sum_{i=1}^{M} \alpha_i S_i\right)^{+} - P(\mathbf{S}, \tau)\right) \cdot \left(\left(E - \sum_{i=1}^{M} \alpha_i S_i\right)^{+} - P(\mathbf{S}, \tau)\right), \tag{10.124}$$

where $f^{\lambda}(x)$ is an intensity function and λ is a rationality parameter.

It is required to state that the boundary of a M-dimensional Black-Scholes PDE in option pricing is the solution of the $(M-1)$-dimensional problem while in infinity they approach to zero. Furthermore, at each boundary $S_i = 0$ we have

$$P(S_1, \ldots, S_i \to \infty, \ldots, \tau) = 0. \tag{10.125}$$

There are several approaches to value this option pricing problem in the presence of multi assets using finite difference, finite element schemes and Monte-Carlo method [32]. The most challenging issue in dealing with such nonlinear PDEs is to control the boundedness of the numerical solution, i.e., stability of the numerical scheme when the size of the discretized system gets bigger by considering higher number of assets and nodal points for discretization.

As discussed in the second section of this chapter, another problem is the presence of the cross derivative terms which cause instability and oscillation in the process of solving (10.122) numerically. Thus, the objective of this section is to address a numerically stable finite difference schemes for multi-asset American/European option pricing problems based on the semi-discretization technique.

The matrix involving the second order partial derivative terms, so called the diffusion matrix, can be diagonalized by means of its orthogonal transformation. This technique could be applied to remove the cross derivative terms as it has been done in [43].

But in this section, we follow the suggested $LDL^\top$ factorization given previously in the second section so as to construct a corresponding nonlinear PDE without mixed derivative terms.

In [69] a semi-discretized method has been applied for multi-asset problem under regime-switching. In that work the spatial step sizes are fixed, and so the size of the matrix A in order to obtain L-stability.

To keep going, we first do a same procedure as in the second section by obtaining the corresponding PDE with constant coefficient and then the PDE without cross derivative terms. Thus using the dimensionless logarithmic substitution

$$x_i = \frac{1}{\sigma_i} \ln \frac{S_i}{E}, \ i = 1, \ldots, M, \quad V(\mathbf{x}, \tau) = \frac{P(\mathbf{S}, \tau)}{E}, \tag{10.126}$$

where $\mathbf{x} = [x_1, \ldots, x_M]^\top$, we obtain

$$\begin{gathered} \frac{\partial V}{\partial \tau} = \frac{1}{2} \sum_{i=1, j=1}^{M} \rho_{ij} \frac{\partial^2 V}{\partial x_i \partial x_j} + \sum_{i=1}^{M} \delta_i \frac{\partial V}{\partial x_i} - rV + \frac{1}{E} F(EV), \\ x_i \in \mathbb{R}, \quad i = 1, \ldots, M, \quad 0 < \tau \leq T, \end{gathered} \tag{10.127}$$

where $\delta_i = \frac{r - q_i - \frac{\sigma_i^2}{2}}{\sigma_i}$.

Now by applying the linear transformation discussed before based on the $LDL^\top$ factorization of the correlation matrix [13]

$$\mathbf{y} = [y_1, \ldots, y_M]^\top = C\mathbf{x}, \quad U(\mathbf{y}, \tau) = V(\mathbf{x}, \tau), \tag{10.128}$$

where $C = \left(c_{ij}\right)_{1 \leq i,j \leq M} = L^{-1}$, we can obtain the following simplified transformed nonlinear PDE for multi-asset option pricing problem

$$\frac{\partial U}{\partial \tau} = \frac{1}{2} \sum_{i=1}^{M} D_{ii} \frac{\partial^2 U}{\partial y_i^2} + \sum_{i=1}^{M} \left(\sum_{j=1}^{M} \delta_j c_{ij} \right) \frac{\partial U}{\partial y_i} - rU + \frac{1}{E} F(EU), \tag{10.129}$$

where the cross derivative terms have been removed. Under transformations (10.126) and (10.128) the initial condition (10.121) takes the form

$$U(\mathbf{y}, 0) = \left(1 - \sum_{i=1}^{M} \alpha_i e^{\sigma_i x_i} \right)^+, \tag{10.130}$$

where

$$\mathbf{x} = [x_1, \ldots, x_M]^\top = C^{-1}\mathbf{y}. \tag{10.131}$$

For dealing with the above time-dependent PDEs, one way is the method of lines based on the semi-discretization with respect to spatial variables which results in a system of (linear or nonlinear) ordinary differential equations in time with the corresponding matrix of coefficients A.

Then semi-discretization of Eq. (10.129) is obtained by using the central difference approximation for the spatial derivatives, resulting in the system of (nonlinear) ordinary differential equations (ODEs) of the form

$$\begin{aligned} \frac{du_{j_1,\ldots,j_M}}{d\tau} &= \frac{1}{2}\sum_{i=1}^{M} D_{ii}\frac{u_{j_1,\ldots,j_i-1,\ldots,j_M} - 2u_{j_1,\ldots,j_i,\ldots,j_M} + u_{j_1,\ldots,j_i+1,\ldots,j_M}}{h_i^2} \\ &+ \sum_{i=1}^{M}\left(\sum_{j=1}^{M}\delta_j c_{ij}\right)\frac{u_{j_1,\ldots,j_i+1,\ldots,j_M} - u_{j_1,\ldots,j_i-1,\ldots,j_M}}{2h_i} \\ &- r u_{j_1,\ldots,j_M} + \frac{1}{E}F(Eu_{j_1,\ldots,j_M}), \end{aligned} \tag{10.132}$$

which its stencil has only $2M + 1$ mesh points in contrast to $M^2 + M + 1$ mesh points based on the recent finite difference method given in [69].

To construct conditions for finding stable solutions, we first consider that for $i = 1, \ldots, M$:

$$h_i = \beta_i h, \tag{10.133}$$

$$d_i = \frac{D_{ii}}{\beta_i^2}, \quad d = \sum_{i=1}^{M} d_i, \tag{10.134}$$

$$c_i = \sum_{j=1}^{M} \delta_j c_{ij}, \tag{10.135}$$

$$a_0 = -\frac{1}{h^2}\left(d + rh^2\right), \tag{10.136}$$

$$a_{+i} = \frac{1}{2h^2}\left(d_i + \frac{h}{\beta_i}c_i\right), \tag{10.137}$$

$$a_{-i} = \frac{1}{2h^2}\left(d_i - \frac{h}{\beta_i}c_i\right). \tag{10.138}$$

The nonlinear system (10.132) with the boundary and initial conditions can be presented in the following vector form

$$\begin{cases} \frac{d\mathbf{u}}{d\tau}(\tau) = A\mathbf{u}(\tau) + \lambda\left(\mathbf{u}(0) - \mathbf{u}(\tau)\right)^{+}, \\ \mathbf{u}(0) = [u_0(0), \ldots, u_N(0)]^{\top}, \end{cases} \tag{10.139}$$

where $u_j(0) = U(\boldsymbol{\xi}_j, 0) = \left(1 - \sum_{i=1}^{M} \alpha_i e^{\sigma_i x_i(\boldsymbol{\xi}_j)}\right)^{+}$, wherein $x_i(\boldsymbol{\xi}_j) = \left(C^{-1}\boldsymbol{\xi}_j\right)_i$ is the i-th entry of $C^{-1}\boldsymbol{\xi}_j$.

The entries of the matrix A are given by:

$$a_{ij} = \begin{cases} a_0, & \boldsymbol{\xi}_i \in \Omega,\ j = i, \\ a_{\pm 1}, & \boldsymbol{\xi}_i \in \Omega,\ j = i \pm 1, \\ a_{\pm m} & \boldsymbol{\xi}_i \in \Omega,\ j = i \pm \prod_{n=1}^{m-1}(N_n + 1),\ 2 \le m \le M, \\ 0, & \text{otherwise.} \end{cases} \tag{10.140}$$

Note that as the chosen artificial boundary conditions do not change with τ, then their derivative with respect to τ are zero which motivates the appearance of zeros in the corresponding rows of A.

If $k = \frac{T}{N_\tau}$, so $\tau^n = nk$, $n = 0, \ldots, N_\tau$. Thus for full discretization we have [16]:

$$\mathbf{u}(\tau^{n+1}) = e^{Ak}\mathbf{u}(\tau^n) + \lambda \int_0^k e^{As}\left(\mathbf{u}(0) - \mathbf{u}(\tau^{n+1} - s)\right)^{+} ds. \tag{10.141}$$

Now, by replacing $\mathbf{u}(\tau^{n+1} - s)$ by the known value $\mathbf{u}(\tau^n)$ corresponding to $s = k$, we attain

$$\int_0^k e^{As}\left(\mathbf{u}(0) - \mathbf{u}(\tau^{n+1} - s)\right)^{+} ds \approx \left(\int_0^k e^{As} ds\right)\left(\mathbf{u}(0) - \mathbf{u}(\tau^n)\right)^{+}. \tag{10.142}$$

We use the accurate Simpson's rule

$$\int_0^k e^{As} ds \approx k\varphi(A, k) \tag{10.143}$$

where $\varphi(A, k) = \frac{1}{6}\left(I + 4e^{A\frac{k}{2}} + e^{Ak}\right)$.

Denoting $\mathbf{u}^n = \mathbf{u}(\tau^n)$, we get the final explicit scheme

$$\mathbf{u}^{n+1} = e^{Ak}\mathbf{u}^n + k\lambda\varphi(A, k)\left(\mathbf{u}^0 - \mathbf{u}^n\right)^{+}, \quad \tau^n = nk,\ n = 0, \ldots, N_\tau - 1. \tag{10.144}$$

This is the proposed explicit full-discretized FD method for solving multi-asset option pricing problem which is stable under two conditions along spatial and temporal variables.

Coefficients a_{-i} and a_{+i}, $i = 1, \ldots, M$, depend on d_i and c_i, see (10.134) and (10.135) respectively. If step size h is chosen as

$$h \leq \min_{1 \leq i \leq M} \frac{d_i}{|c_i|}, \tag{10.145}$$

then a_{-i} and a_{+i} are non-negative. This is the first condition on the step size along spatial variable which could result in the positivity of the numerical schemes after several investigations on the structure of the schemes using bounds on matrix exponential and Metzler matrices.

Subsequently we may prove that $u_i^n \leq 1$, $0 \leq i \leq N$, $0 \leq n \leq N_\tau$ using the induction principle. We remark that

$$u_i^0 \leq 1, \tag{10.146}$$

and from (10.144) u_i^{n+1} is a function g_i on the arguments $u_0^n, \ldots, u_N^n$, given by

$$u_i^{n+1} = g_i(u_0^n, \ldots, u_N^n) = \left(e^{Ak}\right)_i \mathbf{u}^n + k\lambda \left(\varphi(A,k)\right)_i \left(\mathbf{u}^0 - \mathbf{u}^n\right)^+ . \tag{10.147}$$

And furthermore by the non-negativity of e^{Ak} and $\varphi(A,k)$ one gets

$$\frac{\partial g_i}{\partial u_j^n} \geq \left(e^{Ak}\right)_{ij} - k\lambda \left(\varphi(A,k)\right)_{ij}, \quad 0 \leq i,j \leq N. \tag{10.148}$$

Finally, we attain the following bound for the temporal step size

$$k < \frac{h^2}{d + (r+\lambda)h^2}. \tag{10.149}$$

Theorem 10.5 *With previous notation under conditions* (10.145) *and* (10.149) *the numerical solution* $\mathbf{u}^n$ *of the scheme* (10.144) *is non-negative and* $\|\cdot\|_\infty$*-stable, with* $\|\mathbf{u}^n\|_\infty \leq 1$ *for all values of* $\lambda \geq 0$ *and any time level* $0 \leq n \leq N_\tau$.

In what follows, we try to investigate the robustness of the proposed approach for solving several experiments in the presence of multi assets.

Example 10.6 The American basket call option of two assets is considered with the following parameters [51]

$$\sigma_1 = 0.12,\ \sigma_2 = 0.14,\ r = 0.03,\ \rho = 0.3,\ q_1 = 0.01,\ q_2 = 0.01,\ T = 0.5,\ E = 100. \tag{10.150}$$

Table 10.10 American basket call option price comparisons

Nodes	Proposed method	HOC
12 × 12	3.18982	2.86247
24 × 24	3.35338	3.27894
48 × 48	3.41344	3.35094

Table 10.11 Option price on an equidistant grid of $n \times n \times n$ nodes

n	P_h	P_l	KM (with rationality)
8	11.4957	12.862	12.394
16	13.3457	13.150	13.055
32	13.3272	13.221	13.235
64	13.2470	13.239	13.241
Reference value (P)	13.245		

In the following table, we include the results at $\mathbf{S} = (100, 100)$ for $\lambda = 100$, various spatial step sizes h and corresponding k under the discussed conditions. The numerical solution by high-order computational method of [51] is denoted by HOC (Table 10.10).

Example 10.7 As a numerical example we consider the European basket call option with no dividends and the following parameters (see [43, p. 76])

$$\sigma_1 = 0.3,\ \sigma_2 = 0.35,\ \sigma_4 = 0.4,\ r = 0.04,\ \rho_{ij} = 0.5, \alpha_i = \frac{1}{3},\ T = 1,\ E = 100. \tag{10.151}$$

The spot price is chosen to be $S_1 = S_2 = S_3 = E$. The reference value $P_{ref} = 13.245$ is computed by using an accurate Fast Fourier Transform technique (see [43, Chap. 4]). Since the considered option is of European style, penalty term is not necessary and λ is chosen to be zero.

The numerical results of the proposed method P_h are presented in the following table and compared with the sparse grid solution technique P_l on an equidistant grid of [43] and the method of [69] denoted by KM with rationality approach [30] (Table 10.11).

As could be seen from the numerical experiments, the theoretical bounds for the temporal and spatial variables are quite useful and necessary in solving real-life problems. The transformations made the process of solving this type of options quite easier and much more efficient. After spatial semi-discretization, the problem is fully discretized. We could handle American/European put/call options.

References

1. Abramowitz, M., Stegun, I.A.: Handbook of Mathematical Functions: With Formulas, Graphs, and Mathematical Tables. Dover Books on Mathematics. National Bureau of Standard, USA (1961)
2. Bates, D.S.: Jumps and stochastic volatility: exchange rate processes implicit Deutsche Mark options. Rev. Financ. Stud. **9**, 69–107 (1996)
3. Black, F., Scholes, M.: The pricing of options and corporate liabilities. J. Polit. Econ. **81**, 637–654 (1973)
4. Boyarchenko, S.I., Levendorskii, S.Z.: Option pricing for truncated Lévy processes. Int. J. Theor. Appl. Finance **3**(3), 549–552 (2000)
5. Bunch, J.R., Kaufman, L.: Some stable methods for calculating inertia and solving symmetric linear systems. Math. Comput. **137**, 163–179 (1977)
6. Carr, P., Geman, H., Madan, D.B., Yor, M.: The fine structure of asset returns: an empirical investigation. J. Bus. **75**(2), 305–332 (2002)
7. Casabán, M.C., Company, R., Jódar, L., Romero, J.V.: Double discretization difference schemes for partial integro-differential option pricing jump diffusion models. Abstr. Appl. Anal. **2012**, 1–20 (2012)
8. Clift, S.S., Forsyth, P.A.: Numerical solution of two asset jump diffusion models for option valuation. Appl. Numer. Math. **58**, 743–782 (2008)
9. Company, R., Jódar, L., Fakharany, M.: Positive solutions of European option pricing with CGMY process models using double discretization difference schemes. Abstr. Appl. Anal. **2013**, 1–11 (2013)
10. Company, R., Jódar, L., Fakharany, M., Casabán, M.-C.: Removing the correlation term in the option pricing Heston model: numerical analysis and computing. Abstr. Appl. Anal. **2013**, 1–11 (2013)
11. Company, R., Egorova, V.N., Jódar, L.: Solving American option pricing models by the front-fixing method: numerical analysis and computing. Abstr. Appl. Anal. 9 pp. (2014). Article ID 146745
12. Company, R., Egorova, V., Jódar, L.: Constructing positive reliable numerical solution for American call options: a new front-fixing approach. J. Comput. Appl. Math. **291**, 422–431 (2016)
13. Company, R., Egorova, V., Jodar, L., Soleymani, F.: A mixed derivative terms removing method in multi-asset option pricing problems. Appl. Math. Lett. **60**, 108–114 (2016)
14. Cont, R., Tankov, P.: Financial Modelling with Jump Processes. Chapman & Hall/CRC, Boca Raton, FL (2004)
15. Cont, R., Voltchkova, E.: A finite difference scheme for option pricing in jump diffusion and exponential Lévy models. SIAM J. Numer. Anal. **43**(4), 1596–1626 (2005)
16. Cox, S., Matthews, P.: Exponential time differencing for stiff systems. J. Comput. Phys. **176**(2), 430–455 (2002)
17. Cox, J.C., Ross, S.A.: The valuation of options for alternative stochastic processes. J. Financ. Economet. **3**, 145–166 (1976)
18. Crank, J.: Free and Moving Boundary problems. Oxford University Press, Oxford (1984)
19. Davis, P., Rabinowitz, P.: Methods of Numerical Integration, 2nd edn. Academic, New York (1984)
20. Duffy, D.J.: Finite Difference Methods in Financial Engineering: A Partial Differential Approach. Wiley, Chichester (2006)
21. Duffy, D.: Unconditionally stable and second-order accurate explicit finite difference schemes using domain transformation: part I one-factor equity problems. Technical report, SSRN, 2009. Available at SSRN: http://ssrn.com/abstract=1552926
22. Düring, B., Fournié, M.: High-order compact finite difference scheme for option pricing in stochastic volatility models. J. Comput. Appl. Math. **236**, 4462–4473 (2012)

23. Düring, B., Heuer, C.: High-order compact schemes for parabolic problems with mixed derivatives in multiple space dimensions. SIAM J. Numer. Anal. **53**, 2113–2134 (2015)
24. Düring, B., Fournié, M., Jüngel, A.: Convergence of a high-order compact finite difference scheme for a nonlinear Black-Scholes equation. ESAIM: Mathematical Modelling and Numerical Analysis - Modélisation Mathématique et Analyse Numérique **38**(2), 359–369 (2004)
25. Egorova, V.N., Company, R., Jódar, L.: A new efficient numerical method for solving American option under regime switching model. Comput. Math. Appl. **71**(1), 224–237 (2016)
26. Egorova, V., Tan, S.-H., Lai, C.-H., Company, R., Jódar, L.: Moving boundary transformation for American call options with transaction cost: finite difference methods and computing. Int. J. Comp. Math. **94**, 345–362 (2017)
27. Ehrhardt, M., Mickens, R.: A fast, stable and accurate numerical method for the Black-Scholes equation of American options. Int. J. Theor. Appl. Financ. **11**, 471–501 (2008)
28. Fakharany, M., Company, R., Jódar, L.: Positive finite difference schemes for a partial integro-differential option pricing model. Appl. Math. Comp. **249**, 320–332 (2014)
29. Fakharany, M., Company, R., Jódar, L.: Solving partial integro-differential option pricing problems for a wide class of infinite activity Lévy processes. J. Comput. Appl. Math. **296**, 739–752 (2016)
30. Gad, K.S.T., Pedersen, J.L.: Rationality parameter for exercising American put. Risks **3**(2), 103 (2015)
31. Garabedian, P.R.: Partial Differential Equations. AMS Chelsea Publishing, USA (1998)
32. Glasserman, P.: Monte Carlo Methods in Financial Engineering. Springer, New York (2003)
33. Golub, G.H., van Loan, C.F.: Matrix Computations, 4th edn. Johns Hopkins University Press, Baltimore (2012)
34. Gustafsson, B., Kreiss, H.O., Oliger, J.: Time-Dependent Problems and Difference Methods, 2nd edn. Wiley, New York (2013)
35. Heston, S.L.: A closed-form solution for options with stochastic volatility with applications to bond and currency options. Rev. Financ. Stud. **6**, 327–343 (1993)
36. Higham, N.J.: Stability of the diagonal pivoting method with partial pivoting. SIAM J. Matrix Anal. Appl. **18**, 52–65 (1997)
37. Higham, N.J.: Accuracy and Stability of Numerical Algorithms, 2nd edn. SIAM, Philadelphia (2002)
38. Hull, J., White, A.: The pricing of options with stochastic volatilities. J. Finan. **42**, 281–300 (1987)
39. in 't Hout, K.J.: Stability of ADI schemes for multidimensional diffusion equations with mixed derivative terms. Appl. Numer. Math. **74**, 83–94 (2013)
40. Kwok, Y.K.: Mathematical Models of Financial Derivatives. Springer, Berlin (2008)
41. Landau, H.: Heat conduction in a melting solid. Quart. Appl. Math. **8**, 81–95 (1950)
42. Lapidus, L., Pinder, G.F.: Numerical Solution of Partial Differential Equations in Science and Engineering. Wiley-Interscience, New York (1982)
43. Leentvaar, C.C.W.: Pricing multi-asset options with sparse grids. PhD thesis, TU Delft (2008)
44. Madan, D., Yor, M.: Representing the CGMY and Meixner Lévy processes as time changed Brownian motions. J. Comput. Finance **12**(1), 27–47 (2008)
45. Madan, D.B., Carr, P., Chang, E.C.: The Variance Gamma process and option pricing. Eur. Financ. Rev. **2**, 79–105 (1998)
46. Merton, R.C.: Option pricing when underlying stock returns are discontinuous. J. Financ. Econ. **3**, 125–144 (1976)
47. Nielsen, B., Skavhaug, O., Tvelto, A.: Penalty and front-fixing methods for the numerical solution of American option problems. J. Comput. Finance **5**, 69–97 (2002)
48. Nielsen, B.F., Skavhaug, O., Tveito, A.: Penalty methods for the numerical solution of American multi-asset option problems. J. Comp. Appl. Math. **222**(1), 3–16 (2008)
49. Pealat, G., Duffy, D.: The alternating direction explicit (ADE) method for one-factor problems. Wilmott Mag. **2011**(54), 54–60 (2011)
50. Pooley, D.M., Forsyth, P.A., Vetzal, K.R.: Numerical convergence properties of option pricing PDEs with uncertain volatility. IMA J. Numer. Anal. **23**, 241–267 (2003)

51. Rambeerich, N., Tangman, D., Lollchund, M., Bhuruth, M.: High-order computational methods for option valuation under multifactor models. Eur. J. Oper. Res. **224**(1), 219–226 (2013)
52. Reisinger, C., Wittum, G.: Efficient hierarchial approximation of high-dimensional option pricing problems. SIAM J. Sci. Comput. **29**, 440–458 (2007)
53. Saib, A., Tangman, Y., Thakoor, N., Bhuruth, M.: On some finite difference algorithms for pricing American options and their implementation in Mathematica. In: Proceedings of the 11th International Conference on Computational and Mathematical Methods in Science and Engineering, 2011
54. Salmi, S., Toivanen, J., Von Sydow, L.: Iterative methods for pricing American options under the bates model. Procedia Comput. Sci. **18**, 1136–1144 (2013)
55. Sauer, T.: Computational solution of stochastic differential equations. WIREs Comput. Stat. **5**, 362–371 (2013)
56. Schoutens, W.: Lévy Processes in Finance: Pricing Financial Derivatives. Wiley, New York (2003)
57. Schoutens, W., Teugels, J.L.: Lévy processes, polynomials and martingales. Commun. Statist.-Stoch. Models **14**(1, 2), 335–349 (1998)
58. Ševčovič, D.: Analysis of the free boundary for the pricing of an American call option. Eur. J. Appl. Math. **2**, 25–37 (2001)
59. Ševčovič, D.: Transformation methods for evaluating approximations to the optimal exercise boundary for linear and nonlinear Black-Scholes equations. In: Nonlinear Models in Mathematical Finance: New Research Trends in Option Pricing, pp. 173–218. Nova Science Publishers, Inc., New York (2008)
60. Smith, G.D.: Numerical Solution of Partial Differential Equations: Finite Difference Methods, 3rd edn. Clarendon Press, Oxford (1985)
61. Strikwerda, J.: Finite Difference Schemes and Partial Differential Equations, 2nd edn. Society for Industrial and Applied Mathematics, Philadelphia (2004)
62. Tavella, D.: Quantitative Methods in Derivatives Pricing. Wiley, New York (2012)
63. Tavella, D., Randall, C.: Pricing Financial Instruments: The Finite Difference Method. Wiley, New York (2007)
64. Toivanen, J.: Numerical valuation of European and American options under Kou's jump-diffusion model. SIAM J. Sci. Comput. **30**(4), 1949–1970 (2008)
65. Wang, I.R., Wan, J.W., Forsyth, P.A.: Robust numerical valuation of European and American options under the CGMY process. J. Comput. Finance **10**(4), 31–69 (2007)
66. Wilmott, P., Dewynne, J., Howison, S.: Option Pricing: Mathematical Models and Computation. Oxford Financial Press, Oxford (1998)
67. Witelski, T.P., Bowen, M.: ADI schemes for higher-order nonlinear diffusion equations. Appl. Numer. Math. **45**, 331–351 (2003)
68. Wu, L., Kwok, Y.-K.: A front-fixing method for the valuation of American option. J. Financ. Eng. **6**, 83–97 (1997)
69. Yousuf, M., Khaliq, A.Q.M., Liu, R.: Pricing American options under multi-state regime switching with an efficient L-stable method. Int. J. Comput. Math. **92**, 2530–2550 (2015)
70. Zvan, R., Forsyth, P.A., Vetzal, K.R.: Negative coefficients in two-factor option pricing models. J. Comput. Finance **7**, 37–73 (2003)

Chapter 11
Modified Barrier Penalization Method for Pricing American Options

Miglena N. Koleva and Radoslav L. Valkov

Abstract We propose a modified interior penalization method which is applicable to different types of American options. Further, we develop an efficient numerical approach for solving the resulting nonlinear parabolic partial differential problem. Numerical experiments illustrate the performance of the method.

11.1 Introduction

The pricing of early-exercise securities is important in quantitative finance because these are the most widely-traded type of instruments on the derivative market. The American-style option valuation is an illustrative example of an optimal stopping time problem which could be further formulated as a parabolic variational inequality.

Let S stand for the underlying asset price process, following a standard geometric Brownian motion with volatility σ and drift equal to the interest rate r while t is the time to maturity. For computational purposes one must truncate the spatial domain $S \in [0, \infty)$ and introduce the far field boundary location $S_{\max}$. We consider pricing with the following conditions on the parabolic boundary:

$$V(S, 0) = V^*(S), \quad V(0, t) = V_L, \quad V(S_{\max}, t) = V_R,$$

where $V_L \geq 0$ and $V_R \geq 0$ are given constants. The American put is a classical Stefan problem where the payoff is convex, continuous but nonsmooth.

M.N. Koleva (✉)
Faculty of Natural Science and Education, Ruse University Angel Kanchev, Ruse, Bulgaria
e-mail: mkoleva@uni-ruse.bg

R.L. Valkov
Department of Mathematics - Computer Sciences, University of Antwerp, Antwerp, Belgium
e-mail: radoslav.valkov@uantwerpen.be

M. Ehrhardt et al. (eds.), *Novel Methods in Computational Finance*,
Mathematics in Industry 25, DOI 10.1007/978-3-319-61282-9_11

The option value with maturity T satisfies the parabolic variational inequality, cf. e.g. [4]

$$LV(S,t) := V_t - \frac{1}{2}\sigma^2 S^2 V_{SS} - rV_S + rV \geq 0 \perp V(S,t) \geq V^*(S) \text{ in } (0,\infty) \times [0,T),$$

which could be written down as the following linear complementarity problem

$$\text{LCP}: \begin{cases} LV(S,t) = \lambda \geq 0, \\ V(S,t) - V^*(S) \geq 0, \;\; LV(S,t) \cdot (V(S,t) - V^*(S)) = 0. \end{cases}$$

From the complementarity condition

$$V \geq V^*, \; \lambda \geq 0, \; \lambda \cdot (V(S,t) - V^*(S)) = 0$$

we infer for the Lagrange multiplier

$$\lambda = \max\left(0, \lambda + \epsilon^{-1}(V^* - V)\right) \tag{11.1}$$

for any sufficiently small (penalty) parameter $\epsilon > 0$. Thus, we get to solve the following nonlinear equation, equipped with the complementarity condition and drawing analogies with the augmented Lagrangian method:

$$LV(S,t) - \max\left(0, \lambda + \epsilon^{-1}(V^* - V)\right) = 0.$$

Superimposing infinite penalty when violating the constraint $V(S,t) - V^*(S) \geq 0$ we may embed the LCP [1] in the family of the nonlinear equations

$$LV^{\epsilon}(S,t) - \max\left(0, \epsilon^{-1}(V^* - V^{\epsilon})\right) = 0. \tag{11.2}$$

The penalty method guarantees in an asymptotic sense the fulfilment of constraints by including in the objective function an additional penalty term. If we consider the upper bound on the multiplier $\Lambda_{\max}$ we may state the following approximation, see [4]:

$$LV^{\epsilon}(S,t) - \max\left(0, \Lambda_{\max} + \epsilon^{-1}(V^* - V_{\epsilon})\right) = 0.$$

There are, however, some issues with this approach since the early exercise constraint is not strictly satisfied by the solution for fixed small ϵ while the penalty term is nonsmooth and unbounded. The following interior approximation aims to tackle these drawbacks with $\mathrm{C} \geq rK$ for pricing the American put, cf. [6, 11],

$$LV^{\epsilon}(S,t) - \frac{\epsilon \mathrm{C}}{V^{\epsilon} + \epsilon - (K - S)} = 0 \tag{11.3}$$

where the general interior penalty method, applicable to any type of payoff, is

$$LV^{\epsilon}(S,t) - \frac{\epsilon C}{V^{\epsilon} + \epsilon - V^{*}} = 0. \tag{11.4}$$

Zhang and Wang [14] prove convergence of the penalized solution V^{ϵ} to the solution of the underlying variational inequality V. However, a major issue with this approach is its dependence on ϵ and some vague parameter C, resulting in overall lower accuracy and more Newton iterations per time level.

We therefore consider modifying this interior barrier method in order to enhance the performance [8]. Let us set up the fully-discrete LCP in order to present our considerations in a clear and concise manner. First, by the method of lines, we define a smooth nonuniform spatial grid and approximate the spatial derivatives by second-order finite difference formulas. After backward Euler time discretization with step Δt we have to solve the following discrete linear complementarity problem for $U^n \in \mathbb{R}^{m-1}$

$$\begin{cases} (I - \Delta t A)U^n = U^{n-1} + \Delta t g + \Delta t \lambda^n \\ \lambda^n \geq 0, \;\; U^n \geq U^0, \;\; \lambda^n \cdot \left(U^n - U^0\right) = 0, \end{cases}$$

where $A \in \mathbb{R}^{(m-1)\times(m-1)}$ stands for the spatial discretization matrix, $g \in \mathbb{R}^{m-1}$ is the boundary information (assuming Dirichlet conditions on the elliptic boundary) and $\lambda^n \in \mathbb{R}^{m-1}$ is the nonnegative auxiliary (multiplier) vector which satisfies

$$\lambda^n = \max\left(0, \lambda^n + \frac{1}{\Delta t}(U^0 - U^n)\right). \tag{11.5}$$

The solution U^n of the discrete LCP is the saddle point of the Lagrange functional

$$\Lambda(U^n, \lambda^n) = \frac{1}{2}(I - \Delta t A)U^n \cdot U^n - b^n \cdot U^n - \Delta t \lambda^n \cdot (U^n - U^0), \; b^n := U^{n-1} + \Delta t g.$$

Let us now observe the following equivalence [12]

$$U^n - U^0 \geq 0 \Leftrightarrow \epsilon \log\left(1 + \epsilon^{-1}(U^n - U^0)\right) \geq 0$$

and further we shall modify the Lagrange functional accordingly

$$\Lambda(U^n, \lambda^n) = \frac{1}{2}(I - \Delta t A)U^n \cdot U^n - b^n \cdot U^n - \Delta t \lambda^n \cdot (\epsilon \log\left(1 + \epsilon^{-1}(U^n - U^0)\right)).$$

From the Karush-Kuhn-Tucker conditions we get the discrete LCP:

$$\begin{cases} (I - \Delta t A)U^n - \Delta t \lambda^n \frac{\epsilon}{U^n - U^0 + \epsilon} = b^n \\ \lambda^n \geq 0, \;\; U^n \geq U^0, \;\; \lambda^n \cdot \left(U^n - U^0\right) = 0. \end{cases} \tag{11.6}$$

As a matter of fact, if we consider the fairly rough estimate for the multiplier $\lambda^n \leq rK$ in the put case we get the discrete interior penalty method (11.4):

$$(I - \Delta t A)U^n - \Delta t \frac{rK\epsilon}{U^n - U^0 + \epsilon} = b^n.$$

Substituting $U^0 = \max(K-S, 0)$ with $K-S$ as in Eq. (11.3) is a band-aid for the case of put option to fix the accuracy and minimize the penalty term in the continuation region where $S > K$, far away from the free boundary.

11.2 The Finite Difference Method

There are many papers using the penalty method for solving American options, see [1, 2, 5–7, 9–11, 14]. In this section we present a simplified version of barrier penalty method (11.6), introduced in Sect. 11.1.

We consider the penalized problem which approximates the LCP for some sufficiently small positive parameter ϵ

$$V_t^\epsilon - \frac{1}{2}\sigma^2 S^2 V_{SS}^\epsilon - rSV_S^\epsilon + rV^\epsilon - g(S, V^\epsilon) = 0, \; (S,t) \in (0, S_{\max}) \times [0, T), \tag{11.7}$$
$$V^\epsilon(S,0) = V^*(S), \;\; V^\epsilon(0,t) = V_L, \;\; V^\epsilon(S_{\max}, t) = V_R$$

with the penalty term

$$g(S, V^\epsilon) = \frac{\epsilon\lambda}{V^\epsilon + \epsilon - V^*}. \tag{11.8}$$

For given integers m and N we define $\Delta t = T/N$, $t^n = n\Delta t$ and the nonuniform spatial grid

$$\overline{\omega} = \{S_0 = 0, \; S_{i+1} = S_i + h_i, \; i = 0, \dots, m-1, \; S_m = S_{\max}\},$$

where the discrete solution, computed on the mesh $\overline{\omega}$ is denoted by $U_i^n = V^\epsilon(S_i, t^n)$. Let us now write down the considered finite difference approximations of the first derivative for $h_i = S_{i+1} - S_i$, $\hbar_i = 0.5(h_i + h_{i-1})$

$$(U_{\check{S}})_i^n = \frac{U_{i+1}^n - U_i^n}{h_i}, \;\; (U_{\hat{S}})_i^n = \frac{U_i^n - U_{i-1}^n}{h_{i-1}}, \;\; (U_{\mathring{S}})_i^n = \frac{h_{i-1}U_{\check{S}i}^{\,n} + h_i U_{\hat{S}i}^{\,n}}{2\hbar_i},$$

where $(U_{\check{S}})_i^n$, $(U_{\hat{S}})_i^n$ are of first order and $(U_{\mathring{S}})_i^n$ of second on a smooth grid. The second derivative is further approximated as

$$(U_{SS})_i^n = \left((U_{\check{S}})_i^n - (U_{\hat{S}})_i^n\right)/\hbar_i.$$

After backward Euler time discretization of (11.7), (11.8) and application of the maximal use of central differencing with flag $\chi := \mathrm{H}(\sigma^2 S_i - rh_i)$ (H stands for the Heaviside function), see Wang and Forsyth [13], we get the following system of nonlinear equations for $n = 0, \ldots, N$ and $i = 2, \ldots, m-1$:

$$L^h U_i^{n+1} - \frac{\epsilon\lambda_i}{U_i^{n+1} + \epsilon - U^0} = 0,$$

$$U(0, t^{n+1}) = V_L,\ U(S_m, t^{n+1}) = V_R,\ U(S_i, 0) = U^0(S_i) = V^*(S_i), \tag{11.9}$$

$$L^h U_i^n := \frac{U_i^{n+1} - U_i^n}{\Delta t} - \frac{\sigma^2 S_i^2}{2}(U_{SS})_i^n - rS_i\left(\chi(U_{\overset{\circ}{S}})_i^n + (1-\chi)(U_{\check{S}})_i^n\right) + rU_i^n.$$

Next, on the base of (11.1), (11.5) we set the simple version of λ:

$$\lambda_i = \max\{0, L^h U_i^0\},\ \ i = 1, \ldots, m-1. \tag{11.10}$$

Numerical experiments show that with this choice of λ we attain similar precision as with λ^n, computed by (11.5), but for smaller computational cost.

We find the solution U^{n+1} by initiating a Newton's iteration process with initial guess $U^{(0)} = U^n$, where the Newton increment on the $(k+1)$th step $\Delta^{(k+1)} = U^{(k+1)} - U^{(k)}$ is the solution of the following tridiagonal system of linear equations

$$\begin{aligned} &-A_i\Delta_{i-1}^{(k+1)} + C_i^{(k)}\Delta_i^{(k+1)} - B_i\Delta_{i+1}^{(k+1)} \\ &\qquad = U_i^n + A_i P_{i-1}^{(k)} - \widetilde{C}_i U_i^{(k)} + B_i U_{i+1}^{(k)} + F_i^{(k)}, \end{aligned} \tag{11.11}$$

where $A_0 = A_m = B_0 = B_m = 0$, $C_1^{(k)} = C_m^{(k)} = 1$, $F_1^{(k)} = V_L$, $F_m^{(k)} = V_R$ and

$$A_i = \tfrac{S_i\Delta t}{2\hbar_i h_{i-1}}\left(\sigma^2 S_i - \chi r h_i\right),\ \ B_i = \tfrac{S_i\Delta t}{2\hbar_i h_i}\left(\sigma^2 S_i + \chi r h_{i-1}\right) + (1-\chi)\tfrac{rS_i\Delta t}{h_i},$$

$$C_i^{(k)} = \widetilde{C}_i + \tfrac{\epsilon\lambda_i\Delta t}{(U_i^{(k)} + \epsilon - U_i^0)^2},\ \ \widetilde{C}_i = 1 + A_i + B_i + r\Delta t,\ \ F_i^{(k)} = \tfrac{\epsilon\lambda_i\Delta t}{U_i^{(k)} + \epsilon - U_i^0},$$

The iteration process is terminated when reaching the desired tolerance i.e. we set $U^{n+1} := U^{(k+1)}$ when $\max_i\{|\Delta_i^{(k+1)}|/(\max\{1, U_i^{(k+1)}\})\} < \mathrm{tol}$.

At each iteration k, in view of the definition of χ, the coefficient matrix $M^{(k)} = \mathrm{tridiag}[-A_i, C_i^{(k)}, -B_i]$, being strictly diagonally dominant and $A_i, C_i^{(k)}, B_i > 0$ it is an M-matrix.

Theorem 11.1 *The approximate option value* U_i^n*, obtained by the scheme* (11.9), (11.10) *satisfy*

$$U_i^n \geq U_i^0,\ \ i = 0, \ldots, m,\ \ n = 0, \ldots, N.$$

Proof We follow the same line of consideration as in [11]. Rewrite the scheme (11.9), (11.10) in the following equivalent form

$$(1+A_i+B_i+r\Delta t)U_i^{n+1}-B_iU_{i+1}^{n+1}-A_iU_{i-1}^{n+1}=U_i^n+\frac{\epsilon\lambda\Delta t}{U_i^{n+1}+\epsilon-U^0}. \tag{11.12}$$

Let $w_i^n=U_i^n-U_i^0$. Thus, from (11.12) we obtain

$$(1+A_i+B_i+r\Delta t)w_i^{n+1}-B_iw_{i+1}^{n+1}-A_iw_{i-1}^{n+1}=w_i^n+\frac{\epsilon\lambda\Delta t}{w_i^{n+1}+\epsilon}-\Delta tL^hU_i^0. \tag{11.13}$$

Define $w^{n+1}=\min\limits_i w_i^{n+1}$ and let j be an index, such that $w_j^{n+1}=w^{n+1}$. For $i=j$, from (11.13) we have

$$(1+A_i+B_i+r\Delta t)w^{n+1}\geq w_j^n+B_iw^{n+1}+A_iw^{n+1}+\frac{\epsilon\lambda\Delta t}{w_j^{n+1}+\epsilon}-\Delta tL^hU_j^0$$

and therefore

$$(1+r\Delta t)w^{n+1}\geq w_j^n+\frac{\epsilon\lambda\Delta t}{w_j^{n+1}+\epsilon}-\Delta tL^hU_j^0.$$

Rearranging the above inequality, we obtain

$$(1+r\Delta t)w^{n+1}-\frac{\epsilon\lambda\Delta t}{w^{n+1}+\epsilon}+\Delta tL^hU_j^0\geq w_j^n\geq w^n.$$

We use induction method on n: taking into account that $w^0\geq 0$, assume that $w^n\geq 0$ and prove $w^{n+1}\geq 0$. Now we have

$$\mathcal{F}(w^{n+1})\geq 0,\ \text{ where }\ \mathcal{F}(w):=(1+r\Delta t)w-\frac{\epsilon\lambda\Delta t}{w+\epsilon}+\Delta tL^hU_i^0.$$

Observe that

$$\mathcal{F}(0)=-\Delta t(\lambda-L^hU_i^0)=\Delta t(L^hU_i^0-\max\{0,L^hU_i^0\})=$$
$$\Delta t\begin{cases}0, & \text{if } L^hU_i^0\geq 0,\\ L^hU_i^0, & \text{if } L^hU_i^0<0,\end{cases}\quad \text{i.e. } \mathcal{F}(0)\leq 0.$$

Further, from

$$\mathscr{F}'(w) := 1 + r\Delta t + \frac{\epsilon\lambda\Delta t}{(w+\epsilon)^2} \geq 0$$

and $\mathscr{F}(w^{n+1}) \geq 0$ we conclude that $w^{n+1} \geq 0$.

For the discrete scheme (11.9), (11.10) we develop a two-grid algorithm TGA [7].

Let us define two non-uniform spatial grids—a coarse mesh $\overline{\omega}_c$ and a fine grid $\overline{\omega}_f$

$$\overline{\omega}_c = \{S_0 = 0,\ S_{i+1} = S_i + h_i^c,\ i = 1, \ldots, m_c - 1,\ S_{m_c} = S_{\max}\},$$

$$\overline{\omega}_f = \{S_0 = 0,\ S_{i+1} = S_i + h_i^f,\ i = 1, \ldots, m_f - 1,\ S_{m_f} = S_{\max}\},$$

where $m_f \gg m_c$ and the discrete solution, computed on the mesh $\overline{\omega}_*$ is denoted by $U_{i,*}^n = U(S_i, t^n)$.

Algorithm 2 (TGA)

At each time level $n = 0, 1, \ldots$ we perform the two steps:

1: Set $U_c^{(0)} := U_c^n$ and compute U_c^{n+1} by (11.9), (11.10) through Newton's iterations (11.11) on the coarse mesh $\overline{\omega}_c$.

2: Set $U_f^{(0)} := I(U_c^{n+1})$, where $I(U_c)$ is the interpolant of P_c on the fine grid, perform *only one* Newton's iteration (11.11) on the fine mesh $\overline{\omega}_f$ and get U_f^{n+1}.

11.3 Numerical Experiments

We consider an American butterfly option with the payoff

$$V^*(S) = \max\{S - K_1, 0\} - 2\max\{S - K\} + \max\{S - K_2, 0\},$$

where K_1, $K = (K_1 + K_2)/2$, K_2 are the strikes and $V_L = V_R = 0$. The model parameters are: $K_1 = 90$, $K_2 = 100$, $\sigma = 0.2$, $r = 0.1$, $\epsilon = 1.\text{e}{-6}$. We will test the relevance of the modified penalty method (11.9), (11.10) and the accuracy, order of convergence and efficiency of the constructed TGA.

The linearized system (11.11) is solved by BiConjugate gradients stabilized method using preconditioning with upper and lower triangular matrix. For stopping criteria, we chose tol=1.e−6.

In the computational domain for $S_{\max} = 200$, we use a smooth non-uniform grid, cf. in 't Hout et al. [3]—uniform inside the region $[S_l, S_r] = [1/2K, 3/2K]$, and non-uniform outside with stretching parameter $c = K/10$:

$$S_i := \phi(\xi_i) = \begin{cases} S_l + c\sinh(\xi_i), & \xi_{\min} \le \xi_i < 0, \\ S_l + c\xi_i, & 0 \le \xi_i \le \xi_{\text{int}}, \\ S_r + c\sinh(\xi_i - \xi_{\text{int}}), & \xi_{\text{int}} \le \xi_i < \xi_{\max}. \end{cases}$$

The uniform partition of $[\xi_{\min}, \xi_{\max}]$ is defined through $\xi_{\min} = \xi_0 < \cdots < \xi_M = \xi_{\max}$:

$$\xi_{\min} = \sinh^{-1}\left(\frac{-S_l}{c}\right), \quad \xi_{\text{int}} = \frac{S_r - S_l}{c}, \quad \xi_{\max} = \xi_{\text{int}} + \sinh^{-1}\left(\frac{S_{\max} - S_r}{c}\right).$$

Example 11.1 (Early Exercise Constraint) We compare the different penalty methods—interior penalty (11.3), exterior penalty (11.2) and modified barrier penalty (11.8), (11.10) in maintaining the condition $V^* < U$. On Fig. 11.1 we plot the corresponding solutions at $T = 1$ and payoff, while on Fig. 11.2 we plot the difference $U^n - V^*$ for exterior penalty (11.2) and modified penalty (11.8), (11.10). We observe that for butterfly option, only with modified barrier penalty (11.8), the numerical solution satisfy the early exercise constraint. Thus, the statement of Theorem 11.1 is verified.

Example 11.2 (One-Grid Computations) We perform computations only on one mesh $\overline{\omega}$, i.e. step 1 of TGA with time steps $\Delta t = h$ and $\Delta t = h^2$, $h = \min_i h_i$. The results are listed in Tables 11.1 and 11.2. We give the values of the solution at strike points K_1 and K at final time T, diff—the absolute value of the difference in the solution from the coarser grid, *CR*—computed as $\log_2$ from the ratio of the changes on successive grids, iter—the averaging number of iterations k at each time level and CPU time (in seconds). We observe that the order of convergence in space at strike points is closed to two and the computational process is more efficient for smaller time step.

Example 11.3 (TGA) For the numerical tests, we set $\Delta t = h^f$, $\Delta t = (h^f)^2$, $h^f = \min_i h_i^f$ and $m_f = (m_c)^2/S_{\max}$, i.e. $h^f = (h^c)^2$ in the case of uniform meshes. The results are given in Tables 11.3 and 11.4. We observe that the order of convergence on the coarse mesh, tested at strike points K_1 and K is closed to four, i.e. $\mathcal{O}(\Delta t + |h^c|^4 + |h^f|^2)$, $|h| = \max_i h_i$. Also, the TGA accelerate the computational efficiency. Comparable values of the solution in Tables 11.1, 11.2, 11.3, and 11.4 are highlighted.

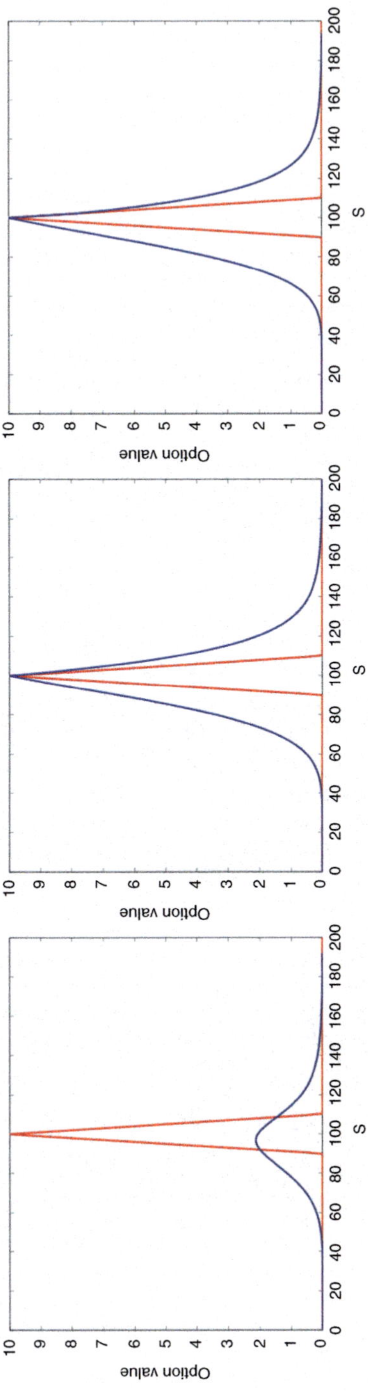

Fig. 11.1 Payoff and option value. *Left*: interior penalty (11.3); *Center*: exterior penalty (11.2); *Right*: modified penalty (11.8)

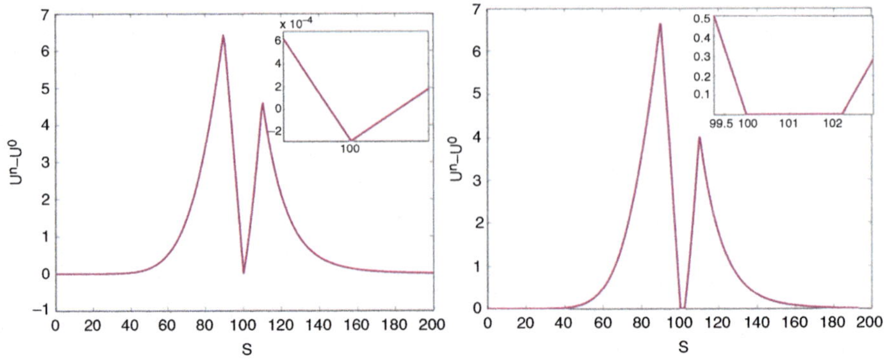

Fig. 11.2 $U^n - V^*$, *Left*: exterior penalty (11.2); *Right*: modified penalty (11.8)

Table 11.1 Values of $U(K_1, T)$, $U(K, T) = 10 + \widetilde{U}(K, T)$, convergence rate (*CR*), averaging number of iterations (iter) and CPU time, *one-grid computations*, $\Delta t = h$

m	$U(K_1, T)$	diff	CR	$\widetilde{U}(K, T)$	diff	CR	iter	CPU
400	6.70216			5.3352e−7			10	0.05
800	6.90697	2.048e−1		5.7003e−7	3.99e−8		10	0.09
1600	7.00017	2.980e−1		9.8146e−7	4.10e−7		9.54	0.23
3200	7.05562	5.545e−2	2.426	1.1000e−6	1.20e−7	1.773	7.46	0.56
6400	7.08314	2.752e−2	1.011	1.1481e−6	4.81e−8	1.319	5.77	1.44
12,800	7.09683	1.370e−2	1.006	1.1750e−6	2.69e−8	0.839	4.39	3.85
25,600	7.10366	6.831e−3	1.004	1.1886e−6	1.35e−8	0.986	3.49	11.91
51,200	7.10706	3.401e−3	1.006	1.1944e−6	5.75e−9	1.241	2.99	47.72
102,400	7.10877	1.707e−3	0.995	1.1966e−6	2.23e−9	1.364	2.57	208.75
204,800	7.10962	8.548e−4	0.998	1.1975e−6	8.58e−10	1.378	2.27	1175.56

Table 11.2 Values of $U(K_1, T)$, $U(K, T) = 10 + \widetilde{U}(K, T)$, convergence rate (*CR*), averaging number of iterations (iter) and CPU time, *one-grid computations*, $\Delta t = h^2$

m	$U(K_1, T)$	diff	CR	$\widetilde{U}(K, T)$	diff	CR	iter	CPU
400	6.95822			1.0960e−6			7.35	0.08
800	7.07028	1.121e−1		1.1724e−6	7.63e−8		4.47	0.26
1600	7.10047	3.019e−2	1.892	1.1918e−6	1.94e−8	1.977	3.28	0.93
3200	7.10797	7.505e−3	2.008	1.1967e−6	4.97e−9	1.963	2.60	4.49
6400	7.10985	1.880e−3	1.997	1.1978e−6	1.09e−9	2.190	2.17	23.31
12,800	7.11032	4.696e−4	2.001	1.1980e−6	2.14e−10	2.350	2.03	143.96

Table 11.3 Values of $U(K_1, T)$, $U(K, T) = 10 + \widetilde{U}(K, T)$, convergence rate ($CR$), averaging number of iterations (iter) and CPU time, *two-grid computations*, $\Delta t = (h^f)$

m_c	m_f	$U(K_1, T)$	diff	CR	$\widetilde{U}(K, T)$	diff	CR	iter	CPU
400	800	6.90690			1.0501e−6			7.83	0.18
800	3200	7.05559	1.49e−1		1.1623e−6	1.10e−7		4.72	0.36
1600	12,800	7.09682	4.12e−2	1.855	1.1885e−6	2.85e−8	1.949	3.39	2.53
3200	51,200	7.10706	1.02e−2	2.014	1.1960e−6	7.51e−9	1.925	2.73	30.82
6400	204,800	7.10962	2.56e−3	1.994	1.1977e−6	1.65e−9	2.189	2.25	538.10

Table 11.4 Values of $U(K_1, T)$, $U(K, T) = 10 + \widetilde{U}(K, T)$, convergence rate ($CR$), averaging number of iterations (iter) and CPU time, *two-grid computations*, $\Delta t = (h^f)^2$

m_c	m_f	$U(K_1, T)$	diff	CR	$\widetilde{U}(K, T)$	diff	CR	iter	CPU
400	800	7.07020			1.1782e−6			5.77	0.38
800	3200	7.10795	3.77e−2		1.1978e−6	1.96e−8		3.74	6.94
1600	12,800	7.11032	2.37e−3	3.992	1.1985e−6	6.71e−10	4.868	2.58	203.55

11.4 Conclusions

In contrast to the interior (11.3) and exterior (11.2) penalty methods, the modified penalty method guarantees that the solution always satisfy the early exercise constraint, independently of the type of the option.

The two-grid algorithm attains fourth order convergence in space on the coarse mesh. We observe very fast performance of the presented TGA, independently of the choice of the time step size. One and the same accuracy as with one-grid computations, can be obtained by TGA, saving computational time.

Acknowledgements This research was supported by the European Union under Grant Agreement number 304617 (FP7 Marie Curie Action Project Multi-ITN STRIKE—Novel Methods in Computational Finance and the Bulgarian National Fund of Science under Project I02/20-2014.

References

1. Forsyth, P.A., Vetzal, K.R.: Quadratic convergence for valuing American options using a penalty method. SIAM J. Sci. Comput. **23**(6), 2095–2122 (2002)
2. Gyulov, T.B., Valkov, R.L.: American option pricing problem transformed on finite interval. Int. J. Comput. Math. **93**(5), 821–836 (2016)
3. Haentjens, T., in 't Hout, K.J.: Alternating direction implicit finite difference schemes for the Heston-Hull-White partial differential equation. J. Comput. Financ. **16**(1), 83–110 (2012)
4. Ito, K., Kunisch, L.: Parabolic variational inequalities: the Lagrange multiplier approach. J. Math. Pures Appl. **85**, 415–449 (2006)
5. Jaillet, P., Lamberton, D., Lapeyre, B.: Variational inequalities and the pricing of American options. Acta Appl. Math. **21**(3), 263–289 (1990)

6. Khaliq, A.Q.M., Voss, D.A., Kazmi, S.H.K.: A linearly implicit predictor-corrector scheme for pricing American option using a penalty method approach. J. Bank. Financ. **30**, 489–502 (2006)
7. Koleva, M.N., Valkov, R.L.: Two-grid algorithms for pricing American options by a penalty method. In: ALGORITMY 2016, Slovakia, Publishing House of Slovak University of Technology in Bratislava, pp. 275–284 (2016). ISBN 978-80-227-4454-4
8. Koleva, M.N., Valkov, R.L.: Numerical penalization algorithms for pricing American options. In: Proceedings of the International Conference on Numerical Methods for Scientific Computations and Advanced Applications, Sofia, Fastumprint, pp. 45–48 (2016)
9. Kovalor, R., Linetsky, V., Marcozzi, M.: Pricing multi-asset American options: a finite element method-of-lines with smooth penalty. J. Sci. Comput. **33**, 209–237 (2007)
10. Mikula, K., Ševčovič, D., Stehlíková, B.: Analytical and Numerical Methods for Pricing Financial Derivatives. Nova Science, Hauppauge (2011)
11. Nielsen, B.F., Skavhaug, O., Tveito, A.: Penalty and front-fixing methods for the numerical solution of American option problems. J. Comput. Finance **5**(4), 69–97 (2001)
12. Polyak, R.A.: Modified barrier functions (theory and methods). Math. Program. **54**, 177–222 (1992)
13. Wang, J., Forsyth, P.A.: Maximal use of central differencing for Hamilton-Jacobi-Bellman PDEs in finance. SIAM J. Numer. Anal. **46**(3), 1580–1601 (2008)
14. Zhang, K., Wang, S.: Convergence property of an interior penalty approach to pricing American option. J. Ind. Manage. Optim. **7**, 435–447 (2011)

Part IV
Numerical Methods in Finance

Chapter 12
Newton-Based Solvers for Nonlinear PDEs in Finance

Shih-Hau Tan and Choi-Hong Lai

Abstract In this chapter, different Newton-based solvers are introduced to solve fully nonlinear PDEs generated from financial problems. The first one concentrates on solving the root-finding problem from the nonlinear system after applying the standard finite difference method with implicit scheme. The second one addresses to solve the deferred correction problem which is transformed from the original PDE. Different numerical experiments in terms of accuracy and efficiency are compared and some improvements using Newton-like methods are also discussed.

12.1 Introduction

The ideal market assumptions, including frictionless, perfect liquid, zero transaction costs and constant volatility, used in the classical linear Black-Scholes PDE model for option pricing [2] are not always true in the real market. In particular factors like transaction costs, illiquid market effects or other market feedback need to be considered. Modification to the classical Black-Scholes model usually comes into the form of a nonlinear Black-Scholes equation in which the volatility function depends on the option price itself and/or its derivatives. Over the last two decades there were various models aimed to take account of the above situations as discussed in the literature [3, 4, 8, 10, 13, 15, 23].

The generalized nonlinear Black-Scholes equation to be considered may be written as

$$\frac{\partial V}{\partial t} + \frac{1}{2}\sigma^2 S^2 \frac{\partial^2 V}{\partial S^2} + rS\frac{\partial V}{\partial S} - rV = 0, \tag{12.1}$$

where $V(S,t) \in \{S : S \geq 0\} \times [0,T)$ is the price of the option, r is the risk-free interest rate, t is the time, T is the maturity date, S is the current spot price of the underlying asset and σ is the nonlinear volatility which depends on V, V_t, V_S, and

S.-H. Tan (✉) • C.-H. Lai
Department of Mathematical Sciences, University of Greenwich, London, UK
e-mail: shihhau.tan@gmail.com; C.H.Lai@gre.ac.uk

M. Ehrhardt et al. (eds.), *Novel Methods in Computational Finance*,
Mathematics in Industry 25, DOI 10.1007/978-3-319-61282-9_12

V_{SS}, where

$$V_t = \frac{\partial V}{\partial t}, V_S = \frac{\partial V}{\partial S}, V_{SS} = \frac{\partial^2 V}{\partial S^2},$$

and the boundary and terminal conditions depend on the type of option to be considered. For example, in the case of an European call option these conditions are

$$\begin{cases} V(S,T) = (S-K)^+, & \text{for} \quad 0 \le S < S_{\max}, \\ V(0,t) = 0, & \text{for} \quad 0 \le t \le T, \\ V(S,t) = S - Ke^{-r(T-t)}, & \text{when} \quad S = S_{\max}. \end{cases}$$

Here $S_{\max}$ is usually chosen as an integer multiple of K.

In this section, the aim is to solve Eq. (12.1) with nonlinear volatility $\sigma = \sigma(S, V_{SS})$ which depends on the underlying asset and the second order derivative V_{SS}. For instances:

I. $\sigma = \sigma_0 \sqrt{1 + e^{r(T-t)} a S^2 V_{SS}}$ (A Simplified Barles and Soner Model [1]),
II. $\sigma = \sigma_0 \sqrt{(1 + \mu(S V_{SS})^{\frac{1}{3}})}$ (Risk Adjusted Pricing Methodology Model [15]),
III. $\sigma = \sigma_0 (1 - \rho\lambda(S) S V_{SS})^{-1}$ (Frey-Patie Model [11]).

Here σ_0 is a constant historical volatility; a, μ, ρ and $\lambda(S)$ are suitable constants and function which present the financial factors under different considerations.

To solve the nonlinear parabolic partial differential equation, an implicit finite difference scheme with standard notations and the transformation $\tau = T - t$ can be adopted which transforms Eq. (12.1) to

$$\frac{V_i^{n+1} - V_i^n}{\Delta\tau} - \frac{1}{2}(\sigma_i^{n+1})^2 S_i^2 \frac{V_{i+1}^{n+1} - 2V_i^{n+1} + V_{i-1}^{n+1}}{(\Delta S)^2} - rS_i \frac{V_{i+1}^{n+1} - V_{i-1}^{n+1}}{2\Delta S} + rV_i^{n+1} = 0, \tag{12.2}$$

where $\Delta\tau = T/N$ and $\Delta S = S_{\max}/(M+1)$ are the sizes of the temporal and spatial discretisation, respectively, with number of grid points $N + 1$ (time) and $M + 2$ (space), $\tau^n = n\Delta\tau, n = 0, 1, \ldots, N$; $S_i = i\Delta S, i = 0, 1, \ldots, M+1$; and V_i^n denotes the finite difference approximation of $V(S_i, n\Delta\tau)$, $n = 0, 1, \ldots, N$. Note that for each n, V_0^n and V_{M+1}^n are given by the boundary conditions. Equation (12.2) can be simplified to

$$a_i V_{i-1}^{n+1} + b_i V_i^{n+1} + c_i V_{i+1}^{n+1} = V_i^n,$$

where the tri-diagonal coefficients,

$$a_i = \Delta\tau\left(-\frac{S_i^2(\sigma_i^{n+1})^2}{2(\Delta S)^2} + \frac{rS_i}{2(\Delta S)}\right), \quad b_i = 1 + \Delta\tau\left(\frac{(\sigma_i^{n+1})^2 S_i^2}{(\Delta S)^2} + r\right),$$

$$c_i = \Delta\tau\left(-\frac{S_i^2(\sigma_i^{n+1})^2}{2(\Delta S)^2} - \frac{rS_i}{2(\Delta S)}\right),$$

depend on the volatility itself. The discretisation leads to the nonlinear system of equations

$$H(V^{n+1})V^{n+1} = V^n, \tag{12.3}$$

where

$$H(V^{n+1}) = \begin{pmatrix} b_1 & c_1 & 0 & \cdots & & \cdots & 0 \\ a_2 & b_2 & c_2 & 0 & \cdots & \cdots & 0 \\ \cdots & & \cdots & & \cdots & & \cdots \\ \cdots & & \cdots & & \cdots & & \cdots \\ 0 & \cdots\cdots & 0 & a_{M-1} & b_{M-1} & c_{M-1} \\ 0 & \cdots\cdots & & 0 & a_M & b_M \end{pmatrix}$$

and $V^n = (V_1^n, V_2^n, \ldots, V_M^n)^T$.

Equation (12.3) will need numerical solvers to obtain a solution. Frozen coefficient method is a common technique and the main idea is to keep the coefficient $H(V^{n+1})$ lagging behind when iteratively solving for a new V^{n+1} until the solution converges. Algorithm 3, shown below, outlines the steps of the method. The drawback of the method is that the number of iterations can become large in cases without proper initial guess (or may even diverge).

In the following sections, some solvers based on Newton's linearisation are introduced [6, 14] to show other approaches for solving Eq. (12.1). These approaches

---

Algorithm 3 Frozen coefficient method

Input: M, N, r, σ_0, tol, initial condition $V^0 = V(S, \tau = 0)$, initial guess V^*
Output: $V^N = V(S, \tau = T)$
1: **for** n = 0 to N-1 **do**
2: $\quad V^{n+1} = H(V^*)^{-1}V^n$
3: $\quad$ **if** $\|V^{n+1} - V^*\| < tol$ **then**
4: $\quad\quad V^{n+1} = V^*$
5: $\quad\quad$ **break**
6: $\quad$ **else**
7: $\quad\quad V^* = V^{n+1}$ and go back to 2.
8: $\quad$ **end if**
9: **end for**

---

concentrate on different linearisation techniques and methodologies of using direct or iterative solver for the nonlinear system (12.3). The developed Newton-based solvers can provide an appropriate direction in order to achieve a faster convergence to the solution, but need to afford the evaluation of updating terms which is like a trade-off. Different numerical results are shown to discuss more details.

12.2 Root-Finding Problem

There are different approaches in implementing Newton's method, and the most common way is to change the original problem to a root-finding algorithm and to update the numerical approximate solution by using the Jacobian matrix. This idea can be applied to solve Eq. (12.3) by defining

$$G(V^{n+1}) = H(V^{n+1})V^{n+1} - V^n = 0, \tag{12.4}$$

and by calculating $[Jac(G(V^{n+1}))]^{-1}$, which is the inverse of the Jacobian matrix of G for the update as stated in Algorithm 4. In order to obtain the Jacobian matrix efficiently, a decomposition of the nonlinear matrix $H(V^{n+1})$ as the one below may be used,

$$H(V^{n+1}) = \Sigma^{n+1} H_1 + H_2, \quad \text{where} \quad \Sigma^{n+1} = \text{diag}((\sigma_i^{n+1})^2).$$

Note that H_1 and H_2 are constant tri-diagonal matrices. By using this decomposition, the Jacobian matrix of G becomes

$$Jac(G(V^{n+1})) = \frac{\partial[H(V^{n+1})V^{n+1}]}{\partial V^{n+1}} = H(V^{n+1}) + \text{diag}(H_1\, V^{n+1})\nabla(\Sigma^{n+1}),$$

Algorithm 4 Root-finding problem (NM1)

Input: M, N, r, σ_0, tol, initial condition $V^0 = V(S, \tau = 0)$, initial guess V^*
Output: $V^N = V(S, \tau = T)$
1: **for** n = 0 to N-1 **do**
2: $\quad V^{n+1} = V^*$
3: $\quad G^{n+1} = H(V^{n+1})V^{n+1} - V^n$
4: $\quad$ **if** $\|G^{n+1}\| < tol$ **then**
5: $\quad\quad V^{n+1} = V^*$
6: $\quad\quad$ **break**
7: $\quad$ **else**
8: $\quad\quad V^* = V^* - (Jac(G^{n+1}))^{-1}G^{n+1}$ and go back to 3.
9: $\quad$ **end if**
10: **end for**

where $\nabla(\Sigma^{n+1}) = ((\nabla(\sigma_1^{n+1})^2)^T, (\nabla(\sigma_2^{n+1})^2)^T, \ldots, (\nabla(\sigma_M^{n+1})^2)^T)^T$ (each $\nabla(\sigma_k^{n+1})^2$ is a row vector) simplifies the computing of the Jacobian matrix in terms of the nonlinear volatility. In the concerning models (Model I,II and III) $\nabla(\Sigma^{n+1})$ is at most a tri-diagonal matrix. Each $\nabla(\sigma_k^{n+1})^2$ can be obtained by either deriving the exact formula or by using a finite difference replacement of the spatial derivatives.

Algorithm 4 stops on the norm of the function G or could have been extended by also including a check on the correction of the Newton's iteration. A more appropriate size of performing the updating is discussed in Sect. 12.3.

12.3 Newton-Like Methods

Solving the root-finding problem with Algorithm 4 can obtain the numerical solution of the original nonlinear partial differential equation (12.1). But there are some drawbacks of this method, for example, the oscillations of root-finding function might occur which affects the convergence of the solution. Also the cost of evaluating Jacobian matrix is computationally expensive when analytic formula doesn't exist. Therefore in this section, the aim is to introduce several improvements for Algorithm 4 to make it more efficient and robust. Different strategies are introduced here and some of them may be merged or combined in actual implementation to optimise the performance of the iterative method. To simply the notation, Eq. (12.4) is rewritten as

$$G(v) = 0,$$

with the updating formula

$$v^{k+1} = v^k + \delta v^k, \quad \delta v^k = -[Jac(G(v^k))]^{-1} G(v^k),$$

where k represents the iteration step.

- ***Damped Newton's Method***

The updating direction δv^k sometimes may become too large resulting into an unnecessary large change to the current approximate solution leading to a lost in accuracy. Also this large change may occur oscillations of the root-finding function which affects the convergence of solution a lots. In order to avoid this situation, a damping factor s is usually inserted to scale the direction in order to provide a smaller and safer correction in the updating process. This leads to

$$v^{k+1} = v^k + s\delta v^k,$$

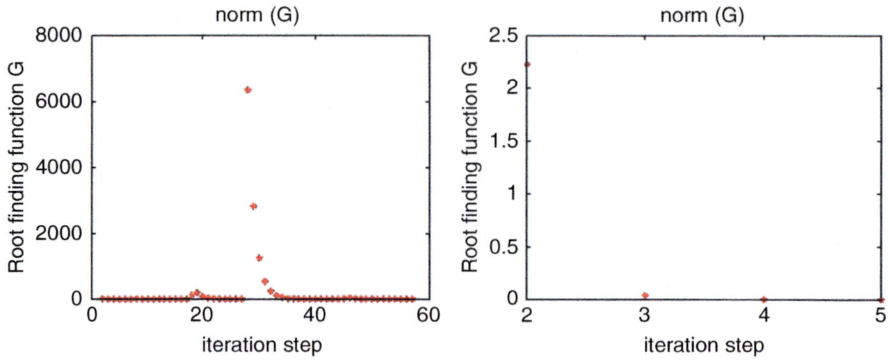

Fig. 12.1 Results of Frey-Patie model with $\Delta\tau = 2.833\text{e}{-04}$, $\Delta S = 0.75$. The *left plot* is without damped updating and the *right plot* is using damped updating

in which s may be chosen as 2^{-m} (see [16]) such that m is the smallest integer satisfying

$$\left\|G(v^k + 2^{-m}\delta v^k)\right\| < (1 - \alpha 2^{-m}) \left\|G(v^k)\right\|,$$

where α is usually chosen as a small number.

Figure 12.1 gives an example of using damped Newton's method to achieve an adaptive updating. It can be observed that using damped Newton's method can avoid the oscillations of the root-finding function, though it might require to evaluate the root-finding function at least twice within a Newton's iteration. However this extra cost can guarantee a more robust result of applying the Newton-based solver.

- ***Inexact Newton's method***

One main problem when applying Newton's method is that it is expensive to calculate the updating direction δv^k, which requires the inversion of the Jacobian matrix $[Jac(G)]^{-1}$. Therefore reducing the number of iterations would be a way to shorten the overall computation time. The idea is to do the whole Newton's iteration approximately as shown in [5] by finding an update direction which satisfies the inexact Newton condition

$$\left\|G(v^k) + Jac[G(v^k)]\delta v^k\right\| < \eta \left\|G(v^k + \delta v^k)\right\|,$$

where η is known as the forcing term.

In terms of implementation there are two loops, one being the inner loop for finding the updating direction and the other being the outer Newton's iterative loop. The choice of η is important. Small values of η reduce the iteration to become simply Newton's method. Other choices of η may not improve the result, but rather lead to a poorer one. Discussions of suitable choices can be found from [16, 18].

- ***Jacobian-free Newton-Krylov (JFNK) method***

Once the procedure for regulating the direction in the updating formula and the inexact Newton condition are applied, the most challenging problem is the evaluation of the Jacobian matrix, or, alternatively, the efficient multiplication of the Jacobian matrix and a vector in the inexact Newton condition when an exact formula is not readily available. To take on this challenge the so-called Jacobian-free Newton-Krylov technique, which is one of the variants from the class of Newton-Krylov methods, may be used. In the Jacobian-free Newton-Krylov method, in which the evaluation of the Jacobian matrix is not necessary, the input for the inexact Newton condition can be approximated as

$$Jac(G)u \approx \frac{G(v+\epsilon u) - G(v)}{\epsilon},$$

where ϵ is a small number. The choice of ϵ again is important which can not be too large. For more detailed discussion and numerical analysis one can check [18].

- ***Broyden-type method***

Another approach to avoid the excessive computation of the Jacobian matrix is Broyden's method. From experiences through numerical experiments, a highly accurate Jacobian matrix in each iteration is often unnecessary in achieving fast convergence. Broyden's method relies on the concept of a generalised secant method which leads to an iterative scheme for approximating the Jacobian matrix,

$$Jac(G)_n = Jac(G)_{n-1} + \frac{\Delta G - Jac(G)_{n-1}\Delta v}{\|\Delta v\|^2}\Delta v^T.$$

One shortcoming of Broyden's method is that the matrix structure of each new Broyden's approximation to the Jacobian matrix easily changes from one iteration to another. For example, suppose that the original Jacobian matrix is a tri-diagonal matrix. Then it might become a full matrix after applying one step of Broyden's method. This means that it might perturb the approximate solution in the Newton's iteration in the wrong direction. There are several modifications of Broyden's method [12, 19, 20], for example, one is to preserve the matrix structure by using the sparse Broyden method in [22]. In some cases the implementation of this sparsity preservation method does not save much time. Therefore a simple trick adopted in the numerical tests is to pick up only the tri-diagonal entries after performing each iteration in Broyden's method. Comparisons of using different Broyden-type methods can be found in [7].

- ***Fundamental algorithm***

Finally, a fundamental algorithm in combining these Newton-like techniques with Algorithm 4, as given below in Algorithm 5, in an attempt to improve the performance of root-finding approach in solving nonlinear parabolic equations.

Algorithm 5 Newton-like method (NLM)

Input: M, N, r, σ_0, tol, initial condition $V^0 = V(S, \tau = 0)$, initial guess V^*
Output: $V^N = V(S, \tau = T)$

1: **for** n = 0 to N-1 **do**
2: $\quad V^{n+1} = V^*$
3: $\quad G^{n+1} = H(V^{n+1})V^{n+1} - V^n$
4: $\quad$ **if** $\|G^{n+1}\| < tol$ **then**
5: $\qquad V^{n+1} = V^*$
6: $\qquad$ **break**
7: $\quad$ **else**
8: $\qquad$ • Evaluate $Jac(G)d$ by the *JFNK* method or evaluate $Jac(G)$ by Broyden's method,
9: $\qquad$ • Decide d to satisfy the Inexact Newton condition,
10: $\qquad$ • Use d and apply Damped Newton method to do update V^*, and go back to 3.
11: $\quad$ **end if**
12: **end for**

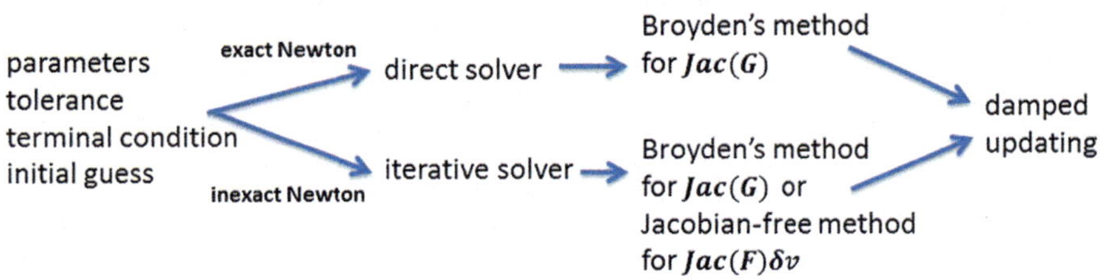

Fig. 12.2 Different strategies to implement Newton-like method

Figure 12.2 also illustrate different strategies of implementing Newton-like methods in exact or inexact sense.

12.4 Deferred Correction Problem

Instead of solving the root-finding problem as shown in Eq. (12.4), another approach of applying Newton's linearisation is to consider a smooth function F representing the nonlinear Black-Scholes equation (12.1), i.e.

$$F(V_\tau, V_S, V_{SS}, V) = V_\tau - \frac{1}{2}\sigma^2 S^2 V_{SS} - rSV_S + rV = 0,$$

and the linearisation of the function F at $(V_\tau^*, V_S^*, V_{SS}^*, V^*)$ in direction (e_τ, e_S, e_{SS}, e) reads as follows:

$$\begin{aligned} &F(V_\tau^* + e_\tau, V_S^* + e_S, V_{SS}^* + e_{SS}, V^* + e) \\ &= F(V_\tau^*, V_S^*, V_{SS}^*, V^*) + \frac{\partial F}{\partial V_\tau^*} e_\tau + \frac{\partial F}{\partial V_S^*} e_S + \frac{\partial F}{\partial V_{SS}^*} e_{SS} + \frac{\partial F}{\partial V^*} e + O(D^2), \end{aligned} \tag{12.5}$$

Algorithm 6 Deferred correction problem (NM2)

Input: M, N, r, σ_0, tol, initial guess V^*
Output: $V^N = V(S, \tau = T)$

```
1: for n = 0 to N-1 do
2:     Evaluate ∂F/∂V*_τ, ∂F/∂V*_S, ∂F/∂V*_SS, ∂F/∂V* by V*
3:     Solve Eq. (12.5) to obtain e
4:     if ||e|| < tol then
5:         V^{n+1} = V*
6:         break
7:     else
8:         V* = V* + e and go back to 2.
9:     end if
10: end for
```

where D^2 represents all higher order terms and the partial derivatives are evaluated at $(V_\tau^*, V_S^*, V_{SS}^*, V^*)$.

The original nonlinear differential equation becomes a linear parabolic partial differential equation of the correction term e with zero boundary and initial conditions. Equation (12.5) is easier to solve once all the coefficients like $\frac{\partial F}{\partial V_\tau^*}$, $\frac{\partial F}{\partial V_S^*}$, $\frac{\partial F}{\partial V_{SS}^*}$, $\frac{\partial F}{\partial V^*}$ are computed and these coefficients again can be evaluated by either deriving the exact formula or by using a finite difference replacement of the derivatives. This approach is also known as Waveform-Newton [9, 17, 21] and is described in Algorithm 6.

12.5 Numerical Experiments

In this section different comparisons of solving Eq. (12.1) are shown and discussed in terms of the accuracy, number of iterations and computation time with boundary and terminal conditions given as introduced in Sect. 12.1. Some common model parameters were chosen as

$$\sigma_0 = 0.4, \ K = 100, \ r = 0.03, \ S_{\min} = 0, \ S_{\max} = 300, \ T = 1/12$$

and the tolerance of Newton-based solvers was set as 10^{-8}. The initial guess V^* for Newton-based solvers was selected to be the approximate solution at the previous time level (for the first iteration the $V^* = V(S, \tau = 0)$). M and N are the grid points used for the spatial and temporal discretisation respectively. The special parameters used in Model I, II, III are $a = 0.05, \rho = 0.03, \lambda(S) \equiv 1, \mu = 0.05$. For exact Newton's method the Thomas algorithm was used as the tri-diagonal solver to do the inversion of Jacobian matrix. For the inexact Newton's method, Jacobi method was applied as an iterative solver. The computations were implemented using MATLAB R2013b.

12.5.1 Comparisons of Numerical Methods with Explicit Invariant Solution

In order to ensure all the Newton-based solvers are accurate, the explicit invariant solutions for the Frey-Patie model (model III) derived by Bordag in [3, (86), (87)] with parameters $c = -0.05$, $d_1 = 0$, $d_2 = 30$ were computed and taken as reference solutions for evaluating the experimental order of convergence (EOC). The boundary conditions and initial conditions were generated from these invariant solutions.

The experimental order of convergence (or convergence ratio) is constructed from the convergence rate of the error defined as follows:

$$\alpha = \frac{\log((Err)_{m+1}/(Err)_m)}{\log((\Delta S)_{m+1}/(\Delta S)_m)}.$$

Here the error E is defined as $E = \|V(S,\tau) - \hat{V}(S,\tau)\|/\|\hat{V}(S,\tau)\|$ for $S \in [0.5E, 1.5E]$, where $V(S,\tau)$ is the solution from numerical solver, and $\hat{V}(S,\tau)$ is from the invariant solution. The ratio $(\Delta S)^2/\Delta\tau$ is fixed to be 108,000, and $(\Delta S)_{m+1}/(\Delta S)_m = 0.5$. Tables 12.1 and 12.2 show results for the l_∞ maximum norm and l_2 integral norm. Both results demonstrate that all the solvers converge to the same solution which converges to the explicit invariant solution with refined grid points.

Table 12.1 EOC for the Frey-Patie Model with the l_∞ maximum norm

$\Delta\tau$	ΔS	E_{NM1}	α_{NM1}	E_{NM2}	α_{NM2}	E_{Frozen}	α_{Frozen}
0.00833	30	2.93e−05	–	2.93e−05	–	2.93e−05	
0.00208	15	1.72e−06	4.09	1.72e−06	4.09	1.72e−06	4.09
5.21e−04	7.5	1.02e−07	4.08	1.02e−07	4.08	1.02e−07	4.08
1.30e−04	3.75	2.50e−08	2.02	2.50e−08	2.02	2.50e−08	2.02
3.26e−05	1.875	5.00e−09	2.32	5.00e−09	2.32	5.00e−09	2.32
8.14e−06	0.9375	1.25e−09	2.00	1.25e−09	2.00	1.25e−09	2.00

Table 12.2 EOC for the Frey-Patie Model with the l_2 integral norm

$\Delta\tau$	ΔS	E_{NM1}	α_{NM1}	E_{NM2}	α_{NM2}	E_{Frozen}	α_{Frozen}
0.00833	30	2.93e−05	–	2.93e−05	–	2.93e−05	
0.00208	15	1.79e−06	4.03	1.79e−06	4.03	1.79e−06	4.03
5.21e−04	7.5	1.39e−07	3.68	1.39e−07	3.68	1.39e−07	3.68
1.30e−04	3.75	4.46e−08	1.64	4.46e−08	1.64	4.46e−08	1.64
3.26e−05	1.875	1.25e−08	1.83	1.25e−08	1.83	1.25e−08	1.83
8.14e−06	0.9375	4.32e−09	1.53	4.32e−09	1.53	4.32e−09	1.53

Table 12.3 Average number of iterations for Model I

M	N	NM1	NM1*	NM2	NM2*	Frozen	Frozen*
40	40	4.02	4.20	4.07	4.60	10.1	27.6
80	80	4.02	4.40	5.03	5.60	9.28	53.4
160	160	5.01	5.4	–	–	–	–

Table 12.4 Average number of iterations for Model II

M	N	NM1	NM1*	NM2	NM2*	Frozen	Frozen*
40	40	5.00	5.00	14.2	14.0	4.32	5.00
80	80	3.53	4.20	21.2	21.0	4.22	5.00
160	160	3.28	5.00	32.6	32.0	4.12	5.20

Table 12.5 Average number of iterations for Model III

M	N	NM1	NM1*	NM2	NM2*	Frozen	Frozen*
40	40	3.15	4.00	3.05	3.40	5.20	6.60
80	80	3.21	5.40	3.10	4.20	5.28	12.8
160	160	3.21	8.6	3.15	7.40	–	–

12.5.2 Comparisons of Number of Iterations

Tables 12.3, 12.4 and 12.5 show the number of iterations of using frozen coefficient method and Newton-based solvers to different models. The numbers 'NM1, NM2, Frozen' are the average of number of iterations in all time steps and the numbers 'NM1*, NM2*, Frozen*' are the average of number of iteration just for the first five time steps. The numerical experiments show that the root-finding approach (NM1) with adaptive updating (damped Newton's method) is the most robust one as the deferred correction (NM2) is sensitive to the nonlinearity of problems and sometimes also doesn't converge. Frozen coefficient method requires more number of iterations and as shown in the tables some results diverge when grid sizes become small.

An observation shown in Fig. 12.3 is that after several time steps, the number of iterations for frozen coefficient method seems to reduce and not very far from the ones in Newton-based solvers. This can help to create hybrid solvers, namely using Newton-based solvers in the beginning of several time steps, and switch to frozen coefficient method which the computation is cheaper.

12.5.3 Comparisons of Computation Time

Tables 12.6, 12.7 and 12.8 show the computation time of using frozen coefficient method, Newton-based and the hybrid solvers to different models. The Jacobian

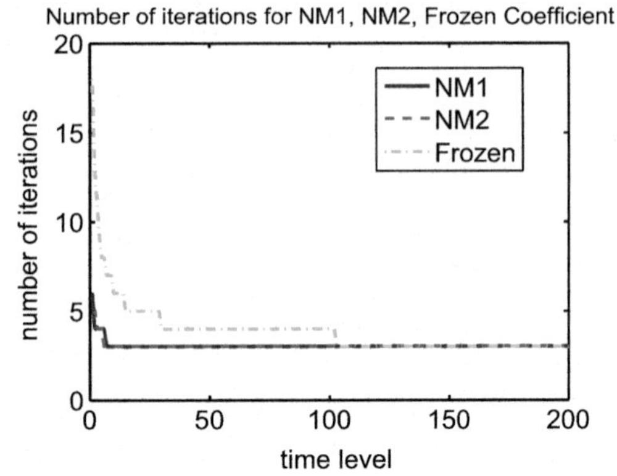

Fig. 12.3 Number of iterations in the first 240 time steps for Model III with $\Delta\tau = 8.14\text{e}{-06}$, $\Delta S = 0.9375$. The *red line* is for NM1; the *blue* one is for NM2; the *black* one is for the frozen coefficient method

Table 12.6 Computation time (sec) for Model I

M	N	NM1	NM2	Frozen	Hybrid
40	40	0.08	0.08	0.10	0.10
80	80	0.19	0.17	0.23	0.19
160	160	0.88	–	–	0.59

Table 12.7 Computation time (sec) for Model II

M	N	NM1	NM2	Frozen	Hybrid
40	40	0.08	0.16	0.07	0.08
80	80	0.17	0.64	0.14	0.14
160	160	0.51	3.56	0.40	0.37

Table 12.8 Computation time (sec) for Model III

M	N	NM1	NM2	Frozen	Hybrid
40	40	0.08	0.07	0.05	0.10
80	80	0.16	0.12	0.07	0.17
160	160	0.50	0.33	–	0.46

matrix is evaluated by derived formula. The hybrid solver is a merged version of root-finding approach and frozen coefficient method which in the first 6 time steps, Algorithm 4 is applied to have a faster convergent result, and after that Algorithm 3 is used to get the benefit of easier calculation. The results show that the hybrid solver can take advantages from both solvers and is also robust.

Another comparison in Fig. 12.4 shows using different Newton-like methods to solve Frey-Patie Model. The aim is to compare using exact and inexact Newton's methods, combined with finite difference method and Broyden-type method to approximate the Jacobian matrix. The grid points of temporal discretisation is fixed to be $N = 1000$ and of spatial discretisation were chosen as $M = 50, 100, 150, 200, 250$. Damped Newton's method was also used to guarantee

Fig. 12.4 Different strategies to implement Newton-like method

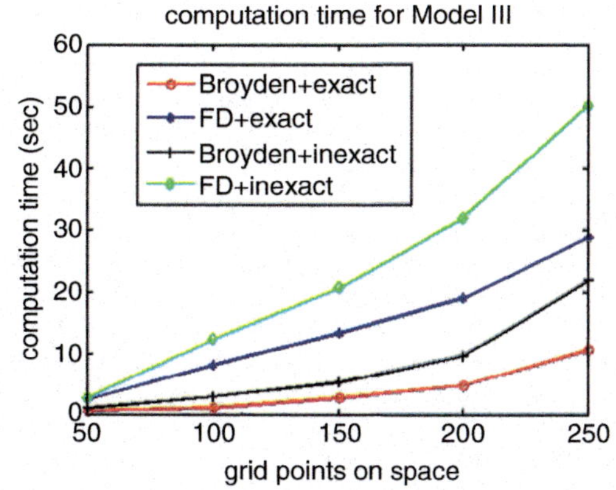

an adaptive updating. It is observed that Broyden-type method can replace the finite difference method to evaluate the Jacobian matrix and reduce the computation time. Also direct solver performs well than iterative solver, as the systems to be solved are always with simple tri-diagonal structure. It demonstrates that if there is no exact formula for Jacobian matrix, then a better strategy of solving nonlinear Black-Scholes equation in one-dimensional case is to choose exact Newton's method with some approximation of Jacobian matrix like Broyden-type methods.

12.6 Conclusions

Some important algorithms based on Newton's method and its variants are introduced to solve nonlinear parabolic partial differential equations from financial market. The main idea is to use Newton's method with invariant approaches, and to improve the method based on various techniques in order to tackle drawbacks. Essentially these Newton-based solvers are efficient and robust, and can be adopted to different kinds of nonlinear Black-Scholes equations. In the end some improvements like using hybrid solver, or applying Newton-like methods to do some approximation are also addressed and compared with numerical experiments.

Acknowledgements The authors thank Prof. Matthias Ehrhardt, Dr. E. Jan W. ter Maten from Bergische Universität Wuppertal, Prof. Daniel Ševčovič from Comenius University Bratislava, Prof. Kevin Parrott and Dr. André Ribeiro from University of Greenwich, for the fruitful discussions.

References

1. Barles, G., Soner, H.M.: Option pricing with transaction costs and a nonlinear Black-Scholes equation. Finance Stochast. **2**, 369–397 (1998). https://people.math.ethz.ch/~hmsoner/pdfs/38-Soner-FS-98.pdf
2. Black, F., Scholes, M.S.: The pricing of options and corporate liabilities. J. Polit. Econ. **81**, 637–654 (1973)
3. Bordag, L.A., Frey, R.: Pricing options in illiquid markets: symmetry reductions and exact solutions, in Nonlinear Models in Mathematical Finance: New Research Trends in Option Pricing, pp. 103–130. Nova Science, New York (2008)
4. Company, R., Jódar, L., Pintos, J.: A consistent stable numerical scheme for a nonlinear option pricing model in illiquid markets. Math. Comput. Simul. **82**(10), 1972–1985 (2012)
5. Dembo, R.S., Eisenstat, S.C., Steihaug,T.: Inexact Newton methods. SIAM J. Numer. Anal. **19**(2), 400–408 (1982)
6. Ďuriš, K., Tan, S.-H., Lai, C.-H., Ševčovič, D.: Comparison of the analytical approximation formula and Newton's method for solving a class of nonlinear Black-Scholes parabolic equations, Comput. Methods Appl. Math. **16**(1), 35–50 (2016)
7. Egorova, V., Tan, S.-H., Lai, C.-H., Company, R., Jódar, L.: Moving boundary transformation for American call options with transaction cost: finite difference methods and computing. Int. J. Comput. Math. **94**, 1–18 (2015). Published online: 08 Dec 2015
8. Ehrhardt, M. (ed.): Nonlinear Models in Mathematical Finance. Nova Science, New York (2008)
9. Erdman, D.J., Rose, D.J.: Newton waveform relaxation techniques for tightly coupled systems. IEEE Trans. Comput. Aided Des. **11**(5), 598–606 (1992)
10. Frey, R.: Market illiquidity as a source of model risk in dynamic hedging. In: Gilbson, E. (ed.) Model Risk. RISK Publication, London (2000)
11. Frey, R., Patie, P.: Risk management for derivatives in illiquid markets: a simulation study. In: Sandmann, K., Schönbucher, P.J. (eds.) Advances in Finance and Stochastics, pp. 137–160. Springer, Berlin (2002). http://www.math.uni-leipzig.de/~frey/frey-patie-simulation-study.pdf
12. Griewank, A.: Broyden updating, the good and the bad. Documenta Mathematica - Extra Volume ISMP 2012, pp. 301–315. http://www.math.uiuc.edu/documenta/vol-ismp/45_griewank-andreas-broyden.pdf.
13. Guyon, J., Henry-Labordère, P.: Nonlinear Option Pricing. CRC Press, Boca Raton (2013)
14. Heider, P.: Numerical methods for nonlinear Black-Scholes equations. Appl. Math. Finance **17**, 59–81 (2010)
15. Jandačka, M., Ševčovič, D.: On the risk-adjusted pricing-methodology-based valuation of vanilla options and explanation of the volatility smile. J. Appl. Math. **3**, 235–258 (2005)
16. Kelley, C.T.: Solving Nonlinear Equation with Newton's Method. SIAM, Philadelphia (1987)
17. Kevenaar, T.A.M.: Periodic steady state analysis using shooting and waveform-Newton. Int. J. Circuit Theory Appl. **22**, 51–60 (1994)
18. Knoll, D.A., Keyes, D.E.: Jacobian-free Newton-Krylov methods: a survey of approaches and applications, J. Comput. Phys. **193**, 357–397 (2004)
19. Martínez, J.M.: Practical quasi-Newton methods for solving nonlinear systems, J. Comput. Appl. Math. **124**(1), 97–121 (2000)
20. Rheinboldt,W.C.: Quasi-Newton methods. Lecture Notes TU Munich (2000). https://www-m2.ma.tum.de/foswiki/pub/M2/Allgemeines/SemWs09/quasi-newt.pdf
21. Saleh, R.A., White,J.: Accelerating relaxation algorithms for circuit simulation using waveform-Newton and step-size refinement. IEEE Trans. Comput. Aided Des. **9**(9), 951–958 (1990)
22. Schubert, L.K.: Modification of a quasi-Newton method for nonlinear equations with a sparse Jacobian. J. Math. Comput. **24**, 27–23 (1970)
23. Ševčovič, D., Stehlíková, B., Mikula,K.: Analytical and Numerical Methods for Pricing Financial Derivatives. Nova Science, New York (2011)

Chapter 13
Implicit-Explicit Schemes for European Option Pricing with Liquidity Shocks

Walter Mudzimbabwe and Lubin Vulkov

Abstract We consider the numerical valuation of European options with liquidity shocks. We propose two implicit-explicit schemes that preserve the positivity of the differential problem. Comparison principle and convergence for the difference schemes are provided. Numerical tests illustrate the theoretical results and show second order accuracy after Richardson extrapolation in time. (This chapter is a summary of the paper (Mudzimbabwe and Vulkov, J Comp Appl Math 299:245–256, 2016); all theoretical statements in this summary are proved in that reference.)

13.1 Introduction

Ludkovski and Shen [5] proposed a nonlinear pricing mechanism based on utility maximization. Their model is an example of regime switching models. They consider a investor whose utility is described by an exponential utility function $\mathscr{U}(x) = -e^{-\gamma x}$ where $\gamma > 0$ is the coefficient of risk aversion.

The investor seeks to maximise utility of both terminal wealth X_T and option payoff h at time horizon $T < \infty$, which is chosen to coincide with the expiration date of all securities in market model. The prices $R^i(t,S)$ are also related to the indifference prices. Then the pair $\{R^i(t,S),\ i = 0,1\}$ is the unique viscosity solutions of the coupled semi-linear system,

$$\begin{aligned} &R_t^0 + \tfrac{1}{2}\sigma^2 S^2 R_{SS}^0 - \tfrac{\nu_{01}}{\gamma} e^{-\gamma(R^1-R^0)} + \tfrac{(d_0+\nu_{01})}{\gamma} = 0, \\ &R_t^1 - \tfrac{\nu_{10}}{\gamma} e^{-\gamma(R^0-R^1)} + \tfrac{\nu_{10}}{\gamma} = 0. \end{aligned} \tag{13.1}$$

The terminal conditions are:

$$R^i(T,S) = h(S), \quad i = 0,1.$$

W. Mudzimbabwe • L. Vulkov (✉)
Department of Numerical Analysis and Statistics, Ruse University Angel Kanchev, Ruse, Bulgaria
e-mail: wmudzimbabwe@uni-ruse.bg; lvalkov@uni-ruse.bg

M. Ehrhardt et al. (eds.), *Novel Methods in Computational Finance*, Mathematics in Industry 25, DOI 10.1007/978-3-319-61282-9_13

Here σ is the volatility of the underlying, ν_{01}, ν_{10} are the transition intensities from state (0) to state (1) and vice versa, respectively, μ denotes the drift of the underlying and $d_0 = \mu^2/2\sigma^2$, see [4, 5] for more details.

The indifference prices

$$p = R^0 + \gamma^{-1} \ln F_0(t), \quad q = R^1 + \gamma^{-1} \ln F_1(t),$$

can be shown to satisfy the parabolic-ordinary system

$$\begin{aligned} &p_t + \tfrac{1}{2}\sigma^2 S^2 p_{SS} - \tfrac{v_{01}}{\gamma}\tfrac{F_1}{F_0} e^{-\gamma(q-p)} + \tfrac{(d_0+v_{01})}{\gamma} - \tfrac{1}{\gamma}\tfrac{F_0'}{F_0} = 0, \\ &q_t - \tfrac{v_{10}}{\gamma}\tfrac{F_0}{F_1} e^{-\gamma(p-q)} + \tfrac{v_{10}}{\gamma} - \tfrac{1}{\gamma}\tfrac{F_1'}{F_1} = 0 \end{aligned} \tag{13.2}$$

with terminal conditions

$$p(T, S) = q(T, S) = h(S). \tag{13.3}$$

The numerical solution of the system (13.2) is the main object of this section. A possible way to build an efficient numerical solution of (13.2), (13.3) is to implement an IMEX method [1–3, 9]. In this procedure, the diffusion term is discretized implicitly in time and the reaction terms are discretized explicitly. By making the substitutions $\tau = T - t$, $u = \gamma R^0$ and $v = \gamma R^1$, the system (13.1) becomes

$$\begin{aligned} L^p(u, v) &\equiv u_\tau - \tfrac{1}{2}\sigma^2 S^2 u_{SS} + a e^u e^{-v} - b = 0, \\ L^0(u, v) &\equiv v_\tau + c e^v e^{-u} - c = 0, \end{aligned} \tag{13.4}$$

where $a = \nu_{01}$, $b = d_0 + \nu_{01}$, $c = \nu_{10}$. In accordance with (13.3) we take the initial conditions to be

$$u(0, S) = u_0(S) = \gamma h(S), \quad v(0, S) = v_0(S) = \gamma h(S). \tag{13.5}$$

13.2 IMEX Linear Scheme

Here we develop a linear IMEX scheme to solve the coupled semi-linear parabolic-ordinary system problem (13.4). We consider a call option with boundary conditions [10, 11]

$$u(\tau, 0) = \varphi_l(\tau) = 0, \quad u(\tau, S) = \varphi_r(\tau) \approx S_{\max} \text{ for large } S.$$

The left natural boundary condition for u reads as follows

$$u_\tau(\tau, 0) = -a e^{-(v(\tau,0) - u(\tau,0))} + b.$$

On the domain $Q_T = \Omega \times [0, T]$ we introduce the uniform mesh $w_{S\tau} = w_S \times w_\tau$:

$$\overline{w}_S = \{S_i = i\Delta S,\ \ \Delta S > 0,\ \ i = 0, 1, \ldots, I;\ \ I\Delta S = S_{max}\},\ \ \overline{w}_S = w_S \cup \{S_0, S_I\};$$

$$\overline{w}_\tau = \{\tau_j = j\Delta\tau,\ \ \Delta\tau > 0,\ \ j = 0, 1, \ldots, J;\ \ J\Delta\tau = T\}.,\ \ \overline{w}_\tau = w_\tau \cup \{\tau_0, \tau_J\}$$

Next, on the discrete domain $w_{S\tau}$ we approximate the problem (13.4)–(13.5) by the finite difference scheme [7, 8]

$$L^p(U, V) = \frac{U_i^{j+1} - U_i^j}{\Delta\tau} - \frac{1}{2}\sigma^2 S_i^2 \frac{U_{i-1}^{j+1} - 2U_i^{j+1} + U_{i+1}^{j+1}}{(\Delta S)^2} +$$

$$ae^{-V_i^j} e^{U_i^j} - b = 0, \tag{13.6}$$

$$L^0(U, V) = \frac{V_i^{j+1} - V_i^j}{\Delta\tau} + ce^{-U_i^j} e^{V_i^j} - c = 0, \tag{13.7}$$

for $i = 1, \ldots, I-1, j = 0, 1, \ldots, J-1.$

$$U_i^0 = U_0(S_i),\ \ i = 0, 1, \ldots, I; \tag{13.8}$$

$$U_0^j = \varphi_l(\tau_j),\ \ U_i^j = \varphi_r(\tau_j),\ \ j = 0, 1, \ldots, J; \tag{13.9}$$

$$V_i^0 = V_0(S_i),\ \ i = 1, \ldots, I. \tag{13.10}$$

We approximate the natural boundary conditions explicitly

$$U_0^{j+1} = U_0^j - \Delta\tau(ae^{-V_0^j} e^{U_0^j} - b).$$

In algebraic form of the on the $(j+1)$-th time level $j = 0, 1, \ldots, J-1$, the scheme (13.6) and (13.7) reads as follows

$$-A_i U_{i-1}^{j+1} + C_i U_i^{j+1} - B_i U_{i+1}^{j+1} = F_i,$$

$$V_i^{j+1} = V_i^j - \Delta\tau c e^{-U_i^j + V_i^j} + c,$$

where

$$A_i = B_i = \frac{1}{2}\sigma^2 \frac{S_i^2}{(\Delta S)^2},\ \ C_i = \frac{1}{\Delta\tau} + A_i + B_i,$$

$$F_i = \frac{1}{\Delta\tau} U_i^j - ae^{-V_i^j} e^{U_i^j} + b,\ \ i = 1, \ldots, I-1;$$

Theorem 13.1 *Let* $(u, v) \in C^{2,4}(Q_T)$ *is classical solution of problem* (13.4)–(13.5). *Then the error estimate holds*

$$\|u - U\|_{C(w_{S\tau})} + \|v - V\|_{C(w_{S\tau})} \le C(\Delta\tau + (\Delta S)^2),$$

for sufficiently small ΔS *and* $\Delta\tau$, *where the constant* C *doesn't depend on* ΔS *and* $\Delta\tau$.

The discrete comparison principle for (U, V) is important for the positivity of the numerical indifference prices p and q.

Theorem 13.2 *Suppose that the assumptions of Theorem 13.1 hold and* $(\overline{U}, \overline{V})$, $(\underline{U}, \underline{V})$ *be grid functions defined on* $\overline{w}_{S\tau}$. *Let the following inequalities*

$$L^p(\overline{U}, \overline{V}) \ge L^p(\underline{U}, \underline{V}), \quad L^0(\overline{U}, \overline{V}) \ge L^0(\underline{U}, \underline{V}),$$
$$\overline{U}_i^0 \ge \underline{U}_i^0, \quad \overline{V}_i^0 \ge \underline{V}_i^0, \quad i = 0, \dots, I,$$
$$\overline{V}_0^j \ge \underline{V}_0^i, \quad \overline{U}_M^j \ge \underline{U}_M^j, \quad j = 1, \dots, J.$$

hold and let ΔS *and* $\Delta\tau$ *are sufficiently small such that* $\Delta\tau$ *satisfies*

$$\Delta\tau < \min(a, c)e^{2C_u}e^{2C_v}.$$

Then

$$\overline{U}_i^j \ge \underline{U}_i^j, \quad \overline{V}_i^j \ge \underline{V}_i^j, \quad i = 0, 1 \dots, I, \quad j = 0, 1, \dots, J.$$

13.3 IMEX Linearised Scheme

Let us start with the fully implicit scheme:

$$\frac{U_i^{j+1} - U_i^j}{\Delta\tau} - \frac{1}{2}\sigma^2 S_i^2 \frac{U_{i-1}^{j+1} - 2U_i^{j+1} + U_{i+1}^{j+1}}{(\Delta S)^2} + ae^{V_i^{j+1}}e^{U_i^{j+1}} - b = 0,$$
$$\frac{V_i^{j+1} - V_i^j}{\Delta\tau} + ce^{-U_i^{j+1}}e^{V_i^{j+1}} - c = 0,$$

for $i = 1, 2, \dots, I-1, j = 0, 1, \dots, J-1$ with boundary and initial approximations given by (13.8)–(13.10). By Taylor expansion the exponential nonlinear terms can be linearised to obtain the IMEX linearised scheme:

$$\frac{U_i^{j+1}-U_i^j}{\Delta\tau}-\frac{1}{2}\sigma^2 S_i^2\frac{U_{i-1}^{j+1}-2U_i^{j+1}+U_{i+1}^{j+1}}{(\Delta S)^2}+a(e^{-V_i^j}e^{U_i^j}(1+V_i^j-U_i^j)$$

$$+e^{-V_i^j}e^{U_i^j}(U_i^{j+1}-V_i^{j+1}))-b=0, \tag{13.11}$$

$$\frac{V_i^{j+1}-V_i^j}{\Delta\tau}+c(e^{-U_i^j}e^{V_i^j}(1-V_i^j+U_i^j)+e^{-U_i^j}e^{V_i^j}(V_i^{j+1}-U_i^{j+1}))-c=0, \tag{13.12}$$

which simplifies to

$$-\hat{A}_iU_{i-1}^{j+1}+\hat{C}_iU_i^{j+1}-\hat{B}_iU_{i+1}^{j+1}+\hat{D}_iV_i^{j+1}=\hat{F}_i, \tag{13.13}$$

$$\hat{E}_iU_i^{j+1}+\hat{K}_iV_i^{j+1}=G_i, \tag{13.14}$$

where

$$\hat{A}_i=\hat{B}_i=\frac{1}{2}\sigma^2\frac{S_i^2}{(\Delta S)^2},\ \hat{C}_i=\frac{1}{\Delta\tau}+\hat{A}_i+\hat{B}_i+ae^{U_i^j-V_i^j},$$

$$\hat{D}_i=-a\tau e^{U_i^j-V_i^j},\ \hat{F}_i=\frac{1}{\Delta\tau}U_i^j-a\tau e^{U_i^j-V_i^j}(1+V_i^j-U_i^j)+b\Delta\tau,$$

$$\hat{E}_i=-c\Delta\tau e^{V_i^j-U_i^j},\ \hat{K}_i=\frac{1}{\Delta\tau}+ce^{V_i^j-U_i^j},$$

$$G_i=\frac{1}{\Delta\tau}V_i^j-c\tau e^{V_i^j-U_i^j}(1-V_i^j+U_i^j)+c.$$

Substituting V_i^{j+1} from (13.14) into (13.13) we get

$$-\hat{A}_iU_{i-1}^{j+1}+\left(\hat{C}_i-\frac{\hat{D}_i\hat{E}_i}{\hat{K}_i}\right)U_i^{j+1}-\hat{B}_iU_{i+1}^{j+1}=\hat{F}_i-\frac{\hat{D}_i}{\hat{K}_i}\tilde{F}_i,$$

$$V_i^{j+1}=\frac{G_i}{\hat{K}_i}-\frac{\hat{E}_i}{\hat{K}_i}U_i^{j+1},\quad i=1,\ldots,I-1$$

with U_i^0, $i=0,1,\ldots,I$ and U_0^j, U_I^j, $j=0,1,\ldots,J$ given by (13.8)–(13.10).

Since $ae^{U_i^j-V_i^j}>0$, the diagonal domination can significantly increase in comparison with IMEX linear scheme (13.6) and (13.7).

Theorem 13.3 *Suppose that the assumptions of Theorem 13.1 hold and that there exists a classical solution $(u,v)\in C^{2,4}(Q_T)$ of problem (13.4)–(13.5). Then for sufficiently small ΔS and $\Delta\tau$ the following error estimate holds:*

$$\|u-U\|_{C(w_{S\tau})}+\|v-V\|_{C(w_{S\tau})}\le C(\Delta\tau+(\Delta S)^2),$$

where the constant C doesn't depend on ΔS and $\Delta\tau$.

13.4 Numerical Experiments

In what follows, we try to investigate the accuracy, effectiveness and convergence of the implicit-explicit linear scheme (13.6)–(13.10) (Scheme 1) and implicit-explicit linearised scheme (13.11) and (13.12) (Scheme 2). We perform numerical experiments both with uniform and non-uniform meshes. Also, we present results of numerical experiments using Richardson extrapolation in time.

The following two tables show the accuracy in maximal discrete norm $\|\cdot\|_\infty$ and convergence rate at initial time $t = 0$, using two consecutive meshes with formulas

$$\text{Order} = \log_2(E^w_{I/2}/E^w_I), \quad E^w_I = \|W^{I/2} - W^I\|_\infty,$$

where W is R^0 or R^1.

We improve the convergence in time by applying Richardson extrapolation [3]. To this end we use the formula

$$Y_n = \frac{2^p W_n - Z_n}{2^p - 1},$$

where p is order of numerical solution (1 in our case) and W_n is the solution obtained using time step $\Delta\tau/2$ and Z_n is the solution obtained using time step $\Delta\tau$. The resulting solution Y_n has order of accuracy $p + 1$.

Table 13.1 shows the result of applying this technique to the Scheme 1. The order of accuracy in time is now two. Similarly this technique is applied to Scheme 2, see Table 13.2. The convergence is much slower but smoother compared to the explicit based Scheme 1 due the error of linearisation. Tables 13.1 and 13.2 shows second order in time.

In Fig. 13.1 we compare option values p and q at issue and maturity in the liquid and illiquid states using the parameters $\mu = 0.06$, σ=0.3, $\nu_{01} = 1$, $\nu_{10} = 12$, $K = 2$, $T = 1$, $S_{\max} = 5$ and $\gamma = 1$ using Scheme 1. These values are taken arbitrary but

Table 13.1 Convergence results for at the money ($S = 2$, $K = 2$, $S_{\min} = 0$ and $S_{\max} = 5$) and taking $\Delta\tau = \Delta S/2$ based on Scheme 1 using Richardson extrapolation

I	Z_n, W_n	Y_n	Difference	Ratio (order)
10	0.2451080			
20	0.2465578	0.2480075		
40	0.2472811	0.2480045	3.02e−6	
80	0.2476431	0.2480051	5.79e−7	5.22 (2.38)
160	0.2478242	0.2480053	1.96e−7	2.96 (1.56)
320	0.2479148	0.2480053	5.13e−8	3.82 (1.93)
640	0.2479600	0.2480053	1.27e−8	4.05 (2.02)
1280	0.2479827	0.2480053	3.08e−9	4.12 (2.04)
2560	0.2479940	0.2480053	7.45e−10	4.13 (2.05)

Table 13.2 Convergence results for at the money ($S = 2$, $K = 2$, $S_{\min} = 0$ and $S_{\max} = 5$) and taking $\Delta\tau = \Delta S/2$ based on Scheme 2 using Richardson extrapolation

I	Z_n, W_n	Y_n	Difference	Ratio (order)
10	0.2451717			
20	0.2465832	0.2479947		
40	0.2472928	0.2480023	7.64e−6	
80	0.2476486	0.2480045	2.14e−6	3.57 (1.84)
160	0.2478269	0.2480051	6.22e−7	3.44 (1.78)
320	0.2479161	0.2480053	1.78e−7	3.49 (1.81)
640	0.2479607	0.2480053	4.93e−8	3.62 (1.85)
1280	0.2479830	0.2480053	1.32e−8	3.73 (1.90)
2560	0.2479942	0.2480053	3.46e−9	3.81 (1.93)
5120	0.2479998	0.2480053	8.95e−10	3.87 (1.95)
10240	0.2480026	0.2480053	2.29e−10	3.91 (1.97)

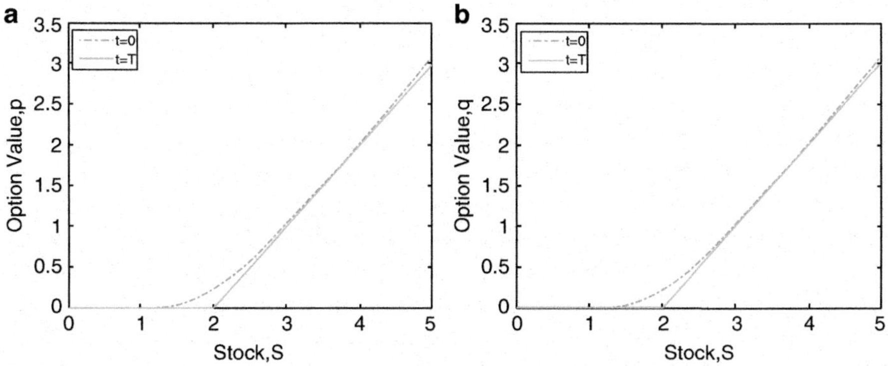

Fig. 13.1 Comparing European option values at issue and maturity in the liquid and illiquid states for the IMEX Linear scheme. (**a**) p at $t = 0$ and $t = T$. (**b**) q at $t = 0$ and $t = T$

are justified e.g. $\nu_{10} = 12$ is much larger than ν_{01} and means that the market will be illiquid for about a month during a year.

Figure 13.2 illustrate the solution from the linearised scheme, using the same parameters. Figures 13.1 and 13.2 show that for this set of parameters, the solution (p, q) is positive using both schemes. This is in line with our theoretical results that $p, q \geq 0$ when $h \geq 0$.

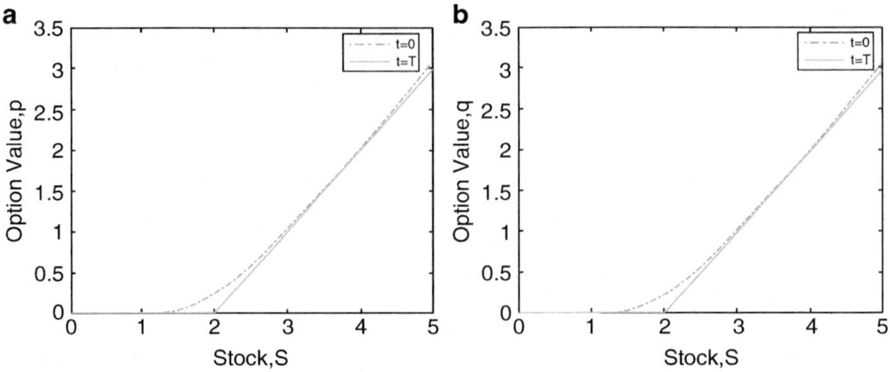

Fig. 13.2 Comparing European option values at issue and maturity in the liquid and illiquid states for the IMEX linearised scheme. (**a**) p at $t = 0$ and $t = T$. (**b**) q at $t = 0$ and $t = T$

13.5 Conclusions

In this chapter, we have considered a one-dimensional problem of European options with liquidity shocks. We have constructed and analyzed two IMEX finite difference schemes and for both schemes they preserve the positivity property of the differential solution. The second one (the IMEX linearized scheme) has better diagonal domination, hence it is monotone. It would be interesting to consider extensions of the IMEX schemes to the American options with liquidity shocks. In this case one has to solve a free boundary problem. It could be written as a linear complementary problem which could be discretized using the schemes given here. The extension is beyond the scope of this chapter, and we leave it for further work.

Acknowledgements This research was supported by the European Union under Grant Agreement number 304617 (FP7 Marie Curie Action Project Multi-ITN STRIKE—Novel Methods in Computational Finance) and the Bulgarian National Fund of Science under Project I02/20-2014.

Tables 13.1, 13.2 and Figs. 13.1, 13.2 are reproduction from [6] with permission of Elsevier, License number 4050190237741/February 15, 2017.

References

1. Asher, U.M., Ruuth, S.J., Wetton, B.T.: Implicit-explicit methods for time dependent partial differential equations. SIAM J. Numer. Anal. **32**(3), 797–823 (1995)
2. Briani, M., Natalini, R., Russo, G.: Implicit-explicit numerical schemes for jump-diffusion processes. CALCOLO **44**, 33–57 (2007)
3. Farago, I., Izsak, F., Szabo, T., Kriston, A.: An IMEX scheme for reaction-diffusion equations: applications for a PEM fuel cell model. Cent. Eur. J. Math. **11**(4), 746–759 (2013)
4. Gyulov, T., Vulkov, L.: Well-posedness and comparison principle for option pricing with switching liquidity (2015). arXiv: 1502.07622

5. Ludkovski, M., Shen, Q.: European option pricing with liquidity shocks. Int. J. Theor. Appl. Fin. **16**(7), 135–143 (2013)
6. Mudzimbabwe, W., Vulkov, L.: IMEX schemes for a parabolic-ODE system of European options with liquidity shock. J. Comp. Appl. Math. **299**, 245–256 (2016)
7. Samarskii, A.A.: Theory of Difference Scheme. Marcel Dekker, Inc., New York (2001)
8. Tavella, D., Randall, C.: Pricing Financial Instruments: The Finite Difference Method. Wiley, New York (2000)
9. Verwer, J.G., Blom, J.G., Hundsdorfer, W.: An implicit-explicit approach for atmosphere transport-chemistry problems. Appl. Numer. Math. **20**, 191–209 (1996)
10. Wilmott, P., Dewynne, J., Howison, S.: Option Pricing: Mathematical Models and Computations. Wiley, New York (1998)
11. Windcliff, H., Forsyth, P.A., Vetzal, K.R.: Analysis of the stability of the linear boundary condition for the Black-Scholes equation. J. Comp. Fin. **8**(1), 65–92 (2004)

Chapter 14
A Highly Efficient Numerical Method for the SABR Model

Álvaro Leitao, Lech A. Grzelak, and Cornelis W. Oosterlee

Abstract In Leitao et al. (Appl Math Comput 293:461–479, 2017), we have presented a one time-step Monte Carlo simulation of the SABR model (Hagan et al. Wilmott Mag 84–108, 2002). The technique is based on an efficient simulation of SABR's time-integrated variance process. We base our approach on the derivation of the cumulative distribution function of the integrated variance by means of Fourier techniques and the use of a copula to approximate the conditional distribution (integrated variance conditional on the SABR volatility process). Resulting is a fast simulation algorithm which can be employed to price European options under the SABR dynamics. This converts our approach into an alternative to Hagan analytic formula for short maturities, where some known issues of the implied volatility expression for small strike values are overcome. A generalization of this technique to the multiple time-step case has been presented in Leitao et al. (On an efficient multiple time-step Monte Carlo simulation of the SABR model 2016, submitted for publication. Available at SSRN: http://ssrn.com/abstract=2764908).

14.1 Introduction

The *Stochastic Alpha Beta Rho (SABR)* model [6] is an established SDE system which is often used for interest rates and FX modeling in practice. The model belongs to the so-called *stochastic local volatility (SLV)* models. The idea behind SLV models is that the modeling of volatility is partly done by a local volatility and partly by a stochastic volatility contribution, aiming to preserve the advantages and minimize the disadvantages of the individual models.

Á. Leitao (✉) • C.W. Oosterlee
TU Delft, Delft Institute of Applied Mathematics, Delft, The Netherlands

CWI-Centrum Wiskunde & Informatica, Amsterdam, The Netherlands
e-mail: A.LeitaoRodriguez@tudelft.nl; c.w.oosterlee@cwi.nl

L.A. Grzelak
ING, Quantitative Analytics, Amsterdam, The Netherlands

TU Delft, Delft Institute of Applied Mathematics, Delft, The Netherlands
e-mail: L.A.Grzelak@tudelft.nl

M. Ehrhardt et al. (eds.), *Novel Methods in Computational Finance*,
Mathematics in Industry 25, DOI 10.1007/978-3-319-61282-9_14

In the original paper [6], the authors have provided a closed-form approximation formula for the implied volatility under SABR dynamics. This is important for practitioners, as it can be used highly efficiently within the calibration exercise. However, the closed-form expression is derived by perturbation theory and its applicability is thus not general. The formula is for example not always accurate for small strike values and/or high volatility.

In [10], we have proposed a one time-step Monte Carlo method for the SABR model. We based our approach on an accurate approximation of the cumulative distribution function of the so-called *time-integrated variance* (conditional on the SABR volatility), using Fourier techniques and copulas. Resulting is a fast simulation algorithm which can be employed to price European derivative contracts up to 2 years under the SABR dynamics. This kind of contract is often traded in foreign-exchange (FX) markets. Our approach can thus be seen as an alternative to Hagan's analytic formula for short maturities that may be employed for model calibration purposes. In this chapter, this work is summarized.

14.2 The SABR Model

The SABR model [6] is defined by the following SDE system, with independent Brownian motions $\mathrm{d}\hat{W}_S(t)$ and $\mathrm{d}\hat{W}_\sigma(t)$,

$$\begin{aligned}
\mathrm{d}S(t) &= \sigma(t)S^\beta(t)\left(\rho \mathrm{d}\hat{W}_\sigma(t) + \sqrt{1-\rho^2}\mathrm{d}\hat{W}_S(t)\right), \quad & S(0) &= S_0 \exp(rT), \\
\mathrm{d}\sigma(t) &= \alpha\sigma(t)\mathrm{d}\hat{W}_\sigma(t), & \sigma(0) &= \sigma_0.
\end{aligned} \tag{14.1}$$

Here $S(t) = \bar{S}(t)\exp(r(T-t))$ denotes the forward price of the asset $\bar{S}(t)$, with r the interest rate, S_0 the spot price, T the maturity and $\sigma(t)$ a stochastic volatility process, with $\sigma(0) = \sigma_0$. The model parameters are $\alpha > 0$ (the volatility of volatility), $0 \leq \beta \leq 1$ (the variance elasticity) and ρ (the correlation coefficient).

Based on the work by Islah [7], an analytic approximation for the *cumulative distribution function* (CDF) of the SABR conditional process has been obtained. For some $S(0) > 0$, the conditional CDF of $S(t)$ with an absorbing boundary at $S(t) = 0$, and given the volatility, $\sigma(t)$, and the conditional time-integrated variance, $\int_0^t \sigma^2(s)\mathrm{d}s|\sigma(t)$, reads

$$Pr\left(S(t) \leq X|S(0) > 0, \sigma(t), \int_0^t \sigma^2(s)\mathrm{d}s\right) = 1 - \chi^2(a; b, c), \tag{14.2}$$

where

$$a = \frac{1}{\nu(t)}\left(\frac{S(0)^{1-\beta}}{(1-\beta)} + \frac{\rho}{\alpha}\left(\sigma(t) - \sigma(0)\right)\right)^2, \qquad c = \frac{X^{2(1-\beta)}}{(1-\beta)^2\nu(t)},$$

$$b = 2 - \frac{1 - 2\beta - \rho^2(1-\beta)}{(1-\beta)(1-\rho^2)}, \qquad \nu(t) = (1-\rho^2)\int_0^t \sigma^2(s)\mathrm{d}s,$$

and $\chi^2(x;\delta,\lambda)$ is the non-central chi-square CDF. This formula is exact in the case of $\rho = 0$ and constitutes an *approximation* otherwise.

Therefore, to perform an "almost exact" one time-step Monte Carlo method for the SABR model, several steps need to be performed:

- **Simulation of the SABR volatility process.** By Eq. (14.1), the stochastic volatility process of the SABR model exhibits a log-normal distribution.
- **Simulation of the time-integrated variance process, $\int_0^t \sigma^2(s)\mathrm{d}s|\sigma(t)$.** This conditional distribution is not available in closed-form and needs to be approximated.
- **Simulation of the SABR forward price process.** The forward price $S(t)$ can be simulated by inverting the CDF in Eq. (14.2).

In the above steps of the SABR model almost exact simulation, the challenging part is the simulation of the time-integrated variance, $\int_0^T \sigma^2(s)\mathrm{d}s$, conditional on $\sigma(T)$ [and $\sigma(0)$]. We propose a computationally efficient approximation, based on a copula multi-variate distribution, to simulate the conditional distribution of $\int_0^T \sigma^2(s)\mathrm{d}s$ given the volatility $\sigma(T)$. Simulation by means of a copula technique requires the CDF of the involved marginal distributions. In Sect. 14.3, the derivation of the CDF of $\int_0^T \sigma^2(s)\mathrm{d}s$ is presented in detail. Hereafter, for notational convenience, we will use $Y(T) := \int_0^T \sigma^2(s)\mathrm{d}s$.

14.3 CDF of SABR's Time-Integrated Variance

In this section, we present a procedure to approximate the CDF of the time-integrated variance, $Y(T)$. Since we will work in the log-space, an approximation of the CDF of $\log Y(T)$, $F_{\log Y(T)}$, will be derived. We approximate $Y(T)$ by its discrete analogue, i.e.

$$\int_0^T \sigma^2(s)\mathrm{d}s \approx \sum_{j=1}^{M} \sigma^2(t_j)\Delta t =: \hat{Y}(T),$$

where M is the number of discrete time points,[1] i.e., $t_j = j\Delta t$, $j = 1, \ldots, M$ and $\Delta t = \frac{T}{M}$. $\hat{Y}(T)$ is subsequently transformed to the logarithmic domain, being $f_{\log \hat{Y}(T)}$ the *probability density function* (PDF) of $\log \hat{Y}(T)$. This PDF is found by approximating the associated characteristic function, $\phi_{\log \hat{Y}(T)}$, and applying a Fourier inversion procedure. The characteristic function and the desired PDF of $\log \hat{Y}(T)$ form a so-called Fourier pair. Based on the work in [11], we develop a recursive procedure to recover the characteristic function of $f_{\log \hat{Y}(T)}$. We start by defining the sequence,

$$R_j = \log\left(\frac{\sigma^2(t_j)}{\sigma^2(t_{j-1})}\right) = \log\left(\sigma^2(t_j)\right) - \log\left(\sigma^2(t_{j-1})\right), \tag{14.3}$$

where R_j is the logarithmic increment of $\sigma^2(t)$ between t_j and t_{j-1}. As the volatility process follows log-normal dynamics, the R_j are independent and identically distributed, i.e. $R_j \stackrel{d}{=} R$. By the definition of R_j in Eq. (14.3), we write $\sigma^2(t_j)$ as

$$\sigma^2(t_j) = \sigma^2(t_0)\exp(R_1 + R_2 + \cdots + R_j). \tag{14.4}$$

At this point, a backward recursion procedure in terms of R_j will be set up by which we will recover $\phi_{\log \hat{Y}(T)}$. We define

$$Y_1 = R_M, \quad Y_j = R_{M+1-j} + Z_{j-1}, \quad j = 2, \ldots, M. \tag{14.5}$$

with $Z_j = \log(1 + \exp(Y_j))$.

By Eqs. (14.4) and (14.5), the discrete time-integrated variance can be expressed as

$$\hat{Y}(T) = \sum_{i=1}^{M} \sigma^2(t_i)\Delta t = \Delta t \sigma_0^2 \exp(Y_M). \tag{14.6}$$

From Eq. (14.6) and by applying the definition of the characteristic function, we determine $\phi_{\log \hat{Y}(T)}$, as follows

$$\phi_{\log \hat{Y}(T)}(u) = \mathrm{e}[\exp(iu \log \hat{Y}(T))] = \exp\left(iu \log\left(\Delta t \sigma_0^2\right)\right) \phi_{Y_M}(u).$$

We have reduced the computation of $\phi_{\log \hat{Y}(T)}$ to the computation of ϕ_{Y_M}. An accurate and efficient way of approximating ϕ_{Y_M} was derived in [10], which is employed also here. Once the approximation of ϕ_{Y_M}, $\hat{\phi}_{Y_M}$, has been derived, we

[1] These time points are not to be confused with the Monte Carlo time steps. We will have only one Monte Carlo time-step. M is the number of points for the discrete approximation of $Y(T)$.

recover $f_{\log \hat{Y}(T)}$ from $\phi_{\log \hat{Y}(T)}$ by employing the COS method [4], as follows

$$f_{\log \hat{Y}(T)}(x) \approx \frac{2}{\hat{b}-\hat{a}} \sum_{k=0}^{N-1}{}' C_k \cos\left((x-\hat{a})\frac{k\pi}{\hat{b}-\hat{a}}\right),$$

with

$$C_k = \Re\left(\phi_{\log \hat{Y}(T)}\left(\frac{k\pi}{\hat{b}-\hat{a}}\right)\exp\left(-i\frac{\hat{a}k\pi}{\hat{b}-\hat{a}}\right)\right),$$

where N is the number of COS expansion terms, $[\hat{a}, \hat{b}]$ is the support of $\log \hat{Y}(T)$ and the prime $'$ and $\Re$ symbols indicate division of the first term in the summation by two and taking the real part of the complex-valued expressions in the brackets, respectively. The CDF $F_{\log \hat{Y}(T)}$ is then obtained by integration.

14.4 Simulation of $Y(T)|\sigma(T)$: Copula Approach

The CDF, $F_{\log \hat{Y}(T)}$ derived in Sect. 14.3 is now employed to simulate $Y(T)|\sigma(T)$ by means of a *copula*. In order to define any copula, a measure for the correlation between the involved distributions needs to be determined. Here, we choose the Pearson's correlation coefficient, $\mathscr{P}$, which is directly employed in many copulas (Gaussian and Student **t** copulas, for example) and a relation with Kendall's τ exists for *Archimedean* copulas. By definition, the Pearson coefficient for $\log Y(T)$ and $\log \sigma(T)$ is given by

$$\mathscr{P}_{\log Y(T), \log \sigma(T)} = \frac{\operatorname{cov}\left[\log \int_0^T \sigma^2(s)\mathrm{d}s, \log \sigma(T)\right]}{\sqrt{\operatorname{var}\left[\log \int_0^T \sigma^2(s)\mathrm{d}s\right]\operatorname{var}\left[\log \sigma(T)\right]}}.$$

We employ the following approximation

$$\log \int_0^T \sigma^2(s)\mathrm{d}s \approx \int_0^T \log \sigma^2(s)\mathrm{d}s = 2\int_0^T \log \sigma(s)\mathrm{d}s.$$

where the logarithm and the integral are interchanged. Since the log function is concave, this approximation forms a lower bound (Jensen's inequality) for the true value. After some algebraic manipulations, an approximation of the Pearson's correlation coefficient is then obtained as

$$\mathscr{P}_{\log Y(T), \log \sigma(T)} \approx \frac{\frac{1}{2}\alpha^2 T^2}{\sqrt{\frac{1}{3}\alpha^4 T^4}} = \frac{\sqrt{3}}{2}. \tag{14.7}$$

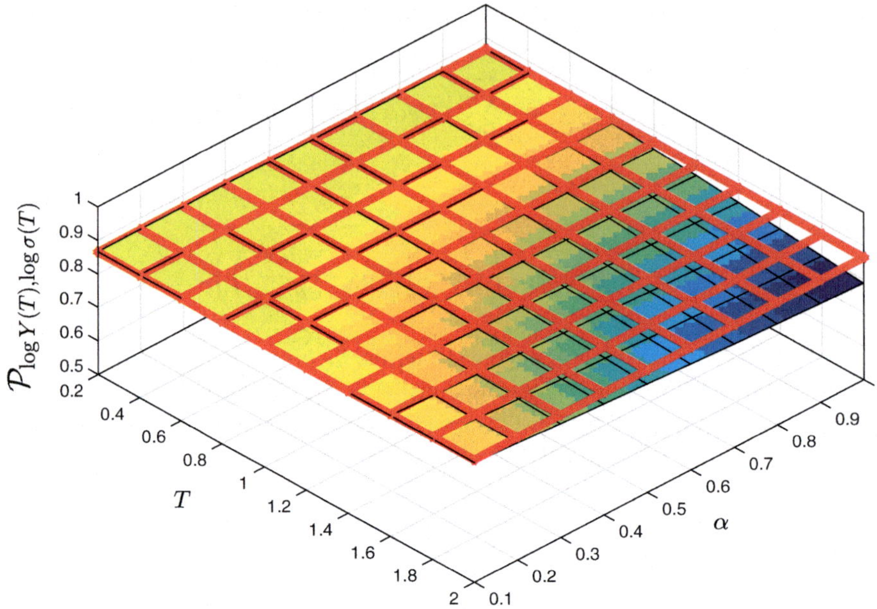

Fig. 14.1 Pearson's correlation coefficient: Empirical (surface) vs. approximation (red grid)

The approximation obtained is a *constant value*. We can show numerically that, for the problems at hand, this appears to be a very reasonable approximation. The correlation between $\log Y(T)$ and $\log \sigma(T)$ is affected by the maturity time T, and by the *volatility-of-volatility* parameter α. In Fig. 14.1, the empirical and approximated (red grid) correlations are depicted for typical values of T and α. Because we focus on short maturity options, we restrict $T \in [0, 2]$ and $\alpha \in (0, 1]$. The experiment shows that, only in rather extreme cases ($\alpha > 0.7$), the differences between the empirical and approximated correlation increase, but remain within five basis points and do not affect the performed simulation significantly.

The following steps describe the complete procedure for the simulation of $Y(T)$ given $\sigma(T)$ by using a bivariate copula approach:

1. Determine $F_{\log \sigma(T)}$ (analytically) and $F_{\log \hat{Y}(T)}$ (Sect. 14.3).
2. Determine the correlation by Eq. (14.7).
3. Define a bivariate copula distribution.
4. Generate correlated uniforms, $U_{\log \sigma(T)}$ and $U_{\log \hat{Y}(T)}$ from the bivariate copula.
5. From $U_{\log \sigma(T)}$ and $U_{\log \hat{Y}(T)}$ invert the original marginals, $F_{\log \sigma(T)}$ and $F_{\log \hat{Y}(T)}$.
6. Finally, the samples of $Y(T)|\sigma(T)$ are obtained by taking exponentials.

14.5 Simulation of $S(T)$ Given $S(0)$, $\sigma(T)$ and $\int_0^T \sigma^2(s)\mathrm{d}s$

We complete the one time-step SABR simulation by the conditional sampling of $S(t)$. The most commonly used techniques can be classified in two categories: direct inversion of the SABR distribution function given in Eq. (14.2) and moment-matching approaches. The direct inversion procedure has a higher computational cost because of the evaluation of the non-central χ^2 distribution. However some recent developments make this computation affordable. In [3], the authors proposed a forward asset simulation based on a combination of moment-matching (Quadratic Gaussian) and enhanced direct inversion procedures. We employ this technique also here.

As already pointed out by Andersen [1], the almost exact simulation of the asset price in some stochastic volatility models can result in loss of the martingale property, due to the approximation of a continuous process by its discrete equivalent. This is especially seen when the size of time time-step is large and for certain configurations of the SABR parameters, like small β values, close-to-zero initial asset values S_0, high *vol-vol* parameter α or large initial volatility σ_0. We employ a simple but effective numerical martingale correction, as follows

$$S(t) = S(t) - \frac{1}{n}\sum_{i=1}^{n} S_i(t) + \mathrm{e}[S(t)] = S(t) - \frac{1}{n}\sum_{i=1}^{n} S_i(t) + S_0,$$

where $S_i(t)$ represents the i-th Monte Carlo sample.

14.6 Numerical Experiments

Different numerical experiments have been carried out. In Sect. 14.6.1, we compare different copulas for the simulation of $Y(T)|\sigma(T)$. After that, in Sect. 14.6.2, we employ the best fitting copulas in a SABR pricing experiment. We consider several representative SABR data sets with special characteristics, like a zero correlation (Set I), a normal SABR model (Set II) and a high *volatility-of-volatility* SABR set (Set III). The parameter values are shown in Table 14.1. Other parameters in our one-step SABR method include the number of discrete time points, $M = 1000$ and the number of COS elements: $N = 150$.

Table 14.1 Data sets

	S_0	σ_0	α	β	ρ	T
Set I	1.0	0.5	0.4	0.7	0.0	2
Set II	0.05	0.1	0.4	0.0	−0.8	0.5
Set III	0.04	0.4	0.8	1.0	−0.5	2

14.6.1 Copula Approach Analysis

We consider here Gaussian and Student **t** copulas (Gaussian-based), and Clayton, Frank and Gumbel copulas (Archimedean). In order to select an optimal copula, we assess a so-called *goodness-of-fit (GOF)* criterion for copulas. We split the analysis, first evaluating the Archimedean copulas and subsequently, after choosing the optimal copula from this class, performing an analysis of the remaining copulas. The GOF testing for Archimedean copulas is a graphic procedure proposed by Genest and Rivest [5]. Given a function $\hat{K}$ based on based on Kendall's processes, the graphical GOF test consists in comparing the distance $\hat{\lambda}(u) := u - \hat{K}(u)$ and the *empirical* distance, $\lambda(u)$. In Fig. 14.2, the distances $\hat{\lambda}(u)$ between three Archimedean copulas (Clayton, Frank and Gumbel) are depicted. The experiment is performed for each data set in Table 14.1. As the measurable quantity, the mean squared error (MSE) of $\hat{\lambda}(u) - \lambda(u)$ is presented in Table 14.2.

From the GOF results for the Archimedean copulas, we find that the Gumbel copula fits best in our framework. Thus, we perform a GOF test including the Gaussian, Student **t** and Gumbel copulas. Hence, we perform a new GOF test which is based on the distances between the *empirical Deheuvels copula*, $\mathscr{C}_d$, and the analyzed copula $\mathscr{C}$. By using the discrete L^2 norm, this GOF measure reads

$$D_d\left(\mathscr{C}_d, \mathscr{C}\right) = \|\mathscr{C}_d - \mathscr{C}\|_{L^2},$$

where d is the number of random variables to build the copula, $d = 2$. In Table 14.3, the distances, D_2, between the tested copulas and the Deheuvels copula are shown.

According to the GOF results, the three copulas perform very similarly. When longer maturities are considered, the Gumbel copula exhibits smaller errors. In terms of speed, the Gaussian copula is around three times faster than the Gumbel

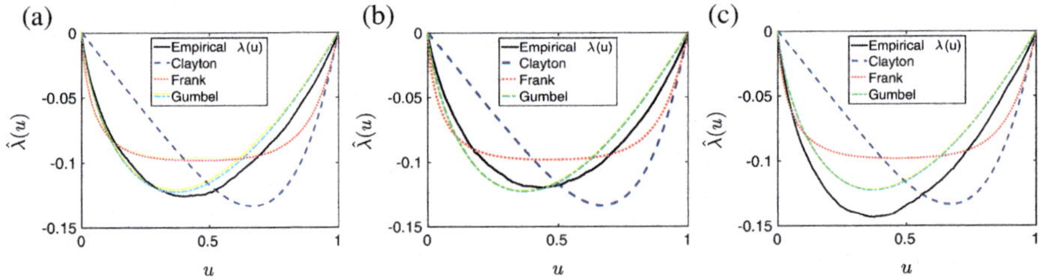

Fig. 14.2 Archimedean GOF test: $\hat{\lambda}(u)$ vs. empirical $\lambda(u)$. (**a**) Set I. (**b**) Set II. (**c**) Set III

Table 14.2 MSE of $\hat{\lambda}(u) - \lambda(u)$

	Clayton	Frank	Gumbel
Set I	1.3469×10^{-3}	2.9909×10^{-4}	5.1723×10^{-5}
Set II	1.0885×10^{-3}	2.1249×10^{-4}	8.4834×10^{-5}
Set III	2.1151×10^{-3}	7.5271×10^{-4}	2.6664×10^{-4}

Table 14.3 Generic GOF: D_2

	Gaussian	Student **t** ($\nu = 5$)	Gumbel
Set I	5.0323×10^{-3}	5.0242×10^{-3}	3.8063×10^{-3}
Set II	3.1049×10^{-3}	3.0659×10^{-3}	4.5703×10^{-3}
Set III	5.9439×10^{-3}	6.0041×10^{-3}	4.3210×10^{-3}

Table 14.4 Convergence in n: mean and standard deviation of the error (basis points) and time (s)

	$n = 1000$	$n = 10,000$	$n = 100,000$	$n = 1,000,000$
Gaussian (Set I, X_1)				
Error	519.58 (204.02)	132.39 (68.03)	37.42 (16.55)	16.23 (7.66)
Time	0.3386	0.3440	0.3857	0.5733
Gumbel (Set I, X_1)				
Error	151.44 (199.36)	−123.76 (86.33)	34.14 (17.03)	11.59 (6.58)
Time	0.3492	0.3561	0.3874	0.6663

copula, although the impact in the overall method is very small. The Student **t** copula is discarded since its accuracy and performance are very similar to the Gaussian copula and the calibration of the ν parameter adds extra complexity. As a general strategy, we conclude that the Gumbel copula is the most robust choice. When short maturities are considered, the Gaussian copula may be a satisfactory alternative.

14.6.2 Pricing European Options by the One-Step SABR Method

For the pricing experiment, the strike values X_i are chosen by the expression $X_i(T) = S(0)\exp(0.1 \times T \times \delta_i)$, $\delta_i = -1.5, -1.0, \ldots, 1.0, 1.5$.

First of all, the accuracy and the performance of the one-step SABR method are analyzed, considering the Gaussian and Gumbel copulas for the simulation of the time-integrated variance. In Table 14.4, the convergence of our method when the number of samples, n, is increased is empirically shown. We present the mean and the standard deviation of the error in basis points for the *implied volatilities* given by our one-step method and the reference price (Antonov et al. [2]) when Set I and X_1 are employed. We observe a reduction in the error (both mean and standard deviation) according to the expected Monte Carlo ratio ($1/\sqrt{n}$). Also in Table 14.4, the execution times of the one-step SABR method are shown. We can see that the number of paths hardly affects the performance.

To further test the one-step SABR method, in Table 14.5, the differences (in basis points) between the obtained implied volatilities with Hagan's formula, Monte Carlo simulation with a Milstein discretization and the one-step SABR method and

Table 14.5 One-step SABR method with Gaussian and Gumbel copulas—implied volatilities: differences in basis points

Strikes	X_1	X_2	X_3	X_4	X_5	X_6	X_7
Set I (reference: [2])							
Hagan	55.07	52.34	50.08	N/A	47.04	46.26	45.97
MC	23.50	21.41	19.38	N/A	16.59	15.58	14.63
Gaussian	16.23	20.79	24.95	N/A	33.40	37.03	40.72
Gumbel	11.59	15.57	19.12	N/A	25.41	28.66	31.79
Set II (reference: [8])							
Hagan	−558.82	−492.37	−432.11	−377.47	−327.92	−282.98	−242.22
MC	5.30	6.50	7.85	9.32	10.82	12.25	13.66
Gaussian	9.93	9.98	10.02	10.20	10.57	10.73	11.04
Gumbel	−9.93	−9.38	−8.94	−8.35	−7.69	−6.83	−5.79
Set III (MC Milstein)							
Hagan	287.05	252.91	220.39	190.36	163.87	141.88	126.39
Gaussian	16.10	16.76	16.62	15.22	13.85	12.29	10.67
Gumbel	6.99	3.79	0.67	−2.27	−5.57	−9.79	−14.06

several strikes are presented. Our copula-based one-step method achieves a very high accuracy.

14.7 Conclusions

In this chapter an efficient method to obtain samples of the SABR dynamics based on the time-integrated variance has been developed. The technique employs a Fourier method to derive the CDF of the time-integrated variance. By means of a copula, the conditional sampling technique is obtained. Its application gives us a fast and accurate *one time-step* Monte Carlo method for the SABR model simulation. By numerical experiments, we have shown that our method does not suffer from well-known problems of the Hagan formula in the case of small strike values and higher volatilities.

The use of the proposed technique is restricted to option maturities up to 2 years and European-type options. In [9], we have generalized the methodology to the *multiple time-step* case. This generalization allows us to deal with problems with longer maturities (more than 2 years) and also with more involved exotic options (early-exercise and path-dependent options). We need to introduce new method components to deal with the increasing computational complexity in that case.

Acknowledgements The first author is supported by the European Union in the FP7-PEOPLE-2012-ITN Program under Grant Agreement Number 304617 (FP7 Marie Curie Action, Project Multi-ITN STRIKE—Novel Methods in Computational Finance).

References

1. Andersen, L.B.G.: Efficient simulation of the Heston stochastic volatility model. J. Comput. Finance **11**(3), 1–22 (2008)
2. Antonov, A., Konikov, M., Spector, M.: SABR spreads its wings. Risk Mag. **26**, 58–63 (2013)
3. Chen, B., Oosterlee, C.W., van der Weide, H.: A low-bias simulation scheme for the SABR stochastic volatility model. Int. J. Theor. Appl. Finance **15**(2), 1250016-1–1250016-37 (2012)
4. Fang, F., Oosterlee, C.W.: A novel pricing method for European options based on Fourier-cosine series expansions. SIAM J. Sci. Comput. **31**, 826–848 (2008)
5. Genest, C., Rivest, L.-P.: Statistical inference procedures for bivariate Archimedean copulas. J. Am. Stat. Assoc. **88**(423), 1034–1043 (1993)
6. Hagan, P.S., Kumar, D., Lesniewski, A.S., Woodward, D.E.: Managing smile risk. Wilmott Mag. **1**, 84–108 (2002)
7. Islah, O.: Solving SABR in exact form and unifying it with LIBOR market model (2009). Available at SSRN: http://ssrn.com/abstract=1489428
8. Korn, R., Tang, S.: Exact analytical solution for the normal SABR model. Wilmott Mag. **2013**(66), 64–69 (2013)
9. Leitao, Á., Grzelak, L.A., Oosterlee, C.W.: On an efficient multiple time-step Monte Carlo simulation of the SABR model. (2016, submitted for publication). Available at SSRN: http://ssrn.com/abstract=2764908
10. Leitao, Á., Grzelak, L.A., Oosterlee, C.W.: On a one time-step Monte Carlo simulation approach of the SABR model: Application to European options. Appl. Math. Comput. **293**, 461–479 (2017)
11. Zhang, B., Oosterlee, C.W.: Efficient pricing of European-style Asian options under exponential Lévy processes based on Fourier cosine expansions. SIAM J. Financ. Math. **4**(1), 399–426 (2013)

Chapter 15
PDE Methods for SABR

Jörg Kienitz, Thomas McWalter, and Roelof Sheppard

Abstract In this chapter we consider the general SABR model which includes the Free Boundary SABR model considered in Chap. 4. We summarize as well as further illustrate the results from Kienitz et al. (Clipping the wings of SABR—stable and efficient numerics for sticky SABR. Preprint, 2016, submitted), Kienitz (Approximate and PDE Solution to the Boundary Free SABR Model—Applications to Pricing and Calibration. SSRN, http://papers.ssrn.com/sol3/papers.cfm?abstract_id=2647344, 2015) and Kienitz and Caspers (Interest rates explained volume 2: term structure models. Palgrave MacMillan, Basingstoke, 2017). The dynamic is given by the system of SDEs:

$$
\begin{aligned}
dF(t) &= \alpha(t)C(F(t))dW_1(t) \\
d\alpha(t) &= \nu\alpha(t)dW_2(t) \\
\langle dW_1(t), dW_2(t)\rangle &= \rho dt \\
F(0) &= f_0 \\
\alpha(0) &= \alpha_0
\end{aligned}
$$

$(W_1(t))_t$ and $(W_2(t))_t$ are correlated Brownian motions. f_0 and $\alpha_0 \geq 0$ are the starting values called the current forward rate and the initial volatility. The parameter $\nu > 0$ is the volatility of volatility and $\rho \in [-1, 1]$ is the correlation parameter. The function C is a local volatility parametrization.

J. Kienitz (✉)
Department of Actuarial Science, University of Cape Town, Cape Town, South Africa

Applied Mathematics and Numerical Analysis, University of Wuppertal, Wuppertal, Germany
e-mail: joerg.kienitz@gmx.de

T. McWalter
University of Cape Town, Cape Town, South Africa

University of Johannesburg, Johannesburg, South Africa
e-mail: tom@analytical.co.za

R. Sheppard
Standard Bank, Johannesburg, South Africa
e-mail: roelof.sheppard@standardbank.co.za

M. Ehrhardt et al. (eds.), *Novel Methods in Computational Finance*,
Mathematics in Industry 25, DOI 10.1007/978-3-319-61282-9_15

The case $C(F) = F^{\beta}$ is the original SABR model which was introduced in Hagan et al. (Wilmott Mag 1:84–108, 2002). Special versions of the model are known as the log-normal SABR model, $C(F) = F$, the Normal or Bachelier SABR model, $C(F) = 1$ and the shifted SABR model $C(F) = (F + b)^{\beta}$, with $b > 0$. Recently, Antonov et al. (Free boundary SABR. RISK, 2015) introduced the Free Boundary SABR model. They proposed to choose $C(F) = |F|^{\beta}$. The choices of the local volatility reflects the area of application of the model. For instance $C(F) = (F + b)^{\beta}$ or $C(F) = 1$ and $C(F) = |F|^{\beta}$ makes it possible to cope with negative rates prevailing in some interest rate markets including the EUR and CHF.

We consider a stable numerical scheme for the Free Boundary SABR model and show that it is stable for all types of parameters.

15.1 Introduction

The SABR model gained its popularity from the ease of use due to the asymptotic expansion formula derived in [9] and its flexibility to fit market observed volatility data. This solution covers the case of the classic SABR as well as the Bachelier and the shifted SABR model. The log-normal, $\sigma_{\mathrm{BS}}(K, T)$, or the Bachelier volatility, $\sigma_{\mathrm{B}}(K, T)$, can approximately be expressed in terms of the SABR parameters, the strike K and the maturity T. These values can then be used with the Black76, resp. Bachelier formula to price European Call and Put options. The approximation formula for the Bachelier model was considered in Chap. 4. Improvements to the asymptotic expansion such as [22] extend the area of application. For further improvements and their numerical implementation see [17].

The known deficiencies of the asymptotic expansions, namely the problems stemming from using it for very long dated options, large values of volatility of volatility or very small strikes were solved by practitioners using ad hoc methods. Especially during and in the aftermath of the financial crisis with very low or even negative rates and high values for implied volatilities revealed the deficiencies even more. It was realized that the density obtained from Call prices derived from using implied volatilities calculated by the asymptotic expansion formula led to negative values and, thus, to arbitrage. Again practitioners developed ad hoc methods for instance to be use for replication methods to price CMS based contracts—called wing extensions, see [14] or [5]. The first theoretical solution for this problem can be found in [6] who considered the density as the sum of two components that are always positive. These components reflect the decomposition into a continuous part and to an absorbing part. The latter is a Dirac measure in 0. In addition to that result an upper bound on the time to maturity was derived. Beyond that bound the standard asymptotic expansion cannot be applied safely. For related recent papers on the SABR model we refer the reader to [15, 16, 18].

Recent work of [10] suggests to consider the cumulative probability density

$$\mathbb{P}\left[F < F_T < F + dF | F_t = f_0, \alpha_t = \alpha_0\right]$$

The equation for the forward density is then

$$\frac{\partial q}{\partial T} = \frac{\alpha^2}{2} \frac{\partial^2 \left[\left(1 + 2\rho\nu y + \nu^2 y^2\right) \exp\left(\rho\nu v \Gamma(F)(T-t)\right) C^2(F) q\right]}{\partial F^2}$$

with $y(F) = \int_{f_0}^{F} \frac{dg}{C(g)}$ and $\Gamma(F) = \frac{C(F) - C(f_0)}{F - f_0}$. Subsuming all the notation into a function $D(F) := \sqrt{\alpha^2 + 2\alpha\rho\nu y(F) + \nu^2 y(F)^2} F^{\beta}$ we have

$$\frac{\partial q}{\partial T} = \frac{v^2}{2} \frac{\partial^2 \left[D^2(F) q\right]}{\partial F^2}$$

To keep track of the lower boundary of F called F_{Min} at 0 for the standard SABR model or $-b$ for the shifted model as well as for very large values, called F_{Max}, two additional equations are derived based on conservation laws. Summarizing, the numerical method is designed to keep the martingale property as well as the fact that the cumulative density sums up to 1. We finally have to consider the following set of equations for q, q^L and q^R:

$$\frac{\partial q}{\partial T} = \frac{\alpha^2}{2} \frac{\partial^2 \left[D^2(F) q\right]}{\partial F^2}, \quad q(0) = \delta(F - f_0) \tag{15.1}$$

$$\frac{dq^L}{dT} = \lim_{F \downarrow F_{\text{Min}}} \frac{v}{2} \left[D^2(F) q\right]_F, \quad q^L(0) = 0 \tag{15.2}$$

$$\frac{dq^R}{dT} = - \lim_{F \uparrow F_{\text{Max}}} \frac{v}{2} \left[D^2(F) q\right]_F, \quad q^R(0) = 0 \tag{15.3}$$

While Hagan et al. [10] suggest to use a Crank-Nicholson discretization to solve the system of equations, LeFloch and Kennedy [21] showed that this is not efficient and even leads to erratic effects. This was earlier remarked by Duffy [7] in a different context. Thus, different numerical schemes for efficiently solve the system of equations to produce reliable and stable results. For our considerations we take the Lawson-Swayne scheme, see [20], they propose to apply as one of the schemes leading to stable results for SABR. We extend the applicability of the effective PDE solution to the case of the Free Boundary SABR model as well by applying a suitable transformation.

In a series of publications on the SABR model [2–4] choose to split the Call option price $C(K, T)$ and consider the sum of the intrinsic value $(F - K)^+$ and the time value denoted by $\mathcal{O}_F^{SABR}(T, K)$. Then, the authors derive an integral representation for $\mathscr{O}(T, K)$ for the case $\rho = 0$ which is close to the one given in [12]. The main idea from [2, 3] which was extended to the free SABR model in [4] was to find an efficient approximation to the integral expression (15.4). The corresponding formula for $\mathscr{O}_F^{SABR}(T, K)$ is given in Chap. 4 (4.11),

$$\mathscr{O}_F^{SABR}(T, K) = \frac{1}{\pi}\sqrt{|Kf_0|}\{1_{K\geq 0}A_1 + \sin(\gamma\pi)A_2\} \tag{15.4}$$

with $\gamma = \left|-\frac{1}{2}(1-\beta)^{-1}\right|$ and the corresponding values

$$A_1 = \int_0^\pi \frac{\sin(\phi)\sin(\gamma\phi)}{b - \cos(\phi)} \frac{G(T\nu^2, s(\phi))}{\cosh(s(\phi))} d\phi$$

$$A_2 = \int_0^\infty \frac{\sinh(\psi)\left(1_{K\geq 0}\cosh(\gamma\psi) + 1_{K<0}\sinh(\gamma\psi)\right)}{b + \cosh(\psi)} \frac{G(T\nu^2, s(\psi))}{\cosh(s(\psi))} d\psi$$

The functions appearing in the above equations can be found in Chap. 4. While theoretically appealing this approach can lead to numerical instabilities for certain values of correlation and far out of the money strikes, especially for $|\rho|$ large, i.e. near 1 the proposed projection is not applicable. This will not be addressed further but we remark that the proposed PDE as well as the approximate solutions can be applied for $-1 \leq \rho \leq 1$. This is essential for calibration.

The aim of the current chapter is twofold. First, we extend numerical methods to be applicable to the Free Boundary SABR model—denoted by fSABR from now on. This results in a PDE approach similar to that of [10] for the standard SABR model. Second, we derive Finite Difference schemes based on [23] that are applied to the fSABR model. Finally, we wish to fill a gap in the current literature namely the comparison of the proposed numerical schemes in terms of accuracy, efficient pricing, stability for extreme parameters and the calculation of hedge sensitivities. To this end we consider the different numerical methods for different versions of the general SABR model and for different values of the SABR parameters. Then, we derive European Call option prices, implied volatilities and risk-neutral densities.

We have organized this chapter as follows: We use the results from [10] and [21] to adapt the techniques to work for the fSABR model. To this end we briefly review the numerical scheme which we apply. Then, we make the necessary changes to apply the methods in case of the fSABR model.

The following sections are dedicated to the benchmark solution. This is a finite difference method (FDM) in two dimensions based on Soviet splitting that extends work of [23] to the more general setting of fSABR. This method serves as a benchmark for all the other numerical techniques we considered.

Finally, we compare all the proposed numerical method in terms of accuracy for different types of SABR models and for different parameter values. From the numerical studies we reach the conclusion that the performance of the numerical techniques, the applicability for extreme model parameters and the calculation of hedge sensitivities are in favour of the PDE solution.

15.2 (Free Boundary Boundary) SABR: PDE Solution

Before we present our numerical approach to the fSABR model we outline some of the possible pitfalls of the integration approach:

- Inapplicability of projection method Inaccuracy for the projected solution for ITM and OTM options
  - Inaccuracy for ITM and OTM options
  - Arbitrage possibilities
  - Constraints on the SABR parameters
- Stability of the numerical integration with respect to the used number of quadrature points and for choosing the upper bound for numerical integration

For more information on the model and pitfalls see also Chap. 4

15.2.1 Effective PDE

In [21] many Finite-Difference schemes for calculating the density for the (No-Arbitrage) SABR model are proposed. Here we take the Lawson-Swayne scheme, [20], and modify it in a way such that it can be applied to the fSABR model.

First, we derive the effective PDE for the fSABR model. To this end we use the techniques from [10] and adapt it to the our setting. We use (15.1)–(15.3) together with the local volatility function $C(F) = |F|^{\beta}$. The function $D(F)$ is given by

$$D(F) = \sqrt{\alpha^2 + 2\alpha\rho\nu\left(\int_{f_0}^{F}|g|^{-\beta}dg\right) + \nu^2\left(\int_{f_0}^{F}|g|^{-\beta}dg\right)^2}\,|F|^{\beta} \tag{15.5}$$

To efficiently solve this equation we fix the notation and for $j = 1,\ldots,N_F,\quad n = 0,\ldots,N_T - 1$ we consider $z_j = z^- + jh$, $y_j = y(z_j - \frac{h}{2})$, $F_j = F(y_j)$, $C_j = D(F_j)$ and

$$\Gamma_j = \frac{|F_j|^{\beta} - f^{\beta}}{F_j - f_0}$$

$$E_j(T) = \exp(\rho\nu\alpha\Gamma_j T).$$

Take $t_n = nT/N$ and $q_j^n = q(z_j, t_n)$. To define the discrete operator governing the evolution let

$$\begin{aligned} L_j^n q(z_j, t_n) &= \frac{1}{\Delta} \frac{C_{j-1}}{F_j - F_{j-1}} E_{j-1}(t_n) q(z_{j-1}, t_n) \\ &\quad - \frac{1}{\Delta} \left(\frac{C_j}{F_{j+1} - F_j} + \frac{C_j}{F_j - F_{j-1}} \right) E_j(t_n) q(z_j, t_n) \\ &\quad + \frac{1}{\Delta} \frac{C_{j+1}}{F_{j+1} - F_j} E_{j+1} q(z_{j+1}, t_n) \end{aligned}$$

For the boundaries we update according to

$$\begin{aligned} \frac{C_0}{F_1 - F_0} E_0(T) q(t_0, T) &= -\frac{C_1}{F_1 - F_0} E_1(T) q(z_1, T) \\ \frac{C_{J+1}}{F_{J+1} - F_J} E_{J+1} q(t_{J+1}, T) &= -\frac{C_J}{F_{J+1} - F_J} E_J(T) q(z_J, T) \end{aligned}$$

The discretization is done using the Lawson-Swayne scheme which for $b = 1 - \frac{\sqrt{2}}{2}$ is

$$\begin{aligned} q_j^{n+b} - q_j^n &= b \Delta L_j^{n+b} q_j^{n+b} \\ P^L(t_{n+b}) - P^L(t_n) &= b\Delta \frac{C_1}{F_1 - F_0} E_1(t_{n+b}) q_1^{n+b} \\ P^R(t_{n+b}) - P^R(t_n) &= b\Delta \frac{C_J}{F_{J+1} - F_J} E_J(t_{n+b}) q_J^{n+b}. \end{aligned}$$

The step from $n + b$ we step forward to $n + 2b$ by

$$\begin{aligned} q_j^{n+2b} - q_j^{n+b} &= b \Delta L_j^{n+2b} q_j^{n+2b} \\ P^L(t_{n+2b}) - P^L(t_{n+b}) &= b\Delta \frac{C_1}{F_1 - F_0} E_1(t_{n+2b}) q_1^{n+2b} \\ P^R(t_{n+2b}) - P^R(t_{n+b}) &= b\Delta \frac{C_J}{F_{J+1} - F_J} E_J(t_{n+2b}) q_J^{n+2b}. \end{aligned}$$

The final step computes the new values at t_{n+1} and is given by

$$\begin{aligned} q_j^{n+1} &= (\sqrt{2} + 1) q_j^{n+2b} - \sqrt{2} q_j^{n+b} \\ P^L(t_{n+1}) &= (\sqrt{2} + 1) P^L(t_{n+2b}) - \sqrt{2} P^L(t_{n+b}) \\ P^R(t_{n+1}) &= (\sqrt{2} + 1) P^R(t_{n+2b}) - \sqrt{2} P^R(t_{n+b}) \end{aligned}$$

The final outcome are the density q for $F_{\text{Min}} < F < F_{\text{Max}}$ and the values at F_{Min} and F_{Max}.

Within the described set up the choice of the coordinate transformation $z \mapsto y \mapsto f$ is the crucial part. To account for the local volatility structure of the fSABR model we take the transformation given as follows:

$$f(y) = p(p(f^{1-\beta} + (1-\beta)y))^{1/(1-\beta)} \tag{15.6}$$

with

$$p := \begin{cases} 1 & , y \geq -\frac{f^{1-\beta}}{1-\beta} \\ -1 & , \text{else} \end{cases}$$

with $y(z) = \frac{\alpha}{\nu}\sinh(\nu z) + \rho(\cosh(\nu z) - 1)$. For illustration, we consider the values given in [4], $\alpha = 0.005095939$, $\beta = 0.1$, $\nu = 0.3$, $\rho = -0.3$ with forward $f_0 = 0.005$ and $T = 3$. The grid consists of $N_F = 160$ and $N_T = 160$ points. First, we show the transformed coordinates in Fig. 15.1.

In fact this transformation is the only thing that has to be changed to solve the system of equations necessary to numerically approximate the probability density function in the fSABR setting. The resulting probability density is then used to price European Call options. This is done by numerically integrating the density. Using this approach for a given maturity options on different strikes can be computed in one sweep. Next, we use the integration and the PDE approach for pricing European Call options. Figure 15.2 show densities obtained by the PDE method using different values for T.

Take $\alpha_0 = 0.014703823$, $\beta = 0.3$, $\nu = 0.3$, $\rho = -0.3$, $f_0 = 0.005$ and $T = 3$, we consider the effects of introducing a displacement into the fSABR model as well as an artificial lower bound F_{Min}. Using this set-up it is possible to control the stickyness. The results are shown in Fig. 15.3. The stickyness controls the peak of the distribution which is at 0 for the standard fSABR model. Furthermore, the lower bound controls the region where we wish to have contributions of the smile. This proves useful for the replication method used for CMS pricing for instance.

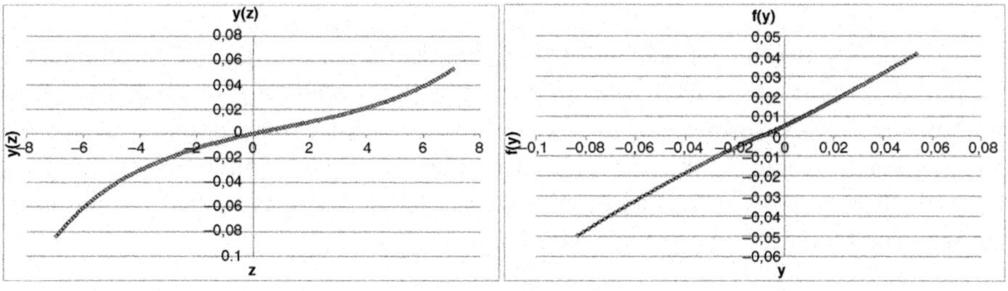

Fig. 15.1 Coordinate transformation for the fSABR using $\alpha_0 = 0.005095939$, $\beta = 0.1$, $\nu = 0.3$, $\rho = -0.3$ with forward $f_0 = 0.005$. We plotted the function $y(\cdot)$ (*left*) and the function $f(y(\cdot))$ (*right*)

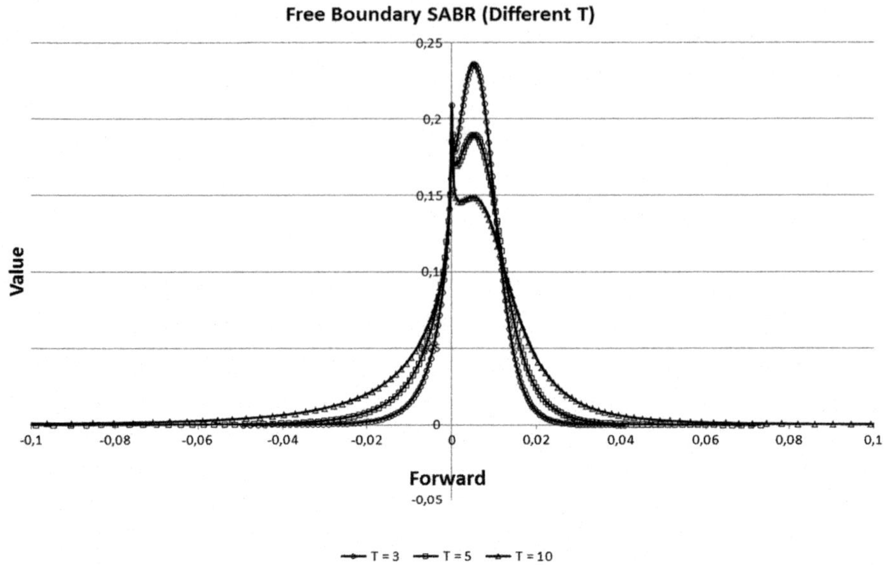

Fig. 15.2 Densities obtained using different values for T

To use the effective PDE solution we need to price financial instruments. First, we remark that we constructed the solution in a way that

$$hq_j(T) = \int_{zj-1}^{z_j} q(T,z)dz = \int_{F(z_{j-1})}^{F(z_j)} Q(T,F)dF \tag{15.7}$$

$$hF_{j-\frac{1}{2}}q_j(T) = \int_{zj-1}^{z_j} F(z)q(T,z)dz = \int_{F(z_{j-1})}^{F(z_j)} FQ(T,F)dF \tag{15.8}$$

If a strike value K is such that $F_j < K < F_{j+1}$ for some grid points F_j and F_{j+1} we need to take two things into account. First, we need to account for the integration from K to F_{j+1} and, second, we need to handle the boundary $F_{\max}$. To this end we regard the probability distribution obtained by the PDE method as a discrete cumulative probability on each grid cell. To account for strikes not lying on the grid we use for the ease of implementation a linear interpolation to get the corresponding density within a single grid. This can be accomplished by keeping the overall probability (15.7) and the forward (15.8). Then, the density to be used for numerical integration can be used on the subgrid (15.9).

$$Q(T,F) = \frac{hq_j}{F_j - F_{j-1}}\left(1 + 3b\frac{2F - F_{j-1} - F_j}{F_j - F_{j-1}}\right) \tag{15.9}$$

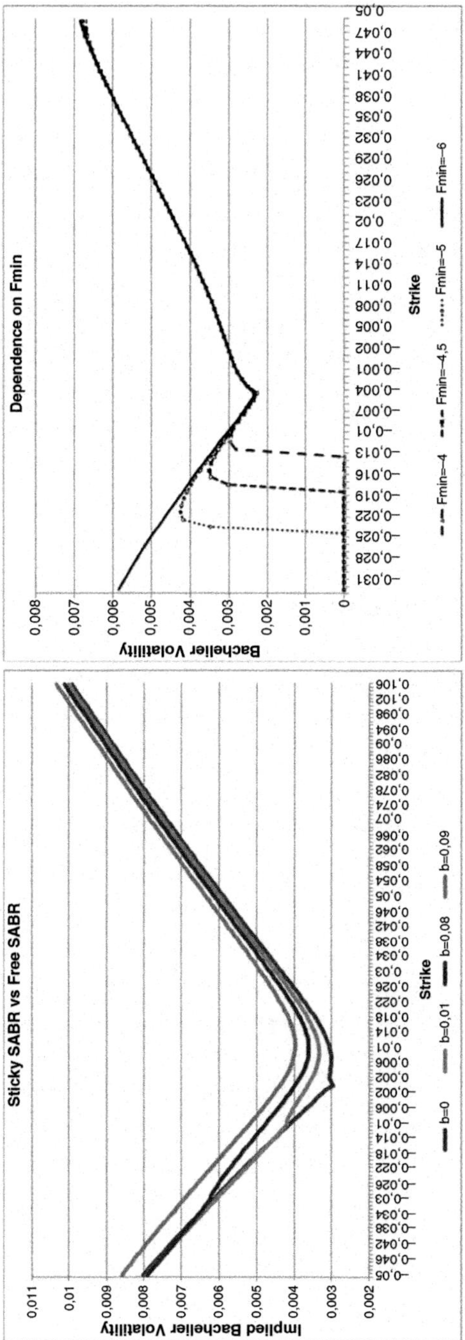

Fig. 15.3 Effects of a variation of the Sticky SABR model controlling the stickiness (*left*) using a displacement and the shape of the smile on the left wing (*right*) using an artificial lower bound

The value of b can be calculated since we know the cumulative density within each grid. Thus, the price of a European Call is given by (15.10)

$$\begin{aligned} C(T,K) = &\frac{1}{2}hq_k\frac{(F_k-K)^2}{F_k-F_{k-1}}\cdot\left(1+b_k\frac{F_k+2K-3F_{k-1}}{F_k-F_{k-1}}\right) \\ &+h\sum_{j=k+1}^{J}(F_{j-\frac{1}{2}}-K)q_j(T)+(F_{\text{Max}}-K)Q^R(T) \end{aligned} \tag{15.10}$$

The same can be done for pricing European Put options. We have

$$\begin{aligned} P(T,K) = &\frac{1}{2}hq_k\frac{(K-F_k)^2}{F_k-F_{k-1}}\cdot\left(1+b_k\frac{3F_k-2K-F_{k-1}}{F_k-F_{k-1}}\right) \\ &+h\sum_{j=k+1}^{J}(K-F_{j-\frac{1}{2}})q_j(T)+(K-F_{\text{Min}})Q^L(T). \end{aligned}$$

This extends to arbitrary payoff functions on the forward $F(T)$.

15.3 Benchmark PDE Solution

In the general SABR framework the value of a European Call option, $V(t_0, F_0, \alpha_0)$, with strike K and maturity T, can be obtained by solving the following PDE arising using the Feynman-Kac Theorem

$$\frac{\partial V}{\partial t}+\frac{1}{2}\alpha^2C^2(F)\frac{\partial^2 V}{\partial F^2}+\rho\nu\alpha^2C(F)\frac{\partial^2 V}{\partial F\partial\alpha}+\frac{1}{2}\nu^2\alpha^2\frac{\partial^2 V}{\partial\alpha^2}=0 \tag{15.11}$$

Equation (15.11) is on the domain $(t, F, \alpha) \in \Omega = [t_0, T] \times [F_{min}, F_{max}] \times [0, \alpha_{max}]$ subject to the following boundary conditions

$$V(T,F,\alpha) = \max(F-K,0) \tag{15.12}$$

$$V(t,F_{min},\alpha) = 0 \tag{15.13}$$

$$\frac{\partial V}{\partial F}(t,F_{max},\alpha) = 1 \tag{15.14}$$

$$\frac{\partial V}{\partial t}(t,F,0) = 0 \tag{15.15}$$

$$\frac{\partial V}{\partial t}(t,F,\alpha_{max})+\frac{1}{2}\alpha_{max}^2C^2(F)\frac{\partial^2 V}{\partial F^2} = 0 \tag{15.16}$$

For Eqs. (15.12)–(15.16) t_0 is the valuation date, F_0 the spot forward rate and α_0 the calibrated volatility parameter. The parameters F_{min}, F_{max} and α_{max} of the domain is chosen such that perturbations on the boundaries have minimal impact on the value of the contingent claim.

The terminal condition of the boundary value problem is given by the payoff function, see Eq. (15.12). Equation (15.13) reflects that the Call Option will be worthless when the forward, F, is very small and Eq. (15.14) specifies a von Neumann boundary condition. The lower boundary condition is obtained by letting $\alpha = 0$ in Eq. (15.11) whereas the upper boundary follows from the assumption that the value of a Call Option will not be very sensitive to changes in volatility (vega) if volatility is very large.

The Finite Difference Method used to solve this boundary value problem is based on techniques described in [23] and can be summarized as follows:

1. Partition the domain, $[F_{min}, F_{max}] \times [0, \alpha_{max}]$, into a non-uniform grid. This grid is chosen such that the number of grid points is concentrated at F_0 and α_0 and can be obtained my applying a grid generating function to uniform partitions of $[F_{min}, F_{max}]$ and $[0, \alpha_{max}]$. Let $x_{min} = x_0 < x_1 < \cdots < x_m = x_{max}$ denote a uniform partition for an arbitrary interval $[x_{min}, x_{max}]$ where $x_i - x_{i-1} = x_{i+1} - x_i$ for $i = 1, \ldots, m-1$. To obtain the benchmark solution we made use of the following grid generating function

$$g(x) = \hat{x} + \frac{c}{p}\sinh\left[p(x - x_{min}) + \sinh^{-1}\left(-\frac{p}{c}(\hat{x} - x_{min})\right)\right]$$

where $\hat{x}$ is the concentration point, c defines the density of the grid at $\hat{x}$ and p is chosen such that $g(x_{max}) = x_{max}$. By applying this grid generating function we obtain the following non-uniform partitions for $[F_{min}, F_{max}]$ and $[0, \alpha_{max}]$

$$F_{min} = f_0 < f_1 < \cdots < f_m = F_{max}$$

$$0 = a_0 < a_1 < \cdots < a_n = \alpha_{max}$$

Let $0 = t_0 < t_1 < \cdots < t_l = T$ be a uniform partition for the time interval $[t_0, T]$ where $t_i - t_{i-1} = \Delta t$ for $i = 1, \ldots, l$ and denote the finite difference approximation at each node point by

$$v_{ij}^k \approx V(t_k, f_i, a_j)$$

for $k = 0, \ldots, l$,$i = 0, \ldots, m$ and $j = 0, \ldots, n$.

2. The solution at time t_0 can be obtained by marching backward trough time with the following Yanenko scheme

$$\frac{\widetilde{v}_{ij} - v_{ij}^k}{\Delta t} = \frac{1}{2}a_j^2 C^2(F_i)\Delta_F^2 \widetilde{v}_{ij} + \frac{1}{2}\rho\nu a_j^2 C(F_i)\Delta_{F\alpha} v_{ij}^k$$

$$\frac{v_{ij}^{k-1} - \widetilde{v}_{ij}}{\Delta t} = \frac{1}{2}\nu a_j^2 \Delta_\alpha^2 v_{ij}^{k-1} + \frac{1}{2}\rho\nu a_j^2 C(F_i)\Delta_{F\alpha}\widetilde{v}_{ij}$$

where Δ_F^2, Δ_α^2 and $\Delta_{F\alpha}$ are second order finite difference approximations for $\frac{\partial^2}{\partial F^2}$, $\frac{\partial^2}{\partial \alpha^2}$ and $\frac{\partial^2}{\partial F \alpha}$. This scheme has better stability properties than traditional ADI schemes for non-smooth payoff functions and is better with approximating mixed derivatives, see [8].

3. The major drawback of this Yanenko scheme is that it is only first order accurate in time. Accuracy can be drastically improved by making use of extrapolation techniques as described in [13]. The benchmark solution was obtained by applying similar extrapolation techniques to the above-mentioned Yanenko scheme as described in [23].

15.4 Numerical Results

In this section we consider some examples from the literature and apply the proposed numerical techniques.

First, we consider the standard SABR model on some challenging parameter sets proposed in [10]. Here we consider all the numerical methods and display the results. Second, we apply a numerical study to the fSABR case. Here we consider the parameter sets from [4] and take different time to maturities, extreme parameters and, furthermore, consider the Greeks Δ and Γ.

15.4.1 SABR

For the numerical study we consider parameter sets based on [4] and [10]. Thus, we take the parameters $\alpha_0 = 0.318$, $\beta = 0$, $\rho = -0.183$ and $\nu = 0.777$ for Set 1 and $\alpha = 0.329$, $\beta = 0.5$, $\rho = -0.455$ and $\nu = 0.867$ for Set 2. To analyze the accuracy of the different numerical methods we chose to calculate the implied Bachelier volatility. Table 15.1 displays the results:

15.4.2 Free SABR

For the free SABR model we take the following sets of parameters, $f_0 = 0.005$, $\alpha_0 = 0.6F_0^{1-\beta}$, $\beta = 0.1$, $\nu = 0.3$ and $\rho = -0.3$ for FSet 1 and $f_0 = 0.01$, $\alpha = 0.3f_0^{1-\beta}$, $\nu = 0.3$, $\beta = 0.25$ and $\rho = -0.3$ for FSet 2. We consider the fSABR model with the parameter set FSet 1 with $T = 3$. The benchmark price is obtained by applying the 2d FDM scheme introduced in Sect. 15.3. Then, the PDE approach from Sect. 15.2 and the two asymptotic expansions from Chap. 4 as well

Table 15.1 Numerical results for all the considered numerical methods on the parameter Set1 and Set2

PDE B Vol	Int B Vol	FDM B Vol	Approx B Vol	
−0.012	0.00479340553283691	0.00480592250823975	0.00477105379104614	0.00554412603378296
−0.011	0.0046190619468689	0.00463113188743591	0.00460058450698853	0.00533342361450195
−0.01	0.00444278120994568	0.00445470213890076	0.00442653894424438	0.00512048602104187
−0.009	0.00426456332206726	0.0042763352394104	0.00425055623054504	0.00490501523017883
−0.008	0.0040842592716217	0.00409618020057678	0.00407233834266663	0.00468693673610687
−0.007	0.00390186905860901	0.00391408801078796	0.00389143824577332	0.00446595251560211
−0.006	0.00371716916561127	0.00372976064682007	0.00370830297470093	0.00424191355705261
−0.005	0.00352993607521057	0.00354334712028503	0.00352218747138977	0.0040145218372345
−0.004	0.00334024429321289	0.00335469841957092	0.0033336877822876	0.00378362834453583
−0.003	0.00314779579639435	0.00316381454467773	0.00314243137836456	0.00354886054992676
−0.002	0.00295259058475494	0.00297091901302338	0.00294826924800873	0.00331006944179535
−0.001	0.00275485217571259	0.00277623534202576	0.00275105237960815	0.00306710600852966
0	0.00255495309829712	0.00257380306720734	0.00255174934864044	0.00281855463981628
0.001	0.00235404819250107	0.00238310545682907	0.00235170125961304	0.00257000327110291
0.002	0.00215500593185425	0.0021817535161972	0.00215306878089905	0.00231925398111343
0.003	0.00196471810340881	0.00198621302843094	0.00196304172277451	0.00207517296075821
0.004	0.00179983675479889	0.00181056559085846	0.00179864466190338	0.00185763463377953
0.005	0.00169601291418076	0.00171840190887451	0.00169478356838226	0.00171113759279251
0.006	0.00169519335031509	0.00166384503245354	0.00169405713677406	0.00170985236763954
0.007	0.00179000198841095	0.00174239277839661	0.00178895890712738	0.00183224678039551
0.008	0.00193700194358826	0.00187985599040985	0.00193577259778976	0.00201921910047531
0.009	0.00210627913475037	0.00204306095838547	0.00210452824831009	0.00223066657781601
0.01	0.00228419899940491	0.00221662223339081	0.00228255987167358	0.00245008617639542
0.011	0.00246487557888031	0.00239372253417969	0.00246234238147736	0.00267095863819122
0.012	0.00264555215835571	0.00257134437561035	0.00264406204223633	0.00289052724838257

(continued)

Table 15.1 (continued)

PDE B Vol	Int B Vol	FDM B Vol	Approx B Vol	
−0.012	0.00272750854492187	0.00272750854492187	0.00272750854492187	0.00278472900390625
−0.011	0.00263214111328125	0.00263214111328125	0.00263214111328125	0.0026702880859375
−0.01	0.0025177001953125	0.0025177001953125	0.0025177001953125	0.00256538391113281
−0.009	0.00241279602050781	0.00241279602050781	0.00241756439208984	0.00245571136474609
−0.008	0.00230789184570312	0.00230312347412109	0.00230789184570312	0.00234603881835937
−0.007	0.00219583511352539	0.00219345092773437	0.00219821929931641	0.00223159790039062
−0.006	0.00208377838134766	0.00208377838134766	0.00208377838134766	0.00211715698242187
−0.005	0.00197410583496094	0.00197172164916992	0.00197649002075195	0.00200271606445312
−0.004	0.0018608570098877	0.00185847282409668	0.00186324119567871	0.00188827514648437
−0.003	0.00174880027770996	0.0017470121383667	0.00175118446350098	0.00177383422851562
−0.002	0.00163793563842773	0.00163793563842773	0.00163853168487549	0.00166118144989014
−0.001	0.0015336275100708	0.00153601169586182	0.00153183937072754	0.00155508518218994
0	0.00146269798278809	0.00147446990013123	0.00146239995956421	0.00155910849571228
0.001	0.00152632594108582	0.0015345960855484	0.00151947140693665	0.00156305730342865
0.002	0.00153969973325729	0.00154629349708557	0.00153250992298126	0.0015612319111824
0.003	0.00153172761201859	0.00153768807649612	0.00152546912431717	0.00154387205839157
0.004	0.00151943415403366	0.00152435153722763	0.00151420012116432	0.00152558088302612
0.005	0.00151220709085464	0.00149086117744446	0.00150773674249649	0.00151393935084343
0.006	0.0015159510076046	0.00151660293340683	0.00151187181472778	0.0015152245759964
0.007	0.00153381377458572	0.0015316903591156	0.00152982771396637	0.00153161585330963
0.008	0.0015663355588913	0.00156145542860031	0.00156223773956299	0.00156380236148834
0.009	0.00161200761795044	0.00160463154315948	0.00160753726959229	0.00160999596118927
0.01	0.00166833400726318	0.00165864825248718	0.00166326761245728	0.00166751444339752
0.011	0.00173255801200867	0.0017210841178894	0.001726895570755	0.0017334520816803
0.012	0.00180274248123169	0.0017896294593811	0.0017961859703064	0.00180542469024658

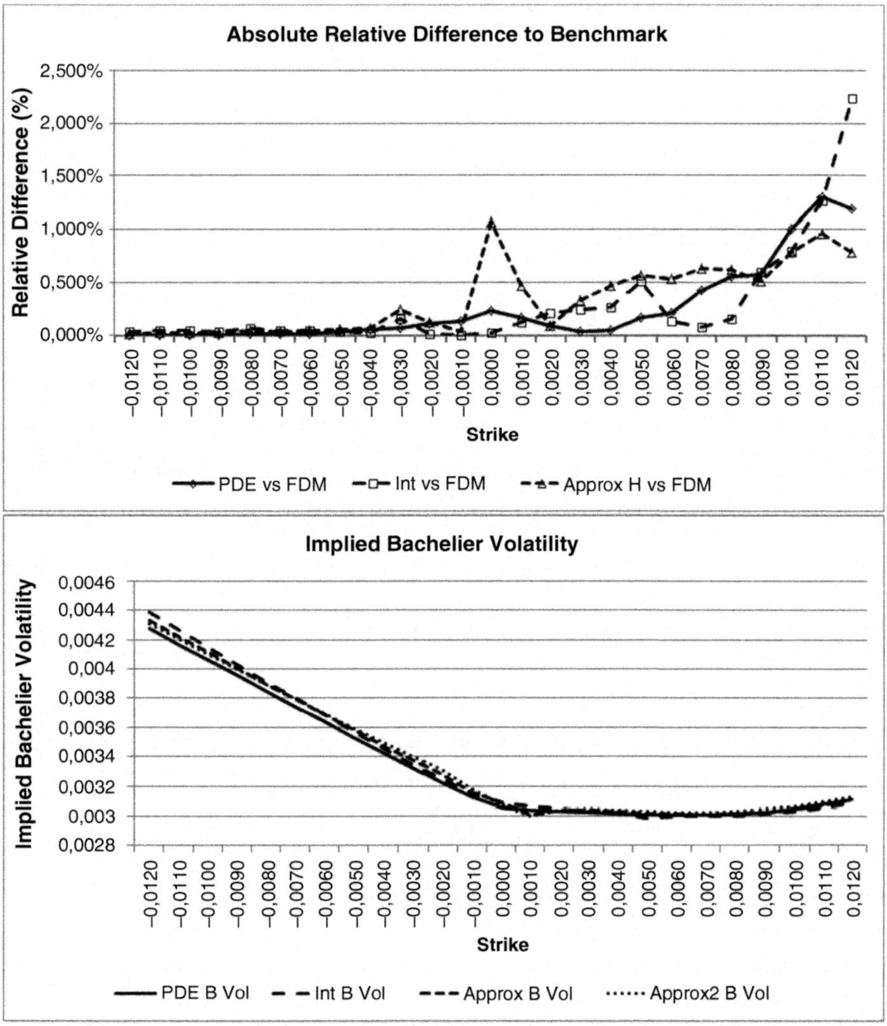

Fig. 15.4 Performance of the Numerical Techniques against the Benchmark on FSet1 in terms of prices (*top*) and implied Bachelier volatility (*bottom*)

as the integration method from [4] are applied. In Fig. 15.4 we show the relative error with respect to the benchmark price. For this set of parameters the PDE as well as the Approximation based on the Hagan approach show the best accuracy. The integration is the worst.

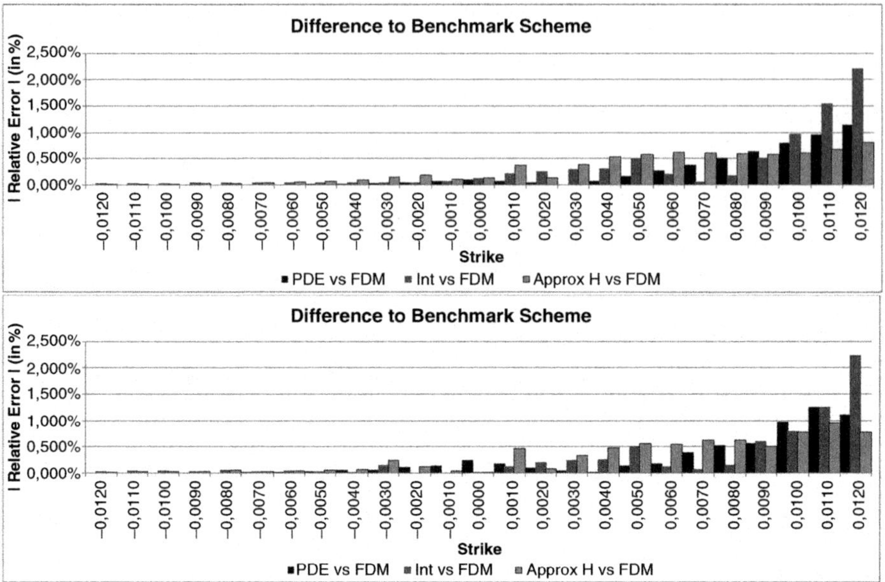

Fig. 15.5 Difference $|Price/Price_{\text{Benchmark}} - 1|$ of the different numerical methods. We consider FSet1 (*top*) and FSet2 (*bottom*)

Further illustrations on the performance of the numerical methods can be found in Fig. 15.5 which shows the difference to our benchmark implementation.

15.4.3 Stability for Extreme Parameters

Let us consider the parameter set FSet1. Keeping the values of this parameter set but varying one parameter the time to maturity T, the correlation ρ and the volatility of volatility ν. These parameters are known to challenge a numerical method or even make it inapplicable. We consider:

- $T = 1, 5, 10, 20, 50$

  The finding here is that except of the PDE solution none of the other numerical techniques is capable to accurately price Call options for all maturities and, thus, give accurate values for the implied Bachelier volatilities. As expected the quality of the approximation formulas deteriorate. The results can be seen in Fig. 15.6. But if we include higher order terms in the standard expansion from [9] this effect can be accounted for.

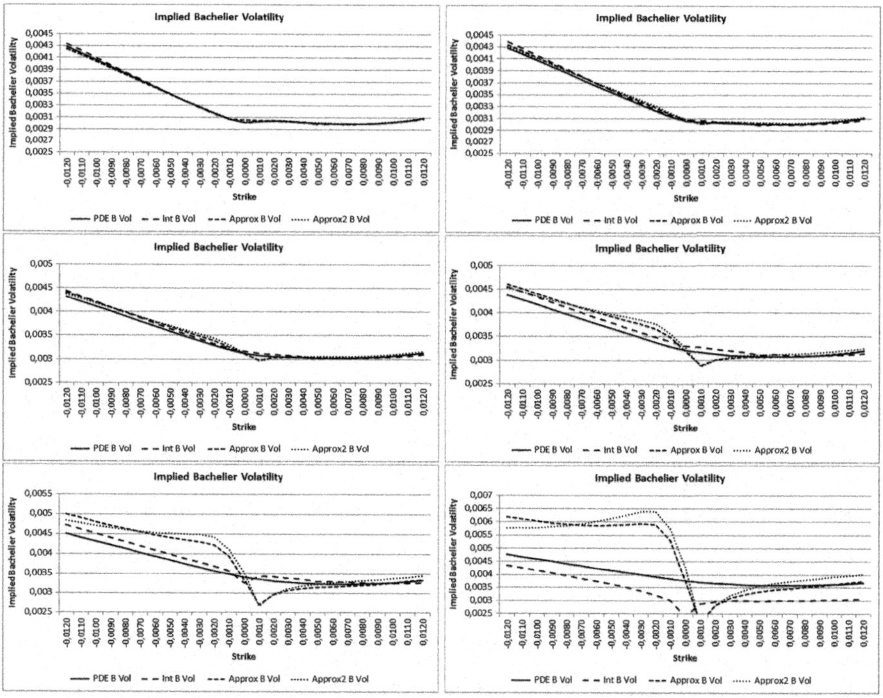

Fig. 15.6 Implied Bachelier Volatility for FSet1 and different $T = 1Y, 3Y, 5Y, 10Y, 20Y, 50Y$ from *top left* to *bottom right*. The different numerical techniques are shown

- $\rho = -0.9, 0.9$
  We find that the integration technique cannot cope with all values for the correlation parameter. It is well known that for certain market parameters to recover the prices, resp. volatilities the model parameters have to be chosen near the lowest/largest possible values. Thus, care has to be taken by choosing the appropriate domain for the parameters. For values of the correlation parameter ρ near 1 even the PDE method does not seem to work properly. But it is possible to adjust the method by choosing a larger grid and increasing the number of space steps. Especially, we observe that the method based on integration and projection breaks down here. The results can be seen in Fig. 15.7.
- $\nu = 0.5, 1, 2$
  The approximation formulas, especially the formula based on the Hagan asymptotic expansion is not suitable for large values of volatility of volatility. The calculated values for the implied Bachelier volatilities are far off the correct results. But we observe that the approximation method based on Approx 2 from

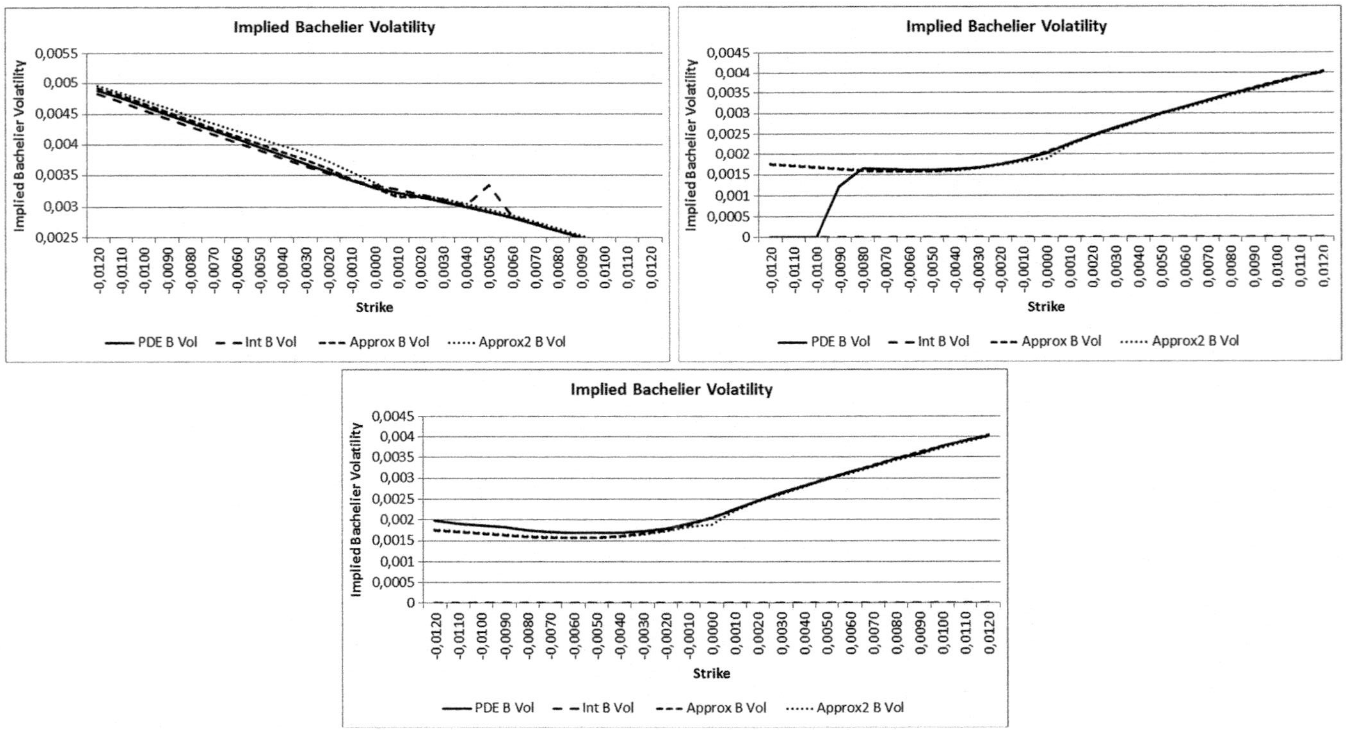

Fig. 15.7 Implied Bachelier Volatility for FSet1 and $\rho = -0.9, 0.9$ from *left* to *right*. Adjusting the size of the grid allows to apply the PDE method for all correlation values. The graph with the increased grid is shown on the *bottom*. The method based on integration and projection breaks down for $\rho = 0.9$

Chap. 4 is better suited here. The results can be seen in Fig. 15.8. Again, higher order terms in the approximate solution can be used to cope with high values of the volatility of volatility.

We show the effect by plotting the implied Bachelier volatility in Figs. 15.6, 15.7 and 15.8. It shows that the PDE method is stable for all times to maturity, especially very long dated options can be priced safely.

15.4.4 The Greeks

In this subsection we analyze the Greeks, $\Delta = \frac{\partial C(K,T)}{\partial f_0}$ and $\Gamma = \frac{\partial^2 C(K,T)}{\partial f_0^2}$ for all the suggested methods in terms of accuracy. To this end we consider the Call prices $C(K,T,\pm\epsilon)$ depending on a parameter $\epsilon > 0$ where this expression means that we calculate the Call option price taking $f_{0,\pm\epsilon} = f_0 \pm \epsilon$.

We consider

$$\Delta = \frac{C(K,T,\epsilon) - C(K,T,-\epsilon)}{2\epsilon}$$

$$\Gamma = \frac{C(K,T,-\epsilon) + C(K,T,\epsilon) - 2C(K,T,0)}{2\epsilon}$$

We observe the following:

- The accuracy and the stability depends on the choice of the parameter ϵ, see for instance [19] on a systematic account of numerical differentiation. For instance the PDE method cannot be applied with too small values for ϵ. For the integration approach the value of ϵ should not be too large. Figure 15.9 illustrates this result. The asymptotic approximation result is not sensitive to value of ϵ. In practice people often wish to choose not the theoretical optimal value but 1% or $1bp$.
- For accurately calculating the Greeks the number of discretization steps applied for the PDE method is crucial. Too few space steps give inaccurate results but we also observe that the choice of time steps does not effect that much.

Finally, we show the performance of all methods with respect to the benchmark method. Tables 15.2 and 15.3 summarize our findings.

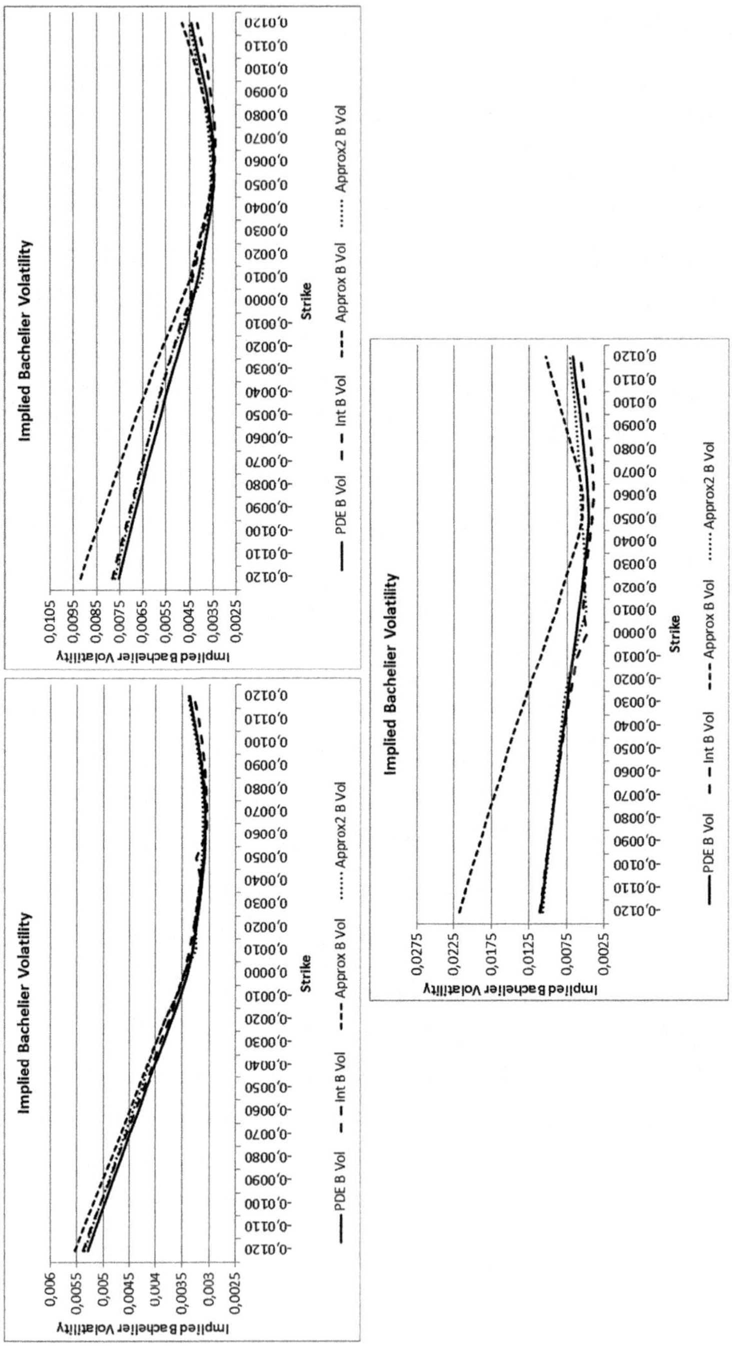

Fig. 15.8 Implied Bachelier Volatility for FSet1 and $\nu = 0.5, 1, 2$ from *top left* to *bottom*

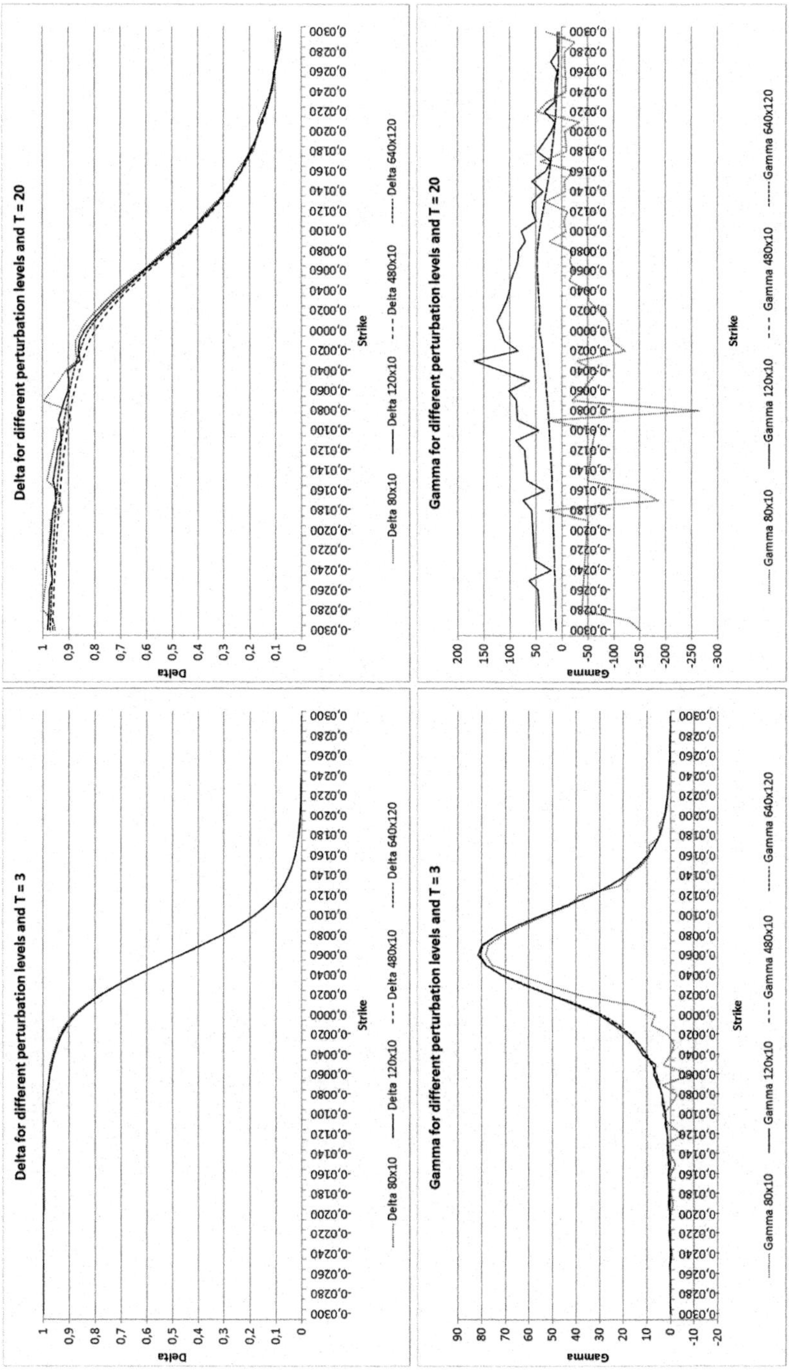

Fig. 15.9 The Δ (*top*) for $T = 3$ (*left*) and $T = 20$ (*right*) as well as Γ (*bottom*) for $T = 3$ (*left*) and $T = 20$ (*right*) for the parameter set FSet1 and different values for the space and time discretization for the Lawson-Swayne scheme

Table 15.2 Δ for the parameter sets FSet 1 and FSet 2

Strike	Delta PDE	Delta I	Delta FDM	Delta Approx
−0.012	0.993698275262026	0.990447197031438	0.99378527605855	0.988285348032612
−0.011	0.992137796694447	0.988700601795237	0.992142364617401	0.985639069806445
−0.01	0.990153391781245	0.98655348149697	0.9901582624028	0.982322004796417
−0.009	0.987608012445636	0.983892718777633	0.98759819708645	0.978145327374573
−0.008	0.98433473046346	0.980566747909271	0.98431575184855	0.972862166531479
−0.007	0.98010028290457	0.976370225519909	0.98002818615155	0.966148871469532
−0.006	0.974605112652844	0.971021706056485	0.9743450131457	0.957580819520697
−0.005	0.967425175182187	0.964130715210617	0.967264991482955	0.946602948614858
−0.004	0.95800399732016	0.955148420371758	0.95790960370744	0.932499955284393
−0.003	0.945575034048376	0.943291903752431	0.94526181836995	0.914390608727274
−0.002	0.929055409014991	0.927421795930724	0.92927371199052	0.891363807905389
−0.001	0.906868593180497	0.90581431766313	0.906239998656895	0.863544997591736
0	0.876191952082227	0.875288571969262	0.875523789468095	0.835715150727823
0.001	0.832036000513688	0.831444752687396	0.831222491696075	0.778508719086273
0.002	0.775984725637476	0.775975577281972	0.77496280258135	0.715526611194037
0.003	0.709575876760165	0.710361034886996	0.708618257441765	0.647950690593902
0.004	0.634529313712575	0.63601399582319	0.63365369167202	0.575531663988991
0.005	0.553451601333911	0.555310044542572	0.5529497164859	0.500000000000001
0.006	0.469830232251657	0.471621514341837	0.469383794872715	0.424027499676016
0.007	0.387661609034369	0.38896549399256	0.387191462432928	0.350677893550867
0.008	0.310821940119526	0.311362056858625	0.310427076971173	0.282850158773035
0.009	0.242412377371175	0.24211414432084	0.241989008786594	0.22277741159748
0.01	0.184290467902872	0.18327678296186	0.183888923924435	0.171719942087015
0.011	0.137001024230099	0.135495938119273	0.136667094993404	0.129926386744704
0.012	0.0999616519945859	0.0982147713233497	0.0996121930412275	0.0968222689339598

−0.012	0.999940349880034	0.0267836736157767	0.999924655720793	0.0395977976103135
−0.011	0.999920068514712	0.0361576888863874	0.999901630970978	0.0513504187088509
−0.01	0.999891607509963	0.0496652013785848	0.999869883597775	0.0674972595177126
−0.009	0.999850930417581	0.0691212440079997	0.999825342666866	0.0900907373008186
−0.008	0.999791651009468	0.0977511050437307	0.999761603116857	0.1223738756162
−0.007	0.999703333656564	0.14149750876026	0.999668279508259	0.169636280716867
−0.006	0.999569039905457	0.208727935309058	0.999527922384714	0.240834937062506
−0.005	0.999357513349764	0.315913324172232	0.99930992287332	0.351837467146843
−0.004	0.9990121986084	0.494988756002609	0.998957616604925	0.532376850242988
−0.003	0.998421917354974	0.805960194239647	0.998358417982148	0.842387549125634
−0.002	0.997340979528987	1.38613861597366	0.99726499693537	1.41571699831949
−0.001	0.995138880748623	2.5866997919741	0.995035696164351	2.60561617516572
0	0.989143578914452	5.94750694256178	0.988689475696133	6.07241278788306
0.001	0.961345762286021	20.7616702692995	0.962725578734294	20.1958434472432
0.002	0.909753463914061	47.147902671412	0.91371091701171	45.7190970668247
0.003	0.82565732701215	85.1995600694463	0.832434206380994	83.28953901072
0.004	0.704014075105699	127.902401925663	0.712262458265014	126.809833785502
0.005	0.552428762388186	158.442301739893	0.558655810328055	145.084470003481
0.006	0.393305109346701	160.219210443305	0.394181878408048	163.35910622146
0.007	0.253585480165132	132.874583931795	0.249294228610074	135.559981193388
0.008	0.149856043830698	92.7819423250735	0.143517730257981	93.6555820197043
0.009	0.0829986142764157	56.8739093000416	0.077373305550558	56.4270018360078
0.01	0.044214552709342	32.0011482204636	0.0402649760455638	31.1705766647596
0.011	0.0231836799854041	17.1789496178206	0.0207303385137513	16.4559809421021
0.012	0.0121740571395168	9.05313778964605	0.0107381548829	8.5486602326492

Table 15.3 Γ for the parameter sets FSet 1 and FSet 2

Strike	Gamma PDE	Gamma I	Gamma FDM	Gamma Approx
−0.012	1.33730476153948	3.62531683215095	1.45031914769791	4.07740047940475
−0.011	1.69670215333095	4.16067703797041	2.04499811859958	4.98617098894372
−0.01	2.16502452330888	4.79914686966204	2.57651565839685	6.114567536325
−0.009	2.76648058123927	5.56703721923263	3.2575211480982	7.51827010109923
−0.008	3.55105676489581	6.49943000981257	4.0971504044994	9.26667701548617
−0.007	4.57884335959731	7.64379055884223	5.35874842209905	11.4453414391813
−0.006	5.9063639536068	9.0652688319523	6.91458640340002	14.1577784825569
−0.005	7.65094556814082	10.8544879376393	8.78829794348848	17.5252385457847
−0.004	9.91986797147827	13.1389137868199	11.2238007262989	21.6815493076644
−0.003	12.8822062808974	16.0991800021368	14.8419801781607	26.7575392228631
−0.002	16.7435413121293	19.9919875176237	17.9457921258997	32.8471787883894
−0.001	21.7542727512562	25.1832601560337	24.9340076712299	39.9665927196952
0	28.2677222046236	32.2653580187016	32.1339819687313	48.3683101727
0.001	38.6314110472018	41.6872347842878	41.6511551885302	57.0086894022444
0.002	50.0429971762259	52.3175503388835	51.9426843723797	64.4044949304427
0.003	61.287971588427	62.9261428039627	62.6173246869501	70.4886602153048
0.004	71.0491453277391	72.2524874622807	71.9878171115997	74.6248117527645
0.005	77.9597748578182	77.0597905998781	79.0933619721397	76.2634198783405
0.006	80.9309688681514	81.8670937374755	81.8639345570699	75.0496028005228
0.007	79.4569733436718	80.4223877126121	80.0324819284449	71.0745364358029
0.008	73.8422258880618	74.8161178463114	74.3364817165049	64.7685436683102
0.009	65.108119111702	66.0129117358824	65.5565573557779	56.8753320529815
0.01	54.7030053709516	55.4345529218839	55.005250084239	48.2579220901602
0.011	44.0179741629545	44.5312135154163	44.130797116188	39.7103648901165
0.012	34.1405462278334	34.4328861398503	34.315033503083	31.8275319340533

−0.012	0.999930496028749	0.0239614781002984	0.99979054360445	0.164293929502679
−0.011	0.999920727453399	0.0183560775988967	0.999724060523153	0.220994291963428
−0.01	0.999885108256649	0.0431187870998662	0.999632092728037	0.301009911093941
−0.009	0.9998344952124	0.057343407799948	0.999502847572921	0.415740760220662
−0.008	0.9997826963946	0.0752357976006174	0.999317935470451	0.583209220847425
−0.007	0.999691161199801	0.127301413001268	0.999047891825705	0.832632786156029
−0.006	0.9995378774541	0.251238138600576	0.998643986248647	1.21275671536442
−0.005	0.99937544356994	0.190382500521447	0.998022520930855	1.80764078462625
−0.004	0.99898738735324	0.396179864801602	0.997032953979791	2.76795169684013
−0.003	0.99827759094282	0.49877924867929	0.995388354264664	4.37636303151495
−0.002	0.99729530432217	1.09616271351965	0.992499034567417	7.19281047830561
−0.001	0.996081952719805	0.536528474971015	0.987032083024034	12.3929643358264
0	0.98875213282111	5.8826886700598	0.972613580465403	24.2206142733216
0.001	0.96183318557447	20.1379231561999	0.930182186593882	49.4781276176755
0.002	0.910601547898735	46.7805046636899	0.866317554649707	79.7319000596201
0.003	0.826595681411665	85.0008759937702	0.772695681259515	112.782826837974
0.004	0.704746317106863	127.921129204914	0.647418213200163	140.532098305058
0.005	0.552849242305563	158.586156739254	0.499999999999997	152.122043301124
0.006	0.393400723932104	160.442320545953	0.351622917949465	141.35300649221
0.007	0.253469768397669	133.049285319355	0.225503644149104	113.175703708539
0.008	0.149654824515295	92.8553522006568	0.134070641036936	79.7614661813119
0.009	0.0827948485854845	56.8865081194633	0.0757713266391107	51.1435740046268
0.01	0.0440330750681649	31.9987998460744	0.041740717750978	30.8731511608453
0.011	0.0230384808333123	17.1630598844725	0.0228597858560535	18.0460977743406
0.012	0.0120641677565168	9.02887777284111	0.0126123771691592	10.4214543626257

15.5 Conclusions

Our conclusion is that the PDE method based on the effective solution of the general SABR model should be applied in practice. For the standard SABR model this is described in [10] whereas we extended it to the case of the fSABR model here. This method is the most reliable and gives accurate results for extreme parameters and for the Greeks. If efficiently implemented, e.g. using the method proposed here it is also reasonable fast.

All other methods have their advantages mainly in terms of speed but this often comes with a decreasing range of applicability. For instance the very fast approximation methods cannot handle extreme parameters as large values for the maturity for instance. Let us summarize our findings:

- PDE and asymptotic expansions can be carried over to the fSABR model
- All the numerical methods have been benchmarked against 2d FDM scheme
- PDE method is the most accurate approximation to the benchmark solution
- Approximation formulas efficient near ATM and small time to maturity
- Integration formula not applicable for all values of correlation (PROBLEMS for calibration)
- Greeks can be calculated and it is shown that the approximation formula leads to values different than the PDE and Integration approach
- Accuracy of PDE scheme relies of ϵ and mainly on the number of space steps

Thus, we conclude that an appropriate implementation of the No-Arbitrage SABR as well as the extension to the fSABR model should apply the PDE solution approach. Comparison to our 2d FDM benchmark approach have shown that it is accurate, stable and due to the fact the efficient numerical schemes exist is reasonable fast.

We expect that the method of the effective PDE is applicable to a larger set of stochastic volatility models. The authors applied the model successfully to the case of the ZABR model, see [1].

Furthermore, [11] showed that for a variety of stochastic volatility models effective PDE can be derived. This makes the described technique applicable to a variety of models.

References

1. Andreasen, J., Huge, B.N.: Expanded forward volatility. RISK **1**, 101–107 (2013)
2. Antonov, A., Spector, M.: Advanced analytics for the SABR model, SSRN (2012)
3. Antonov, A., Spector, M.: SABR spreads its wings. RISK (2013)
4. Antonov, A., Konikov, M., Spector, M.: Free boundary SABR. RISK (2015)
5. Benaim, S., Dodgson, M., Kainth, D.: An arbitrage free method for smile extrapolation. Preprint (2010). http://www.quarchhome.org/RiskTailsPaper_v5.pdf
6. Doust, P.: No-arbitrage SABR. J. Comput. Finance **15**, 3–31 (2012)
7. Duffy, D.: A critique of the Crank-Nicolson scheme – strengths and weaknesses for financial instrument pricing. Wilmott Mag. **4**, 68–76 (2004)

8. Duffy, D.: Finite Difference Methods in Financial Engineering: A Partial Differential Equation Approach. Wiley, New York (2006). ISBN: 0470858826
9. Hagan, P.S., Kumar, D., Lesniewski, A.S., Woodward, D.E.: Managing smile risk. Wilmott Mag. **1**, 84–108 (2002)
10. Hagan, P., Kumar, D., Lesniewski, A.S., Woodward, D.E.: Arbitrage free SABR. Wilmott (2014)
11. Hagan, P., Kumar, D., Lesniewski, A.S., Woodward, D.E.: Universal smiles. Wilmott (2016)
12. Islah, O.: Soving SABR in Exact Form and Unifying it with LIBOR Market Model. SSRN eLibrary (2009)
13. Khaliq, A.Q.M., Twizell, E.H.: L-stable splitting methods for the simple heat equation in two space dimensions with homogeneous boundary conditions. SIAM J. Numer. Anal. **3**(23), 473–484 (1986)
14. Kienitz, J.: Pricing CMS spread options. Talk on ICBI Global Derivatives, Paris (2011)
15. Kienitz, J.: Approximate and PDE solution to the boundary free SABR model – applications to pricing and calibration. SSRN (2015). http://papers.ssrn.com/sol3/papers.cfm?abstract_id=2647344
16. Kienitz, J., Caspers, P.: Interest Rates Explained Volume 2: Term Structure Models. Palgrave MacMillan, Basingstoke (2017)
17. Kienitz, J., Wetterau, D.: Financial Modeling – Theory, Implementation and Practice – (with Matlab source). Wiley, New York (2012)
18. Kienitz, J., McWalter, T., Sheppard, R.: Clipping the Wings of SABR – Stable and Effcient Numerics for Sticky SABR – (2016, submitted). Preprint
19. Kopecky, K.A.: Advanced Macroeconomics I & II: Computational Methods for Macroeconomics and Applications. University of Western Ontario Lectures Notes ECO 613/614 Fall 2007
20. Lawson, J.D., Swayne, D.A.: A simple efficient algorithm for the solution of heat conduction problems. In: Proceedings of the 6th Manitoba Conference on Numerical Mathematics, pp. 239–250 (1976)
21. LeFloch, F., Kennedy, G.: Finite difference techniques for arbitrage free SABR (2013). Unpublished
22. Obloj, J.: Fine tune your smile. Wilmott Magazine **3**, 84–108 (2007)
23. Sheppard, R.: Pricing equity derivatives under stochastic volatility: a partial differential equation approach. Master's thesis, University of the Witwatersrand, Johannesburg (2007)

Part V
Compact FDMs and Splitting Schemes

Chapter 16
Sparse Grid High-Order ADI Scheme for Option Pricing in Stochastic Volatility Models

Bertram Düring, Christian Hendricks, and James Miles

Abstract We present a sparse grid high-order alternating direction implicit (ADI) scheme for option pricing in stochastic volatility models. The scheme is second-order in time and fourth-order in space. Numerical experiments confirm the computational efficiency gains achieved by the sparse grid combination technique.

16.1 Introduction

Stochastic volatility models such as the Heston model [22] have become one of the standard approaches in financial option pricing. For some stochastic volatility models and under additional restrictions, closed-form solutions can be obtained by Fourier methods (e.g. [9, 22]). Another approach is to derive approximate analytic expressions, see e.g. [2] and the literature cited therein. In general, however,—even in the Heston model [22] when the parameters in it are non constant—the partial differential equations (PDEs) arising from stochastic volatility models have to be solved numerically.

In the mathematical literature, there are many papers on numerical methods for option pricing, mostly addressing the one-dimensional case of a single risk factor and using standard, second order finite difference methods (see, e.g., [40] and the references therein). More recently, high-order finite difference schemes (fourth order in space) were proposed [19, 36, 39] that use a compact stencil (three points in space). In the option pricing context, see e.g. [11, 12, 30].

There are less works considering numerical methods for option pricing in stochastic volatility models, i.e., for two spatial dimensions. Finite difference approaches that are used are often standard, second-order methods, e.g. in [28] where different efficient methods for solving the American option pricing problem

B. Düring (✉) • J. Miles
Department of Mathematics, University of Sussex, Pevensey II, BN1 9QH Brighton, UK
e-mail: bd80@sussex.ac.uk; james.miles@sussex.ac.uk

C. Hendricks
Chair of Applied Mathematics/Numerical Analysis, Bergische Universität Wuppertal, Gaußstraße 20, 42097 Wuppertal, Germany
e-mail: hendricks@uni-wuppertal.de

M. Ehrhardt et al. (eds.), *Novel Methods in Computational Finance*,
Mathematics in Industry 25, DOI 10.1007/978-3-319-61282-9_16

for the Heston model are proposed. In [10] a high-order compact finite difference scheme for option pricing in the Heston model is derived and this approach is extended to non-uniform grids in [14]. Other approaches include finite element-finite volume [43], multigrid [5], sparse wavelet [23], FFT-based [35] or spectral methods [42].

The classical alternating direction implicit (ADI) method, introduced by Peaceman and Rachford [33], Douglas [6, 7], Fairweather and Mitchell [32], is a very powerful method that is especially useful for solving parabolic equations (*without* mixed derivative terms) on rectangular domains. Beam and Warming [1], however, have shown that no simple ADI scheme involving only discrete solutions at time levels n and $n+1$ can be second-order accurate in time in the presence of mixed derivatives. To overcome this limitation, unconditionally stable ADI schemes which are second order in time have been proposed by Hundsdorfer and Verwer [26, 27] and more recently by in't Hout and Welfert [24]. These schemes are second-order accurate in time and space. In [25] different second-order ADI schemes of this type are applied to the Heston model. In [13] this approach is combined with different high-order discretisations in space, using high-order compact schemes for two-dimensional convection-diffusion problems *with mixed derivatives and constant coefficients*. In [21] this approach is combined with sparse grids and applied to multi-dimensional diffusion equations, again *with constant coefficients*. Building on the ideas in [13, 26, 27], a high-order (second-order accurate in time and fourth-order accurate in space) ADI method for option pricing in stochastic volatility models which involve the solution of two-dimensional convection-diffusion equations *with mixed derivative terms and space-dependent coefficients* is derived in [16].

In this chapter we combine the approaches from [21] and [16], to obtain a *sparse grid high-order ADI scheme* for option pricing in stochastic volatility models. In the next section we recall stochastic volatility models for option pricing and the related convection-diffusion partial differential equations. Section 16.3 is devoted to the Hundsdorfer-Verwer ADI splitting in time. The spatial discretisation is introduced in Sect. 16.4 for the implicit steps, and in Sect. 16.5 for the explicit steps. The solution of the resulting scheme and the discretisation of boundary conditions are discussed in Sects. 16.6 and 16.7. The sparse grid combination technique is explained in Sect. 16.8. We present numerical convergence results in Sect. 16.9.

16.2 Stochastic Volatility Models

We consider the following class of stochastic volatility models: assume that asset spot price $0 \leq S(t) < \infty$ and variance $0 \leq \sigma(t) < \infty$ follow two stochastic diffusive processes for $t \in [0, T]$,

$$dS(t) = \mu S(t)dt + \sqrt{\sigma(t)}S(t)dW^{(1)}(t), \tag{16.1a}$$

$$d\sigma(t) = \kappa(\sigma(t))^{\alpha}(\theta - \sigma(t))dt + v(\sigma(t))^{\beta}dW^{(2)}(t), \tag{16.1b}$$

which are characterised by two Brownian motions, $dW^{(1)}(t)$ and $dW^{(2)}(t)$, with constant correlation parameter $dW^{(1)}(t)dW^{(2)}(t) = \rho dt$. The drift coefficient for stochastic asset returns is given by the mean return of the asset where $\mu \in \mathbb{R}$ and the diffusion coefficient is given by $\sqrt{\sigma(t)}S(t)$.

The drift coefficient of the asset variance is given by $\kappa(\sigma(t))^{\alpha}(\widetilde{\theta} - \sigma(t))$, where constants $\kappa \geq 0$ and $\theta \geq 0$ are the mean reversion speed of $\sigma(t)$ and the long run mean of $\sigma(t)$, respectively. The diffusion coefficient is given by $v(\sigma(t))^{\beta}$ where constant $v \geq 0$ is the volatility of volatility. The constant riskless interest rate is denoted by $r \geq 0$. The constants α, β determine the stochastic volatility model used.

The class of stochastic volatility models (16.1) includes a number of known stochastic volatility models: The most prominent stochastic volatility model, the *Heston model* [22] (also called *square root (SQR) model*) specifies the variance by

$$d\sigma(t) = \kappa\,(\theta - \sigma(t))\,\mathrm{d}t + v\sqrt{\sigma(t)}\mathrm{d}W^{(2)}(t).$$

Other known stochastic volatility models include the *GARCH* (or *VAR model*) model, see [8], where the stochastic variance is modelled by

$$d\sigma(t) = \kappa\,(\theta - \sigma(t))\,\mathrm{d}t + v\sigma(t)\mathrm{d}W^{(2)}(t),$$

and the *3/2 model* [31] in which the variance follows the process

$$d\sigma(t) = \kappa\,(\theta - \sigma(t))\,\mathrm{d}t + v\sigma^{\frac{3}{2}}(t)\mathrm{d}W^{(2)}(t).$$

All of the three stochastic volatility models mentioned above use a linear mean-reverting drift for the stochastic process of the variance $v(t)$, but there are also models, in which the drift is mean reverting in a non-linear fashion. Following [4], we denote these models with an additional "N": in the *SQRN model* the stochastic variance follows

$$d\sigma(t) = \kappa\sigma(t)\,(\theta - \sigma(t))\,\mathrm{d}t + v\sqrt{\sigma(t)}\mathrm{d}W^{(2)}(t),$$

in the *VARN model*

$$d\sigma(t) = \kappa\sigma(t)\,(\theta - \sigma(t))\,\mathrm{d}t + v\sigma(t)\mathrm{d}W^{(2)}(t),$$

and in the $3/2$-*N model*

$$d\sigma(t) = \kappa\sigma(t)\,(\theta - \sigma(t))\,\mathrm{d}t + v\sigma^{\frac{3}{2}}(t)\mathrm{d}W^{(2)}(t),$$

see [4].

Applying standards arbitrage arguments and Itos lemma to the class of stochastic volatility models (16.1), we can derive the following second order partial differential equation for any financial derivative $V(S, \sigma, t)$, to be solved backwards in time with $0 < S < \infty, 0 < \sigma < \infty, t \in [0, T)$:

$$V_t + \frac{S^2\sigma}{2} V_{SS} + \rho v \sigma^{\beta + \frac{1}{2}} S V_{S\sigma} + \frac{v^2 \sigma^{2\beta}}{2} V_{\sigma\sigma} + rSV_s + [\kappa \sigma^{\alpha}(\theta - \sigma) - \lambda_0 \sigma] V_{\sigma} - rV = 0. \tag{16.2}$$

Here, $\lambda_0 \sigma(t)$ is the market price of volatility risk, where $\lambda_0 \in \mathbb{R}$, which is usually assumed to be proportional to the variance. In the following we assume $\lambda_0 = 0$ for streamlining the presentation. The generalisation to the case $\lambda_0 \neq 0$ is straightforward by consistently adding in the additional term in the coefficient of V_σ. The boundary conditions and final condition are determined by the type of financial derivative $V(S, \sigma, t)$ we are solving for. The boundary conditions of any European option will depend on a prescribed exercise price, denoted here by $E > 0$. For example, in the case of the European Put Option:

$$\begin{aligned}
V(S, \sigma, T) &= \max(E - S, 0), && 0 < S < \infty,\ 0 < \sigma < \infty,\\
\lim_{S \to \infty} V(S, \sigma, t) &= 0, && 0 < \sigma < \infty,\ 0 < t < T,\\
V(0, \sigma, t) &= E \exp(-r(T - t)), && 0 < \sigma < \infty,\ 0 < t < T,\\
\lim_{\sigma \to \infty} V_\sigma(S, \sigma, t) &= 0, && 0 < S < \infty,\ 0 < t < T,
\end{aligned}$$

The remaining boundary condition at $\sigma = 0$ can be obtained by looking at the formal limit $\sigma \to 0$ in (16.2), i.e.,

$$V_t + rSV_S + \kappa\theta V_\sigma - rV = 0, \quad T > t \geq 0,\ S > 0,\ \text{as } \sigma \to 0. \tag{16.3}$$

This boundary condition is used frequently, e.g. in [28, 43]. Alternatively, one can use a homogeneous Neumann condition [5], i.e.,

$$V_\sigma(S, 0, t) = 0, \quad 0 < S < \infty,\ 0 < t < T. \tag{16.4}$$

By using a change of variables:

$$x = \ln\frac{S}{E}, \quad y = \frac{\sigma}{v}, \quad \tau = T - t, \quad u = \exp(r\tau)\frac{V}{E}$$

we transform the partial differential equation to an convection-diffusion equation in two spatial dimensions with a mixed derivative term. The transformed partial differential equation and boundary/initial conditions are now satisfied by $u(x, y, \tau)$, where $x \in \mathbb{R}$, $y > 0$, $\tau \in (0, T]$:

$$u_\tau = \frac{vy}{2}u_{xx} + \frac{(vy)^{2\beta}}{2}u_{yy} + \rho(vy)^{\beta+\frac{1}{2}}u_{xy} + \left(r - \frac{vy}{2}\right)u_x + \left(\kappa\,(vy)^\alpha\,\frac{\theta - vy}{v}\right)u_y, \tag{16.5}$$

$$u(x,y,0) = \max(1-\exp(x),0), \quad -\infty < x < \infty, 0 < y < \infty, \tag{16.6a}$$

$$\lim_{x\to\infty} u(x,y,\tau) = 0, \quad 0 < y < \infty, 0 \le \tau < T, \tag{16.6b}$$

$$\lim_{x\to-\infty} u(x,y,\tau) = 1, \quad 0 < y < \infty, 0 \le \tau < T, \tag{16.6c}$$

$$\lim_{y\to\infty} u_y(x,y,\tau) = 0, \quad -\infty < x < \infty, 0 < \tau \le T, \tag{16.6d}$$

$$\lim_{y\to 0} u_y(x,y,\tau) = 0, \quad -\infty < x < \infty, 0 < \tau \le T. \tag{16.6e}$$

In order to discretise the problem and solve numerically, we truncate our spatial boundaries to finite values. Take $L_1 \le x \le K_1$, where $L_1 < K_1$, and $L_2 \le y \le K_2$, where $0 < L_2 < K_2$, so that the spatial domain forms a closed rectangle in $\mathbb{R}^2$ of $M \times N$ points with uniform spacing of Δ_x in the x-direction and Δ_y in the y-direction:

$$x_i = L_1 + (i-1)\Delta_x,\ i = 1,2,\ldots,M, \qquad y_j = L_2 + (j-1)\Delta_y,\ j = 1,2,\ldots,N.$$

The lower y-boundary is truncated to $L_2 > 0$ to ensure non-degeneracy of the partial differential equation for all values of y. We assume cell aspect ratios to be moderate. We also take a uniform partition of $\tau \in [0,T]$ into P points such that $\tau_k = (k-1)\Delta_\tau$, where $k = 1,2,\ldots,P$. We denote the discrete approximation of $u((i-1)\Delta_x, (j-1)\Delta_y, (k-1)\Delta_\tau)$ by $u^k_{i,j}$ and $U^n = (u^n_{i,j})_{i,j}$.

16.3 Hundsdorfer-Verwer ADI Splitting Scheme

We consider the Alternating Direction Implicit (ADI) time-stepping numerical method proposed by Hundsdorfer and Verwer [26, 27]. Our partial differential Eq. (16.5) takes the form $u_\tau = F(u)$. We employ the splitting $F(u) = F_0(u) + F_1(u) + F_2(u)$ where unidirectional and mixed derivative differential operators are given by:

$$F_0(u) = \rho(vy)^{\beta+\frac{1}{2}}u_{xy},\ F_1(u) = \frac{vy}{2}u_{xx} + \left(r - \frac{vy}{2}\right)u_x,$$

$$F_2(u) = \frac{(vy)^{2\beta}}{2}u_{yy} + \left(\kappa\,(vy)^\alpha\,\frac{\theta - vy}{v}\right)u_y. \tag{16.7}$$

We consider (16.5) with the splitting (16.7) and look for a semi-discrete approximation $U^n \approx u(\tau_n)$ at time $n\Delta_\tau$. Given an approximation U^{n-1} we can calculate an approximation for U^n at time $n\Delta_\tau$ using the differential operators from (16.7):

$$Y_0 = U^{n-1} + \Delta_t F(U^{n-1}), \tag{16.8a}$$

$$Y_1 = Y_0 + \phi\Delta_t(F_1(Y_1) - F_1(U^{n-1})), \tag{16.8b}$$

$$Y_2 = Y_1 + \phi\Delta_t(F_2(Y_2) - F_2(U^{n-1})), \tag{16.8c}$$

$$\widetilde{Y}_0 = Y_0 + \psi\Delta_t(F(Y_2) - F(U^{n-1})), \tag{16.8d}$$

$$\widetilde{Y}_1 = \widetilde{Y}_0 + \phi\Delta_t(F_1(\widetilde{Y}_1) - F_1(Y_2)), \tag{16.8e}$$

$$\widetilde{Y}_2 = \widetilde{Y}_1 + \phi\Delta_t(F_2(\widetilde{Y}_2) - F_2(Y_2)), \tag{16.8f}$$

$$U^n = \widetilde{Y}_2. \tag{16.8g}$$

The parameter ψ is taken to be $\psi = 1/2$ to ensure second-order accuracy in time. The parameter ϕ is typically fixed to $\phi = 1/2$. Larger values give stronger damping of the implicit terms while lower values return better accuracy. The role of ϕ is discussed in [26]. Its influence in the connection with high-order spatial approximations is investigated numerically in [16].

The first and fourth step in (16.8) can be solved explicitly, while the remaining steps are solved implicitly. Our aim is to derive high-order spatial discretisations of the differential operators. Following [13] we combine high-order compact finite difference methods for the implicit steps with a (classical, non-compact) high-order stencil for the explicit steps.

16.4 High-Order Compact Scheme for Implicit Steps

For $F_1(u)$, consider the one-dimensional convection-diffusion equation

$$u_{xx} + c_1 u_x = c_2 g \tag{16.9}$$

with constants $c_1 = 2r/(vy) - 1$ and $c_2 = 2/(vy)$. To discretise the partial derivatives in (16.9), we employ standard, centered second-order finite difference operators, denoted by δ_{x0} and δ_x^2. The second-order terms in the truncation error involve higher-order partial derivatives, u_{xxx} and u_{xxxx}. Hence, if we can find second-order accurate expressions for u_{xxx} and u_{xxxx}, using only information on the compact stencil, then it will be possible to approximate u_x and u_{xx} with fourth order accuracy on the compact stencil. By differentiating (16.9) once and twice with respect to x, respectively, it is possible to express u_{xxx} and u_{xxxx} in terms of first- and second-order derivatives of u and g with respect to x. We obtain the following relations, concisely written in matrix form,

$$\begin{pmatrix} 1 & 0 & \frac{1}{6} & 0 \\ 0 & 1 & 0 & \frac{1}{12} \\ 0 & c_1\Delta_x^2 & 1 & 0 \\ 0 & 0 & c_1 & 1 \end{pmatrix} \begin{pmatrix} u_x \\ u_{xx} \\ \Delta_x^2 u_{xxx} \\ \Delta_x^2 u_{xxxx} \end{pmatrix} = \begin{pmatrix} \delta_{x0}u_{i,j} \\ \delta_x^2 u_{i,j} \\ c_2\Delta_x^2 g_x \\ c_2\Delta_x^2 g_{xx} \end{pmatrix} + \begin{pmatrix} \mathcal{O}(\Delta_x^4) \\ \mathcal{O}(\Delta_x^4) \\ 0 \\ 0 \end{pmatrix} = \begin{pmatrix} \delta_{x0}u_{i,j} \\ \delta_x^2 u_{i,j} \\ c_2\Delta_x^2\delta_{x0}g_{i,j} \\ c_2\Delta_x^2\delta_x^2 g_{i,j} \end{pmatrix} + \mathcal{O}(\Delta_x^4).$$

This shows that only second-order approximations for u_x, u_{xx}, g_x and g_{xx} are needed. Using these relations to discretise (16.9) and to replace the partial derivatives u_{xxx} and u_{xxxx} in the truncation error, yields a fourth-order compact approximation for (16.9) at all points of the spatial grid except those that lie on the x- and y-boundaries. We refer to [16] for more details of the derivation of the compact high-order spatial discretisation.

To approximate $F_1(u)$ at points along the x boundaries of the inner grid of the spatial domain, we will require a contribution from the Dirichlet values at the x-boundaries of the spatial domain. We collect these separately in a vector d. Details on the boundary conditions are given in Sect. 16.7. The resulting linear system to be solved can be written in matrix form:

$$A_x u = B_x g + d,$$

where $u = (u_{2,2}, u_{2,3}, \ldots, u_{M-1,N-1})$, $g = (g_{2,2}, g_{2,3}, \ldots, g_{M-1,N-1})$. The coefficient matrices A_x and B_x are block diagonal matrices, with the following structure:

$$A_x = \begin{pmatrix} A_x^{1,1} & 0 & 0 & 0 \\ 0 & A_x^{2,2} & 0 & 0 \\ 0 & 0 & \ddots & 0 \\ 0 & 0 & 0 & A_x^{N-2,N-2} \end{pmatrix}, \quad B_x = \begin{pmatrix} B_x^{1,1} & 0 & 0 & 0 \\ 0 & B_x^{2,2} & 0 & 0 \\ 0 & 0 & \ddots & 0 \\ 0 & 0 & 0 & B_x^{N-2,N-2} \end{pmatrix},$$

where each $A_x^{j,j} = \text{diag}[a_{-1}^{j,j}, a_0^{j,j}, a_1^{j,j}]$ and $B_x^{j,j} = \text{diag}[b_{-1}^{j,j}, b_0^{j,j}, b_1^{j,j}]$ are tri-diagonal matrices. Explicit expression for all coefficients are given in [16].

For $F_2(u)$ the derivation can be presented in a concise form, similar as for $F_1(u)$, again we refer to [16] for additional details. Consider the one-dimensional convection-diffusion equation

$$u_{yy} + c_1 u_y = c_2 g \tag{16.10}$$

with $c_1(y) = 2\kappa(vy)^{\alpha-2\beta}(\theta - vy)/v$ and $c_2(y) = 2/(vy)^{2\beta}$, the necessary relations can be concisely written in matrix form,

$$\begin{pmatrix} 1 & 0 & \frac{1}{6} & 0 \\ 0 & 1 & 0 & \frac{1}{12} \\ c_1'\Delta_y^2 & c_1\Delta_y^2 & 1 & 0 \\ c_1''\Delta_y^2 & 2c_1'\Delta_y^2 & c_1 & 1 \end{pmatrix} \begin{pmatrix} u_y \\ u_{yy} \\ \Delta_y^2 u_{yyy} \\ \Delta_y^2 u_{yyyy} \end{pmatrix}$$
$$= \begin{pmatrix} \delta_{y0} u_{i,j} \\ \delta_y^2 u_{i,j} \\ \Delta_y^2(\delta_{y0} c_{2,j} g_{i,j} + c_{2,j}\delta_{y0} g_{i,j}) \\ \Delta_y^2(\delta_y^2 c_{2,j} g_{i,j} + 2\delta_{y0} c_{2,j}\delta_{y0} g_{i,j} + c_{2,j}\delta_y^2 g_{i,j}) \end{pmatrix} + \mathcal{O}(\Delta_y^4),$$

where the first two lines of the system correspond to standard, central second-order difference approximations, while the third and fourth are obtained from the repeated differentiation of (16.10). Using these relations to discretise (16.10) and to replace the partial derivatives u_{yyy} and u_{yyyy} in the truncation error, yields a fourth-order compact approximation for (16.10).

We obtain a linear system which can be represented in matrix form:

$$A_y u = B_y g$$

where $u = (u_{2,2}, u_{2,3}, \ldots, u_{M-1,N-1})$, $g = (g_{2,2}, g_{2,3}, \ldots, g_{M-1,N-1})$. We do not impose any boundary conditions in y-direction, but discretise the boundary grid points with the same scheme, and handle resulting ghost points via extrapolation; details on the boundary conditions are given in Sect. 16.7. The coefficient matrices A_y and B_y are block tri-diagonal matrices with the following structures:

$$\mathbf{A_y} = \begin{pmatrix} A_y^{1,1} & A_y^{1,2} & 0 & 0 & 0 \\ A_y^{2,1} & A_y^{2,2} & A_y^{2,3} & 0 & 0 \\ 0 & \ddots & \ddots & \ddots & 0 \\ 0 & 0 & A_y^{N-3,N-4} & A_y^{N-3,N-3} & A_y^{N-3,N-2} \\ 0 & 0 & 0 & A_y^{N-2,N-3} & A_y^{N-2,N-2} \end{pmatrix},$$

$$\mathbf{B_y} = \begin{pmatrix} B_y^{1,1} & B_y^{1,2} & 0 & 0 & 0 \\ B_y^{2,1} & B_y^{2,2} & B_y^{2,3} & 0 & 0 \\ 0 & \ddots & \ddots & \ddots & 0 \\ 0 & 0 & B_y^{N-3,N-4} & B_y^{N-3,N-3} & B_y^{N-3,N-2} \\ 0 & 0 & 0 & B_y^{N-2,N-3} & B_y^{N-2,N-2} \end{pmatrix},$$

where each $A_y^{j,j} = \text{diag}[a^{i,j}]$ and $B_y^{j,j} = \text{diag}[b^{i,j}]$ are diagonal matrices. Explicit expression for all coefficients are given in [16].

16.5 High-Order Scheme for Explicit Steps

The first and fourth steps of the ADI scheme (16.8) operate only on previous approximations to explicitly calculate an updated approximation. The differential operator in these steps takes the form of the right hand side of (16.5). For the mixed derivative term it seems not to be possible to exploit the structure of the differential operator to obtain a fourth-order approximation on a compact computational stencil. Hence, in order to maintain fourth-order accuracy of the scheme in the explicit steps of (16.8), the derivatives in each differential operator F_0, F_1 and F_2 are approximated using classical, fourth-order central difference operators which operate on a larger 5×5-stencil in the spatial domain. Here we use the shift operator defined by:

$$s_x = e^{\Delta_x \delta_x} \text{ where } (s_x u)_{i,j} = u_{i+1,j}, \quad s_y = e^{\Delta_y \delta_y} \text{ where } (s_y u)_{i,j} = u_{i,j+1}.$$

For $F_1(u) = \frac{vy}{2} u_{xx} - (\frac{vy}{2} - r) u_x$, we have the following scheme:

$$\left[\frac{vy}{2} u_{xx} + \left(r - \frac{vy}{2}\right) u_x\right]_{i,j} = \frac{vy_j}{2} \left(\frac{-s_x^{-2} + 16 s_x^{-1} - 30 + 16 s_x - s_x^2}{12 \Delta_x^2}\right) u_{i,j}$$
$$+ \left(r - \frac{vy_j}{2}\right) \left(\frac{s_x^{-2} - 8 s_x^{-1} + 8 s_x - s_x^2}{12 \Delta_x}\right) u_{i,j} + \mathcal{O}(\Delta_x^4).$$

For $F_2(u) = \frac{(vy)^{2\beta}}{2} u_{yy} + \frac{\kappa (vy)^{\alpha} (\theta - vy)}{v} u_y$, we have:

$$\left[\frac{(vy)^{2\beta}}{2} u_{yy} + \frac{\kappa (vy)^{\alpha} (\theta - vy)}{v} u_y\right]_{i,j}$$
$$= \frac{(vy_j)^{2\beta}}{2} \left(\frac{-s_y^{-2} + 16 s_y^{-1} - 30 + 16 s_y - s_y^2}{12 \Delta_y^2}\right) u_{i,j}$$
$$+ \frac{\kappa (vy_j)^{\alpha} (\theta - vy_j)}{v} \left(\frac{s_y^{-2} - 8 s_y^{-1} + 8 s_y - s_y^2}{12 \Delta_y}\right) u_{i,j} + \mathcal{O}(\Delta_y^4).$$

Finally, for the mixed derivative term $F_0 = \rho (vy)^{\beta + \frac{1}{2}} u_{xy}$, the following computational stencil is used:

$$\left[\rho (vy)^{\beta + \frac{1}{2}} u_{xy}\right]_{i,j}$$
$$= \rho (vy_j)^{\beta + \frac{1}{2}} \left(\frac{s_x^{-2} - 8 s_x^{-1} + 8 s_x - s_x^2}{12 \Delta_x}\right) \left(\frac{s_y^{-2} - 8 s_y^{-1} + 8 s_y - s_y^2}{12 \Delta_y}\right) u_{i,j}$$
$$+ \mathcal{O}(\Delta_x^4 \Delta_y^4) + \mathcal{O}(\Delta_x^4) + \mathcal{O}(\Delta_y^4).$$

•	$\mathrm{u}_{4,1}$	$u_{4,2}$	$u_{4,3}$	$u_{4,4}$
•	$\mathrm{u}_{3,1}$	$u_{3,2}$	$u_{3,3}$	$u_{3,4}$
•	$\mathrm{u}_{2,1}$	$u_{2,2}$	$u_{2,3}$	$u_{2,4}$
•	$\mathrm{u}_{1,1}$	$\mathrm{u}_{1,2}$	$\mathrm{u}_{1,3}$	$\mathrm{u}_{1,4}$
⊙	○	○	○	○

Fig. 16.1 Example: evaluation of $F(u_{2,2})$ using the 5×5-point computational stencil in the lower left corner of the computational domain; ghost points outside the computational domain at which values are extrapolated from the interior of the domain are marked by bullets (•,○,⊙), grid points on the boundary are set in Roman

Using these fourth-order approximations, the first and fourth step in (16.8) can be computed directly. The values at the spatial boundaries for each solution of the ADI scheme are determined by the boundary conditions, the computational stencil is required for all remaining points in the spatial domain. For the explicit steps, the 5×5-point computational stencil exceeds the spatial boundary when we wish to approximate differential operator $F(u)$ at any point along the boundary of the spatial domain's inner grid. For example if we wish to evaluate $F(u_{2,2})$, we will require contributions from ghost points which fall outside the spatial domain, as marked by bullet points in Fig. 16.1. We extrapolate information from grid points $u(x_i, y_j)$, where $i = 1, \ldots, M-1, j = 1, \ldots, N-1$ to establish values at these ghost points for the purpose of evaluating the differential operator $F(u)$ at any point along the boundary of the inner grid of the spatial domain. To calculate the values at these ghost points, we use the following five-point extrapolation formulae for three cases:

$$
\begin{aligned}
x = L_1(\bullet): \quad & u_{i,0} = 5u_{i,1} - 10u_{i,2} + 10u_{i,3} - 5u_{i,4} + u_{i,5} + \mathcal{O}(\Delta_x^5),\\
y = L_2(\circ): \quad & u_{0,j} = 5u_{1,j} - 10u_{2,j} + 10u_{3,j} - 5u_{4,j} + u_{5,j} + \mathcal{O}(\Delta_y^5),\\
x = L_1, y = L_2(\odot): \quad & u_{0,0} = 5u_{1,1} - 10u_{2,2} + 10u_{3,3} - 5u_{4,4} + u_{5,5} + \mathcal{O}(\Delta_x^5)\\
& + \mathcal{O}(\Delta_x^4\Delta_y) + \mathcal{O}(\Delta_x^3\Delta_y^2) + \mathcal{O}(\Delta_x^2\Delta_y^3) + \mathcal{O}(\Delta_x\Delta_y^4) + \mathcal{O}(\Delta_y^5).
\end{aligned}
$$

The extrapolation at the $x = K_1$ and $y = K_2$ boundaries and the remaining three corners is handled analogously.

16.6 Solving the High-Order ADI Scheme

Starting from a given U^{n-1}, the ADI scheme (16.8) involves six approximation steps to obtain U^n, the solution at the next time level. The first approximation Y_0 can be solved for explicitly using the 5×5-point computational stencil derived in Sect. 16.5. The second approximation for our solution, denoted by Y_1, has to be solved for implicitly:

$$Y_1 = Y_0 + \phi\Delta_t(F_1(Y_1) - F_1(U^{n-1})) \iff F_1(Y_1 - U^{n-1}) = \frac{1}{\phi\Delta_t}(Y_1 - Y_0). \tag{16.11}$$

We apply the fourth-order compact scheme established in Sect. 16.4 to solve (16.11). In matrix form we obtain

$$A_x(Y_1 - U^{n-1}) = B_x\Big(\frac{1}{\phi\Delta_t}(Y_1 - Y_0)\Big) + d.$$

Collecting unknown Y_1 terms on the left hand side and known terms Y_0, U^{n-1} and d on the right hand side we get

$$(B_x - \phi\Delta_t A_x)\, Y_1 = B_x Y_0 - \phi\Delta_t A_x U^{n-1} - \phi\Delta_t d.$$

To solve, we invert the tri-diagonal matrix $(B_x - \phi\Delta_t A_x)$. For the third step of the ADI scheme, we proceed analogously, and use the high-order compact scheme presented in Sect. 16.4 to solve for Y_2 implicitly. The fourth, fifth and sixth step of the ADI scheme are performed analogously as the first, second and third steps, respectively.

Note that the matrix $(B_x - \phi\Delta_t A_x)$ appears twice in the scheme (16.8), in the second and fifth step. Similarly, $\big(B_y - \phi\Delta_t A_y\big)$ appears in the third and the sixth step. Hence, using LU-factorisation, only two matrix inversions are necessary in each time step of scheme (16.8). Moreover, since the coefficients in the partial differential Eq. (16.5) do not depend on time, and the matrices are therefore constant, they can be LU-factorised before iterating in time to obtain a highly efficient algorithm.

The combination of the fourth-order spatial discretisation presented in Sect. 16.4 and 16.5 with the second-order time splitting (16.8) yields a high-order ADI scheme with order of consistency two in time and four in space.

16.7 Boundary Conditions

For the case of the Dirichlet conditions at $x = L_1$ and $x = K_1$ we impose

$$u(L_1, y_j, \tau_k) = 1 - e^{r\tau + L_1}, \qquad j = 1, 2, \ldots, N,\ k = 1, 2, \ldots,$$
$$u(K_1, y_j, \tau_k) = 0, \qquad j = 1, 2, \ldots, N,\ k = 1, 2, \ldots.$$

Using the homogeneous Neumann conditions (16.6d) and (16.6e) which are correct in the limit $y \to \infty$ and $y \to 0$, respectively, at the (finite) boundaries $y = L_2 > 0$ and $y = K_2$ would result in a dominant error along these boundaries. Hence, we do not impose any boundary condition at these two boundaries but discretise the partial differential equation using the computational stencil from the interior. The values of the unknown on the boundaries are set by extrapolation from values in the interior. This introduces a numerical error, and it needs to be considered that the order of extrapolation should be high enough not to affect the overall order of accuracy. We refer to Gustafsson [20] to discuss the influence of the order of the approximation on the global convergence rate. We use the following extrapolation formulae:

$$u_{i,1}^k = 5u_{i,2}^k - 10u_{i,3}^k + 10u_{i,4}^k - 5u_{i,5}^k + u_{i,6}^k + \mathcal{O}(\Delta_y^6),$$
$$u_{i,N}^k = 5u_{i,N-1}^k - 10u_{i,N-2}^k + 10u_{i,N-3}^k - 5u_{i,N-4}^k + u_{i,N-5}^k + \mathcal{O}(\Delta_y^6).$$

16.8 Sparse Grid Combination Technique

Due to the ADI splitting and the compactness of the finite difference discretisation in the implicit steps, the computational effort grows linearly with the number of unknowns, namely $\mathcal{O}(N \cdot M)$. In the following we use the so-called *sparse grid combination technique* to reduce the number of grid nodes and thus also the computational effort. Sparse grids go back to Smolyak [38], who used them for numerical integration. Zenger [41], Bungartz et al. [3] and Schiekofer [37] extended his idea and applied sparse grids to solve PDEs with finite element, finite volume and finite difference methods. These methods in general require hierarchical, tree-like data structures, which makes the data structure management more complicated than in the full grid case. With the help of the sparse grid combination technique [18] this problem can be overcome. Here, full tensor-based solutions are linearly combined to construct the sparse grid solution. This allows us to use standard full grid PDE solvers. Hence, this approach is very versatile and broadly applicable. Furthermore, each sub-solution can be computed independently, which makes it easily parallelisable.

The combination technique is based on the error splitting structure of the underlying numerical scheme. Let the numerical solution of the HO-ADI scheme be given by u_l with multi-index $l = (l_1, l_2)$ and mesh widths $\Delta_x = 2^{-l_1}(K_1 - L_1)$, $\Delta_y = 2^{-l_2}(K_2 - L_2)$. We assume that our numerical scheme satisfies an error splitting structure of the form

$$u - u_l = \Delta_x^4 w_1(\Delta_x) + \Delta_y^4 w_2(\Delta_y) + \Delta_x^4 \Delta_y^4 w_{1,2}(\Delta_x, \Delta_y),$$

with functions $w_1, w_2, w_{1,2}$ bounded by some constant $C \in \mathbb{R}^+$. The mesh widths Δ_x and Δ_y are independent of one another. Since the error functions w_1 and w_2 only depend on either Δ_x or Δ_y, we can subtract two solutions with the same mesh width in one coordinate direction, such that the error term cancels out. Exploiting this idea further leads to the combination technique

$$u_n^s = \sum_{|l|_1=n+1} u_l - \sum_{|l|_1=n} u_l. \tag{16.12}$$

Applying the error splitting from above, the lower order terms cancel out and we obtain

$$\begin{aligned} u_n^s = u &+ 2^{-4(n+1)} R_1 w_1(2^{-(n+1)} R_1) + 2^{-2(n+1)(R_2)} w_2(2^{-(n+1)} R_2) \\ &+ 2^{-4(n+1)} R_1 R_2 \sum_{i=0}^{n+1} w_{1,2}(2^{-i} R_1, 2^{-(n+1-i)} R_2) \\ &- 2^{-4n} R_1 R_2 \sum_{i=0}^{n} w_{1,2}(2^{-i} R_1, 2^{-(n-i)} R_2), \end{aligned}$$

where $R_1 = K_1 - L_1$ and $R_2 = K_2 - L_2$. As w_1, w_2 and $w_{1,2}$ are bounded by C the pointwise error is given by

$$|u_n^s - u| = \mathcal{O}(n 2^{-4n}),$$

which is equivalent to

$$|u_n^s - u| = \mathcal{O}(\Delta^4 \log_2(\Delta^{-1})) \tag{16.13}$$

for $\Delta = 2^{-n}$. We observe that the error of the sparse grid combination technique is deteriorated by a factor of $\log_2(\Delta^{-1})$ compared to the fourth-order full grid solution.

Figure 16.2 shows the two-dimensional grid hierarchy at levels $n = 0, \ldots, 4$. The sparse grid in two dimensions at level n consists of sub-grids, whose sum of refinement levels fulfils $|l|_1 = n$. Hence the number of grid points on each sub-grid grows with $\mathcal{O}(2^n)$. As the number of grids increases with $\mathcal{O}(n)$, this leads to $\mathcal{O}(n2^n)$ nodes in the sparse grid. Let $\Delta = 2^{-n}$, then this results in $\mathcal{O}(\Delta^{-1} \log_2(\Delta^{-1}))$ grid points compared to $\mathcal{O}(\Delta^{-2})$ nodes in the full grid. Thus we are able to reduce the number of grid nodes significantly while maintaining a high accuracy.

It should be noted that for larger n the combination technique as introduced above involves solutions on grids which violate the assumption of moderate cell aspect ratios which may lead to reduced accuracy and potential instability of the scheme due to the extreme distortion of the grid. This aspect of the combination technique is of general nature and not specific to our scheme. A usual remedy would be to exclude solutions on extremely distorted grids in (16.12). For further details we refer to the pertinent literature on sparse grids.

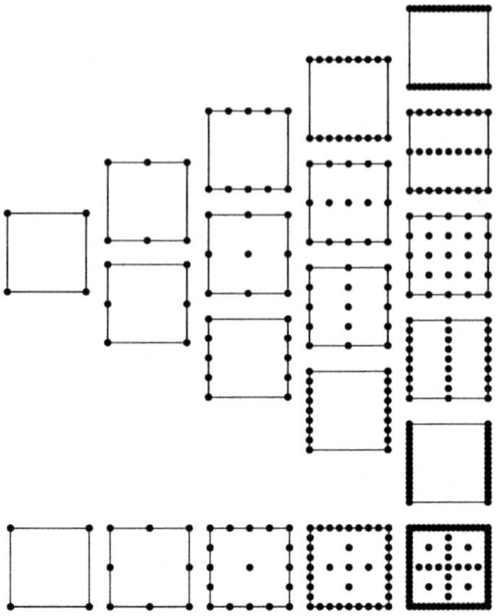

Fig. 16.2 Sub-grids and sparse grid for $n = 0, \ldots, 4$

16.9 Numerical Experiments

In this section we test the proposed sparse grid high-order ADI scheme. Beside the accuracy of the full grid solution we are also interested in the efficiency of the combined sparse grid solution.

It is well known that due to the non-smooth nature of the payoff function in option pricing problems one cannot expect to see higher-order in practice [34]. Some form of smoothing has to be applied to the initial condition. In [29] suitable smoothing operators are identified in Fourier space. Since the order of convergence of our high-order compact scheme is four, we could use the smoothing operator Φ_4 as in [15], given by its Fourier transformation

$$\hat{\Phi}_4(\omega) = \left(\frac{\sin\left(\frac{\omega}{2}\right)}{\frac{\omega}{2}} \right)^4 \left[1 + \frac{2}{3} \sin^2\left(\frac{\omega}{2}\right) \right].$$

This leads to the smooth initial condition determined by

$$\tilde{u}_0(x,y) = \int_{-3h}^{3h} \int_{-3h}^{3h} \Phi_4\left(\frac{\tilde{x}}{h}\right) \Phi_4\left(\frac{\tilde{y}}{h}\right) u_0(x - \tilde{x}, y - \tilde{y}) \; d\tilde{x}\, d\tilde{y}$$

for any stepsize $h > 0$, where u_0 is the original initial condition and $\Phi_4(x)$ denotes the Fourier inverse of $\hat{\Phi}_4(\omega)$, see [29]. As $h \to 0$, the smooth initial condition $\tilde{u}_0$ tends towards the original initial condition u_0 and the approximation of the smoothed problem tends towards the true solution. For our numerical experiments we use this smoothing operator which has already been applied successfully to option pricing problems in [15].

A numerical solution computed on a grid with $\Delta_x = \Delta \cdot (K_1 - L_1)$, $\Delta_y = \Delta \cdot (K_2 - L_2)$ and time step $\Delta_t = 5 \cdot \Delta^2$ serves as a reference solution, where $\Delta = 2^{-8}$. Since the accuracy of option prices close to the strike price is of highest interest from a practitioner's point of view, we compute the maximum absolute error in the region $[0.5E, 2E] \times [0.05, 1]$. The grid parameters of the computational domain are chosen to be $L_1 = -5$, $K_1 = 1.5$, $L_2 = 0.05$ and $K_2 = 2.5$. The parameters of the ADI method are $\psi = 1/2$ and $\phi = 1/2$, cf. Sect. 16.3. The full grid solution is computed with step sizes $\Delta_x = \Delta \cdot (K_1 - L_1)$, $\Delta_y = \Delta \cdot (K_2 - L_2)$ and $\Delta_t = 5 \cdot \Delta^2$ with $\Delta = 2^{-n}$, while the sparse grid solution u_n^s is constructed according to definition (16.12). In order to avoid instabilities due to the extreme distortion of the grid we neglect grids within the combination technique, where $l_i \leq 2$ for $i = 1, 2$. Thus, the finest resolution in one of the sub-grids along one coordinate direction is given by $\Delta = 2^{-(n-3)}$.

We compare the performance of the high-order ADI scheme in the full and sparse grid case for a European put option with the parameters given in table in Fig. 16.3. Figure 16.4 shows the maximum error plotted versus the grid resolution Δ for both

Parameter	Value
Strike price	$E = 100$
Time to maturity	$T = 1$
Interest rate	$r = 0.05$
Volatility of volatility	$v = 0.1$
Mean reversion speed	$\kappa = 2$
Long run mean of volatility	$\theta = 0.1$
Correlation	$\rho = -0.5$
Stochastic volatility drift parameter	$\alpha = 0.5$
Stochastic volatility diffusion parameter	$\beta = 0.5$

Fig. 16.3 Parameters used in the numerical experiments

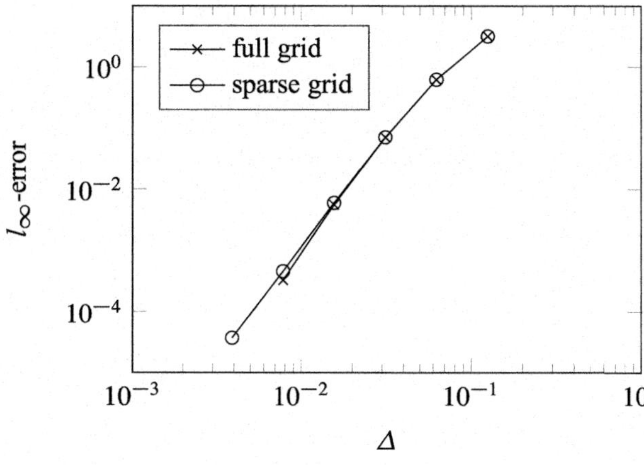

Fig. 16.4 Error decay of the full grid for $n = 3, 4, \ldots, 7$ and sparse grid combination technique for $n = 6, 7, \ldots, 11$

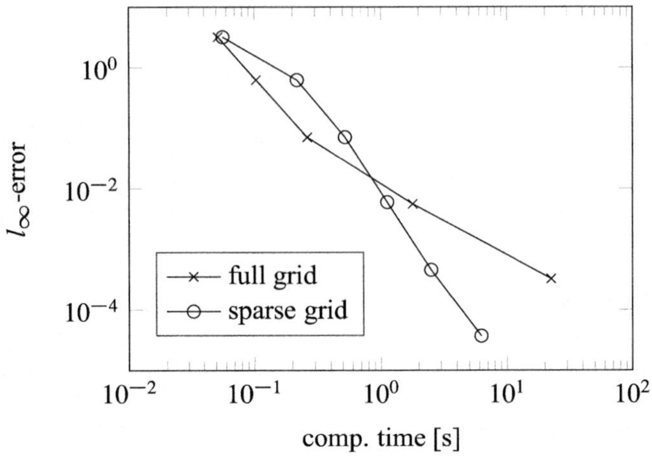

Fig. 16.5 Error versus computational time for the full grid for $n = 3, 4, \ldots, 7$ and sparse grid combination technique for $n = 6, 7, \ldots, 11$

cases. The fourth-order compact finite difference scheme achieves an estimated numerical convergence order of 3.33, the error of the sparse grid solution decays slightly slower due to the logarithmic factor in (16.13).

To illustrate the computational efficiency we compare the run-time to the accuracy in Fig. 16.5 for both approaches. We confirm that, as the mesh width decreases, the lower number of employed grid nodes in the sparse grid method outweighs its slightly lower convergence rate. The serial implementation of the combination technique outperforms the full grid solver in the high accuracy region, reducing the computational time by about an order of magnitude, while achieving a similar accuracy.

Acknowledgements BD acknowledges support by the Leverhulme Trust research project grant 'Novel discretisations for higher-order nonlinear PDE' (RPG-2015-69). CH is supported by the European Union in the FP7-PEOPLE-2012-ITN Program under Grant Agreement Number 304617 (FP7 Marie Curie Action, Project Multi-ITN STRIKE—Novel Methods in Computational Finance). Further CH acknowledges partial support from the bilateral German-Spanish Project HiPeCa—High Performance Calibration and Computation in Finance, Programme Acciones Conjuntas Hispano-Alemanas financed by DAAD. JM has been supported in part by a studentship under the EPSRC Doctoral Training Partnership (DTP) scheme (grant number EP/L505109/1).

References

1. Beam, R.M., Warming, R.F.: Alternating direction implicit methods for parabolic equations with a mixed derivative. SIAM J. Sci. Stat. Comput. **1**(1), 131–159 (1980)
2. Benhamou, E., Gobet, E., Miri, M.: Time dependent Heston model. SIAM J. Finan. Math. **1**, 289–325 (2010)

3. Bungartz, H., Griebel, M.: Sparse grids. Acta Numer. **13**, 1–123 (2004)
4. Christoffersen, P., Jacobs, K., Mimouni, K.: Volatility dynamics for the S&P500: evidence from realized volatility, daily returns, and option prices. Rev. Financ. Stud. **23**, 3141–3189 (2010)
5. Clarke, N., Parrott, K.: Multigrid for American option pricing with stochastic volatility. Appl. Math. Financ. **6**(3), 177–195 (1999)
6. Douglas, J.: Alternating direction methods for three space variables. Numer. Math. **4**, 41–63 (1962)
7. Douglas, J., Gunn, J.E.: A general formulation of alternating direction methods. I. Parabolic and hyperbolic problems. Numer. Math. **6**, 428–453 (1964)
8. Duan, J.: The GARCH option pricing model. Math. Financ. **5**(1), 13–32 (1995)
9. Düring, B.: Asset pricing under information with stochastic volatility. Rev. Deriv. Res. **12**(2), 141–167 (2009)
10. Düring, B., Fournié, M.: High-order compact finite difference scheme for option pricing in stochastic volatility models. J. Comput. Appl. Math. **236**(17), 4462–4473 (2012)
11. Düring, B., Fournié, M., Jüngel, A.: Convergence of a high-order compact finite difference scheme for a nonlinear Black-Scholes equation. Math. Model Numer. Anal. **38**(2), 359–369 (2004)
12. Düring, B., Fournié, M., Jüngel, A.: High-order compact finite difference schemes for a nonlinear Black-Scholes equation. Int. J. Theor. Appl. Financ. **6**(7), 767–789 (2003)
13. Düring, B., Fournié, M., Rigal, A.: High-order ADI schemes for convection-diffusion equations with mixed derivative terms. In: Azaïez, M. et al. (eds.) Spectral and High Order Methods for Partial Differential Equations - ICOSAHOM'12. Lecture Notes in Computational Science and Engineering 95, pp. 217–226. Springer, Berlin/Heidelberg (2013)
14. Düring, B., Fournié, M., Heuer, C.: High-order compact finite difference schemes for option pricing in stochastic volatility models on non-uniform grids. J. Comput. Appl. Math. **271**(18), 247–266 (2014)
15. Düring, B., Heuer, C.: High-order compact schemes for parabolic problems with mixed derivatives in multiple space dimensions. SIAM J. Numer. Anal. **53**(5), 2113–2134 (2015)
16. Düring, B., Miles, J.: High-order ADI scheme for option pricing in stochastic volatility models. J. Comput. Appl. Math. **316**, 109–121 (2017)
17. Fournié, M., Rigal, A.: High order compact schemes in projection methods for incompressible viscous flows. Commun. Comput. Phys. **9**(4), 994–1019 (2011)
18. Griebel, M., Schneider, M., Zenger, C.: A combination technique for the solution of sparse grid problems. IMACS Elsevier, Iterative Methods Linear Algebra **16**, 263–281 (1992)
19. Gupta, M.M., Manohar, R.P., Stephenson, J.W.: A single cell high-order scheme for the convection-diffusion equation with variable coefficients. Int. J. Numer. Methods Fluids **4**, 641–651 (1984)
20. Gustafsson, B.: The convergence rate for difference approximation to general mixed initial-boundary value problems. SIAM J. Numer. Anal. **18**(2), 179–190 (1981)
21. Hendricks, C., Ehrhardt, M., Günther, M.: High-order ADI schemes for diffusion equations with mixed derivatives in the combination technique. Appl. Numer. Math. **101**, 36–52 (2016)
22. Heston, S.L.: A closed-form solution for options with stochastic volatility with applications to bond and currency options. Rev. Financ. Stud. **6**(2), 327–343 (1993)
23. Hilber, N., Matache, A., Schwab, C.: Sparse wavelet methods for option pricing under stochastic volatility. J. Comput. Financ. **8**(4), 1–42 (2005)
24. in't Hout, K.J., Welfert, B.D.: Stability of ADI schemes applied to convection-diffusion equations with mixed derivative terms. Appl. Numer. Math. **57**, 19–35 (2007)
25. in't Hout, K.J., Foulon, S.: ADI finite difference schemes for option pricing in the Heston model with correlation. Int. J. Numer. Anal. Mod. **7**, 303–320 (2010)
26. Hundsdorfer, W.: Accuracy and stability of splitting with stabilizing corrections. Appl. Numer. Math. **42**, 213–233 (2002)
27. Hundsdorfer, W., Verwer, J.G.: Numerical Solution of Time-Dependent Advection-Diffusion-Reaction Equations. Springer Series in Computational Mathematics, vol. 33. Springer, Berlin (2003)

28. Ikonen, S., Toivanen, J.: Efficient numerical methods for pricing American options under stochastic volatility. Numer. Methods Partial Differ. Equ. **24**(1), 104–126 (2008)
29. Kreiss, H.O., Thomée, V., Widlund, O.: Smoothing of initial data and rates of convergence for parabolic difference equations. Commun. Pure Appl. Math. **23**, 241–259 (1970)
30. Liao, W., Khaliq, A.Q.M.: High-order compact scheme for solving nonlinear Black-Scholes equation with transaction cost. Int. J. Comput. Math. **86**(6), 1009–1023 (2009)
31. Lewis, A.L.: Option Valuation Under Stochastic Volatility. Finance Press, Newport Beach, (2000)
32. Mitchell, A.R., Fairweather, G.: Improved forms of the alternating direction methods of Douglas, Peaceman, and Rachford for solving parabolic and elliptic equations. Numer. Math. **6**, 285–292 (1964)
33. Peaceman, D.W., Rachford Jr., H.H.: The numerical solution of parabolic and elliptic differential equations. J. Soc. Ind. Appl. Math. **3**, 28–41 (1959)
34. Pooley, D.M., Vetzal, K.R., Forsyth, P.A.: Convergence remedies for non-smooth payoffs in option pricing. J. Comput. Financ. **6**(4), 25–40 (2003)
35. Ruijter, M.J., Oosterlee, C.W.: Two-dimensional Fourier cosine series expansion method for pricing financial options. SIAM J. Sci. Comput. **34**(5), 642–671 (2012)
36. Rigal, A.: Schémas compacts d'ordre élevé: application aux problèmes bidimensionnels de diffusion-convection instationnaire I. C.R. Acad. Sci. Paris. Sr. I Math. **328**, 535–538 (1999)
37. Schiekofer, T.: Die Methode der Finiten Differenzen auf dünnen Gittern zur Lösung elliptischer und parabolischer partieller Differentialgleichungen. PhD thesis, Universität Bonn (1999)
38. Smolyak, S.: Quadrature and interpolation formulas for tensor products of certain classes of functions. Dokl. Akad. Nauk SSSR **148**, 1042–1045 (1963)
39. Spotz, W.F., Carey, C.F.: Extension of high-order compact schemes to time-dependent problems. Numer. Methods Partial Differ. Equ. **17**(6), 657–672 (2001)
40. Tavella, D., Randall, C.: Pricing Financial Instruments: The Finite Difference Method. John Wiley & Sons, New York (2000)
41. Zenger, C.: Sparse grids. Technical report, Institut für Informatik, Technische Universität München, Oct. 1990
42. Zhu, W., Kopriva, D.A.: A spectral element approximation to price European options with one asset and stochastic volatility. J. Sci. Comput. **42**(3), 426–446 (2010)
43. Zvan, R., Forsyth, P.A., Vetzal, K.R.: Penalty methods for American options with stochastic volatility. J. Comput. Appl. Math. **91**(2), 199–218 (1998)

Chapter 17
Essentially High-Order Compact Schemes with Application to Stochastic Volatility Models on Non-Uniform Grids

Bertram Düring and Christof Heuer

Abstract We present high-order compact schemes for a linear second-order parabolic partial differential equation (PDE) with mixed second-order derivative terms in two spatial dimensions. The schemes are applied to option pricing PDE for a family of stochastic volatility models. We use a non-uniform grid with more grid-points around the strike price. The schemes are fourth-order accurate in space and second-order accurate in time for vanishing correlation. In our numerical convergence study we achieve fourth-order accuracy also for non-zero correlation. A combination of Crank-Nicolson and BDF-4 discretisation is applied in time. Numerical examples confirm that a standard, second-order finite difference scheme is significantly outperformed.

17.1 Introduction

We consider the following parabolic partial differential equation for $u = u(x_1, x_2, t)$ in two spatial dimensions and time,

$$du_\tau + a_1 u_{x_1x_1} + a_2 u_{x_2x_2} + b_{12} u_{x_1x_2} + c_1 u_{x_1} + c_2 u_{x_2} = 0 \tag{17.1}$$

in $\Omega\times]0,T] =: Q_T$, subject to suitable boundary conditions and initial condition $u(x_1, x_2, 0) = u_0(x_1, x_2)$ with $T > 0$ and $\Omega = \left[x_{\min}^{(1)}, x_{\max}^{(1)}\right] \times \left[x_{\min}^{(2)}, x_{\max}^{(2)}\right] \subset \mathbb{R}^2$ with $x_{\min}^{(i)} < x_{\max}^{(i)}$ for $i = 1, 2$. The functions $a_i = a_i(x_1, x_2, \tau) < 0$, $b_{12} = b_{12}(x_1, x_2, \tau)$, $c_i = c(x_1, x_2, \tau)$, $d = d(x_1, x_2, \tau)$ map Q_T to $\mathbb{R}$, and $a_i(\cdot, \tau)$, $b(\cdot, \tau)$, $c_i(\cdot, \tau)$, and $d(\cdot, \tau)$ are assumed to be in $C^2(\Omega)$ and $u(\cdot, t) \in C^6(\Omega)$ for all $\tau \in]0, T]$. We define

B. Düring (✉)
Department of Mathematics, University of Sussex, Pevensey II, BN1 9QH Brighton, UK
e-mail: bd80@sussex.ac.uk

C. Heuer
Chair of Applied Mathematics/Numerical Analysis, Bergische Universität Wuppertal, Gaußstraße 20, 42097 Wuppertal, Germany
e-mail: heuer.chr@googlemail.com

M. Ehrhardt et al. (eds.), *Novel Methods in Computational Finance*, Mathematics in Industry 25, DOI 10.1007/978-3-319-61282-9_17

a uniform spatial grid G with step size Δx_k in x_k direction for $k = 1, 2$. Setting $f = -du_\tau$ and applying a standard, second-order central difference approximation leads to the elliptic problem

$$\begin{aligned} f =& A_0 - \frac{a_1(\Delta x_1)^2}{12}\frac{\partial^4 u}{\partial x_1^4} - \frac{a_2(\Delta x_2)^2}{12}\frac{\partial^4 u}{\partial x_2^4} - \frac{b_{12}(\Delta x_1)^2}{6}\frac{\partial^4 u}{\partial x_1^3 \partial x_2} \\ &- \frac{b_{12}(\Delta x_2)^2}{6}\frac{\partial^4 u}{\partial x_1 \partial x_2^3} - \frac{c_1(\Delta x_1)^2}{6}\frac{\partial^3 u}{\partial x_1^3} - \frac{c_2(\Delta x_2)^2}{6}\frac{\partial^3 u}{\partial x_2^3} + \varepsilon, \end{aligned} \tag{17.2}$$

with $A_0 := a_1 D_1^c D_1^c U_{i_1,i_2} + a_2 D_2^c D_2^c U_{i_1,i_2} + b_{12} D_1^c D_2^c U_{i_1,i_2} + c_1 D_1^c U_{i_1,i_2} + c_2 D_2^c U_{i_1,i_2}$, where D_k^c denotes the central difference operator in x_k direction, and $\varepsilon \in \mathcal{O}\left(h^4\right)$ if $\Delta x_k \in \mathcal{O}(h)$ for $h > 0$. We call a finite difference scheme high-order compact (HOC) if its consistency error is of order $\mathcal{O}\left(h^4\right)$ for $\Delta x_1, \Delta x_2 \in \mathcal{O}(h)$ for $h > 0$, and it uses only points on the compact stencil, $U_{k,p}$ with $k \in \{i_1 - 1, i_1, i_1 + 1\}$ and $p \in \{i_2 - 1, i_2, i_2 + 1\}$, to approximate the solution at $(x_{i_1}, x_{i_2}) \in \mathring{G}$.

17.2 Auxiliary Relations for Higher Derivatives

Our aim is to replace the third- and fourth-order derivatives in (17.2) which are multiplied by second-order terms by equivalent expressions which can be approximated with second order on the compact stencil. Indeed, if we differentiate (17.1) (using $f = -du_\tau$) once with respect to x_k ($k = 1, 2$), we obtain relations

$$\frac{\partial^3 u}{\partial x_1^3} = A_1, \qquad \frac{\partial^3 u}{\partial x_2^3} = A_2, \tag{17.3}$$

where we can discretise A_i with second order on the compact stencil using the central difference operator. Analogously, we obtain

$$\begin{aligned} \frac{\partial^4 u}{\partial x_1^4} =& B_1 - \frac{b_{12}}{a_1}\frac{\partial^4 u}{\partial x_1^3 \partial x_2} &\iff&& \frac{\partial^4 u}{\partial x_1^3 \partial x_2} &= \frac{a_1}{b_{12}} B_1 - \frac{a_1}{b_{12}}\frac{\partial^4 u}{\partial x_1^4}, \\ \frac{\partial^4 u}{\partial x_2^4} =& B_2 - \frac{b_{12}}{a_2}\frac{\partial^4 u}{\partial x_1 \partial x_2^3} &\iff&& \frac{\partial^4 u}{\partial x_1 \partial x_2^3} &= \frac{a_2}{b_{12}} B_2 - \frac{a_2}{b_{12}}\frac{\partial^4 u}{\partial x_2^4}, \\ \frac{\partial^4 u}{\partial x_1^3 \partial x_2} =& C_1 - \frac{a_2}{a_1}\frac{\partial^4 u}{\partial x_1 \partial x_2^3} &\iff&& \frac{\partial^4 u}{\partial x_1 \partial x_2^3} &= C_2 - \frac{a_1}{a_2}\frac{\partial^4 u}{\partial x_1^3 \partial x_2}, \end{aligned} \tag{17.4}$$

where we can approximate B_k and C_k with second order on the compact stencil using the central difference operator. A detailed derivation can be found in [3, 5].

17.3 Derivation of High-Order Compact Schemes

In general it is not possible to obtain a HOC scheme for (17.1), since there are four fourth-order derivatives in (17.2), but only three auxiliary equations for these in (17.4). Hence, we propose four different versions of the numerical schemes, where only one of the fourth-order derivatives in (17.2) is left as a second-order remainder term. Using (17.3) and (17.4) in (17.2) we obtain as *Version 1* scheme

$$\begin{aligned} f =& A_0 - \frac{c_1(\Delta x_1)^2}{6}A_1 - \frac{c_2(\Delta x_2)^2}{6}A_2 - \frac{a_2(\Delta x_2)^2}{12}B_2 - \frac{b_{12}(\Delta x_2)^2}{12}C_2 \\ & - \frac{a_1\left(2a_2(\Delta x_1)^2 - a_1(\Delta x_2)^2\right)}{12a_2}B_1 + \frac{a_1\left(a_2(\Delta x_1)^2 - a_1(\Delta x_2)^2\right)}{12a_2}\frac{\partial^4 u}{\partial x_1^4} + \varepsilon, \end{aligned} \tag{17.5}$$

as *Version 2* scheme

$$\begin{aligned} f =& A_0 - \frac{c_1(\Delta x_1)^2}{6}A_1 - \frac{c_2(\Delta x_2)^2}{6}A_2 - \frac{a_1(\Delta x_1)^2}{12}B_1 - \frac{b_{12}(\Delta x_1)^2}{12}C_1 \\ & - \frac{a_2\left(2a_1(\Delta x_2)^2 - a_2(\Delta x_1)^2\right)}{12a_1}B_2 + \frac{a_2\left(a_1(\Delta x_2)^2 - a_2(\Delta x_1)^2\right)}{12a_1}\frac{\partial^4 u}{\partial x_2^4} + \varepsilon, \end{aligned} \tag{17.6}$$

as *Version 3* scheme

$$\begin{aligned} f =& A_0 - \frac{c_1(\Delta x_1)^2}{6}A_1 - \frac{c_2(\Delta x_2)^2}{6}A_2 - \frac{a_1(\Delta x_1)^2}{12}B_1 - \frac{a_2(\Delta x_2)^2}{12}B_2 \\ & - \frac{b_{12}(\Delta x_2)^2}{12}C_2 + \frac{b_{12}\left(a_1(\Delta x_2)^2 - a_2(\Delta x_1)^2\right)}{12a_2}\frac{\partial^4 u}{\partial x_1^3 \partial x_2} + \varepsilon, \end{aligned} \tag{17.7}$$

and, finally, as *Version 4* scheme

$$\begin{aligned} f =& A_0 - \frac{c_1(\Delta x_1)^2}{6}A_1 - \frac{c_2(\Delta x_2)^2}{6}A_2 - \frac{a_1(\Delta x_1)^2}{12}B_1 - \frac{a_2(\Delta x_2)^2}{12}B_2 \\ & - \frac{b_{12}(\Delta x_1)^2}{12}C_1 + \frac{b_{12}\left(a_2(\Delta x_1)^2 - a_1(\Delta x_2)^2\right)}{12a_1}\frac{\partial^4 u}{\partial x_1 \partial x_2^3} + \varepsilon. \end{aligned} \tag{17.8}$$

Employing the central difference operator with $\Delta x = \Delta y = h$ for $h > 0$ to discretise A_i, B_i, C_i, in (17.5)–(17.8) and neglecting the remaining lower-order term leads to four semi-discrete (in space) schemes. A more detailed description of this approach can be found in [3, 5]. When $a_1 \equiv a_2$ or $b_{12} \equiv 0$ these schemes are fourth-order consistent in space, otherwise second-order.

In time, we apply the implicit BDF4 method on an equidistant time grid with stepsize $k \in \mathcal{O}(h)$. The necessary starting values are obtained using a Crank-Nicolson time discretisation, where we subdivide the first time steps with a step size $k' \in \mathcal{O}\left(h^2\right)$ to ensure the fourth-order time discretisation in terms of h.

With additional information on the solution of (17.1) even better results are possible. If the specific combination of pre-factors in (17.1) and the higher derivatives in the second-order terms is sufficiently small, the second-order term dominates the computational error only for very small step-sizes h. Before this error term becomes dominant one can observe a fourth-order numerical convergence. In this case we call the scheme essentially high-order compact (EHOC).

17.4 Application to Option Pricing

In this section we apply our numerical schemes to an option pricing PDE in a family of stochastic volatility models, with a generalised square root process for the variance with nonlinear drift term,

$$dS_t = \mu S_t \mathrm{d}t + \sqrt{v_t} S_t \mathrm{d}W_t^{(1)}, \qquad dv_t = \kappa v_t^{\alpha} \left(\theta - v_t\right) \mathrm{d}t + \sigma \sqrt{v_t} \mathrm{d}W_t^{(2)},$$

with $\alpha \geq 0$, a correlated, two-dimensional Brownian motion, $dW_t^{(1)} dW_t^{(2)} = \rho dt$, as well as drift $\mu \in \mathbb{R}$ of the stock price S, long run mean $\theta > 0$, mean reversion speed $\kappa > 0$, and volatility of volatility $\sigma > 0$. For $\alpha = 0$ one obtains the standard Heston model, for $\alpha = 1$ the SQRN model, see [1]. Using Itô's lemma and standard arbitrage arguments, the option price $V = V(S, v, t)$ solves

$$\frac{\partial V}{\partial t} + \frac{vS^2}{2}\frac{\partial^2 V}{\partial S^2} + \rho\sigma v S \frac{\partial^2 V}{\partial S \partial v} + \frac{\sigma^2 v}{2}\frac{\partial^2 V}{\partial v^2} + rS\frac{\partial V}{\partial S} + \kappa v^{\alpha}\left(\theta - v\right)\frac{\partial V}{\partial v} - rV = 0, \tag{17.9}$$

where $S, \sigma > 0$ and $t \in [0, T[$ with $T > 0$. For a European Put with exercise price K we have the final condition $V(S, T) = \max\left(K - S, 0\right)$. The transformations $\tau = T - t$, $u = e^{r\tau} V/K$, $\hat{S} = \ln(S/K)$, $y = v/\sigma$ as well as $\hat{S} = \varphi\left(x\right)$ [2], lead to

$$\varphi_x^3 u_\tau + \frac{\sigma y}{2}\left[\varphi_x u_{xx} + \varphi_x^3 u_{yy}\right] - \rho\sigma y \varphi_x^2 u_{xy}$$
$$+ \left[\frac{\sigma y \varphi_{xx}}{2} + \left(\frac{\sigma y}{2} - r\right)\varphi_x^2\right] u_x - \kappa\sigma^{\alpha} y^{\alpha}\frac{\theta - \sigma y}{\sigma}\varphi_x^3 u_y = 0,$$

with initial condition $u(x, y, 0) = \max\left(1 - e^{\varphi(x)}, 0\right)$. The function φ is considered to be four times differentiable and strictly monotone. It is chosen in such a way that grid points are concentrated around the exercise price K in the S–v plane when using a uniform grid in the x–y plane.

Dirichlet boundary conditions are imposed at $x = x_{\min}$ and $x = x_{\max}$ similarly as in [2],

$$u(x_{\min}, y, \tau) = u(x_{\min}, y, 0), \quad u(x_{\max}, y, \tau) = u(x_{\max}, y, 0),$$

for all $\tau \in [0, \tau_{\max}]$ and $y \in [y_{\min}, y_{\max}]$. At the boundaries $y = y_{\min}$ and $y = y_{\max}$ we employ the discretisation of the interior spatial domain and extrapolate the resulting ghost-points using

$$U_{i,-1} = 3U_{i,0} - 3U_{i,1} + U_{i,2} + \mathcal{O}(h^3),$$
$$U_{i,M+1} = 3U_{i,M} - 3U_{i,M-1} + U_{i,M-2} + \mathcal{O}(h^3),$$

for $i = 0, \dots, N$. Third-order extrapolation is sufficient here to ensure overall fourth-order convergence [4].

17.5 Numerical Experiments

We employ the function $\varphi(x) = \sinh(c_2 x + c_1(1-x))/\zeta$, where $c_1 = \operatorname{asinh}(\zeta \hat{S}_{\min})$, $c_2 = \operatorname{asinh}(\zeta \hat{S}_{\max})$ and $\zeta > 0$. We use $\kappa = 1.1$, $\theta = 0.2$, $v = 0.3$, $r = 0.05$, $K = 100$, $T = 0.25$, $v_{\min} = 0.1$, $v_{\max} = 0.3$, $S_{\min} = 1.5$, $S_{\max} = 250$, $\rho = 0, -0.4$ and $\zeta = 7.5$. Hence, $x_{\max} - x_{\min} = y_{\max} - y_{\min} = 1$. For the Crank-Nicolson method we use $k'/h^2 = 0.4$, for the BDF4 method $k/h = 0.1$. We smooth the initial condition according to [3, 6], so that the smoothed initial condition tends towards the original initial condition for $h \to 0$. We neglect the case $\alpha = 0$ (Heston model), since a numerical study of that case has been performed in [2]. In the numerical convergence plots we use a reference solution U_{ref} on a fine grid ($h = 1/320$) and report the absolute l^2-error compared to U_{ref}. The numerical convergence order is computed from the slope of the linear least square fit of the points in the log-log plot.

Figure 17.1a shows the transformation from x to S. The transformation focuses on the region around the strike price. Figures 17.1b–e show that the HOC schemes lead to a numerical convergence order of about 3.5, whereas the standard, second-order central difference discretisation (SD) leads to convergence orders of about 2.3, in the case of vanishing correlation. In all cases with non-vanishing correlation ($\rho \neq 0$) we observe only slightly improved convergence for Version 1 (V1) when comparing it to the standard discretisation. Version 2 (V2) and Version 3 (V3), however, lead to similar convergence orders as the HOC scheme, even for non-vanishing correlation. Results of Version 4 are not shown as this scheme shows instable behaviour in this example.

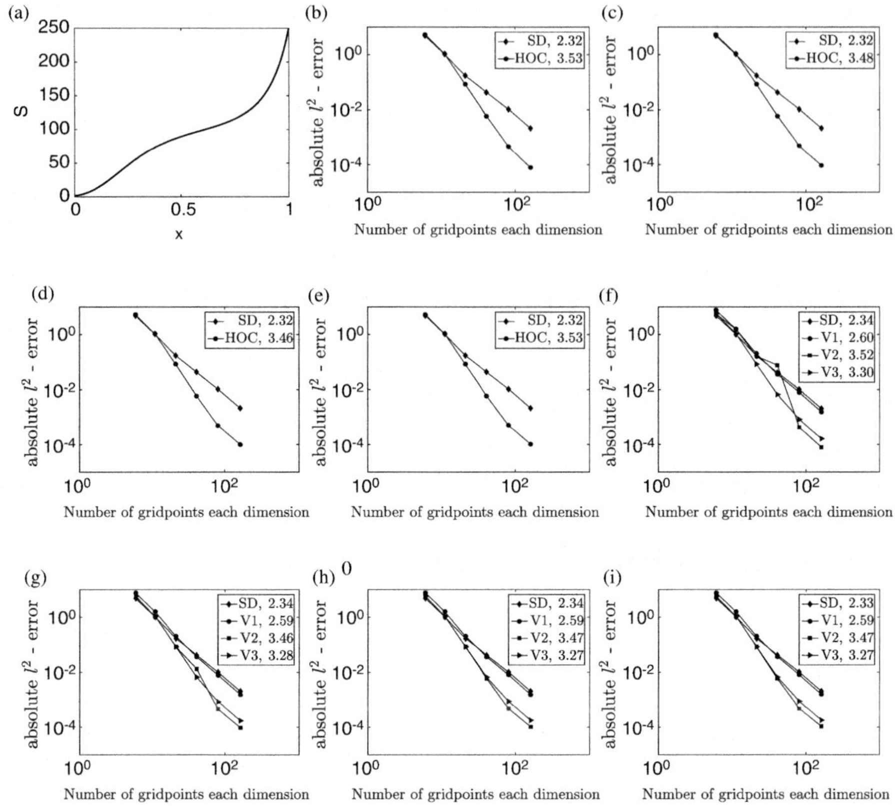

Fig. 17.1 Transformation of the spatial grid and numerical convergence plots. (**a**) Transformation with $\zeta = 7.5$. (**b**) HOC: $\alpha = 0.25$, $\rho = 0$. (**c**) HOC: $\alpha = 0.5$, $\rho = 0$. (**d**) HOC: $\alpha = 0.75$, $\rho = 0$. (**e**) HOC: SQRN model, $\rho = 0$. (**f**) EHOC: $\alpha = 0.25$, $\rho = -0.4$. (**g**) EHOC: $\alpha = 0.5$, $\rho = -0.4$. (**h**) EHOC: $\alpha = 0.75$, $\rho = -0.4$. (**i**) EHOC: SQRN model, $\rho = -0.4$

In summary, we obtain high-order compact schemes for vanishing correlation and achieve high-order convergence also for non-vanishing correlation for the family (17.9) of stochastic volatility model. A standard, second-order discretisation is significantly outperformed in all cases.

Acknowledgements BD acknowledges support by the Leverhulme Trust research project grant 'Novel discretisations for higher-order nonlinear PDE' (RPG-2015-69). CH was partially supported by the European Union in the FP7-PEOPLE-2012-ITN Program under Grant Agreement Number 304617 (FP7 Marie Curie Action, Project Multi-ITN STRIKE—Novel Methods in Computational Finance).

References

1. Christoffersen, P., Jacobs, K., Mimouni, K.: Models for S&P500 dynamics: evidence from realized volatility, daily returns, and option prices. Rev. Financ. Stud. **23**(8), 3141–3189 (2010)
2. Düring, B., Fournié, M., Heuer, C.: High-order compact finite difference schemes for option pricing in stochastic volatility models on non-uniform grids. J. Comput. Appl. Math. **271**(18), 247–266 (2014)
3. Düring, B., Heuer, C.: High-order compact schemes for parabolic problems with mixed derivatives in multiple space dimensions. SIAM J. Numer. Anal. **53**(5), 2113–2134 (2015)
4. Gustafsson, B.: The convergence rate for difference approximations to general mixed initial-boundary value problems. SIAM J. Numer. Anal. **18**(2), 179–190 (1981)
5. Heuer, C.: High-order compact finite difference schemes for parabolic partial differential equations with mixed derivative terms and applications in computational finance. PhD thesis, University of Sussex (Aug. 2014). http://sro.sussex.ac.uk/49800/
6. Kreiss, H.O., Thomee, V., Widlund, O.: Smoothing of initial data and rates of convergence for parabolic difference equations. Commun. Pure Appl. Math. **23**, 241–259 (1970)

Chapter 18
High Order Compact Schemes for Option Pricing with Liquidity Shocks

Miglena N. Koleva, Walter Mudzimbabwe, and Lubin G. Vulkov

Abstract This chapter concerns the numerical pricing of European options for markets with liquidity shocks. We derive and analyze high-order weighted compact finite difference schemes (WCFDS). Numerical simulations for the price and Greeks, using WCFDS combined with Richardson extrapolation in time are presented.

18.1 Introduction

We consider the market model of Ludkovski and Shen [8], described by the parabolic-ordinary system

$$\left| \begin{array}{l} R_t^0 + \dfrac{1}{2}\sigma^2 S^2 R_{SS}^0 - \dfrac{v_{01}}{\gamma} e^{-\gamma(R^1 - R^0)} + \dfrac{(d_0 + v_{01})}{\gamma} = 0, \\ R_t^1 - \dfrac{v_{10}}{\gamma} e^{-\gamma(R^0 - R^1)} + \dfrac{v_{10}}{\gamma} = 0, \end{array} \right. \tag{18.1}$$

subject with the following terminal conditions

$$R^i(S,T) = H(S), \qquad i = 0, 1.$$

The process is backward in time ($T \geq t \geq 0$) and $S \geq 0$ is the underlying asset. Next, σ is the volatility of the stock, ν_{01} is the transition rate from state $0 \to 1$, ν_{10} is the transition rate from state $1 \to 0$ (assuming that the market has two states: liquid - 0 and illiquid - 1), $d_0 = \mu/(2\sigma^2)$, while μ is the drift.

Numerical solution of this problem is considered in [5, 9]. In this chapter we will extend the results in [5] and construct high-order weighted compact finite difference scheme (WCFDS) for the model problem.

M.N. Koleva (✉) • W. Mudzimbabwe • L.G. Vulkov
Faculty of Natural Science and Education, Ruse University Angel Kanchev, Ruse, Bulgaria
e-mail: mkoleva@uni-ruse.bg; wmudzimbabwe@uni-ruse.bg; lvalkov@uni-ruse.bg

M. Ehrhardt et al. (eds.), *Novel Methods in Computational Finance*,
Mathematics in Industry 25, DOI 10.1007/978-3-319-61282-9_18

Just as in [5, 8, 9] we are interesting in buyer indifference prices p, q, which satisfy the semi-parabolic system

$$\left| \begin{array}{l} p_t + \frac{1}{2}\sigma^2 S^2 p_{SS} - \frac{\upsilon_{01}}{\gamma}\frac{F_1}{F_0}e^{-\gamma(q-p)} + \frac{(d_0+\upsilon_{01})}{\gamma} - \frac{1}{\gamma}\frac{F_0'}{F_0} = 0, \\ q_t - \frac{\upsilon_{10}}{\gamma}\frac{F_0}{F_1}e^{-\gamma(p-q)} + \frac{\upsilon_{10}}{\gamma} - \frac{1}{\gamma}\frac{F_1'}{F_1} = 0, \end{array} \right. \qquad (18.2)$$

where $S \in \mathbb{R}^+$, $0 \le t \le T$ and $p(S,T) = q(S,T) = H(S)$.

We transform this system in a suitable for our numerical method form [5, 8, 9]. For this purpose, we set $\widetilde{u} := \gamma R^0$, $\widetilde{\upsilon} = \gamma R^1$. Then, taking into account that $e^{\gamma(p-R^0)} = F_0(t)$ and $e^{\gamma(q-R^1)} = F_1(t)$, i.e.

$$p = R^0 + \gamma^{-1}\ln F_0(t) = \gamma^{-1}\left(\widetilde{u} + \ln F_0(t)\right),$$
$$q = R^1 + \gamma^{-1}\ln F_1(t) = \gamma^{-1}\left(\widetilde{\upsilon} + \ln F_1(t)\right),$$

a comparison principle in (p,q) solutions will be equivalent to a comparison principle for the $(\widetilde{u},\widetilde{\upsilon})$ variables.

It is easily to check that the pair $(\widetilde{u},\widetilde{\upsilon})$ satisfies the problem:

$$\left| \begin{array}{l} \widetilde{u}_t + \frac{1}{2}\sigma^2 S^2 \widetilde{u}_{SS} = ae^{-\left(\widetilde{\upsilon}-\widetilde{u}\right)} - b, \\ \widetilde{\upsilon}_t = ce^{-\left(\widetilde{u}-\widetilde{\upsilon}\right)} - c, \end{array} \right. \qquad (18.3)$$

where $a = \upsilon_{01}$, $b = \upsilon_{01} + d_0$, $c = \upsilon_{10}$ and $\widetilde{u}(S,T) = \gamma H(S)$, $\widetilde{\upsilon}(S,T) = \gamma H(S)$.

As in many models in financial mathematics [2, 10], in order to to formulate our model as initial value problem, to overcome the degeneration and simultaneously to remove the reaction term, we use the substitution

$$\tau = T - t, \quad x = \ln S, \quad u = e^{-\frac{x}{2}-\frac{\sigma^2\tau}{8}}\widetilde{u}, \quad \upsilon = \widetilde{\upsilon}. \qquad (18.4)$$

Thus, from (18.3) we obtain the following initial value parabolic-ordinary system

$$\left| \begin{array}{l} u_\tau - \frac{1}{2}\sigma^2 u_{xx} = e^{-\frac{x}{2}-\frac{\sigma^2\tau}{8}}\left(-ae^{ue^{\frac{x}{2}+\frac{\sigma^2\tau}{8}}-\upsilon} + b\right), \;\; x \in \mathbb{R}, \; 0 \le \tau \le T, \\ \upsilon_\tau = -ce^{-ue^{\frac{x}{2}+\frac{\sigma^2\tau}{8}}+\upsilon} + c, \;\; x \in \mathbb{R}, \; 0 \le \tau \le T, \end{array} \right. \qquad (18.5)$$

$$u(x,0) = \gamma e^{-\frac{x}{2}}H(e^x), \quad \upsilon(x,0) = \gamma H(e^x), \;\; x \in \mathbb{R}.$$

Also, the writer's indifference price $p^w(S,t)$ and $q^w(S,t)$ are solutions of the problem, see [8, Remark 4.6]

$$\left| \begin{array}{l} p_t^w + \frac{1}{2}\sigma^2 S^2 p_{SS}^w + \frac{\upsilon_{01}}{\gamma}\frac{F_1}{F_0}e^{\gamma(q^w-p^w)} - \frac{(d_0+\upsilon_{01})}{\gamma} + \frac{1}{\gamma}\frac{F_0'}{F_0} = 0, \\ q_t^w + \frac{\upsilon_{10}}{\gamma}\frac{F_0}{F_1}e^{\gamma(p^w-q^w)} - \frac{\upsilon_{10}}{\gamma} + \frac{1}{\gamma}\frac{F_1'}{F_1} = 0, \end{array} \right.$$

where $S \in \mathbb{R}^+, 0 \le t \le T$ and $p^w(S,T) = q^w(S,T) = H(S)$.

Next, applying the transformation

$$p^w = R^0 - \gamma^{-1}\ln F_0(t) = \gamma^{-1}\left(\widetilde{u} - \ln F_0(t)\right),$$
$$q^w = R^1 - \gamma^{-1}\ln F_1(t) = \gamma^{-1}\left(\widetilde{\upsilon} - \ln F_1(t)\right),$$

in combination with (18.4) we have

$$\left| \begin{array}{l} u_\tau - \frac{1}{2}\sigma^2 u_{xx} = e^{-\frac{x}{2}-\frac{\sigma^2\tau}{8}}\left(ae^{-ue^{\frac{x}{2}+\frac{\sigma^2\tau}{8}}+\upsilon} - b\right), \;\; x\in\mathbb{R},\; 0\le\tau\le T, \\ \upsilon_\tau = ce^{ue^{\frac{x}{2}+\frac{\sigma^2\tau}{8}}-\upsilon} - c, \;\; x\in\mathbb{R},\; 0\le\tau\le T, \end{array} \right. \tag{18.6}$$

$$u(x,0) = \gamma e^{-\frac{x}{2}}H(e^x), \quad \upsilon(x,0) = \gamma H(e^x), \;\; x\in\mathbb{R}.$$

In what follows, we will solve the problem (18.5), corresponding to option buyer's price model (18.2).

18.2 Construction and Analysis of WCFDS

There are a number of methods in the literature for solving non-linear parabolic and hyperbolic problems, but the results in the case of exponential non-linear term are scarce [11]. Efficient finite difference schemes for model (18.2) are constructed and analyzed in [5, 9]. The numerical scheme, presented in [9] is of second order in space, while in [5] is developed implicit-explicit compact finite difference schemes (CFDS). For a single PDEs in finance, CFDS are proposed in many papers, see e.g. [2, 3, 7].

In this section we construct a weighted fourth-order in space compact finite difference scheme for problem (18.5). In the same manner can be obtained WCFDS for the problem (18.6).

The discretization of problem (18.5) is based on the Padé approximation. Let $[L^-, L^+]$ is the computational interval. We consider Dirichlet boundary conditions $U_0(L^-,t)$, $U_N(L^+,t)$, which corresponds to boundary conditions $(u^-(e^{L^-},t)$, $u^+(e^{L^+},t))$ of the model (18.2).

First, we approximate the system (18.5) in time. To this aim, we introduce uniform mesh in time, with step size $\triangle\tau$ and $\tau^{n+1} = n\triangle\tau$, $n = 0, 1, \ldots$. Further we use the following notations

$$\nu := \nu(x) = \nu(x, \tau^n), \quad \widehat{\nu} := \widehat{\nu}(x) = \widetilde{\nu}(x, \tau^{n+1}),$$

$$g_1(u, v, x, \tau) := e^{-\frac{x}{2}-\frac{\sigma^2\tau}{8}}(-ae^{ue^{\frac{x}{2}+\frac{\sigma^2\tau}{8}}-v} + b),$$

$$g_2(u, v, x, \tau) := -ce^{-ue^{\frac{x}{2}+\frac{\sigma^2\tau}{8}}+v} + c.$$

Thus, for $\theta_1 \in [0, 1]$ and $\theta_2 \in [0, 1]$, the semi-discretization in time of (18.5) reads as follows

$$\left| \begin{array}{l} \dfrac{\widehat{u} - u}{\triangle\tau} - \dfrac{1}{2}\theta_1\sigma^2\widehat{u}'' - \dfrac{1}{2}(1-\theta_1)\sigma^2 u'' = \theta_2\widehat{g}_1 + (1-\theta_2)g_1, \quad x \in \mathbb{R}, \\ \dfrac{\widehat{v} - v}{\triangle\tau} = \theta_2\widehat{g}_2 + (1-\theta_2)g_2, \quad x \in \mathbb{R}, \end{array} \right. \tag{18.7}$$

where $\widehat{g}_1 = g_1(\widehat{u}, \widehat{v}, x, \tau^{n+1})$, $g_1 = g_1(u, v, x, \tau^n)$.

Let us consider the uniform partition of the space interval and define the mesh $\omega_h = \{x_i : x_i = L^- + ih,\ i = 0, \ldots, N,\ h = (L^+ - L^-)/N\}$. Using the notation $u_i = u(x_i)$, we define the second-order central finite difference approximation by

$$\delta_x^2 u_i = \frac{u_{i+1} - 2u_i + u_{i-1}}{h^2}.$$

Applying Taylor series expansion for the second derivative approximation of the solution u in the first equation of (18.7), we obtain

$$\begin{aligned} \theta_1\delta_x^2\widehat{u}_i + (1-\theta_1)\delta_x^2 u_i &= \theta_1\widehat{u}_i'' + (1-\theta_1)u_i'' \\ &+ \frac{h^2}{12}\left(\delta_x^2\theta_1\widehat{u}_i'' + \delta_x^2(1-\theta_1)u_i'' + \mathcal{O}(h^2)\right) + \mathcal{O}(h^4). \end{aligned} \tag{18.8}$$

Substituting $\theta_1\widehat{u}_i'' + (1-\theta_1)u_i''$ from (18.7) in (18.8) and multiplying the resulting equation by $-\sigma^2/2$, for the first equation in (18.7) we find

$$\begin{aligned} &-\frac{\sigma^2}{2}\left(\theta_1\delta_x^2\widehat{u} + (1-\theta_1)\delta_x^2 u\right) \\ &= \left(1 + \frac{h^2}{12}\delta_x^2\right)\left(\frac{u - \widehat{u}}{\triangle\tau} + \theta_2\widehat{g}_1 + (1-\theta_2)g_1\right) + \mathcal{O}(h^4). \end{aligned} \tag{18.9}$$

This formula enable to construct different difference schemes for problem (18.6).

Let U_i and V_i are discrete approximations of $u(x_i)$ and $v(x_i)$, respectively and $U = (U_0, U_1, \ldots, U_N)^T$, $V = (V_0, V_1, \ldots, V_N)^T$. In view of (18.7) and (18.9), e.g. see [1] we derive the full discretization of (18.5)

$$\left|\begin{array}{l} -\dfrac{\sigma^2}{2}\left(\theta_1\delta_x^2\widehat{U} + (1-\theta_1)\delta_x^2 U\right) \\ \qquad\qquad = \left(1 + \dfrac{h^2}{12}\delta_x^2\right)\left(\dfrac{U-\widehat{U}}{\Delta\tau} + \theta_2\widehat{g}_1 + (1-\theta_2)g_1\right), \\ \dfrac{\widehat{V}-V}{\Delta\tau} = \theta_2\widehat{g}_2 + (1-\theta_2)g_2, \end{array}\right. \quad (18.10)$$

$$U^0 = u(x,0), \;\; V^0 = v(x,0),$$

$$\widehat{U}_0 = \gamma e^{-\frac{L^-}{2}-\frac{\sigma^2\tau}{8}} u^-(e^{L^-}, \tau^{n+1}), \;\; \widehat{U}_N = \gamma e^{-\frac{L^+}{2}-\frac{\sigma^2\tau}{8}} u^+(e^{L^+}, \tau^{n+1}).$$

The local truncation errors of the first and second equations in (18.10) are

$$\begin{aligned} T_r(U) &= \frac{1}{2}\left(\frac{\sigma^2}{2}(2\theta_1 - 1)u''_\tau + (2\theta_2 - 1)u_{\tau\tau}\right)\Delta\tau \\ &+ \frac{1}{4}\left(\frac{\sigma^2}{4}u''_{\tau\tau} + \frac{1}{3}u_{\tau\tau\tau}\right)(\Delta\tau)^2 \\ &+ \frac{1}{24}\left(\frac{\sigma^2}{2}(2\theta_1 - 1)u^{iv}_\tau + (2\theta_2 - 1)u''_{\tau\tau}\right)\Delta\tau h^2 \\ &+ \frac{1}{48}\left(\frac{\sigma^2}{4}u^{iv}_{\tau\tau} - \frac{1}{3}u''_{\tau\tau\tau}\right)(\Delta\tau)^2 h^2 - \frac{\sigma^2}{480}u^{vi}h^4, \\ T_r(V) &= (2\theta_2 - 1)\frac{v_{\tau\tau}}{2}\Delta\tau + \frac{v_{\tau\tau\tau}}{12}(\Delta\tau)^2, \end{aligned}$$

where $(\cdot)_\tau := \frac{\partial(\cdot)}{\partial\tau}$, $(\cdot)' := \frac{\partial(\cdot)}{\partial x}$ at point $(\overline{x}, \overline{\tau})$, $\overline{x} \in (x_i - h, x_i + h)$, $\overline{\tau} \in (\tau^n, \tau^n + \Delta\tau)$.

For the solution of the numerical scheme (18.10), $\theta_1 = 1$ and $\theta_2 = 0$ in [5] is proved the following convergence result.

Theorem 18.1 ([5]) *Suppose that there exists a classical solution $(u, v) \in C^{6,2}(Q_T)$ of problem (18.5) on $Q_T = (-L, L) \times (0, T)$. Then for sufficiently small h and $\Delta\tau$ the following error estimate holds:*

$$\|u - U\|_\infty + \|v - V\|_\infty \le C(\Delta\tau + \Delta\tau h^2 + h^4),$$

where the constant $C > 0$ doesn't depend on h and $\Delta\tau$.

The same result can be established for WCFDS, $\theta_1 = \theta_2 = 1$ and $\theta_1 = \theta_2 = 0$. For Crank-Nicolson scheme, i.e. WCFDS, $\theta_1 = \theta_2 = 0.5$ we establish a high-order of convergence

$$\|u - U\|_\infty + \|v - V\|_\infty \le C(\Delta\tau^2 + \Delta\tau^2 h^2 + h^4). \quad (18.11)$$

In the case $\theta_2 > 0$, we solve the system of non-linear equations (18.10) by classical Newton method. The iteration process continues until the difference (in maximal discrete norm) between two consecutive iterations, for both solutions U and V is less than given tolerance tol.

18.3 Numerical Simulations

We test the efficiency, accuracy and the order of convergence in maximal discrete norm ($\|\cdot\|$) of the numerical solution of the model (18.1), $\sigma = 0.2$, $\gamma = 1$, $\mu_0 = 0.06$, $T = 0.5$, $\nu_{01} = 1$, $\nu_{10} = 12$. Also, we verify the convergence rate of the numerical buyer's and writer's indifference prices, computed by WCFDS (18.10) and the corresponding WCFDS for writer's price. To this aim we use the following formula

$$CR^{\mathscr{W}} = \log_2 \frac{\|\mathscr{W}_{N/2} - \mathscr{W}_N\|}{\|\mathscr{W}_N - \mathscr{W}_{2N}\|},$$

where $\mathscr{W}_N := [W_0, \dots, W_N]^\top$ is the computed numerical solution for buyer's price or writer's price at final time T.

For the test example we choose the payoff function $H(S) = \max(0, S-E)$, where E is the strike price. Thus, the compatibility with initial data, boundary conditions are $u^- = H(L^-)$ and $u^+ = H(L^+)$.

The computations are performed in two domains: $\Omega^1 = [-5, 5]$ and $\Omega^2 = [-5\ln 2, 5\ln 2]$ for $E = 2$. Note that $E = 2$ is a grid node of the original mesh S in Ω^2. In the case $\theta_2 > 0$, we choose the tol $= 1.e - 6$ for the iteration process.

To improve the order of convergence in time, we implement active Richardson extrapolation (RE) [4]

$$\widetilde{\mathscr{W}}_N = \frac{2^\rho \mathscr{W}_N(\triangle\tau/2) - \mathscr{W}_N(\triangle\tau)}{2^\rho - 1}, \tag{18.12}$$

where $\mathscr{W}_N(\triangle\tau)$ is the numerical solution computed for time step $\triangle\tau$ and ρ is the order of convergence of the underlying numerical method. Then, $\widetilde{W}_N$ is of order $\rho + 1$.

Regarding to [6], we cannot obtain fourth order convergence in space for the given pay off function, because it is not smooth. We smooth the function $H(x) = \max(0, e^x - E)$ and therefore the initial conditions $u^0(x) = u(x, 0)$ and $v^0(x) = v(x, 0)$ of (18.5). Following the results in [6], we use the smoothing operators Φ_4, given by its Fourier transform $\widehat{\Phi}_4$

$$\widehat{\Phi}_4 = \left(\frac{\sin(w/2)}{w/2}\right)^4 \left[1 + \frac{2}{3}\sin^2(w/2)\right].$$

The resulting smooth function is

$$\widetilde{H}(x^*) = \frac{1}{h} \int_{-3h}^{3h} \Phi_4\left(\frac{x}{h}\right) H(x^* - x)dx,$$

where Φ_4 is the Fourier inverse of $\widehat{\Phi}_4$. This technique is used only for grid nodes near by point $\ln E$. In smooth regions remains the original initial function. Similar smoothing technique is applied in [3].

The resulting smooth function, converges to the original function H as $h \to 0$. Moreover, the solution of the smoothed problem converges (with high-order rate) to the solution of (18.5), see [6].

Taking into account Theorem 18.1, if we apply active Richardson extrapolation (18.12), $\rho = 1$, the likely order of convergence of the numerical solution is $\mathcal{O}(\Delta\tau^2 + \Delta\tau^2 h^2 + h^4)$. Further, to verify this convergence rate, both in space and time, we perform the numerical test with fixed ratio $\Delta\tau = h^2$. The same is for the fully implicit CFDS ($\theta_1 = \theta_2 = 1$) and the explicit CFDS ($\theta_1 = \theta_2 = 0$).

To verify the estimate (18.11), for $\theta_1 = \theta_2 = 0.5$ we perform the computations with time step $\Delta\tau = h^2$. If the Richardson extrapolation (18.12), $\rho = 2$ is used, then the expected convergence is $\mathcal{O}(\Delta\tau^3 + \Delta\tau^3 h^2 + h^4)$ and it is naturally to choose $\Delta\tau = h^{4/3}$ to confirm this order, both in space and time.

We give the results for *indifference writer's price per contract* p^w, *indifference buyer's price per contract* p and the corresponding option values R^0.

In Table 18.1 is presented the order of convergence of the numerical solutions p and p^w, obtained from U, solving (18.10) (for (18.5) and (18.6)) with WCFDS, $\theta_1 = 1$, $\theta_2 = 0$ and applying Richardson extrapolation, $\rho = 1$. We observe that the order of convergence in space is closed to four, and because of the fixed time-space space size ratio ($\Delta\tau = h^2$), we can conclude that the convergence rate in time is not less than two.

In Table 18.2 are listed the buyer's and writer' values of R^0 at point $S = E$ in Ω^2, obtained as before from U with WCFDS, $\theta_1 = 1$, $\theta_2 = 0$. Here 'diff' is the absolute value of the difference between solutions, computed on two consecutive space meshes. The given order of convergence in space is at strike point $S = E$.

In Table 18.3 we give the buyer's values of R^0 at point $S = E$ in Ω^2 and corresponding order of convergence, obtained from U with WCFDS, $\theta_1 = \theta_2 = 0.5$ for $\Delta t = h^2$ without Richardson extrapolation and for $\Delta t = h^{4/3}$ with Richardson

Table 18.1 CR^p and CR^{p^w} in discrete maximum norm, corresponding to buyer's and writer's prices

	Numerical buyer's price		Numerical writer's price	
N	Ω^1	Ω^2	Ω^1	Ω^2
40	3.3125	2.8886	3.3352	2.8423
80	3.1405	3.3780	3.1413	3.3958
160	3.5548	3.6732	3.5936	3.7403
320	3.7489	3.8577	3.8427	4.0178

Table 18.2 Values of R^0 and convergence rate (CR) at point E, corresponding to buyer's price

	Buyer's price			Writer's price		
N	R^0	diff	CR	R^0	diff	CR
20	0.1569522			0.1137874		
40	0.1337066	2.32456e-2	2.8886	0.0915802	2.22072e-2	2.8423
80	0.1305678	3.13885e-3	3.3780	0.0884836	3.09657e-3	3.3958
160	0.1302659	3.01911e-4	3.6732	0.0881894	2.94203e-4	3.7403
320	0.1302422	2.36673e-5	3.8577	0.0881674	2.20150e-5	4.0183
640	0.1302406	1.63251e-6		0.0881660	1.35863e-6	

Table 18.3 Values of R^0 for buyer's price, convergence rate (CR) at point E and number of iterations (it(Δt)), WCFDS, $\theta_1 = \theta_2 = 0.5$

	$\Delta t = h^2$			$\Delta t = h^{4/3}$ with RE, $\rho = 2$		
N	R^0	CR	it(Δt)	R^0	CR	it(Δt)+it($\Delta t/2$)
20	0.1555562		4	0.1555440		5.50+4
40	0.1336810	2.8127	3	0.1336764	2.8141	4+3.80
80	0.1305675	3.3674	2.97	0.1305669	3.3686	3+3
160	0.1302658	3.6719	2	0.1302658	3.6713	3+2.98
320	0.1302422	3.8576	2	0.1302422	3.8602	2.99+2
640	0.1302406		2	0.1302406		2+2

Table 18.4 Values of R^0 for buyer's price, convergence rate (CR) at point E and number of iterations (it(Δt)), WCFDS, $\theta_1 = \theta_2 = 1$, Richardson extrapolation, $\rho = 1$

N	R^0	diff	CR	it(Δt)+it($\Delta t/2$)
20	0.1555382			5+4
40	0.1336685	2.18695e-2	2.8172	3.76+3
80	0.1305656	3.10292e-3	3.3711	3+2.95
160	0.1302657	2.99894e-4	3.6708	3+2.65
320	0.1302422	2.35480e-5	3.8571	2+2
640	0.1302406	1.62501e-6		2+2

extrapolation with $\rho = 2$. We show the average number of iterations it(Δt) at each time level, resulting from the mesh with time step size Δt. Similar results for WCFDS, $\theta_1 = \theta_2 = 1$, combined with Richardson extrapolation, $\rho = 1$ are given in Table 18.4.

In Table 18.5 we compare CPU times (in seconds) of WCFDS for different values of θ_1 and θ_2, with or without Richardson extrapolation, reaching one and the same value $R^0(E, T) = 0.13024056$ for buyer's price.

Table 18.5 CPU times of WCFDS for different values of θ_1 and θ_2, with or without RE

$\theta_1 = \theta_2 = 0.5$	$\theta_1 = \theta_2 = 0$	$\theta_1 = 1, \theta_2 = 0$	$\theta_1 = \theta_2 = 0.5$	$\theta_1 = \theta_2 = 1$
RE, $\rho = 2$	RE, $\rho = 1$	RE, $\rho = 1$		RE, $\rho = 1$
22.02	109.74	112.24	149.36	449.70

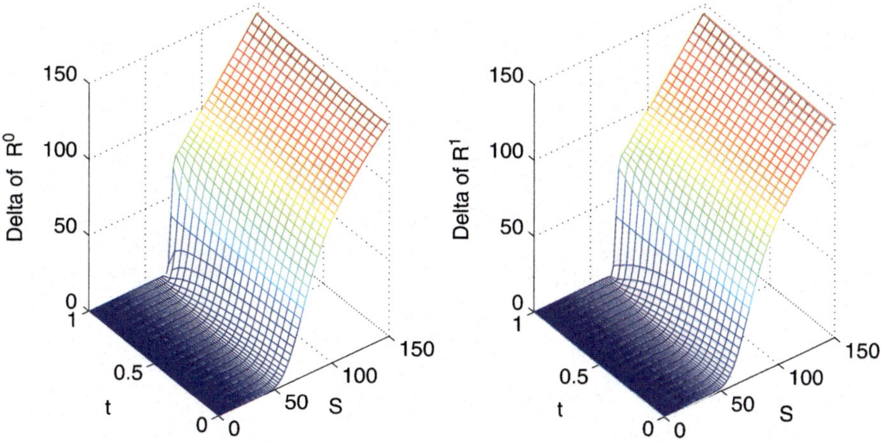

Fig. 18.1 Numerical Greek Delta of R^0 (*left*) and R^1 (*right*) for buyer's price, $0 \leq t \leq 1$

It is not a surprise that, the highest-order Crank-Nicolson CFDS ($\theta_1 = \theta_2 = 0.5$), combined with Richardson extrapolation, $\rho = 2$ is faster than the other methods, although for coarse meshes, the algorithm requires a little bit more iterations (in comparison with fully implicit scheme) to reach the desired precision.

We observe a very good efficiency of the semi-implicit CFDS ($\theta_1 = 1$, $\theta_2 = 0$). The reason is that the method does not require iterations and computation, respectively updated, of the derivatives of $g_1(\cdot)$, $g_2(\cdot)$ at each new time level/iteration (as is for the schemes with $\theta_2 > 0$), the coefficient matrix is one and the same at each time level (as is for the explicit scheme), so the matrix inversion performs only ones for the whole computational process. Another advantage of the WCFDS, $\theta_1 = 1$, $\theta_2 = 0$ is that the stability region is closed to the one of the fully implicit CFDS ($\theta_1 = 1, \theta_2 = 1$).

On Figs. 18.1, 18.2 we plot the graphics of the numerical Greek Delta (first spatial derivative of the option value) and Greek Gamma (second spatial derivative of the option value), of R^0 and R^1 for buyer's price, computed in Ω^1 by WCFDS, $\theta_1 = 1$, $\theta_2 = 0$ for $E = 70$, $N = 321$, $\triangle\tau = 0.05$.

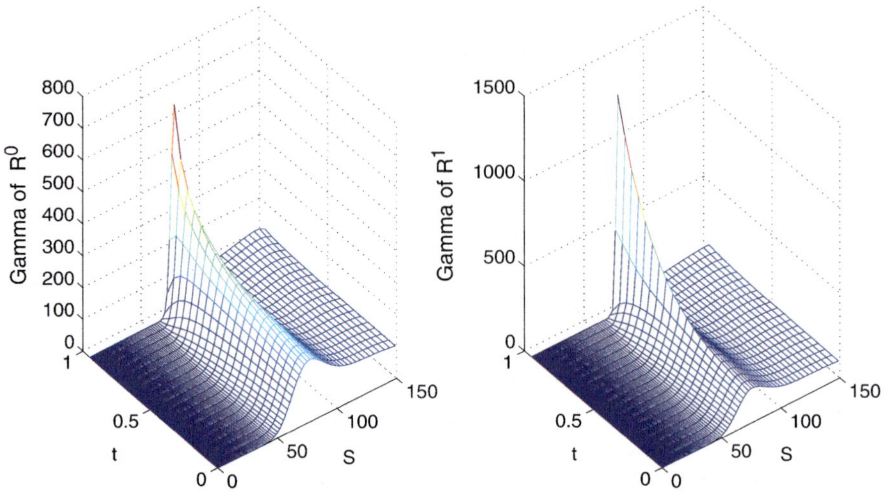

Fig. 18.2 Numerical Greek Gamma of R^0 (*left*) and R^1 (*right*) for buyer's price, $0 \leq t < 1$

18.4 Conclusions

In this chapter we construct high-order weighted compact finite difference schemes for computing options values with liquidity shock. In order to improve the order of convergence in time, we apply Richardson extrapolation. Numerical test confirm the higher order of convergence of the presented WCFDS. We compare the efficiency of the presented method for different values of the weights.

Tables 18.1 and 18.2 are reproduction from [5] with permission of Springer, License number 4045290562481/Feb. 10, 2017.

Acknowledgements This research was supported by the European Union under Grant Agreement number 304617 (FP7 Marie Curie Action Project Multi-ITN STRIKE—Novel Methods in Computational Finance) and the Bulgarian National Fund of Science under Project I02/20-2014.

References

1. Ciment, M., Leventhal, S., Weinberg, B.: The operator compact implicit method for parabolic equations. J. Comput. Phys. **28**(2), 135–166 (1978)
2. Dremkova, E., Ehrhardt, M.: A high-order compact method for nonlinear Black-Scholes option pricing equations of American options. Int. J. Comput. Math. **88**(13), 2782–2797 (2011)
3. Düring, B. Heuer, C.: High-order compact schemes for parabolic problems with mixed derivatives in multiple space dimensions. SIAM J. Numer. Anal. **53**(5), 2113–2134 (2015)
4. Faragó, I., Izsák, F., Szabó, T.: An IMEX scheme combined with Richardson extrapolation methods for some reaction-diffusion equations. Quart. J. Hung. Meteor. Serv. **117**(2), 201–218 (2013)

5. Koleva, M.N., Mudzimbabwe, W., Vulkov, L.G.: Fourth-order compact schemes for a parabolic-ordinary system of European option pricing liquidity shocks model. Numer. Algorithms **74**(1), 59–75 (2016)
6. Kreiss, H.O., Thomee, V., Widlund, O.: Smoothing of initial data and rates of convergence for parabolic difference equations. Commun. Pure Appl. Math. **23**, 241–259 (1970)
7. Liao, W., Khaliq, A.Q.M.: High order compact scheme for solving nonlinear Black-Scholes equation with transaction cost. Int. J. Comput. Math. **86**(6), 1009–1023 (2009)
8. Ludkovski, M., Shen, Q.: European option pricing with liquidity shocks. Int. J. Theory Appl. Financ. **16**(7), 1350043 (2013)
9. Mudzimbabwe, W., Vulkov, L.G.: IMEX schemes for a parabolic-ODE system of European options with liquidity shock. J. Comput. Appl. Math. **299**, 245–256 (2016)
10. Tangman, D.Y., Gopaul, A., Bhuruth, M.: Numerical pricing of options using high-order compact finite difference schemes. J. Comput. Appl. Math. **218**, 270–280 (2008)
11. Wang, L., Chen, W., Wang, C.: An energy-conserving second order numerical scheme for nonlinear hyperbolic equation with exponential nonlinear term. J. Comput. Appl. Math. **280**, 347–366 (2015)

Chapter 19
Alternating Direction Explicit Methods for Linear, Nonlinear and Multi-Dimensional Black-Scholes Models

Zuzana Bučková, Matthias Ehrhardt, Michael Günther, and Pedro Pólvora

Abstract In this chapter we discuss Alternating direction explicit (ADE) methods for one-dimensional convection diffusion equations and their numerical analysis including stability, consistency and convergence results. We propose ADE schemes for the two and three dimensional linear Black-Scholes pricing model. Our implemented methodology can be easily extended to higher dimensions. Extension for nonlinear Frey and Patie model is included at the end of this chapter where the nonlinear part is treated explicitly (This chapter is a shortened version of the papers (Bučková et al., Acta Math Univ Comen 84:309–325, 2015; Bučková et al., AIP Conf Proc 1773:030001, 2016.)).

19.1 Introduction

We focus in this chapter on the ADE methods, as an efficient scheme, which , can be used for a wide range of financial problems. We have considered the ADE method, that strongly uses boundary data in the solution algorithm and hence it is very sensible to incorrect treatment of boundary conditions. We have implemented the ADE scheme for solving linear and nonlinear BS equations by treating the nonlinearity explicitly. ADE scheme consists of two steps (sweeps). In the first step an upward sweeping is used and in the second step on downward sweeping is used and they are combined after each time step. To our knowledge, the ADE scheme has not been applied to nonlinear PDEs before.

Z. Bučková (✉) • M. Ehrhardt • M. Günther
Lehrstuhl für Angewandte Mathematik/Numerische Analysis, Bergische Universität Wuppertal, Wuppertal, Germany
e-mail: zuzana.zikova.buckova@gmail.com; ehrhardt@math.uni-wuppertal.de; guenther@math.uni-wuppertal.de

P. Pólvora
Department of Applied Mathematics and Statistics, Comenius University Bratislava, Bratislava, Slovakia
e-mail: pedro.polvora@gmail.com

M. Ehrhardt et al. (eds.), *Novel Methods in Computational Finance*, Mathematics in Industry 25, DOI 10.1007/978-3-319-61282-9_19

It can compete to the Crank-Nicolson scheme, Alternating Direction Implicit (ADI) and locally one-dimensional LOD splitting method. ADI methods and Splitting methods are examples of the Multiplicative Operator Scheme (MOS), which is difficult to parallelize. Methods from the family of Additive Operator Scheme (AOS) can be parallelized. ADE methods also belong to this group of methods. The ADE scheme consists of two explicit sweeps. The sweeping procedure is done from one boundary to another and vice versa. To our knowledge, the ADE scheme has not been applied to higher dimensional or nonlinear PDEs before.

We consider the Black-Scholes (BS) equation which is a parabolic PDE with space dependent coefficients:

$$v_t = \frac{1}{2}\sigma^2 S^2 v_{SS} + rSv_S - rv, \qquad t \geq 0,\ \forall S \in \mathbb{R}, \tag{19.1}$$

where the solution $v(S,t)$ stands for a European option price. A European call (put) option is a contract between its buyer and holder, to buy (sell) a stock at the maturity time T (final time) for the fixed price K, called also strike price. The solution of the linear equation (19.1) is given in closed form formula and it is known as the Black-Scholes formula. Black-Scholes equation is derived under strict assumptions in the market, such as no transaction costs, illiquidity, etc. Modelling this phenomenon in a more realistic way, it leads to the nonlinear BS model which does not have any more analytical solution.

Since the ADE scheme is explicit, stable and thus efficient, it represents a good candidate to compute the numerical solution of multi-dimensional models in finance.

Here we present the implementation of the ADE schemes to two and three dimensional models appearing in finance, esp. the multi-dimensional linear Black-Scholes model. One of the advantages of this approach is that its fundamental implementation set-up can be transferred to higher dimensions. We study a financial derivative that can be exercised only at a pre-fixed maturity time T (commonly referred as 'European' option) and whose payoff depends on the value of N financial assets with prices $S_1, \ldots, S_N$. We assume a financial market with the standard Black-Scholes assumptions, explained in details e.g. in [15]. Although this is very restrictive from the modelling point of view, it is enough to illustrate the implementation of the ADE schemes in a high-dimensional setting. Under this model the price of a derivative $V(S_1, \ldots, S_N, \tau)$ is given by the following $N-$dimensional linear parabolic PDE:

$$\frac{\partial V}{\partial \tau} = \sum_{i=1}^{N}\sum_{j=1}^{N} \frac{\Gamma_{ij} S_i S_j}{2} \frac{\partial^2 V}{\partial S_i \partial S_j} + \sum_{i=1}^{N} rS_i \frac{\partial V}{\partial S_i} - rV, \qquad \tau \geq 0,\ \forall S \in \mathbb{R}_0^+, \tag{19.2}$$

where r denotes the risk-free interest rate, $\tau = T - t$ is the remaining time to the maturity time T and we have the covariance matrix Γ,

$$\Gamma_{ij} \equiv \rho_{ij}\sigma_i\sigma_j, \qquad i,j = 1, \ldots, N, \tag{19.3}$$

with ρ_{ij} being the correlation between asset i and j and σ_i the standard deviation of the asset i. Additionally we have an initial condition which is defined by the payoff of the option,

$$V(S_1, \ldots, S_N, 0) = \Phi(S_1, \ldots, S_N). \tag{19.4}$$

We obtain different models by choosing different numbers of underlying assets (i.e. the number of spatial variables) and defining different payoff functions with corresponding initial conditions. Here we consider both spread options and call options, which have payoffs given by:

$$2-D \text{ Spread option: } V(S_1, S_2, 0) = \max(S_1 - S_2 - K, 0),$$
$$N-D \text{ Call option: } V(S_1, \ldots, S_N, 0) = \max(\max(S_1, \ldots, S_N) - K, 0)).$$

19.2 Alternating Direction Explicit Schemes

ADE schemes are efficient finite-difference schemes to solve PDEs where the discretization of the spatial derivatives is made using available information of both the current and the previous time-steps such that the solution can be determined without solving a linear system of equations.

ADE schemes were proposed by Saul'ev [12] in 1957, later developed by Larkin [8], Bakarat and Clark [2] in 1964–66. More recently, these schemes have received some attention by Duffy [5, 6] 2013 and Leung and Osher [9] 2005 who have studied and applied these schemes in both financial modelling and other applications.

Some advantages of the ADE methods are that they can be implemented in a parallel framework and are very fast due to their explicitness; for a complete survey on the advantages and the motivation to use them in a wide range of problems we refer the reader to [5, 6].

Numerical analysis results focusing on stability and consistency considerations are described in [9] and [3]. In [3] a numerical analysis of convection-diffusion-reaction equation with constant coefficients and smooth initial data is provided. The authors proved that the ADE method applied to the one-dimensional reaction-diffusion equation on a uniform mesh with the discretization of the diffusion according to Saul'ev [12] and the discretization of the convection term following Towler and Yang [13] is unconditionally stable. If a convection term is added to the equation and upwind discretization for this term is used, the ADE scheme is also unconditionally stable cf. [3].

In the ADE schemes one computes for each time level two different solutions which are referred to as sweeps. Hereby the number of sweeps does not depend on the dimension. It has been shown [3, 5, 9] that for the upward and downward sweep the consistency is of order $O((d\tau)^2 + h^2 + \frac{d\tau}{h})$ where $d\tau$ is the time step and h denotes the space step. An exceptionality of the ADE method is that the average of upward and downward solutions has consistency of order $O((d\tau)^2 + h^2)$. For linear models, unconditional stability results and the $O((d\tau)^2 + h^2)$ order of consistency lead to the $O((d\tau)^2 + h^2)$ convergence order.

Stability, consistency and convergence analysis can be extended to higher dimensional models.

The straightforward implementation to nonlinear cases with preserving good stability and consistency properties of the scheme is also a strong advantage. We show how one can implement this scheme for higher dimensional models by focusing on a linear model. The procedure is implemented for non-linear models as well. One way how to do it is to solve nonlinear equation in each time level, instead of system of nonlinear equations in case of implicit schemes. Another way is to keep nonlinearity in the explicit form and solve it directly. Powerful tool for nonlinear equations represents also the Alternating segment explicit-implicit and the implicit-explicit parallel difference method [16].

19.2.1 The Idea of the ADE scheme

The ADE scheme consists of two explicit sub steps, called sweeps. A sweeping step is constructed from one boundary to another and vice versa. Figure 19.1 is an illustrative example of an upward sweep (analogous to the downward sweep in Fig. 19.2).

Figures 19.1 and 19.2 display the grid for the calculating the price of call option in the Black-Scholes model. The blue line represents the payoff as an initial condition and the green lines are given by Dirichlet boundary conditions for small and big asset values. The calculation is provided backward in time.

To calculate the value of the yellow point we use the black values. We can see that we do not use only values from the previous time level but also already known values from the current time level, which preserve explicitness of the scheme. After

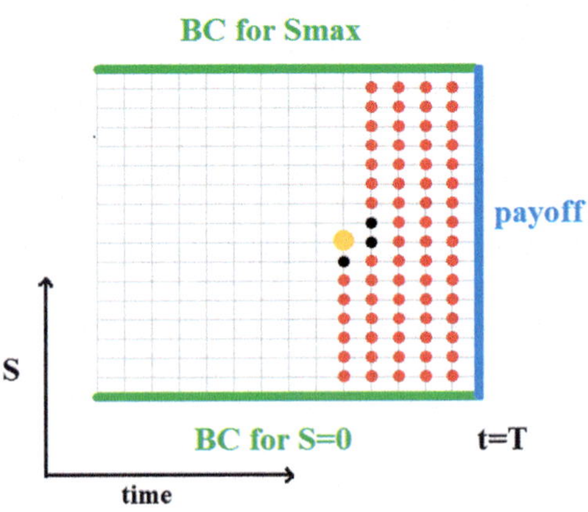

Fig. 19.1 Upward sweep

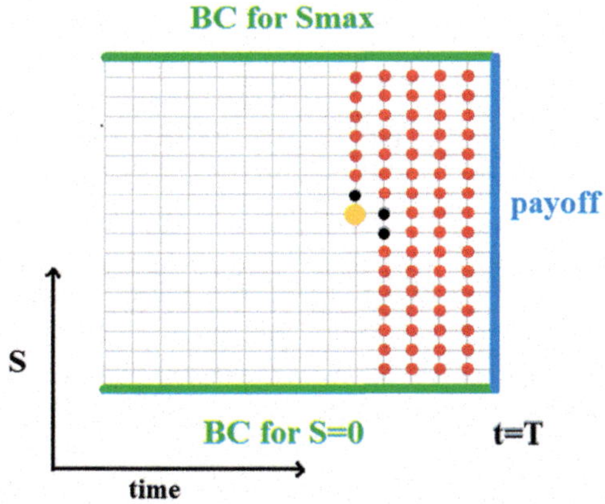

Fig. 19.2 Downward sweep

each time level we combine the solutions from the upward and downward sweep by averaging.

To introduce the ADE method systematically we follow the lines of Leung and Osher [9], and Duffy [5]. The computational spatial interval $(x_{\min}, x_{\max})$, or $(0, S_{\max})$, respectively, is divided into J subintervals, i.e. the space step is $h = (x_{\max} - x_{\min})/J$ and the grid points $x_j = jh$, or $h = S_{\max}/J$, $S_j = jh$, respectively. Thus we get for the coefficients of the BS equation (19.1) $a(S_j) = \frac{1}{2}\sigma^2(jh)^2$, $b(S_j) = rjh$, $c(S_j) = r$.

We consider the resulting spatial semidiscretization to the PDE (19.1), i.e. the following system of ODEs

$$v' = A(v)v, \qquad t > 0, \tag{19.5}$$

with $v(t) \in \mathbb{R}^{J-1}$. Let us consider for simplicity a uniform grid; the time interval $[0, T]$ is divided uniformly into N sub-intervals, with the step size $k = T/N$, i.e. we have the grid points $t_n = nk$. Applying the trapezoidal rule to (19.5) leads to the Crank-Nicolson scheme

$$v^{n+1} = \big[I - kA(v^n)\big]^{-1}\big[I + kA(v^n)\big]v^n, \tag{19.6}$$

where $v^n \approx v(t_n)$. While this classical scheme (19.6) is unconditionally stable and of second order in time and space, it becomes computationally expensive to invert the operator $I - kA(v^n)$ especially in higher space dimensions. In order to obtain an efficient scheme while keeping the other desirable properties, this operator is split additively by the matrix decomposition $A = L + D + U$, where L is lower diagonal, D is diagonal and U denotes an upper-diagonal matrix. Next, following the notation of [9] we further define the *symmetric splitting*

$$B = L + \frac{1}{2}D, \qquad C = U + \frac{1}{2}D. \tag{19.7}$$

Then we can formulate the three steps of the ADE scheme with its upward/downward sweeps and the combination (also for higher dimensions) as

$$\text{UP} \qquad u^{n+1} = \big[I - kB(v^n)\big]^{-1}\big[I + kC(v^n)\big]v^n, \tag{19.8}$$

$$\text{DOWN} \qquad d^{n+1} = \big[I - kC(v^n)\big]^{-1}\big[I + kB(v^n)\big]v^n, \tag{19.9}$$

$$\text{COMB} \qquad v^{n+1} = \frac{1}{2}\big[u^{n+1} + d^{n+1}\big]. \tag{19.10}$$

In other words, in the two sweeps above we assign the solution values that are already computed on the new time level to the operator to be inverted. Hence, the resulting scheme is *explicit*, i.e. efficient. There remain the questions, if we could preserve the unconditional stability and second order accuracy. This will be our main topic in the sequel.

Let us summarize the procedure for one space dimension. The approximation to the solution $v(x,t)$ at the grid point (x_j, t_n) is $c(x_j, t_n) =: c_j^n$ given as an average of upward sweep u_j^n and downward sweep d_j^n. This combination c_j^n contains the initial data at the beginning. For $n = 0, 1, \dots, N-1$ we repeat the following steps:

1. Initialization: $u_j^n = c_j^n, \quad d_j^n = c_j^n, \qquad j = 1, \dots, J-1$
2. **U**pward sweep: $u_j^{n+1}, \qquad j = 1, \dots, J-1$
3. **D**ownward sweep: $d_j^{n+1}, \qquad j = J-1, \dots, 1$
4. **C**ombination: $\quad c^{n+1} = (u^{n+1} + d^{n+1})/2$

Using different approximation strategies for the convection, diffusion and reaction terms we obtain different variations of the ADE schemes, which were proposed by Saul'ev [12].

19.2.2 Solving PDEs with the ADE Method

We start considering the partial differential equation (PDE)

$$v_t = a\, v_{xx} + b\, v_x - c\, v, \qquad t \ge 0,\ \forall x \in \mathbb{R}, \tag{19.11}$$

with the constant coefficients $a \ge const. > 0$, $b \ge 0$, $c \ge 0$ and supplied with smooth initial data. We denote the analytical classical solution of (19.11) by $v := v(x,t)$ and use subscripts to abbreviate partial differentiation, e.g. $v_{xx} := \partial^2 v / \partial x^2$.

Secondly, we will consider the classical linear Black-Scholes (BS) equation

$$v_t = \frac{1}{2}\sigma^2 S^2 v_{SS} + rSv_S - rv, \qquad t \ge 0,\ \forall S \in \mathbb{R}, \tag{19.12}$$

which is a generalization of the PDE (19.11) to space dependent coefficients. In computational finance a solution $v(S,t)$ of the PDE (19.12) represents a European option price. A European option is a contract between the holder of the option and the future buyer, that at a time instance T, the expiration time, the underlying asset (stock) can be sold or bought (call or put option) for a fixed strike price K. Using the Black-Scholes formula the option price is calculated for the corresponding underlying asset price S (stock price) in a time interval $t \in (0,T)$.

Let us note that the BS equation (19.12) is derived under quite restrictive market assumptions, which are not very realistic. Relaxing these assumptions leads to new models (e.g. including transaction costs, illiquidity on the market) that are strongly nonlinear BS equations that can only be solved analytically in very simple cases.

While there exist analytical tools to solve explicitly (19.11) and (19.12), the interest in studying the ADE method for these simple 1D cases is the fact that we want to extend this approach in a subsequent work to nonlinear PDEs and to higher dimensions. Applying the ADE to the nonlinear BS equations we need to solve only a scalar nonlinear equation (instead of a nonlinear system of equations for a standard implicit method). Thus, the computational effort using ADE instead of an implicit scheme is highly reduced. Also, for higher space dimensions the number of ADE sweeps does not increase, it remains two. These facts make the ADE methods an attractive candidate to study them in more detail.

19.2.3 The Modified Difference Quotients for the ADE Method

In this subsection we want to illustrate the outcome of the previous Sect. 19.2.1. Thus, we select some spatial discretization and investigate which ADE scheme will result.

For the discretization of the *diffusion term* we use, cf. [12]

$$\begin{aligned} \frac{\partial^2 v(x_j,t_n)}{\partial x^2} &\approx \frac{u_{j+1}^n - u_j^n - u_j^{n+1} + u_{j-1}^{n+1}}{h^2}, && j = 1,\dots,J-1 \\ \frac{\partial^2 v(x_j,t_n)}{\partial x^2} &\approx \frac{d_{j+1}^{n+1} - d_j^{n+1} - d_j^n + d_{j-1}^n}{h^2}, && j = J-1,\dots,1. \end{aligned} \tag{19.13}$$

In order to obtain a symmetric scheme we use the following approximations of the *reaction term*, the same for the upward and downward sweep

$$\begin{aligned} v(x_j,t_n) &\approx \frac{u_j^{n+1} + u_j^n}{2}, && j = 1,\dots,J-1, \\ v(x_j,t_n) &\approx \frac{d_j^{n+1} + d_j^n}{2}, && j = J-1,\dots,1. \end{aligned} \tag{19.14}$$

Different approximations of the *convection term* are possible [4, 9]. In the following we state three of them. First, Towler and Yang [13] used special kind of centered differences

$$\begin{aligned} \frac{\partial v(x_j,t_n)}{\partial x} &\approx \frac{u_{j+1}^n - u_{j-1}^{n+1}}{2h}, && j = 1,\dots,J-1, \\ \frac{\partial v(x_j,t_n)}{\partial x} &\approx \frac{d_{j+1}^{n+1} - d_{j-1}^n}{2h}, && j = J-1,\dots,1. \end{aligned} \tag{19.15}$$

More accurate approximations were proposed by Roberts and Weiss [11], Piacsek and Williams [10]

$$\begin{aligned} \frac{\partial v(x_j,t_n)}{\partial x} &\approx \frac{u_{j+1}^n - u_j^n + u_j^{n+1} - u_{j-1}^{n+1}}{2h}, && j = 1,\dots,J-1, \\ \frac{\partial v(x_j,t_n)}{\partial x} &\approx \frac{d_{j+1}^{n+1} - d_j^{n+1} + d_j^n - d_{j-1}^n}{2h}, && j = J-1,\dots,1. \end{aligned} \tag{19.16}$$

As a third option we will use *upwind* approximations combined with the ADE technique. Since we have in mind financial applications we will focus on left going waves, i.e. $b > 0$ in (19.11). Right going waves $b < 0$ are treated analogously.

The well-known first order approximation reads

$$\frac{\partial v(x_j,t)}{\partial x} \approx \frac{v_{j+1}(t) - v_j(t)}{h} \qquad j = J-1,\dots,1, \tag{19.17}$$

and the forward difference of second order [14]

$$\frac{\partial v(x_j,t)}{\partial x} \approx \frac{-v_{j+2}(t) + 4v_{j+1}(t) - 3v_j(t)}{2h}, \qquad j = J-1,\dots,1. \tag{19.18}$$

Applying the ADE time splitting idea of Sect. 19.2.1 we obtain for the upwind strategy (19.17)

$$\begin{aligned} \frac{\partial v(x_j,t_{n+1})}{\partial x} &\approx \frac{u_{j+1}^n - u_j^n}{h}, && j = 1,\dots,J-1, \\ \frac{\partial v(x_j,t_{n+1})}{\partial x} &\approx \frac{d_{j+1}^n - d_j^n + d_{j+1}^{n+1} - d_j^{n+1}}{2h}, && j = J-1,\dots,1, \end{aligned} \tag{19.19}$$

and for the second order approximation

$$\begin{aligned} \frac{\partial v(x_j,t_{n+1})}{\partial x} &\approx \frac{-u_{j+2}^n + 4u_{j+1}^n - 3u_j^n}{2h}, && j = 1,\dots,J-1, \\ \frac{\partial v(x_j,t_{n+1})}{\partial x} &\approx \frac{-d_{j+2}^n + 4d_{j+1}^n - 3d_j^n - d_{j+2}^{n+1} + 4d_{j+1}^{n+1} - 3d_j^{n+1}}{4h}, && j = J-1,\dots,1. \end{aligned} \tag{19.20}$$

We will show that this upwind approximation (19.19) leads to a stable scheme.

19.3 Stability of the ADE Method

In this section we investigate the stability of the proposed ADE method using the matrix approach in Sect. 19.3.1 and the classical von Neumann method in Sect. 19.3.2. For the convection-diffusion-reaction equation (19.11) we obtain unconditional stability using the matrix approach. This stability analysis can be extended by adding homogeneous BCs, without affecting the stability results. This is our motivation to deal with the matrix approach.

19.3.1 Stability Analysis Using the Matrix Approach

We are motivated by [9], where the authors claim and proof that "if A is symmetric negative definite, the ADE scheme is unconditionally stable". We have to define symmetric discretization quotients to get symmetric discrete operators. For reaction-diffusion equation applying central difference quotients we get symmetric operator A and we can follow the ideas for the proof for the heat equation from [9].

Using upwind discretization formulas instead of central differencing leads also to an unconditionally stable scheme. "If A is lower-triangular with all diagonal elements negative, the ADE scheme is unconditionally stable" is generally claimed and proved in [9]. In the following we choose suitable differentiating approximations, we formulate theorems about stability properties and prove it.

Theorem 19.1 *The ADE scheme applied to the reaction-diffusion PDE* (19.11) *(with $b = 0$) is unconditionally stable.*

Proof Without loss of generality we focus on the upward sweep

$$\frac{u_j^{n+1} - u_j^n}{k} = a\frac{u_{j+1}^n - u_j^n - u_j^{n+1} + u_{j-1}^{n+1}}{h^2} - c\frac{u_j^{n+1} + u_j^n}{2}.$$

Let us denote the parabolic mesh ratio $\alpha := a\frac{k}{h^2}$, $\gamma := ck$; where a, c are constants.

$$u_j^{n+1} = u_j^n + \alpha\left(u_{j+1}^n - u_j^n - u_j^{n+1} + u_{j-1}^{n+1}\right) - \frac{\gamma}{2}\left(u_j^{n+1} + u_j^n\right)$$

$$\left(1 + \alpha + \frac{\gamma}{2}\right)u_j^{n+1} + (-\alpha)u_{j-1}^{n+1} = \left(1 - \alpha - \frac{\gamma}{2}\right)u_j^n + \alpha u_{j+1}^n \qquad (19.21)$$

We follow roughly the train of thoughts of Leung and Osher [9] and write the upward sweep (19.21) with homogeneous BCs in matrix notation

$$A_u u^{n+1} = B_u u^n, \qquad n \geq 0,$$

with $A_u, B_u \in R^{(J-1)\times(J-1)}$ given by

$$A_u = \begin{pmatrix} 1+\alpha+\frac{\gamma}{2} & 0 & \dots & 0 \\ -\alpha & \ddots & & \vdots \\ \vdots & \ddots & \ddots & 0 \\ 0 & \dots & -\alpha & 1+\alpha+\frac{\gamma}{2} \end{pmatrix} = I + \begin{pmatrix} \alpha+\frac{\gamma}{2} & 0 & \dots & 0 \\ -\alpha & \ddots & & \vdots \\ \vdots & \ddots & \ddots & 0 \\ 0 & \dots & -\alpha & \alpha+\frac{\gamma}{2} \end{pmatrix}$$

$$A_u =: I + E,$$

$$B_u = \begin{pmatrix} 1-\alpha-\frac{\gamma}{2} & \alpha & \dots & 0 \\ 0 & \ddots & \ddots & \vdots \\ \vdots & & \ddots & \alpha \\ 0 & \dots & 0 & 1-\alpha-\frac{\gamma}{2} \end{pmatrix} = I - \begin{pmatrix} \alpha+\frac{\gamma}{2} & -\alpha & \dots & 0 \\ 0 & \ddots & \ddots & \vdots \\ \vdots & & \ddots & -\alpha \\ 0 & \dots & 0 & \alpha+\frac{\gamma}{2} \end{pmatrix}$$

$$B_u =: I - E^\top.$$

Next, we consider the matrices

$$A_u^\top + A_u = 2I + D,$$

$$\text{where} \qquad D := E + E^\top = \begin{pmatrix} 2\alpha+\gamma & -\alpha & \dots & 0 \\ -\alpha & \ddots & \ddots & \vdots \\ \vdots & & \ddots & -\alpha \\ 0 & \dots & -\alpha & 2\alpha+\gamma \end{pmatrix}.$$

The matrix D is positive definite and thus we can define the induced D-norm as

$$||C||_D^2 := \sup_{x\neq 0} \frac{||C_x||_D^2}{||x||_D^2} = \sup_{x\neq 0} \frac{x^\top C^\top DCx}{x^\top Dx},$$

and the upward sweep can be written as

$$U^{n+1} = A_u^{-1} B_u \, U^n.$$

Next, we consider the D-norm for the upward sweep matrix $A_u^{-1}B_u$

$$||A_u^{-1}B_u||_D^2 := \sup_{x\neq 0} \frac{x^\top B_u^\top A_u^{-\top} DA_u^{-1} B_u x}{x^\top Dx}$$

The numerator $B_u^\top A_u^{-\top} DA_u^{-1} B_u$ can be easily rewritten after a few algebraic steps as $D - 2\gamma(A_u^{-1}D)^\top(A_u^{-1}D)$. From our notation $A_u = I + E$ and $B_u = I - E^\top$ follows

$$B_u^\top A_u^{-\top} DA_u^{-1} B_u = (I - E^\top)^\top A_u^{-\top} DA_u^{-1}(I - E^\top)$$

where $E^\top = D - E$. An expression in terms of matrices A_u and D gets the following form:

$$\begin{aligned}
(A_u^\top - D)^\top A_u^{-\top} D A_u^{-1}(A_u^\top - D) \\
&= D - DA_u^{-1}D - D^\top A_u^{-\top} D + DA_u^{-\top} DA_u^{-1} D \\
&= D - DA_u^{-\top} A_u^\top A_u^{-1} D - DA_u^{-\top} A_u A_u^{-1} D + DA_u^{-\top} DA_u^{-1} D \\
&= D + DA_u^{-\top}\big[- A_u^{-\top} - A_u + D\big] A_u^{-1} D \\
&= D - 2(A_u^{-1}D)^\top (A_u^{-1}D)
\end{aligned}$$

and hence it follows

$$||A_u^{-1}B_u||_D^2 = 1 - 2 \sup_{x \neq 0} \frac{||A_u^{-1}Dx||_2^2}{||x||_D^2}.$$

Thus the spectral radius of the upward sweep matrix $A_u^{-1}B_u$ reads

$$\rho(A_u^{-1}B_u) \leq ||A_u^{-1}B_u||_D < 1$$

and we can conclude that the upward sweep is unconditionally stable.

An analogous result holds for the downward step. In the corresponding equation

$$A_d d^{n+1} = B_d d^n, \qquad n \geq 0 \tag{19.22}$$

the matrices A_d and B_d are defined as $A_d = A_u^\top$ and $B_d = B_u^\top$. The analysis is done analogously: we can define a positive definite matrix and follow again the steps from the previous proof of the Theorem 19.1. Consequently also the combination, as an arithmetic average of these two sub steps, is also unconditionally stable.

The stability analysis using the matrix approach according to [9] worked for reaction-diffusion equations with constant coefficients. However, this proof is not transferable for the stability analysis of methods with non-symmetric terms, e.g. the difference quotients for the convection term proposed by Towler and Yang (Eq. 19.15), or Roberts and Weiss (Eq. 19.16), cf. Sect. 19.3.2.

As a remedy we can apply a modified upwind discretization of the convection term. The resulting structure of the matrices A_u, B_u is different but we can perform a similar proof.

Theorem 19.2 *ADE scheme, using upwind discretization in convection term, applied to the reaction-diffusion-convection equation* (19.11) *is unconditionally stable in the upward sweep and unconditionally stable in the downward one.*

Proof Again, without loss of generality, we focus on the upward sweep and consider an upwind discretization for a left-going wave, i.e. $b \geq 0$ (since later we would like to extend this approach for the Black-Scholes model, where $b \geq 0$). In the upward sweep we use difference quotients using values just from the old time level (19.17)

$$\frac{u_j^{n+1}-u_j^n}{k} = a\frac{u_{j+1}^n - u_j^n - u_j^{n+1} + u_{j-1}^{n+1}}{h^2} + b\frac{u_{j+1}^n - u_j^n}{h} - c\frac{u_j^{n+1}+u_j^n}{2}.$$

Using the abbreviations $\alpha := a\frac{k}{h^2}$, $\beta := b\frac{k}{h} \geq 0$, $\gamma := ck$, we can write

$$-\alpha u_{j-1}^{n+1} + \Big(1+\alpha+\frac{\gamma}{2}\Big)u_j^{n+1} = \Big(1-\alpha-\beta-\frac{\gamma}{2}\Big)u_j^n + \Big(\alpha+\beta\Big)u_{j+1}^n \qquad (19.23)$$

We follow again roughly the ideas of Leung and Osher [9] and consider the upward sweep (19.23) with homogeneous BCs

$$A_u u^{n+1} = B_u u^n, \qquad n \geq 0,$$

with the system matrices $A_u, B_u \in R^{(J-1)\times(J-1)}$ given by

$$A_u = \begin{pmatrix} 1+\alpha+\frac{\gamma}{2} & 0 & \dots & 0 \\ -\alpha & & \ddots & \vdots \\ \vdots & & \ddots \ \ddots & 0 \\ 0 & & \dots\ -\alpha & 1+\alpha+\frac{\gamma}{2} \end{pmatrix} =: I+E,$$

$$B_u = \begin{pmatrix} 1-\alpha-\beta-\frac{\gamma}{2} & \alpha+\beta & \dots & 0 \\ 0 & \ddots & \ddots & \vdots \\ \vdots & & \ddots & \alpha+\beta \\ 0 & \dots & 0 & 1-\alpha-\beta-\frac{\gamma}{2} \end{pmatrix} =: I-F.$$

$$\text{where} \qquad D := E+F = \begin{pmatrix} 2\alpha+\beta+\gamma & -\alpha-\beta & \dots & 0 \\ -\alpha & \ddots & \ddots & \vdots \\ \vdots & & \ddots & -\alpha-\beta \\ 0 & \dots & -\alpha & 2\alpha+\beta+\gamma \end{pmatrix}.$$

The matrix D is not symmetric but obviously positive definite.

In the sequel we have just outlined the steps which differ from the previous proof. The numerator $B_u^\top A_u^{-\top} D A_u^{-1} B_u$ can be easily rewritten after a few algebraic steps as $D - 2\gamma(A_u^{-1}D)^\top(A_u^{-1}D)$.

From our notation $A_u = I+E$ and $B_u = I-F$ follows

$$B_u^\top A_u^{-\top} D A_u^{-1} B_u = (I-F)^\top A_u^{-\top} D A_u^{-1}(I-F)$$

where $F := D - E$. An expression in terms of matrices A_u and D gets the following form:

$$(I + E - D)^\top A_u^{-\top} D A_u^{-1} (I + E - D) = (A_u - D)^\top A_u^{-\top} D A_u^{-1} (A_u - D)$$

and we proceed the same way as in the previous proof.

For the *downward sweep* we have:

$$\frac{d_j^{n+1} - d_j^n}{k} = a\frac{d_{j+1}^{n+1} - d_j^{n+1} - d_j^n + d_{j-1}^n}{h^2} + b\frac{d_{j+1}^n + d_{j+1}^{n+1} - d_j^n - d_j^{n+1}}{2h} - c\frac{d_j^{n+1} + d_j^n}{2}.$$

Using the abbreviations $\alpha := a\frac{k}{h^2}$, $\beta := b\frac{k}{h} \le 0$, $\gamma := ck$, we can write

$$\left(1 + \alpha + \frac{\beta}{2} + \frac{\gamma}{2}\right) d_j^{n+1} + \left(-\alpha - \frac{\beta}{2}\right) d_{j+1}^{n+1} = \alpha d_{j-1}^n + \left(1 - \alpha - \frac{\beta}{2} - \frac{\gamma}{2}\right) d_j^n + \frac{\beta}{2} d_{j+1}^n \tag{19.24}$$

$$A_D d^{n+1} = B_D d^n, \qquad n \ge 0,$$

with A_D, $B_D \in R^{(J-1)\times(J-1)}$ given by matrices A_D, B_D. The matrix A_D is upper-diagonal $A_D = \text{diag}(1 + \alpha + \frac{\beta}{2} + \frac{\gamma}{2}, -\alpha - \frac{\beta}{2})$. The matrix B_D is tridiagonal with diagonal terms: $B_D = \text{diag}(\alpha, 1 - \alpha - \frac{\beta}{2} - \frac{\gamma}{2}, \frac{\beta}{2})$. Likewise we construct matrices $D = \text{diag}(-\alpha, 2\alpha + \beta + \gamma, -\alpha - \beta)$ as a tridiagonal positive definite matrix. We can follow the same way of proof and thus we conclude the unconditional stability of the downward sweep.

19.3.2 Von Neumann Stability Analysis for the Convection-Diffusion-Reaction Equation

Since analysis using matrix approach was suitable for upwind kind of approximation in convection term, here we investigate stability properties of the ADE schemes, where discretization of convection term is provided according to [13] and [11].

We consider the convection-diffusion-reaction equation (19.11) and focus on the sequel on the upward sweep of the ADE procedure. An appropriate choice for the approximation of the convection term is the one due to Roberts and Weiss [11], since performing just a downward sweep leads to the unconditionally stable solution.

Theorem 19.3 *The ADE scheme with the Roberts and Weiss approximation in the convection term, applied to the PDE* (19.11) *is conditionally stable in the upward sweep and unconditionally stable for the downward one.*

Proof Using Roberts and Weiss discretization in convection term we get

$$\frac{u_j^{n+1}-u_j^n}{k} = a\frac{u_{j+1}^n - u_j^n - u_j^{n+1} + u_{j-1}^{n+1}}{h^2} + b\frac{u_{j+1}^n - u_j^n + u_j^{n+1} - u_{j-1}^{n+1}}{2h} - c\frac{u_j^{n+1}+u_j^n}{2}.$$

Let us denote the parabolic mesh ratio $\alpha := a\frac{k}{h^2}$, the hyperbolic mesh ratio $\beta := b\frac{k}{h}$ and $\gamma := ck$; where a, b, c are nonnegative constants.

$$u_j^{n+1} = u_j^n + \alpha\left(u_{j+1}^n - u_j^n - u_j^{n+1} + u_{j-1}^{n+1}\right) + \frac{\beta}{2}\left(u_{j+1}^n - u_j^n + u_j^{n+1} - u_{j-1}^{n+1}\right) - \frac{\gamma}{2}\left(u_j^{n+1} + u_j^n\right)$$

Applying von Neumann ansatz $u_j^n := e^{\xi tn} e^{i\lambda xj}$ the amplification factor A_1 reads:

$$A_1 = \frac{A + Be^{i\lambda h}}{C + De^{-i\lambda h}}$$

where $A = 1-\alpha-\beta/2-\gamma/2; B = \alpha+\beta/2; C = 1+\alpha-\beta/2+\gamma/2; D = -\alpha+\beta/2$.

For stability we require $|A_1| \le 1$, i.e.

$$|A_1|^2 = A_1\overline{A_1} = \frac{\left(A + Be^{i\lambda h}\right)\left(A + Be^{-i\lambda h}\right)}{\left(C + De^{-i\lambda h}\right)\left(C + De^{i\lambda h}\right)} \le 1$$

$$A^2 + B^2 + 2AB\cos(\lambda h) \le C^2 + D^2 + 2CD\cos(\lambda h)$$

$$2\,(AB - CD)\cos(\lambda h) \le C^2 + D^2 - A^2 - B^2$$

$$(4\alpha - 4\alpha\beta - \beta\gamma)\cos(\lambda h) \le 4\alpha - 4\alpha\beta - \beta\gamma + 2\gamma. \tag{19.25}$$

We need to check two cases with respect to the sign of $(4\alpha - 4\alpha\beta - \beta\gamma)$.

- **Case 1:** By substituting α, β, γ into $4\alpha - 4\alpha\beta - \beta\gamma > 0$ we get following condition:

$$\alpha < \frac{a}{2Pe} - \frac{ck}{4} \tag{19.26}$$

where $Pe = \frac{bh}{2}$ is the so-called Péclet number. In this case equation (19.25) can be rewritten as

$$\cos(\lambda h) \le \frac{4\alpha - 4\alpha\beta - \beta\gamma + 2\gamma}{4\alpha - 4\alpha\beta - \beta\gamma} \qquad \forall \lambda h \tag{19.27}$$

i.e.

$$1 \le 1 + \frac{2\gamma}{4\alpha - 4\alpha\beta - \beta\gamma}$$

or

$$0 \le \frac{2\gamma}{4\alpha - 4\alpha\beta - \beta\gamma}. \tag{19.28}$$

We can notice that condition (19.28) is satisfied for all the possible values of parameters, since $\gamma > 0$ and $4\alpha - 4\alpha\beta - \beta\gamma > 0$.

- **Case 2:** We consider $4\alpha - 4\alpha\beta - \beta\gamma < 0$, what is equivalent with the condition

$$\alpha > \frac{a}{2Pe} - \frac{ck}{4}. \tag{19.29}$$

In this case equation (19.25) can be rewritten as

$$\cos(\lambda h) \ge \frac{4\alpha - 4\alpha\beta - \beta\gamma + 2\gamma}{4\alpha - 4\alpha\beta - \beta\gamma} \qquad \forall \lambda h \tag{19.30}$$

i.e.

$$-1 \ge 1 + \frac{2\gamma}{4\alpha - 4\alpha\beta - \beta\gamma},$$

or

$$2 \le \frac{-2\gamma}{4\alpha - 4\alpha\beta - \beta\gamma} \tag{19.31}$$

or

$$\frac{1}{2}\beta\gamma + 2\alpha\beta - 2\alpha \le \gamma \tag{19.32}$$

or

$$\alpha \le \frac{\alpha}{\beta} - \frac{\gamma}{4} + \frac{\gamma}{2\beta} \tag{19.33}$$

After substituting α, β, γ and after elementary algebraic steps we get

$$\alpha \le \frac{a}{2Pe} - \frac{ck}{4} + \frac{ch}{4b}. \tag{19.34}$$

Case 2 leads to conditions (19.29) and (19.34) what means that

$$\alpha \in \left(\frac{a}{2Pe} - \frac{ck}{4}, \frac{a}{2Pe} - \frac{ck}{4} + \frac{ch}{4b} \right] \tag{19.35}$$

To sum up case 1 and case 2 we can claim that conditions (19.26) and (19.35) and also considering the situation where $(4\alpha - 4\alpha\beta - \beta\gamma) = 0$ we get

$$\alpha \leq \frac{a}{2Pe} - \frac{ck}{4} + \frac{ch}{4b}. \tag{19.36}$$

For the downward sweep we get the following amplification factor:

$$A_2 = \frac{\left[1 - \alpha + \frac{\beta}{2} - \frac{\gamma}{2}\right] + [\alpha - \frac{\beta}{2}]e^{-i\lambda h}}{\left[1 + \alpha + \frac{\beta}{2} + \frac{\gamma}{2}\right] + [-\alpha - \frac{\beta}{2}]e^{i\lambda h}} \tag{19.37}$$

Stability condition $|A_2|^2 \leq 1$ leads to the formula:

$$\cos(\lambda h) \leq \frac{4\alpha + 4\alpha\beta + \beta\gamma + 2\gamma}{4\alpha + 4\alpha\beta + \beta\gamma}.$$

Let us note that the last condition can be simplified to the condition:

$$\frac{2\gamma}{4\alpha + 4\alpha\beta + \beta\gamma} \geq 0. \tag{19.38}$$

The coefficients α, β, γ are positive, i.e. the condition (19.38) is satisfied and thus we have the unconditional stability for the downward sweep using the Roberts and Weiss approximation, which completes the proof.

In case of the Roberts and Weiss approximation we propose to use only the unconditional stable downward sweep.

Theorem 19.4 *The ADE scheme, using Towler and Yang approximation in the convection term, applied to the PDE* (19.11) *is conditionally stable in both sweeps.*

Proof For the Towler and Yang approximation the stability condition for the upward sweep reads

$$(4\alpha - 2\alpha\beta - \beta\gamma)\cos(\lambda h) \leq 4\alpha - 2\alpha\beta + 2\gamma, \tag{19.39}$$

where again we can distinguish 2 cases with respect to the sign of left hand side of the Eq. (19.39).

- **Case 1:** If $(4\alpha - 2\alpha\beta - \beta\gamma) > 0$, it means

$$\alpha < \frac{a}{Pe} - \frac{ck}{2}. \tag{19.40}$$

In this case equation (19.39) can be rewritten as

$$\cos(\lambda h) \leq \frac{4\alpha - 4\alpha\beta - \beta\gamma + 2\gamma}{4\alpha - 2\alpha\beta - \beta\gamma} \tag{19.41}$$

$$1 \leq 1 + \frac{\gamma(2+\beta)}{4\alpha - 2\alpha\beta - \beta\gamma}$$

$$0 \leq \frac{\gamma(2+\beta)}{4\alpha - 2\alpha\beta - \beta\gamma}. \tag{19.42}$$

We can notice that condition (19.42) is satisfied for all the possible values of parameters, since $\gamma \geq 0$ and $(2+\beta) > 0$ and $4\alpha - 2\alpha\beta - \beta\gamma > 0$.

- **Case 2:** We consider $(4\alpha - 2\alpha\beta - \beta\gamma) < 0$, what is equivalent with the condition

$$\alpha > \frac{a}{Pe} - \frac{ck}{2}. \tag{19.43}$$

In this case equation (19.39) can be rewritten as

$$\cos(\lambda h) \geq \frac{4\alpha - 4\alpha\beta - \beta\gamma + 2\gamma}{4\alpha - 2\alpha\beta - \beta\gamma} \tag{19.44}$$

$$-2 \geq \frac{\gamma(2+\beta)}{4\alpha - 2\alpha\beta - \beta\gamma}$$

After substituting α, β, γ and simplification it leads to the condition

$$\alpha \leq \frac{a}{Pe} - \frac{ck}{2} + \frac{ch}{2b} + \frac{1}{2}. \tag{19.45}$$

In case 2 we obtain two conditions (19.43) and (19.45), namely:

$$\alpha \in \left(\frac{a}{2Pe} - \frac{ck}{2}, \frac{a}{Pe} - \frac{ck}{2} + \frac{ch}{2b} + \frac{1}{2}\right] \tag{19.46}$$

From case 1 condition (19.40) and case 2 condition (19.46) in Towler and Yang case and considering also possibility of $(4\alpha - 2\alpha\beta - \beta\gamma) = 0$ we can sum up

$$\alpha \leq \frac{a}{Pe} - \frac{ck}{2} + \frac{ch}{2b} + \frac{1}{2}. \tag{19.47}$$

For the downward sweep the stability condition is

$$\cos(\lambda h) \leq \frac{4\alpha + 2\alpha\beta + 2\gamma}{4\alpha + 2\alpha\beta + \beta\gamma},$$

which leads to the condition:

$$\frac{k}{h^2} \leq \frac{1}{Pe} \tag{19.48}$$

Both sweeps in the Towler and Yang discretization of the convection term in the reaction-diffusion-convection equation are conditionally stable under the conditions (19.47) and (19.48)

19.4 Consistency Analysis of the ADE Methods

In this section we provide a consistency analysis of the ADE methods for solving the convection-diffusion-reaction equation (19.11) and for the BS model.

19.4.1 Consistency of the ADE Scheme for Convection-Diffusion-Reaction Equations

We study the consistency of the following ADE discretization

$$\frac{u_j^{n+1}-u_j^n}{k} = a\frac{u_{j+1}^n - u_j^n - u_j^{n+1} + u_{j-1}^{n+1}}{h^2} + b\frac{u_{j+1}^n - u_j^n + u_j^{n+1} - u_{j-1}^{n+1}}{2h} - c\frac{u_j^{n+1}+u_j^n}{2}$$

to the convection-diffusion-reaction equation (19.11). The local truncation error (LTE) of the upward sweep is given by

$$\begin{aligned} LTE_{\text{up}} &= k\Big(-\frac{1}{2}v_{tt} + \frac{1}{2}av_{xxt} + \frac{1}{2}bv_{xt}\Big) \\ &\quad + k^2\Big(-\frac{1}{6}v_{ttt} + \frac{1}{4}av_{xxtt} + \frac{1}{4}bv_{xtt}\Big) + h^2\Big(\frac{1}{12}av_{xxxx}\frac{1}{6}bv_{xxx}\Big) \\ &\quad - kh\Big(\frac{1}{6}av_{xxxt} + \frac{1}{4}bv_{xxt}\Big) - \frac{k}{h}av_{xt} - \frac{k^2}{h}\Big(\frac{1}{2}av_{xtt}\Big) - \frac{k^3}{h}\Big(\frac{1}{6}av_{xttt}\Big), \end{aligned}$$

and analogously the LTE for the downward sweep reads

$$\begin{aligned} LTE_{\text{down}} &= k\Big(-\frac{1}{2}v_{tt} + \frac{1}{2}av_{xxt} + \frac{1}{2}bv_{xt}\Big) \\ &\quad + k^2\Big(-\frac{1}{6}v_{ttt} + \frac{1}{4}av_{xxtt} + \frac{1}{4}bv_{xtt}\Big) + h^2\Big(\frac{1}{12}av_{xxxx}\frac{1}{6}bv_{xxx}\Big) \\ &\quad + kh\Big(\frac{1}{6}av_{xxxt} + \frac{1}{4}bv_{xxt}\Big) + \frac{k}{h}av_{xt} + \frac{k^2}{h}\Big(\frac{1}{2}av_{xtt}\Big) + \frac{k^3}{h}\Big(\frac{1}{6}av_{xttt}\Big). \end{aligned}$$

Thus we end up for the LTE for the combined sweep

$$LTE_{\text{ADE}} = k\Big(-\frac{1}{2}v_{tt} + \frac{1}{2}av_{xxt} + \frac{1}{2}bv_{xt}\Big) + k^2\Big(-\frac{1}{6}v_{ttt} + \frac{1}{4}av_{xxtt} + \frac{1}{4}bv_{xtt}\Big) + h^2\Big(\frac{1}{12}av_{xxxx}\frac{1}{6}bv_{xxx}\Big)$$

Assuming a constant parabolic mesh ratio k/h^2, the first order term in k can be written in the form $O(k) = O(h^2)$ and hence we get

$$LTE_{\text{ADE}} = k^2\Big(-\frac{1}{6}v_{ttt} + \frac{1}{4}av_{xxtt} + \frac{1}{4}bv_{xtt}\Big) + h^2\Big(\frac{1}{12}av_{xxxx}\frac{1}{6}bv_{xxx} - \frac{1}{2}v_{tt} + \frac{1}{2}av_{xxt} + \frac{1}{2}bv_{xt}\Big)$$

Hence, the order of consistency of the ADE method for the PDE (19.11) is $O(k^2 + h^2)$.

19.4.2 The Consistency of the ADE Method for the Linear BS Model

As an extension of the PDE (19.11) we consider now the linear BS equation.

Theorem 19.5 *The order of consistency of the ADE method for the linear BS equation is $O(k^2 + h^2)$ in both sweeps and in the final combined solution.*

Proof The linear BS PDE is a special case of (19.11) with the space-dependent coefficients $a(S) = \frac{1}{2}\sigma^2 S^2$, $b(S) = rS$, $c(S) = r$. The LTE for the upward sweep reads:

$$\begin{aligned} LTE_{\text{BS}} = {} & k\Big(-\frac{1}{2}v_{tt} + \frac{1}{2}av_{xxt} + \frac{1}{2}bv_{xt}\Big) \\ & + k^2\Big(-\frac{1}{6}v_{ttt} + \frac{1}{4}av_{xxtt} + \frac{1}{4}bv_{xtt}\Big) + h^2\Big(\frac{1}{12}av_{xxxx}\frac{1}{6}bv_{xxx}\Big) \\ & + kh\Big(-\frac{1}{6}av_{xxxt} - \frac{1}{4}bv_{xxt}\Big) + \frac{k}{h}\Big(-av_{xt}\Big) \\ & + \frac{k^2}{h}\Big(-\frac{1}{2}av_{xtt}\Big) + \frac{k^3}{h}\Big(-\frac{1}{6}av_{xttt}\Big) \end{aligned}$$

If we assume a constant parabolic mesh ratio $\alpha = k/h^2$, then we get

$$LTE = k\Big(-\frac{1}{2}v_{tt}\Big) + k^2\Big(-\frac{1}{6}v_{ttt}\Big) = \alpha h^2\Big(-\frac{1}{2}v_{tt}\Big) + k^2\Big(-\frac{1}{6}v_{ttt}\Big),$$

where we neglected higher order terms. A similar result holds for the downward sweep. We have shown that consistency for the linear BS model is $O(k^2 + h^2)$ in downward, upward and hence also in the combination.

19.4.3 Application and Numerical Experiments with the Linear Model

We apply the ADE method and calculate a price for a vanilla European call option in a classic linear BS model with constant coefficients. Choosing the following set of parameters $r = 0.03$ (interest rate); $q = 0$ (continuous dividend yield); $\sigma = 0.2$ (volatility); $T = 1$ (maturity time in years); $S_{\max} = 90$ (maximal stock price); $K = 30$ (strike price); and defining a grid with $N = 50$ time steps; $J = 200$ space steps we get an option price, which is shown in Fig. 19.3 (Fig. 19.4).

In this subsection we analyze the computational and theoretical order of convergence. In table in Fig. 19.5 it is recorded an error as a difference between numerical solution using ADE method and the closed form BS formula for different meshes with fixed mesh ratio 0.23. In table in Fig. 19.6 ratios of errors from the table in Fig. 19.5 are calculated. One can observe that using double space steps, ratio of errors converges to the number 4, what confirms that the theoretical order of convergence is 2.

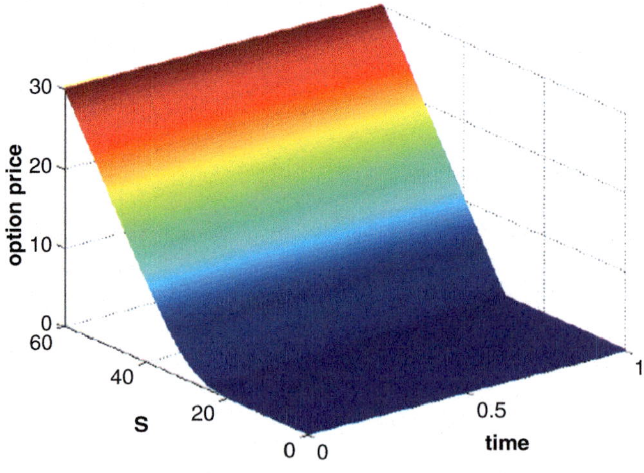

Fig. 19.3 Option price

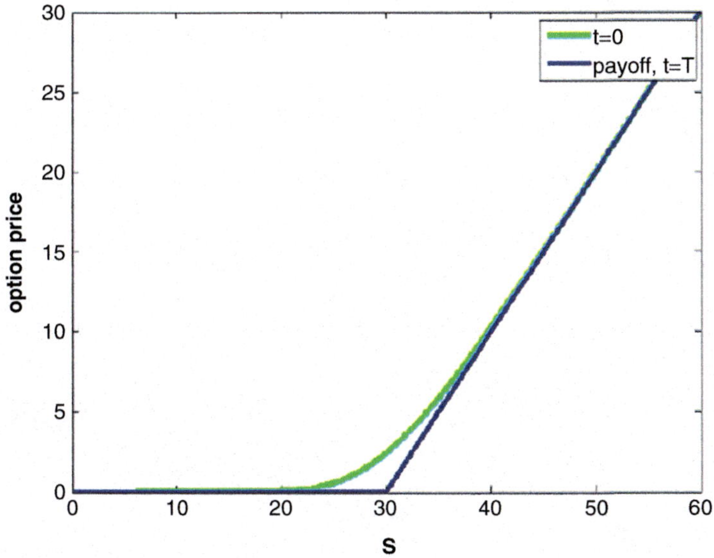

Fig. 19.4 Solution at time $t = 0$ and $t = T$

N	J	mesh ratio	error
3	50	0.23	0.2458
12	100	0.23	0.0855
50	200	0.23	0.0208
200	400	0.23	0.0052
800	800	0.23	0.0013

Fig. 19.5 Error as a difference between exact solution and approximation

ratio of errors	
error50/error100	2.87
error100/error200	4.11
error200/error400	4
error400/error800	4

Fig. 19.6 Ratio of errors

Figures 19.7, 19.8, 19.9, 19.10, and 19.11 show an error on different grids, as a difference between numerical solution and the exact one (from the BS formula). Table in Fig. 19.5 records the maximum value of the error from the time $t = 0$, it means that we observe the maximal value of the errors whole calculation in the current time. At the beginning of the calculation (nearby maturity time) we can observe the highest error, which is caused by the non-smooth initial data. This error decreases during the calculation. The finer the mesh, the faster the decrease of the error (19.7)–(19.11).

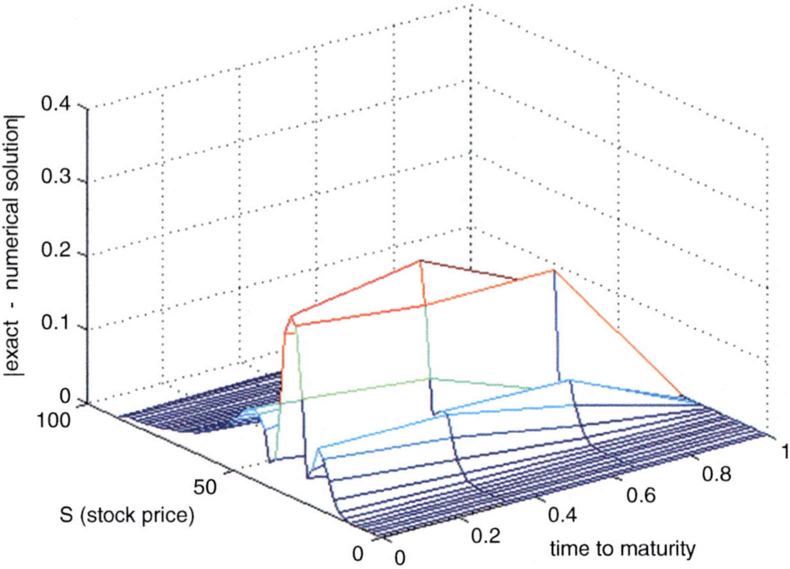

Fig. 19.7 Error, $N = 3, J = 50$

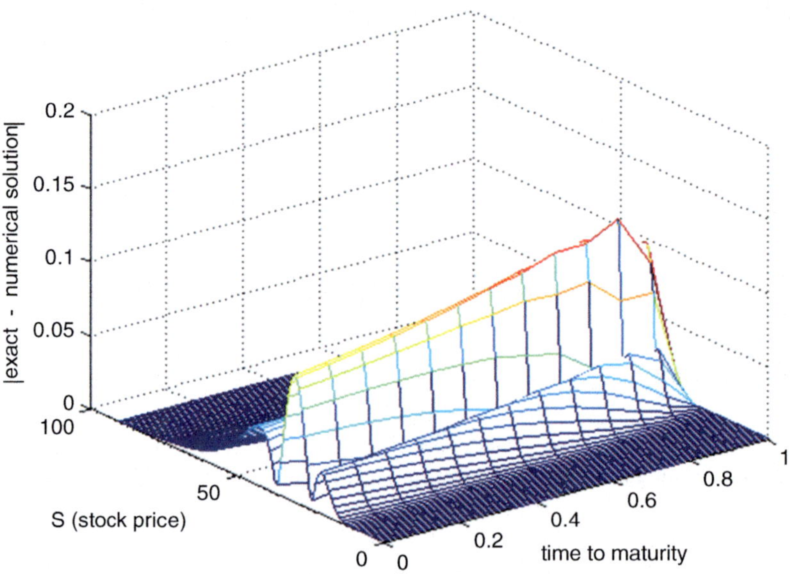

Fig. 19.8 Error, $N = 12, J = 100$

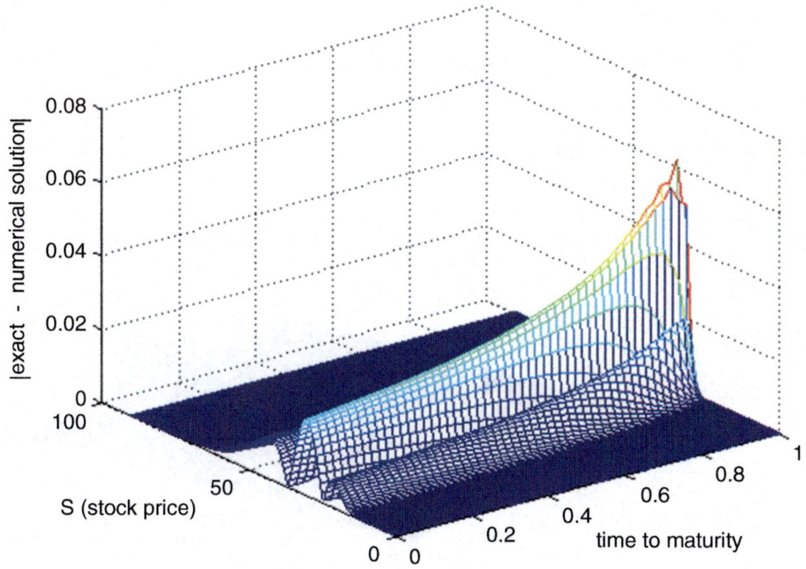

Fig. 19.9 Error, $N = 50, J = 200$

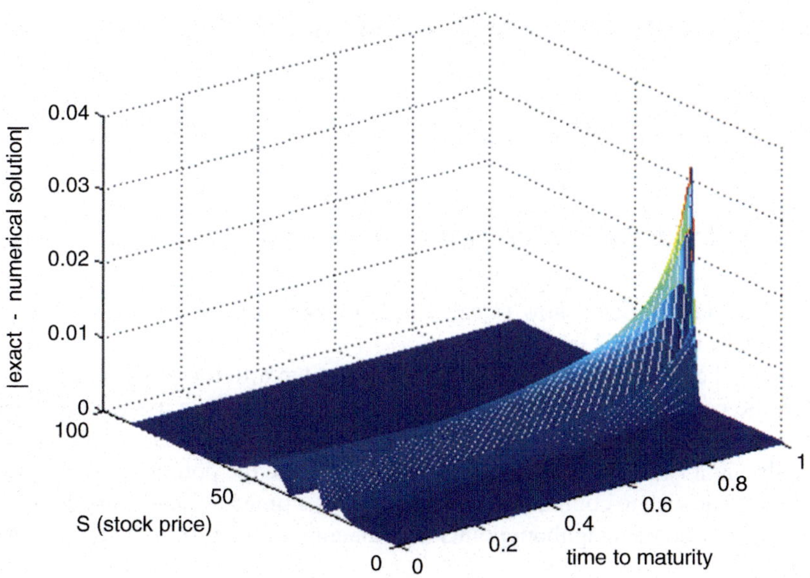

Fig. 19.10 Error, $N = 200, J = 400$

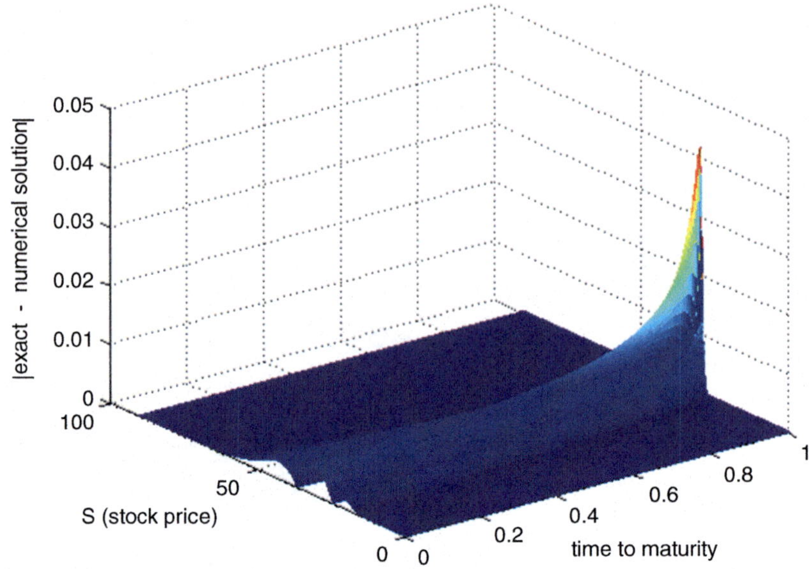

Fig. 19.11 Error, $N = 800, J = 800$

19.5 ADE Schemes for Multi-Dimensional Models

In this section we introduce the ADE scheme for multi-dimensional PDE models. We first consider the 2D case and then we proceed to higher dimensional cases.

19.5.1 ADE Schemes for Two-Dimensional Models

We now explain in detail how to construct the ADE scheme for two-dimensional PDE models, i.e. $N = 2$ in (19.2).

The first key aspect of this scheme is choosing the difference quotients approximating the partial derivatives of our equation in a way that we use the information from both time levels without the need to solve a linear system of equations. In particular, in a two dimensional setting, we would use the points as exemplified in Fig. 19.12: we wish to compute the value in black, at time level $n+1$, and we use the information from the neighbour points with an empty filling, from both time level n and $n + 1$.

The second key aspect is that in order to improve the accuracy of this scheme, for each time-level two different calculations of the grid points are done using different difference quotients, these are referred to as the *downward sweep* and the *upward sweep*. Then, the solution at that time level is taken as the average of both sweeps.

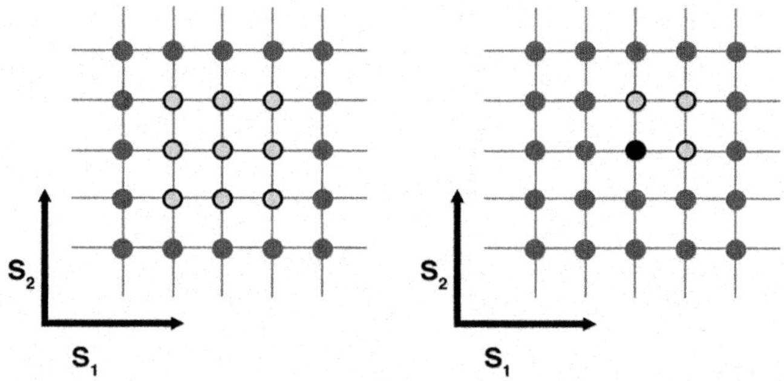

Fig. 19.12 Downward sweep. *Left figure*: time level n. *Right figure*: time level $n + 1$. We depict the spatial grid for two different time-steps. The *empty circles* represent the points used in the computation of the value at the location of the *black circle*. S_1 and S_2 denote the spatial dimensions

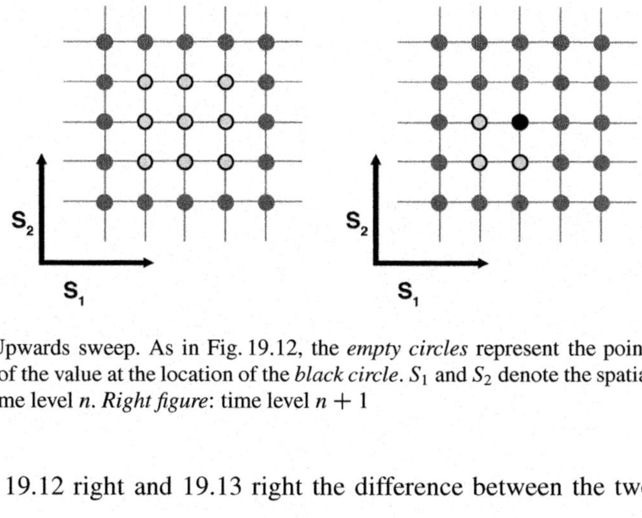

Fig. 19.13 Upwards sweep. As in Fig. 19.12, the *empty circles* represent the points used in the computation of the value at the location of the *black circle*. S_1 and S_2 denote the spatial dimensions. *Left figure*: time level n. *Right figure*: time level $n + 1$

From Figs. 19.12 right and 19.13 right the difference between the two sweeps is apparent.

The final key aspect is that the structure imposed by the stencil illustrated in Figs. 19.12 and 19.13 is not by itself enough to guarantee that the scheme is explicit, we must make sure that the empty filling points in the time level $n + 1$ have been computed before we compute the black point. This imposes a structure on the algorithm to compute the points as illustrated in Fig. 19.14.

For a fixed time level and starting from the boundary we see that in the first step we can only compute the points numbered as 1, since, our stencil is as described in the Figs. 19.12 right, 19.13 right. After computing these points we have a total of four points that can be now computed, these are numbered as 2. Hence, we chose any of those points which in turn allows new points to be computed, and so forth. As long as we respect this order, our algorithm is fully explicit.

As we can see in the second step, we have more than one possibility per step as to what point to compute, hence, there are different sequences of points. A natural choice is to choose the sequence of points as shown in Fig. 19.15. We called this

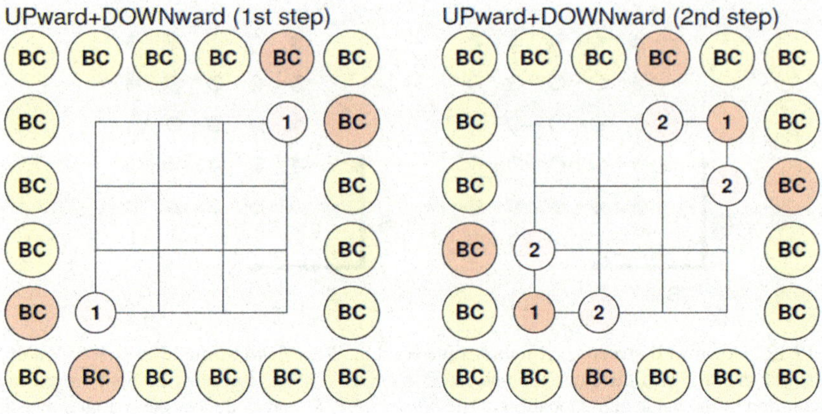

Fig. 19.14 First steps of the algorithm in the 2D case. Elements numbered 1 correspond to the step 1 from both sweeps UP and DOWN. Elements numbered 2 correspond to the elements that can be computed as the second step also for both sweeps

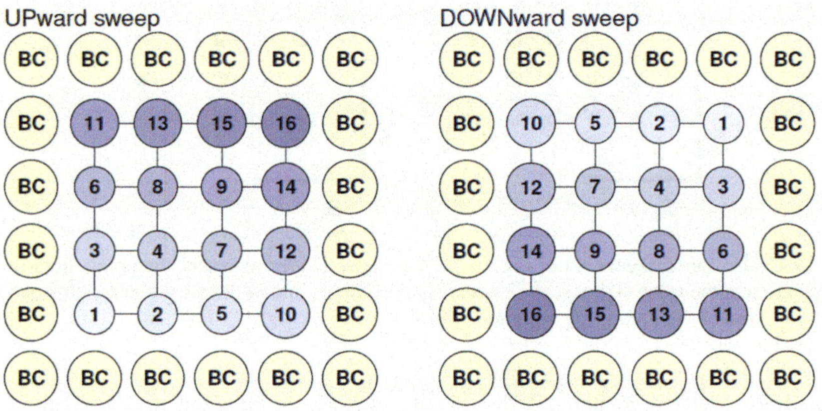

Fig. 19.15 The complete algorithm in the 2D case. The points are computed in the order of the numbering. The left part of the figure refers to the UP sweep and the right to the DOWN sweep. Approach of numbering is called jumping approach or house numbering

approach of numbering as a *jumping approach* or *house numbering approach*. We are moving from one corner of the square to another where diagonal points are computed and the others. We could do the same strategy in higher dimensions, but it is not straightforward and yields no advantage in comparison with the next approach. The approach we have implemented is a *row-wise ordering* and it is displayed in Fig. 19.16. It is just more straightforward way of ordering grid points. It is also more convenient to use this approach in hypercubes.

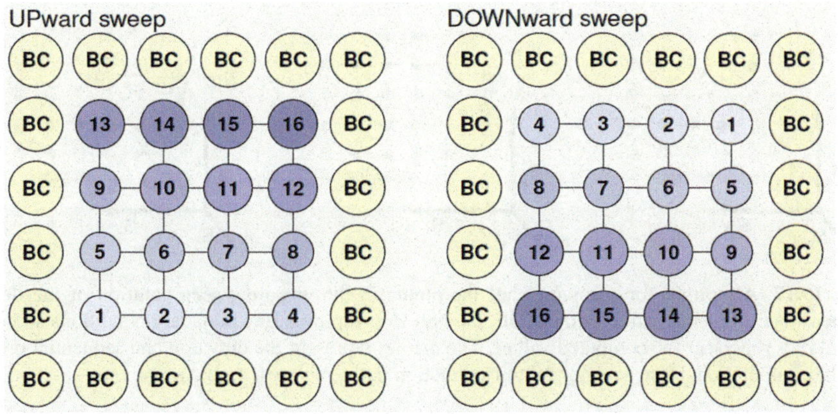

Fig. 19.16 The complete algorithm in the 2D case. The points are computed in the order of the numbering. The left part of the figure refers to the UP sweep and the right to the DOWN sweep. Approach of numbering is called line approach or sequence approach

19.5.2 ADE Schemes for Three and Higher Dimensional Models

In this section we describe how to extend the two-dimensional ADE scheme introduced before to three and higher dimensional models. We suggest an algorithm which can be extended to higher dimensional models quite easily.

As for the two dimensional case, a key part of the ADE in higher dimensions is to choose the proper difference quotients such that we keep good stability and consistency properties and explicitness of the scheme. Solely for simplicity we will use a uniform grid.

Consider the three dimensional case where we are solving the PDE of the price of an option under the linear Black-Scholes model introduced before, with three underlying assets. The PDE's solution will be a four-dimensional function where one dimension represents time and the other three are spatial dimensions representing the values of the underlying assets. For each time level, we have a three dimensional solution which can be illustrated as a three-dimensional grid. Recall that the initial condition is given for $V(S_1, S_2, S_3, 0)$ and step by step we calculate the values for the new time layer.

As before, we retain the explicitness of the scheme by using only values that have already been computed at the current time level. Specifically, this explicit (as in the lower dimensional case) is obtained by computing the value of points in a particular sequence that only uses points that either arise from the previous time level or that have been already computed for the current time level.

For illustration purposes we depict a two-dimensional slice of the domain in Fig. 19.17. We see that we move in a straight line in one dimension until we hit the boundary and then we proceed to the next point in the second dimension and

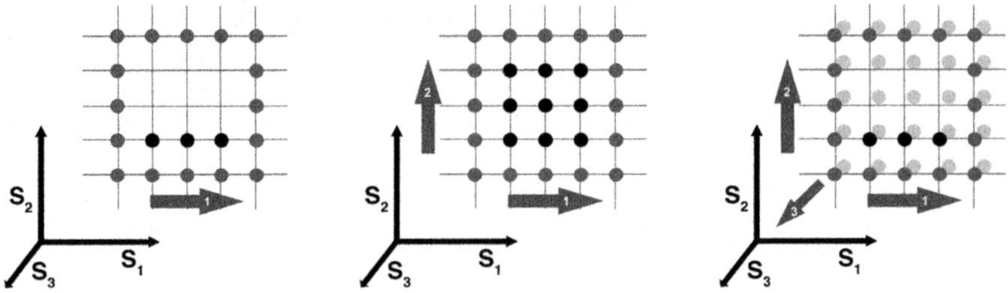

Fig. 19.17 Algorithm for computing all the points in the upward sweep solution of the three dimensional implementation of the ADE, the *grey dots* represent the boundary conditions and the *black dots* represent the computed values. The *arrows* represent the direction and sequential order of the computation. First, second and third direction in the pictures, respectively

so forth. By using this approach, the extension to higher dimensional models is straightforward.

19.5.3 Boundary Conditions

In higher dimensional models we also have to deal with the issue of boundary conditions. Just as in the three dimensional model 8 boundary conditions are required (each edge of the cube), for a N-dimensional model 2^N boundary conditions have to be prescribed. In an ideal case we prescribe values for the maximum values of the assets prices (truncated values) as Dirichlet boundary conditions. Alternatively one could also consider Neumann boundary conditions or Robin type boundary conditions.

19.6 The Numerical Scheme

The discretization of the PDE (19.2) is done on a uniform grid. In the time domain we have N_τ subintervals of the interval $[0, T]$, thus the time step size is defined as $d\tau = T/N_t$. As we have N different underlying assets our spatial space is N-dimensional. In our numerical studies we consider both $N = 2$ and $N = 3$.

For the 3-dimensional model we have 3 spatial intervals $[x_{min}, x_{max}]$, $[y_{min}, y_{max}]$, $[z_{min}, z_{max}]$, specifically $[0, S_{1max}]$, $[0, S_{2max}]$, $[0, S_{3max}]$ as all stocks have non-negative values.

The space steps on the uniform grid are defined by the following $h_\alpha = S_{\alpha max}/N_\alpha$ for $\alpha = 1, \ldots, 3$, where $S_{\alpha max}$ denotes the maximal value for the asset α and N_α denotes the number of points for the direction of the α asset.

A point on the spatial grid is then given by $[x_i, y_j, z_k]$ with $x_i = (i-1)h_1, y_j = (j-1)h_2, z_k = (k-1)h_3$; where $i = 1, \ldots, N_1+1, j = 1, \ldots, N_2+1, k = 1, \ldots, N_3+1$.

The discrete numerical solution of the 3-dimensional Black-Scholes equation at $[x_i, y_j, z_k]$ and time level n for the upward sweep is denoted by $u^n_{ijk} = u(x_i, y_j, z_k, n)$ and for the downward sweep is denoted by $u^n_{ijk} = d(x_i, y_j, z_k, n)$,

Since this notation would easily become very cumbersome we will introduce some abbreviations: $u(x_i, y_j, z_k, n)$ and $d(x_i, y_j, z_k, n)$ will be shortened to u^n and d^n. When we consider u at a point shifted from the point indexed by (i, j, k) we will introduce a subscript $u^n_{\beta+}$ where β denotes the direction where we're performing the shift. For example,

$$u(x_i, y_{j+1}, z_k, n) =: u^n_{2+} \qquad u(x_i, y_j, z_{k-1}, n) =: u^n_{3-}. \tag{19.49}$$

In the case that we have shifts in multiple directions we simply introduce another subscript, for example,

$$u(x_{i-1}, y_{j+1}, z_k, n) =: u^n_{1-2+} \qquad u(x_i, y_{j+1}, z_{k-1}, n) =: u^n_{3-2+}. \tag{19.50}$$

This notation would not be suitable if we denote a point such as $u(x_i, y_{j+3}, z_k, n)$ but since we are considering only a first-order scheme we will not have shifts of more than 1 unit, therefore this notation is appropriate.

19.6.1 Algorithm of the Scheme

We can construct the upward sweep and the downward sweep separately for each time step and then combine them, this bring opportunities for the parallelization of the scheme. The upward sweep is calculated in a way that we are moving from one corner of the hypercube to the opposite. The downward sweep is constructed in the opposite way. This procedure can be done in different ways, but it is important to keep the explicitness of the scheme in each of the sweeps. In the following we outline the algorithms. As an illustration the upward sweep of this algorithm is represented in Fig. 19.17.

According to the described procedure we construct upward and downward sweep of the solution and after each time level we calculated its average. This way we get final numerical solution c^n.

For $n = 0, 1, \ldots, N_t - 1$ we repeat

1. Initialization: $u^n = c^n; \quad d^n = c^n$
2. Upward: $u^{n+1}_{ijk}; \qquad i = 1, \ldots, N_1 - 1; j = 1, \ldots, N_2 - 1; k = 1, \ldots, N_3 - 1$
3. Downward: $d^{n+1}_{ijk}; \quad i = N_1 - 1, \ldots, 1; j = N_2 - 1, \ldots, 1; k = N_3 - 1, \ldots, 1$
4. $c^n = (u^{n+1} + d^{n+1})/2$

19.6.2 Upward Finite Difference Quotients and Its Numerical Scheme

Finite difference quotients using the upward sweep in the ADE scheme are introduced. Exact continuous solution of the PDE (19.2) in the point $x_i, y_j, z_k, \tau_{n+\frac{1}{2}}$ is denoted as: $V := V(x,y,z,\tau)|_{(x_i,y_j,z_k,\tau_{n+\frac{1}{2}})}$ and e.g. in the time level n it is denoted as: $V^n := V(x,y,z,\tau)|_{(x_i,y_j,z_k,\tau_n)}$, For derivatives it holds as follow: $\frac{\partial V}{\partial \tau} := \frac{\partial V(x,y,z,\tau)}{\partial \tau}|_{(x_i,y_j,z_k,\tau_{n+\frac{1}{2}})}$. Approximation of the V^n is denoted as u^n for an upward sweep.

$$V \simeq \frac{V^n + V^{n+1}}{2}. \tag{19.51}$$

For the time derivative the explicit Euler discretization is used:

$$\frac{\partial V}{\partial \tau} = \frac{V^{n+1} - V^n}{d\tau} + \mathcal{O}(\tau_n^2). \tag{19.52}$$

In the convection term we choose the Robert and Weiss approximation [11]

$$\frac{\partial V}{\partial S_\alpha} = \frac{V^n_{\alpha+} - V^n + V^{n+1} - V^{n+1}_{\alpha-}}{2h_\alpha} + \mathcal{O}(h_\alpha^2), \qquad \forall \alpha = 1,2,3 \tag{19.53}$$

and the diffusion term is approximated by a special kind of central difference,

$$\frac{\partial^2 V}{\partial S_\alpha^2} = \frac{V^n_{\alpha+} - V^n - V^{n+1} + V^{n+1}_{\alpha-}}{h_\alpha^2} + \mathcal{O}(h_\alpha^2), \qquad \forall \alpha = 1,2,3. \tag{19.54}$$

Note that in all the above mentioned difference quotients we use values from two time layers in the fashion that we can use all the values from the previous time layer, but due to the algorithm explained in Fig. 19.16 only known values from the current time layer are used to keep the explicitness of the algorithm.

We approximate mixed term derivatives in an explicit way, as well:

$$\frac{\partial^2 V}{\partial S_\alpha S_\beta} = \frac{V^n_{\alpha+\beta+} - V^n_{\alpha+\beta-} - V^n_{\alpha-\beta+} + V^n_{\alpha-\beta-}}{4h_\alpha h_\beta} + \mathcal{O}(h_\alpha^2 + h_\beta^2) \qquad \forall \alpha, \beta = 1,2,3. \tag{19.55}$$

We now use the difference quotients introduced above to discretize the 3-dimensional Black-Scholes PDE (19.2). Let us define,

$$\gamma_1^{ij}(x_1, x_2) \equiv \frac{dt}{2h_i h_j} \Gamma_{ij} S_i(x_1) S_j(x_2), \qquad \gamma_2^{i}(x_1) \equiv \frac{dt}{2h_i} r S_i(x_1),$$

with $S_i(p) = (p-1)h_i$. The discretized equation for the 3D model becomes,

$$u^{n+1} - u^n = \sum_{i=1}^{3} \gamma_1^{ii} \left[u_{i+}^n - u^n - u^{n+1} + u_{i-}^{n+1} \right] \tag{19.56}$$
$$+ \sum_{i=1}^{3} \sum_{j=1, i \neq j}^{3} \frac{\gamma_1^{ij}}{4} \left[u_{i+j+}^n - u_{i+j-}^n - u_{i-j+}^n + u_{i-j-}^n \right]$$
$$+ \sum_{i=1}^{3} \gamma_2^i \left[u_{i+}^n - u^n + u^{n+1} - u_{i-}^{n+1} \right] - r \frac{u^n + u^{n+1}}{2}$$

The resulting algorithm is fully explicit, if we follow the procedure illustrated in Fig. 19.14. From Eq. (19.56) we express u^{n+1} and we realize an explicit formula for the scheme.

19.6.3 Difference Quotients and Numerical Scheme for the Downward Sweep

Let V be the exact continuous solution in point $x_i, y_j, z_k, \tau_{n+\frac{1}{2}}$. The approximation of the V^n obtained by the downward sweep is d^n, where the following difference quotients are used:

$$V \simeq \frac{V^n + V^{n+1}}{2}, \tag{19.57}$$

$$\frac{\partial V}{\partial \tau} = \frac{V^{n+1} - V^n}{dt} + \mathcal{O}(\tau_n^2), \tag{19.58}$$

$$\frac{\partial V}{\partial S_\alpha} = \frac{V_{\alpha+}^{n+1} - V^n + V^{n+1} - V_{\alpha-}^n}{2h_\alpha} + \mathcal{O}(h_\alpha^2), \qquad \forall \alpha = 1, 2, 3, \tag{19.59}$$

$$\frac{\partial^2 V}{\partial S_\alpha^2} = \frac{V_{\alpha+}^{n+1} - V^n - V^{n+1} + V_{\alpha-}^n}{h_\alpha^2} + \mathcal{O}(h_\alpha^2), \qquad \forall \alpha = 1, 2, 3, \tag{19.60}$$

$$\frac{\partial^2 V}{\partial S_\alpha S_\beta} = \frac{V_{\alpha+\beta+}^n - V_{\alpha+\beta-}^n - V_{\alpha-\beta+}^n + V_{\alpha-\beta-}^n}{4h_\alpha h_\beta} + \mathcal{O}(h_\alpha^2 + h_\beta^2), \qquad \forall \alpha, \beta = 1, 2, 3. \tag{19.61}$$

In the same manner we get the discretized equation for the downward sweep,

$$d^{n+1} - d^n = \sum_{i=1}^{3} \gamma_1^{ii} \left[d_{i+}^{n+1} - d^n - d^{n+1} + d_{i-}^n \right] \tag{19.62}$$

$$+ \sum_{i=1}^{3} \sum_{j=1, i \neq j}^{3} \frac{\gamma_1^{ij}}{4} \left[d_{i+j+}^n - d_{i+j-}^n - d_{i-j+}^n + d_{i-j-}^n \right]$$

$$+ \sum_{i=1}^{3} \gamma_2^i \left[d_{i+}^{n+1} - d^n + d^{n+1} - d_{i-}^n \right] - r \frac{d^n + d^{n+1}}{2}.$$

19.7 Numerical Results and Experimental Study of Convergence

We now present numerical results for two particular cases of the implementation of the ADE scheme to Black-Scholes pricing models. In particular, we show the results for the price of a Spread option depending on two underlying assets S_1 and S_2 and a three-dimensional European Call Option on three underlying assets S_1, S_2 and S_3. For both cases we show illustrations of the obtained price surfaces and experimental convergence rates.

19.7.1 Two Dimensional Black-Scholes Model

We denote the Black-Scholes price for a spread option by $V(S_1, S_2, \tau)$ where $\tau = T - t$ is the time to maturity T. Recall that the payoff of a spread option is

$$V(S_1, S_2, 0) = \max(S_1 - S_2 - K, 0)$$

where $K \in \mathbb{R}^+$ denotes the strike price. The boundary conditions are given by:

$$V(S_1, 0, \tau) = BS_{1d}(S_1, \tau), \quad S_1, \tau \in \mathbb{R}^+,$$

$$V(0, S_2, \tau) = 0, \quad S_2, \tau \in \mathbb{R}^+,$$

$$V(S_1^{max}, S_2, \tau) = e^{-q_1 \tau} S_1 - e^{-r\tau}(S_2 + K), \quad S_1^{max} := S_1 \gg S_2 + K,$$

$$V(S_1, S_2^{max}, \tau) = V_{kirk}(S_1, S_2^{max}, \tau),$$

where $BS_{1d}(S_1, \tau)$ denotes the Black-Scholes price formula for a call option on a stock with price S and time to maturity τ and $V_{kirk}(S_1, S_2^{max}, \tau)$ denotes the approximation in [1].

We choose the different grid configurations displayed in Table 19.1 and we use following parameters: volatility of S_2 σ_2 is 0.1; volatility of S_1, σ_1 is 0.25; correlation of S_1 and S_2, ρ is -0.33; maturity time T is 1 year; strike price K is 3; maximal stock price for S_1 $S_{1\max}$ is 60; maximal stock price for S_2 $S_{2\max}$ is 225. As an example we display the numerical solution for the option price at $\tau = T$ (or equivalently $t = 0$) with a grid of $N_1 = N_2 = 20$ spatial points and $N_t = 50$ temporal points in Fig. 19.18.

In the Fig. 19.19 we display a log-log plot of the errors in the L_2 norm (blue line) and the theoretical second order of convergence (red line).

Table 19.1 Specifications of different grids

	N_1	N_2	N_t	$d\tau/h_1^2$	$d\tau/h_2^2$
Solution 1	5	5	3	0.0578	0.004
Solution 2	10	10	12	0.0578	0.004
Solution 3	20	20	50	0.0578	0.004
Solution 4	40	40	200	0.0578	0.004
Solution 5	80	80	800	0.0578	0.004
Solution 6	160	160	3200	0.0578	0.004

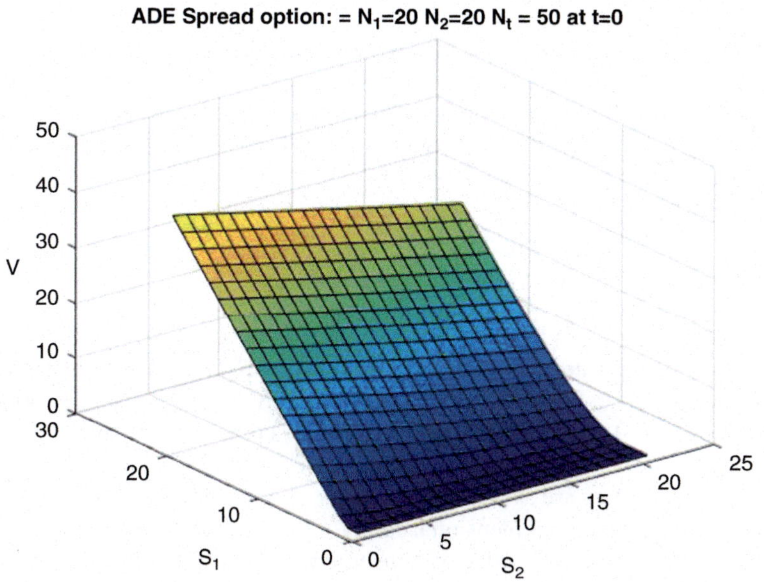

Fig. 19.18 Numerical solution at the final time for two dimensional spread option on the grid with $N_1 = N_2 = 20$ space steps and $N_t = 50$ time steps

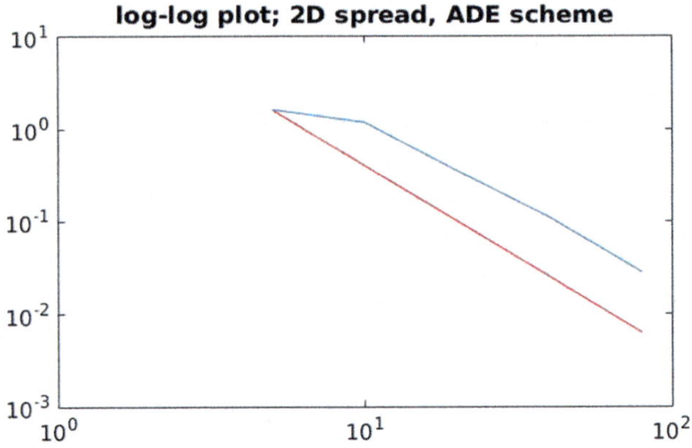

Fig. 19.19 Experimental convergence analysis

19.7.2 Three Dimensional Black-Scholes Model

We now show the results of the implementation of the ADE to the three dimensional Black-Scholes model for the price $V(S_1, S_2, S_3, \tau)$ of a call option, where $\tau = T - t$ denotes the time to maturity T and S_i denotes the value of the underlying asset i. Recall the payoff for a call option:

$$V(S_1, S_2, S_3, 0) = \max\left((\max(S_1, S_2, S_3) - K, 0)\right).$$

with $K \in \mathbb{R}^+$ denoting the strike price. The boundary conditions are taken from the numerical solution of the 2D Black-Scholes model, BS_{2d}, implemented as outlined in Sect. 19.7.1 but for a call-option payoff,

$$V(S_i = 0, t) = BS_{2d}(S_j, S_k, t), \quad i, j, k = 1, 2, 3, \quad i \neq j \neq k$$

$$V(S_i = S_i^{max}, t) = \max\left(S_i^{max} - K, 0\right), \quad i, j, k = 1, 2, 3.$$

The model parameters are following: volatility of S_1 σ_1 is 0.11; volatility of S_2 σ_2 is 0.5; volatility of S_3, σ_3 is 0.25; correlation of S_1 and S_2, ρ is 0.3; correlation of S_2 and S_3, ρ is -0.2; correlation of S_1 and S_3, ρ is 0.5; maturity time T is 1 year; strike price K is 30.

The grid parameters are as follows: $N_1 = N_2 = N_3 = 20$; $N_t = 50$.

Analogously to the two dimensional case, for the three dimensional case we have computed the experimental order of convergence using different grid settings cf. Table 19.2. Numerical results (Fig. 19.20) confirm that we keep second order of convergence also in the three dimensional model.

Table 19.2 Usage of different grids

N_1	N_2	N_3	N_t	$d\tau/h_1^2$	$d\tau/h_2^2$	$d\tau/h_3^2$
5	5	5	3	0.004	0.004	0.004
10	10	10	12	0.004	0.004	0.004
20	20	20	50	0.004	0.004	0.004
40	40	40	200	0.004	0.004	0.004
80	80	80	800	0.004	0.004	0.004
160	160	160	3200	0.004	0.004	0.004

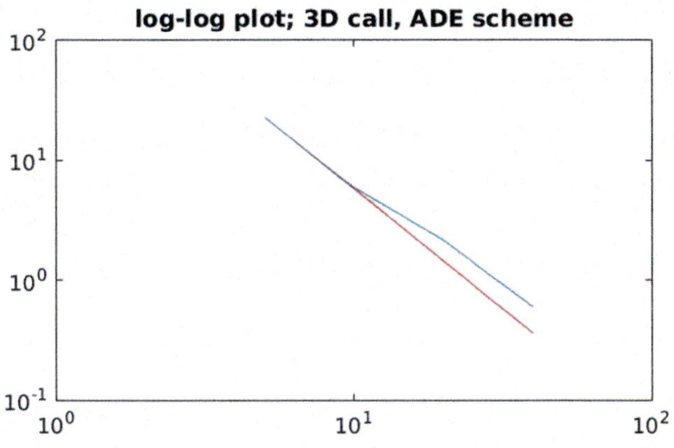

Fig. 19.20 Experimental order of convergence in three dimensional call option model

19.8 Influence of the Dimensionality on the Computational Complexity of the Scheme

In this section we highlight the fact, where the ADE scheme has a good potential to be an effective scheme in higher dimensions. We compare it with the behaviour of the classical Crank-Nicolson scheme.

The solution of the option price for the Crank-Nicolson (CN) scheme is implemented with a lot of optimization steps, so we do not compare the real time for the calculation. We focus on the fact observed in the Fig. 19.21 for the CN scheme is growing with dimension. It means for the same number of total points in a grid we need more time in the 3D model as in the 2D model. The explanation is coming from the construction of the scheme. Although for the same number of total points in a grid the size of the matrix is the same, but its structure is different. For 3D more non-diagonal terms are present and to compute solution in the implicit scheme is becoming costly for higher-dimensional models.

The costs for the ADE schemes in Fig. 19.22 for higher dimensions are not increasing, even opposite, since the calculation of the explicit scheme depends only on the total number of grid points and size of the stencil.

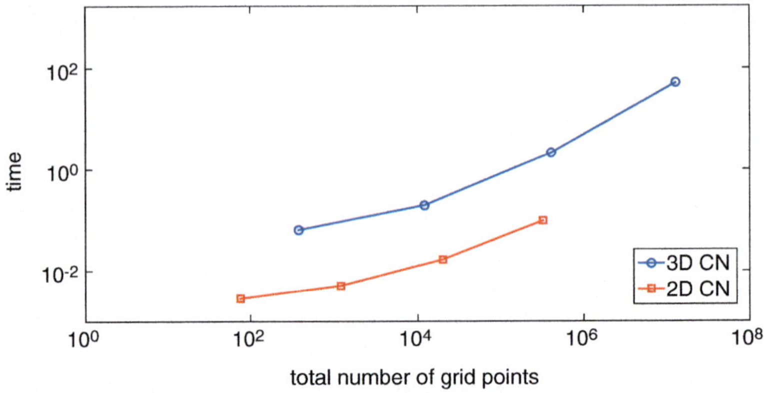

Fig. 19.21 Computational complexity with respect to the total number of points in the grid for Crank-Nicolson scheme

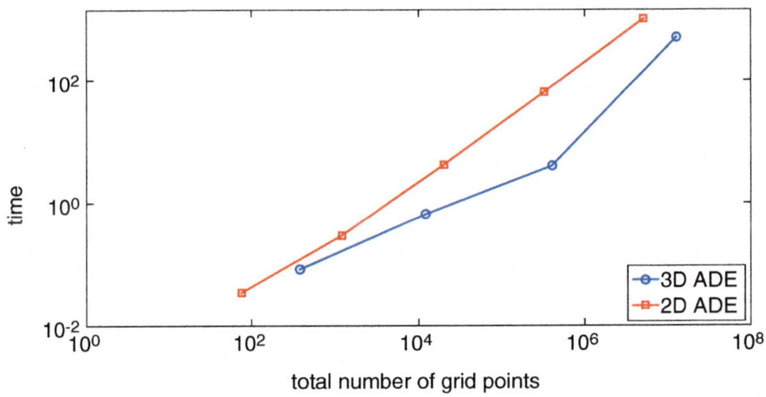

Fig. 19.22 Computational complexity with respect to the total number of points in the grid for ADE scheme

19.9 Application of the ADE Scheme to the Frey and Patie Model

We provide numerical experiments on the nonlinear Frey and Patie model [7], which considers the nonlinear Black-Scholes equation

$$v_t + \frac{\sigma^2 S^2}{2} \frac{v_{SS}}{(1 - \rho S v_{SS})^2} = 0. \tag{19.63}$$

This is a backward-in-time equation with a terminal condition, the so-called pay-off function. For the call option it is $\max(S - K, 0)$. To obtain an initial condition we apply the time reversal $\tau = T - t$, where τ is a new variable representing time remaining to the maturity:

$$-v_\tau + \frac{\sigma^2 S^2}{2} \frac{v_{SS}}{(1-\rho S v_{SS})^2} = 0, \tag{19.64}$$

with the initial condition: $v(S,0) = (S-K)^+$.

A uniform grid was used in time and in space. The time $\tau = 0, \ldots, T$ is divided into N sub-intervals, with the step length $k = \frac{T}{N}$ and time points $t^n = nk$, for $k = 0, \ldots, N$.

The space variable (stock price) $S \in (0, S_{max})$ is divided into M subintervals. The step size in space is $h = \frac{S_{max}}{M}$. The corresponding index to the stock price S_j at jh is j, for $j = 0, \ldots, M$.

We have chosen a boundary condition from the asymptotic behaviour of the exact analytical solution of the linear BS model. For small values of the stock price the option price $C(S,t)$ satisfies $C(0,t_n) = 0, \quad \forall n = 0, \ldots, N$. For $S \to S_{max}$ it is $C(S_{max}, t) = S_{max} - Ke^{-r\tau}$.

The scheme consists of two independent steps (sweeps). In the first step the upward sweeping is used. The particular solution at the grid point $(j, n+1)$ is U_j^{n+1} (marked in bold to highlight the unknown in the equation)

$$-\left(\frac{\mathbf{U_j^{n+1}} - C_j^n}{k}\right) + S_j^2 \frac{\sigma^2}{2} \frac{\frac{C_{j+1}^n - C_j^n - \mathbf{U_j^{n+1}} + U_{j-1}^{n+1}}{h^2}}{\left(1 - \rho S_j \frac{C_{j+1}^n - C_j^n - \mathbf{U_j^{n+1}} + U_{j-1}^{n+1}}{h^2}\right)^2} = 0. \tag{19.65}$$

The second step is downward sweeping. At the grid point $(j, n+1)$ the particular solution V_j^{n+1} is described by the following equation:

$$-\left(\frac{\mathbf{V_j^{n+1}} - C_j^n}{k}\right) + S_j^2 \frac{\sigma^2}{2} \frac{\frac{V_{j+1}^{n+1} - \mathbf{V_j^{n+1}} - C_j^n + C_{j-1}^n}{h^2}}{\left(1 - \rho S_j \frac{V_{j+1}^{n+1} - \mathbf{V_j^{n+1}} - C_j^n + C_{j-1}^n}{h^2}\right)^2} = 0. \tag{19.66}$$

After solving the Eqs. (19.65) and (19.66) we calculate the combination of the upward and downward sweeps as an average of it:

$$C_j^{n+1} = \frac{U_j^{n+1} + V_j^{n+1}}{2} \quad \text{for} \quad j = 1, \ldots M. \tag{19.67}$$

We repeat this procedure for each time n. The obtained solution C_j^n is the numerical approximation of the solution $v(S,t)$ of the Eq. (19.64) at each grid point (Table 19.3).

In Table 19.4 we can see the effect that with a finer mesh (in time and space) and constant mesh ratio the error decreases; here, by halving the time step size, with a factor close to 4, what is also the result from the theoretical (quadratic) order.

Table 19.3 Absolute error between numerical solution and reference solution on fine mesh

N	J	Mesh ratio	Error
3	50	0.23	0.1911
12	100	0.23	0.0987
48	200	0.23	0.0340
192	400	0.23	0.0094
768	800	0.23	0.0024

Table 19.4 Ratio of errors with different number of space steps

	Ratio of errors
Error50/error100	1.93
Error100/error200	2.89
Error200/error400	3.59
Error400/error800	3.88

$Error50$ corresponds to the absolute error with $J = 50$

19.10 Conclusions

We have considered the alternating direction explicit (ADE) method, that strongly uses boundary data in the solution algorithm and hence is very sensible to incorrect treatment of boundary conditions. We have implemented the ADE scheme for solving linear and nonlinear BS equations by reducing the nonlinearity to scalar equations. For linear equations the ADE scheme consists of two explicit sweeps. The sweeping procedure is done from one boundary to another and in the opposite way. The final solution is defined as an average of these two sweeps after each time step. Numerical analysis in the sense of studying stability and consistency of this method has been provided. For the linear heat equation with the constant coefficients we get unconditional stability using the matrix analysis approach. The consistency of the upward and downward sweeps is of order $O\left(k^2 + h^2 + \left(\frac{k}{h}\right)\right)$, and for the final (averaged) solution we obtain a consistency of order $O\left(k^2 + h^2\right)$.

The ADE can compete with the Crank-Nicolson scheme, ADI and LOD splitting methods. Applying the ADE method to linear models leads to an explicit scheme with unconditional stability. Applying ADE to the nonlinear model it does not lead to the explicit scheme any more. Now, at each sweep we need to solve a series of scalar nonlinear equations, but not any more a full nonlinear system of equations. Thus, the computational effort using the ADE instead of the full implicit scheme is highly reduced. For nonlinear cases we obtain only conditional stability. To our knowledge, the ADE scheme has not been applied to nonlinear PDEs before.

Alternating Direction Implicit methods and Splitting methods are examples of the Multiplicative Operator Scheme (MOS), which is difficult to parallelise. Methods from the family of the Additive Operator Scheme (AOS) can be parallelised. The ADE method belongs to this last group of methods.

Acknowledgements The authors were partially supported by the European Union in the FP7-PEOPLE-2012-ITN Program under Grant Agreement Number 304617 (FP7 Marie Curie Action, Project Multi-ITN STRIKE—Novel Methods in Computational Finance).

References

1. Alexander, C., Venkatramanan, A.: Analytic approximations for spread options. Available at SSRN 1012521, 2007
2. Barakat, H.Z., Clark, J.A.: On the solution of the diffusion equations by numerical methods. J. Heat Transf. **88**(4), 421–427 (1966)
3. Bučková, Z., Ehrhardt, M., Günther, M.: Alternating direction explicit methods for convection diffusion equations. Acta Math. Univ. Comen. **84**(2), 309–325 (2015)
4. Campbell, L.J., Yin, B.: On the stability of alternating-direction explicit methods for advection-diffusion equations. Numer. Methods Partial Differ. Equ. **23**(6), 1429–1444 (2007)
5. Duffy, D.J.: Finite Difference Methods in Financial Engineering: A Partial Differential Equation Approach. John Wiley & Sons, Chichester (2013)
6. Duffy, D.J., Andrea, G.: C# for Financial Markets. John Wiley & Sons, Chichester (2013)
7. Frey, R., Patie, P.: Risk management for derivatives in illiquid markets: a simulation study. In: Advances in Finance and Stochastics, pp. 137–159. Springer, Berlin/Heidelberg (2002)
8. Larkin, B.K.: Some stable explicit difference approximations to the diffusion equation. Math. Comput. **18**(86), 196–202 (1964)
9. Leung, S., Osher, S.: An alternating direction explicit (ADE) scheme for time-dependent evolution equations. Preprint UCLA June, 9:2005, 2005
10. Piacsek, S.A., Williams, G.P.: Conservation properties of convection difference schemes. J. Comput. Phys. **6**(3), 392–405 (1970)
11. Roberts, K.V., Weiss, N.O.: Convective difference schemes. Math. Comput. **20**, 272–299 (1966)
12. Saul'ev, V.K.: A method of numerical integration of diffusion equations. Dokl. Akad. Nauk SSSR **115**, 1077–1079 (1957)
13. Towler, B.F., Yang, R.Y.K.: Numerical stability of the classical and the modified Saul'yev's finite-difference methods. Comput. Chem. Eng. **2**(1), 45–51 (1978)
14. Vichnevetsky, R., Bowles, J.B.: Fourier Analysis of Numerical Approximations of Hyperbolic Equations. Society for Industrial & Applied Mathmatics, Philadelphia (1982)
15. Wilmott, P., Dewynne, J., Howison, S.: Option Pricing: Mathematical Models and Computation. Oxford financial press, Oxford (1993)
16. Zhao, W., Yang, X., Wu, L.: Alternating segment explicit-implicit and implicit-explicit parallel difference method for the nonlinear Leland equation. Adv. Differ. Equ. **2016**(1), 1–18 (2016)

Chapter 20
Numerical Study of Splitting Methods for American Option Valuation

Karel in 't Hout and Radoslav L. Valkov

Abstract This chapter deals with the numerical approximation of American-style option values governed by partial differential complementarity problems. For a variety of one- and two-asset American options we investigate by ample numerical experiments the temporal convergence behaviour of three modern splitting methods: the explicit payoff approach, the Ikonen-Toivanen approach and the Peaceman-Rachford method. In addition, the temporal accuracy of these splitting methods is compared to that of the penalty approach.

20.1 Introduction

American-style options are one of the most common instruments on the derivative markets and their valuation is of major interest to the financial industry. In this chapter we investigate by ample numerical experiments the accuracy and convergence of a collection of recent splitting methods that are employed in the numerical valuation of one- and two-asset American options.

Let $u(s_1, s_2, t)$ denote the fair value of a two-asset American-style option under the Black-Scholes framework if at t time units before the given maturity time T the underlying asset prices equal $s_1 \geq 0$ and $s_2 \geq 0$. Let $\phi(s_1, s_2)$ denote the payoff of the option and define the spatial differential operator

$$\mathscr{A} = \tfrac{1}{2}\sigma_1^2 s_1^2 \frac{\partial^2}{\partial s_1^2} + \rho\sigma_1\sigma_2 s_1 s_2 \frac{\partial^2}{\partial s_1 \partial s_2} + \tfrac{1}{2}\sigma_2^2 s_2^2 \frac{\partial^2}{\partial s_2^2} + r s_1 \frac{\partial}{\partial s_1} + r s_2 \frac{\partial}{\partial s_2} - r. \tag{20.1}$$

Here constant r is the risk-free interest rate, the positive constant σ_i denotes the volatility of the price of asset i for $i = 1, 2$ and the constant $\rho \in [-1, 1]$ stands for the correlation factor pertinent to the two underlying asset price processes.

K. in 't Hout (✉) • R.L. Valkov
Department of Mathematics and Computer Science, University of Antwerp, Middelheimlaan 1, B-2020 Antwerp, Belgium
e-mail: karel.inthout@uantwerpen.be; radoslav.valkov@uantwerpen.be

M. Ehrhardt et al. (eds.), *Novel Methods in Computational Finance*, Mathematics in Industry 25, DOI 10.1007/978-3-319-61282-9_20

It is well-known that the function u satisfies the partial differential complementarity problem (PDCP)

$$u \geq \phi, \quad \frac{\partial u}{\partial t} \geq \mathscr{A} u, \quad (u-\phi)\left(\frac{\partial u}{\partial t} - \mathscr{A} u\right) = 0, \tag{20.2}$$

valid pointwise for (s_1, s_2, t) with $s_1 > 0$, $s_2 > 0$, $0 < t \leq T$. The initial condition is prescribed by the payoff,

$$u(s_1, s_2, 0) = \phi(s_1, s_2) \tag{20.3}$$

for $s_1 \geq 0$, $s_2 \geq 0$ and boundary conditions are given by imposing (20.2) for $s_1 = 0$ and $s_2 = 0$, respectively. The three conditions in (20.2) naturally induce a decomposition of the (s_1, s_2, t)-domain: the early exercise region is the set of all points (s_1, s_2, t) where $u = \phi$ holds and the continuation region is the set of all points (s_1, s_2, t) where $\partial u / \partial t = \mathscr{A} u$ holds. The joint boundary of these two regions is referred to as the early exercise boundary or free boundary.

For most American-style options both the option value function and the early exercise boundary are unknown in (semi-)closed analytical form. Accordingly, one resorts to numerical methods for their approximation. Following the method of lines, one first discretizes the PDCP (20.2) in the spatial variables (s_1, s_2) and next discretizes in the temporal variable t. This leads to a linear complementarity problem (LCP) in each time step.

Various approaches have been proposed in the literature to handle these LCPs. In this chapter we study three modern splitting methods: the explicit payoff (EP) approach, the Ikonen-Toivanen (IT) approach and the Peaceman-Rachford (PR) method. The IT splitting approach was introduced in [12, 13] and has recently been combined in [5] with Alternating Direction Implicit (ADI) schemes for the temporal discretization. The PR method was proposed in [14].

At present the convergence theory pertinent to the IT splitting approach appears to be limited in the literature. The main goal of this chapter is to gain more insight into its convergence behaviour by numerically studying temporal discretization errors. In addition, all splitting approaches above are compared to the popular penalty (P) approach, which was introduced for American option valuation in [3, 17, 18]. An outline of our chapter is as follows.

In Sect. 20.2 a suitable spatial discretization of the PDCP (20.2) is formulated. In Sect. 20.3 the temporal discretization methods under consideration are described: the θ-EP method, the θ-IT method, the PR method, the θ-P method and three families of ADI-IT methods. Section 20.4 provides an illuminating interpretation of the IT approach and the PR method. Subsequently, an ample numerical study of the temporal discretization errors for all methods above is performed in Sect. 20.5, where five American-style options are considered. Conclusions are presented in Sect. 20.6.

20.2 Spatial Discretization

The numerical solution of the PDCP (20.2) commences with the discretization of the spatial differential operator $\mathscr{A}$ defined by (20.1). To this purpose, the unbounded spatial domain is first truncated to a square $(0, S_{\max}) \times (0, S_{\max})$ with given value $S_{\max}$ chosen sufficiently large. We prescribe homogeneous Neumann conditions at the far-field boundaries $s_1 = S_{\max}$ and $s_2 = S_{\max}$, consistent with the payoffs under consideration.

For the spatial discretization a suitable nonuniform Cartesian grid is taken,

$$(s_{1,i}\,,\, s_{2,j}) \quad \text{for } 0 \le i \le m_1,\ 0 \le j \le m_2,$$

where $0 = s_{k,0} < s_{k,1} < \ldots < s_{k,m_k} = S_{\max}$ is the mesh in the k-th spatial direction for $k = 1, 2$. The use of nonuniform grids, instead of uniform ones, can yield a substantial improvement in efficiency. We consider here the type of grid used in [4, 5]. Let $k \in \{1, 2\}$ and let $[S_{\text{left}}, S_{\text{right}}]$ be any given fixed subinterval of $[0, S_{\max}]$ that is of practical interest in the k-th direction. Let parameter $d > 0$ and let equidistant points $\xi_{\min} = \xi_0 < \xi_1 < \ldots < \xi_{m_k} = \xi_{\max}$ be given with

$$\xi_{\min} = \sinh^{-1}\left(\frac{-S_{\text{left}}}{d}\right),$$

$$\xi_{\text{int}} = \frac{S_{\text{right}} - S_{\text{left}}}{d},$$

$$\xi_{\max} = \xi_{\text{int}} + \sinh^{-1}\left(\frac{S_{\max} - S_{\text{right}}}{d}\right).$$

Note that $\xi_{\min} < 0 < \xi_{\text{int}} < \xi_{\max}$. The mesh $0 = s_{k,0} < s_{k,1} < \ldots < s_{k,m_k} = S_{\max}$ is then defined through the transformation

$$s_{k,l} = \varphi(\xi_l) \quad (0 \le l \le m_k),$$

where

$$\varphi(\xi) = \begin{cases} S_{\text{left}} + d\sinh(\xi) & (\text{for } \xi_{\min} \le \xi \le 0), \\ S_{\text{left}} + d\xi & (\text{for } 0 < \xi \le \xi_{\text{int}}), \\ S_{\text{right}} + d\sinh(\xi - \xi_{\text{int}}) & (\text{for } \xi_{\text{int}} < \xi \le \xi_{\max}). \end{cases}$$

By construction, this mesh is uniform inside the interval $[S_{\text{left}}, S_{\text{right}}]$ and nonuniform outside, where the mesh widths outside the interval are always larger than the mesh width inside. The parameter d controls the fraction of mesh points that lie inside. Let $\Delta\xi = \xi_1 - \xi_0$. It is readily shown that the above mesh is smooth, in the sense that there exist real constants $C_0, C_1, C_2 > 0$ such that the mesh widths $h_{k,l} = s_{k,l} - s_{k,l-1}$ satisfy

$$C_0\,\Delta\xi \le h_{k,l} \le C_1\,\Delta\xi \;\text{ and }\; |h_{k,l+1} - h_{k,l}| \le C_2\,(\Delta\xi)^2 \;\text{ uniformly in } l,\ m_k.$$

For our applications it turns out to be beneficial for accuracy if the given strike price $K > 0$ of an option is located exactly midway between two successive mesh points. This can be achieved with the above type of mesh as follows. Fix an interval $[S_{\text{left}}, S_{\text{right}}]$ such that K lies in the middle. Then $K = \varphi(\xi_{\text{int}}/2)$. Let $\nu \ge 1$ be any given integer and take

$$\Delta\xi = \frac{\xi_{\text{int}} - 2\,\xi_{\text{min}}}{\nu},$$

so that $\xi_{\text{min}} = \xi_0 < \xi_1 < \ldots < \xi_\nu = \xi_{\text{int}} - \xi_{\text{min}}$. It holds that $s_{k,\nu} = \varphi(\xi_{\text{int}} - \xi_{\text{min}}) = S_{\text{left}} + S_{\text{right}}$. The point $\xi_{\text{int}}/2$ is the middle of the interval $[\xi_{\text{min}}, \xi_{\text{int}} - \xi_{\text{min}}]$ and lies exactly midway between two successive ξ-mesh points whenever ν is odd. This implies that K lies exactly midway between two successive s_k-mesh points whenever ν is odd. Then let $m_k = m_k(\nu)$ be the smallest integer such that $m_k\Delta\xi \ge \xi_{\text{max}} - \xi_{\text{min}}$ and reset ξ_{max} to $\xi_{\text{min}} + m_k\Delta\xi$.

The discretization of the operator $\mathscr{A}$ is performed using finite differences. Let $f : \mathbb{R} \to \mathbb{R}$ be any given smooth function. Let $\{x_l\}_{l\in\mathbb{Z}}$ be any given increasing sequence of mesh points and $\Delta x_l = x_l - x_{l-1}$ for all l. For approximating the first and second derivatives of f we consider the following well-known finite difference formulas:

$$f'(x_l) \approx \alpha_0 f(x_l) + \alpha_1 f(x_{l+1}), \tag{20.4a}$$

$$f'(x_l) \approx \beta_{-1} f(x_{l-1}) + \beta_0 f(x_l) + \beta_1 f(x_{l+1}), \tag{20.4b}$$

$$f''(x_l) \approx \delta_{-1} f(x_{l-1}) + \delta_0 f(x_l) + \delta_1 f(x_{l+1}) \tag{20.4c}$$

with

$$\alpha_0 = \frac{-1}{\Delta x_{l+1}}, \qquad \alpha_1 = \frac{1}{\Delta x_{l+1}},$$

$$\beta_{-1} = \frac{-\Delta x_{l+1}}{\Delta x_l(\Delta x_l + \Delta x_{l+1})}, \qquad \beta_0 = \frac{\Delta x_{l+1} - \Delta x_l}{\Delta x_l \Delta x_{l+1}}, \qquad \beta_1 = \frac{\Delta x_l}{\Delta x_{l+1}(\Delta x_l + \Delta x_{l+1})},$$

$$\delta_{-1} = \frac{2}{\Delta x_l(\Delta x_l + \Delta x_{l+1})}, \qquad \delta_0 = \frac{-2}{\Delta x_l \Delta x_{l+1}}, \qquad \delta_1 = \frac{2}{\Delta x_{l+1}(\Delta x_l + \Delta x_{l+1})}.$$

The approximation (20.4a) is the first-order forward formula. The approximations (20.4b), (20.4c) are both central formulas that are second-order whenever the mesh is smooth.

For the discretization of the terms $\partial^2 u/\partial s_k^2$ ($k = 1, 2$) in $\mathscr{A}u$, formula (20.4c) is taken. For the terms $\partial u/\partial s_k$ ($k = 1, 2$) formula (20.4b) is applied at each mesh point

$s_{k,l}$ such that the corresponding coefficient

$$r s_{k,l} \beta_{-1} + \tfrac{1}{2}\sigma_k^2 s_{k,l}^2 \delta_{-1}$$

is nonnegative, otherwise formula (20.4a) is used. This mixed central/forward discretization of the convection terms is often employed in the literature. For the cross derivative term $\partial^2 u/\partial s_1 \partial s_2$ the finite difference formulas used for $\partial u/\partial s_1$, $\partial u/\partial s_2$ are successively applied. Concerning the boundary of the spatial domain, at $s_k = 0$ all spatial derivative terms involving the k-th direction vanish, so that this part of the boundary is trivially dealt with ($k = 1, 2$). At $s_k = S_{\max}$ the Neumann condition directly yields $\partial u/\partial s_k$ and $\partial^2 u/\partial s_k^2$ is approximated using a virtual point $s_{k,m_k+1} > S_{\max}$, where the value at this point is defined by linear extrapolation ($k = 1, 2$).

The given spatial discretization leads to a semidiscrete PDCP system

$$U(t) \geq U_0, \quad U'(t) \geq AU(t), \quad (U(t) - U_0)^{\mathrm{T}}(U'(t) - AU(t)) = 0 \tag{20.5}$$

for $0 < t \leq T$ with $U(0) = U_0$. Here $U(t)$ denotes the $M \times 1$ vector representing the semidiscrete approximation to the option value function $u(\cdot, \cdot, t)$ on the spatial grid, where $M = (m_1 + 1)(m_2 + 1)$. The $M \times M$ matrix A and the initial $M \times 1$ vector U_0 are given, where the latter represents the payoff function ϕ on the spatial grid. The vector inequalities are to be interpreted componentwise and the symbol ^{T} denotes taking the transpose.

Taking into account the selection of the finite difference formulas and the boundary conditions, it readily follows that $-A$ is always an M-matrix (see e.g. [2]) whenever the correlation $\rho = 0$. This feature is widely used in the computational finance literature to prove favourable properties of numerical methods. If $\rho \neq 0$, then $-A$ is usually not an M-matrix anymore when standard finite difference formulas for the mixed derivative are applied. Advanced techniques exist to overcome this, but in the present chapter we shall adhere to the standard discretization above.

20.3 Temporal Discretization

For the temporal discretization of the obtained semidiscrete PDCP system (20.5) we deal in Sect. 20.3.1 with the well-known family of θ-methods. Special instances of this are the Crank-Nicolson (CN) method for $\theta = \frac{1}{2}$, and the backward Euler (BE) method for $\theta = 1$. Next, in Sect. 20.3.2 three prominent families of Alternating Direction Implicit (ADI) schemes are considered.

20.3.1 θ-Methods

Let parameter $\theta > 0$. Let I denote the $M \times M$ identity matrix. Let $\Delta t = T/N$ with integer $N \geq 1$ be a given time step and let temporal grid points $t_n = n\Delta t$ for integers $0 \leq n \leq N$. The θ-method applied to (20.5) defines approximations $U_n \approx U(t_n)$ successively for $n = 1, 2, \ldots, N$ by

$$U_n \geq U_0, \tag{20.6a}$$

$$(I - \theta\Delta tA)U_n \geq (I + (1-\theta)\Delta tA)U_{n-1}, \tag{20.6b}$$

$$(U_n - U_0)^{\mathrm{T}}((I - \theta\Delta tA)U_n - (I + (1-\theta)\Delta tA)U_{n-1}) = 0. \tag{20.6c}$$

The fully discrete PDCP (20.6) forms a linear complementarity problem (LCP) for the vector U_n. Much attention has been paid in the literature to the solution of LCPs. We consider in the following several approximation approaches that are popular in the present time-dependent context.

The explicit payoff (EP) approach is arguably the most commonly used in financial practice. It yields the simple method (20.7), generating for $n = 1, 2, \ldots, N$ approximations $\widehat{U}_n \approx U(t_n)$.

θ-EP method:

$$(I - \theta\Delta tA)\bar{U}_n = (I + (1-\theta)\Delta tA)\widehat{U}_{n-1}, \tag{20.7a}$$

$$\widehat{U}_n = \max\{\bar{U}_n\,,\, U_0\} \tag{20.7b}$$

with $\widehat{U}_0 = U_0$ where the maximum of any two vectors is to be taken componentwise. Method (20.7) can be regarded as a fractional step splitting technique in which one first performs a time step by ignoring the American constraint and next applies this constraint explicitly, compare [1]. More precisely, the latter means projecting $\bar{U}_n$ onto the closed convex subspace of vectors $V \in \mathbb{R}^M$ satisfying $V \geq U_0$. The computational cost per time step of the θ-EP method is essentially the same as that in the case of the European counterpart of the option, which is very favourable.

The Ikonen-Toivanen (IT) operator splitting approach [12, 13] has the same computational cost. It yields the

θ-IT method:

$$(I - \theta\Delta tA)\bar{U}_n = (I + (1-\theta)\Delta tA)\widehat{U}_{n-1} + \Delta t\,\widehat{\lambda}_{n-1}, \tag{20.8a}$$

$$\begin{cases} \widehat{U}_n - \bar{U}_n - \Delta t\,(\widehat{\lambda}_n - \widehat{\lambda}_{n-1}) = 0, \\ \widehat{U}_n \geq U_0,\;\; \widehat{\lambda}_n \geq 0,\;\; (\widehat{U}_n - U_0)^{\mathrm{T}}\,\widehat{\lambda}_n = 0 \end{cases} \tag{20.8b}$$

with $\widehat{\lambda}_0 = 0$. The vector $\widehat{U}_n$ and the auxiliary vector $\widehat{\lambda}_n$ are computed in two stages. In the first stage, an intermediate approximation $\bar{U}_n$ is defined by the linear system (20.8a). In the second stage, $\bar{U}_n$ and $\widehat{\lambda}_{n-1}$ are updated to $\widehat{U}_n$ and $\widehat{\lambda}_n$ by (20.8b). It is easily seen that for these updates one has the simple formula

$$\begin{cases} \widehat{U}_n = \max\left\{\bar{U}_n - \Delta t\,\widehat{\lambda}_{n-1}\,,\; U_0\right\}, \\ \widehat{\lambda}_n = \max\left\{0\,,\; \widehat{\lambda}_{n-1} + (U_0 - \bar{U}_n)/\Delta t\right\}. \end{cases} \tag{20.9}$$

A related approach has been considered by Lions & Mercier [14], which was inspired by the original Peaceman-Rachford (PR) directional splitting scheme [15]. It can be formulated as the
PR method:

$$(I - \tfrac{1}{2}\Delta t A)\bar{U}_n = \widehat{U}_{n-1} + \tfrac{1}{2}\Delta t\,\widehat{\lambda}_{n-1}, \tag{20.10a}$$

$$\begin{cases} \widehat{U}_n = \max\left\{(I + \tfrac{1}{2}\Delta t A)\bar{U}_n\,,\; U_0\right\}, \\ \widehat{\lambda}_n = \max\left\{0\,,\; U_0 - (I + \tfrac{1}{2}\Delta t A)\bar{U}_n\right\}/(\tfrac{1}{2}\Delta t). \end{cases} \tag{20.10b}$$

A useful interpretation of the IT splitting approach and the PR method shall be given in Sect. 20.4.

Let *Large* > 0 be any fixed large number and integer $\kappa \geq 1$. The penalty approach has been proposed for American option valuation in [3, 17, 18]. It yields
θ-P method:

$$\begin{gathered} \left(I - \theta\Delta t A + P_n^{(k)}\right)\bar{U}_n^{(k+1)} = (I + (1-\theta)\Delta t A)\widehat{U}_{n-1} + P_n^{(k)} U_0 \\ \text{for } k = 0, 1, \ldots, \kappa - 1 \text{ and } \widehat{U}_n = \bar{U}_n^{(\kappa)}. \end{gathered} \tag{20.11}$$

This forms an iteration in each time step. Here $\bar{U}_n^{(0)} = \widehat{U}_{n-1}$ and $P_n^{(k)}$ (for $0 \leq k < \kappa$) is defined as the diagonal matrix with l-th diagonal entry equal to *Large* whenever $\bar{U}_{n,l}^{(k)} < U_{0,l}$ and zero otherwise. In each time step, κ linear systems have to be solved, involving different matrices. Accordingly, the penalty method is computationally more expensive per time step than the three foregoing methods. A natural convergence criterion is

$$\max_l \frac{|\bar{U}_{n,l}^{(k+1)} - \bar{U}_{n,l}^{(k)}|}{\max\{1, |\bar{U}_{n,l}^{(k+1)}|\}} < tol \quad \text{or} \quad P_n^{(k+1)} = P_n^{(k)}, \tag{20.12}$$

with given sufficiently small tolerance $tol > 0$. Let ϵ denote the machine precision of the computer. A rule of thumb[1] for the choice of penalty factor is then

$$Large \approx \alpha \frac{tol}{\epsilon} \quad \text{with } \alpha = 10^{-2}. \tag{20.13}$$

We have $\epsilon \approx 10^{-16}$ and choose $tol = 10^{-7}$ and $Large = 10^7$. In our applications, the average number of iterations κ per time step lies between 1 and 2.

20.3.2 ADI Schemes

ADI schemes are attractive for the temporal discretization of semidiscrete multidimensional PDEs as their computational cost per time step is directly proportional to the number of spatial grid points M from the semidiscretization, which is optimal. For these schemes, the matrix A is split into

$$A = A_0 + A_1 + A_2. \tag{20.14}$$

Here A_0 is the part of A that corresponds to the semidiscretization of the mixed derivative term. This matrix is nonzero whenever the correlation ρ is nonzero. Next, A_1 and A_2 are the parts of A that correspond to the semidiscretization of all spatial derivative terms in the s_1- and s_2-directions, respectively, and also contain an equal part of $-rI$. These two matrices are essentially tridiagonal (that is, up to a possible permutation).

In the literature on the numerical valuation of European-style options, three prominent families of ADI schemes have been considered: the Douglas (Do) scheme, the Modified Craig-Sneyd (MCS) scheme and the Hundsdorfer-Verwer (HV) scheme, compare e.g. [7]. In [5, 6] these schemes have been adapted for the numerical valuation of American-style options by combining them with the IT splitting approach, leading to so-called ADI-IT methods:
Do-IT method:

$$\begin{cases} Y_0 = (I + \Delta t A)\widehat{U}_{n-1} + \Delta t\,\widehat{\lambda}_{n-1}, \\ Y_j = Y_{j-1} + \theta \Delta t A_j \left(Y_j - \widehat{U}_{n-1}\right) \qquad (j = 1, 2), \\ \bar{U}_n = Y_2, \\ \\ \widehat{U}_n = \max\left\{\bar{U}_n - \Delta t\,\widehat{\lambda}_{n-1}\,,\, U_0\right\}, \\ \widehat{\lambda}_n = \max\left\{0\,,\, \widehat{\lambda}_{n-1} + (U_0 - \bar{U}_n)/\Delta t\right\}. \end{cases} \tag{20.15}$$

[1] Suggested to the authors by P.A. Forsyth.

MCS-IT method:

$$
\begin{cases}
Y_0 = (I + \Delta t A)\widehat{U}_{n-1} + \Delta t\, \widehat{\lambda}_{n-1}, \\
Y_j = Y_{j-1} + \theta \Delta t A_j \left(Y_j - \widehat{U}_{n-1}\right) & (j = 1, 2), \\
\widetilde{Y}_0 = Y_0 + \left(\theta \Delta t\, A_0 + (\tfrac{1}{2} - \theta)\Delta t A\right)\left(Y_2 - \widehat{U}_{n-1}\right), \\
\widetilde{Y}_j = \widetilde{Y}_{j-1} + \theta \Delta t A_j \left(\widetilde{Y}_j - \widehat{U}_{n-1}\right) & (j = 1, 2), \\
\bar{U}_n = \widetilde{Y}_2, \\
\\
\widehat{U}_n = \max\left\{\bar{U}_n - \Delta t\, \widehat{\lambda}_{n-1}\,,\, U_0\right\}, \\
\widehat{\lambda}_n = \max\left\{0\,,\, \widehat{\lambda}_{n-1} + (U_0 - \bar{U}_n)/\Delta t\right\}.
\end{cases}
\tag{20.16}
$$

HV-IT method:

$$
\begin{cases}
Y_0 = (I + \Delta t A)\widehat{U}_{n-1} + \Delta t\, \widehat{\lambda}_{n-1}, \\
Y_j = Y_{j-1} + \theta \Delta t A_j \left(Y_j - \widehat{U}_{n-1}\right) & (j = 1, 2), \\
\widetilde{Y}_0 = Y_0 + \tfrac{1}{2}\Delta t A \left(Y_2 - \widehat{U}_{n-1}\right), \\
\widetilde{Y}_j = \widetilde{Y}_{j-1} + \theta \Delta t A_j \left(\widetilde{Y}_j - Y_2\right) & (j = 1, 2), \\
\bar{U}_n = \widetilde{Y}_2, \\
\\
\widehat{U}_n = \max\left\{\bar{U}_n - \Delta t\, \widehat{\lambda}_{n-1}\,,\, U_0\right\}, \\
\widehat{\lambda}_n = \max\left\{0\,,\, \widehat{\lambda}_{n-1} + (U_0 - \bar{U}_n)/\Delta t\right\}.
\end{cases}
\tag{20.17}
$$

The Do-IT method constitutes the basic ADI-IT method. The MCS-IT and HV-IT methods form different extensions to this method, which require about twice the amount of computational work per time step.

In the ADI-IT methods above, the A_0 part is always treated in an explicit manner and the A_1 and A_2 parts in an implicit manner. In each time step, linear systems have to be solved with the two matrices $I - \theta \Delta t A_j$ for $j = 1, 2$. Since these matrices are both tridiagonal, the solution can be done very efficiently by computing once, upfront, their *LU* factorizations and then employ these in all time steps. It thus follows that the computational cost per time step of each ADI-IT method is directly proportional to the number of spatial grid points M, which is very favourable.

Concerning the underlying ADI schemes it holds that the MCS and HV schemes both have a classical order of consistency (that is, for fixed nonstiff ODE systems) equal to two for any value θ. We mention that the MCS scheme with $\theta = \frac{1}{2}$ is the so-called Craig-Sneyd (CS) scheme. For the Do scheme, if A_0 is nonzero, then the classical order of consistency is only equal to one. This lower order is due to the fact that in this scheme the A_0 part is treated in a simple, forward Euler fashion.

20.4 An Interpretation of the IT Approach and the PR Method

In this section we present an illuminating interpretation of the IT splitting approach and the PR method. It is obtained upon rewriting the semidiscrete PDCP (20.5) by means of an auxiliary variable $\lambda(t)$, often called a Lagrange multiplier:

$$U'(t) = AU(t) + \lambda(t), \tag{20.18a}$$

$$U(t) \geq U_0, \ \ \lambda(t) \geq 0, \ \ (U(t) - U_0)^{\mathrm{T}}\lambda(t) = 0. \tag{20.18b}$$

Suppose for the moment that $\lambda(t)$ is known and write the ODE system (20.18a) in splitted form as

$$U'(t) = F(t, U(t)) + G(t, U(t))$$

with

$$F(t, V) = AV \ \text{ and } \ G(t, V) = \lambda(t) \quad (0 \leq t \leq T, \ V \in \mathbb{R}^M).$$

Assume $\widehat{U}_{n-1} \approx U(t_{n-1})$ is given and consider $\widehat{U}_n \approx U(t_n)$ defined by

$$\begin{cases} Y_0 = \widehat{U}_{n-1} + \Delta t\, F(t_{n-1}, \widehat{U}_{n-1}) + \Delta t\, G(t_{n-1}, \widehat{U}_{n-1}), \\ Y \ = Y_0 + \theta_1 \Delta t \left(F(t_n, Y) - F(t_{n-1}, \widehat{U}_{n-1})\right), \\ Z \ = Y + \theta_2 \Delta t \left(G(t_n, Z) - G(t_{n-1}, \widehat{U}_{n-1})\right), \\ \widehat{U}_n = Z. \end{cases} \tag{20.19}$$

The above can be viewed as a *Douglas type splitting* scheme for (20.18a) involving two parameters θ_1, θ_2, compare e.g. [11]. Note that the splitting here is different from the directional splitting considered in Sect. 20.3.2. A simple relation holds between the scheme (20.19) and the θ-IT method (20.8): upon taking $\theta_1 = \theta$ and $\theta_2 = 1$, writing $Y = \bar{U}_n$ and replacing $\lambda(t_q)$ by an approximation $\widehat{\lambda}_q$ for $q \in \{n-1, n\}$, it is easily seen that (20.19) becomes (20.8a) together with the first line of (20.8b). The second line of (20.8b), which complements the θ-IT method, forms a discrete analogue of the complementarity condition (20.18b) at $t = t_n$. We mention that a related, operator-theoretic derivation was given in [14] if $\theta = 1$, where it was called the Douglas-Rachford scheme.

The above interpretation of the θ-IT method is directly extended to all ADI-IT methods (20.15), (20.16), (20.17) upon nesting into (20.19) the directional splitting of the function F induced by (20.14).

Consider next a *Peaceman-Rachford type splitting* scheme for (20.18a),

$$\begin{cases} Y &= \widehat{U}_{n-1} + \tfrac{1}{2}\Delta t\, F(t_{n-1/2}, Y) + \tfrac{1}{2}\Delta t\, G(t_{n-1}, \widehat{U}_{n-1}), \\ Z &= Y \quad + \tfrac{1}{2}\Delta t\, F(t_{n-1/2}, Y) + \tfrac{1}{2}\Delta t\, G(t_n, Z), \\ \widehat{U}_n &= Z. \end{cases} \tag{20.20}$$

Elaborating (20.20), and next replacing $\lambda(t_q)$ by an approximation $\widehat{\lambda}_q$ for $q \in \{n-1, n\}$, gives

$$\widehat{U}_n = (I + \tfrac{1}{2}\Delta t A)\bar{U}_n + \tfrac{1}{2}\Delta t\,\widehat{\lambda}_n \quad \text{with } \bar{U}_n \text{ defined by (20.10a).}$$

The discrete analogue of the complementarity condition (20.18b) at $t = t_n$ reads

$$\widehat{U}_n \geq U_0, \;\; \widehat{\lambda}_n \geq 0, \;\; (\widehat{U}_n - U_0)^{\mathrm{T}}\,\widehat{\lambda}_n = 0.$$

This is equivalent, for any given $\varepsilon > 0$, to

$$\widehat{U}_n - U_0 = \max\left\{0\,,\, \widehat{U}_n - U_0 - \varepsilon\widehat{\lambda}_n\right\} \;\text{ and }\; \widehat{\lambda}_n = \max\left\{0\,,\, \widehat{\lambda}_n - (\widehat{U}_n - U_0)/\varepsilon\right\}.$$

Selecting $\varepsilon = \frac{1}{2}\Delta t$ and inserting the above expression for $\widehat{U}_n$ yields (20.10b). Hence, the PR method (20.10) can be viewed as obtained from a Peaceman-Rachford type splitting scheme, with the comment that the pertinent splitting is not directional. This interpretation corresponds to the operator-theoretic exposition given in [14].

We remark that a natural variant to the θ-IT method is obtained by selecting $\theta_1 = \theta_2 = \theta$ in (20.19). This leads to (20.8) except that in the first line of the update (20.8b) the step size Δt is replaced by $\theta\Delta t$. Accordingly, the same replacement occurs in (20.9). As it turns out, for $\theta = \frac{1}{2}$ this variant of the θ-IT method is equivalent to the PR method.

20.5 Numerical Study

In the following we present extensive numerical experiments for the temporal discretization methods described in Sect. 20.3. Our main objectives are to study their actual convergence behaviour in the numerical solution of (20.5) and to assess their relative performance.

To this purpose, we study the *temporal discretization error* at $t = T = N\Delta t$, on a natural region of interest, defined by

$$\widehat{e}(\Delta t; m) = \max\{|U_l(T) - \widehat{U}_{N,l}| \,:\, 0 \leq i,j \leq m,\; \tfrac{1}{2}K < s_{1,i}\,,\, s_{2,j} < \tfrac{3}{2}K\}. \tag{20.21}$$

Here $U(T)$ represents the exact solution to the semidiscrete PDCP (20.5) for $t = T$ and $l = l(i,j)$ denotes the index such that the components $U_l(T)$ and $\widehat{U}_{N,l}$ correspond to the spatial grid point $(s_{1,i}, s_{2,j})$. In our experiments always the same number of mesh points in the two spatial directions is taken, $m_1 = m_2 = m$. For each given m, a reference solution for $U(T)$ is computed by applying the θ-P method with $\theta = \frac{1}{2}$ and $N = 10m$ time steps.

Clearly, (20.21) measures the temporal error in the maximum norm, which is the most relevant norm in financial practice. Note that the spatial discretization error is not contained in (20.21). We investigate here in detail the error due to the temporal discretization itself. This will lead to important new insights. We take the number of time steps N directly proportional to m, which forms the common situation in applications. The following methods are considered:

- BE-EP: (20.7) with $\theta = 1$
- BE-IT: (20.8) with $\theta = 1$
- BE-P: (20.11) with $\theta = 1$
- CN-EP: (20.7) with $\theta = 1/2$
- CN-IT: (20.8) with $\theta = 1/2$
- CN-P: (20.11) with $\theta = 1/2$
- PR: (20.10)

and

- Do-IT: (20.15) with $\theta = 1/2$
- CS-IT: (20.16) with $\theta = 1/2$
- MCS-IT: (20.16) with $\theta = 1/3$
- HV-IT: (20.17) with $\theta = 1/(2 + \sqrt{2})$

The selected values of θ for methods (20.15), (20.16), (20.17) are motivated by the favourable unconditional stability results obtained for the underlying ADI schemes in [8, 9].

It is well-known that in financial applications the payoff function ϕ is usually nonsmooth at one or more given points, which has an adverse effect on the accuracy of numerical solution methods. For the spatial discretization, this effect can be alleviated by constructing a spatial grid such that the coordinates of these points of nonsmoothness always lie exactly midway between two successive mesh points. Such a construction has been considered in Sect. 20.2. Subsequently, for the temporal discretization, a common approach is to apply backward Euler damping, also known as Rannacher time stepping. In the case of European options, this means that the first few time steps are all replaced by two substeps with step size $\Delta t/2$ of the backward Euler method. In analogy to this, we always replace each of the first two time steps of any of the θ-EP, θ-IT and θ-P methods by two substeps with step size $\Delta t/2$ of the same method using $\theta = 1$. Next, for damping the PR method and all ADI-IT methods, the θ-IT method is employed with $\theta = 1$.

20.5.1 One-Asset American Options

We begin with the special case of one-asset American options under the Black-Scholes framework. The pertinent (one-dimensional) spatial differential operator is

$$\mathscr{A} = \tfrac{1}{2}\sigma^2 s^2 \frac{\partial^2}{\partial s^2} + rs\frac{\partial}{\partial s} - r. \tag{20.22}$$

The spatial discretization is performed as in Sect. 20.2, where for the nonuniform mesh the following parameter values are taken,

$$d = K/3, \;\; S_{\text{left}} = 0.8K, \;\; S_{\text{right}} = 1.2K, \;\; S_{\max} = 5K. \tag{20.23}$$

As a first example we consider an American put option, which has payoff $\phi(s) = \max(K - s, 0)$ (for $s \geq 0$), and choose financial parameter values

$$r = 0.02, \;\; \sigma = 0.40, \;\; T = 0.5, \;\; K = 100. \tag{20.24}$$

Figure 20.1 displays, for all methods listed above except the ADI-IT methods, their temporal discretization errors $\widehat{e}(\Delta t; m)$ for $N = m$ and 20 different values m between

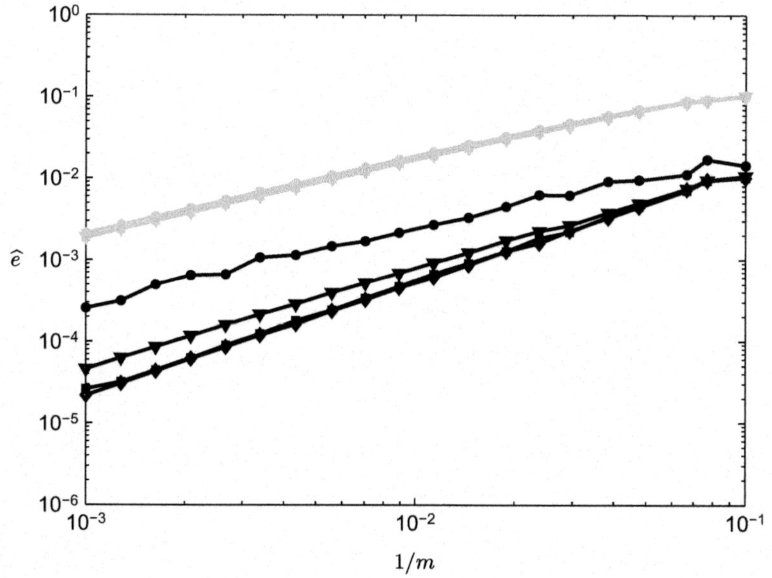

Fig. 20.1 American put option and parameters (20.24). Temporal error $\widehat{e}(\Delta t; m)$ versus $1/m$ with $N = m$ for $10 \leq m \leq 1000$. Constant step sizes. BE-EP: *light bullets*, BE-IT: *light squares*, BE-P: *light triangles*, CN-EP: *dark bullets*, CN-IT: *dark squares*, CN-P: *dark triangles*, PR: *dark diamonds*

10 and 1000, given by an equal number of appropriate odd values ν (see Sect. 20.2). One observes that the errors obtained with the three methods BE-EP, BE-IT, BE-P are very close to each other. They show a first-order convergence behaviour, as might be expected. The errors obtained with the four methods CN-EP, CN-IT, CN-P, PR are substantially smaller. Of these four, the CN-EP method is the least accurate. The errors for the CN-IT, CN-P, PR methods are relatively close to each other and a convergence order approximately equal to 1.3 is observed for these. Clearly, this order is significantly lower than two, which is attributed to the nonsmoothness of the option value function near the early exercise boundary, see e.g. [3].

We next choose the more challenging example of an American butterfly option, see [10]. It has the nonconvex payoff

$$\phi(s) = \max(s - K_1, 0) - 2\max(s - K, 0) + \max(s - K_2, 0)$$

with strikes $K_1 < K_2$ and $K = (K_1 + K_2)/2$. Figure 20.2 displays, analogously to the above, the temporal discretization errors for all methods under consideration with financial parameter values

$$r = 0.02, \ \ \sigma = 0.40, \ \ T = 0.5, \ \ K_1 = 80, \ \ K_2 = 120. \tag{20.25}$$

The BE-IT, BE-P, CN-IT, CN-P, PR methods reveal a neat first-order convergence behaviour. The explicit payoff methods, BE-EP and CN-EP, invariably yield large errors and appear to converge only very slowly as $N = m$ increases. We mention

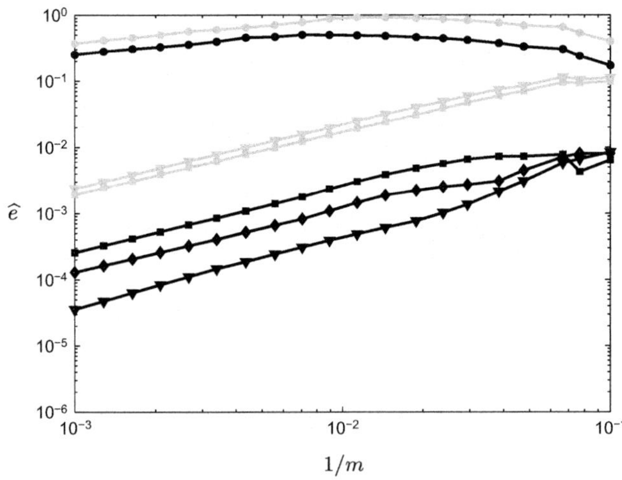

Fig. 20.2 American butterfly option and parameters (20.25). Temporal error $\widehat{e}(\Delta t; m)$ versus $1/m$ with $N = m$ for $10 \le m \le 1000$. Constant step sizes. BE-EP: *light bullets*, BE-IT: *light squares*, BE-P: *light triangles*, CN-EP: *dark bullets*, CN-IT: *dark squares*, CN-P: *dark triangles*, PR: *dark diamonds*

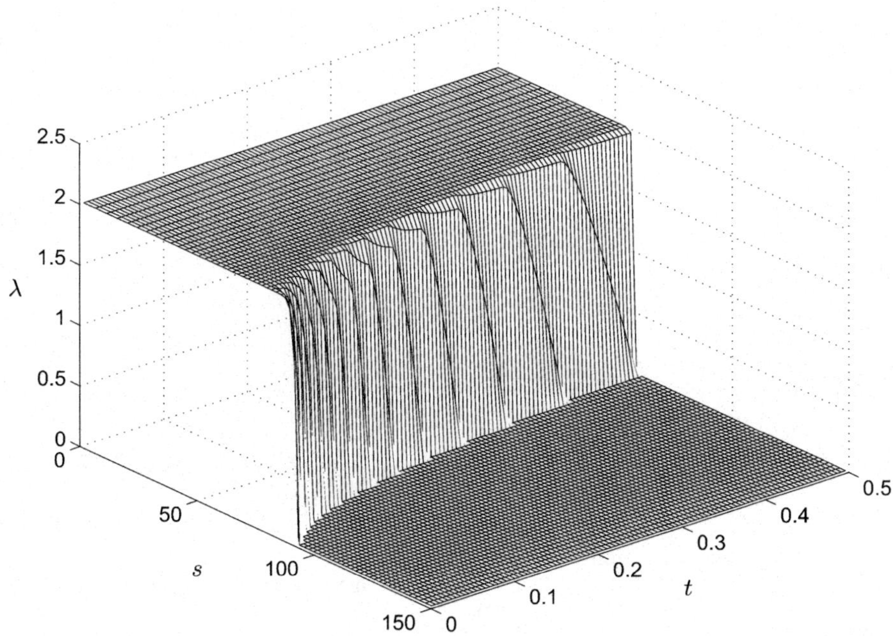

Fig. 20.3 American put option and parameters (20.24). Lagrange multipliers $\widehat{\lambda}_{n,i}$ versus $(s_{1,i}, t_n)$ in $[0, \frac{3}{2}K] \times (0, T]$ for BE-IT method and $m = 100$

that for the latter methods the temporal errors are largest near the strike K (which is always an early exercise point for the butterfly option).

Ample additional experiments in the case of one-asset American put and butterfly options support our above conclusions. The CN-based methods are in general more accurate than the BE-based methods, as could be expected. Further, it is found that the PR method is often somewhat more accurate than the CN-IT method.

Examining the Lagrange multiplier vectors $\widehat{\lambda}_n$ provides further useful insight. Figures 20.3 and 20.4 display for the American put and butterfly options, respectively, the Lagrange multiplier $\widehat{\lambda}_{n,i}$ versus $(s_{1,i}, t_n)$ in the domain $[0, \frac{3}{2}K] \times (0, T]$ for the BE-IT method and $m = 100$. The subdomain where the multiplier is nonzero represents the early exercise region. Clearly, for the butterfly option this region forms a narrow neighbourhood of the strike K. Replacing the BE-IT method by the CN-IT method or the PR method yields essentially the same outcome as in Figs. 20.3 and 20.4. Upon increasing m, the outcome for the American put remains approximately the same, but for the American butterfly the maximum grows, in a manner directly proportional to m. The latter phenomenon can be explained from the nonsmoothness (kink) of the exact butterfly option value function at the strike K at all times, which renders the numerical valuation of the American butterfly much more challenging than that of the American put.

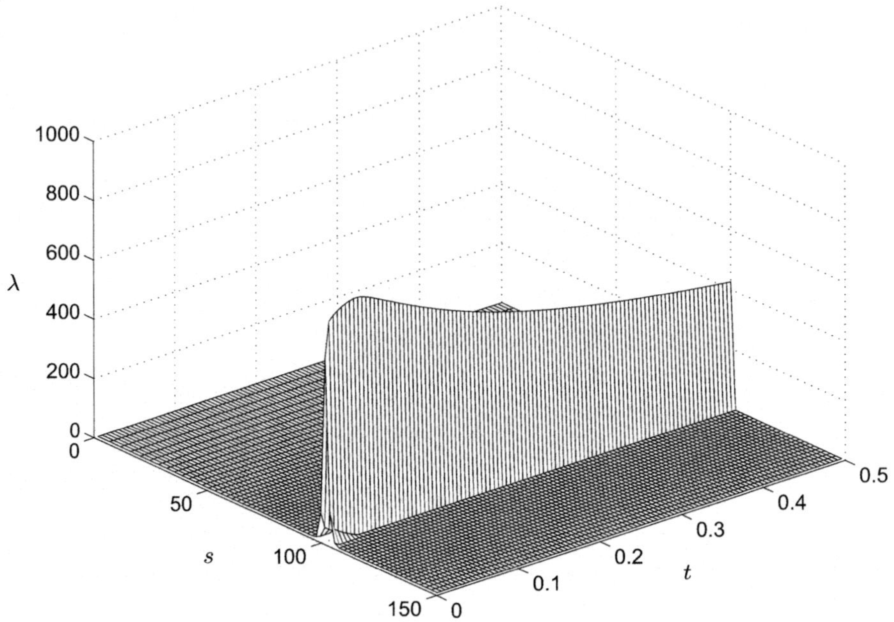

Fig. 20.4 American butterfly option and parameters (20.25). Lagrange multipliers $\widehat{\lambda}_{n,i}$ versus $(s_{1,i}, t_n)$ in $[0, \frac{3}{2}K] \times (0, T]$ for BE-IT method and $m = 100$

It was demonstrated in [3] that by employing suitable adaptive variable step sizes, instead of constant step sizes, one can recover second-order convergence for the CN-P method. All temporal discretization methods from Sect. 20.3 are extended straightforwardly to variable step sizes. We consider here temporal grid points defined upfront by (compare also e.g. [13, 16])

$$t_n = \left(\frac{n}{N}\right)^2 T \quad \text{for } n = 0, 1, 2, \ldots, N. \tag{20.26}$$

The corresponding step sizes are smallest near $t = 0$ (which is where the option value and early exercise boundary vary strongest) and they grow linearly with n. Figures 20.5 and 20.6 are the analogues of Figs. 20.1 and 20.2, respectively, obtained in the case of these variable step sizes. Indeed, for the CN-P method a favourable second-order convergence behaviour is observed. We note that relatively larger temporal errors can occur near the early exercise boundary in the case of the put option, resulting in the occasional "peaks" in Fig. 20.5, see also [3]. With variable step sizes, however, the CN-P method is in general substantially more accurate than all other methods under consideration. For the other methods, employing variable step sizes does not lead to a significant improvement in accuracy compared to constant step sizes. Because with the CN-P method the pertinent LCP in each time step is essentially solved exactly, we conclude that for the other CN-based methods

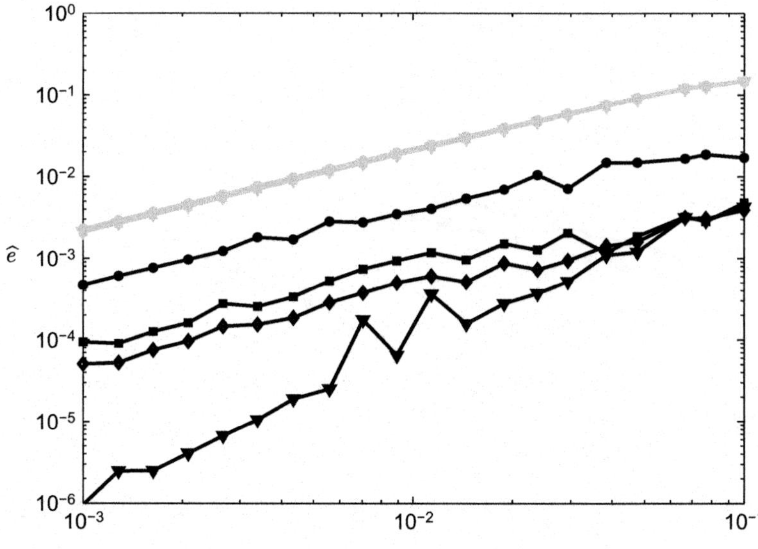

Fig. 20.5 American put option and parameters (20.24). Temporal error $\widehat{e}(\Delta t; m)$ versus $1/m$ with $N = m$ for $10 \leq m \leq 1000$. Variable step sizes. BE-EP: *light bullets*, BE-IT: *light squares*, BE-P: *light triangles*, CN-EP: *dark bullets*, CN-IT: *dark squares*, CN-P: *dark triangles*, PR: *dark diamonds*

(including PR) the error due to the approximate solution of the LCP in each time step dominates the error due to the CN time stepping; notice that the temporal discretization error (20.21) can be viewed as the sum of these two errors, since $U(T) - \widehat{U}_N = (U(T) - U_N) + (U_N - \widehat{U}_N)$ with U_N defined by (20.6).

20.5.2 Two-Asset American Options

We next consider numerical experiments for several two-asset American options. The spatial discretization of the PDCP (20.2) is done on the nonuniform grid described in Sect. 20.2 with parameter values (20.23). For the temporal discretization we apply all methods listed at the beginning of this section except those using the explicit payoff approach.

As a first example an American put option on the minimum of two asset prices is taken. It has payoff $\phi(s_1, s_2) = \max(K - s, 0)$ with $s = \min(s_1, s_2)$. We choose financial parameter values from [18],

$$r = 0.05, \ \ \sigma_1 = 0.30, \ \ \sigma_2 = 0.30, \ \ \rho = 0.50, \ \ T = 0.5, \ \ K = 40. \tag{20.27}$$

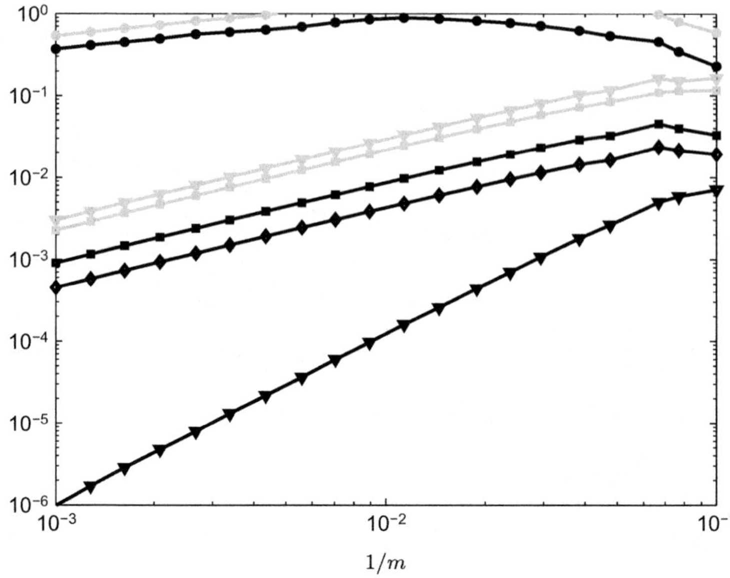

Fig. 20.6 American butterfly option and parameters (20.25). Temporal error $\widehat{e}(\Delta t; m)$ versus $1/m$ with $N = m$ for $10 \le m \le 1000$. Variable step sizes. BE-EP: *light bullets*, BE-IT: *light squares*, BE-P: *light triangles*, CN-EP: *dark bullets*, CN-IT: *dark squares*, CN-P: *dark triangles*, PR: *dark diamonds*

The numerically approximated early exercise region for $t = T$ is shown in Fig. 20.7. We compute the temporal discretization errors $\widehat{e}(\Delta t; m)$ for $N = m$ constant step sizes and 15 different values m between 10 and 200, corresponding to an equal number of odd values ν. Figure 20.9 displays the obtained results for the θ-based methods (Fig. 20.8 will be discussed on the following page). Similar to the one-asset American put option case, the errors for the BE-IT and BE-P methods are very close to each other and show an approximate first-order convergence behaviour. Also, as before, the errors for the CN-based methods are substantially smaller and relatively close to each other and show a convergence order approximately equal to 1.3. In Fig. 20.10 the results are displayed for the four ADI-IT methods under consideration. The obtained accuracies with the CS-IT, MCS-IT and HV-IT methods are about the same and close to those for the CN-based methods. The observed convergence orders for these three ADI-IT methods are thus also approximately equal to 1.3. The Do-IT method is substantially less accurate than the three more advanced ADI-IT methods, but it is somewhat more accurate than the BE-IT and BE-P methods. The observed convergence order for the Do-IT method is slightly smaller than one.

As a second example we consider an American put option on the arithmetic average of two asset prices, which has the payoff $\phi(s_1, s_2) = \max(K - s, 0)$ with $s = (s_1 + s_2)/2$. The numerically approximated early exercise region when $t = T$ is displayed in Fig. 20.8. Figures 20.11 and 20.12 form the analogues of Figs. 20.9 and 20.10, respectively, for this option. Comparing the achieved accuracies of the

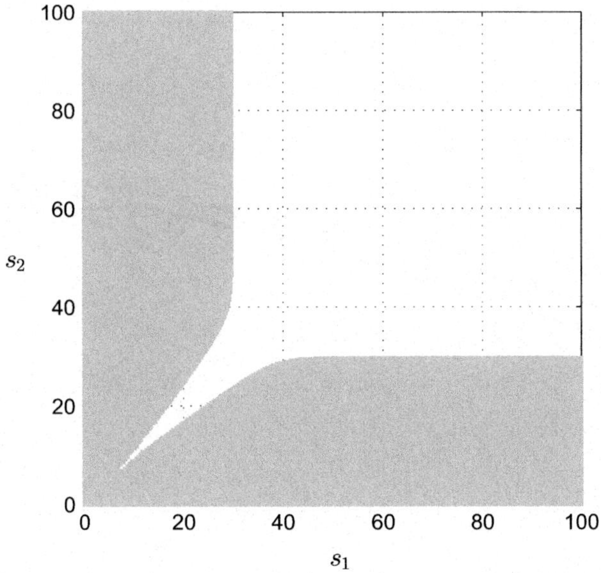

Fig. 20.7 Early exercise region if $t = T$ for two-asset American put on the minimum and parameters (20.27)

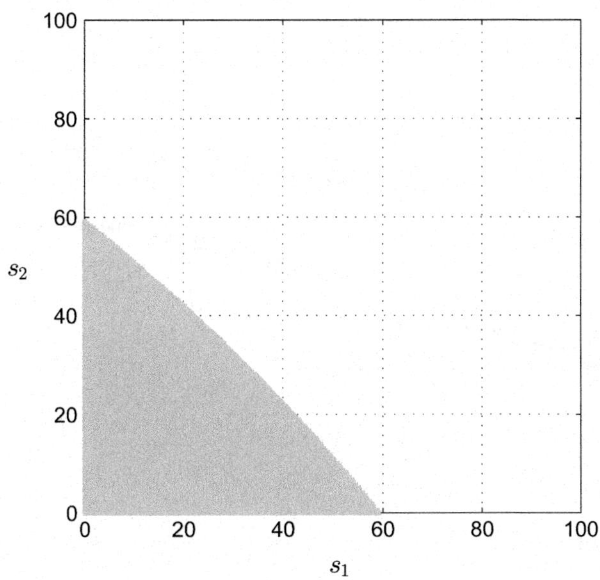

Fig. 20.8 Early exercise region if $t = T$ for two-asset American put on the average and parameters (20.27)

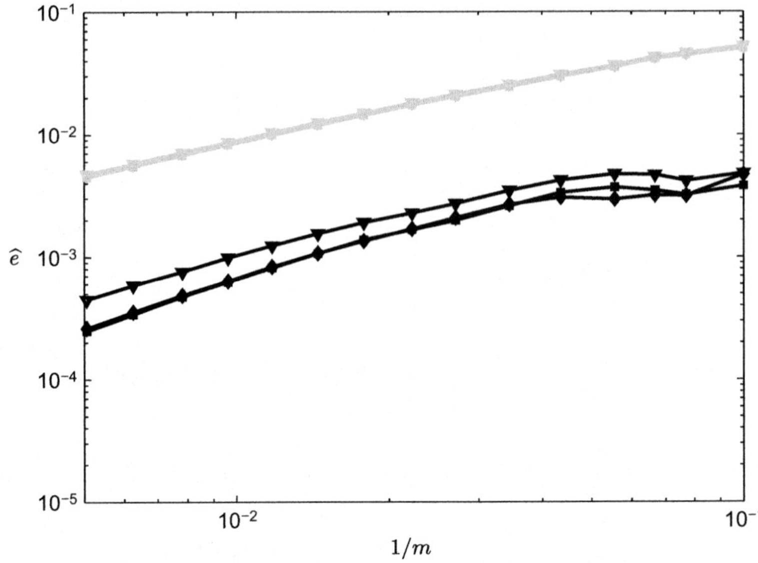

Fig. 20.9 Two-asset American put on the minimum and parameters (20.27). Temporal error $\widehat{e}(\Delta t; m)$ versus $1/m$ with $N = m$ for $10 \leq m \leq 200$. Constant step sizes. BE-IT: *light squares*, BE-P: *light triangles*, CN-IT: *dark squares*, CN-P: *dark triangles*, PR: *dark diamonds*

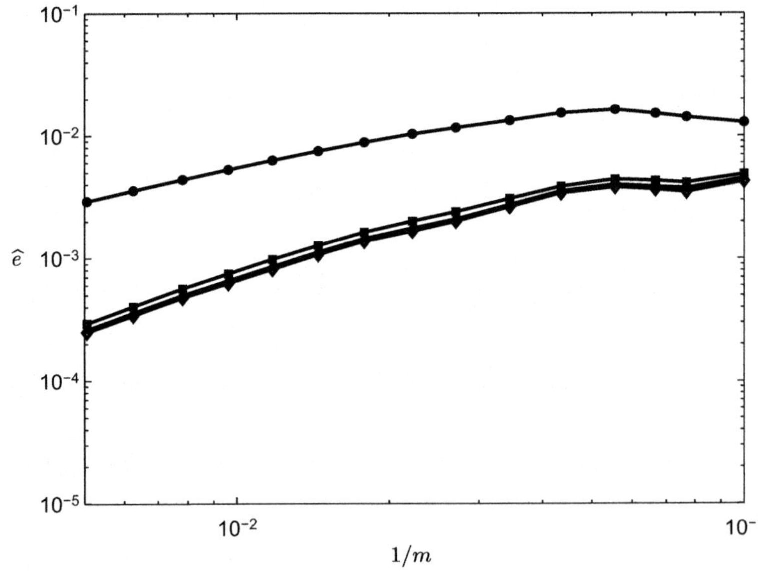

Fig. 20.10 Two-asset American put on the minimum and parameters (20.27). Temporal error $\widehat{e}(\Delta t; m)$ versus $1/m$ with $N = m$ for $10 \leq m \leq 200$. Constant step sizes. Do-IT: *dark bullets*, CS-IT: *dark squares*, MCS-IT: *dark stars*, HV-IT: *dark diamonds*

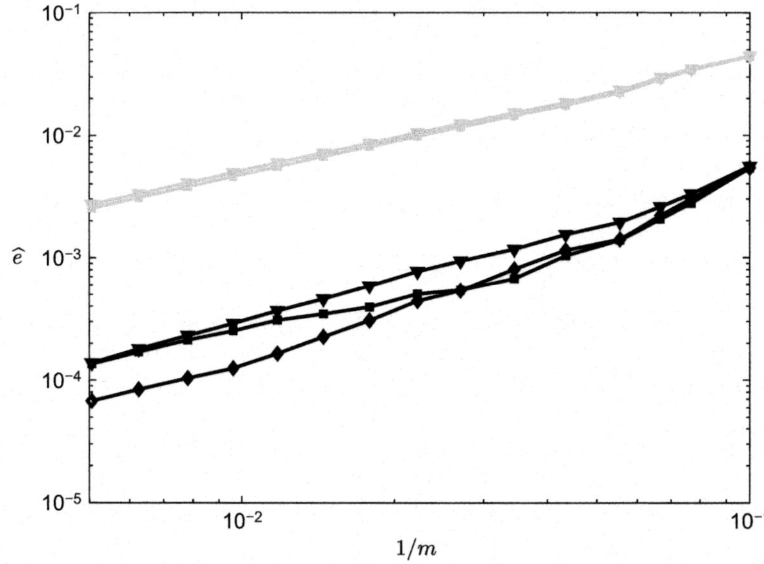

Fig. 20.11 Two-asset American put on the average and parameters (20.27). Temporal error $\widehat{e}(\Delta t; m)$ versus $1/m$ with $N = m$ for $10 \leq m \leq 200$. Constant step sizes. BE-IT: *light squares*, BE-P: *light triangles*, CN-IT: *dark squares*, CN-P: *dark triangles*, PR: *dark diamonds*

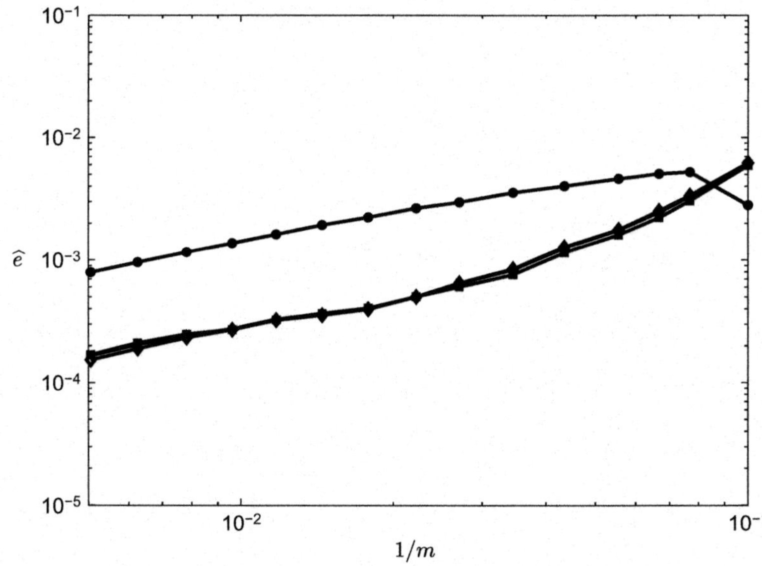

Fig. 20.12 Two-asset American put on the average and parameters (20.27). Temporal error $\widehat{e}(\Delta t; m)$ versus $1/m$ with $N = m$ for $10 \leq m \leq 200$. Constant step sizes. Do-IT: *dark bullets*, CS-IT: *dark squares*, MCS-IT: *dark stars*, HV-IT: *dark diamonds*

different methods, the same conclusions are obtained as in the case of the put option on the minimum, with the exception of the PR method, which is often somewhat more accurate than the other methods. The observed convergence orders for the CN-P and PR methods are approximately equal to 1.1 and 1.3, respectively, and for all other methods they are slightly smaller than one.

As a third example we select an American butterfly option on the maximum of two asset prices with payoff

$$\phi(s_1, s_2) = \max(s - K_1, 0) - 2\max(s - K, 0) + \max(s - K_2, 0)$$

with $s = \max(s_1, s_2)$ and $K = (K_1 + K_2)/2$. For this option the early exercise region encompasses all points (s_1, K) and (K, s_2) with $0 \leq s_1, s_2 \leq K$. We choose financial parameter values

$$r = 0.05,\ \ \sigma_1 = \sigma_2 = 0.30,\ \ \rho = 0.50,\ \ T = 0.5,\ \ K_1 = 32,\ \ K_2 = 48. \tag{20.28}$$

The obtained temporal errors for the θ-based methods and the ADI-IT methods are displayed in Figs. 20.13 and 20.14, respectively. The outcome is quite distinct from, and less favourable than, that in all foregoing (one- and two-asset) American option examples. For the BE-IT, BE-P and CN-P methods a neat first-order convergence is observed. Of these, the CN-P method is by far the most accurate. For all other

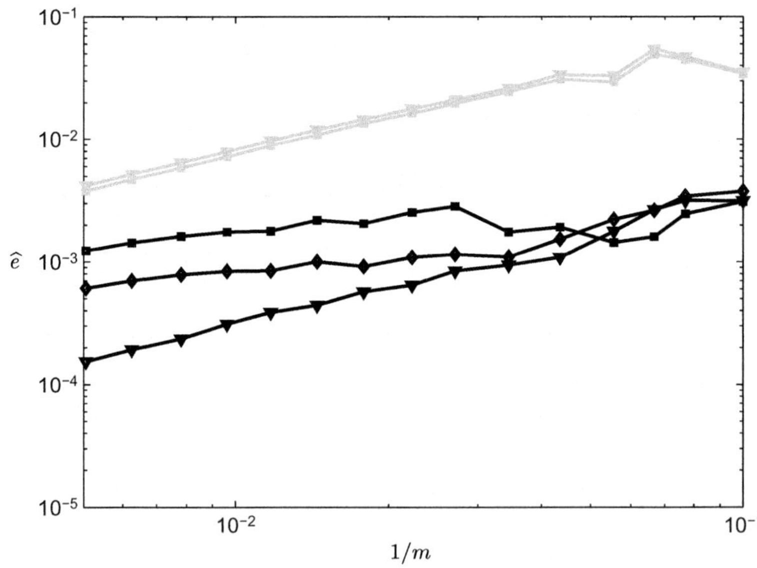

Fig. 20.13 Two-asset American butterfly and parameters (20.28). Temporal error $\widehat{e}(\Delta t; m)$ versus $1/m$ with $N = m$ for $10 \leq m \leq 200$. Constant step sizes. BE-IT: *light squares*, BE-P: *light triangles*, CN-IT: *dark squares*, CN-P: *dark triangles*, PR: *dark diamonds*

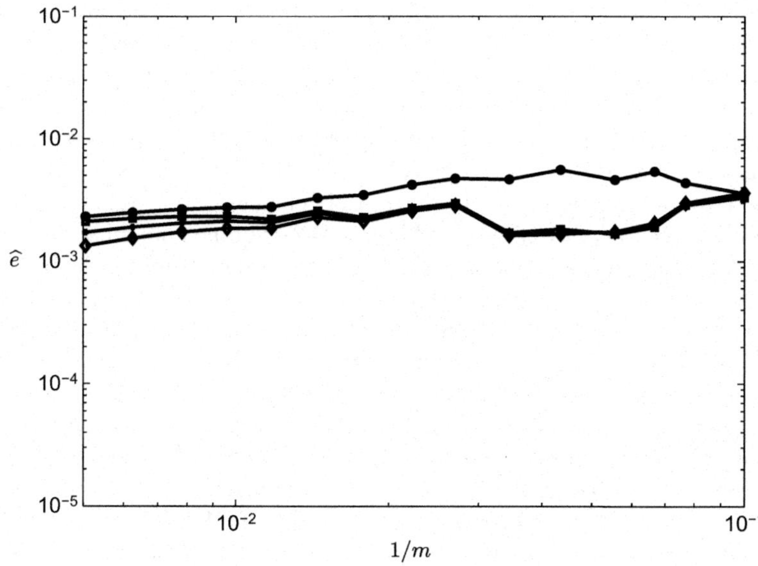

Fig. 20.14 Two-asset American butterfly and parameters (20.28). Temporal error $\widehat{e}(\Delta t; m)$ versus $1/m$ with $N = m$ for $10 \leq m \leq 200$. Constant step sizes. Do-IT: *dark bullets*, CS-IT: *dark squares*, MCS-IT: *dark stars*, HV-IT: *dark diamonds*

methods the converge behaviour is unclear. We note that setting the correlation $\rho = 0$, or computing the reference solution by the PR method instead of the CN-P method, does not change this conclusion. Subsequent experiments for the two-asset American butterfly option up to the value $m = 500$ suggest that for the CN-IT and PR methods there is convergence in $m = N$ of order 0.5. The converge behaviour of the ADI-IT methods is difficult to assimilate in this example and further research is required.

Employing the temporal grid points (20.26), corresponding to variable step sizes, leads in general again to a substantial improvement in accuracy for the CN-P method. In the case of the two-asset butterfly option, a smooth second-order convergence behaviour is observed. For the two-asset put options above, such a favourable result is obtained when the region of interest for the temporal error (20.21) does not intersect with the early exercise boundary.

20.6 Conclusions

In this chapter an ample numerical study has been performed for a collection of contemporary temporal discretization methods for PDCPs modelling the fair values of one- and two-asset American-style options. To this purpose, a detailed numerical investigation has been carried out of the temporal discretization error (20.21). Here the maximum norm is considered and the number of time steps N has been taken directly proportional to the number of mesh points m in each spatial direction.

Five American-style options are chosen for the numerical experiments: the one-asset put, the one-asset butterfly, the two-asset put on the minimum, the two-asset put on the arithmetic average, and the two-asset butterfly on the maximum. For the temporal discretization, the backward Euler (BE) and Crank-Nicolson (CN) methods are selected together with three ADI schemes: Douglas (Do), Modified Craig-Sneyd (MCS) and Hundsdorfer-Verwer (HV). For the numerical treatment of the LCPs that occur in each time step, the explicit payoff (EP) approach, the Ikonen-Toivanen (IT) splitting approach and the penalty (P) approach are considered. In addition to this, the Peaceman-Rachford (PR) method has been selected, which is related to the CN-IT method. For the ADI schemes, only the combination with the IT splitting approach is studied in the present chapter.

The two explicit payoff methods, BE-EP and CN-EP, have been considered just for one-asset options. They show a temporal convergence order equal to 1.0 for the one-asset put option, but in the case of the one-asset butterfly option their temporal errors turn out to be large and convergence appears to be slow.

In contrast, for all five options above, the BE-IT and BE-P methods always show a temporal convergence order close to 1.0 and the CN-P method a convergence order between 1.0 and 1.3. By employing suitable variable step sizes, defining the temporal grid points (20.26), the CN-P method reveals a favourable convergence order close to 2.0 whenever the early exercise boundary is not contained in the region of interest. For all other methods under consideration, using these variable step sizes does unfortunately not lead to an improvement in their convergence behaviour.

The CN-IT and PR methods always show a convergence order between approximately 1.0 and 1.3, except for the two-asset butterfly option, where it appears to reduce to 0.5.

Concerning the ADI-IT methods, the Do-IT method with $\theta = 1/2$ shows a convergence order about equal to 1.0 for both two-asset put options. The MCS-IT methods with $\theta = 1/3$ and $\theta = 1/2$ and the HV-IT method with $\theta = 1/(2 + \sqrt{2})$ show convergence orders approximately equal to 1.3 and 1.0 for these two options, respectively. The convergence behaviour of the ADI-IT methods is unclear in the case of the two-asset butterfly option.

The above observations on the temporal convergence behaviour of the methods employing the IT splitting approach appear to be largely new in the literature. Only for the BE-IT method a directly related theoretical result is known to us, see [5]. The numerical results in this chapter are in agreement with this theoretical result. For the methods using the penalty approach, our observations agree well with the (theoretical and practical) findings in [3]. Clearly, a temporal convergence order close to or equal to two is only observed in the experiments in this chapter for the CN-P method applied with suitable variable step sizes.

Comparing the size of the temporal errors of the different methods with constant step size, the experiments suggest that for the two methods BE-IT, BE-P these are always very similar, and the same is valid for the three methods CN-IT, CN-P, PR.

The latter group was found to be always significantly more accurate than the former group. Further, the PR method proved to be often somewhat more accurate than the CN-IT method. The ADI-based methods MCS-IT and HV-IT revealed a similar accuracy to the CN-based methods in the case of the two-asset put options. The Do-IT method was significantly less accurate than these. With the pertinent variable step sizes, the CN-P method has been found to be the most accurate in general among all methods under consideration.

Based on the numerical experiments discussed in this chapter, and taking into account the amount of computational work per time step, the MCS-IT and HV-IT methods are recommended in the numerical solution of (20.2) for two-asset American-style options whenever the payoff function and the financial parameters are standard. If the payoff function is more advanced, such as the (nonconvex) two-asset butterfly on the maximum, we recommend the CN-P method with variable step sizes. The BE-IT and BE-P methods are advocated if general applicability is important, which goes at the expense of temporal accuracy. As a comprise between general applicability and temporal accuracy, the PR method forms a good candidate.

Acknowledgements This work has been supported by the European Union in the FP7-PEOPLE-2012-ITN Program under Grant Agreement Number 304617 (FP7 Marie Curie Action, Project Multi-ITN STRIKE—Novel Methods in Computational Finance).

References

1. Barles, G., Daher, Ch., Romano, M.: Convergence of numerical schemes for parabolic equations arising in finance theory. Math. Models Methods Appl. Sci. **5**, 125–143 (1995)
2. Berman, A., Plemmons, R.J.: Nonnegative Matrices in the Mathematical Sciences. SIAM, Philadelphia (1994)
3. Forsyth, P.A., Vetzal, K.R.: Quadratic convergence for valuing American options using a penalty method. SIAM J. Sci. Comput. **23**, 2095–2122 (2002)
4. Haentjens, T., in 't Hout, K.J.: Alternating direction implicit finite difference schemes for the Heston-Hull-White partial differential equation. J. Comput. Financ. **16**, 83–110 (2012)
5. Haentjens, T., in 't Hout, K.J.: ADI schemes for pricing American options under the Heston model. Appl. Math. Financ. **22**, 207–237 (2015)
6. Haentjens, T., in 't Hout, K.J., Volders, K.: ADI schemes with Ikonen-Toivanen splitting for pricing American put options in the Heston model. In: Simos, T.E., et al. (eds.) Numerical Analysis and Applied Mathematics. AIP Conference Proceedings, vol. 1281, 231–234 (2010)
7. in 't Hout, K.J., Foulon, S.: ADI finite difference schemes for option pricing in the Heston model with correlation. Int. J. Numer. Anal. Model. **7**, 303–320 (2010)
8. in 't Hout, K.J., Welfert, B.D.: Stability of ADI schemes applied to convection-diffusion equations with mixed derivative terms. Appl. Numer. Math. **57**, 19–35 (2007)
9. in 't Hout, K.J., Welfert, B.D.: Unconditional stability of second-order ADI schemes applied to multi-dimensional diffusion equations with mixed derivative terms. Appl. Numer. Math. **59**, 677–692 (2009)
10. Howison, S.D., Reisinger, C., Witte, J.H.: The effect of nonsmooth payoffs on the penalty approximation of American options. SIAM J. Financ. Math. **4**, 539–574 (2013)
11. Hundsdorfer, W., Verwer, J.G.: Numerical Solution of Time-Dependent Advection-Diffusion-Reaction Equations. Springer, Berlin (2003)

12. Ikonen, S., Toivanen, J.: Operator splitting methods for American option pricing. Appl. Math. Lett. **17**, 809–814 (2004)
13. Ikonen, S., Toivanen, J.: Operator splitting methods for pricing American options under stochastic volatility. Numer. Math. **113**, 299–324 (2009)
14. Lions, P.L., Mercier, B.: Splitting algorithms for the sum of two nonlinear operators. SIAM J. Numer. Anal. **16**, 964–979 (1979)
15. Peaceman, D.W., Rachford, H.H.: The numerical solution of parabolic and elliptic differential equations. J. Soc. Ind. Appl. Math. **3**, 28–41 (1955)
16. Reisinger, C., Whitley, A.: The impact of a natural time change on the convergence of the Crank–Nicolson scheme. IMA J. Numer. Anal. **34**, 1156–1192 (2014)
17. Zvan, R., Forsyth, P.A., Vetzal, K.R.: Penalty methods for American options with stochastic volatility. J. Comp. Appl. Math. **91**, 199–218 (1998)
18. Zvan, R., Forsyth, P.A., Vetzal, K.R.: A finite volume approach for contingent claims valuation. IMA J. Numer. Anal. **21**, 703–731 (2001)

Chapter 21
High-Order-Compact ADI Schemes for Pricing Basket Options in the Combination Technique

Christian Hendricks, Christof Heuer, Matthias Ehrhardt, and Michael Günther

Abstract In this chapter we combine high-order-compact (HOC) and alternating-direction-implicit (ADI) schemes for pricing basket options in a sparse grid setting HOC schemes exploit the structure of the underlying partial differential equation to obtain a high order of consistency while employing a compact stencil. As time discretisation we propose an efficient ADI splitting to derive a stable scheme. The combination technique is used to construct the so called sparse grid solution, which leads to a significant reduction of necessary grid points and thus to a lower computational effort.

21.1 Introduction

We consider the d-dimensional Black-Scholes partial differential equation (PDE)

$$\frac{\partial V}{\partial t} + \frac{1}{2}\sum_{i,j=1}^{d} \rho_{ij}\sigma_i\sigma_j S_i S_j \frac{\partial^2 V}{\partial S_i \partial S_j} + r\sum_{i=1}^{d} S_i \frac{\partial V}{\partial S_i} - rV = 0$$

in $\Omega \times \Omega_t$ with $\Omega = [0, S_1^{\max}] \times \cdots \times [0, S_d^{\max}]$ and $\Omega_t = [0, T]$. The volatility of the single assets S_i is denoted by $\sigma_i > 0$, their correlation is given by ρ_{ij} for $i, j = 1, \ldots, d$. The risk-free interest rate is given by r. At maturity $t = T$ the option value is given by its payoff

$$g(S_1, \ldots, S_d) = (K - S_1 - \cdots - S_d)^+ \quad \text{(Put)},$$

$$g(S_1, \ldots, S_d) = (S_1 + \cdots + S_d - K)^+ \quad \text{(Call)}$$

C. Hendricks (✉) • C. Heuer • M. Ehrhardt • M. Günther
Lehrstuhl für Angewandte Mathematik/Numerische Analysis, Bergische Universität Wuppertal, Wuppertal, Germany
e-mail: hendricks@math.uni-wuppertal.de; heuer.chr@googlemail.com; ehrhardt@math.uni-wuppertal.de; guenther@math.uni-wuppertal.de

M. Ehrhardt et al. (eds.), *Novel Methods in Computational Finance*, Mathematics in Industry 25, DOI 10.1007/978-3-319-61282-9_21

with the strike price $K > 0$. We apply the transformations $x_i = \log(S_i)$ for $i = 1, \ldots, d$, $\tau = T - t$ and $u = e^{r\tau}V$, which leads to the transformed PDE

$$\frac{\partial u}{\partial \tau} - \frac{1}{2}\sum_{i,j=1}^{d} \rho_{ij}\sigma_i\sigma_j \frac{\partial^2 u}{\partial x_i \partial x_j} - \sum_{i=1}^{d}\left(r - \tfrac{1}{2}\sigma_i^2\right)\frac{\partial u}{\partial x_i} = 0, \tag{21.1}$$

with space-independent coefficients. In order to solve the PDE (21.1) numerically we use the method-of-lines approach and end up with a semi-discretisation in space

$$\frac{\partial u}{\partial \tau} = F(u(\tau)), \qquad 0 \le \tau \le T, \qquad u(0) = g.$$

We consider ADI splitting schemes in the time domain with the decomposition of the spatial discretisation

$$F(u) = F_0(u) + F_1(u) + \cdots + F_d(u),$$

where F_0 stems from all mixed derivatives and F_i for $i = 1, \ldots, d$ belongs to the unidirectional contribution of the i-th coordinate in the PDE (21.1). Within the ADI framework the F_0 part will always be treated explicitly. We propose a HOC finite difference discretisation of the F_i-terms to compute a highly accurate solution while employing a compact stencil. To reduce the number of grid points we use the combination technique to compute the so called sparse grid solution. Compared to a tensor-based full grid with $\mathcal{O}(h^{-d})$ points in space, the sparse grid consists of only $\mathcal{O}(h^{-1}\log(h^{-1})^{d-1})$ nodes. Under suitable regularity assumptions the pointwise rate of convergence is $\mathcal{O}(h^4\log(h^{-1})^{d-1})$ if a fourth order scheme is used to compute the sub solutions.

21.2 HOC Finite Differences

We derive a HOC approximation of the single F_i arising in the decomposition of F. Throughout this chapter we use standard finite difference operators to approximate the derivatives. A central discretisation to the first and second derivative of order two is given by

$$\delta_{x_i}^2 u_k = \tfrac{1}{h_i^2}\left(u_{k+1} - 2u_k + u_{k-1}\right) = \frac{\partial^2 u}{\partial x_i^2} + \mathcal{O}(h_i^2),$$

$$\delta_{x_i}^0 u_k = \tfrac{1}{2h_i}\left(u_{k+1} - u_{k-1}\right) = \frac{\partial u}{\partial x_i} + \mathcal{O}(h_i^2).$$

The mixed derivative term F_0 is approximated with the help of fourth order stencils

$$\tilde{\delta}^0_{x_i} u_k = \tfrac{1}{12h_i}\left(-u_{k+2} + 8u_{k+1} - 8u_{k-1} + u_{k-2}\right) = \frac{\partial u}{\partial x_i} + \mathscr{O}(h_i^4).$$

Thus we can approximate F_0 via

$$F_0(u) = \sum_{\substack{i,j=1\\ i\neq j}}^{d} \tfrac{1}{2}\rho_{ij}\sigma_i\sigma_j\tilde{\delta}^0_{x_i}\tilde{\delta}^0_{x_j} u_{kl} + \sum_{i,j} \mathscr{O}(h_i^4 h_j^4).$$

As F_0 is always treated explicitly we do not expect any significant adverse effects incorporating these large stencils regarding the computational effort. The unidirectional contributions are given by

$$F_i(u) = \tfrac{1}{2}\sigma_i^2 \frac{\partial^2 u}{\partial x_i^2} + \left(r - \tfrac{1}{2}\sigma_i^2\right)\frac{\partial u}{\partial x_i} = f, \tag{21.2}$$

for $i = 1, \ldots, d$ and some arbitrary right hand side f. Inserting the finite difference operators we obtain

$$\begin{aligned} F_i(u_k) = \tfrac{1}{2}\sigma_i^2\delta^2_{x_i}u_k - \tfrac{1}{2}\sigma_i^2\tfrac{h_i^2}{12}\frac{\partial^4 u}{\partial x_i^4} \\ + \left(r - \tfrac{1}{2}\sigma_i^2\right)\delta^0_{x_i}u_k - \left(r - \tfrac{1}{2}\sigma_i^2\right)\tfrac{h_i^2}{6}\frac{\partial^3 u}{\partial x_i^3} + \mathscr{O}(h_i^4) = f_k. \end{aligned} \tag{21.3}$$

Since the truncation error in (21.3) is of order two, we can derive a fourth order approximation if the third and fourth derivative are approximated with second order accuracy. In order to derive these approximations, we differentiate Eq. (21.2) once with respect to x_i and get

$$\frac{\partial^3 u}{\partial x_i^3} = \frac{2}{\sigma_i^2}\frac{\partial f}{\partial x_i} - \left(\frac{2r}{\sigma_i^2} - 1\right)\frac{\partial^2 u}{\partial x_i^2}. \tag{21.4}$$

Differentiating (21.2) twice with respect to x_i gives

$$\frac{\partial^4 u}{\partial x_i^4} = \frac{2}{\sigma_i^2}\frac{\partial f}{\partial x_i} - \left(\frac{2r}{\sigma_i^2} - 1\right)\left(\frac{2}{\sigma_i^2}\frac{\partial f}{\partial x_i} - \left(\frac{2r}{\sigma_i^2} - 1\right)\frac{\partial^2 u}{\partial x_i^2}\right). \tag{21.5}$$

The derivatives (21.4) and (21.5) can be approximated via central discretisation on the compact stencil with second order. Hence using (21.4) and (21.5) in (21.3) leads to a fourth order accurate approximation

$$\left(\frac{h_i^2\left(r-\frac{\sigma_i^2}{2}\right)^2}{6\sigma_i^2}+\frac{\sigma_i^2}{2}\right)\delta_{x_i}^2 u_k+\left(r-\frac{\sigma_i^2}{2}\right)\delta_{x_i}^0 u_k$$

$$= f_k + \frac{h_i^2}{12}\delta_{x_i}^2 f_k + \frac{h_i^2\left(r-\frac{\sigma_i^2}{2}\right)}{6\sigma_i^2}\delta_{x_i}^0 f_k. \tag{21.6}$$

Rewriting this scheme in terms of matrices or symbolic operators gives

$$A_{x_i}U = B_{x_i}F$$

for vectors U and F, where A_{x_i} corresponds to the left hand side of (21.6) and B_{x_i} to its right hand side. The semi-discrete scheme can thus be written as

$$\frac{\partial u}{\partial t} = F_0(u) + B_{x_1}^{-1}A_{x_1}u + \cdots + B_{x_d}^{-1}A_{x_d}u + \mathcal{O}(h_1^4) + \cdots + \mathcal{O}(h_d^4) + \sum_{i,j}\mathcal{O}(h_i^4 h_j^4).$$

21.3 HOC-ADI Schemes

We now apply three well known ADI schemes to the spatial discretisation given in the previous section, namely
HOC Douglas scheme:

$$\begin{cases} Z_0 &= \prod_{j=1}^d B_{x_j}u_n + \Delta_t\left(\prod_{j=1}^d B_{x_j}F_0(u_n) + \sum_{i=1}^d \prod_{\substack{j=1\\ j\neq i}}^d B_{x_j}A_{x_i}u_n\right)\\ (B_{x_i}-\theta\Delta_t A_{x_i})Z_i &= Z_{i-1} - \theta\Delta_t \prod_{j=i+1}^d B_{x_j}A_{x_i}u_n \text{ for } i=1,\ldots,d\\ u_{n+1} &= Z_d, \end{cases} \tag{21.7}$$

HOC Craig-Sneyd scheme:

$$\begin{cases} Z_0 &= \prod_{j=1}^d B_{x_j}u_n + \Delta_t\left(\prod_{j=1}^d B_{x_j}F_0(u_n) + \sum_{i=1}^d \prod_{\substack{j=1\\ j\neq i}}^d B_{x_j}A_{x_i}u_n\right)\\ (B_{x_i}-\theta\Delta_t A_{x_i})Z_i &= Z_{i-1} - \theta\Delta_t \prod_{j=i+1}^d B_{x_j}A_{x_i}u_n \text{ for } i=1,\ldots,d\\ \tilde{Z}_0 &= Z_0 + \frac{1}{2}\Delta_t\left(\prod_{j=1}^d B_{x_j}F_0(Z_d) - \prod_{j=1}^d B_{x_j}F_0(u_n)\right)\\ (B_{x_i}-\theta\Delta_t A_{x_i})\tilde{Z}_i &= \tilde{Z}_{i-1} - \theta\Delta_t \prod_{j=i+1} B_{x_j}A_{x_i}u_n \text{ for } i=1,\ldots,d\\ u_{n+1} &= \tilde{Z}_d. \end{cases} \tag{21.8}$$

HOC Modified Craig-Sneyd scheme:

$$\begin{cases} Z_0 & = \prod_{j=1}^{d} B_{x_j} u_n + \Delta_t \left(\prod_{j=1}^{d} B_{x_j} F_0(u_n) + \sum_{i=1}^{d} \prod_{\substack{j=1 \\ j\neq i}}^{d} B_{x_j} A_{x_i} u_n \right) \\ (B_{x_i} - \theta \Delta_t A_{x_i})\, Z_i & = Z_{i-1} - \theta \Delta_t \prod_{j=i+1}^{d} B_{x_j} A_{x_i} u_n \text{ for } i = 1, \ldots, d \\ \hat{Z}_0 & = Z_0 + \theta \Delta_t \left(\prod_{j=1}^{d} B_{x_j} F_0(Z_d) - \prod_{j=1}^{d} B_{x_j} F_0(u_n) \right) \\ \tilde{Z}_0 & = \hat{Z}_0 + (\tfrac{1}{2} - \theta) \Delta_t \left(\prod_{j=1}^{d} B_{x_j} F(Z_d) - \prod_{j=1}^{d} B_{x_j} F(u_n) \right) \\ (B_{x_i} - \theta \Delta_t A_{x_i})\, \tilde{Z}_i & = \tilde{Z}_{i-1} - \theta \Delta_t \prod_{j=i+1} B_{x_j} A_{x_i} u_n \text{ for } i = 1, \ldots, d \\ u_{n+1} & = \tilde{Z}_d . \end{cases} \tag{21.9}$$

The Douglas scheme, see [2], exhibits a consistency order 2 in time if $\theta = \frac{1}{2}$ and $F_0 = 0$, order 1 otherwise. The consistency order in time of the Craig-Sneyd scheme, see [1], is given by 2 if and only if $\theta = \frac{1}{2}$. The modified Craig-Sneyd scheme, see [5], exhibits consistency order 2 in time for any θ. The Craig-Sneyd and the Modified Craig-Sneyd scheme can be seen as an extension of the Douglas scheme.

21.4 Combination Technique

In order to construct the solution on the sparse grid we use the combination technique, which exploits the error splitting structure to linearly combine an anisotropic sequence of solutions in such a way that low order error terms cancel out. We assume

$$u(\mathbf{x_h}) - u_{\mathbf{l}} = \sum_{k=1}^{d} \sum_{\substack{\{j_1,\ldots,j_k\} \\ \subseteq \{1,\ldots,d\}}} w_{j_1,\ldots,j_k}(.; h_{j_1}, \ldots, h_{j_k}) h_{j_1}^4 \cdots h_{j_k}^4,$$

as error with bounded coefficient functions w. The analytical solution on the discrete grid $\mathbf{x_h}$ is denoted by $u(\mathbf{x_h})$. Note that such an splitting structure can be shown for a wide class of PDEs and linear finite difference schemes [7]. Combining the solutions according to

$$u_n^s = \sum_{q=0}^{d-1} (-1)^q \binom{d-1}{q} \sum_{|\mathbf{l}|_1 = n-q} u_{\mathbf{l}},$$

we can expect a pointwise rate of convergence of $\mathcal{O}(h^4 \log(h^{-1})^{d-1})$. Here u_n^s denotes the sparse grid solution on level n. The numerical sub solutions $u_{\mathbf{l}}$ are computed on a grid with step sizes $(h_1, h_2, \ldots, h_d) = \left(2^{-l_1} \cdot c_1, 2^{-l_2} \cdot c_2, \ldots, 2^{-l_d} \cdot c_d\right)$ with multi-index $\mathbf{l} = (l_1, l_2, \ldots, l_d)$ and grid length c_i in the coordinate direction i for $i = 1, \ldots, d$.

21.5 Numerical Experiments

In this section we apply our numerical schemes to a European basket put option with two underlyings with parameters

$$T = 1, \quad K = 20, \quad \sigma_1 = 0.4, \quad \sigma_2 = 0.3, \quad \rho_{12} = 0.5,$$
$$x_i^{\min} = -5 \quad \text{and} \quad x_i^{\max} = \log(5K)$$

for $i = 1, 2$. Figure 21.1 shows the results of our numerical tests. In the time domain we use the lowest θ value ensuring unconditionally stability in the case of standard second order finite differences [3, 4]. All three schemes show a stable behaviour, see Fig. 21.1a and lead to their expected convergence order. Figure 21.1b and c

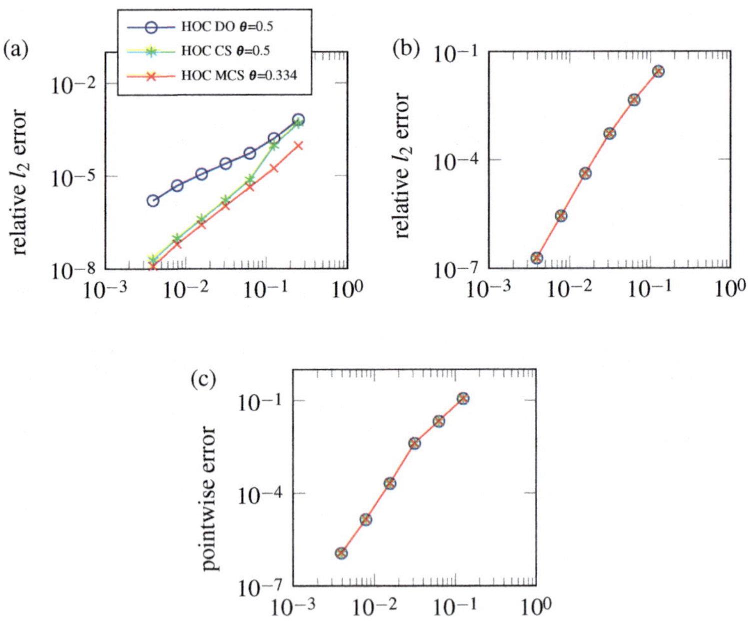

Fig. 21.1 Numerical convergence plots. (**a**) Error in time, $\Delta_t \to 0$. (**b**) Spatial error, $h \to 0$. (**c**) Error on sparse grid for $h \to 0$

show the evolution of the error on the full and sparse grid. We compute the sparse grid error at the central grid node, which is the only point that belongs to all sub grids and is therefore not influenced by the interpolation technique used to combine the solutions. The convergence in both plots is in line with the theoretical findings. Please note that the initial value has been smoothed according to Kreiss et al. [6] in order to overcome the deteriorations from the non-smooth option's payoff.

21.6 Conclusions

In this chapter we introduced HOC-ADI schemes to price basket options. The number of grid points could be reduced significantly using sparse grids and the combination technique. In a forthcoming paper we will generalise these schemes to problems settings with space-dependent coefficients. Furthermore we analyse the stability in the von Neumann framework.

Acknowledgements The authors were partially supported by the European Union in the FP7-PEOPLE-2012-ITN Program under Grant Agreement Number 304617 (FP7 Marie Curie Action, Project Multi-ITN STRIKE—Novel Methods in Computational Finance)

References

1. Craig, I.J.D., Sneyd, A.D.: An alternating-direction implicit scheme for parabolic equations with mixed derivatives. Comput. Math. Appl. **16**(4), 341–350 (1988)
2. Douglas, J., Rachford, H.: On the numerical solution of heat conduction problems in two and three space variables. Trans. Am. Math. Soc. **82**, 421–439 (1956)
3. in't Hout, K.J., Mishra, C.: Stability of the modified Craig-Sneyd scheme for two-dimensional convection-diffusion equations with mixed derivative terms. Math. Comput. Simul. **81**, 2540–2548 (2011)
4. in't Hout, K.J., Welfert, B.D.: Stability of ADI schemes applied to convection-diffusion equations with mixed derivative terms. Appl. Numer. Math. **57**(1), 19–35 (2007)
5. in't Hout, K.J., Welfert, B.D.: Unconditional stability of second-order ADI schemes applied to multi-dimensional diffusion equations with mixed derivative terms. Appl. Numer. Math. **59**(3-4), 677–692 (2009)
6. Kreiss, H.O., Thomée, V., Widlund, O.: Smoothing of initial data and rates of convergence for parabolic difference equations. Commun. Pure Appl. Math. **23**(2), 241–259 (1970)
7. Reisinger, C.: Analysis of linear difference schemes in sparse grid combination technique. IMA J. Numer. Anal. **33**(2), 544–581 (2013)

Part VI
Scientific Computing

Chapter 22
Splitting Methods for Fokker-Planck Equations Related to Jump-Diffusion Processes

Beatrice Gaviraghi, Mario Annunziato, and Alfio Borzì

Abstract A splitting implicit-explicit (SIMEX) scheme for solving a partial integro-differential Fokker-Planck equation related to a jump-diffusion process is investigated. This scheme combines the method of Chang-Cooper for spatial discretization with the Strang-Marchuk splitting and first- and second-order time discretization methods. It is proven that the SIMEX scheme is second-order accurate, positive preserving, and conservative. Results of numerical experiments that validate the theoretical results are presented. (This chapter is a summary of the paper Gaviraghi et al. (Appl Math Comput, 2017); all theoretical statements in this summary are proved in that reference.)

22.1 Introduction

The Fokker-Planck (FP) equation (also known as the forward Kolmogorov equation) governs the time evolution of the probability density function (PDF) of a Markovian stochastic processes and plays a fundamental role in any problem involving random quantities [8, 19], including stochastic processes with jumps. Within this class of problems, Lévy processes, that have stationary and independent increments, are increasing in popularity from the need of modeling the market behavior beyond the Black-Scholes framework [5] and in statistical physics [17].

We consider the numerical discretization of the FP equation corresponding to a jump-diffusion Markov process $X_t \in \mathbb{R}^d$, for $t \in I = [t_0, t_f]$, which is modelled by the following stochastic differential problem

$$\begin{cases} dX_t = b(X_t)dt + \sigma(X_t)dW_t + dP_t, \\ X_{t_0} = X_0, \end{cases} \tag{22.1}$$

B. Gaviraghi • M. Annunziato • A. Borzì (✉)
Lehrstuhl für Mathematik IX (Chair Scientific Computing), University of Würzburg, Würzburg, Germany
e-mail: beatrice.gaviraghi@mathematik.uni-wuerzburg.de; alfio.borzi@mathematik.uni-wuerzburg.de; mannunzi@unisa.it

M. Ehrhardt et al. (eds.), *Novel Methods in Computational Finance*, Mathematics in Industry 25, DOI 10.1007/978-3-319-61282-9_22

where $X_0 \in \mathbb{R}^d$ is a given initial random data. This stochastic differential equation (SDE) relates the infinitesimal increments of the stochastic process X_t to both deterministic and random increments, given by the multidimensional Wiener process $W_t \in \mathbb{R}^m$ and the compound Poisson process $P_t \in \mathbb{R}^d$. We denote with $\lambda \in \mathbb{R}^+$ the rate of the time events of a compound Poisson process and with $g(y)$ the PDF of the size of its jumps. The density $g(y)$ is nonnegative and normalized, $\int_{\mathbb{R}^d} g(y)dy = 1$. The deterministic functions $b : \mathbb{R}^d \to \mathbb{R}^d$ and $\sigma : \mathbb{R}^d \to \mathbb{R}^{d\times m}$ represent the drift and the diffusion coefficients, respectively. We assume that the matrix σ is full-rank. The solvability of (22.1) follows under growth and regularity conditions on b and σ; see, e.g. [2, 13].

The PDF $f(x,t)$ of X_t is defined in $\mathbb{R}^d \times I$ and it is governed by the following FP partial integro-differential equation (PIDE)

$$\partial_t f(x,t) = \mathscr{L} f(x,t) + \mathscr{I} f(x,t), \tag{22.2}$$

where the two linear operators $\mathscr{L}$ and $\mathscr{I}$ are defined as follows

$$\begin{aligned} \mathscr{L} f(x,t) &:= -\sum_{i=1}^{d} \partial_{x_i} \left(b_i(x) f(x,t)\right) + \sum_{i,j=1}^{d} \partial^2_{x_i x_j} \left(a_{ij}(x) f(x,t)\right) \\ \mathscr{I} f(x,t) &:= \lambda \int_{\mathbb{R}^d} f(x-y,t) g(y) dy - \lambda f(x,t), \end{aligned} \tag{22.3}$$

where $a_{ij}(x)$ are the elements of a matrix defined as $a_{ij}(x) := \frac{1}{2}\sum_{k=1}^{d} \sigma_{ik}(x)\sigma_{jk}(x)$. Since the diffusion coefficient σ is full rank, the matrix a is positive definite. We note that the operator $\mathscr{L}$ is related to the drift-diffusion component of the process and $\mathscr{I}$ to the jump term.

In application, PIDEs naturally arise when option prices in jump diffusion models have to be computed [3, 5, 6, 16]. We recall that in [1, 4, 15] a conservative and positive preserving finite difference scheme was proposed and analyzed for the case of Itô SDEs without jumps. Conservative numerical schemes for our FP PIDE problem have been less investigated.

The purpose of this work is to implement and analyse a conservative scheme that solves the FP problem for a jump-diffusion process, with appropriate initial and boundary conditions.

In the next section, we define our FP problem. In Sect. 22.3, we illustrate our discretization method and the resulting SIMEX schemes. Section 22.4 is devoted to the convergence analysis of our SIMEX1 and SIMEX2 schemes, i.e. the stability in time and second-order accuracy in space. In particular, we discuss first- and second-order accuracy in time for the SIMEX1 and SIMEX2 schemes, respectively. In Sect. 22.5, we show that our numerical schemes guarantee the same structural properties of the PDF solution, i.e. the non-negativity and conservation of total probability of the PDF solution. Section 22.6 presents results of numerical experiments that validate the theoretical results.

22.2 The PIDE Fokker-Planck Problem

We focus on the following FP problem

$$\begin{cases} \partial_t f(x,t) = \mathscr{L} f(x,t) + \mathscr{I} f(x,t) + \psi(x,t) & \text{for } (x,t) \in \Omega \times I \\ f(x,0) = f_0(x) \text{ for } x \in \Omega \\ \mathscr{F}(x,t) = 0 \text{ for } (x,t) \in \partial\Omega \times I, \end{cases} \tag{22.4}$$

where f_0 is the PDF of X_0 in (22.1) and $\mathscr{F}$ is defined below in (22.7). The source term ψ is included for the purpose of error analysis. Notice that the space variable x has the same domain of the considered stochastic process X_t in (22.1).

Our main focus is (22.4) with $\psi = 0$, whose solution is the PDF of the process X_t governed by (22.1). Therefore, in the case of zero source term ψ, f satisfies the following conditions

$$\begin{aligned} &1) f(x,t) \geq 0 \text{ for each } (x,t) \in \Omega \times I, \\ &2) \int_\Omega f(x,t)dx = 1 \text{ for each time } t \in I, \end{aligned} \tag{22.5}$$

provided that the initial data f_0 is the PDF of the initial data X_0. The positivity is ensured by standard maximum principle arguments [9].

Notice that the differential operator $\mathscr{L}$ in (22.3) can be written in divergence form as follows

$$\mathscr{L} f(x,t) = \nabla \cdot \mathscr{F}(x,t), \tag{22.6}$$

where $\mathscr{F}(x,t)$ is defined as follows

$$\begin{aligned} \mathscr{F}(x,t) &:= B(x)f(x,t) + C(x)\nabla f(x,t) \\ B_i(x) &:= \sum_{i=1}^{d} \partial_{x_j} a_{ij}(x) - b_i(x) \\ C_{ij}(x) &:= a_{ij}(x). \end{aligned} \tag{22.7}$$

Notice that (22.7) defines the flux of the drift-diffusion process and the condition $\mathscr{F} = 0$ (zero flux) on the boundary corresponds to that process evolving in a domain with reflecting barriers at the boundary. Moreover, when $g(x)$ is also set for the reflecting boundary conditions, the following conservation property holds

Proposition 22.1 *Let us consider our FP problem with $\psi = 0$, and assume that the density g satisfies*

$$\int_\Omega g(x)dx = 1.$$

Then

$$\int_\Omega f(x,t) = \int_\Omega f_0(x)dx \qquad \forall t \in I.$$

22.3 Discretization Scheme of the PIDE

For simplicity, we discuss our FP problem in one dimension. Consider the domain $\Omega = (r, s)$ and the time interval $I = [t_0, t_f]$. We set the mesh sizes h and δt as follows

$$h := \frac{s-r}{N-1} \text{ and } \delta t := \frac{t_f - t_0}{M}, \text{ with } N, M \in \mathbb{N}.$$

We consider uniform meshes in space and time. We have

$$\begin{aligned} \Omega_h &:= \{x_j = r + (j-1)h,\, j = 1, \ldots, N\} \subset \bar{\Omega} \\ I_{\delta t} &:= \{t_n = t_0 + n\delta t,\, n = 0, \ldots, M\} \subset I. \end{aligned} \tag{22.8}$$

First, we carry out the spatial discretization of the integro-differential terms of $\mathcal{L}$ and $\mathcal{I}$ defined in (22.3), that together lead to a large system of ODEs. Exploiting the divergence form (22.6) of the differential operator $\mathcal{L}$, we discretize the spatial derivative of $\mathcal{F}$ defined by (22.7) using the Chang-Cooper (CC) scheme [4, 15]. This is a cell-centered finite-volume scheme with an exponential fitting technique to determine the fluxes at the cell boundaries, $x_{j\pm\frac{1}{2}}, j = 1, \ldots, N$, and the unknown variables are considered on the grid points $x_j, j = 1, \ldots, N$. In the following, $\Phi_j(t)$ and $\Phi_{j\pm\frac{1}{2}}(t)$ denote the time-continuous restrictions of a generic function $\Phi(x,t)$ to the grid points x_j and $x_{j\pm\frac{1}{2}} := \frac{x_j + x_{j\pm 1}}{2}$, respectively, accordingly to the method of the lines.

Using the CC scheme, we have the following discretization formula

$$\partial_x \mathcal{F}(x_j, t) \approx \frac{\mathcal{F}_{j+\frac{1}{2}}(t) - \mathcal{F}_{j-\frac{1}{2}}(t)}{h},$$

where

$$\mathcal{F}_{j+\frac{1}{2}}(t) = \left((1-\delta_j)B_{j+\frac{1}{2}} + \frac{1}{h}C_{j+\frac{1}{2}}\right) f_{j+1}(t) - \left(\frac{1}{h}C_{j+\frac{1}{2}} - \delta_j B_{j+\frac{1}{2}}\right) f_j(t).$$

The parameter $\delta_j \in [0, 1]$ is defined as follows

$$\delta_j := \frac{1}{hB_{j+\frac{1}{2}}/C_{j+\frac{1}{2}}} - \frac{1}{\exp\{hB_{j+\frac{1}{2}}/C_{j+\frac{1}{2}}\} - 1}. \tag{22.9}$$

The integral over Ω in the PIDE is approximated by the midpoint rule [7]. Hence the discretization of the integral operator takes the following form

$$\mathscr{I}f(x_j,t) \approx \lambda\left(h\sum_{i=1}^{N} f(x_i,t)g(x_j-x_i) - f(x_j,t)\right).$$

The space discretization above gives the following approximation

$$\begin{cases} f'_{\mathrm{SD}}(t) = (\mathscr{A}+\mathscr{G})f_{\mathrm{SD}}(t) + \Psi(t) \\ f_{\mathrm{SD}}(t_0) = f_{\mathrm{SD}}(0), \end{cases} \tag{22.10}$$

where $\mathscr{A}$ and $\mathscr{G}$, denoting the matrices related to the operators $\mathscr{A}$ and $\mathscr{G}$ respectively, are defined below. Notice that (22.10) is a system of ordinary differential equations parametrized by the space mesh size h in $\mathscr{A}$ and $\mathscr{G}$. In other words, $f_{\mathrm{SD}}(t) = \{f_{SD,1}(t),\dots,f_{SD,N}(t)\} \in \mathbb{R}^N$ can be viewed as a grid function, where each component describes the time evolution of f_{SD} on the correspondent grid point of Ω_h. The initial value $f_{\mathrm{SD}}(0)$ and the source term Ψ represent the restriction on the grid Ω_h of the sufficiently smooth initial data f_0 and of the source term ψ in (22.4), respectively.

The matrix $\mathscr{A}$ in (22.10) follows from the CC spatial discretization and the boundary condition on $\mathscr{F}$. By setting $w_j = \exp\{hB_{j+\frac{1}{2}}/C_{j+\frac{1}{2}}\}$, $\beta_j = C_{j+\frac{1}{2}}/h - \delta_j B_{j+\frac{1}{2}} = B_{j+\frac{1}{2}}/(w_j-1)$, the tridiagonal matrix $\mathscr{A}$ is defined as follows

$$[\mathscr{A}]_{ij} = \begin{cases} \beta_{i-1}/h & j=i-1,\ 2\le i\le N, \\ -(\beta_{i-1}w_{i-1}+\beta_i)/h & j=i, \qquad 1\le i\le N, \quad \beta_0=\beta_N=0, \\ \beta_i w_i/h & j=i+1,\ 1\le i\le N-1, \\ 0 & \text{otherwise.} \end{cases} \tag{22.11}$$

The matrix $\mathscr{G}$ in (22.10) is defined as follows

$$\mathscr{G} := \lambda(G-I),$$

where I denotes the N–dimensional identity matrix and G is the matrix with normalized columns as follows

$$[G]_{ij} := \frac{h\,g(x_i-x_j)}{\sum_{k=1}^{N} h\,g(x_k-x_j)}. \tag{22.12}$$

The choice of the normalization in (22.12) is discussed in Sect. 22.5.

The next step of our discretization procedure is to consider the semi-discrete system (22.10), for which we use an operator splitting method. Setting δt also as the splitting time step, we apply the Strang-Marchuk (SM) splitting scheme [11, 12, 14, 20]. In the following, we refer to the time-continuous solution of the splitting scheme as $f_{SP}(t)$. The idea to process the computation for $\mathscr{A}$ and $\mathscr{G}$ at separate time steps. The initial data is set as follows, $f_{SP}(t_0) := f_{SD}(t_0)$, where f_{SP} is the splitting solution and f_{SD} is the solution of the semi-discretized system (22.10) without splitting.

In each time interval $[t_n, t_{n+1}]$, given the splitting solution $f_{SP}(t_n)$, the following subproblems, connected via the initial conditions, are solved

$$\begin{aligned}
&1. \begin{cases} \phi_1'(t) = \mathscr{A}\phi_1(t) \\ \phi_1(t_n) = f_{SP}(t_n) \end{cases} && t \in [t_n, t_{n+\frac{1}{2}}] \\
&2. \begin{cases} \phi_2'(t) = \mathscr{G}\phi_2(t) + \Psi(t) \\ \phi_2(t_n) = \phi_1(t_{n+\frac{1}{2}}) \end{cases} && t \in [t_n, t_{n+1}] \\
&3. \begin{cases} \phi_3'(t) = \mathscr{A}\phi_3(t) \\ \phi_3(t_{n+\frac{1}{2}}) = \phi_2(t_{n+1}) \end{cases} && t \in [t_{n+\frac{1}{2}}, t_{n+1}]. \\
&4. \begin{cases} f_{SP}(t_{n+1}) := \phi_3(t_{n+1}) \end{cases}
\end{aligned} \tag{22.13}$$

This system of continuous-time equations is approximated by time discretization. The fully discrete numerical solution will be referred to as $\hat{f} = (f_j^n)$, $j = 1, \ldots, N$, $n = 0, \ldots, M$. We propose to use two different time discretization methods that together with the space discretization give the schemes named SIMEX1 and SIMEX2.

In SIMEX1, the solution of the first and third step of (22.13) is carried out with the implicit Euler method, while the second step is explicit, in order to avoid the drawback of inverting dense matrices. Given f^n at time t_n, the three initial value problems in (22.13) read as follows

$$\begin{aligned}
&1. \frac{f^{n+\frac{1}{2}} - f^n}{\delta t/2} = \mathscr{A} f^{n+\frac{1}{2}} \\
&2. \frac{f^{(n+\frac{1}{2})^*} - f^{n+\frac{1}{2}}}{\delta t} = \mathscr{G} f^{n+\frac{1}{2}} + \Psi(t_n) \\
&3. \frac{f^{n+1} - f^{(n+\frac{1}{2})^*}}{\delta t/2} = \mathscr{A} f^{n+1},
\end{aligned} \tag{22.14}$$

where the unknown are sequentially solved: $f^n \to f^{n+\frac{1}{2}} \to f^{(n+\frac{1}{2})^*} \to f^{n+1}$.

The time discretization of (22.13) in SIMEX2 is carried out with the predictor corrector method. Given f^n at time t_n, the discretization of the three initial value problems in (22.13) take the following form

$$
\begin{aligned}
&1.\ \begin{cases} \bar{f}^{n+\frac{1}{2}} = f^n + \frac{\delta t}{2}\mathscr{A} f^n \\ f^{n+\frac{1}{2}} = f^n + \frac{\delta t}{4}\left[\mathscr{A} f^n + \mathscr{A}\bar{f}^{n+\frac{1}{2}}\right] \end{cases} \\
&2.\ \begin{cases} \bar{f}^{n+\frac{1}{2}^*} = f^{n+\frac{1}{2}} + \delta t[\mathscr{G} f^{n+\frac{1}{2}} + \Psi(t_n)] \\ f^{n+\frac{1}{2}^*} = f^{n+\frac{1}{2}} + \frac{\delta t}{2}\left[\mathscr{G} f^{n+\frac{1}{2}} + \Psi(t_n) + \mathscr{G}\bar{f}^{n+\frac{1}{2}^*} + \Psi(t_{n+1})\right] \end{cases} \\
&3.\ \begin{cases} \bar{f}^{n+1} = f^{n+\frac{1}{2}^*} + \frac{\delta t}{2}\mathscr{A} f^{n+\frac{1}{2}^*} \\ f^{n+1} = f^{n+\frac{1}{2}^*} + \frac{\delta t}{4}\left[\mathscr{A} f^{n+\frac{1}{2}^*} + \mathscr{A}\bar{f}^{n+1}\right]. \end{cases}
\end{aligned}
\tag{22.15}
$$

22.4 Convergence Analysis

In this section, we investigate stability and accuracy properties of the SIMEX1 and SIMEX2 schemes. After determining the order of convergence of the spatial discretization, we focus on the rate of convergence in time.

The discrete L^2-scalar product of two grid functions u and v on $\Omega_h \times I_{\delta t}$ is defined as follows

$$(u, v)_{L^2_{h,\delta t}} := h\delta t \sum_{j=1}^{N}\sum_{n=0}^{M} u_j^n v_j^n,$$

with associated norm $\|u\|_{L^2_{h,\delta t}} := \sqrt{(u,u)_{L^2_{h,\delta t}}}$. In a similar fashion, the discrete L^2_h inner product and norm are defined for functions w, z on the spatial grid Ω_h as follows

$$(w, z)_{L^2_h} := h\sum_{j=1}^{N} w_j z_j \text{ and } \|w\|_{L^2_h} := \sqrt{(w,w)_{L^2_h}}.$$

We aim at comparing the continuous PIDE solution f of (22.4) and the numerical solution $\hat{f}$, which is defined on the grid points of $\Omega_h \times I_{\delta t}$. Central to our analysis is the following inequality

$$\|f - \hat{f}\|_{L^2_{h,\delta t}} \le \|f_h - f_{SD}\|_{L^2_{h,\delta t}} + \|f_{SD} - f_{SP}\|_{L^2_{h,\delta t}} + \|f_{SP} - \hat{f}\|_{L^2_{h,\delta t}}, \tag{22.16}$$

where $f_h(t) = f(\bar{x}, t) \in \mathbb{R}^N$ is the PIDE solution restricted to $\bar{x} \in \Omega_h$, f_{SD} solves (22.10) and f_{SP} is obtained as in (22.13). In (22.16), the $L^2_{h,\delta t}$ norms are computed after evaluating the continuous functions at the points of the meshes Ω_h and $I_{\delta t}$ defined in (22.8).

In the following, we provide bounds for each of the three norms of (22.16). Specifically, we can prove that $\|f_h - f_{\mathrm{SD}}\|_{L^2_{h,\delta t}} = \mathcal{O}(h^2)$. In Proposition 22.3, we obtain $\|f_{\mathrm{SD}} - f_{\mathrm{SP}}\|_{L^2_{h,\delta t}} = \mathcal{O}(\delta t^2)$. Further, for the SIMEX1 scheme, we report in Proposition 22.5 the following, $\|f_{\mathrm{SP}} - \hat{f}\|_{L^2_{h,\delta t}} = \mathcal{O}(\delta t)$; and in Proposition 22.7, we obtain $\|f_{\mathrm{SP}} - \hat{f}\|_{L^2_{h,\delta t}} = \mathcal{O}(\delta t^2)$ for the SIMEX2 scheme.

Proposition 22.2 *If the solution f of* (22.4) *has continuous space derivatives up to the 4th order and g in* (22.3) *is twice continuously differentiable, the spatial truncation error α, related to the semi-discrete Eq. (22.13), is consistent of order 2 as follows*

$$\|\alpha(t)\|_{L^2_h} = \mathcal{O}(h^2).$$

We now discuss a bound for the second addend in (22.16) related to the splitting method. Consider the matrices $\mathscr{A}$ and $\mathscr{G}$ as in (22.11) and (22.12), respectively, and define the operator $\mathscr{S}$ as follows

$$\mathscr{S} := e^{\frac{\delta t}{2}\mathscr{A}}\, e^{\delta t \mathscr{G}}\, e^{\frac{\delta t}{2}\mathscr{A}}. \tag{22.17}$$

Consider the time interval $[t_n, t_{n+1}]$. We apply four times the variation of constants formula for ODEs [12] to integrate the ODE systems (22.13). Therefore, the splitting solution can be formally written as follows

$$\begin{aligned} f_{\mathrm{SP}}(t_{n+1}) =& \mathscr{S} f_{\mathrm{SP}}(t_n) + e^{\frac{\delta t}{2}\mathscr{A}} e^{\frac{\delta t}{2}\mathscr{G}} \int_0^{\frac{\delta t}{2}} e^{(\frac{\delta t}{2}-s)\mathscr{G}} \Psi(t_n + s)ds \\ &+ e^{\frac{\delta t}{2}\mathscr{A}} \int_0^{\frac{\delta t}{2}} e^{(\frac{\delta t}{2}-s)\mathscr{G}} \Psi(t_{n+\frac{1}{2}} + s)ds. \end{aligned} \tag{22.18}$$

We define with d_n the local truncation splitting error for each $n = 0, \ldots, M-1$, which is the residual obtained at time t_n by inserting the exact solution of the semidiscretized system (22.10) in the formal expression of the splitting solution (22.18) as follows

$$\begin{aligned} d_n :=& f_{\mathrm{SD}}(t_{n+1}) - \mathscr{S} f_{\mathrm{SD}}(t_n) - e^{\frac{\delta t}{2}\mathscr{A}} e^{\frac{\delta t}{2}\mathscr{G}} \int_0^{\frac{\delta t}{2}} e^{(\frac{\delta t}{2}-s)\mathscr{G}} \Psi(t_n + s)ds \\ &- e^{\frac{\delta t}{2}\mathscr{A}} \int_0^{\frac{\delta t}{2}} e^{(\frac{\delta t}{2}-s)\mathscr{G}} \Psi(t_{n+\frac{1}{2}} + s)ds. \end{aligned} \tag{22.19}$$

Define the global splitting error at time t_n as $E_n := f_{SD}(t_n) - f_{SP}(t_n)$. Subtracting (22.18) from (22.19), we obtain the following relation

$$E_{n+1} = \mathscr{S} E_n + d_n \tag{22.20}$$

Exploiting the linearity of the solution operator $\mathscr{S}$ and the fact that $E_0 = 0$, we recursively apply (22.20), and obtain

$$E_n = \sum_{k=0}^{n} \mathscr{S}^{n-k} d_k. \tag{22.21}$$

Proposition 22.3 *Let Ψ in* (22.10) *be of class $\mathscr{C}^1([t_0, t_f])$. Then $\|f_{SD} - f_{SP}\|_{L^2_{h,\delta t}} = \mathscr{O}(\delta t^2)$.*

Next, we discuss a bound for the third term in (22.16), where the full numerical solution $\hat{f} = f_j^n$, $j = 1, \ldots, N$, $n = 0, \ldots, M$, is given by either the Euler discretization or by the predictor corrector scheme. By the definition of the operator $\mathscr{S}$ in (22.17), we write the solution of (22.13) in a convenient form as follows

$$f_{SP}(t_{n+1}) = \mathscr{S} f_{SP}(t_n) + e^{\frac{\delta t}{2}\mathscr{A}} \int_0^{\delta t} e^{(\delta t - s)\mathscr{G}} \Psi(t_n + s) ds, \tag{22.22}$$

where we applied three times the variation of constants formula [12] to (22.13).

Next, we write (22.14) in a compact form. Given f^n, the computation of f^{n+1} is carried out as follows

$$\begin{aligned} f^{n+1} &= \left(I - \frac{\delta t}{2}\mathscr{A}\right)^{-1} \left((I + \delta t \mathscr{G})(I - \frac{\delta t}{2}\mathscr{A})^{-1} f^n + \delta t \Psi(t_n)\right) \\ &= R_1(\mathscr{A}, \mathscr{G}, \delta t) f^n + \delta t \left(I - \frac{\delta t}{2}\mathscr{A}\right)^{-1} \Psi(t_n), \end{aligned} \tag{22.23}$$

where $R_1(\mathscr{A}, \mathscr{G}, \delta t) := \left(I - \frac{\delta t}{2}\mathscr{A}\right)^{-1} (I + \delta t \mathscr{G})(I - \frac{\delta t}{2}\mathscr{A})^{-1}$ is the amplification factor. The non singularity of the matrix $I - \frac{\delta t}{2}\mathscr{A}$ is discussed in [15].

For each time window $[t_n, t_{n+1}]$ we define the respective time truncation error T_n, obtained by inserting the formal splitting solution f_{SP} defined in (22.22) in the numerical approximation given by (22.23). We have

$$T_n := f_{SP}(t_{n+1}) - R_1(\mathscr{A}, \mathscr{G}, \delta t) f_{SP}(t_n) - \delta t \left(I - \frac{\delta t}{2}\mathscr{A}\right)^{-1} \Psi(t_n) \tag{22.24}$$

Proposition 22.4 (Consistency of SIMEX1) *The truncation error* (22.24) *is of order $\mathscr{O}(\delta t^2)$.*

Next, we define, for each $n = 1, \ldots, M$, the time discretization error e_n as follows

$$e_n := f_{\text{SP}}(t_n) - f^n,$$

such that by subtracting (22.24) from (22.23), we obtain the following relation

$$e_{n+1} = R_1(\mathscr{A}, \mathscr{G}, \delta t)e_n + T_n, \tag{22.25}$$

Proposition 22.5 (Accuracy of SIMEX1) *If the time step* δt *is chosen such that* $\|R_1(\mathscr{A}, \mathscr{G}, \delta t)\|_{L_h^2} \leq 1$, *then* $\|f_{SP} - \hat{f}\|_{L^2_{h,\delta t}} = \mathscr{O}(\delta t)$.

Next, we write (22.15) in a compact form. Given f^n, the computation of f^{n+1} is carried out as follows

$$f^{n+1} = R\left(\frac{\delta t}{2}\mathscr{A}\right) R(\delta t \mathscr{G}) R\left(\frac{\delta t}{2}\mathscr{A}\right) f^n + R\left(\frac{\delta t}{2}\mathscr{A}\right) \bar{\Psi}_n, \tag{22.26}$$

where

$$\bar{\Psi}_n := \frac{\delta t}{2}\left[(I + \delta t \mathscr{G})\Psi(t_n) + \Psi(t_{n+1})\right]$$

and the function R is the amplification factor; given a matrix M and $z \in \mathbb{R}$, R is defined as $R(zM) := I + zM + \frac{z^2}{2}M^2$.

For each time window $[t_n, t_{n+1}]$, we define the time truncation error T_n, obtained by inserting the formal splitting solution f_{SP} defined in (22.22) in the numerical approximation given by (22.26)

$$T_n := f_{\text{SP}}(t_{n+1}) - R\left(\frac{\delta t}{2}\mathscr{A}\right) R(\delta t \mathscr{G}) R\left(\frac{\delta t}{2}\mathscr{A}\right) f_{\text{SP}}(t_n) - R\left(\frac{\delta t}{2}\mathscr{A}\right) \bar{\Psi}_n. \tag{22.27}$$

Proposition 22.6 (Consistency of SIMEX2) *The truncation error* (22.27) *is of order* $\mathscr{O}(\delta t^3)$.

Now, we introduce $R_2(\mathscr{A}, \mathscr{G}, \delta t) := R\left(\frac{\delta t}{2}\mathscr{A}\right) R(\delta t \mathscr{G}) R\left(\frac{\delta t}{2}\mathscr{A}\right)$ and state the following.

Proposition 22.7 (Accuracy of SIMEX2) *If the time step* δt *is chosen such that* $\|R_2(\mathscr{A}, \mathscr{G}, \delta t)\|_{L_h^2} \leq 1$, *then* $\|f_{SP} - \hat{f}\|_{L^2_{h,\delta t}} = \mathscr{O}(\delta t^2)$.

22.5 Positivity and Conservativeness of the SIMEX Schemes

In this section, we show that the SIMEX1 and SIMEX2 schemes are conservative and conditionally positive preserving. First, we focus on the SIMEX1 scheme, where the time discretization is given by the Euler scheme. Given the numerical solution f^n at time t_n, we compute f^{n+1} as follows

$$\begin{aligned} &1. \frac{f^{n+\frac{1}{2}} - f^n}{\frac{\delta t}{2}} = \mathscr{A} f^{n+\frac{1}{2}} \\ &2. \frac{f^{n+\frac{1}{2}^*} - f^{n+\frac{1}{2}}}{\delta t} = \mathscr{G} f^{n+\frac{1}{2}} \\ &3. \frac{f^{n+1} - f^{n+\frac{1}{2}^*}}{\frac{\delta t}{2}} = \mathscr{A} f^{n+1}. \end{aligned} \tag{22.28}$$

Proposition 22.8 *Let us consider* (22.28). *Assume that* $\delta t \le \min\{\frac{1}{\lambda}, \frac{2}{L}\}$, *with* λ *rate of jumps of the compound Poisson process P in* (22.1) *and let L be the Lipschitz constant of the function B that defines* $\mathscr{F}$ *in* (22.7). *If* $f_j^n \ge 0$ *for each* $j = 1, \ldots, N$, *then* $f_j^{n+1} \ge 0$ *for each* $j = 1, \ldots, N$.

Proposition 22.9 *Let us consider* (22.28). *The total probability is preserved, in the sense that*

$$\sum_{j=1}^{N} f_j^n = \sum_{j=1}^{N} f_j^{n+1}.$$

Next, we focus on the SIMEX2 scheme, where the time discretization is given by the predictor-corrector scheme. Given the numerical solution f^n at time t_n, the three steps required to compute f^{n+1} are as follows

$$\begin{aligned} &1. \begin{cases} \bar{f}^{n+\frac{1}{2}} = f^n + \frac{\delta t}{2} \mathscr{A} f^n \\ f^{n+\frac{1}{2}} = f^n + \frac{\delta t}{4} \left[\mathscr{A} f^n + \mathscr{A} \bar{f}^{n+\frac{1}{2}} \right] \end{cases} \\ &2. \begin{cases} \bar{f}^{n+\frac{1}{2}^*} = f^{n+\frac{1}{2}} + \delta t \mathscr{G} f^{n+\frac{1}{2}} \\ f^{n+\frac{1}{2}^*} = f^{n+\frac{1}{2}} + \frac{\delta t}{2} \left[\mathscr{G} f^{n+\frac{1}{2}} + \mathscr{G} \bar{f}^{n+\frac{1}{2}} \right] \end{cases} \\ &3. \begin{cases} \bar{f}^{n+1} = f^{n+\frac{1}{2}^*} + \frac{\delta t}{2} \mathscr{A} f^{n+\frac{1}{2}^*} \\ f^{n+1} = f^{n+\frac{1}{2}^*} + \frac{\delta t}{4} \left[\mathscr{A} f^{n+\frac{1}{2}^*} + \mathscr{A} \bar{f}^{n+1} \right]. \end{cases} \end{aligned} \tag{22.29}$$

Proposition 22.10 *Let us consider* (22.29) *and suppose that* $\delta t \le \min\{\frac{1}{\lambda}, \frac{2}{\max_j |\mathscr{A}_{jj}|}\}$, *where* λ *is the rate of the compound Poisson process P in* (22.1), *and* $\mathscr{A}_{jj}$ *are the diagonal elements of* $\mathscr{A}$ *as defined in* (22.11). *If* $f_j^n \ge 0$ *for each* $j = 1, \ldots, N$, *then* $f_j^{n+1} \ge 0$ *for each* $j = 1, \ldots, N$.

Remark 22.1 We obtain an estimate for $\max_j |\mathscr{A}_{jj}|$ as follows.

$$|\mathscr{A}_{jj}| = (\beta_i + \beta_{i-1} w_{i-1})/h = \frac{C_{j+1/2}}{h^2} \frac{r_{j+1/2}}{e^{r_{j+1/2}} - 1} + \frac{C_{j-1/2}}{h^2} \frac{-r_{j-1/2}}{e^{-r_{j-1/2}} - 1}$$

where $r(x) = hB(x)/C(x)$ is the Péclet number, and $r_{j+1/2} = r(x_{j+1/2})$. Therefore,

$$\max_j |\mathscr{A}_{jj}| \leq \frac{\max_x(C(x))}{h^2}(\max(1, 1 - r(x)) + \max(1, 1 + r(x))) \leq$$
$$\frac{\max_x(C(x))}{h^2}(2 + \max_x(|r(x)|)).$$

Hence, the positivity bound to δt in the previous theorem becomes

$$\delta t \leq \min\{\frac{1}{\lambda}, \frac{2h^2}{\max_x\{C(x)\}(2 + h \max_x\{|B(x)|/C(x)\})}\}.$$

We observe that for vanishing space step size h, the time step size must vanishes with order 2. In small diffusion regime, i.e. $C(x) \simeq 0$, it scales linearly as $\delta t < 2h/\max_x\{B(x)\}$.

Proposition 22.11 *Let us consider* (22.29). *The total probability is conserved, in the sense that*

$$\sum_{j=1}^{N} f_j^n = \sum_{j=1}^{N} f_j^{n+1}.$$

22.6 Numerical Experiments

In this section, we present results of numerical experiments for the FP problem (22.4). We set $b(x) = -x$ and $\sigma = \sqrt{2}$, i.e. $a(x) = 1$ (Ornstein-Uhlenbeck process, e.g. see [18] for applications). The jumps of the compound Poisson process are chosen to have rate $\lambda = 5$ and the distribution g is chosen to be $g \sim \mathscr{N}(3, 0.2^2)$. We consider the domain $\Omega = (-15, 30)$ and $I = [0, 1]$. In order to test the performance of the SIMEX schemes, we set the solution to (22.4) as the following a moving Gaussian

$$f(x,t) = \frac{1}{\sqrt{2\pi}\tilde{\sigma}} \exp\left\{-\frac{(x - \mu t)^2}{2\tilde{\sigma}^2}\right\},$$

with $\mu = 10$ and $\tilde{\sigma} = 3$, hence compute analytically the corresponding source term $\psi(x,t)$, as discussed in Sect. 22.4.

Table 22.1 $L^2_{h,\delta t}$-error of the scheme SIMEX1

N	M	$\|f-\hat{f}\|_{h,\delta t}$
100	100	$5.39 \cdot 10^{-3}$
200	400	$1.09 \cdot 10^{-3}$
400	800	$2.90 \cdot 10^{-4}$
800	1600	$7.37 \cdot 10^{-5}$

Table 22.2 $L^2_{h,\delta t}$-error of the scheme SIMEX2

N	M	$\|f-\hat{f}\|_{h,\delta t}$
200	200	$1.10 \cdot 10^{-3}$
400	400	$2.93 \cdot 10^{-4}$
800	800	$7.45 \cdot 10^{-5}$
1600	1600	$1.87 \cdot 10^{-5}$

In Table 22.1, we report the norm of the SIMEX1 solution error as a function of the mesh size. We see that the scheme is first-order accurate in time and second-order accurate in space, as proved in Sect. 22.4.

In Table 22.2, we present results for the same test case, obtained with the SIMEX2 method. We have second-order convergence in time and space.

References

1. Annunziato, M., Borzì, A.: A Fokker-Planck control framework for multidimensional stochastic processes. J. Comput. Appl. Math. **237**, 487–507 (2013)
2. Applebaum, D.: Lévy Processes and Stochastic Calculus. Cambridge University Press, Cambridge (2004)
3. Briani, M., Natalini, R.: Asymptotic high-order schemes for integro-differential problems arising in markets with jumps. Commun. Math. Sci. **4**, 81–96 (2006)
4. Chang, J.S., Cooper, G.: A practical difference scheme for Fokker-Planck equations. J. Comput. Phys. **6**, 1–16 (1970)
5. Cont, R., Tankov, P.: Financial Modeling with Jump Processes. Chapman and Hall, London (2004)
6. Cont, R., Voltchkova, E.: Integro-differential equations for option prices in exponential Lévy models. Finance Stochast. **9**, 299–325 (2005)
7. Dahlquist, G., Björck, A.: Numerical Methods in Scientific Computing, vol. 1. Society for Industrial and Applied Mathematics, Philadelphia (2008)
8. Gardiner, C.: Stochastic Methods: A Handbook for the Natural and Social Sciences. Springer, Berlin (2009)
9. Garroni, M.G., Menaldi, J.L.: Green Functions for Second-Order Parabolic Integro-Differential Problems. Longman, Harlow (1992)
10. Gaviraghi, B., Annunziato, M., Borzì, A.: Analysis of splitting methods for solving a partial-integro differential Fokker-Planck equation. Appl. Math. Comput. **294**, 1–17 (2017)
11. Geiser, J.: Decomposition Methods for Differential Equations: Theory and Applications. Chapman and Hall, London (2009)
12. Hundsdorfer, W., Verwer, J.G.: Numerical Solutions of Time-Dependent Advection-Diffusion-Reaction Equations. Springer, Berlin (2010)

13. Kloeden, P.E., Platen, E.: Numerical Solutions of Stochastic Differential Equations. Springer, Berlin (1999)
14. Marchuk, G.I.: Methods of Numerical Mathematics. Springer, New York (1981)
15. Mohammadi, M., Borzì, A.: Analysis of the Chang-Cooper discretization scheme for a class of Fokker-Planck equations. J. Numer. Math. **23**, 271–288 (2015)
16. Pascucci, A.: PDE and Martingale Methods in Option Pricing. Springer, New York (2011)
17. Paul, W., Baschnagel, J.: Stochastic Processes, from Physics to Finance. Springer, New York (2010)
18. Priola, E., Zabczyk, J.: Densities for Ornstein-Uhlenbeck processes with jumps. Bull. Lond. Math. Soc. **41**, 41–50 (2009)
19. Risken, R.: The Fokker-Planck Equation: Methods of Solution and Applications. Springer, New York (1996)
20. Strang, G.: On the construction and comparison of difference schemes. SIAM J. Numer. Anal. **5**, 506–517 (1968)

Chapter 23
A Fokker-Planck Based Approach to Control Jump Processes

Beatrice Gaviraghi, Mario Annunziato, and Alfio Borzì

Abstract A framework for the optimal sparse-control of the probability density function of a jump-diffusion process is presented. This framework is based on the partial integro-differential Fokker-Planck (FP) equation that governs the time evolution of the probability density function of this process. In the stochastic process and, correspondingly, in the FP model the control function enters as a time-dependent coefficient. The objectives of the control are to minimize a discrete-in-time, resp. continuous-in-time, tracking functionals and its L^2- and L^1-costs, where the latter is considered to promote control sparsity. An efficient proximal scheme for solving these optimal control problems is considered. Results of numerical experiments are presented to validate the theoretical results and the computational effectiveness of the proposed control framework. (This chapter is a summary of the paper Gaviraghi et al. (Appl Math 7:1978–2004, 2016); all theoretical statements in this summary are proved in that reference.)

23.1 Introduction

In this chapter, we focus on control problems related to stochastic jump-diffusion (JD) processes evolving in a bounded domain with reflecting barriers. These problems are very important in the modeling of the dynamics of stock prices [6, 18] and of market models; see, e.g., [14]. In both cases, the need arises to control these models involving random quantities and, correspondingly, stochastic optimization problems must be solved. Such problems have largely been examined in the scientific literature [9, 19, 27]. In this work, we consider the problem of controlling a stochastic process by following an alternative approach already proposed in [1, 2, 20], where the problem is reformulated from stochastic to deterministic. The key idea of this strategy is to assume the probability density function (PDF) be the

B. Gaviraghi • M. Annunziato • A. Borzì (✉)
Lehrstuhl für Mathematik IX (Chair Scientific Computing), University of Würzburg, Würzburg, Germany
e-mail: beatrice.gaviraghi@mathematik.uni-wuerzburg.de; alfio.borzi@mathematik.uni-wuerzburg.de; mannunzi@unisa.it

M. Ehrhardt et al. (eds.), *Novel Methods in Computational Finance*, Mathematics in Industry 25, DOI 10.1007/978-3-319-61282-9_23

state function of the considered process. The time evolution of the PDF is modeled by the Fokker-Planck (FP) equation, also known as forward Kolmogorov equation.

In the case of JD process, the FP equation takes the form of a partial-integro differential equation (PIDE), which endowed with initial and boundary conditions, gives rise to a FP initial-boundary value problem. While the Cauchy data must be the initial distribution of the given random variable, the boundary conditions of a FP problem depend on the properties of the barriers on the considered stochastic model [7, 23]. Furthermore, starting from a controlled stochastic differential equation, a controlled FP equation is obtained, leading to the formulation of optimal control problems governed by the FP PIDE. For a general reference on PDE-based optimal control problems see [4, 16, 25]. The strategy of designing control functions for stochastic models based on the FP formulation was considered in [1–3, 20] for different stochastic problems other than JD processes. In the present work, FP sparse control problems are considered for the first time.

For control purposes, we focus on tracking objectives that include mean expectation values as in [15] and costs of the control that include sparsity promoting L^1-cost functionals as in, e.g., [22, 24]. With the aim of solving these optimization problems, we use a very recent proximal iterative schemes developed in [17, 21].

In the next section, we discuss the functional setting of the FP problem modeling the evolution of the PDF of a JD stochastic process. In Sect. 23.3, we formulate our optimal control problems. Section 23.4 is devoted to the formulation of the corresponding first-order optimality systems. In Sect. 23.5, we illustrate a proximal method for solving our optimal control problems. Section 23.6 is devoted to the illustration of our numerical tests, including a discussion on the robustness of the algorithm to the choice of the parameters of the optimization problem.

23.2 The Fokker-Planck Equation of a Jump-Diffusion Process

Consider a time interval $I := [t_0, t_f]$ and a stochastic process $X = \{X_t\}_{t\in I}$ with range in a bounded domain $\Omega \subset \mathbb{R}^n$. We assume that the set Ω is convex with Lipschitz boundary. The dynamic of X is governed by the following stochastic differential problem

$$\begin{cases} dX_t = b(X_t, t)dt + \sigma(X_t)dW_t + dP_t \\ X_{t_0} = X_0, \end{cases} \tag{23.1}$$

where $X_0 \in \mathbb{R}^n$ is a random variable with known distribution. The functions $b : \Omega \times I \to \mathbb{R}^n$ and $\sigma : \Omega \to \mathbb{R}^{n\times d}$ represent the drift and the diffusion coefficients, respectively. We assume that σ is full rank. Random increments to the process X in (23.1) are given by a Wiener process $W \in \mathbb{R}^d$ and a compound Poisson process $P \in \mathbb{R}^n$. The rate of jumps and the jump distribution are denoted with $\lambda \in \mathbb{R}^+$ and

$\hat{g}$, respectively. Define $a : \Omega \to \mathbb{R}^{n\times n}$, $a_{ij}(x) := \sum_{k=1}^{d} \frac{1}{2}\sigma_{ik}(x)\sigma_{jk}(x)$. Since σ is full rank, a is positive definite, and hence there exists $c_a > 0$ such that

$$v^T a(x) v \geq c_a \|v\|^2_{\mathbb{R}^n}, \qquad \text{for each } v \in \mathbb{R}^n, \text{ a.e. in } \Omega. \tag{23.2}$$

In this work, we consider a stochastic process with reflecting barriers. This assumption determines the boundary conditions for the FP equation corresponding to (23.1), see below. Define $Q := \Omega \times I$ and $\Sigma := \partial\Omega \times I$, and denote with f the PDF of the process given by (23.1). It is known [7, 23] that the time evolution of f is modeled by the following FP equation of PIDE type

$$\partial_t f(x,t) = \mathcal{L}f(x,t) + \mathcal{I}f(x,t), \quad (x,t) \in Q, \tag{23.3}$$

where the differential operator $\mathcal{L}$ and the integral operator $\mathcal{I}$ are defined as follows

$$\mathcal{L}f(x,t) := -\sum_{i=1}^{n} \partial_i \left(b(x,t)f(x,t)\right) + \sum_{i,j=1}^{n} \partial^2_{ij}\left((a(x)f(x,t)\right), \tag{23.4}$$

and

$$\mathcal{I}f(x,t) := \lambda \left[\int_{\mathbb{R}^n} f(y,t) g(x,y) dy - f(x,t) \right], \tag{23.5}$$

respectively. The definition of g in (23.5) takes into account the presence of reflecting barriers and the dependence on the jump amplitude $\hat{g}$, as we discuss later.

Notice that the differential operator $\mathcal{L}$ can be rewritten as follows

$$\mathcal{L}f(x,t) := \nabla \cdot F(x,t),$$

where the flux of the differential part is given by

$$F(x,t) := B(x,t)f(x,t) + a(x)\nabla f(x,t). \tag{23.6}$$

and

$$B_i(x) := \sum_{j=1}^{n} \partial_j a_{ij}(x) - b_i(x,t), \tag{23.7}$$

for each $i = 1, \ldots, n$.

The PDF f of X in (23.1) in the bounded domain Ω is obtained by solving (23.3), endowed by suitable initial and boundary conditions. In our setting, the initial data f_0 represents the PDF of the initial random variable X_0. The choice of a bounded domain with reflecting barriers results in the following zero-flux boundary conditions for the FP model

$$F(x,t) \cdot \hat{n}(x,t) = 0 \text{ for } (x,t) \in \Sigma, \tag{23.8}$$

where $\hat{n}$ is the unit outward normal on $\partial\Omega$.

Notice that the flux F corresponds to the differential part of the FP equation, that is, to the drift and diffusion components of the stochastic process. In order to take into account the action of a reflecting barrier on the jumps, we consider a suitable definition of the kernel g, which can be conveniently illustrated in the one-dimensional case as follows.

Consider $n = 1$ and $\Omega = (r, s)$. We consider a jump density $\hat{g}$ with compact support $(r-s, s-r)$. With this choice, multiple reflections in the dynamic of the JD process are avoided. The kernel g in (23.5) takes the following form

$$g(x, y) := \hat{g}(x-y) + \hat{g}(2r-x-y)\Theta(r+s-x-y) + \hat{g}(2s-x-y)\Theta(x+y-r-s), \tag{23.9}$$

where Θ is the step function defined by

$$\Theta(z) := \begin{cases} 1, & z > 0 \\ 0, & z < 0 \\ 1/2, & z = 0. \end{cases}$$

Since $\hat{g}$ is a PDF, we have that $\int_{r-s}^{s-r} \hat{g}(z)\, dz = 1$; from (23.9), it results that the function g satisfies

$$\int_{\mathbb{R}} g(x, y)dx = 1 \text{ for all } y \in \Omega. \tag{23.10}$$

Our FP problem is stated as follows

$$\begin{cases} \partial_t f(x, t) = \nabla \cdot F(x, t) + \mathcal{I} f(x, t) & \text{for } (x, t) \in Q \\ f(x, 0) = f_0(x) & \text{for } x \in \Omega \\ F(x, t) \cdot n(x, t) = 0 & \text{for } (x, t) \in \Sigma. \end{cases} \tag{23.11}$$

The choice of the boundary conditions (23.8) together with (23.10) ensures that the total probability is preserved over time.

Next, we recall some definitions concerning the functional spaces needed to state the existence and uniqueness of solutions to (23.11). The space $C^0(\Omega)$ refers to the functions that are continuous in Ω and it is endowed with the supremum norm. Let α be a constant, $\alpha > 1/2$. The space $C^\alpha(\Sigma)$ refers to the functions that are Hölder continuous on Σ, with Hölder exponent α with respect to the space variable. The space $L^\infty(\Omega)$ denotes all the functions that are bounded on Ω, up to a set of zero measure. The spaces $H^1(\Omega)$ and $H^{2,1}(Q)$ are defined as follows

$$\begin{aligned} H^1(\Omega) :=& \{v \in L^2(\Omega) \mid \partial_i v \in L^2(\Omega), i = 1, \ldots, n\}, \\ H^{2,1}(Q) :=& \{v \in L^2(Q) \mid \partial_t v, \partial_i v, \partial^2_{ij} v \in L^2(Q), i, j = 1, \ldots, n\}. \end{aligned} \tag{23.12}$$

These spaces are endowed with the following norms

$$\|v\|_{H^1(\Omega)} := \sum_{|i|=0}^{1} \|\partial^i v\|_{L^2(\Omega)}, \quad \|v\|_{H^{2,1}(Q)} := \sum_{2j+|i|=0}^{2} \|\partial_t^j \partial_x^i v\|_{L^2(Q)},$$

where $i, j \in \mathbb{N}^n$ denote multi-indices. We also consider the Gelfand triple $V \subset H \subset V^*$, with $V := H^1(\Omega)$ and $H := L^2(\Omega)$. The space V^* denotes the dual space of V. This allows us to define the following space $W := \{y \in L^2(I; V)$ with $\partial_t y \in L^2(I; V^*)\}$.

We assume that the coefficients a and b in (23.4) satisfy the following conditions

$$\begin{aligned} &a(x) \in C^0(\Omega) \cap C^\alpha(\partial\Omega), \partial_i a(x) \in L^\infty(\Omega) \cap C^\alpha(\partial\Omega), \partial_{ij}^2 a(x) \in L^\infty(\Omega) \\ &b_i(x,t) \in L^\infty(Q) \cap C^\alpha(\Sigma), \partial_i b_j(x,t) \in L^\infty(Q) \end{aligned} \tag{23.13}$$

for each $i, j = 1, \ldots, n$.

Proposition 23.1 *Assume* (23.13) *and* $f_0 \in H^1(\Omega)$. *Then the unique solution to* (23.11) *belongs to* $L^2(I; V)$, *with* $\partial_t f \in L^2(I; V^*)$. *Therefore,* $f \in W$. *Moreover,* $f \in C(I; H)$.

We define $\mathscr{F} := \{f \in W \,|\, f(0) = f_0\}$, where f_0 is the initial data in (23.11). The initial-boundary value problem (23.11) can be stated as $\mathscr{K}(f) = 0$, where the map $\mathscr{K}$ is defined as follows

$$\mathscr{K}(f) := \begin{cases} \partial_t f - \nabla \cdot F(f) - \mathscr{I} f & \text{in } Q \\ F(f), & \text{on } \Sigma, \end{cases} \tag{23.14}$$

with F and $\mathscr{I}$ defined in (23.6) and (23.5), respectively.

23.3 Two Fokker-Planck Optimal Control Problems

In this section we discuss existence of solutions to our optimal control problems governed by (23.14). We consider a control function u that acts through the drift function $b = b(x, t; u)$ by means of a time-dependent control $u = u(t) \in \mathbb{R}$. Therefore we refer to (23.14) as $\mathscr{K}(f, u) = 0$. We assume that b is a smooth function of its arguments and that assumption (23.13) is fulfilled. We remark that a time-dependent control function is a natural choice considering that it originates from the stochastic differential model where the time is the only independent variable.

We assume the presence of control constraints given by $u_a, u_b \in \mathbb{R}$, with $u_a < 0 < u_b$. We denote

$$\mathscr{U}_{ad} := \{u \in \mathscr{U} : u_a \le u \le u_b\} \subset \mathscr{U} := L^2(I). \tag{23.15}$$

Let ν and γ be positive constants. We consider the following objective

$$\mathscr{J}(f,u) := D(f) + \frac{\nu}{2}\|u\|_2^2 + \gamma\|u\|_1. \tag{23.16}$$

The term $D(f)$ in (23.16) represents a tracking objective that involves the expectation value of X_t, $\mathbb{E}[X_t] := \int_\Omega xf(x,t)dx$, and a desired trajectory or a discrete set of values (e.g. measurements). We investigate the following two cases.

1. Discrete time tracking. Given a set of values $\{\xi_k\}_{k=1}^K$ at different times $t_k \in (t_0, t_f)$, $\forall k =, 1, \ldots, K$, we have

$$D(f) := \frac{1}{K}\sum_{k=1}^{K}\left(\xi_k - \int_\Omega xf(x,t_k)dx\right)^2. \tag{23.17}$$

2. Continuous time tracking. Given a square-integrable function $\xi : I \to \mathbb{R}$, we have

$$D(f) := \int_{t_0}^{t_f}\left(\xi(t) - \int_\Omega xf(x,t)dx\right)^2 dt. \tag{23.18}$$

The norms in (23.16) are defined as follows

$$\|u\|_2 := \left(\int_I |u(t)|^2 dt\right)^{1/2} \text{ and } \|u\|_1 := \int_I |u(t)|dt.$$

We investigate the following optimal control problem(s)

$$\begin{aligned} &\min \mathscr{J}(f,u) \\ &\text{such that } \mathscr{K}(f,u) = 0, \quad (f,u) \in \mathscr{F} \times \mathscr{U}_{ad}. \end{aligned} \tag{23.19}$$

In order to discuss the existence of solutions to (23.19), we consider the control-to-state operator $\mathscr{S} : \mathscr{U} \to \mathscr{F}$, that maps a given $u \in \mathscr{U}$ into $\mathscr{S}(u) := f \in \mathscr{F}$, where (f,u) satisfies $\mathscr{K}(f,u) = 0$. Note that the definition of $\mathscr{U}_{ad}$ in (23.15) ensures that b satisfies (23.13). Because of Proposition 23.1, the operator $\mathscr{S}$ is well defined.

Therefore the constrained optimization problem (23.19) can be transformed into an unconstrained one as follows

$$\min_{u \in \mathscr{U}_{ad}} \hat{\mathscr{J}}(u), \tag{23.20}$$

where $\hat{\mathscr{J}} : u \mapsto \hat{\mathscr{J}}(u) := \mathscr{J}(\mathscr{S}(u), u)$ is the so-called reduced cost functional.

The solvability of (23.20) is ensured by the next theorem, whose proof adapts techniques given in [25] and [10].

Theorem 23.1 *There exists at least one optimal pair* $(\bar{f}, \bar{u})$ *that solves* (23.19)*, such that* $\bar{u}$ *solves* (23.20) *and* $\bar{f} = \mathscr{S}(\bar{u}) \in \mathscr{F}$.

The next proposition can be proved by using standard arguments [2, 25].

Proposition 23.2 *The mapping* $\mathscr{S} : \mathscr{U} \to C(I; H), u \mapsto f = \mathscr{S}(u)$ *solution to* (23.11) *is Fréchet differentiable and the directional derivative* $e := \mathscr{S}'(u) \cdot h$ *satisfies the following initial-boundary problem*

$$\begin{cases} \partial_t e = \nabla \cdot F(e) + \mathscr{I}(e) - \nabla \cdot (f b'(u) h) & \text{for } (x,t) \in Q, \\ [F(e) - f b'(u) h] \cdot n = 0 & \text{for } (x,t) \in \Sigma, \\ e(x, t_0) = 0 & \text{for } x \in \Omega, \end{cases} \tag{23.21}$$

where b is the drift in (23.1) *and F is defined in* (23.6).

23.4 Two First-Order Optimality Systems

We follow the standard approach [8, 25, 26] of characterizing the solution of our optimal control problem as the solution to first-order optimality conditions that constitute the optimality system.

Consider the reduced problem (23.20) and write the reduced functional $\hat{\mathscr{J}}$ as $\hat{\mathscr{J}} := \mathscr{J}_1 + \mathscr{J}_2$, $\mathscr{J}_i : \mathscr{U} \to \mathbb{R}^+$, $i = 1, 2$, where

$$\begin{aligned} \mathscr{J}_1(u) &:= D(\mathscr{S}(u)) + \frac{\nu}{2} \|u\|_2^2, \\ \mathscr{J}_2(u) &:= \gamma \|u\|_1. \end{aligned} \tag{23.22}$$

Remark 23.1 The functional $\mathscr{J}_1$ is smooth thanks to Proposition 23.2 and possibly nonconvex, while $\mathscr{J}_2$ is convex and nonsmooth.

The following definitions are needed in order to determine the first-order optimality system. If $\hat{\mathscr{J}}$ is finite at a point u, the Fréchet subdifferential of $\hat{\mathscr{J}}$ at u is defined as follows [8]. We have

$$\partial \hat{\mathscr{J}}(u) := \left\{ \varphi \in \mathscr{U}^* : \liminf_{v \to u} \frac{\hat{\mathscr{J}}(v) - \hat{\mathscr{J}}(u) - \langle \varphi, v - u \rangle}{\|v - u\|_2} \geq 0 \right\}, \tag{23.23}$$

where $\mathscr{U}^*$ is the dual space of $\mathscr{U}$. Any element $\varphi \in \partial \hat{\mathscr{J}}(u)$ is called a subgradient. In our framework, we have

$$\partial \hat{\mathscr{J}}(u) = \nabla \mathscr{J}_1(u) + \partial \mathscr{J}_2(u),$$

since $\mathcal{J}_1$ is Fréchet differentiable at u; this follows from standard arguments [25]. Moreover, for each $\alpha > 0$, it holds that $\partial(\alpha \hat{\mathcal{J}}) = \alpha\partial\hat{\mathcal{J}}$.

The following proposition gives a necessary condition for a local minimum of $\hat{\mathcal{J}}$.

Proposition 23.3 *If $\hat{\mathcal{J}} = \mathcal{J}_1 + \mathcal{J}_2$, with $\mathcal{J}_1$ and $\mathcal{J}_2$ given by* (23.22)*, attains a local minimum in $\bar{u} \in \mathcal{U}_{ad}$, then*

$$0 \in \partial\hat{\mathcal{J}}(\bar{u}),$$

or equivalently

$$-\nabla\mathcal{J}_1(\bar{u}) \in \partial\mathcal{J}_2(\bar{u}).$$

With this preparation and using results in [8, 24], we have that each $\lambda \in \partial\mathcal{J}_2(\bar{u})$, with $\bar{u}$ a local minimum, satisfies the following inequality

$$\langle\nabla\mathcal{J}_1(\bar{u}) + \lambda, v - \bar{u}\rangle \geq 0 \text{ for each } v \in \mathcal{U}_{ad}. \tag{23.24}$$

Moreover, recalling the definition of $\mathcal{J}_2$ in (23.22) and exploiting the isomorphism $\mathcal{U}^* \simeq \mathcal{U}$, the inclusion $\lambda \in \partial\mathcal{J}_2(\bar{u})$ gives the following

$$\lambda \in \Lambda_\gamma := \{l \in L^2(I) : |l| \leq \gamma \text{ a.e. on } I\}. \tag{23.25}$$

A pointwise analysis of (23.24), which takes into account the definition (23.15) of the admissible controls, ensures the existence of two nonnegative functions $\bar{\lambda}_a, \bar{\lambda}_b \in \mathcal{U}^*$ that play the role of Lagrange multipliers [25]. The previous considerations lead to the following proposition, that states the optimality system for the reduced problem (23.20).

Proposition 23.4 *The optimal solution $\bar{u}$ of the minimization problem* (23.20) *with $\hat{\mathcal{J}} = \mathcal{J}_1 + \mathcal{J}_2$ defined in* (23.22)*, is characterized by the existence of $(\bar{\lambda}, \bar{\lambda}_a, \bar{\lambda}_b) \in \Lambda_\gamma \times \mathcal{U}^* \times \mathcal{U}^*$ such that*

$$\begin{cases} \nabla\mathcal{J}_1(\bar{u}) + \bar{\lambda} - \bar{\lambda}_a + \bar{\lambda}_b = 0 \\ \bar{\lambda}_b \geq 0, \qquad u_b - \bar{u} \geq 0, \qquad \langle\bar{\lambda}_b, u_b - \bar{u}\rangle = 0 \\ \bar{\lambda}_a \geq 0, \qquad \bar{u} - u_a \geq 0, \qquad \langle\bar{\lambda}_a, \bar{u} - u_a\rangle = 0 \\ \bar{\lambda} = \gamma \text{ a.e. on } \{t \in I : \bar{u}(t) > 0\} \\ |\bar{\lambda}| \leq \gamma \text{ a.e. on } \{t \in I : \bar{u}(t) = 0\} \\ \bar{\lambda} = -\gamma \text{ a.e. on } \{t \in I : \bar{u}(t) < 0\} \end{cases} \tag{23.26}$$

We refer to the last three conditions in (23.26) *for the pair $(\bar{u}, \bar{\lambda})$ as the complementarity conditions.*

The differentiability of $\mathcal{K}$, $\mathcal{J}_1$ and $\mathcal{S}$ with respect to f and u allows us to compute $\nabla \mathcal{J}_1(u)$ in (23.26) within the adjoint approach. By definition, for each $u \in \mathcal{U}$, we have

$$\nabla \mathcal{J}_1(u) = \nu u + (\mathcal{S}'(u))^* D'(\mathcal{S}(u)).$$

By considering the total derivative of $\mathcal{K}(\mathcal{S}(u), u) = 0$, we have

$$\mathcal{K}_f(\mathcal{S}(u), u)\mathcal{S}'(u) + \mathcal{K}_u(\mathcal{S}(u), u) = 0.$$

Therefore, we obtain

$$\nabla \mathcal{J}_1(u) = \nu u - \mathcal{K}_u(\mathcal{S}(u), u)^* (\mathcal{K}_f(\mathcal{S}(u), u)^*)^{-1} D'(\mathcal{S}(u)).$$

Defining the adjoint variable p as the solution to the following adjoint problem

$$\mathcal{K}_f(\mathcal{S}(u), u)^* p = -D'(\mathcal{S}(u)), \tag{23.27}$$

we obtain the following reduced gradient

$$\nabla \mathcal{J}_1(u) = \nu u + \mathcal{K}_u(f, u)^* p. \tag{23.28}$$

After some calculation, we have that (23.27) can be rewritten as the following adjoint system

$$\begin{cases} -\partial_t p(x,t) = b(x;u)\nabla p(x,t) + \sum_{i,j=1}^n a_{ij}(x)\partial_{ij}^2 p(x,t) + \tilde{\mathcal{I}} p(x,t) + \alpha(x,t) \text{ on } Q \\ p(x,t_f) = 0 \text{ on } \Omega \\ \nabla p(x,t)\cdot n(x,t) = 0 \text{ on } \partial\Omega \times I \\ p(x,t_k^-) = p(x,t_k^+) + \beta(x,k) \text{ on } \Omega, \text{ for each } k = 1,\ldots,K, \end{cases} \tag{23.29}$$

where α and β depend on the choice of D in (23.17) and (23.18). When D is given by (23.17), $\alpha(x,t) = 0$ and $\beta(x,k) = -2x(\xi_k - \int_\Omega sf(s,t_k)ds)$, for each $k = 1,\ldots,K$. On the other hand, when D is given by (23.18), $\alpha(x,t) = 2x(\xi(t) - \int_\Omega sf(s,t)ds)$ and $\beta(x,k) = 0$.

The operator $\tilde{\mathcal{I}}$ is defined as follows

$$\tilde{\mathcal{I}} p(x,t) := \lambda \left[\int_\Omega p(y,t)g(y,x)dy - p(x,t) \right] \text{ for each } (x,t) \in Q.$$

The terminal boundary-value problem (23.29) admits a unique solution $p \in H^{2,1}(Q)$; see [11].

The reduced gradient in (23.28), for given u, f, and p, takes the following form

$$\nabla \hat{\mathscr{J}}_1(u) = \nu u + \int_\Omega f(\nabla p) \cdot (\partial_u b). \tag{23.30}$$

The complementarity conditions in (23.26) can be recast in a more compact form, as follows. We define $\bar{\mu} := \bar{\lambda} - \bar{\lambda}_a + \bar{\lambda}_b$. For each $k \in \mathbb{R}^+$, we define the following quantity

$$\begin{aligned} E(\bar{u}, \bar{\mu}) :=& \bar{u} - \max\{0, \bar{u} + k(\bar{\mu} - \gamma)\} - \min\{0, \bar{u} + k(\bar{\mu} + \gamma)\} + \\ & \max\{0, \bar{u} - u_b + k(\bar{\mu} - \gamma)\} + \min\{0, \bar{u} - u_a + k(\bar{\mu} + \gamma)\}. \end{aligned}$$

The complementarity conditions in (23.26) and the inequalities related to the Lagrange multipliers $\bar{\lambda}_a$ and $\bar{\lambda}_b$, together with the requirement $\bar{\lambda} \in \Lambda_\gamma$, are equivalent to $E(\bar{u}, \bar{\mu}) = 0$; see, e.g., [24].

The previous considerations can be summarized in the following propositions.

Proposition 23.5 (Optimality System for a Discrete-in-Time Tracking Functional)

A local solution $(f, u) \in \mathscr{F} \times \mathscr{U}_{ad}$ of (23.19) with D given by (23.17) is characterized by the existence of $(p, \mu) \in H^{2,1}(Q) \times \mathscr{U}^*$, such that the following system is satisfied

$$\begin{cases} \nu u + \int_\Omega f(\nabla p) \cdot (\partial_u b) + \mu = 0 & \text{a.e. in } I \\ \partial_t f(x,t) = \nabla \cdot F(x,t) + \mathscr{I}(f(x,t)) & \text{for } (x,t) \in Q \\ f(x,0) = f_0(x) & \text{for } x \in \Omega \\ F(x,t) \cdot n(x,t) = 0 & \text{for } (x,t) \in \Sigma \\ -\partial_t p(x,t) = b(x;u)\nabla p(x,t) + \sum_{i,j=1}^n a_{ij}(x)\partial^2_{ij} p(x,t) + \tilde{\mathscr{I}} p(x,t) & \text{for } (x,t) \in Q \\ p(x,t_f) = 0 & \text{for } x \in \Omega \\ \nabla p(x,t) \cdot n(x,t) = 0 & \text{for } (x,t) \in \Sigma \\ p(x,t_k^-) = p(x,t_k^+) - 2x(\xi_k - \int_\Omega s f(s,t_k) ds), \ k = 1, \ldots, K & \text{for } x \in \Omega \\ E(u,\mu) = 0 & \text{a.e. in } I. \end{cases} \tag{23.31}$$

Proposition 23.6 (Optimality System for a Continuous-in-Time Tracking Functional) *A local solution* $(f, u) \in \mathscr{F} \times \mathscr{U}_{ad}$ *of* (23.19) *with* D *given by* (23.18) *is characterized by the existence of* $(p, \mu) \in H^{2,1}(Q) \times \mathscr{U}^*$, *such that the following system is satisfied*

$$\begin{cases} \nu u + \int_\Omega f(\nabla p) \cdot (\partial_u b) + \mu = 0 & a.e.\ in\ I \\ \partial_t f(x,t) = \nabla \cdot F(x,t) + \mathscr{I}(f(x,t)) & for\ (x,t) \in Q \\ f(x,0) = f_0(x) & for\ x \in \Omega \\ F(x,t) \cdot n(x,t) = 0 & for\ (x,t) \in \Sigma \\ -\partial_t p(x,t) = b(x;u)\nabla p(x,t) + \sum_{i,j=1}^n a_{ij}(x)\partial^2_{ij} p(x,t) + \tilde{\mathscr{I}} p(x,)+ & \\ \qquad +2x(\xi(t) - \int_\Omega s f(s,t) ds) & for\ (x,t) \in Q \\ p(x,t_f) = 0 & for\ x \in \Omega \\ \nabla p(x,t) \cdot n(x,t) = 0 & for\ (x,t) \in \Sigma \\ E(u,\mu) = 0 & a.e.\ in\ I. \end{cases} \tag{23.32}$$

Notice that for the discretization of the optimality systems given in (23.31) and (23.32), we use the splitting schemes discussed in [11]; see also [13].

If we follow the optimize-before-discretize (OBD) approach, the optimality system has already been computed on a continuous level as in (23.31) and (23.32) and subsequently discretized. As a consequence, the OBD approach allows one to discretize the forward and adjoint FP problems according to different numerical schemes. However, the OBD procedure might introduce an inconsistency between the discretized objective and the reduced gradient; see [4] and references therein. For this reason, we pursue the DBO (discretize-before-optimize) approach.

23.5 A Proximal Optimization Scheme

In this section, we discuss a proximal optimization scheme for solving (23.20). This scheme and the related theoretical discussion follow the work in [5, 12, 21, 22]. Proximal methods conveniently exploit the additive structure of the reduced objective, and in our framework, we have that the reduced functional $\hat{\mathscr{J}}$ is given by the sum of a nonconvex smooth function $\mathscr{J}_1$ and a convex nonsmooth function $\mathscr{J}_2$ as in (23.22).

For our discussion, we need the following definitions and properties.

Definition 23.1 (Proximity Operator) Let Z be a Hilbert space and l a convex lower semi continuous function, $l : Z \to \mathbb{R}$. The proximity operator $prox_l : Z \to Z$ of l is defined as follows

$$prox_l(z) := \arg\min_{w \in Z} \left\{ l(w) + \frac{1}{2}\|z - w\|_Z^2 \right\}, \quad z \in Z.$$

Proposition 23.7 *Let Z be a Hilbert space and l a convex lower semi continuous function, $l : Z \to \mathbb{R}$, with proximity operator $prox_l$. The following relation holds*

$$p = prox_l(z) \Leftrightarrow z - p \in \partial l(p), \tag{23.33}$$

where ∂l is the subdifferential defined in (23.23).

Proposition 23.8 *The solution $\bar{u}$ of* (23.20) *satisfies*

$$\bar{u} = prox_{\alpha \mathscr{J}_2}(\bar{u} - \alpha \nabla \mathscr{J}_1(\bar{u})). \tag{23.34}$$

for each $\alpha > 0$.

The relation (23.34) suggests that a solution procedure based on a fixed point iteration should be pursued. We discuss how such algorithm can be implemented.

In the following, we assume that $\mathscr{J}_1(u)$ in (23.22) has a locally Lipschitz-continuous gradient $\nabla \mathscr{J}_1$ as follows

$$\|\nabla \mathscr{J}_1(u) - \nabla \mathscr{J}_1(v)\| \leq L\|u - v\|, \tag{23.35}$$

for each $v \in \mathscr{V}$, $\mathscr{V} \subset \mathscr{U}_{ad}$ neighborhood of u, with L a Lipschitz continuity constant. It is shown in [17] that (23.35) implies the following inequality

$$\mathscr{J}_1(u) \leq \mathscr{J}_1(v) + \langle \nabla \mathscr{J}_1(v), u - v \rangle + \frac{L}{2}\|u - v\|^2,$$

for each $v \in \mathscr{V}$, and hence

$$\min_{u \in \mathscr{U}_{ad}} \left\{ \mathscr{J}_1(u) + \mathscr{J}_2(u) \right\} \leq$$

$$\min_{u \in \mathscr{U}_{ad}} \left\{ \mathscr{J}_1(v) + \mathscr{J}_2(u) + \langle \nabla \mathscr{J}_1(v), u - v \rangle + \frac{L}{2}\|u - v\|^2 \right\}. \tag{23.36}$$

Inequality (23.36) is the starting point for the formulation of a proximal scheme, whose strategy consists of minimizing the right-hand side in (23.36). One can prove the following equality

$$\arg\min_{u \in \mathscr{U}_{ad}} \left\{ \mathscr{J}_1(v) + \mathscr{J}_2(u) + \langle \nabla \mathscr{J}_1(v), u - v \rangle + \frac{L}{2}\|u - v\|^2 \right\} =$$

$$\arg\min_{u \in \mathscr{U}_{ad}} \left\{ \mathscr{J}_2(u) + \frac{L}{2} \left\| u - \left(v - \frac{1}{L}\nabla \mathscr{J}_1(v) \right) \right\|^2 \right\}. \tag{23.37}$$

Recall the definition of $\mathscr{J}_2$ in (23.22). The following lemma gives an explicit expression for the right-hand side in (23.37).

Lemma 23.1 *Let $\mathcal{U}_{ad}$ be as in (23.15). Then*

$$\operatorname*{arg\,min}_{u\in\mathcal{U}_{ad}} \left\{ \tau\|u\|_1 + \frac{1}{2}\|u-w\|^2 \right\} = \mathbb{S}_\tau^{\mathcal{U}_{ad}}(w) \text{ for each } w \in \mathcal{U},$$

where the projected soft thresholding function $\mathbb{S}_\tau^{\mathcal{U}_{ad}}$ is defined as follows

$$\mathbb{S}_\tau^{\mathcal{U}_{ad}}(w) := \begin{cases} \min\{w-\tau, u_b\} & \text{on } \{t \in I : w(t) > \tau\} \\ 0 & \text{on } \{t \in I : |w(t)| \le \tau\} \\ \max\{w+\tau, u_a\} & \text{on } \{t \in I : w(t) < -\tau\}. \end{cases}$$

Based on this lemma, we conclude the following

$$\operatorname*{arg\,min}_{u\in\mathcal{U}_{ad}} \left\{ \mathcal{J}_2(u) + \frac{L}{2}\left\| u - \left(v - \frac{1}{L}\nabla\mathcal{J}_1(v)\right)\right\|^2 \right\} = \mathbb{S}_{\frac{\gamma}{L}}^{\mathcal{U}_{ad}}\left(v - \frac{1}{L}\nabla\mathcal{J}_1(v)\right),$$

which can be taken as starting point for a fixed-point algorithm as follows

$$u_{k+1} = \mathbb{S}_{\frac{\gamma}{L_k}}^{\mathcal{U}_{ad}}\left(u_k - \frac{1}{L_k}\nabla\mathcal{J}_1(v_k)\right), \tag{23.38}$$

where L_k is the local Lipschitz continuity constant defined in (23.35). Such method has been investigated in, e.g., [5, 22]. In this chapter, we apply an extension of (23.38), which takes for each iteration k the following form

$$u_{k+1} = \mathbb{S}_{\gamma/L_k}^{\mathcal{U}_{ad}}\left(u_k - \frac{1}{L_k}\nabla\mathcal{J}_1(u_k) + \theta_k(u_k - u_{k-1})\right), \tag{23.39}$$

with $\theta_k \in (0,1)$. This method has been proposed in [17]. Our inertial proximal method is summarized in the following algorithm.

Next, we discuss the convergence of our algorithm, using some existing results [17, 21].

Proposition 23.9 *The sequence $\{u_i\}_{i\in\mathbb{N}}$ generated by (23.39) satisfies the following properties.*

- *The sequence $\{\hat{\mathcal{J}}(u_i)\}_{i\in\mathbb{N}}$ converges in $\mathbb{R}$.*
- *There exists a weakly convergent subsequence $\{u_{i_j}\}_{j\in\mathbb{N}} \subset \{u_i\}_{i\in\mathbb{N}}$.*

Definition 23.2 (Proximal Residual) The proximal residual r is defined as follows

$$r(u) := u - \mathbb{S}_\gamma^{\mathcal{U}_{ad}}\left(u - \nabla\mathcal{J}_1(u)\right). \tag{23.40}$$

Algorithm 7 Inertial proximal method

Input: initial guess u_0, $i = 0$, $i_{\max}$, $\theta_i \in (0, 1)$, tolerance *tol*, initial guess of the Lipschitz constant $L = L_0$.

1: **while** $i \leq i_{\max}$, do: **do**

2: Evaluate $\nabla \mathscr{J}_1(u_i)$ according to Algorithm 8.

3: Update $L_i = \eta L_i$ until

$$\mathscr{J}_1(\tilde{u}) \leq \mathscr{J}_1(u_i) + \langle \nabla_u \mathscr{J}_1(u_i), \tilde{u} - u_i \rangle + \frac{L_i}{2} \|\tilde{u} - u_i\|^2$$

where

$$\tilde{u} = \mathbb{S}^{\mathscr{U}_{ad}}_{\gamma/L_i}(u_i - \frac{1}{L_i} \nabla \mathscr{J}_1(u_i) + \theta_i(u_i - u_{i-1})).$$

4: Set $u_{i+1} = \tilde{u}$.

5: Compute E according to (23.31) or (23.32).

6: If $E < tol$, break.

7: **if** $E < tol$ **then**

8: **break**

9: **end if**

10: $i = i + 1$.

11: **end while**

Algorithm 8 Evaluation of the gradient

Input: u_i, initial value f_0 at time t_0, terminal value p_f at time t_f.

1: Compute f_i, given f_0 and u_i.

2: Compute p_i.

3: Evaluate $\nabla \mathscr{J}_1(u_i)$ according to (23.30).

Proposition 23.8 tells us that $r(u) = 0$ in $L^2(I)$ whenever u solves (23.20). The next proposition establish a connection between the condition $r(u) = 0$ and the solution provided by Algorithm 7; see, e.g., [22].

Proposition 23.10 *Let* $\{u_i\}_{i\in\mathbb{N}}$ *be the sequence generated by Algorithm 7. Then the following holds*

$$\min_{0\leq i\leq i_{max}} \|r(u_i)\|_2^2 = \mathscr{O}(1/i_{max}).$$

23.6 Numerical Experiments

We present results of numerical experiments to validate our optimal control framework. Our purpose is to determine a sparse control $u = u(t)$ such that the expected value of the process X defined by (23.1) minimizes the quantity defined by (23.17) and (23.18).

We take $\Omega = (-100, 100)$ and $[t_0, t_f] = [0, 1]$, and assume that the initial probability density function f_0 is given, $f_0 \sim \mathcal{N}(0, 10)$. The compound Poisson process corresponds to the choice $\lambda = 3$ and $g \sim \mathcal{N}(0, 0.04)$. We take $b(x, u) := x - u(t)$ and $\sigma(x) = 50$. In case of (23.17), we consider $\xi = [-5, 20, 0, 10, -10, 0, -5, 20]$. In the case of (23.18), we take $\xi(t) = 20\sin(10t)$. We choose $N = 300$ and $M = 200$.

We investigate the behavior of the optimal solution considering the full optimization setting including the L^1-cost, i.e. $\gamma \geq 0$, and the control constraints given by the bounds $u_a = -20$, $u_b = 20$ in (23.15). For simplicity, we discuss only the case with $\nu = 10^{-2}$.

In Fig. 23.1, we depict the optimal controls for three different choices of values of γ and considering the discrete-in-time tracking functional given by (23.17). In Fig. 23.2, we show the optimal controls for three different choices of values of γ and considering the continuous-in-time tracking functional given by (23.18). In both cases, we can clearly see that increasing the value of the parameter γ significantly increases the sparsity of the solution, as expected.

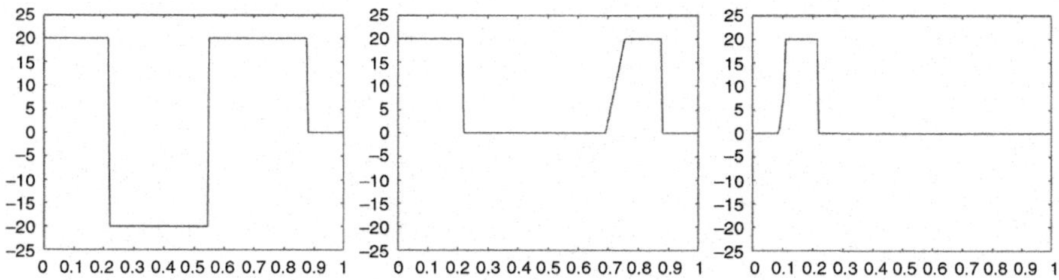

Fig. 23.1 *From left to right*: optimal controls with $\gamma = 0.25, 3, 5$; tracking objective given by (23.17)

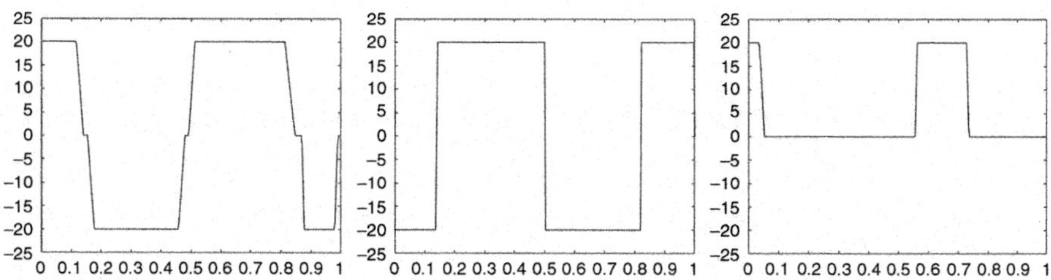

Fig. 23.2 *From left to right*: optimal controls with $\gamma = 1/2, 3, 5$; tracking objective given by (23.18)

References

1. Annunziato, M., Borzì, A.: Optimal control of probability density functions of stochastic processes. Math. Model. Anal. **15**(4), 393–407 (2010)
2. Annunziato, M., Borzì, A.: A Fokker-Planck control framework for multidimensional stochastic processes. J. Comput. Appl. Math. **237**, 487–507 (2013)
3. Annunziato, M., Borzì, A., Magdziarz, M., Weron, A.: A fractional Fokker-Planck control framework for subdiffusion processes. Opt. Control Appl. Methods **37**, 290–304 (2016)
4. Borzì, A., Schulz, V.: Computational Optimization of Systems Governed by Partial Differential Equations. Society for Industrial and Applied Mathematics, Philadelphia (2012)
5. Combettes, P.L., Wajs, V.R.: Signal recovery by proximal forward-backward splitting. Multiscale Model. Simul. **4**(4), 1168–1200 (2005)
6. Cont, R., Tankov, P.: Financial Modeling with Jump Processes. Chapman and Hall, London (2004)
7. Cox, D.R., Miller, H.D.: The Theory of Stochastic Processes. CRC Press, Boca Raton (1977)
8. Ekeland, I., Témam, R.: Convex Analysis and Variational Problems. Society for Industrial and Applied Mathematics, Philadelphia (1999)
9. Fleming, W.H., Soner, M.: Controled Markov Processes and Viscosity Solutions. Springer, New York (1993)
10. Fursikov, A.V.: Optimal Control of Distributed Systems. Theory and Applications. American Mathematical Society, Providence, RI (2000)
11. Gaviraghi, B., Annunziato, M., Borzì, A.: Analysis of splitting methods for solving a partial-integro Fokker-Planck equation. Appl. Math. Comput. **294**, 1–17 (2017)
12. Gaviraghi, B., Schindele, A., Annunziato, M., Borzì, A.: On optimal sparse-control problems governed by jump-diffusion processes. Appl. Math. **7**, 1978–2004 (2016)
13. Geiser, J., Decomposition Methods for Differential Equations: Theory and Applications. Chapman and Hall, London (2009)
14. Harrison, J.M., Pliska, S.R.: A stochastic calculus model of continuous trading: complete markets. Stoch. Process. Appl. **15**, 313–316 (1983)
15. Jäger, S., Kostina, E.A.: Parameter estimation for forward Kolmogorov equation with application to nonlinear exchange rate dynamics. Proc. Appl. Math. Mech. **5**(1), 745–746 (2005)
16. Lions, J.L.: Optimal Control of Systems Governed by Partial Differential Equations. Springer, Berlin (1971)
17. Ochs, P., Chen, Y., Brox, Th., Pock, Th.: iPiano: Inertial proximal algorithm for nonconvex optimization. SIAM J. Imag. Sci. **7**(2), 1388–1419 (2014)
18. Pascucci, A.: PDE and Martingale Methods in Option Pricing. Springer, New York (2011)
19. Pham, H.: Continuous-Time Stochastic Control and Optimization with Financial Applications. Springer, New York (2009)
20. Roy, S., Annunziato, M., Borzì, A.: A Fokker-Planck feedback control-constrained approach for modelling crowd motion. J. Comput. Theor. Transp. (2016). http://dx.doi.org/10.1080/23324309.2016.1189435
21. Schindele, A., Borzì, A.: Proximal schemes for parabolic optimal control problems with sparsity promoting cost functionals. Int. J. Control 1–19 (2016). http://dx.doi.org/10.1080/00207179.2016.1245870
22. Schindele, A., Borzì, A.: Proximal methods for elliptic optimal control problems with sparsity cost functional. Appl. Math. **7**(9), 967–992 (2016)
23. Schuss, Z.: Theory and Applications of Stochastic Processes: An Analytical Approach. Springer, New York (2010)
24. Stadler, G.: Elliptic optimal control problems with $\mathscr{L}^1$-control costs and applications for the placement of control devices. Comput. Optim. Appl. **44**(2), 159–181 (2009)
25. Tröltzsch, F.: Optimal Control of Partial Differential Equations: Theory, Methods and Applications. American Mathematical Society, Providence (2010)

26. Ulbrich, M.: Semismooth Newton Methods for Variational Inequalities and Constrained Optimization Problems in Function Spaces. Society for Industrial and Applied Mathematics, Philadelphia (2011)
27. Yong, J., Zhou, X.Y.: Stochastic Controls, Hamiltonian Systems and Hamilton-Jacobi-Bellman Equations. Springer, New Jersey (2000)

Chapter 24
Proper Orthogonal Decomposition in Option Pricing

José P. Silva, E. Jan W. ter Maten, Michael Günther, and Matthias Ehrhardt

Abstract In this chapter model order reduction (MOR) and the forward-backward duality are combined to generate forward and backward reduced models. We show that both resulting models are numerically efficient models and can in most situations reduce the computational effort in comparison with the full order models, when applying ADI and BDF2 time discretization schemes on a centered second-order and Chang-Cooper spatial discretizations, respectively. For the MOR part, a Proper Orthogonal Decomposition approach was taken.

24.1 Model Order Reduction

Model Order Reduction (MOR) [1] emerged at the end of the twentieth century as an answer to the increasing complexity of the models being developed. Higher and higher resolution schemes led to bigger problems, which, in turn, led to the development of new, more accurate schemes (non-uniform and refined grids, higher-order schemes, sparse schemes, parallelization, problem-specific hardware, etc.). The goal of MOR is to generate smaller models, faster to solve and, if not with similar, with high enough precision with respect to the original Full Order Model (FOM). The Reduced Order Model (ROM) is then a cheaper and faster proxy of the FOM, making it ideal for multi-query problems: parameter studies, parameter optimization, inverse problems, control problems. In finance, and particularly option pricing, inverse problems arise when calibrating model parameters to market data, with volatility being one of the parameters, for example.

Among the different MOR techniques, cf. [1], Proper Orthogonal Decomposition (POD) stands out as a fairly robust technique as it is one of the few techniques able to tackle general non-linear problems. Due to its data-driven approach, it generates ROM in a tailored way. The POD approach consists essentially of two steps.

J.P. Silva (✉) • E.J.W. ter Maten • M. Günther • M. Ehrhardt
Lehrstuhl für Angewandte Mathematik/Numerische Analysis, Bergische Universität Wuppertal, Wuppertal, Germany
e-mail: silva@math.uni-wuppertal.de; termaten@math.uni-wuppertal.de; guenther@math.uni-wuppertal.de; ehrhardt@math.uni-wuppertal.de

M. Ehrhardt et al. (eds.), *Novel Methods in Computational Finance*,
Mathematics in Industry 25, DOI 10.1007/978-3-319-61282-9_24

In the first one, we collect information on the original problem by solving the full order model(s) and then proceed to generate a basis for a (best approximating) subspace from that information.

In the second step, the original model is projected (solution is looked for in this subspace) onto this subspace, a procedure known as Galerkin projection or as Petrov-Galerkin projection, depending whether it is projected on the same or on a different subspace, respectively.

POD can in its essence be described as a mathematical technique that, given an ensemble of data, constructs a basis for the ensemble that is optimal in the least-squares sense. Let X be a real Hilbert space, with inner product $(\cdot,\cdot)_X$, and $Y = \left[y_1\ y_2 \ldots y_n\right]$ an ensemble of n snapshots $y_i \in X$. The snapshots contain the solution for different configurations of the problem, i.e., it may contain the solution at different time instances for an evolution problem, it may contain the solution for different parameter values or any other configuration, which will need to be approximated by the ROM. Then, for some $l \ll n$ a POD basis is an optimal orthonormal basis $\psi_j, j = 1, \ldots, l$ such that the square error between the elements y_i and its l-partial sum of the decomposition of y_i in the space spanned by ψ_j, is minimized, i.e.

$$\min_{\{\psi_k\}_{k=1}^{l}} \mathscr{J}(\boldsymbol{\psi}) = \min_{\{\psi_k\}_{k=1}^{l}} \sum_{i=1}^{n} \left\| y_i - \sum_{j=1}^{l} \left(y_i, \psi_j\right)_X \psi_j \right\|_X^2, \tag{24.1}$$

subject to $\left(\psi_i, \psi_j\right)_X = \delta_{ij}$. It can be proven [4] that the above minimization problem is equivalent to the eigenvalue problem

$$YY^\top \psi = \lambda \psi.$$

Factorizing Y using a singular value decomposition (SVD), it can be proven [4] that the resulting left-singular vectors form a POD basis, where $\lambda = \sigma^2$, with σ the singular values of Y. For the POD basis, $\mathscr{J}(\psi) = \sum_{i=l+1}^{n} \lambda_i = \sum_{i=l+1}^{n} \sigma_i^2$. The size of the basis l necessary for a good approximation is problem dependent, but we can take the relative error as a good indicator. As we are minimizing a sum of squares, this criterion guarantees that we are maximizing the information on the reconstructed snapshots in the least-squares sense,

$$\mathscr{E}(l) = \frac{\sum_{i=1}^{l} \sigma_i^2}{\sum_{i=1}^{n} \sigma_i^2}. \tag{24.2}$$

As the singular values are ordered and reflect the relevance of each dimension in the state space, (24.2) is sometimes called *relative information measure*. It is important to note that this basis will originate a reduced model, which will approximate optimally the existing snapshots and but not necessarily the model as a whole.

When in possession of the POD basis, we proceed to process the existing partial differential equation (PDE) onto the space spanned by the POD Basis. Rewriting the PDE as

$$\frac{\partial}{\partial t}u = \mathscr{L}u, \tag{24.3}$$

where $\mathscr{L}$ is a differential operator. We will restrict the following derivation to the linear case, with the non-linear case following the same principles but originating a non-linear system to be solved. For more details see [4]. We project the PDE in a Galerkin fashion, i.e.

$$\left(\psi_i, \frac{\partial u}{\partial t}\right)_X = (\psi_i, \mathscr{L}u)_X, \qquad i = 1, \ldots, l.$$

Substituting u by its representation in the POD basis of size l , $u(t, s) = \sum_j^l a_j(t)\, \psi_j(\mathbf{s})$ we obtain the explicit system of ODEs

$$\dot{a}(t)_i = \sum_{j=1}^{l} a_j(t)\,(\psi_i, \mathscr{L}\psi_l)_X \qquad i = 1, \ldots, l,$$

making use of the orthonormality property of the basis.

Before proceeding further, we should call attention to the fact that there is an implicit dependence of the basis on the PDE. Let us assume that our differential operator $\mathscr{L}$ has a parametric dependence on a parameter vector $\boldsymbol{\mu}$.[1] The parametric dependence will affect the solution in a non-linear way, which in turn will affect the resulting basis, which ultimately will introduce a parametric dependence through the basis on the reduced model. As a result of POD being a technique based on the SVD of the solution, i.e. a numerical technique is employed to calculate the basis, the parametric dependence of the basis on the parameters cannot be determined. For that reason, we will omit its representation, having in mind that the basis is not parameter independent. This dependence will be more obvious in Sect. 24.3, where a parametric ROM is generated and the basis keeps changing with each addition of snapshots from a different parameter vector at each iteration.

When solving the PDE a decision has to be made if the discretization step comes before or after the projection one. We can discretize the PDE and perform model order reduction on the system of matrices or we can project the PDE and then discretize. We will use the former variant, which in spite of not guaranteeing better results, it stems from a more fundamental approach. We solve the FOM numerically, generate the POD Basis and then project the discretized system of ODE, obtained as a result of applying the method of lines (MOL) to discretize our PDE in space.

[1]The previous example can be extended to the case where the dependence is on an initial condition, geometry or boundary condition.

24.2 Forward-Backward Approach

The price of derivatives usually satisfies deterministic equations, as in general it depends on deterministic quantities of Markov Processes, i.e., they are a function of an expectation of a random variable associated with a stochastic process. One way in which we can obtain a deterministic equation, which can then be of use in determining the price is through the Kolmogorov equations. The Kolmogorov equations are the evolution equations for the transition probability density function $p(t,x;t_0,x_0)$ of the statistics of the stochastic process X_t. The transition probability density function $p(t,x;t_0,x_0)$ represents the probability of the stochastic process X_t transitioning from the state x_0 at t_0 to the state x at t. The PDE for $p(t,x)$ reads

$$\frac{\partial p}{\partial t} = \frac{1}{2}\sum_{i,j}\frac{\partial^2}{\partial x_i \partial x_j}\left(\sum_k B_{ik}(t,\mathbf{x})\,B_{jk}(t,\mathbf{x})\,p\right) - \sum_i \frac{\partial}{\partial x_i}\left(A_i(t,\mathbf{x})\,p\right) \tag{24.4}$$

where B_{ij} and A_i are the diffusion and drift coefficients of the diffusion process, respectively.

We will call this the forward approach and the price is calculated by discounting the expected value of the payoff at maturity,

$$u(t,x) = \mathbb{E}\left[e^{-r(T-t)}u(T,X_T)\,|X_t = x\right] = e^{-r(T-t)}\int p(T,y;t,x)\,\phi_T(y;K)\,dy, \tag{24.5}$$

with $\phi_T(x;K)$ a function of the underlying x representing the payoff at maturity. Although it has no dependence on time, to make it more clear that it represents the payoff at $t = T$, we add T as index.

On the other hand, we also obtain an evolution equation for a deterministic equation through Itô's Lemma. In this case, which we will call the backward approach, the parabolic equation resulting from applying Itô's Lemma to our quantity of interest will tell us how to evolve the payoff from maturity back to today.

In the forward approach, we want to calculate the transition probability density function $p(t,x;t_0,x_0)$ from the initial time t_0 until time T for all the possible values of x, such that the integral in (24.5) can be evaluated for all the European payoffs $\phi_T(\cdot;K)$ traded in the market, as well as for all the maturities up to $T-t_0$. Once the strike price K, the maturity T, the payoff function ϕ_T and the current time t_0 are known, calculating the price consists of

1. Start with a Dirac delta initial condition $p(t_0,x;t_0,x_0) = \delta(x-x_0)$
2. Integrate (24.4) from $t = t_0 < T$ up to $t = T$
3. Obtain the expected future payoff from (24.5)
4. Discount to present time $t = t_0$

In point 3. a quadrature involving the product of the numerically determined transition pdf $p(t,x)$ and the payoff ϕ_T is performed, in a step which can also be

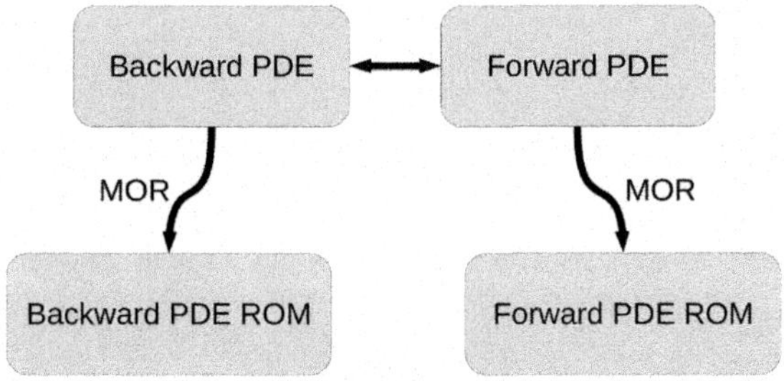

Fig. 24.1 Diagram of FOM ROM relations

described as a post-processing step. Note that ϕ_T is usually continuous, piecewise linear and $p(t,x;t_0,x_0)$ is a smooth function. Therefore, in some particular cases of ϕ_T some special quadrature methods may be applicable.

For comparison, the backward approach follows similar steps

1. Start from the payoff $u(T,x) = \phi_T(x;K)$ (the same in step 3. for the forward approach)
2. Integrate (24.3) backwards from $t = T$ down to $t = t_0 < T$
3. Price is $u(t_0,x)$ at $x = x_0$

The above approaches result in the following rules of thumb:

Backward Equation For calculating the price of one option $u(t,x = X;K_0,T_0)$ for different t,x and given pair K_0,T_0.

Forward Equation For calculating the price of different options $u(t_0,x = X_0;K,T)$ for different K,T and given pair t_0,X_0. We have two possible PDEs, which we'll need to solve and in both cases there is nothing which prohibits us from applying MOR. Figure 24.1 represents the four ways in which we can calculate the price and their interconnections. In Sect. 24.4 we present results for the four cases.

24.3 Speed-up

In the last couple of years, splitting methods have become the go-to choice for the efficient time integration step of the system of ODE resulting from the space discretization of the PDE (24.3) [5]. In particular Alternate Direction Implicit schemes, which result in very efficient schemes to solve the sparse multidimensional PDE, have now seen its stability and consistent properties proved. The system matrix A resulting from the spatial discretization of a d-dimensional problem has dimensions $A \in \mathbb{R}^{N\times N}, N = \prod_{i=1}^{d} n_i$. The matrix consists mostly of zeros except for 3^d diagonals, in the case of a compact stencil.

In Figs. 24.2 and 24.3, we represent graphically the computational cost of an ADI vs a reduced model generated with basis of different sizes. Note that the reduced

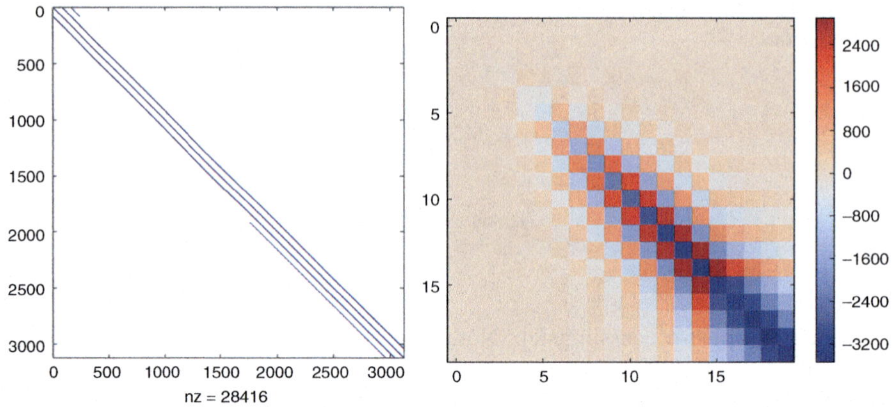

Fig. 24.2 Sparsity of the system matrix from a FOM (a few diagonals) and from a ROM (fully dense matrix)

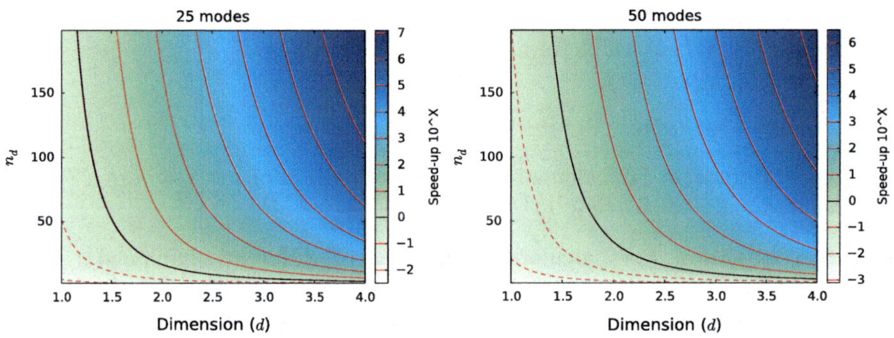

Fig. 24.3 Computational advantage of MOR compared to the Full Order Model when both are solved by an ADI method

model will have much less sparsity than the FOM, as can be seen from the example on the following figure.

The ADI method exploits this sparsity and, therefore, only when the dense ROM can compete with the ADI, it can be of practical use. If we calculate the ratio of total floating point operations in both methods, knowing that the total number of floating point operations for the ADI method will be $f(\cdot) = dn_d\left(\frac{9}{2} + 8n_t + 5dn_t + 2n_tn_d^d\right)$ and using the same scheme for the dense ROM, we obtain the speed-up profile presented in Fig. 24.3. Although splitting the ROM in the system matrices for each dimension will be overkill from the performance point of view, it only adds a multiplier proportional to the number of dimensions of the problem to the computational cost but allows from the practical point of view to use exactly the same scheme as for the FOM, which can simplify its implementation. For a more detailed calculation, see [7].

We see in Fig. 24.3 that we can achieve a significant reduction in the number of operations (speed-up) already for a two-dimensional problem even in the 50 modes case. Due to the exponential dependence on dimension for the ADI method and the respective independence for the reduced model, we can theoretically obtain better and better results the higher the dimension of the problem. Although higher dimensional ADI methods still lack some rigorous proofs of their properties (stability, consistency), in practice they have been applied with success [3] and so, at least for up to four dimensional problems, we can regard Fig. 24.3 as showing realistic cases of application.

24.3.1 Greedy Approach

The greedy approach guides us to select the worst-case scenario. We use the Latin Hypercube Sampling (LHS) for generating a sample of parameter values $\mathscr{P}$ from the parametric space $\mathscr{D}$ covering the whole space. We then compare with the high-fidelity model and select the parameter values which maximize the error and add the snapshots corresponding to this set of parameters. We continue to iterate until the desired number of iterations is reached, or a pre-established error threshold is reached, or the error decrease is smaller than a pre-established threshold.

Input: N_t, $\mathscr{P}$, tol_{sv}
Output: V_{N+1}, μ^*

1: Select mid-point of $\mathscr{D}$ as μ_0
2: Calculate $u(t_i, x; \mu_0)$ for $i = 0, \ldots, N_t$
3: $T_0 = [u(t_0, x; \mu_0) \cdots u(t_{N_t}, x; \mu_0)]$
4: Initialize $N, error$
5: **while** *error* does not satisfy an accuracy condition **do**
6: $\quad \mu^*_{N+1} = \arg\max_{\mu_i \in \mathscr{P}} \Delta_N(\mu_i)$
7: $\quad$ Calculate $u\left(t_i, x; \mu^*_{N+1}\right)$ for $i = 0, \ldots, N_t$
8: $\quad T_{N+1} = \left[T_N \quad u\left(t_0, x; \mu^*_{N+1}\right) \cdots u\left(t_{N_t}, x; \mu^*_{N+1}\right)\right]$
9: $\quad V_{N+1}, S_{N+1} = SVD(T_{N+1})$
10: $\quad size = \min \arg_k \frac{\sum_{i=1}^{k} \sigma_{N+1,i}}{\sum_{i=1}^{n} \sigma_{N+1,i}} \leq tol_{sv}$
11: $\quad V_{N+1} = V_{N+1}(:, 1 : size)$
12: $\quad N = N + 1$
13: $\quad$ update *error*
14: **end while**

In step 5. an error indicator $\Delta_N(\mu_i)$ is used to determine μ^*_{N+1}. Greedy approaches are divided in two types of algorithms, strong greedy and weak greedy. In the strong greedy approach, an exact error is used to calculate the error indicator while in the weak greedy approach an approximation is used for that purpose. A residual approach was tested with very similar results, but we opted to choose the strong greedy approach as in [2]. This step involves the pre-calculation of the FOM solutions for the parameter vectors in $\mathscr{P}$. As this is an offline procedure, which can even be done in parallel, it is not that relevant if the FOM is not prohibitively expensive and it saves time in the online phase by having the snapshots readily available. The error indicator $\Delta(\mu_i)$ is then

$$\Delta(\mu_i) = \|u_{FOM}(0, x; \mu_i) - u_{ROM}(0, x; \mu_i)\|_\infty.$$

The update error in the last step of the algorithm involves calculating the error between the FOM and the ROM for a set of pre-chosen parametric points, which will be different than the ones used for evaluating which snapshots to gather. As with the error indicator Δ_N, we use the max-norm on the solution at maturity in the forward approach and at time t_0 in the backward approach. A more detailed description can be found in [8].

In Sect. 24.4, in Figs. 24.5, 24.6, 24.7, 24.8, we can see the selected parameter values and the corresponding decreasing error.

24.4 Numerical Experiments

For the numerical experiments we select the classical one dimensional Black-Scholes and the Heston models. We select the following parameters for the maturity, strike price and dividend yield for all the models:

$$T = 1, K = 100, q = 0.01.$$

The different type of solutions results in the choice of different computational domains. For the Black-Scholes case, we will select the same computational domain,

$$\Omega = [0, 4K]$$

for both the backward and adjoint (forward) cases. The adjoint equation needs an additional parameter, the current spot x_0, which we choose as $1.2K$. Due to the fact that we selected a call option as the numerical example in both models, this results in the option being in-the-money ($x_0 > K$). In the case of the Heston model we select again $x_0 = 1.2K$ for the current spot price, with the domain

$$\Omega = [0, 30K] \times [0, 15]$$

for the backward equation and

$$\Omega = [0, 10K] \times [0, 1]$$

for the adjoint equation, with the difference laying in the solution profile.

For the numerical parameters, we have used $N_t = 100$ for both models in both variants. For the spatial discretization we choose $N_x = 200$ for the Black-Scholes model and $N_S = 2N_\nu = 64$ for the Heston Model. Regarding the ROM threshold to truncate the basis, we select a ratio between σ_1 and σ_l of 10^{10}. Depending on the range of parameters and accuracy desired, we may decrease the ratio, which will result in a smaller number of basis vectors. For the parametric search, we select the same parameters, r and σ, not forgetting that σ is the volatility in the Black-Scholes model and σ_ν is the volatility (standard deviation) of the stochastic volatility in the Heston Model. In the Black-Scholes model, the parametric space is $\mathscr{D} = [0.02, 0.1] \times [0.1, 0.8]$ and in the Heston Model $\mathscr{D} = [0.025, 0.035] \times [0.35, 0.45]$.

We collect the snapshots at every time step and at a discretized grid for the parametric domain. We discretize the parametric domain $\mathscr{P} = \prod_{i=1}^{n} [\theta_{i,min}, \theta_{i,max}]$ with a uniform grid p_k containing five equally distanced points per parameter, including the extreme values. We will have then 2^5 and 2^5 parameter vectors to generate our snapshots for the Black-Scholes and Heston models, respectively (Table 24.1). For the selection of the parameters for which successively snapshots will be calculated at the specified time levels we perform the greedy search of Sect. 24.3.1 on 100 samples of the parametric domain, sampled using a Latin Hypercube Sampling method. The parameter intervals are centered around values which are common in equity markets.

For the spatial discretization we use a centered second-order scheme for the backward equations, both FOM and ROM, and a Chang-Cooper discretization for the forward ones. This difference results from the fact that, as seen before, the solution of the forward equation represents a probability density function and should, therefore preserve the total probability and be non-negative during the time evolution. The Chang-Cooper scheme is known to preserve those properties [6].

We can observe on the left side of Fig. 24.4 that the first basis function tends to approximate the "average" shape of the solution, with the other ones counting as

Table 24.1 Centers of the parameter ranges

(a) Black-Scholes

r	σ
0.03	0.3

(b) Heston Model

r	θ	κ	σ
0.03	0.16	4	0.3

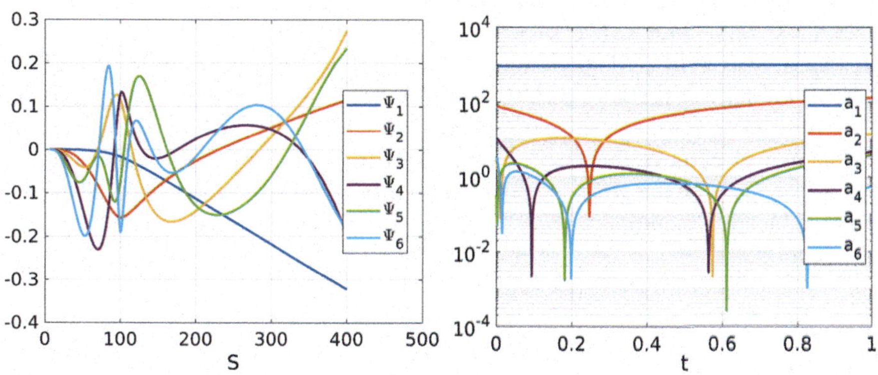

Fig. 24.4 Black-Scholes basis

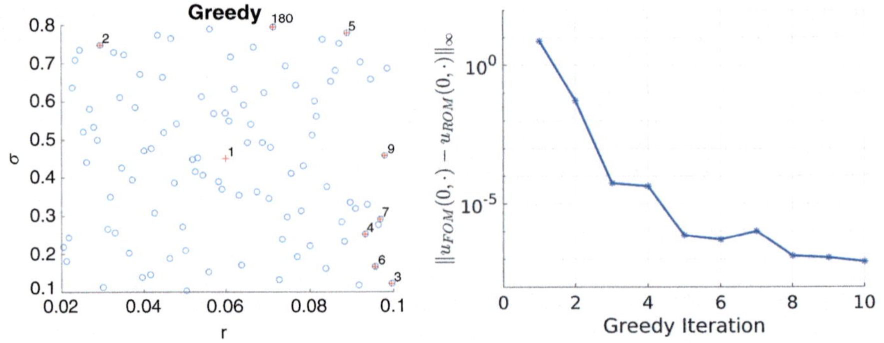

Fig. 24.5 Black-Scholes backward pROM convergence. Note that at the left the worst-cases for mu are at the border of the parameter space. At the right the error at the final time $t = 0$. $\|u_{\text{FOM}}(0,.) - u_{\text{ROM}}(0,.)\|_\infty$ is calculated after injecting u_{ROM} back to the full space

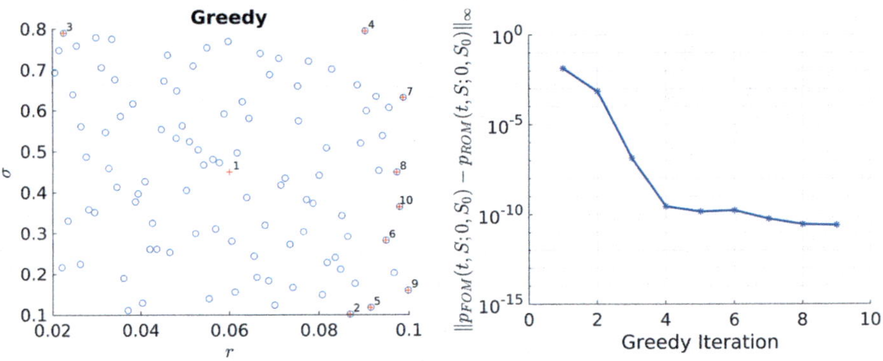

Fig. 24.6 Black-Scholes forward pROM convergence

corrections. Those corrections will have their coefficients $a_i(t)$ changing along time, which can be seen on the right side in Fig. 24.4 for the first six coefficients.

Figures 24.5, 24.6, 24.7, 24.8 show the convergence of the pROM for the backward PDE of the Black-Scholes model, for the forward PDE of the Black-Scholes model, for the backward PDE of the Heston model, and for the forward PDE of the Heston model, respectively. We can see that 10 and 20 greedy iterations are enough to achieve the best possible approximation for the selected ratio of singular values for the Black-Scholes and Heston models, respectively. The size of the reduced system matrix was 42 for the Black-Scholes model and 120 for the Heston model.

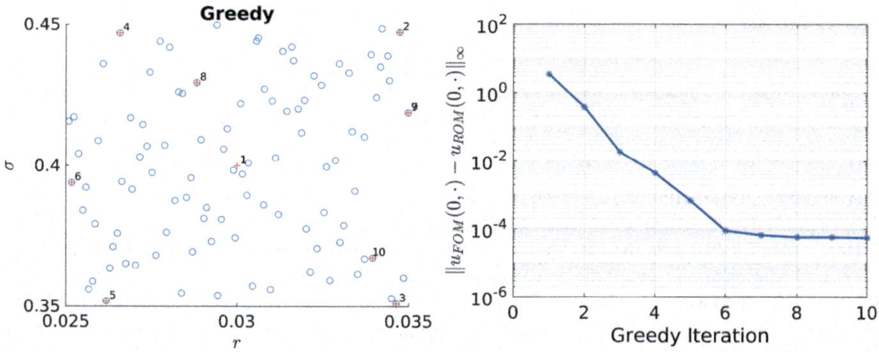

Fig. 24.7 Heston backward pROM convergence

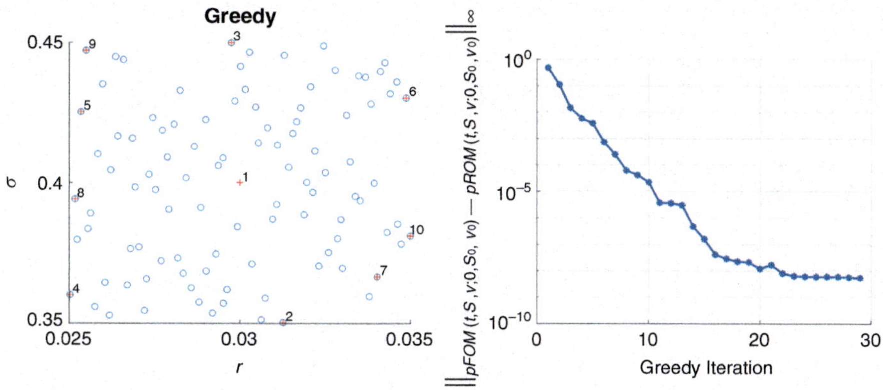

Fig. 24.8 Heston forward pROM convergence

24.5 Conclusions

Having calculated reduced models with the POD for both the 1D Black-Scholes and the Heston models, the level of approximation of the above reduced models has shown to be fit for usage in the Heston case. In the one-dimensional Black-Scholes case, as can be inferred from Fig. 24.3, the reduced model doesn't provide any speed-up, which is not particularly worrying, as the one dimensional model has a closed formula solution. One possible improvement would be combining together the ADI scheme and the Chang-Cooper discretization, as it is still a compact discretization and could potentially benefit from the splitting, especially in 3D or higher dimensions.

Acknowledgements The authors were partially supported by the European Union in the FP7-PEOPLE-2012-ITN Program under Grant Agreement Number 304617 (FP7 Marie Curie Action, Project Multi-ITN STRIKE—Novel Methods in Computational Finance).

References

1. Antoulas, A.C.: Approximation of Large-Scale Dynamical Systems (Advances in Design and Control) (Advances in Design and Control). Society for Industrial and Applied Mathematics, Philadelphia, PA, USA (2005)
2. Haasdonk, B., Salomon, J., Wohlmuth, B.: A reduced basis method for the simulation of American options. In: Numerical Mathematics and Advanced Applications 2011, pp. 821–829. Springer Science + Business Media, Berlin (2012)
3. Haentjens, T.: ADI schemes for the efficient and stable numerical pricing of financial options via multidimensional partial differential equations. Ph.D. thesis, Universiteit Antwerpen (2013)
4. Hinze, M., Volkwein, S.: Proper Orthogonal Decomposition Surrogate Models for Nonlinear Dynamical Systems: Error Estimates and Suboptimal Control, pp. 261–306. Springer, Berlin/Heidelberg (2005)
5. in 't Hout, K., Toivanen, J.: Application of operator splitting methods in finance. arXiv preprint arXiv:1504.01022 (2015)
6. Mohammadi, M., Borzì, A.: Analysis of the Chang-Cooper discretization scheme for a class of Fokker-Planck equations. J. Numer. Math. **23**(3), 271–288 (2015)
7. Silva, J., ter Maten, E.J.W., Günther, M., Ehrhardt, M.: Proper Orthogonal Decomposition in option pricing: Basket options and Heston model. In: G. Russo, V. Capasso, G. Nicosia, V. Romano (Eds): Progress in Industrial Mathematics at ECMI 2014, Series Mathematics in Industry, vol. 22. Springer International Publishing, Cham (2016)
8. Silva, J.P.: Model Order Reduction in Option Pricing. Ph.D. thesis, Bergische Universität Wuppertal (2017)

Part VII
High Performance Computing

Chapter 25
Alternative Parallel Strategies for Linear and Nonlinear PDEs in Option Pricing

Choi-Hong Lai, André M.S. Ribeiro, and Natkunam Kokulan

Abstract This chapter provides an exposition on alternative parallel strategies for pricing options using Black Scholes PDE with nonlinear volatility models. Details of a Laplace transform method and its induced parallel properties when applied to the Black Scholes PDE are discussed. An application is given of the method to the numerical solution of an European option pricing with the Frey-Patie volatility model. An extension to a general transformation technique is explained. Finally a hybrid technique to handle nonlinear volatility is presented.

25.1 The Black-Scholes Equation

Let $V(S,t)$ denote the value of an option, where S is the current value of the underlying asset and t is the time. An European put option pricing problem may be described by the Black-Scholes PDE

$$\frac{\partial V}{\partial t} + \frac{1}{2}\sigma^2 S^2 \frac{\partial^2 V}{\partial S^2} + rS\frac{\partial V}{\partial S} - rV = 0, \tag{25.1}$$

subject to the terminal and boundary conditions

$$\begin{cases} V(S,T) = (K-S)^+, & \text{for} \quad 0 \le S \le S_{max}, \\ V(0,t) = Ke^{-r(T-t)}, & \text{for} \quad 0 \le t < T, \\ V(S_{\max},t) = 0. \end{cases}$$

Here σ is the volatility, r is the interest rate, and K is the strike price. For a perfect market σ may be taken as a constant. For an imperfect market where transaction costs, illiquidity and friction exist σ takes a nonlinear form. In this chapter numerical tests were performed for a hypothetical nonlinear volatility

C.-H. Lai (✉) • A.M.S. Ribeiro • N. Kokulan
Department of Mathematical Sciences, University of Greenwich, Park Row, SE10 9LS, Greenwich, London, UK
e-mail: C.H.Lai@gre.ac.uk; A.M.DaSilvaRibeiro@gre.ac.uk; N.Kokulan@gre.ac.uk

M. Ehrhardt et al. (eds.), *Novel Methods in Computational Finance*, Mathematics in Industry 25, DOI 10.1007/978-3-319-61282-9_25

$\sigma(V) = \sigma_0\sqrt{1+\sin(\frac{V\pi}{K})}$ [1] and the simplified Frey-Patie model $\sigma = \sigma_0(1 - \rho S V_{SS})^{-1}$, where $\rho \neq (SV_{SS})^{-1}$ is a constant which controls the nonlinearity [2].

Note that (25.1) is a backward equation in time. It can be transformed to a forward equation by using $\tau = T - t$ which leads to the forward equation

$$\frac{\partial v}{\partial \tau} = \frac{1}{2}\sigma^2 S^2 \frac{\partial^2 v}{\partial S^2} + rS\frac{\partial v}{\partial S} - rv, \tag{25.2}$$

where $v(S,\tau) \in \{S : S \geq 0\} \times (0,T]$, and supplemented with these conditions

$$\begin{cases} v(S,0) = (K-S)^+, & \text{for} \quad 0 \leq S \leq S_{max}, \\ v(0,\tau) = Ke^{-r\tau}, & \text{for} \quad 0 < \tau \leq T, \\ v(S_{\max},\tau) = 0. \end{cases}$$

In the case when a closed form solution is not possible with certain types of nonlinear volatility, finite difference methods are to be used along the spatial and temporal axes. Typical temporal integration schemes for Eq. (25.2) are forward and backward finite difference methods. These integration schemes lead to an explicit and an implicit scheme, respectively, where data dependence of the current time step cannot be removed from previous time steps. In the latter case where the temporal step size does not require to satisfy a CFL condition does not qualify it to march an arbitrary large step to reach the final solution at $\tau = T$. This leads to the conclusion that concurrent computation of v at many different times is impossible using a temporal marching scheme.

Attempts have been made by several researchers in the direction of modified temporal marching schemes. For examples Miranker and Liniger [3] proposed a family of parallel Runge-Kutta methods; Lions, Maday and Turinici [4] proposed the ‘parareal’ algorithm in time discretisation; and Gander and Vandewalle [5] provided an analysis of the parareal algorithm for parallel temporal integration. The concept relies on partitioning the temporal axis into a number of non-overlapped sections known as time subdomains and the same time-dependent problem is considered in each of these time subdomains. Very often a coarse temporal partitioning is needed to provide corrections using the concept of a multigrid correction method.

The idea of using various transformation methods to induce parallel properties into a given time dependent PDE was first discussed in [6]. It relies on replacing the temporal dependence of a PDE to a simplified form with certain parameters. The technique has been used in many engineering problems in an attempt to obtain closed form solution and in the study of the long term behaviour of many time dependent problems. It is the use of such parametric differential equations that led to the decoupling of problems suitable for parallel processing. Further details may be found from [6]. This chapter provides a detailed exposition of one of the transformation method known as Laplace transform and demonstrates the effectiveness and restrictions of such method in the context of parallel computing.

25.2 The Laplace Transform Method

The forward equation as described in (25.2) is used as the model equation in this section. Let

$$\mathscr{L}(v) \equiv \int_0^\infty e^{-\lambda\tau} v(S,\tau)d\tau = U(\lambda; S)$$

be the Laplace transform of the function $v(S,\tau)$, then the Laplace transform of (25.2) leads to

$$\frac{1}{2}\sigma^2 S^2 \frac{d^2U}{dS^2} + rS\frac{dU}{dS} - (r+\lambda)U = -v(S,0) \tag{25.3}$$

subject to the boundary conditions $U(\lambda;0) = \frac{K}{\lambda+r}$ and $U(\lambda; S_{\max}) = 0$. The above transform is valid only if the volatility σ is a constant. The result of applying the Laplace transform method is to remove the temporal dependence of the problem defined in Eq. (25.2). The transformed problem is now converted to the dependence on the parameter $\lambda \in \{\lambda_j = j\frac{\ln 2}{T} : j = 1, 2, \ldots, m\}$, which is a finite set of transformation parameters. Each element of the set of parametric equations $\{U(x; \lambda_j) \ : j = 1, 2, \ldots, m\}$ satisfies Eq. (25.3) and is decoupled from each other and may be solved independently from the others. Thus an important parallel property is induced to the mathematical model leading to a set of independent parametric equations which may be solved without the knowledge of each other. In terms of distributed computing, this independence amounts to null communication between each computer in the distributed computing environment, assuming there are as many computers as the number of parametric equations.

25.2.1 *Computing the Inverse Laplace*

It is not the intention of this chapter to examine various inverse Laplace transform methods in the retrieval of $v(S,T)$. Instead, without lost of generality, the approximate inverse Laplace transform due to Stehfest [7] was used to compute the weighted sum of the Laplace transformed solutions, i.e.

$$v(S,T) \approx \sum_{j=1}^{m} w_j U(\lambda_j; S) \tag{25.4}$$

where

$$w_j = \frac{\ln 2}{T}(-1)^{m/2+j} \sum_{k=(1+j)/2}^{\min(j,m/2)} \frac{k^{m/2}(2k)!}{(m/2-k)!k!(k-1)!(j-k)!(2k-j)!}$$

are the weighting factors.

Note that m is required to be chosen as an even number if (25.4) is used to compute the inverse Laplace transform. Early experience indicated that the value of m does not require to be larger than 12 and smaller values of m do not lead to good accurate results.

25.2.2 Extension to Nonlinear Volatility

In order to apply the Laplace transform method to Eq. (25.2) the nonlinear volatility should be linearised in a suitably chosen larger time interval. An iterative process is thus required to ensure the nonlinear volatility converges to the correct value. Typically Laplace transform of (25.2) is to be applied in the time interval $\tau \in (T_i, T_{i+1}]$, and the transformed equation in its differential form is given by

$$\frac{1}{2}\sigma(\bar{v})S^2\frac{d^2U}{dS^2} + rS\frac{dU}{dS} - (r+\lambda)U = -v(S,T_i) \tag{25.5}$$

where $U = \mathscr{L}(v)$. An update process is used to renew the value of $\bar{v}$ and an inverse Laplace transform is used to obtain the numerical solution $v^{(n)}(S, T_{i+1})$, using $v^{(n)}(S, T_i)$ as the initial approximation to $\bar{v}$.

Algorithm: Update process - Laplace transform method.
Initial approximation:- $v^{(0)}(S, T_{i+1}) := v^{(n)}(S, T_i)$;
$k := 0$;
Iterate
    $k := k + 1$;
    Compute $\sigma(\bar{v})$ such that $\bar{v} := v^{(k-1)}(S, T_{i+1})$;
    Parallel for $j := 1$ to $m(i)$
        Find $U(\lambda_j; S)$ such that
            $\frac{1}{2}\sigma(\bar{v})S^2\frac{d^2U}{dS^2} + rS\frac{dU}{dS} - (r+\lambda_j)U = -v(S,T_i)$
    End Parallel for
    Compute $v^{(k)}(S, T_{i+1})$ using inverse Laplace in Eq. (25.4)
Until $||v^{(k)}(S, T_{i+1}) - v^{(k-1)}(S, T_{i+1})|| < \epsilon$
$n := k$

Here $m(i)$ is the number of transformation parameters and $T_i = i\Delta\tau$. In order to solve Eq. (25.5) for $U(\lambda_j; S)$, one can employ the same finite difference technique described above. In order to solve (25.2) for $v(S, T)$, this algorithm needs to be iterated through $t_i := T_1, T_2, \ldots, T$ by using suitable values of $m(i)$ in the form of an outer iteration. In essence the actual implementation does not require different values of $m(i)$ for many problems, and the numerical tests shown here used the same number of transformation parameters, denoted as $\bar{m}$, for different values of i during the outer iteration loop. Note that in this case $\Delta\tau$ can be chosen to be much greater than $\delta\tau$ because the fine details of at each time step of a temporal integration is not required in the inverse Laplace transformation calculation.

25.3 Numerical Experiments

Numerical experiments were performed for a hypothetical nonlinear volatility and a simplified Frey-Patie model as described in Sect. 25.1.

25.3.1 Numerical Results for the Hypothetical Nonlinear Volatility

The hypothetical nonlinear volatility was used in an European put option with strike price $K = 100$, $S_{max} = 3.2K$, $r = 0.05$, $\sigma_0 = 0.4$, and the expiry time is one year. An implicit finite difference method was used to solve the nonlinear forward problem defined in Eq. (25.2) with the mesh $\delta S = \frac{S_{max}}{2^{15}}$ and $\delta\tau = \frac{1}{250}$, and a reference solution $V_R(S, 0)$ was obtained for comparison. The computational time for this reference solution on the Stafflinux platform, a virtual machine with two cores from Intel(R) Xeon(R) CPU E5-4650 0 @ 2.70 GHz, 4 GB memory and 16 GB Harddisk, at Greenwich was 3.954 s.

The Laplace transform method was applied to solve the nonlinear forward problem defined in Eq. (25.2) with the same mesh $\delta S = \frac{S_{max}}{2^{15}}$. As the problem is nonlinear a time stepping process is still required in the computation. A series of larger time steps, each defined as $\Delta\tau = T_{i+1} - T_i$, were chosen to be $4\delta\tau$, $8\delta\tau$, $16\delta\tau$, $32\delta\tau$, and $64\delta\tau$. The resulting solution V_L using a $\Delta\tau$ was compared with the reference solution. Table 25.1 records the 2-norm of the point-wise comparison of V_L using different $\Delta\tau$ and m with the reference solution V_R. It can be seen that the effect of $\Delta\tau$ is not significant for this particular hypothetical volatility. The choice of $m \geq 8$ produces similar order of accuracy.

The parallel version was implemented using Fortran with MPI on the Stafflinux platform, as described above, at Greenwich. The parallel run time was simulated on a single processor. Table 25.2 shows the parallel computing time excluding the inverse Laplace transform which is trivial compare with the other equation solvers.

Table 25.1 $||V_L(S,0) - V_R(S,0)||_2$

m / $\Delta\tau$	$64\delta\tau$	$32\delta\tau$	$16\delta\tau$	$8\delta\tau$	$4\delta\tau$
2	0.02190	0.02200	0.02208	0.02208	0.02200
4	0.00173	0.00085	0.00041	0.00035	0.00043
6	0.00104	0.00062	0.00042	0.00031	0.00026
8	0.00107	0.00068	0.00048	0.00037	0.00031
10	0.00107	0.00068	0.00048	0.00036	0.00031
12	0.00107	0.00068	0.00048	0.00036	0.00031
14	0.00107	0.00068	0.00048	0.00036	0.00031

Table 25.2 Parallel computing time for different choices of $\Delta\tau$

m / $\Delta\tau$	$64\delta\tau$	$32\delta\tau$	$16\delta\tau$	$8\delta\tau$	$4\delta\tau$
2	0.1094	0.1735	0.3152	0.5660	0.8877
4	0.1089	0.1600	0.3135	0.4690	0.8418
6	0.0959	0.1610	0.3130	0.4657	0.8393
8	0.0963	0.1610	0.3145	0.4694	0.8683
10	0.1043	0.1744	0.3435	0.5077	0.9159
12	0.0983	0.1645	0.3209	0.4782	0.8597
14	0.0979	0.1647	0.3213	0.4781	0.8581

The results demonstrated that a coarse grain parallel algorithm suits the current purposes. The parallel algorithm is an efficient way of solving the BS PDE with nonlinear volatility. It is observed that the volatility cannot be oscillatory in a given larger time interval, $\Delta\tau$, such as $(T_i, T_{i+1}]$. The solution quality of the Laplace transform method was comparable to the reference solutions V_R.

25.3.2 Numerical Results for Frey-Patie Nonlinear Volatility

The Frey-Patie model was used in the same European put option pricing problem. The same discretisation mesh was used in the reference solution and the Laplace transform method. This set of results was obtained for the case when $\sigma_0 = 0.4$ and $\rho = 0.01$. An implicit finite difference method was used to solve the nonlinear forward problem defined in Eq. (25.2) with the mesh $\delta S = \frac{S_{max}}{100}$ and $\delta\tau = \frac{1}{255}$, and a reference solution $V_R(S,0)$ was obtained for comparison. The computational time for this reference solution on the same Stafflinux platform at Greenwich was 0.019 seconds. Table 25.3 shows that with suitable choice of ρ the choice of larger time stepping is still possible demonstrating the viability of the Laplace transform method for the Frey-Patie model.

The Laplace transform method was implemented using Fortran with MPI on the Stafflinux platform at Greenwich. The parallel run time was simulated on a single processor. Table 25.4 shows the parallel computing time excluding the inverse Laplace transform which is trivial compare with the other equation solvers. Similar conclusion can be drawn as in the case for the hypothetical volatility.

Table 25.3 $||V_L(S,0) - V_R(S,0)||_2$

m / $\Delta\tau$	$64\delta\tau$	$32\delta\tau$	$16\delta\tau$	$8\delta\tau$
2	0.27911	0.28204	0.28361	0.28442
4	0.00728	0.00602	0.00605	0.00630
6	0.00234	0.00238	0.00242	0.00245
8	0.00303	0.00304	0.00305	0.00305
10	0.00302	0.00302	0.00302	0.00302
12	0.00301	0.00301	0.00301	0.00301
14	0.00301	0.00301	0.00301	0.00301

Table 25.4 Parallel computing time for different choices of $\Delta\tau$

m / $\Delta\tau$	$64\delta\tau$	$32\delta\tau$	$16\delta\tau$	$8\delta\tau$
2	0.00015	0.00031	0.00063	0.00125
4	0.00016	0.00031	0.00063	0.00130
6	0.00016	0.00031	0.00063	0.00125
8	0.00015	0.00031	0.00062	0.00124
10	0.00015	0.00031	0.00063	0.00126
12	0.00015	0.00031	0.00063	0.00125
14	0.00015	0.00031	0.00063	0.00125

25.4 A Sequence of Transformations

A generalisation of using the concept of transformation through the use of a sequence of transformation in order to remove the temporal axis was also investigated. The idea is to apply several transformation leading to a differential equation which does not involve time but several parametric differential equations as those described above. The method is derived here.

Consider Eq. (25.1) treating the volatility as a constant in the process of two transformations. First, take $S = e^y$ and $t = T - \tau$ with the stretch transformation $u = e^{r\tau}V$ leads to the forward equation

$$\frac{\partial u}{\partial \tau} - \frac{1}{2}\sigma^2\frac{\partial^2 u}{\partial y^2} - (r - \frac{1}{2}\sigma^2)\frac{\partial u}{\partial y} = 0,$$

in which the first order term can be removed by using $x = y + (r - \frac{1}{2}\sigma^2)\tau$ leading to

$$\frac{\partial u}{\partial \tau} = \frac{1}{2}\sigma^2\frac{\partial^2 u}{\partial x^2}.$$

An invariant transformation using $\eta = \frac{x}{\sqrt{t}}$ and taking $u(x,\tau) = f(\eta)$ leads to the parametric equation

$$\sigma\frac{d^2 f}{d\eta^2} + \frac{1}{2}\frac{df}{d\eta} = 0 \tag{25.6}$$

with suitable boundary conditions easily derived through the same transformation process. Equation (25.6) does not involve temporal axis and the solution $u(x, \tau)$ at many different τ may be evaluated simultaneously. Suppose one requires the numerical solution of u at various different times on a set of fixed grid points along the x-axis defined by the discretised mesh δx, the corresponding mesh size to be used in the transformed system is $\delta\eta = \delta\eta(\delta x, t, \sigma)$. It is possible to simultaneously compute as many time dependent solutions as possible at various different times, i.e.

$$\delta\eta_k = \delta\eta(\delta x, t_k, \sigma)$$

The computations at different times can be performed simultaneously. Each independent calculation depends on the value of t_k which affects the corresponding mesh size $\delta\eta_k$ used only at that particular time. In other words there are as many problems, each defined by Eq. (25.6) and is decoupled from the others, as those times when solutions are required. It is also possible to compute $u(x, \tau)$ using the closed form solution of Eq. (25.6).

The above process may be applied to cases with nonlinear volatility. A linearisation method, such as frozen coefficients, is applied to the forward equation following with the sequence of transformations as described above. This represents one linearisation step of the nonlinear update process which must be repeated until it is converged.

25.5 Conclusions

Two alternative parallel algorithms for nonlinear option pricing were examined. The first one is based on the Laplace transform method in which the temporal dimension is replaced by a set of finite number of parametric equations. This resulted to a coarse grained parallelisation of the numerical Laplace transform. A set of numerical experiments were performed and discussed. The second one is based on the use of a sequence of transformations leading to a differential equation that has no temporal involvement. This technique can also be implemented as a coarse grain parallel algorithm in a similar way as the Laplace transform method.

Experience shown that the nonlinear volatility within a larger time interval, $\Delta\tau$, as the one described in the Laplace transform method is that the rate of change of volatility must not be abrupt. However it is possible to accommodate a steady or constant rate of change of σ which is a necessary condition for an efficient Laplace transform method. This does not apply to the case of the other multiple stretch or similarity transformations.

References

1. Lai, C.-H.: Numerical solutions of certain nonlinear models in European options on a distributed computing environment. In: Ehrhardt, M. (ed.) Nonlinear Models in Mathematical Finance: New Research Trends in Option Pricing, ISBN: 978-1-60456-931-5. Nova Science Publisher, New York, (2008)
2. Frey, R., Patie, P.: Risk management for derivatives in illiquid markets: a simulation study, In: K. Sandmann, P.K. Schönbucher (eds.) Advances in Finance and Stochastics, pp. 137–160. Springer, Berlin (2002)
3. Miranker, W.L., Liniger, W.: Parallel methods for the numerical integration of ordinary differential equations. Math. Comput. **91**, 303–320 (1967)
4. Lions, J.-L., Maday, Y., Turinici, G.: A parareal in time discretisation of PDE's. C.R. Acad. Sci. Paris, Serie I **332**, 661–668 (2001)
5. Gander, M.J., Vandewalle, S.: Analysis of parareal time-parallel time-integration method. SIAM J. Sci. Comput. **29**, 558–578 (2007)
6. Lai, C.-H.: On transformation methods and the induced parallel properties for the temporal domain. In: Magoules, F. (ed.) Substructuring Techniques and Domain Decomposition Methods, ISBN: 1759-3158 (DOI: 10.4203/csets.24.3). Saxe-Coburg, Stirlingshire (2010)
7. Stehfest, H. Numerical inversion of Laplace transforms. Commun. ACM **13**, 47–49 (1970)

Chapter 26
Modern Monte Carlo Methods and GPU Computing

Álvaro Leitao and Cornelis W. Oosterlee

Abstract Pricing early-exercise options under multi-dimensional stochastic processes is a major challenge in the financial sector. In Leitao and Oosterlee (Int J Comput Math 92(12):2433–2454, 2015), a parallel GPU version of the Monte Carlo based Stochastic Grid Bundling Method (SGBM) (Jain and Oosterlee, Appl Math Comput 269:412–431, 2015) for pricing multi-dimensional Bermudan options is presented. The method is based on a combination of simulation, dynamic programming, regression and bundling of Monte Carlo paths. To extend the method's applicability, the problem dimensionality and the number of bundles is increased drastically. This makes SGBM very expensive in terms of computational costs on conventional hardware systems. A parallelization strategy of the method is developed and the GPGPU paradigm is used to reduce the execution time. An improved technique for bundling asset paths, which is more efficient on parallel hardware, is introduced. Thanks to the performance of the GPU version of SGBM, we can fully exploit the method and deal with very high-dimensional problems. Pricing results and comparisons between sequential and GPU parallel versions are presented.

26.1 Introduction

Monte Carlo methods are intensively employed in the financial field. Their simplicity in both interpretation and implementation makes them very attractive for practitioners. However, the main drawback usually attributed to this technique is the high computational cost. Although this fact has been improved in the last decades (due to the rapid evolution of the software and hardware), there still are some particular problems where the application of Monte Carlo methods can

---

Á. Leitao (✉) • C.W. Oosterlee
TU Delft, Delft Institute of Applied Mathematics, Delft, The Netherlands

CWI—Centrum Wiskunde & Informatica, Amsterdam, The Netherlands
e-mail: A.LeitaoRodriguez@tudelft.nl; c.w.oosterlee@cwi.nl

M. Ehrhardt et al. (eds.), *Novel Methods in Computational Finance*,
Mathematics in Industry 25, DOI 10.1007/978-3-319-61282-9_26

be considered "expensive". That is the case when we deal with *multi-* or *high-*dimensional problems. In many cases, the use of Monte Carlo is the only option for this type of problems since other methodologies like PDE- or Fourier-based approaches can not be applied for very high dimensions (more than three), due to the curse of dimensionality. In this chapter, we will focus our analysis on *high-dimensional early-exercise option contracts* and how to combine a Monte Carlo based pricing method with parallel GPU computing resulting in an efficient technology.

In recent years, different Monte Carlo simulation techniques for pricing multi-dimensional early-exercise options were developed. Some representative methods were developed by Longstaff and Schwartz [7] and Tsitsiklis and Van Roy [8]. One of the recent Monte Carlo pricing techniques is the Stochastic Grid Bundling Method (SGBM), proposed by Jain and Oosterlee in [5] for pricing Bermudan options with several underlying assets. The method is a hybrid of *regression-* and *bundling-* based approaches, and uses regressed value functions, together with bundling of the state space to approximate continuation values at different time steps. In [6], the method's applicability has been extended by increasing the number of bundles and the problem dimensionality, which, together, also imply a drastic increase of the number of Monte Carlo paths. As the method becomes much more time-consuming then, we have parallelized the SGBM method taking advantage of the General-Purpose computing on Graphics Processing Units (GPGPU) paradigm. In this chapter, this work is summarized.

26.2 Problem Formulation

This section defines the Bermudan option pricing problem and sets up the notations used in this chapter. A Bermudan option is an option where the buyer has the right to exercise at a set number of times, $t \in [t_0 = 0, \dots, t_m, \dots, t_M = T]$, before the end of the contract, T. $\mathbf{S}_t = (S_t^1, \dots, S_t^d) \in \mathbb{R}^d$ defines the d-dimensional underlying process. Let $h_t := h(\mathbf{S}_t)$ be an adapted process representing the intrinsic value of the option, i.e. the holder of the option receives $\max(h_t, 0)$, if the option is exercised at time t. The value of the option at the terminal time T is equal to the option's payoff, i.e.,

$$V_T(\mathbf{S}_T) = \max(h(\mathbf{S}_T), 0).$$

The conditional continuation value Q_{t_m}, i.e. the expected payoff at time t_m, is

$$Q_{t_m}(\mathbf{S}_{t_m}) = D_{t_m}\mathbb{E}\left[V_{t_{m+1}}(\mathbf{S}_{t_{m+1}})|\mathbf{S}_{t_m}\right],$$

where D_{t_m} is the discount factor. The Bermudan option value at time t_m and state $\mathbf{S}_{t_m}$ is then given by

$$V_{t_m}(\mathbf{S}_{t_m}) = \max(h(\mathbf{S}_{t_m}), Q_{t_m}(\mathbf{S}_{t_m})).$$

We are interested in finding the option value at the initial state $\mathbf{S}_{t_0}$, i.e. $V_{t_0}(\mathbf{S}_{t_0})$.

26.3 Stochastic Grid Bundling Method

SGBM [5] is a simulation-based Monte Carlo method for pricing early-exercise options (such as Bermudan options). SGBM first generates Monte Carlo paths forward in time, which is followed by determining the optimal early-exercise policy, moving backwards in time in a dynamic programming framework, based on the Bellman principle of optimality. The steps involved in the SGBM algorithm are briefly described in the following:

- **Step I: Generation of stochastic grid points.** The grid points in SGBM are generated by Monte Carlo sampling, i.e., by simulating independent copies of sample paths, $\{\mathbf{S}_{t_0}(n), \ldots, \mathbf{S}_{t_M}(n)\}$, $n = 1, \ldots, N$, of the underlying process $\mathbf{S}_t$, all starting from the same initial state $\mathbf{S}_{t_0}$.
- **Step II: Option value at terminal time.** The option value at the terminal time $t_M = T$ is given by

$$V_{t_M}(\mathbf{S}_{t_M}) = \max(h(\mathbf{S}_{t_M}), 0),$$

with $\max(h(\mathbf{S}_{t_M}), 0)$ a multi-dimensional payoff function.

The following steps are subsequently performed for each time step, t_m, $m \leq M$, recursively, moving backwards in time, starting from t_M.

- **Step III: Bundling.** The grid points at t_{m-1} are clustered or *bundled* into $\mathscr{B}_{t_{m-1}}(1), \ldots, \mathscr{B}_{t_{m-1}}(\nu)$ non-overlapping sets or partitions. SGBM employs bundling to approximate the conditional distribution using simulation. It samples this distribution by bundling the grid points at t_{m-1} and then uses those paths that originate from the corresponding bundle to obtain a conditional sample for time t_m.
- **Step IV: Mapping high-dimensional state space to a low-dimensional space.** Corresponding to each bundle $\mathscr{B}_{t_{m-1}}(\beta)$, $\beta = 1, \ldots, \nu$, a parameterized value function $Z : \mathbb{R}^d \times \mathbb{R}^K \mapsto \mathbb{R}$, which assigns values $Z(\mathbf{S}_{t_m}, \alpha^{\beta}_{t_m})$ to states $\mathbf{S}_{t_m}$, is computed. Here $\alpha^{\beta}_{t_m} \in \mathbb{R}^K$ is a vector of free parameters. The objective is then to choose, for each t_m and β, a parameter vector $\alpha^{\beta}_{t_m}$ so that

$$Z(\mathbf{S}_{t_m}, \alpha^{\beta}_{t_m}) \approx V_{t_m}(\mathbf{S}_{t_m}).$$

- **Step V: Computing the continuation and option values at** t_{m-1}. The continuation values for $\mathbf{S}_{t_{m-1}}(n) \in \mathscr{B}_{t_{m-1}}(\beta)$, $n = 1, \ldots, N$, $\beta = 1, \ldots, \nu$, are approximated by

$$\widehat{Q}_{t_{m-1}}(\mathbf{S}_{t_{m-1}}(n)) = \mathbb{E}[Z(\mathbf{S}_{t_m}, \alpha_{t_m}^{\beta})|\mathbf{S}_{t_{m-1}}(n)]. \tag{26.1}$$

The option value is then given by

$$\widehat{V}_{t_{m-1}}(\mathbf{S}_{t_{m-1}}(n)) = \max(h(\mathbf{S}_{t_{m-1}}(n)), \widehat{Q}_{t_{m-1}}(\mathbf{S}_{t_{m-1}}(n))).$$

Due to their importance, the last three steps are more extensively described in the following sections.

26.3.1 Bundling

We propose a bundling technique in the SGBM context which is highly efficient when taking into account our goal of high dimensionality, called *equal-partitioning*. This technique is particularly well-suited for parallel processing: it does not involve an iterative process, distributes the data equally and does not need to store the d-dimensional points. Equal-partitioning has two steps: sorting and splitting. The general idea is to sort the data first under some convenient criterion and then split the sorted data items into sets (i.e. bundles) of equal size. The sorting process is independent of the dimension of the problem, efficient and, furthermore, it is highly parallelizable. In addition, the storage of all Monte Carlo simulation data points can be avoided since only a reduced part is needed in the bundling stage. The split stage assigns directly the portions of data to bundles which will contain the same number of *similar* (following some criterion) data items. Hence, the regression can be performed accurately even though the number of bundles increases in a significant way. Furthermore, the equally sized bundles allow for a better load balancing within the parallel implementation.

26.3.2 Parameterizing the Option Values

The high-dimensional option pricing problem become intractable and requires the approximation of the value function. This can be achieved by introducing a *parameterized value function* $Z : \mathbb{R}^d \times \mathbb{R}^L \mapsto \mathbb{R}$, which assigns a value $Z(\mathbf{S}_{t_m}, \alpha)$ to state $\mathbf{S}_{t_m}$, where $\alpha \in \mathbb{R}^L$ is a vector of free parameters. The goal is to choose, corresponding to each bundle β at time point t_{m-1}, a parameter vector $\alpha_{t_m}^{\beta} := \alpha$ so that,

$$V_{t_m}(\mathbf{S}_{t_m}) \approx Z\left(\mathbf{S}_{t_m}, \alpha_{t_m}^{\beta}\right).$$

SGBM follows the approach of Tsitsiklis and Van Roy [8] and it uses basis functions to approximate the values of the options. Hence, two important decisions have to be made: the form of the function Z and the basis functions. For each particular problem we define several basis functions, $\phi_1, \phi_2, \ldots, \phi_L$, that are typically chosen based on experience, as in the case of the LSM method [7], aiming to represent relevant properties of a given state, $\mathbf{S}_{t_m}$. In our case, the form of $Z\left(\mathbf{S}_{t_m}, \alpha_{t_m}^{\beta}\right)$ depends on $\mathbf{S}_{t_m}$ only through $\phi_k(\mathbf{S}_{t_m})$. Hence, for some function $f : \mathbb{R}^L \times \mathbb{R}^L \mapsto \mathbb{R}$, we can write $Z\left(\mathbf{S}_{t_m}, \alpha_{t_m}^{\beta}\right) = f\left(\phi_k(\mathbf{S}_{t_m}), \alpha_{t_m}^{\beta}\right)$, where

$$Z\left(\mathbf{S}_{t_m}, \alpha_{t_m}^{\beta}\right) = \sum_{k=1}^{L} \alpha_{t_m}^{\beta}(k)\phi_k(\mathbf{S}_{t_m}).$$

An exact computation of the vector of free parameters, $\alpha_{t_m}^{\beta}$, is generally not feasible for the corresponding bundle $\mathscr{B}_{t_{m-1}}(\beta)$. Thus, an approximation $\widehat{\alpha}_{t_m}^{\beta}$ is computed by using ordinary least squares regression.

By using the parameterized option value function $Z\left(\mathbf{S}_{t_m}, \widehat{\alpha}_{t_m}^{\beta}\right)$ for the bundle $\mathscr{B}_{t_{m-1}}(\beta)$, the continuation values in Eq. (26.1) are approximated by

$$\begin{aligned}\widehat{Q}_{t_{m-1}}(\mathbf{S}_{t_{m-1}}(n)) &= D_{t_{m-1}}\mathbb{E}\left[\left(\sum_{k=1}^{L} \widehat{\alpha}_{t_m}^{\beta}(k)\phi_k(\mathbf{S}_{t_m})\right)|\mathbf{S}_{t_{m-1}} = \mathbf{S}_{t_{m-1}}(n)\right] \\ &= D_{t_{m-1}}\sum_{k=1}^{L} \widehat{\alpha}_{t_m}^{\beta}(k)\mathbb{E}\left[\phi_k(\mathbf{S}_{t_m})|\mathbf{S}_{t_{m-1}} = \mathbf{S}_{t_{m-1}}(n)\right],\end{aligned} \tag{26.2}$$

where $\mathbf{S}_{t_{m-1}}(n) \in \mathscr{B}_{t_{m-1}}(\beta)$. The continuation value will give us a reference value to compute the early-exercise policy. The basis functions ϕ_k should be chosen such that the expectations $\mathbb{E}\left[\phi_k(\mathbf{S}_{t_m})|\mathbf{S}_{t_{m-1}} = \mathbf{S}_{t_{m-1}}(n)\right]$ in Eq. (26.2) are easy to calculate, i.e. they are preferably known in closed form or otherwise have analytic approximations. In [6], several choices to determine the basis functions are described, either for particular cases or more general situations.

26.3.3 Estimating the Option Value

The estimation of the option value is the final step in SGBM. We consider the so-called *direct estimator* and *path estimator*. The direct estimator is typically biased high, i.e., it is often an upper bound. The definition of the direct estimator is

$$\widehat{V}_{t_{m-1}}(\mathbf{S}_{t_{m-1}}(n)) = \max\left(h\left(\mathbf{S}_{t_{m-1}}(n)\right), \widehat{Q}_{t_{m-1}}\left(\mathbf{S}_{t_{m-1}}(n)\right)\right),$$

where $n = 1, \ldots, N$. The final option value reads

$$\mathbb{E}[\widehat{V}_{t_0}(\mathbf{S}_{t_0})] = \frac{1}{N}\sum_{n=1}^{N}\widehat{V}_{t_0}(\mathbf{S}_{t_0}(n)).$$

The direct estimator corresponds to Step V in the initial description.

Once the optimal early-exercise policy has been obtained, the path estimator, which is typically biased low, can be developed based on the early-exercise policy. The resulting confidence interval is useful, because, depending on the problem at hand, sometimes the path estimator and sometimes the direct estimator is superior. The obtained confidence intervals are generally small, indicating accurate results. In order to compute the low-biased estimates, we generate a new set of paths, as in common for duality-based Monte Carlo methods, $\mathbf{S}(n) = \{\mathbf{S}_{t_1}(n), \ldots, \mathbf{S}_{t_M}(n)\}$, $n = 1, \ldots, N_{\mathrm{L}}$. Along each path, the approximate optimal policy exercises are

$$\widehat{\tau}^*(\mathbf{S}(n)) = \min\{t_m : h(\mathbf{S}_{t_m}(n)) \geq \widehat{Q}_{t_m}(\mathbf{S}_{t_m}(n)),\ m = 1, \ldots, M\},$$

where $\widehat{Q}_{t_m}(\mathbf{S}_{t_m}(n))$ is previously computed using Eq. (26.2). The path estimator is then defined by $v(n) = h\left(\mathbf{S}_{\widehat{\tau}^*(\mathbf{S}(n))}\right)$. Finally, the low-biased estimate given by the path estimator is

$$\underline{V}_{t_0}(\mathbf{S}_{t_0}) = \lim_{N_{\mathrm{L}}} \frac{1}{N_{\mathrm{L}}}\sum_{n=1}^{N_{\mathrm{L}}} v(n).$$

26.4 Parallel SGBM Method: Implementation Details

The GPU parallel implementation was performed by employing the *Compute Unified Device Architecture*, CUDA, a parallel computing platform and programming model developed by NVIDIA (see [2]).

Since SGBM is based on two clearly separated stages, we parallelize them separately. First of all, the Monte Carlo path generation is parallelized (Step I). As is well-known, Monte Carlo methods are very suitable for parallelization, because of characteristics like a very large number of simulations and data independence. In Fig. 26.1a, we see schematically how the parallelization is done where $p_0, p_1, \ldots, p_{N-1}$ are the CUDA *threads*. The second main stage of SGBM is the regression and the computation of the continuation and option values (Steps IV and V) in each bundle, backwards in time. Due to the data dependency between time steps, the way to parallelize this stage of the method is by parallelizing over the bundles, performing the calculations in each bundle in parallel. Schematic and simplified representation with two bundles is given in Fig. 26.1b. Note that, actually,

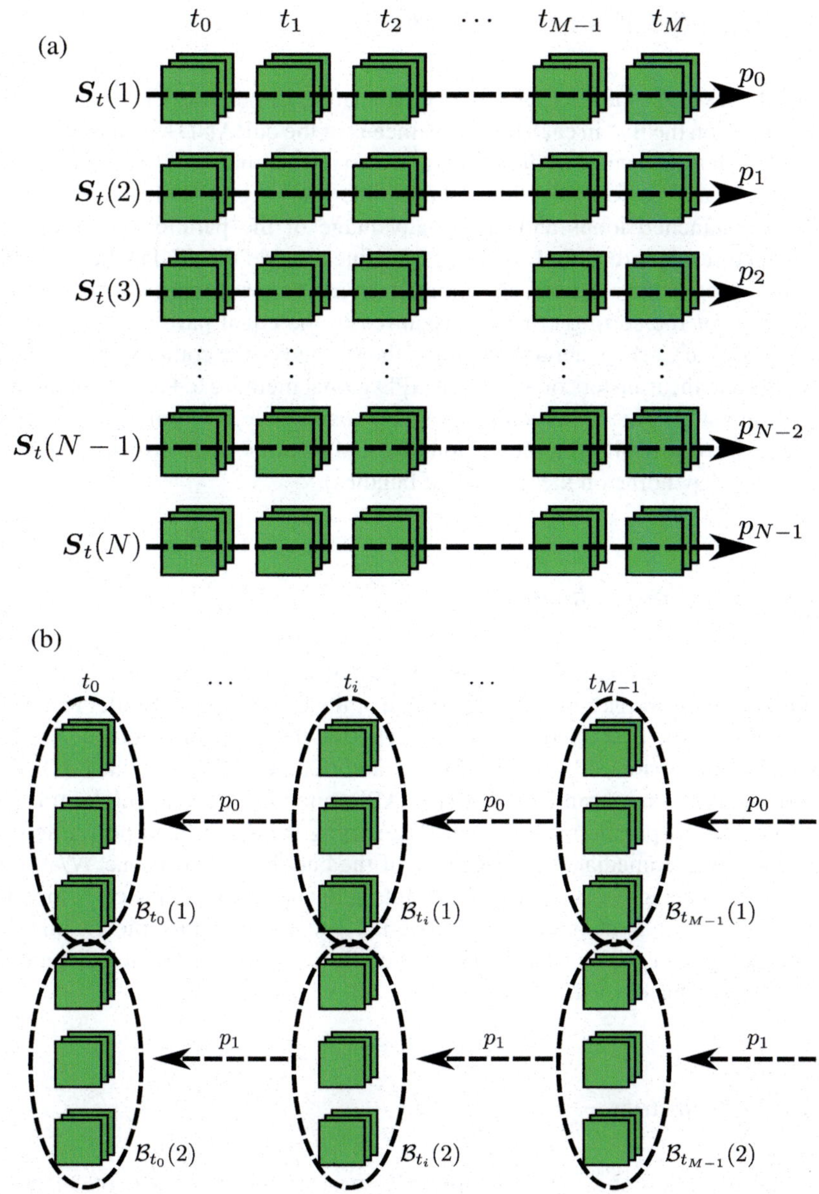

Fig. 26.1 Stages of the parallel SGBM method. (**a**) Monte Carlo stage. (**b**) Bundling stage

several stages of parallelization are performed, one per time step. Between each parallel stage, the bundling (Step III) is carried out.

In the following subsections, we will show more specific details of the CUDA implementation of the parallel SGBM method.

26.4.1 Parallel Monte Carlo paths

We launch one CUDA thread per Monte Carlo path. The necessary random numbers are obtained "on the fly" in each thread by means of the cuRAND standard library. In addition, the intrinsic value of the option and also the computation of the expectation in Eq. (26.2) are performed within the Monte Carlo generator, decreasing the number of launched loops and taking advantage of the parallel execution. The intermediate results are stored in an array defined inside the CUDA kernel which can be allocated in the registers, speeding up the memory accesses. We also perform calculations for the sorting criterion (required in the equal partitioning bundling) inside the Monte Carlo generator, avoiding the storage of the complete Monte Carlo grid points and the transfers of data from GPU global memory to CPU main memory in each time step. This approach gives us a considerable performance and allows us to increase drastically the dimensionality and also the number of Monte Carlo simulations (depending on the number of bundles).

26.4.2 Bundling Scheme

As mentioned, equal-partitioning bundling involves two operations: sorting and splitting. For the sort part, we take advantage of the CUDA Data-Parallel Primitives Library (CUDPP), described in [3]. We choose the parallel Radix sort which is included in version 2.1 of CUDPP. In addition, CUDPP provides a kernel-level Application Programming Interface (API), allowing us to avoid the transfers between stages of parallelization. Once the sorting stage has been performed, the splitting stage is immediate since the size of the bundles is known, i.e. N/ν. Each CUDA thread manages a pointer which points at the beginning of the corresponding region of GPU global memory for each bundle. The memory allocation is made for all bundles together which means that the bundle's memory areas are adjacent and the accesses are faster (*coalescing*).

26.4.3 Estimators

The exercise policy and the final option values can be computed by means of direct and path estimators. For the direct estimator, one CUDA thread per bundle is launched at each time step. For each bundle, the regression and option values are calculated on the GPU. All CUDA threads collaborate in order to compute the continuation value which determines the early-exercise policy. Once the early-exercise policy is determined, the path estimator can be executed. Its parallelization can be done over paths because the early-exercise policy is already known (given by the previous computation of the direct estimator) and is not needed to perform the regression. One CUDA thread per path is launched and it computes the optimal exercise time and the cash flows according to the policy.

26.5 Results

The implementation have been carried out in two programming languages: C and CUDA. This allows us to assess the improvement given by the parallel version compared with the sequential one. Experiments were performed on the Accelerator Island system of the Cartesius Supercomputer (more information in [1]). We consider the d-dimensional problem of pricing basket Bermudan options under the multi-dimensional Geometric Brownian Motion (GBM). The experiments setting is

- Initial state: $\mathbf{S}_{t_0} = (40, 40, \ldots, 40) \in \mathbb{R}^d$.
- Strike: $K = 40$.
- Risk-free interest rate: $r_t = 0.06$.
- Dividend yield rate: $q_\delta = 0.0$, $\delta = 1, 2, \ldots, d$.
- Volatility: $\sigma_\delta = 0.2$, $\delta = 1, 2, \ldots, d$.
- Correlation: $\rho_{i,j} = 0.25$, $j = 2, \ldots, d$, $i = 1, \ldots, j$.
- Maturity: $T = 1.0$.
- Exercise times: $M = 10$.
- Number of basis functions: $L = 3$.

26.5.1 Equal-Partitioning: Convergence Test

First, we wish to test the convergence of the new equal-partitioning technique bundling. Regarding the sorting criterion, we choose the payoff criterion, i.e. we sort the Monte Carlo scenarios following the geometric or arithmetic average of the assets for geometric and arithmetic basket options, respectively. In Fig. 26.2, we show the convergence in option prices for geometric and arithmetic basket Bermudan options with different dimensionalities, i.e. $d = 5$, $d = 10$ and $d = 15$. In the case of the geometric basket option, we can also specify the reference price obtained by the COS method [4].

26.5.2 Parallel SGBM: Performance Test

With the convergence of the equal-partitioning technique shown numerically, we now increase drastically the number of bundles and, hence, the number of Monte Carlo paths, and perform a computational cost experiment, comparing the C and

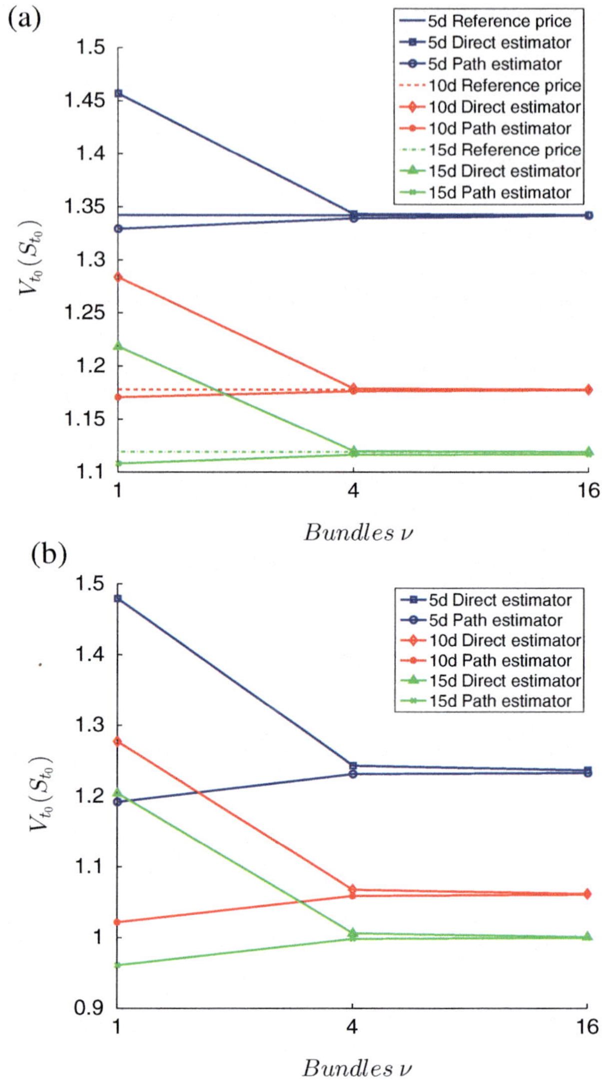

Fig. 26.2 Convergence with equal-partitioning technique. Configuration: $N = 2^{18}$ and $\Delta t = T/M$. (**a**) Geometric basket put option. (**b**) Arithmetic basket put option

CUDA implementations. We now focus on pricing arithmetic basket Bermudan option since it is a more interesting and involved problem. In the first half of the Table 26.1, the execution times for the different SGBM stages, i.e. Monte Carlo path generation (MC), direct estimator computation (DE) and path estimator computation (PE) are shown. The total computational cost for $d = 5$, $d = 10$ and $d = 15$ problems are presented in the second half of the Table 26.1. In both experiments, we observe a significant acceleration of the CUDA versions. The impact of the dimensionality is much less important in the parallel version than in the sequential version, resulting in an increasing speedup in terms of the problem dimension.

Table 26.1 Time (s) for the C and CUDA versions

	Stages time ($d = 5$)			Total time		
	MC	DE	PE	$d = 5$	$d = 10$	$d = 15$
C	79.22	39.64	58.65	256.05	600.09	1143.06
CUDA	0.83	4.14	1.20	8.02	11.23	15.73
Speedup	95.44	9.57	48.87	31.93	53.44	72.67

Configuration: $N = 2^{22}$, $\Delta t = T/M$ and $\nu = 2^{10}$

Table 26.2 Time (s) for a high-dimensional problem

	$\nu = 2^{10}$			$\nu = 2^{14}$		
	$d = 30$	$d = 40$	$d = 50$	$d = 30$	$d = 40$	$d = 50$
C	993.96	1723.79	2631.95	992.29	1724.60	2631.43
CUDA	11.14	17.88	26.99	11.20	17.94	27.07
Speedup	89.22	96.41	97.51	88.60	96.13	97.21

Configuration: $N = 2^{20}$ and $\Delta t = T/M$

This is thank to equal-partitioning bundling technique, since the parallelism can be efficiently exploited.

26.5.3 High-Dimensional Problems

We now test a high-dimensional option pricing problem. In Table 26.2, the execution times for pricing arithmetic basket Bermudan put options in different dimensions and with different numbers of bundles, ν, are presented. Note that the number of bundles hardly influences the execution times and the performance is mainly dependent on the number of paths and the dimensionality. The obtained speedup reaches around 100 times for the 50-dimensional problem.

26.6 Conclusions

In this chapter, we have presented an efficient implementation of the Stochastic Grid Bundling Method on a GPU architecture. Through the GPU parallelism, we could speed up the execution times when the number of bundles and the dimensionality increase drastically. In addition, we have proposed a parallel bundling technique which is more efficient in terms of memory use and more suitable on parallel systems. These two improvements enable us to extend the method's applicability to high-dimensional problems.

Compared with other GPU parallel implementations of early-exercise option pricing methods, our parallel SGBM is very competitive in terms of computational time since we provided a new way to parallelize the backward stage, according to the bundles, which gave us a remarkable performance improvement.

Acknowledgements The first author is supported by the European Union in the FP7-PEOPLE-2012-ITN Program under Grant Agreement Number 304617 (FP7 Marie Curie Action, Project Multi-ITN STRIKE—Novel Methods in Computational Finance). The authors would like to thank Shashi Jain, ING Bank, for providing support and the original codes of the Stochastic Grid Bundling Method.

References

1. Cartesius webpage: https://www.surfsara.nl/systems/cartesius
2. CUDA webpage: http://www.nvidia.com/object/cuda_home_new.html
3. CUDPP webpage: http://cudpp.github.io/
4. Fang, F., Oosterlee, C.W.: Pricing early-exercise and discrete Barrier options by Fourier-cosine series expansions. Numer. Math. **114**(1), 27–62 (2009)
5. Jain, S., Oosterlee, C.W.: The stochastic grid bundling method: efficient pricing of Bermudan options and their Greeks. Appl. Math. Comput. **269**, 412–431 (2015)
6. Leitao, A., Oosterlee, C.W.: GPU acceleration of the stochastic grid bundling method for early-exercise options. Int. J. Comput. Math. **92**(12), 2433–2454 (2015)
7. Longstaff, F.A., Schwartz, E.S.: Valuing American options by simulation: a simple least-squares approach. Rev. Financ. Stud. **14**(1), 113–47 (2001)
8. Tsitsiklis, J.N., Van Roy, B.: Regression methods for pricing complex American-style options. IEEE Trans. Neural Netw. **12**(4), 694–703 (2001)

Chapter 27
Sparse Grid Combination Technique for Hagan SABR/LIBOR Market Model

José Germán López-Salas and Carlos Vázquez Cendón

Abstract SABR models have been used to incorporate stochastic volatility to LIBOR market models (LMM) in order to describe interest rate dynamics and price interest rate derivatives. From the numerical point of view, the pricing of derivatives with SABR/LIBOR market models (SABR/LMMs) is mainly carried out with Monte Carlo simulation. However, this approach could involve excessively long computational times. In the present chapter we propose an alternative pricing based on partial differential equations (PDEs). Thus, we pose the PDE formulation associated to the SABR/LMM proposed by Hagan and Lesniewski (LIBOR market model with SABR style stochastic volatility. Working paper, available at http://lesniewski.us/papers/working/SABRLMM.pdf (2008)). As this PDE is high dimensional in space, traditional full grid methods (like standard finite differences or finite elements) are not able to price derivatives over more than one or two underlying interest rates and their corresponding stochastic volatilities. In order to overcome this curse of dimensionality, a sparse grid combination technique is proposed. So as to assess on the performance of the method a comparison with Monte Carlo is presented.

27.1 Introduction

The LIBOR market model (LMM) [5, 19, 23] has become the most popular interest rate model. The main reason is the agreement between this model and Black's formulas, which are the standard formulas employed in the market [6]. The standard LIBOR market model considers constant volatilities for the forward rates, no volatility smile modeling is taken into account.

Among the different stochastic volatility models offered in the literature, the SABR model proposed by Hagan et al. [17] in the year 2002 stands out for becoming the market standard to reproduce the price of European options. The SABR model

J.G. López-Salas • C.V. Cendón (✉)
Department of Mathematics, Faculty of Informatics, University of A Coruña, Campus Elviña s/n, 15071-A Coruña, Spain
e-mail: jose.lsalas@udc.es; carlosv@udc.es

M. Ehrhardt et al. (eds.), *Novel Methods in Computational Finance*,
Mathematics in Industry 25, DOI 10.1007/978-3-319-61282-9_27

can not be used to price derivatives whose payoff depends on several forward rates. In fact, SABR model works in the terminal measure, under which both the forward rate and its volatility are martingales. This can always be done if we work with one forward rate in isolation at a time. Under this same measure, however, the process for another forward rate and for its volatility would not be driftless.

In order to allow LMM to fit market volatility smiles, different extensions of the LMM that incorporate the volatility smile by means of the SABR model were proposed. These models are known as SABR/LIBOR market models (SABR/LMMs). In this chapter we will deal with the model proposed by Hagan et al. in [16].

While Monte Carlo [12] simulation remains the industry's tool of choice for pricing interest rate derivatives within SABR/LMM setting, several difficulties motivate researchers to address alternative approaches based on partial differential equation (PDE) formulations. The first issue is that the convergence of Monte Carlo methods, although it depends only very weakly on the dimension of the problem, is very slow. The second drawback of Monte Carlo methods is the valuation of options with early-exercise, like in the case of the American options, due to the so-called "Monte Carlo on Monte Carlo" effect. However, the modification of the PDE to a linear complementarity problem is usually straightforward. Finally, the weakest point of Monte Carlo methods appears to be the computation of the sensitivities of the solution with respect to the underlyings, the so-called "Greeks", which are very used by traders, and are directly given by the partial derivatives of the PDE solution.

In view of previous arguments, in the present chapter we pose the equivalent PDE formulation for the SABR/LMM proposed by Hagan. From the numerical point of view, one main difficulty in this PDE formulation lies in its high dimensionality in space-like variables. In order to cope with this so-called *curse of dimensionality* several methods are available in the literature, see [3, 11] for example, which can be put into three categories. The first group uses the Karhunen-Loeve transformation to reduce the stochastic differential equation (SDE) to a lower dimensional equation, therefore this results in a lower dimensional PDE associated to the previously reduced SDE. The second category gathers those methods which try to reduce the dimension of the PDE itself, like for example dimension-wise decomposition algorithms. Finally, the third category groups the methods which reduce the complexity of the problem in the discretization layer, like for example the method of sparse grids, which we use in the present chapter.

The sparse grid method was originally developed by Smolyak [29], who used it for numerical integration. It is mainly based on a hierarchical basis [30, 31], a representation of a discrete function space which is equivalent to the conventional nodal basis, and a sparse tensor product construction. Zenger [33] and Bungartz and Griebel [7] extended this idea and applied sparse grids to solve PDEs with finite elements, finite volumes and finite differences methods. Besides working directly in the hierarchical basis, the sparse grid can also be computed using the combination technique [15] by linearly combining solutions on traditional Cartesian grids with different mesh widths. This is the approach we follow in this chapter. Recently, this technique has been used for a financial application related to the pricing of basket options in [18]. Also in our previous work [21] we have posed the analogous

PDE formulation for the SABR/LMM proposed by Mercurio and Morini [22]. Moreover, we have used the same numerical methodology based on the sparse grids combination technique to solve the resulting high dimensional PDE problem.

The chapter is organized as follows. In Sect. 27.2 we pose the PDE formulation for the Hagan SABR/LMM. In Sect. 27.3 we describe the use of a full grid finite differences scheme for the Hagan model. Numerical results show the limitations of the full grid method when the number of forward rates increases. Therefore, in Sect. 27.4 we describe the sparse grid combination technique applied to the Hagan SABR/LMM and show numerical results that illustrate the behaviour of the method when the number of forward rates increases. For this purpose, a comparison with Monte Carlo simulation results is used.

27.2 The Hagan SABR/LMM PDE

We first consider a set of $N-1$ LIBOR forward rates F_i, $1 \leq i \leq N-1$, $\mathbf{F} = (F_1, \ldots, F_{N-1})$ on the tenor structure $[T_0, T_1, \ldots, T_{N-1}, T_N]$, the accruals being $\tau_i = T_{i+1} - T_i$. Hagan SABR/LMM is defined by the following system of SDEs [16]:

$$
\begin{aligned}
dF_i(t) &= \mu^{F_i}(t) F_i(t)^{\beta_i} dt + \alpha_i V_i(t) F_i(t)^{\beta_i} dW_i^{\mathscr{Q}}(t), \quad F_i(0) \text{ given}, \\
dV_i(t) &= \mu^{V_i}(t) V_i(t) dt + \sigma_i V_i(t) dZ_i^{\mathscr{Q}}(t), \quad V_i(0) = 1,
\end{aligned} \tag{27.1}
$$

which are posed on a probability space $\{\Omega, \mathscr{F}, \mathscr{Q}\}$ with filtration $\{\mathscr{F}_t\}$, $t \in [T_0, T_N]$. On one hand, μ^{F_i} is the drift of the i-th forward rate, $\beta_i \in [0,1]$ is the variance elasticity coefficient, $W_i^{\mathscr{Q}}$ is a standard Brownian motion under the risk neutral measure $\mathscr{Q}$, and $\boldsymbol{\rho}$ is the correlation matrix between the forward rates, i.e.

$$
< dW_i^{\mathscr{Q}}(t), dW_j^{\mathscr{Q}}(t) >= \rho_{ij} dt, \quad \forall i, j \in \{1, \ldots, N-1\}.
$$

On the other hand, V_i is the stochastic volatility of the forward rate F_i, μ^{V_i} is the drift of the i-th stochastic volatility, α_i is a deterministic (constant) instantaneous volatility coefficient used to embed in the model any initial value of the volatility process V_i, $Z_i^{\mathscr{Q}}$ is a standard Brownian motion, and $\boldsymbol{\theta}$ is the correlation matrix between the stochastic volatilities, i.e.

$$
< dZ_i^{\mathscr{Q}}(t), dZ_j^{\mathscr{Q}}(t) >= \theta_{ij} dt, \quad \forall i, j \in \{1, \ldots, N-1\}.
$$

Besides, the Brownian motions driving the forward rates are correlated with those ones driving the stochastic volatilities, $\boldsymbol{\phi}$ will denote the correlation matrix between the forward rates and their stochastic volatilities, i.e.

$$
< dW_i^{\mathscr{Q}}(t), dZ_j^{\mathscr{Q}}(t) >= \phi_{ij} dt, \quad \forall i, j \in \{1, \ldots, N-1\}.
$$

Thus, the correlation structure is given by the block-matrix

$$\boldsymbol{P} = \begin{bmatrix} \boldsymbol{\rho} & \boldsymbol{\phi} \\ \boldsymbol{\phi}^{\mathsf{T}} & \boldsymbol{\theta} \end{bmatrix},$$

which is assumed to be positive definite.

The drifts of the forward rates and their stochastic volatilities are determined by the chosen numeraire. Under the terminal probability measure associated with choosing the bond $P(t, T_N)$ as numeraire, the drifts of the forwards rates are given by

$$\mu^{F_i}(t) = \begin{cases} -\alpha_i V_i(t) \displaystyle\sum_{j=i+1}^{N-1} \frac{\tau_j F_j(t)^{\beta_j}}{1+\tau_j F_j(t)} \rho_{ij}\alpha_j V_j(t) & \text{if } j < N-1, \\ 0 & \text{if } j = N-1, \end{cases}$$

while the drifts of the stochastic volatilities are given by

$$\mu^{V_i}(t) = \begin{cases} -\sigma_i \displaystyle\sum_{j=i+1}^{N-1} \frac{\tau_j F_j(t)^{\beta_j}}{1+\tau_j F_j(t)} \phi_{ij}\alpha_j V_j(t) & \text{if } j < N-1, \\ 0 & \text{if } j = N-1. \end{cases}$$

Our model for the correlation structure is taken from Rebonato [26], who suggests the following functional parameterization:

$$\rho_{ij} = \exp[-\lambda_1 |T_i - T_j|], \tag{27.2}$$

$$\theta_{ij} = \exp[-\lambda_2 |T_i - T_j|], \tag{27.3}$$

$$\phi_{ij} = \operatorname{sign}(\phi_{ii}) \sqrt{|\phi_{ii}\phi_{jj}|} \exp[-\lambda_3 (T_i - T_j)^+ - \lambda_3 (T_j - T_i)^+]. \tag{27.4}$$

So far we have introduced Hagan SABR/LMM. Now suppose we need to price an interest rate product $u(t, \mathbf{F}, \mathbf{V})$ whose payoff at expiry T_N is a function of forward rates from F_1 to F_{N-1}, and also of their stochastic volatilities $\mathbf{V} = (V_1, \ldots, V_{N-1})$. If G is the payoff of the option, then the arbitrage-free value of the option relative to a numeraire $\mathscr{N}$ is given by

$$u(t, \mathbf{F}(t), \mathbf{V}(t)) = \mathbb{E}^{\mathscr{Q}} \left(\frac{G(T, \mathbf{F}(T), \mathbf{V}(T))}{\mathscr{N}(T)} \,\middle|\, \mathscr{F}_t \right). \tag{27.5}$$

Thus, the value u of the option satisfies the PDE

$$\frac{\partial u}{\partial t}+\frac{1}{2}\sum_{i,j=1}^{N-1}\theta_{ij}\sigma_i V_i\sigma_j V_j\frac{\partial^2 u}{\partial V_i\partial V_j}+\frac{1}{2}\sum_{i,j=1}^{N-1}\rho_{ij}\alpha_i V_i F_i^{\beta_i}\alpha_j V_j F_j^{\beta_j}\frac{\partial^2 u}{\partial F_i\partial F_j}+$$
$$\sum_{i,j=1}^{N-1}\phi_{ij}\alpha_i V_i F_i^{\beta_i}\sigma_j V_j\frac{\partial^2 u}{\partial F_i\partial V_j}+\sum_{i=1}^{N-1}\mu^{F_i}(t)F_i^{\beta_i}\frac{\partial u}{\partial F_i}+\sum_{i=1}^{N-1}\mu^{V_i}(t)V_i\frac{\partial u}{\partial V_i}=0, \tag{27.6}$$

with the terminal condition given by the derivative payoff,

$$u(T,\mathbf{F},\mathbf{V})=g(T,\mathbf{F},\mathbf{V}),$$

on $[0,T]\times\mathbb{R}^{N-1}\times\mathbb{R}^{N-1}$. For simplicity of notation, we have used the relative payoff $g(\cdot)=\dfrac{G(\cdot)}{\mathscr{N}(T)}$. This PDE was be derived by applying multi-dimensional Itô's Lemma to u, see [28] for details.

Hereafter, for sake of brevity in the notation, let us consider the following operator:

$$\mathscr{L}[u]=\frac{\partial u}{\partial t}+\frac{1}{2}\sum_{i,j=1}^{N-1}\theta_{ij}\sigma_i V_i\sigma_j V_j\frac{\partial^2 u}{\partial V_i\partial V_j}+\frac{1}{2}\sum_{i,j=1}^{N-1}\rho_{ij}\alpha_i V_i F_i^{\beta_i}\alpha_j V_j F_j^{\beta_j}\frac{\partial^2 u}{\partial F_i\partial F_j}+$$
$$\sum_{i,j=1}^{N-1}\phi_{ij}\alpha_i V_i F_i^{\beta_i}\sigma_j V_j\frac{\partial^2 u}{\partial F_i\partial V_j}+\sum_{i=1}^{N-1}\mu^{F_i}(t)F_i^{\beta_i}\frac{\partial u}{\partial F_i}+\sum_{i=1}^{N-1}\mu^{V_i}(t)V_i\frac{\partial u}{\partial V_i},$$

where u is a function defined on the domain $[0,T]\times\mathbb{R}^{N-1}\times\mathbb{R}^{N-1}$.

27.3 Finite Differences Method with Full Grids

In this section we introduce a full grid finite differences method to solve the problem (27.6). Domain truncation and boundary conditions are proposed. Notice that while the choice of the range of the time variable is totally unambiguous, $[0,T]$, an a priori choice must be made about which values of the space variables are too high or too low to be of interest, so far we will denote them by $[F_i^{min},F_i^{max}]$ and $[V_i^{min},V_i^{max}]$. Selecting boundary values such that the option of interest is too deeply in or out-of-the money is a common and reasonable choice.

We are going to define a $(2N-1)$-dimensional mesh with the time sampled from today (time 0) to the final expiry of the option (time T) at $M+1$ points uniformly spaced by the time step $\Delta t=\dfrac{T}{M}$.

The variables representing the forward rates $\mathbf{F} = (F_1, \ldots, F_{N-1})$ and their stochastic volatilities $\mathbf{V} = (V_1, \ldots, V_{N-1})$, often referred as the "space variables", will be sampled at $R_i + 1$ and $S_i + 1$ points, $i = 1, \ldots, N-1$, spaced by $h_i = \dfrac{F_i^{max} - F_i^{min}}{R_i}$ and $\hat{h}_i = \dfrac{V_i^{max} - V_i^{min}}{S_i}$, respectively.

For a given mesh, each point is uniquely determined by the time level m ($m = 0, \ldots, M$), the index vectors of the $N-1$ forward rates $\mathbf{f} = (f_1, \ldots, f_i, \ldots, f_{N-1})$ and stochastic volatilities $\mathbf{v} = (v_1, \ldots, v_i, \ldots, v_{N-1})$, where $f_i = 0, \ldots, R_i$ and $v_i = 0, \ldots, S_i$. We seek approximations of the solution at these mesh points, which will be denoted by

$$U_{\mathbf{f},\mathbf{v}}^m \approx u(m\Delta t, (f_i h_i)_{1\leq i\leq N-1}, (v_i \hat{h}_i)_{1\leq i\leq N-1}).$$

It is natural for this PDE to be solved backwards in time. We approximate the time derivative by the time-forward approximation

$$\left.\frac{\partial u}{\partial t}\right|_{t=m\Delta t, \mathbf{F}=(f_i h_i)_{1\leq i\leq N-1}, \mathbf{V}=(v_i\hat{h}_i)_{1\leq i\leq N-1}} = \left.\frac{\partial u}{\partial t}\right|_{m,\mathbf{f},\mathbf{v}} \approx \frac{U_{\mathbf{f},\mathbf{v}}^{m+1} - U_{\mathbf{f},\mathbf{v}}^m}{\Delta t}.$$

For the space derivatives we have chosen second-order approximations. We will write $\mathbf{f}_{i\pm1}$ to mean the forward rates index vector $(f_1, \ldots, f_i \pm 1, \ldots, f_{N-1})$ which corresponds to the forward rates point $(f_1 h_1, \ldots, (f_i \pm 1)h_i, \ldots, f_{N-1}h_{N-1})$. The same notation will be used in the case of the stochastic volatilities index vector. The first derivatives are approximated by central differences:

- $\left.\dfrac{\partial u}{\partial F_i}\right|_{m,\mathbf{f},\mathbf{v}} \approx \dfrac{U_{\mathbf{f}_{i+1},\mathbf{v}}^m - U_{\mathbf{f}_{i-1},\mathbf{v}}^m}{2h_i}$,
- $\left.\dfrac{\partial u}{\partial V_i}\right|_{m,\mathbf{f},\mathbf{v}} \approx \dfrac{U_{\mathbf{f},\mathbf{v}_{i+1}}^m - U_{\mathbf{f},\mathbf{v}_{i-1}}^m}{2\hat{h}_i}$.

The second derivatives are approximated by:

- $\left.\dfrac{\partial^2 u}{\partial F_i^2}\right|_{m,\mathbf{f},\mathbf{v}} \approx \dfrac{U_{\mathbf{f}_{i+1},\mathbf{v}}^m - 2U_{\mathbf{f},\mathbf{v}}^m + U_{\mathbf{f}_{i-1},\mathbf{v}}^m}{h_i^2}$,
- $\left.\dfrac{\partial^2 u}{\partial V_i^2}\right|_{m,\mathbf{f},\mathbf{v}} \approx \dfrac{U_{\mathbf{f},\mathbf{v}_{i+1}}^m - 2U_{\mathbf{f},\mathbf{v}}^m + U_{\mathbf{f},\mathbf{v}_{i-1}}^m}{\hat{h}_i^2}$.

The cross derivatives terms are approximated by:

- For $i \neq j$, $\left.\dfrac{\partial^2 u}{\partial F_i \partial F_j}\right|_{m,\mathbf{f},\mathbf{v}} \approx \dfrac{U_{\mathbf{f}_{i+1,j+1},\mathbf{v}}^m + U_{\mathbf{f}_{i-1,j-1},\mathbf{v}}^m - U_{\mathbf{f}_{i+1,j-1},\mathbf{v}}^m - U_{\mathbf{f}_{i-1,j+1},\mathbf{v}}^m}{4h_i h_j}$,

- For $i \neq j$, $\left.\dfrac{\partial^2 u}{\partial V_i \partial V_j}\right|_{m,\mathbf{f},\mathbf{v}} \approx \dfrac{U^m_{\mathbf{f},\mathbf{v}_{i+1,j+1}} + U^m_{\mathbf{f},\mathbf{v}_{i-1,j-1}} - U^m_{\mathbf{f},\mathbf{v}_{i+1,j-1}} - U^m_{\mathbf{f},\mathbf{v}_{i-1,j+1}}}{4\hat{h}_i\hat{h}_j}$,
- $\left.\dfrac{\partial^2 u}{\partial F_i \partial V_j}\right|_{m,\mathbf{f},\mathbf{v}} \approx \dfrac{U^m_{\mathbf{f}_{i+1},\mathbf{v}_{j+1}} + U^m_{\mathbf{f}_{i-1},\mathbf{v}_{j-1}} - U^m_{\mathbf{f}_{i+1},\mathbf{v}_{j-1}} - U^m_{\mathbf{f}_{i-1},\mathbf{v}_{j+1}}}{4h_i\hat{h}_j}$.

The finite differences solution under the so-called θ-scheme satisfies

$$\frac{U^{m+1}_{\mathbf{f},\mathbf{v}} - U^m_{\mathbf{f},\mathbf{v}}}{\Delta t} + \theta W^m_{\mathbf{f},\mathbf{v}} + (1-\theta) W^{m+1}_{\mathbf{f},\mathbf{v}} = 0,$$

where $\theta \in [0, 1]$ and $W^m_{\mathbf{f},\mathbf{v}}$ is the discretization given by

$$\begin{aligned}
W^m_{\mathbf{f},\mathbf{v}} =& \frac{1}{2}\sum_{\substack{i,j=1\\ i\neq j}}^{N-1} \theta_{ij}\sigma_i V_i \sigma_j V_j \frac{U^m_{\mathbf{f},\mathbf{v}_{i+1,j+1}} + U^m_{\mathbf{f},\mathbf{v}_{i-1,j-1}} - U^m_{\mathbf{f},\mathbf{v}_{i+1,j-1}} - U^m_{\mathbf{f},\mathbf{v}_{i-1,j+1}}}{4\hat{h}_i\hat{h}_j} + \\
& \frac{1}{2}\sum_{i=1}^{N-1} \sigma_i^2 V_i^2 \frac{U^m_{\mathbf{f},\mathbf{v}_{i+1}} - 2U^m_{\mathbf{f},\mathbf{v}} + U^m_{\mathbf{f},\mathbf{v}_{i-1}}}{\hat{h}_i^2} + \\
& \frac{1}{2}\sum_{\substack{i,j=1\\ i\neq j}}^{N-1} \rho_{ij}\alpha_i V_i F_i^{\beta_i}\alpha_j V_j F_j^{\beta_j} \frac{U^m_{\mathbf{f}_{i+1,j+1},\mathbf{v}} + U^m_{\mathbf{f}_{i-1,j-1},\mathbf{v}} - U^m_{\mathbf{f}_{i+1,j-1},\mathbf{v}} - U^m_{\mathbf{f}_{i-1,j+1},\mathbf{v}}}{4h_ih_j} + \\
& \frac{1}{2}\sum_{i=1}^{N-1} \alpha_i^2 V_i^2 F_i^{2\beta_i} \frac{U^m_{\mathbf{f}_{i+1},\mathbf{v}} - 2U^m_{\mathbf{f},\mathbf{v}} - U^m_{\mathbf{f}_{i-1},\mathbf{v}}}{h_i^2} + \\
& \sum_{i,j=1}^{N-1} \phi_{ij}\alpha_i V_i F_i^{\beta_i}\sigma_j V_j \frac{U^m_{\mathbf{f}_{i+1},\mathbf{v}_{j+1}} + U^m_{\mathbf{f}_{i-1},\mathbf{v}_{j-1}} - U^m_{\mathbf{f}_{i+1},\mathbf{v}_{j-1}} - U^m_{\mathbf{f}_{i-1},\mathbf{v}_{j+1}}}{4h_i\hat{h}_j} + \\
& \sum_{i=1}^{N-1} \mu^{F_i}(m\Delta t) F_i^{\beta_i} \frac{U^m_{\mathbf{f}_{i+1},\mathbf{v}} - U^m_{\mathbf{f}_{i-1},\mathbf{v}}}{2h_i} + \\
& \sum_{i=1}^{N-1} \mu^{V_i}(m\Delta t) V_i \frac{U^m_{\mathbf{f},\mathbf{v}_{i+1}} - U^m_{\mathbf{f},\mathbf{v}_{i-1}}}{2\hat{h}_i},
\end{aligned} \tag{27.7}$$

and with terminal condition $U^M_{\mathbf{f},\mathbf{v}} = g(T, \mathbf{F}, \mathbf{V})$.

Three different θ values represent three canonical discretization schemes, $\theta = 0$ is the explicit scheme, $\theta = 1$ the fully implicit scheme and $\theta = 0.5$ the Crank-Nicolson scheme. The fully implicit discretization is the best method with respect to stability, whereas the Crank-Nicolson time-stepping provides the best convergence rate. Although the explicit method is the simplest to implement, it has the disadvantage of being conditionally stable.

We shall first discriminate explicit and implicit parts as follows:

$$\frac{U^m_{\mathbf{f},\mathbf{v}}}{\Delta t} - \theta W^m_{\mathbf{f},\mathbf{v}} = \frac{U^{m+1}_{\mathbf{f},\mathbf{v}}}{\Delta t} + (1-\theta) W^{m+1}_{\mathbf{f},\mathbf{v}}. \tag{27.8}$$

As a result of such discretization we arrive to the linear system of equations $\mathbf{A}\mathbf{x} = \mathbf{b}$, where $\mathbf{A}$ is the band matrix of known coefficients, $\mathbf{x}$ is the vector of the unknown solutions $U^m_{\mathbf{f},\mathbf{v}}$ and $\mathbf{b}$ is the vector of known values corresponding to the right-hand side of (27.8).

Equation (27.8) can be rewritten as:

$$\begin{aligned}
&\theta \sum_{i=1}^{N-1} (\hat{b}_i - \hat{r}_i) U^m_{\mathbf{f},\mathbf{v}_{i-1}} + \theta \sum_{i=1}^{N-1} (\hat{b}_i + \hat{r}_i) U^m_{\mathbf{f},\mathbf{v}_{i+1}} + \\
&\theta \sum_{i=1}^{N-1} (b_i - r_i) U^m_{\mathbf{f}_{i-1},\mathbf{v}} + \theta \sum_{i=1}^{N-1} (b_i + r_i) U^m_{\mathbf{f}_{i+1},\mathbf{v}} + \\
&\theta \sum_{ij\in P} a_{ij} \big(U^m_{\mathbf{f}_{i+1},\mathbf{v}_{j+1}} + U^m_{\mathbf{f}_{i-1},\mathbf{v}_{j-1}} - U^m_{\mathbf{f}_{i-1},\mathbf{v}_{j+1}} - U^m_{\mathbf{f}_{i+1},\mathbf{v}_{j-1}} \big) + \\
&\theta \sum_{ij\in C} \hat{\psi}_{ij} \big(U^m_{\mathbf{f},\mathbf{v}_{i+1,j+1}} + U^m_{\mathbf{f},\mathbf{v}_{i-1,j-1}} - U^m_{\mathbf{f},\mathbf{v}_{i-1,j+1}} - U^m_{\mathbf{f},\mathbf{v}_{i+1,j-1}} \big) + \\
&\theta \sum_{ij\in C} \psi_{ij} \big(U^m_{\mathbf{f}_{i+1,j+1},\mathbf{v}} + U^m_{\mathbf{f}_{i-1,j-1},\mathbf{v}} - U^m_{\mathbf{f}_{i-1,j+1},\mathbf{v}} - U^m_{\mathbf{f}_{i+1,j-1},\mathbf{v}} \big) + \\
&\left(-1 - 2\theta \sum_{i=1}^{N-1} (\hat{b}_i + b_i) \right) U^m_{\mathbf{f},\mathbf{v}} = \\
&- \hat{\theta} \sum_{i=1}^{N-1} (\hat{b}_i - \hat{r}_i) U^{m+1}_{\mathbf{f},\mathbf{v}_{i-1}} - \hat{\theta} \sum_{i=1}^{N-1} (\hat{b}_i + \hat{r}_i) U^{m+1}_{\mathbf{f},\mathbf{v}_{i+1}} \\
&- \hat{\theta} \sum_{i=1}^{N-1} (b_i - r_i) U^{m+1}_{\mathbf{f}_{i-1},\mathbf{v}} - \hat{\theta} \sum_{i=1}^{N-1} (b_i + r_i) U^{m+1}_{\mathbf{f}_{i+1},\mathbf{v}} \\
&- \hat{\theta} \sum_{ij\in P} a_{ij} \big(U^{m+1}_{\mathbf{f}_{i+1},\mathbf{v}_{j+1}} + U^{m+1}_{\mathbf{f}_{i-1},\mathbf{v}_{j-1}} - U^{m+1}_{\mathbf{f}_{i-1},\mathbf{v}_{j+1}} - U^{m+1}_{\mathbf{f}_{i+1},\mathbf{v}_{j-1}} \big) \\
&- \hat{\theta} \sum_{ij\in C} \hat{\psi}_{ij} \big(U^{m+1}_{\mathbf{f},\mathbf{v}_{i+1,j+1}} + U^{m+1}_{\mathbf{f},\mathbf{v}_{i-1,j-1}} - U^{m+1}_{\mathbf{f},\mathbf{v}_{i-1,j+1}} - U^{m+1}_{\mathbf{f},\mathbf{v}_{i+1,j-1}} \big) \\
&- \hat{\theta} \sum_{ij\in C} \psi_{ij} \big(U^{m+1}_{\mathbf{f}_{i+1,j+1},\mathbf{v}} + U^{m+1}_{\mathbf{f}_{i-1,j-1},\mathbf{v}} - U^{m+1}_{\mathbf{f}_{i-1,j+1},\mathbf{v}} - U^{m+1}_{\mathbf{f}_{i+1,j-1},\mathbf{v}} \big) + \\
&\left(-1 + 2\hat{\theta} \sum_{i=1}^{N-1} (\hat{b}_i + b_i) \right) U^{m+1}_{\mathbf{f},v},
\end{aligned} \tag{27.9}$$

where $\hat{\theta} = (1-\theta)$, P is the set containing the permutations of the numbers $1,2,\ldots,N-1$ taken two at a time with repetition (the number of elements in P is $(N-1)^2$), C is the set containing the combinations of the numbers $1,2,\ldots,N-1$ taken two at a time without repetition (the number of elements in C is $\binom{N-1}{2} = 2^{-1}(N-1)(N-2)$) and the known coefficients $\hat{b}_i$, b_i, $\hat{r}_i$, r_i, $\hat{\psi}_{ij}$, ψ_{ij} and a_{ij} are defined as

$$\hat{b}_i = \frac{\Delta t \sigma_i^2 V_i^2}{2\hat{h}_i^2}, \quad b_i = \frac{\Delta t \alpha_i^2 V_i^2 F_i^{2\beta_i}}{2h_i^2},$$

$$\hat{r}_i = \frac{\Delta t \mu^{V_i}(t) V_i}{2\hat{h}_i}, \quad r_i = \frac{\Delta t \mu^{F_i}(t) F_i^{\beta_i}}{2h_i},$$

$$\hat{\psi}_{ij} = \frac{\Delta t \theta_{ij} \sigma_i V_i \sigma_j V_j}{4\hat{h}_i \hat{h}_j}, \quad \psi_{ij} = \frac{\Delta t \rho_{ij} \alpha_i V_i F_i^{\beta_i} \alpha_j V_j F_j^{\beta_j}}{4h_i h_j},$$

$$a_{ij} = \frac{\Delta t \phi_{ij} \alpha_i V_i F_i^{\beta_i} \sigma_j V_j}{4h_i \hat{h}_j},$$

where we have denoted $\mathbf{F} = (F_i = f_i h_i)_{1 \le i \le N-1}$ and $\mathbf{V} = (V_i = v_i \hat{h}_i)_{1 \le i \le N-1}$.

27.3.1 Boundary Conditions

In order to specify boundary conditions, a combination of mathematical, financial and heuristic reasoning allows us to find consistent and acceptable ones. There are several possibilities, see [8] for example.

We assume that forward rates and their stochastic volatilities are non negative and hence take values in the range zero to infinity. We first truncate the unbounded interval to a bounded one and then we must specify conditions at the new boundary. Thus we will consider the truncated domain $[F_i^{min}, F_i^{max}] \times [V_i^{min}, V_i^{max}]$, with $F_i^{min} = 0$ and $V_i^{min} = 0$.

For the forward rates we consider Dirichlet boundary conditions. Particularly, the terminal condition holds on the forward rates boundaries, i.e.

$$U^m_{\{\mathbf{f}|\exists f_i = 0\},\mathbf{v}} = U^M_{\mathbf{f},\mathbf{v}}, \quad \forall m = 0,\ldots,M-1,$$

$$U^m_{\{\mathbf{f}|\exists f_i = R_i\},\mathbf{v}} = U^M_{\mathbf{f},\mathbf{v}}, \quad \forall m = 0,\ldots,M-1.$$

At the stochastic volatility boundaries we consider the following conditions:

$$\mathscr{L}[u] = 0, \quad V_k = 0, \tag{27.10}$$

$$\frac{\partial u}{\partial V_k} = 0, \quad V_k = V_{max}. \tag{27.11}$$

Thus, when $V_k = 0$ we require that the PDE itself must be satisfied on this boundary. When V_k approaches to infinity, the price of the derivative becomes independent of V_k. This is reflected by using Neumann conditions instead of the Dirichlet ones used for the forward rates boundaries.

For the boundary $V_k = V_{max}$ in order to maintain the second order accuracy in the discretization of the first derivative the ghost point method is considered. Let us consider the volatility index vector $\mathbf{s} = (v_1, v_2, \ldots, S_k, \ldots, v_{N-1})$. The ghost grid points $U_{\mathbf{f},\mathbf{s}_{k+1}}$ are added. Then, the finite differences scheme of Eq. (27.9) can also be applied at the boundary points $U_{\mathbf{f},\mathbf{s}}$. However, we now have more unknowns than equations. The additional equations come from the central finite differences discretization of the Neumann boundary condition (27.11):

$$\frac{U_{\mathbf{f},\mathbf{s}_{k+1}} - U_{\mathbf{f},\mathbf{s}_{k-1}}}{2\hat{h}_k} = 0,$$

which yields $U_{\mathbf{f},\mathbf{s}_{k+1}} = U_{\mathbf{f},\mathbf{s}_{k-1}}$. Inserting this into the finite differences equation at $V_k = V_{max}$ we achieve

$$\begin{aligned}
&\theta \sum_{\substack{i=1\\ i\neq k}}^{N-1} (\hat{b}_i - \hat{r}_i) U^m_{\mathbf{f},\mathbf{s}_{i-1}} + \theta \sum_{\substack{i=1\\ i\neq k}}^{N-1} (\hat{b}_i + \hat{r}_i) U^m_{\mathbf{f},\mathbf{s}_{i+1}} + 2\theta \hat{b}_k U^m_{\mathbf{f},\mathbf{s}_{k-1}} + \\
&\theta \sum_{i=1}^{N-1} (b_i - r_i) U^m_{\mathbf{f}_{i-1},\mathbf{s}} + \theta \sum_{i=1}^{N-1} (b_i + r_i) U^m_{\mathbf{f}_{i+1},\mathbf{s}} + \\
&\theta \sum_{\substack{ij\in P\\ j\neq k}} a_{ij} \left(U^m_{\mathbf{f}_{i+1},\mathbf{s}_{j+1}} + U^m_{\mathbf{f}_{i-1},\mathbf{s}_{j-1}} - U^m_{\mathbf{f}_{i-1},\mathbf{s}_{j+1}} - U^m_{\mathbf{f}_{i+1},\mathbf{s}_{j-1}} \right) + \\
&\theta \sum_{\substack{ij\in C\\ i\neq k, j\neq k}} \hat{\psi}_{ij} \left(U^m_{\mathbf{f},\mathbf{s}_{i+1,j+1}} + U^m_{\mathbf{f},\mathbf{s}_{i-1,j-1}} - U^m_{\mathbf{f},\mathbf{s}_{i-1,j+1}} - U^m_{\mathbf{f},\mathbf{s}_{i+1,j-1}} \right) + \\
&\theta \sum_{ij\in C} \psi_{ij} \left(U^m_{\mathbf{f}_{i+1,j+1},\mathbf{s}} + U^m_{\mathbf{f}_{i-1,j-1},\mathbf{s}} - U^m_{\mathbf{f}_{i-1,j+1},\mathbf{s}} - U^m_{\mathbf{f}_{i+1,j-1},\mathbf{s}} \right) + \\
&\left(-1 - 2\theta \sum_{i=1}^{N-1} (\hat{b}_i + b_i) \right) U^m_{\mathbf{f},\mathbf{s}} \\
=& -\hat{\theta} \sum_{\substack{i=1\\ i\neq k}}^{N-1} (\hat{b}_i - \hat{r}_i) U^{m+1}_{\mathbf{f},\mathbf{s}_{i-1}} - \hat{\theta} \sum_{\substack{i=1\\ i\neq k}}^{N-1} (\hat{b}_i + \hat{r}_i) U^{m+1}_{\mathbf{f},\mathbf{s}_{i+1}} - 2\hat{\theta} \hat{b}_k U^{m+1}_{\mathbf{f},\mathbf{s}_{k-1}} \\
&- \hat{\theta} \sum_{i=1}^{N-1} (b_i - r_i) U^{m+1}_{\mathbf{f}_{i-1},\mathbf{s}} - \hat{\theta} \sum_{i=1}^{N-1} (b_i + r_i) U^{m+1}_{\mathbf{f}_{i+1},\mathbf{s}}
\end{aligned}$$

$$- \hat{\theta} \sum_{\substack{ij \in P \\ j \neq k}} a_{ij} \left(U^{m+1}_{\mathbf{f}_{i+1},\mathbf{s}_{j+1}} + U^{m+1}_{\mathbf{f}_{i-1},\mathbf{s}_{j-1}} - U^{m+1}_{\mathbf{f}_{i-1},\mathbf{s}_{j+1}} - U^{m+1}_{\mathbf{f}_{i+1},\mathbf{s}_{j-1}} \right)$$

$$- \hat{\theta} \sum_{\substack{ij \in C \\ i \neq k, j \neq k}} \hat{\psi}_{ij} \left(U^{m+1}_{\mathbf{f},\mathbf{s}_{i+1,j+1}} + U^{m+1}_{\mathbf{f},\mathbf{s}_{i-1,j-1}} - U^{m+1}_{\mathbf{f},\mathbf{s}_{i-1,j+1}} - U^{m+1}_{\mathbf{f},\mathbf{s}_{i+1,j-1}} \right)$$

$$- \hat{\theta} \sum_{ij \in C} \psi_{ij} \left(U^{m+1}_{\mathbf{f}_{i+1,j+1},\mathbf{s}} + U^{m+1}_{\mathbf{f}_{i-1,j-1},\mathbf{s}} - U^{m+1}_{\mathbf{f}_{i-1,j+1},\mathbf{s}} - U^{m+1}_{\mathbf{f}_{i+1,j-1},\mathbf{s}} \right) +$$

$$\left(-1 + 2\theta \sum_{i=1}^{N-1} (\hat{b}_i + b_i) \right) U^{m+1}_{\mathbf{f},\mathbf{s}}.$$

27.3.2 Numerical Results

It is not clear where to place F_i^{max} and V_i^{max}. On one hand, it is advantageous to place them far away of the initial forward rates. This reduces the error of the artificial boundary conditions. On the other hand a large computational domain requires a large discretization width. This increases the error of the approximation of the derivatives. In our experiments we will consider $F_i^{max} = 0.1$ and $V_i^{max} = 2.0$, which corresponds to interest rates of 10% and volatilities of 200%.

We are going to value $T_\alpha \times (T_\beta - T_\alpha)$ European swaptions, meaning that the swaption has maturity at time T_α and the length of the underlying swap is $(T_\beta - T_\alpha)$ (also known as the tenor of the swaption).

Some specifications of the financial product are given in Table 27.1 and the employed market data, taken from [4], are shown in Table 27.2. We will consider $\lambda_1 = \lambda_2 = \lambda_3 = 0.1$ in the model for the correlation structure (27.2)–(27.4). Besides, the Crank-Nicolson scheme will be used in (27.8). For solving the system (27.9) the Gauss-Seidel iterative solver has been employed using a tolerance of 10^{-6}.

The numerical experiments have been performed with the following hardware and software configurations: two recent multi-core Intel Xeon CPUs E5-2620 v2 clocked at 2.10 GHz (6 cores per socket) with 62 GBytes of RAM, CentOS Linux, GNU C++ compiler 4.8.2.

First of all, the results from pricing a 1×1 European swaption are discussed. The value ϑ of this swaption is the same as the price of the corresponding caplet, and so

Table 27.1 Specification of the interest rate model

Currency	EUR
Index	EURIBOR
Day count	e30/360
Strike	5.5%

Table 27.2 Market data used in pricing

	Start date	End date	LIBOR rate (%)	Volatility (%)
T_0	29-07-04	29-07-05	2.423306	0
T_1	29-07-05	29-07-06	3.281384	24.73
T_2	29-07-06	29-07-07	3.931690	22.45
T_3	29-07-07	29-07-08	4.364818	19.36
T_4	29-07-08	29-07-09	4.680236	17.43
T_5	29-07-09	29-07-10	4.933085	16.15
T_6	29-07-10	29-07-11	5.135066	15.02
T_7	29-07-11	29-07-12	5.273314	14.24
T_8	29-07-12	29-07-13	5.376115	13.42

Data taken from 27th July 2004

depends only on F_1. Hence, in one dimension a closed form expression for the price of a European swaption can be found by using Black's formula [6]:

$$\vartheta = P(T_0, T_2)\tau_1 \mathrm{Bl}(K, F(T_1, T_2; T_0), \nu_1),$$

where

$$\mathrm{Bl}(K, F, \nu) = F\Phi\big(d_1(K, F, \nu)\big) - K\Phi\big(d_2(K, F, \nu)\big),$$

$$d_1(K, F, \nu) = \frac{\ln(F/K) + \nu^2/2}{\nu},$$

$$d_2(K, F, \nu) = \frac{\ln(F/K) - \nu^2/2}{\nu},$$

$$\nu_i = \sigma_{Black}\sqrt{T_i},$$

where $P(T_0, T_2)$ is the price at time T_0 of a bond with maturity T_2 and σ_{Black} is the constant volatility of the forward rate. This value is equal to 0.659096 basis points (one basis point is one hundredth of one percent, $\frac{1\%}{100} = \frac{1}{10000}$). As Black-Scholes formula for caplets considers constant volatility σ_{Black}, in this first test the volatility of the volatility parameter of Hagan model is considered equal to zero, i.e., $\sigma_1 = 0$, therefore a standard LIBOR market model is used. The solution was found on several levels and Table 27.3 shows the convergence of the model. In all tables of this chapter, *Level* refers to the refinement level n, i.e., the mesh size is $h_i = 2^{-n} \cdot c_i$ in each coordinate direction, where c_i denotes the computational domain length in direction i, which is F_i^{max} in the case of the forward rates and V_i^{max} in the case of their stochastic volatilities. Besides, the solution and the error with respect to the exact solution are also shown in basis points. Additionally, the execution time is measured in seconds and the column labelled as *Grid points* shows the number of

Table 27.3 Convergence of the full grid finite differences solution in basis points in the pricing of a 1×1 swaption, $\sigma_1 = 0$, $V_1(0) = 1$, $\beta_1 = 1$, 128 time steps

Level	Solution	Error	Time	Grid points
3	2.078086	1.418989	0.0024	81
4	1.108211	0.449114	0.0094	289
5	0.779033	0.119936	0.07	1089
6	0.672004	0.012907	0.53	4225
7	0.665176	0.006079	6.34	16,641
8	0.661164	0.002067	84.12	66,049
9	0.659380	0.000283	1122.86	263,169
10	0.659032	0.000064	14,288.34	1,050,625

Exact solution, 0.659096 basis points

grid points employed in the full grid used by the finite differences method without taking into account the time coordinate.

When the volatilities of the volatilities σ_i, $1 \leq i < N$, of the model are non zero or when the length of the underlying swap of the swaption being considered is greater than one, no closed form solutions are available. However, an estimate can be obtained from Monte Carlo simulations. On Table 27.4 Monte Carlo values for the 1×1 European swaption with $\sigma_1 = 0$ are shown for several numbers of paths (*#Paths*). More details about Monte Carlo simulation of SABR/LMMs can be found in the article [9].

In Table 27.5 the pricing of the 1×1 European swaption with $\sigma_1 = 0.3$ for different resolution levels n are shown. In Table 27.6 the results for the 1×2 swaption are given. Note that with this numerical method it was not feasible to price the swaption past refinement level $n = 6$ due to the huge number of required grid points.

Theoretically, it is possible to solve the discrete system (27.9) for a general number of dimensions. However, in computational science, a major problem occurs when the number of dimensions increases. A natural way to reduce the discretization error is to decrease the mesh step in each coordinate direction. However, then the number of grid points in the resulting full grid grows exponentially with the dimension, i.e. the size of the discrete solution increases drastically. This is called the *curse of dimensionality* [2]. Therefore, this procedure of improving the accuracy by decreasing the mesh step is mainly bounded by two factors, the storage and the computational complexity. Due to these limitations, using a full grid discretization method which achieves sufficiently accurate approximations is only possible for problems with up to three or four dimensions, even on the most powerful machines presently available [7].

Table 27.4 Convergence of the Monte Carlo solution in basis points in the pricing of a 1×1 swaption, $\sigma_1 = 0$, $V_1(0) = 1$, $\beta_1 = 1$, 128 time steps

#Paths	Solution
10^5	0.616799
10^7	0.658598
10^9	0.659506

Exact solution, 0.659096 basis points

Table 27.5 Convergence of the full grid finite differences solution in basis points in the pricing of a 1×1 swaption, $\sigma_1 = 0.3$, $\phi_{11} = 0.4$, $V_1(0) = 1$, $\beta_1 = 1$, 128 time steps

Level	Solution	Time	Grid points
3	6.254822	0.0039	81
4	2.501988	0.0122	289
5	1.991646	0.07	1089
6	1.597470	0.62	4225
7	1.526047	7.48	16,641
8	1.519841	98.45	66,049
9	1.519742	1291.76	263,169
10	1.519732	16,238.98	1,050,625

Monte Carlo value using 10^7 paths, 1.657662 basis points

Table 27.6 Convergence of the full grid finite differences solution in basis points in the pricing of a 1×2 swaption, $\sigma_i = 0.3$, $\phi_{ii} = 0.4$, $V_i(0) = 1$, $\beta_i = 1$, 128 time steps

Level	Solution	Time	Grid points
3	5.289644	1.03	6561
4	5.134938	33.84	83,521
5	5.023293	1258.56	1,185,921
6	4.997679	60,396.44	17,850,625

Monte Carlo value using 10^7 paths, 4.564905 basis points

27.4 Sparse Grids and the Combination Technique

Two approaches to try to overcome the curse of dimensionality are increasing the order of accuracy of the applied numerical approximation scheme or reducing the dimension of the problem by choosing suitable coordinates. Both approaches are not always possible for every option pricing problem. In this chapter we will take advantage of the sparse grid combination technique first introduced by Zenger and co-workers [15] in order to try to overcome the curse of dimensionality and allow to use the PDE formulation of SABR/LMM for the pricing problem we are dealing with. The combination technique replicates the structure of a so-called sparse grid by linearly combining solutions on coarser grids of the same dimensionality. This technique reduces the computational effort and the storage space involved with the mentioned traditional finite differences discretization methods. The number of sub-problems to solve increases, while the computational time per problem decreases drastically. This method can be implemented in parallel as each sub-grid is independent of the others. In the next two subsections we give a brief introduction to sparse grids and the combination technique. For a detailed discussion we refer to [7].

27.4.1 Sparse Grids

First, we introduce some notations and definitions. Let $\mathbf{l} = (l_1, l_2, \ldots, l_d) \in \mathbb{N}_0^d$ denote a d-dimensional multi-index. Let $|\mathbf{l}|_1$ and $|\mathbf{l}|_\infty$ denote the discrete L_1−norm and L_∞−norm of the multi-index $\mathbf{l}$, respectively, that are defined as

$$|\mathbf{l}|_1 = \sum_{k=1}^{d} l_k \quad \text{and} \quad |\mathbf{l}|_\infty = \max_{1 \le k \le d} l_k.$$

We define the anisotropic grid $\Omega_\mathbf{l}$ with mesh size $\mathbf{h} = (h_1, h_2, \ldots, h_d) = (2^{-l_1} c_1, 2^{-l_2} c_2, \ldots, 2^{-l_d} c_d)$ with multi-index $\mathbf{l}$ and grid length $\mathbf{c} = (c_1, c_2, \ldots, c_d)$.

Then, the full grid at refinement level $n \in \mathbb{N}$ and mesh size $h_i = 2^{-n} \cdot c_i$ for all i can be defined via the sequence of subgrids

$$\Omega^n = \Omega_{(n,\ldots,n)} = \bigcup_{|\mathbf{l}|_\infty \le n} \Omega_\mathbf{l}.$$

Figure 27.1 visualizes two dimensional full grids for levels $n = 0, \ldots, 4$.

The number of grid points in each coordinate direction of the full grid is $2^n + 1$ and therefore the number of grid nodes in the full grid increases with $O(2^{n \cdot d})$, i.e. grows exponentially with the dimensionality d of the problem.

The sparse grid Ω_s^n at refinement level n consists of all anisotropic Cartesian grids $\Omega_\mathbf{l}$, where the total sum of all refinement factors l_k in each coordinate direction equals the resolution n. Then, the sparse grid Ω_s^n is given by

$$\Omega_s^n = \bigcup_{|\mathbf{l}|_1 \le n} \Omega_\mathbf{l} = \bigcup_{|\mathbf{l}|_1 = n} \Omega_\mathbf{l}.$$

Figure 27.2 shows the two-dimensional grid hierarchy for levels $n = 0, \ldots, 4$.

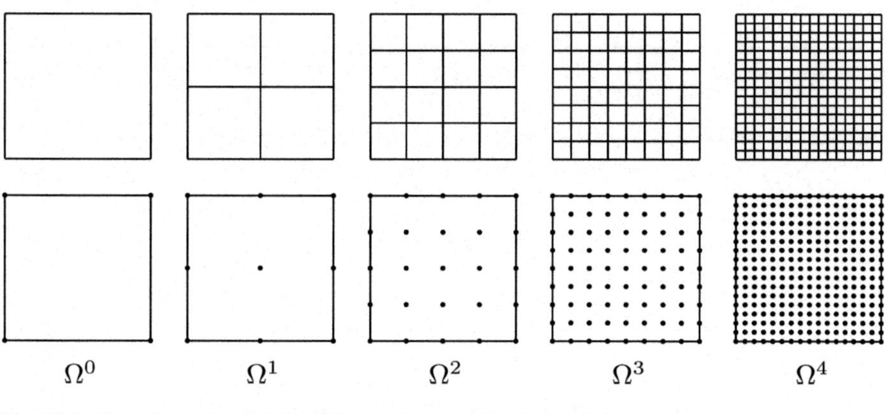

Fig. 27.1 Two-dimensional full grid hierarchy up to level $n = 4$

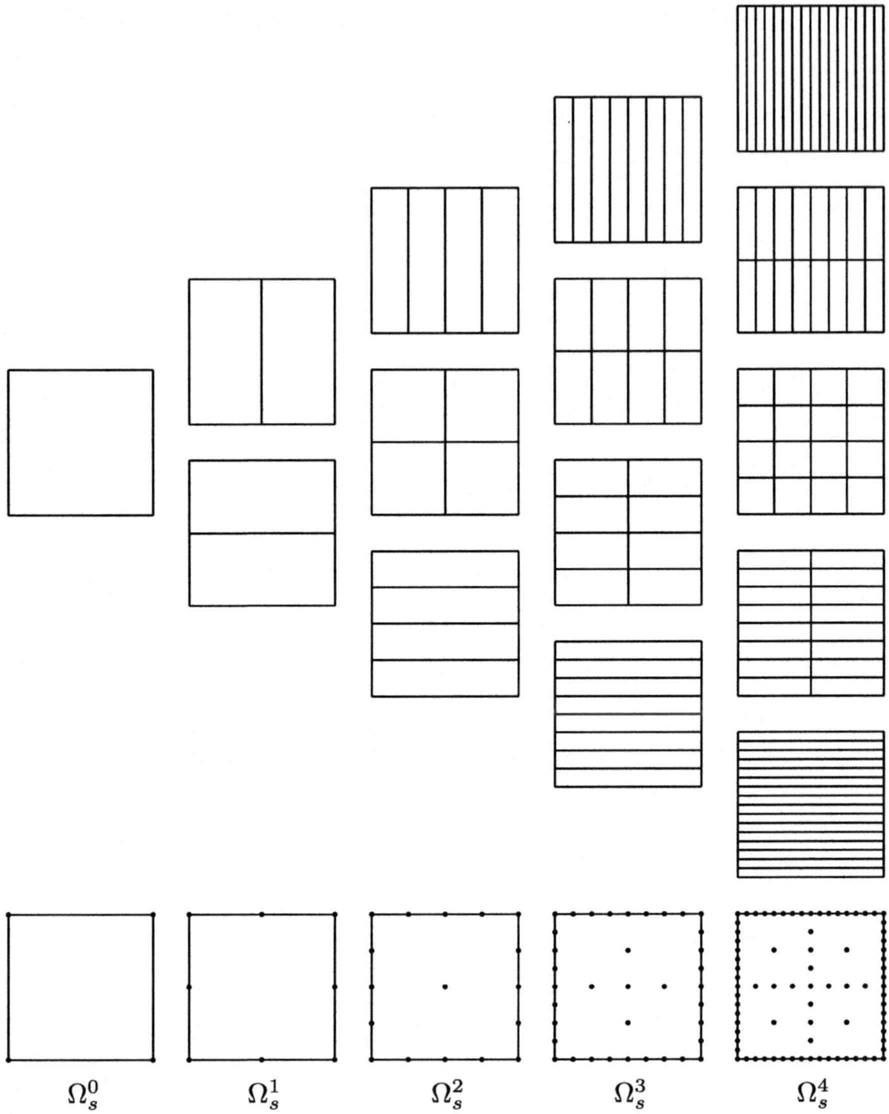

Fig. 27.2 Two-dimensional sparse grid hierarchy up to level $n = 4$

The total number of nodes in the grid $\Omega_{\mathbf{l}}$ is $\prod_{k=1}^{d}(2^{l_k}+1) = O(2^{|\mathbf{l}|_1}) = O(2^n)$. In addition, there exist exactly $\binom{n+d-1}{d-1}$ grids $\Omega_{\mathbf{l}}$ with $|\mathbf{l}|_1 = n$,

$$\begin{aligned}\binom{n+d-1}{d-1} &= \frac{(n+d-1)!}{(d-1)!n!} = \frac{(n+d-1)\cdot\ldots\cdot(n+1)n!}{(d-1)!n!}\\ &= \frac{n+(d-1)}{d-1}\cdot\frac{n+(d-2)}{d-2}\cdot\ldots\cdot\frac{n+(d-(d-1))}{d-(d-1)}\\ &= \left(1+\frac{n}{d-1}\right)\cdot\left(1+\frac{n}{d-2}\right)\cdot\ldots\cdot\left(1+\frac{n}{2}\right)\cdot\left(1+\frac{n}{1}\right)\\ &\leq (1+n)^{d-1} = O(n^{d-1}).\end{aligned}$$

Thus, the total number of grid points of the sparse grid Ω_s^n grows according to

$$\binom{n+d-1}{d-1}\cdot\prod_{k=1}^{d}(2^{l_k}+1) = O(n^{d-1})O(2^n) = O(n^{d-1}2^n), \tag{27.12}$$

which is far less the size of the corresponding full grid with $O(2^{nd})$ grid points. Let $h_n = 2^{-n}$, therefore the sparse grid employs $O(h_n^{-1}\cdot\log_2(h_n^{-1})^{d-1})$ grid points compared to $O(h_n^{-d})$ nodes in the full grid.

Bungartz and Griebel [7] show that the accuracy of the sparse grid using $O(h_n^{-1}\cdot\log_2(h_n^{-1})^{d-1})$ nodes is of order $O(h_n^2\log_2(h_n^{-1})^{d-1}))$ in the case of finite elements discretization and under certain smoothness conditions. Thus, the accuracy of the sparse grid is only slightly deteriorated from the accuracy $O(h_n^2)$ of conventional full grid methods which need $O(h_n^{-d})$ grid points. Therefore, sparse grids need much less points than regular full grids to achieve a similar approximation quality.

However, the structure of a sparse grid is more complicated than the one of a full grid. Common PDE solvers usually manage only full grid solutions. Existing sparse grid methods working directly in the hierarchical basis involve a challenging implementation [1, 32]. This handicap can be circumvented with the help of the sparse grid combination technique which not only exploits the economical structure of the sparse grids but also allows for the use of traditional full grid PDE solvers.

Finally, two and three dimensional sparse grids for several resolution levels n are shown in Figs. 27.3 and 27.4, respectively. Additionally, the growth of the grid points when increasing n can be observed.

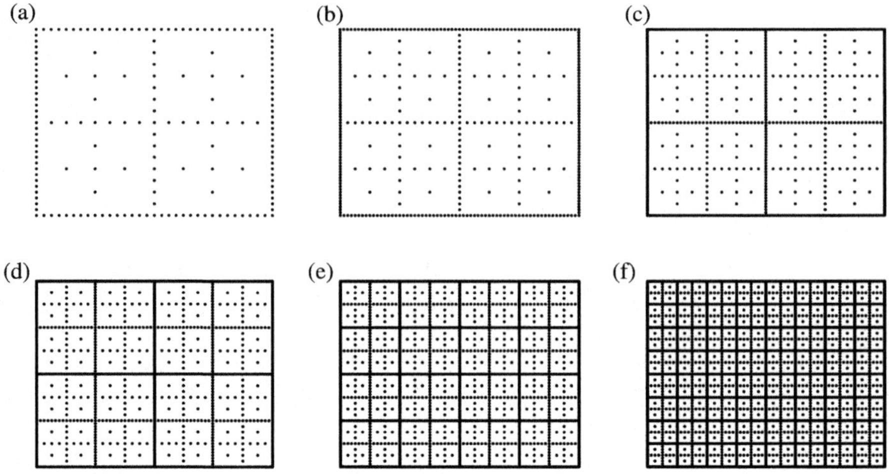

Fig. 27.3 Two dimensional sparse grids for levels $n = 5, \ldots, 10$. (**a**) Ω_s^5, 177 grid points. (**b**) Ω_s^6, 385 grid points. (**c**) Ω_s^7, 833 grid points. (**d**) Ω_s^8, 1793 grid points. (**e**) Ω_s^9, 3841 grid points. (**f**) Ω_s^{10}, 8193 grid points

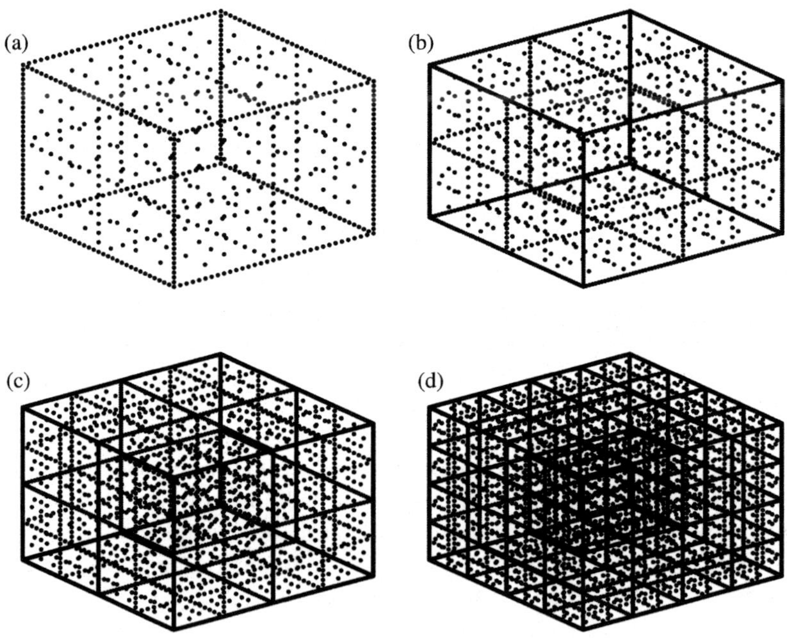

Fig. 27.4 Three dimensional sparse grids for levels $n = 5, 6, 7$ and 8. (**a**) Ω_s^5, 705 grid points. (**b**) Ω_s^6, 1649 grid points. (**c**) Ω_s^7, 3809 grid points. (**d**) Ω_s^8, 8705 grid points

27.4.2 Combination Technique

Similar to the Richardson extrapolation [27], the so-called combination technique linearly combines the numerical solution on the sequence of anisotropic grids $\Omega_{\mathbf{l}}$ where

$$|\mathbf{l}|_1 = n - q, \quad q = 0, \ldots, d-1.$$

The combination technique reads

$$U_s^n = \sum_{q=0}^{d-1} (-1)^q \cdot \binom{d-1}{q} \cdot \sum_{|\mathbf{l}|_1 = n-q} U_{\mathbf{l}}, \quad l_k \geq 0, \quad \forall k = 1, \ldots, d, \tag{27.13}$$

where $U_{\mathbf{l}}$ denotes the numerical solution on the grid $\Omega_{\mathbf{l}}$ and U_s^n the combined solution on the sparse grid Ω_s^n.

The grids employed by the combination technique of level $n = 4$ in two dimensions are shown in Fig. 27.5.

The idea of this technique is that the leading order errors from the discretization on each grid cancel each other out in the combination solution.

The number of grid points involved in the approximation of U_s^n grows according to $O(n^{d-1} \cdot 2^n)$. In fact, from the formula (27.12) we have to solve $\binom{n+d-1}{d-1}$ problems with $O(2^n)$ unknowns, $\binom{n+d-2}{d-1}$ problems with $O(2^{n-1})$ unknowns, ... and $\binom{n}{d-1}$ problems with $O(2^{n-(d-1)})$ unknowns. This results in a total number of $O(n^{d-1} \cdot 2^n)$ grid points which is much less than the $O(2^{n \cdot d})$ grid nodes used by traditional full grid methods. Thus, the efficient use of sparse grids greatly reduces the computing time and the storage requirements which allows for the treatment of problems with ten variables and even more [7].

We have seen that the combination technique linearly combines the numerical solution on several traditional full grids. The solution can be calculated on each of these grids by using any existing PDE numerical method like finite differences, finite volume or finite elements. In addition, since all these sub-problems are independent the combination technique can be parallelized [13].

The combination technique approach presumes the existence of a so-called error splitting. It requires for an associated numerical approximation method on the full grid $\Omega_{\mathbf{l}}$ an error splitting of the form

$$u(\mathbf{x}) - U_{\mathbf{l}}(\mathbf{x}) = \sum_{k=1}^{d} \sum_{\substack{\{j_1,\ldots,j_k\} \\ \subseteq \{1,\ldots,d\}}} C_{j_1,\ldots,j_k}(\mathbf{x}, h_{j_1}, \ldots, h_{j_k}) \cdot h_{j_1}^p \cdot \ldots \cdot h_{j_k}^p, \tag{27.14}$$

at each grid point $\mathbf{x} \in \Omega_{\mathbf{l}}$. Here u denotes the exact solution of the PDE under consideration, $U_{\mathbf{l}}$ the numerical solution on the grid $\Omega_{\mathbf{l}}$, $p > 0$ is the order of accuracy of the numerical approximation method with respect to each

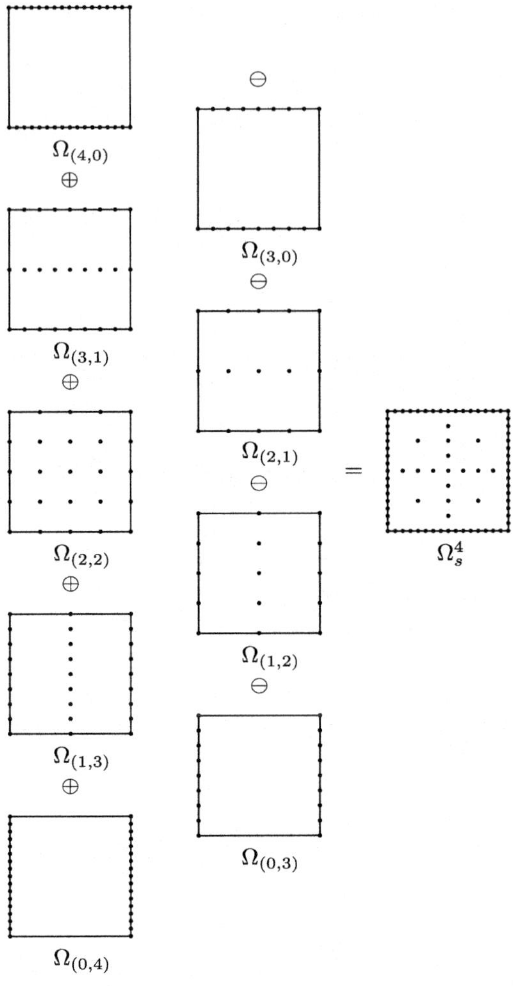

Fig. 27.5 Combination technique with level $n = 4$ in two dimensions

coordinate direction and the coefficient functions $C_{j_1,\dots,j_k}$ of $\mathbf{x}$ and the mesh sizes h_{j_k}, $k = 1, \dots, d$ are required to be bounded by a positive constant K such that

$$|C_{j_1,\dots,j_k}(\mathbf{x}, h_{j_1}, \dots, h_{j_k})| \leq K, \quad \forall k, 1 \leq k \leq d, \quad \forall \{j_1, \dots, j_m\} \subseteq \{1, \dots, d\}.$$

In [14] Griebel and Thurner showed that if the solution of the PDE is sufficiently smooth, the pointwise accuracy of the sparse grid combination technique is $O(n^{d-1} \cdot 2^{-n \cdot p}) = O([\log_2 h_n^{-1}]^{d-1} h_n^p)$, which is only slightly worse than $O(2^{-n \cdot p}) = O(h_n^p)$ obtained by the full grid solution.

The solution at points which do not belong to the sparse grid can be computed through interpolation. The applied interpolation method should provide at least the same order of accuracy of the numerical discretization scheme used to solve the PDE. Otherwise, the accuracy of the numerical scheme will be deteriorated.

27.4.3 Numerical Results

Taking advantage of the previously described sparse grid combination technique, in this section we are pricing the same interest rate derivatives that have been valued in the former Sect. 27.3.2 where traditional full grid finite differences methods were considered. In addition to those products, we are going to price interest rate derivatives with up to four underlying LIBOR rates and their stochastic volatilities, showing that the sparse grid combination technique is able to cope with the curse of dimensionality up to a certain extent. As in the previous Sect. 27.3.2, we will use Crank-Nicolson scheme, we will consider the Gauss-Seidel iterative solver and the same boundary conditions as in Sect. 27.3.1. In the present case, we are interested in the evaluation of the solution at a single point which corresponds with the value of the forward rates at time zero (see Table 27.2) and $V_i(0) = 1$. The numerical solution on each grid handled by the combination technique is interpolated at this point using multilinear interpolation and then added up with the appropriate weights.

The sparse grid combination technique has been implemented to run on multi-core CPUs. The program was optimized and parallelized using OpenMP [25]. CPU times, measured in seconds, correspond to executions using 24 threads, so as to take advantage of Intel Hyperthreading. The speedups of the parallel version with respect to the pure sequential code are around 16. To the best of our knowledge, graphic processor units (GPUs) are not well-suited to parallelize the combination technique, due to the fact that the different grids employed by the combination technique involve memory accesses patterns totally different, therefore, it is not possible to access the device memory in a coalesced way [24], thus GPU global memory can not serve threads in parallel. In this scenario, the GPU code will be ill performing. In the work [10] the authors take advantage of GPUs to parallelize the solver of each full grid considered by the combination technique. However, they do not parallelize the combination technique itself.

In Table 27.7 a 1×1 European swaption is priced. The exact price of this derivative is 0.659096 basis points, as discussed in Sect. 27.3.2. These results are to be compared with those of Table 27.3, where it can be seen how the computational

Table 27.7 Convergence of the sparse grid finite differences solution in basis points in the pricing of a 1×1 swaption, $\sigma_1 = 0$, $V_1(0) = 1$, $\beta_1 = 1$, 128 time steps

Level	Solution	Error	Time	Grid points
3	6.715346	6.056250	0.04	37
4	2.182057	1.522961	0.05	81
5	1.097761	0.438665	0.05	177
6	0.782767	0.123671	0.05	385
7	0.663808	0.004712	0.06	833
8	0.657536	0.001560	0.11	1793
9	0.658183	0.000913	0.46	3841
10	0.659363	0.000267	2.32	8193

Exact solution, 0.659096 basis points

Table 27.8 Convergence of the sparse grid finite differences solution in basis points in the pricing of a 1×1 swaption, $\sigma_1 = 0.3$, $\phi_{11} = 0.4$, $V_1(0) = 1$, $\beta_1 = 1$, 128 time steps

Level	Solution	Time
3	6.818116	0.05
4	2.694770	0.05
5	1.919198	0.05
6	1.596501	0.08
7	1.499332	0.12
8	1.505709	0.14
9	1.515855	0.64
10	1.521027	2.83

Monte Carlo value using 10^7 paths, 1.657662 basis points

Table 27.9 Convergence of the sparse grid finite differences solution in basis points in the pricing of a 1×2 swaption, $\sigma_i = 0.3$, $\phi_{ii} = 0.4$, $V_i(0) = 1$, $\beta_i = 1$, 128 time steps

Level	Solution	Time
7	5.260049	0.21
8	4.951410	0.47
9	4.651916	1.45
10	4.424338	4.10
11	4.463664	17.04
12	4.515542	81.04
13	4.537787	472.07

Monte Carlo value using 10^7 paths, 4.564905 basis points

Table 27.10 Convergence of the sparse grid finite differences solution in basis points in the pricing of a 1×3 swaption, $\sigma_i = 0.3$, $\phi_{ii} = 0.4$, $V_i(0) = 1$, $\beta_i = 1$, 128 time steps

Level	Solution	Time
11	9.177020	151.26
12	8.461583	431.29
13	7.455562	1219.71
14	7.442483	3849.56

Monte Carlo value using 10^7 paths, 7.648443 basis points

times and the grid points employed by the sparse grid combination technique have been substantially reduced.

Next, in Table 27.8 a 1×1 European swaption is priced considering stochastic volatility. These results are to be compared with those of Table 27.5.

In the following Table 27.9, the pricing of a 1×2 European swaption taking into account stochastic volatilities is shown, as in the Table 27.6. For the higher resolution levels, the full grid method became very slow, while the sparse grid combination technique results much faster. Note that the combination technique is able to price successfully the 1×2 European swaption, this was not attainable in Table 27.6.

Finally, in Tables 27.10 and 27.11, 1×3 and 1×4 European swaptions are priced, respectively, taking into account stochastic volatilities. The pricing of these interest rate derivatives was not viable with the full grid approach of Sect. 27.3. In order

Table 27.11 Convergence of the sparse grid finite differences solution in basis points in the pricing of a 1 × 4 swaption, $\sigma_i = 0.3$, $\phi_{ii} = 0.4$, $V_i(0) = 1$, $\beta_i = 1$, 8 time steps

Level	Solution	Time
15	11.316526	16,595.66
16	11.564127	53,184.37

Monte Carlo value using 10^7 paths, 11.674706 basis points

to be able to price derivatives over more than 4 LIBORs and their corresponding stochastic volatilities, the combination technique method should be parallelized to run on a cluster of processors. In [11, Chap. 13] Schröder et al. discuss the parallelization of the combination technique using MPI (*Message Passing Interface*) API. In [20] the authors parallelize the sparse grid combination technique taking advantage of a MapReduce framework, algorithms that are inherently fault tolerant.

Acknowledgements Partially financed by Spanish Grant MTM2013-47800-C2-1-P and by Xunta de Galicia (Grant CN2014/044). First author has also been founded by a FPU Spanish Grant.

References

1. Achatz, S.: Higher order sparse grid methods for elliptic partial differential equations with variable coefficients. Computing **71**(1), 1–15 (2003)
2. Bellmann, R.: Adaptive Control processes: A Guided Tour. Princenton University Press, Princeton (1961)
3. Beylkin, G., Mohlenkamp, M.J.: Algorithms for numerical analysis in high dimensions. SIAM J. Sci. Comput. **26**(6), 2133–2159 (2005)
4. Blackham, J.: Sparse grid solutions to the LIBOR market model. Master's thesis, Magdalen College, University of Oxford (2004)
5. Brace, A., Gatarek, D., Musiela, M.: The market model of interest rate dynamics. Math. Financ. **7**(2), 127–155 (1997)
6. Brigo, D., Mercurio, F.: Interest Rate Models – Theory and Practice. With Smile, Inflation and Credit, 2nd ed. Springer, Berlin/Heidelberg (2007)
7. Bungartz, H.J., Griebel, M.: Sparse grids. Acta Numer. **13**, 147–269 (2004)
8. Duffy, D.J.: Finite Difference Methods in Financial Engineering. A Partial Differential Equation Approach. Wiley Finance Series, Hoboken (2006)
9. Ferreiro, A.M., García, J.A., López-Salas, J.G., Vázquez, C.: SABR/LIBOR market models: pricing and calibration for some interest rate derivatives. Appl. Math. Comput. **242**, 65–89 (2014)
10. Gaikwad, A., Toke, I.M.: GPU based sparse grid technique for solving multidimensional options pricing PDEs. In: Proceedings of the 2nd Workshop on High Performance Computational Finance, WHPCF '09, pp. 6:1–6:9. ACM, New York, NY, USA (2009)
11. Gerstner, T., Kloeden, P. (eds.): Recent Developments in Computational Finance. Foundations, Algorithms and Applications. Interdisciplinary Mathematical Science. World Scientific Publishers, Singapore (2013)
12. Glasserman, P.: Monte Carlo Methods in Financial Engineering. Springer, New York (2003)

13. Griebel, M.: The combination technique for sparse grids solution of PDEs on multiprocessor machines. Parallel Process. Lett. **3**, 66–71 (1992)
14. Griebel, M., Thurner, V.: The efficient solution of fluid dynamics problems by the combination technique. Int. J. Numer. Methods Heat Fluid Flow **5**(3), 251–269 (1995)
15. Griebel, M., Schneider, M., Zenger, C.: A combination technique for the solution of sparse grid problems. In: de Groen, P., Beauwens, R. (eds.) Proceedings of the IMACS International Symposium on Iterative Methods in Linear Albegra, pp. 263–281. Elsevier, Amsterdam, 1992, Brussels (1991)
16. Hagan, P., Lesniewski, A.: LIBOR market model with SABR style stochastic volatility. Working paper, available at http://lesniewski.us/papers/working/SABRLMM.pdf (2008)
17. Hagan, P., Kumar, D., Lesniewski, A., Woodward, D.: Managing Smile Risk. Wilmott Magazine, pp. 84–108 (2002)
18. Hendricks, C., Heuer, C., Ehrhardt, M., Günther, M.: Higher order ADI schemes for parabolic equations in the combinations technique and application to finance. J. Comput. Appl. Math. **316**, 175–194 (2017)
19. Jamshidian, F.: LIBOR and swap market models and measures. Finance Stochast. **1**, 293–330 (1997)
20. Larson, J.W., Hegland, M., Harding, B., Roberts, S., Stals, L., Rendell, A.P., Strazdins, P., Ali, M.M., Kowitz, C., Nobes, R., Southern, J., Wilson, N., Li, M., Oishi, Y.: 2013 International Conference on Computational Science: Fault-Tolerant Grid-Based Solvers: Combining Concepts from Sparse Grids and MapReduce. Procedia Comput. Sci. **18**, 130–139 (2013)
21. López-Salas, J.G., Vázquez, C.: PDE formulation of some SABR/LIBOR market models and its numerical solution with a sparse grid combination technique. Preprint submitted for publication (2016)
22. Mercurio, F., Morini, M.: No-Arbitrage dynamics for a tractable SABR term structure LIBOR Model. Modeling Interest Rates: Advances in Derivatives Pricing, Risk Books, London, UK (2009)
23. Miltersen, K., Sandmann, K., Sondermann, D.: Closed-form solutions for term structure derivatives with lognormal interest rates. J. Financ. **52**(1), 409–430 (1997)
24. Nvidia Corporation: CUDA C Programming guide. Available from www.nvidia.com
25. OpenMP web page: http://www.openmp.org
26. Rebonato, R., Mckay, K., White, R.: The SABR/LIBOR Market Model. Pricing, Calibration and Hedging for Complex Interest-Rate Derivatives, 1st ed. John Wiley & Sons, Hoboken (2009)
27. Richardson, L.F.: The approximate arithmetical solution by finite differences of physical problems including differential equations, with an application to the stresses in a masonry dam. Phil. Trans. R. Soc. A **210**(459-470), 307–357 (1911)
28. Shreve, S.E.: Stochastic Calculus for Finance. Springer, New York (2004)
29. Smolyak, S.: Quadrature and interpolation formulas for tensor products of certain classes of functions. Dokl. Akad. Nauk SSR **148**, 1042–1045 (1963)
30. Yserentant, H.: On the multi-level splitting of finite element spaces. Numer. Math. **49**, 379–412 (1986)
31. Yserentant, H.: Hierarchical bases. In: Proceedings of the 2nd International Conference on Industrial and Applied Mathematics, ICIAM 91, pp. 256–276. Society for Industrial and Applied Mathematics, Philadelphia, PA, USA (1992)
32. Zeiser, A.: Fast matrix-vector multiplication in the sparse-grid Galerkin method. J. Sci. Comput. **47**(3), 328–346 (2010)
33. Zenger, C.: Sparse grids. In: Hackbusch, W. (ed.) Parallel Algorithms for Partial Differential Equations, Proceedings of the 6th GAMM-Seminar, Kiel (1990), vol. 31, pp. 241–251. Vieweg-Verlag (1991)

Part VIII
Software

Chapter 28
Stochastic Filtering Methods in Electronic Trading

Paul Bilokon, James Gwinnutt, and Daniel Jones

Abstract Stochastic filtering methods have found many applications, from Space Shuttles to self-driving cars. In this chapter we shall review some classical and modern filtering algorithms and show how they can be used in finance, especially electronic trading, to estimate and forecast econometric models, stochastic volatility and term structure of risky bonds. We shall discuss the practicalities, such as outlier filtering, parameter estimation, and diagnostics.

28.1 Introduction

In this chapter we shall present the foundations of stochastic filtering from the standpoint of an electronic trading practitioner. We shall not dwell here on the numerous mathematical technicalities and nuances of the subject; these the reader will find elsewhere in the literature, including the selection of texts mentioned in the final section. This selection is, of necessity, incomplete. We do not aim here to present novel research. Instead, we shall refer the reader to examples from a small selection of papers by academics and practitioners as well as tried and tested examples from our own experience as practitioners, omitting many technical details for ease of exposition. All the models discussed in this chapter are implemented in the open-source Python package BayesTSA, which is available at

`https://github.com/thalesians/bayestsa`

The totally ordered set $\mathbb{T}$ will represent the time. Let $(\Omega, \mathscr{F}, \mathbb{P}, (\mathscr{F}_t)_{t\in\mathbb{T}})$ be a filtered probability space satisfying the usual conditions. An $(\mathscr{F}_t)_{t\in\mathbb{T}}$-adapted

P. Bilokon (✉)
Thalesians Ltd, London, UK
e-mail: paul@thalesians.com

J. Gwinnutt
HSBC Bank plc, London, UK
e-mail: james.gwinnutt@hsbc.com

D. Jones
University of Oxford, Oxford, UK
e-mail: Daniel.Jones@maths.ox.ac.uk

M. Ehrhardt et al. (eds.), *Novel Methods in Computational Finance*,
Mathematics in Industry 25, DOI 10.1007/978-3-319-61282-9_28

stochastic process $X = (X_t)_{t\in\mathbb{T}}$, taking values on a complete separable metric space $\mathbb{S}$, will represent the (hidden, latent, unobserved) state of the system at time t. We shall refer to X as the *state process* and to $\mathbb{S}$ as the *state space*. While we cannot observe X directly, we have access to another $(\mathcal{F}_t)_{t\in\mathbb{T}}$-adapted stochastic process $Y = (Y_t)_{t\in\mathbb{T}}$, which is a function of X and a Wiener process $V = (V_t)_{t\in\mathbb{T}}$: $Y_t := h_t(X_t, V_t)$, $t \in \mathbb{T}$. The function h_t is sometimes referred to as the *observation model.*[1] The specification of the dynamics of X, often in the form of an SDE, is called the *process model.*

Let $(\mathcal{Y}_t)_{t\in\mathbb{T}}$ be the σ-algebra generated by the observation process Y. The *filtering problem* consists of computing the *filtering distribution*, $\pi = (\pi_t)_{t\in\mathbb{T}}$, the conditional distribution of X_t given $(\mathcal{Y}_t)_{t\in\mathbb{T}}$. It is defined as a random probability measure, which is—crucially—measurable w.r.t. $(\mathcal{Y}_t)_{t\in\mathbb{T}}$, such that

$$\mathbb{E}\left[\psi(X_t) \mid \mathcal{Y}_t\right] = \int_{\mathbb{S}} \psi(x)\pi_t(dx)$$

for all statistics ψ for which both sides of the above identity make sense. A formal definition of π relies on some technicalities and can be found in [5, Chap. 2].

Interest in the filtering problem dates back to the late 1930s–early 1940s. It was considered in Kolmogorov's (1903–1987) work on time series [75–77] and Wiener's (1894–1964) on improving radar communication during WWII [132],[2] which first appeared in 1942 as a classified memorandum nicknamed "The Yellow Peril" [131], so named after the colour of the paper on which it was printed [5]. In [131, 132], Wiener considered the special case of a stationary state process and additive noise and minimised the mean square error between the estimated and actual state process, working in continuous time. The result is nowadays known as the *Wiener filter.* A discrete-time equivalent of this filter was obtained independently by Kolmogorov and published in earlier in Russian [76]. Due to the contributions by these two scientists, the theory is referred to as the *Wiener–Kolmogorov theory of filtering and prediction* [70].

Kálmán (1930–2016) extended this work to non-stationary processes. This work had military applications, notably the prediction of ballistic missile trajectories. Non-stationary processes were required to realistically model their launch and re-entry phases. Of course, non-stationary processes abound in other fields—even the standard Brownian motion of the basic Bachelier model [4] in finance is non-stationary. Kálmán's fellow electrical engineers initially met his ideas with skepticism, so he ended up publishing in a mechanical engineering journal [67]. In 1960, Kálmán visited the NASA Ames Research Center, where Stanley F. Schmidt

[1]Elsewhere in the literature observations are sometimes referred to as *measurements*, e.g. in [119].

[2]The abstract to the MIT Press edition (ISBN 9780262257190) of this book starts with the following words: "It has been the opinion of many that Wiener will be remembered for his Extrapolation long after Cybernetics is forgotten. Indeed few computer-science students would know today what cybernetics is all about, while every communication student knows what Wiener's filter is."

took interest in this work. This led to its adoption by the Apollo programme and other projects in aerospace and defence.

The discrete-time version of the filter derived in [67] is now known as the *Kálmán filter*. The continuous-time version, known as the *Kálmán–Bucy filter*,[3] was published in a joint paper with Bucy [70]. Before meeting Kálmán, Bucy worked on continuous-time stochastic filtering independently, with Follin, Carlton, and Hanson at John Hopkins Applied Physics Lab in the late 1950s. The Kálmán and Kálmán–Bucy filters address the particularly important special case when the process model and the observation model are linear and both stochastic processes are Gaussian, the so-called *linear-Gaussian* case,[4] where there is an analytic solution for π. The general case of the filtering problem was addressed by Stratonovich (1930–1997) [123, 124], Kushner [78], and Zakai (1926–2015) [136].

The general solutions are, however, infinite-dimensional and not easily applicable. In practice, numerical approximations are employed. *Particle filters* constitute a particularly important class of such approximations. These methods are sometimes referred to as *sequential Monte Carlo (SMC)*, a term coined by Liu and Chen [85]. The Monte Carlo techniques requisite for particle filtering date back to the work of Hammersley and Morton [48]. *Sequential importance sampling (SIS)* dates back to the work of Mayne and Handschin [49, 95]. The important resampling step was added by Gordon, Salmond, and Smith [44], based on an idea by Rubin [112], to obtain the first *sequential importance resampling (SIR)* filter, which, in our experience, remains the most popular particle filtering algorithm used in practice. Other important early contributions to the development of particle filtering include [8, 17, 58, 59, 73, 74, 85].

Our overview of the development of stochastic filtering theory is necessarily brief. We focus on the development of the algorithms to motivate the discussion in the sequel, and merely scratch the surface of advances in stochastic analysis on which these algorithms rely. The reader interested in this important aspect of the theory can examine the work of Cameron (1908–1989), Fujisaki, Girsanov (1934–1967), Kallianpur (1925–2015), Kunita, Liptser, Martin (1911–2004), Shiryaev, Striebel (1929–2014), including the significance of the Cameron-Martin-Girsanov theorem [15, 43], the Fujisaki–Kallianpur–Kunita equation and orthogonal projection in Hilbert spaces [36], the Kallianpur–Striebel formula [64–66, 84], and non-linear filtering of Markov diffusion processes and jump processes [83, 117].

[3]The continuous-time Kálmán–Bucy filter, unlike the Kálmán filter, does not use a predictor-corrector method to update its state estimates. Instead, a differential Riccati equation is integrated over time.

[4]Feynman aptly summarised the reason why linear systems merit special treatment: "Finally, we make some remarks on why *linear* systems are so important. The answer is simple: because we can solve them!" [34, pp. 24–25].

28.2 Special Case: General State-Space Models

We have spoken thus far in considerable generality. Let us turn our attention to a particularly important special case of the filtering problem, *viz.* filtering for discrete-time, *general state-space models*, also known as *hidden Markov models (HMM)* [79]. Such models are prevalent in econometrics due to the relative ease of the estimation of their parameters and forecasting. We shall keep our notation and terminology roughly consistent with [21, Sect. 2.2].

These models are *discrete-time*, i.e. we regard $\mathbb{T}$ as a countable set. Without loss of generality, we shall identify $\mathbb{T}$ with $\mathbb{N}^0$. Since our models consider the evolution of a system over time, they are described as *dynamic*. We shall also assume that the state process $(\boldsymbol{X}_t)_{t\in\mathbb{T}}$ takes values in the Euclidean space $\mathbb{S} = \mathbb{R}^{d_X}$, $d_X \in \mathbb{N}^*$ (so it is now represented by a *state vector*), the observation process $(\boldsymbol{Y}_t)_{t\in\mathbb{T}\setminus 0}$ takes values in the Euclidean space $\mathbb{R}^{d_Y}$, $d_Y \in \mathbb{N}^*$ (so it is now represented by an *observation vector*).

We assume that the sequence of random variables $\boldsymbol{X}_0, \boldsymbol{X}_1, \boldsymbol{X}_2, \ldots$ is a *Markov chain*—i.e., the conditional distribution of $\boldsymbol{X}_t$, $t \in \mathbb{N}^*$, given $\boldsymbol{X}_0, \ldots, \boldsymbol{X}_{t-1}$, depends only upon $\boldsymbol{X}_{t-1}$. Mathematically this means that, for each $A \in \mathscr{B}(R^{d_X})$:

$$\mathbb{P}[\boldsymbol{X}_t \in A \mid \boldsymbol{X}_0 = \boldsymbol{x}_0, \ldots, \boldsymbol{X}_{t-1} = \boldsymbol{x}_{t-1}] = \mathbb{P}[\boldsymbol{X}_t \in A \mid \boldsymbol{X}_{t-1} = \boldsymbol{x}_{t-1}] := \tau_t(A \mid \boldsymbol{x}_{t-1}),$$

where $\tau_t : \mathscr{B}(\mathbb{R}^{d_X}) \times \mathbb{R}^{d_X} \to [0, 1]$ is referred to as the *Markov transition kernel*. The state process $(\boldsymbol{X}_t)_{t\in\mathbb{T}}$ is assumed to be not directly observable, hence the term '*latent* Markov model'.

We regard the sequence of random variables $\boldsymbol{Y}_1, \boldsymbol{Y}_2, \ldots$ as *observations* ordered in time—such a sequence is usually referred to as a *time series*. Each random variable $\boldsymbol{Y}_t$ will be assumed to be conditionally independent of other observations given the state $\boldsymbol{X}_t$, i.e., for $t \in \mathbb{N}^*$,

$$\mathbb{P}[\boldsymbol{Y}_t \in A \mid \boldsymbol{X}_0 = \boldsymbol{x}_0, \ldots, \boldsymbol{X}_t = \boldsymbol{x}_t, (\boldsymbol{Y}_s = \boldsymbol{y}_s)_{s\in\mathbb{N}^*, s\neq t}] = \mathbb{P}[\boldsymbol{Y}_t \in A \mid \boldsymbol{X}_t = \boldsymbol{x}_t]$$

for each $A \in \mathscr{B}(R^{d_Y})$.

For all $t \in \mathbb{N}^*$, $A \in \mathscr{B}(\mathbb{R}^{d_Y})$, the probability measure

$$\gamma_t(A \mid \boldsymbol{x}_t) := \mathbb{P}[\boldsymbol{Y}_t \in A \mid \boldsymbol{X}_t = \boldsymbol{x}_t]$$

is assumed to have a positive density, denoted $p_{\gamma_t}(\boldsymbol{y} \mid \boldsymbol{x})$ and called the *observation density*, w.r.t. the Lebesgue measure, hence

$$\gamma_t(A \mid \boldsymbol{x}_t) = \int_A p_{\gamma_t}(\boldsymbol{y} \mid \boldsymbol{x}_t)\, d\boldsymbol{y}.$$

While $\boldsymbol{X}_t$ is indeed latent, it is related to the observation $\boldsymbol{Y}_t$ via p_{γ_t}.

28.3 Particle Filtering Methods

Particle filtering methods rely on the numerical approximation of π_t with a set of 'particles'. To apply these methods we don't require the Markov transition kernel to have a probability density (the *Markov transition density*), i.e. for all $t \in \mathbb{N}^*$, $A \in \mathscr{B}(\mathbb{R}^{d_x})$, p_{τ_t} in

$$\tau_t(A \mid \boldsymbol{x}_{t-1}) = \int_A p_{\tau_t}(\boldsymbol{x} \mid \boldsymbol{x}_{t-1})\, d\boldsymbol{x}.$$

However, we need to be able to *sample* from the Markov transition kernel.

Here we have given the most common version of the particle filter known as the *sequential importance resampling (SIR)*[5] algorithm [44, 112]. There are many variations on Algorithm 9. For example, in some such variations the number of

---

Algorithm 9 Particle filter: sequential importance resampling (SIR)

1. *Initialisation step*: At time $t = 0$, draw M i.i.d. samples (called *particles*) from the initial distribution $\tau_0 = \pi_0$. Also, initialise M normalised (to 1) weights to an identical value of $\frac{1}{M}$. For $i = 1, 2, \ldots, M$, the samples will be denoted $\hat{\boldsymbol{x}}^{(i)}_{0\mid 0}$ and the normalised weights $\lambda^{(i)}_0$.
2. *Recursive step*: At time $t \in \mathbb{N}^*$, let $(\hat{\boldsymbol{x}}^{(i)}_{t-1\mid t-1})_{i=1,\ldots,M}$ be the particles generated at time $t-1$.

   a. *Importance sampling*:

      i. For $i = 1, \ldots, M$, sample $\hat{\boldsymbol{x}}^{(i)}_{t\mid t-1}$ from the Markov transition kernel $\tau_t(\cdot \mid \hat{\boldsymbol{x}}^{(i)}_{t-1\mid t-1})$.
      ii. For $i = 1, \ldots, M$, use the observation density to compute the non-normalised weights

$$\omega^{(i)}_t := \lambda^{(i)}_{t-1} \cdot p_{\gamma_t}(\boldsymbol{y}_t \mid \hat{\boldsymbol{x}}^{(i)}_{t\mid t-1}) \tag{28.1}$$

   and the values of the normalised weights before resampling ('br')

$${}^{\mathrm{br}}\lambda^{(i)}_t := \frac{\omega^{(i)}_t}{\sum_{k=1}^{M} \omega^{(k)}_t}.$$

   b. *Resampling* (or *selection*): For $i = 1, \ldots, M$, use an appropriate resampling algorithm (such as Algorithm 10) to sample $\hat{\boldsymbol{x}}^{(i)}_{t\mid t}$ from the mixture

$$\sum_{k=1}^{M} {}^{\mathrm{br}}\lambda^{(k)}_t \delta(\boldsymbol{x}_t - \hat{\boldsymbol{x}}^{(k)}_{t\mid t-1}),$$

   where $\delta(\cdot)$ denotes the Dirac delta generalised function, and set the normalised weights after resampling, $\lambda^{(i)}_t$, appropriately (for most common resampling algorithms this means $\lambda^{(i)}_t := \frac{1}{M}$).

---

[5] Alternatively, sometimes this is referred to as *sequential importance sampling with resampling (SISR)* [26].

particles M may vary with time, so instead of M we would have an appropriately defined M_t [82]. In this particular formulation of the filter, all the weights after resampling are set to $\frac{1}{M}$, so we could write (28.1) as

$$\omega_t^{(i)} := p_{Y_t}(\mathbf{y}_t \mid \hat{\mathbf{x}}_{t\mid t-1}^{(i)}),$$

$i = 1, \dots, M$, and omit the initialisation of the normalised weights at time 0, $\lambda_0^{(i)}$, since they wouldn't be used. This isn't the case in all variants of the particle filter. A simpler version of the filter would omit the resampling step altogether; instead, for $i = 1, \dots, M$, setting $\lambda_t^{(i)} := {}^{\text{br}}\lambda_t^{(i)}$ and $\hat{\mathbf{x}}_{t\mid t}^{(i)} := \hat{\mathbf{x}}_{t\mid t-1}^{(i)}$. This version of the algorithm, known as *sequential importance sampling (SIS)* [49, 95], suffers from the *degeneracy problem* [85], [16, Chap. 7] when, in some situations, only a few particles end up with significant weight, and all the other particles have near-zero weights. The resampling step is designed to remedy this. The particular scheme that we give here as Algorithm 10 is known as *multinomial resampling* [31, 44, 112][6] and the filter that uses it is sometimes referred to as the *(weighted) bootstrap filter*. The naïve implementation of multinomial resampling step has the time complexity $\mathcal{O}(M \ln M)$. An implementation based on [17] has the time complexity $\mathcal{O}(M)$. One alternative is to use *stratified sampling* [17, 74, 85], which won't be discussed here. It introduces no computational overhead.

Algorithm 10 Multinomial resampling

Notice that we are working with the *normalised* weights computed before resampling, ${}^{\text{br}}\lambda_t^{(1)}, {}^{\text{br}}\lambda_t^{(2)}, \dots, {}^{\text{br}}\lambda_t^{(M)}$.

1. For $i = 1, 2, \dots, M$, compute the cumulative sums

$${}^{\text{br}}\Lambda_t^{(i)} = \sum_{k=1}^{i} {}^{\text{br}}\lambda_t^{(k)},$$

so that, by construction, $\Lambda_t^{(M)} = 1$.
2. Generate M random samples from $\mathscr{U}(0, 1)$, $u_1, u_2, \dots, u_M$.
3. For each $i = 1, \dots, M$, choose the particle $\hat{\mathbf{x}}_{t\mid t}^{(i)} = \hat{\mathbf{x}}_{t\mid t-1}^{(j)}$ with $j \in \{1, 2, \dots, M-1\}$ such that $u_i \in \left[{}^{\text{br}}\Lambda_t^{(j)}, {}^{\text{br}}\Lambda_t^{(j+1)}\right]$.

Thus we end up with M new particles (*children*), $\hat{\mathbf{x}}_{t\mid t}^{(1)}, \dots, \hat{\mathbf{x}}_{t\mid t}^{(M)}$ sampled from the existing set $\hat{\mathbf{x}}_{t\mid t-1}^{(1)}, \dots, \hat{\mathbf{x}}_{t\mid t-1}^{(M)}$, so that some of the existing particles may disappear, while others may appear multiple times. For each $i = 1, \dots, M$ the number of times $\hat{\mathbf{x}}_{t\mid t-1}^{(i)}$ appears in the resampled set of particles is known as that particle's *replication factor*, $N_t^{(i)}$.
Set the normalised weights after resampling: $\lambda_t^{(i)} := \frac{1}{M}$.
We could view this algorithm as the sampling of the replication factors $N_t^{(1)}, \dots, N_t^{(M)}$ from the multinomial distribution with probabilities ${}^{\text{br}}\lambda_t^{(1)}, \dots, {}^{\text{br}}\lambda_t^{(M)}$, respectively. Hence the name of the method.

[6] See [120] for proofs of convergence.

For each $t \in \mathbb{N}^*$, this algorithm provides the prior and posterior, respectively, approximations to π_t:

$$\pi^M_{t\mid t-1} := \sum_{i=1}^{M} \lambda^{(i)}_{t-1}\delta(\boldsymbol{x}_t - \hat{\boldsymbol{x}}^{(i)}_{t\mid t-1}), \quad \pi^M_{t\mid t} := \sum_{i=1}^{M} \lambda^{(i)}_{t}\delta(\boldsymbol{x}_t - \hat{\boldsymbol{x}}^{(i)}_{t\mid t})$$

(when using the resampling scheme given here the resampled weights are all set to $\frac{1}{M}$, so this constant can be factored out of the sum). The prior state mean and covariance estimates in the particle filter are given by, respectively,

$$\hat{\boldsymbol{x}}_{t\mid t-1} := \sum_{i=1}^{M} \lambda^{(i)}_{t-1}\hat{\boldsymbol{x}}^{(i)}_{t\mid t-1}, \quad \boldsymbol{P}_{t\mid t-1} := \sum_{i=1}^{M} \lambda^{(i)}_{t-1}(\hat{\boldsymbol{x}}^{(i)}_{t\mid t-1} - \hat{\boldsymbol{x}}_{t\mid t-1})^{\mathsf{T}}(\hat{\boldsymbol{x}}^{(i)}_{t\mid t-1} - \hat{\boldsymbol{x}}_{t\mid t-1}).$$

Similarly, one obtains the posterior state mean and covariance before resampling, ${}^{\mathrm{br}}\hat{\boldsymbol{x}}_{t\mid t}$ and ${}^{\mathrm{br}}\boldsymbol{P}_{t\mid t}$, (using the weights ${}^{\mathrm{br}}\lambda^{(i)}_t$ and particles $\hat{\boldsymbol{x}}^{(i)}_{t\mid t-1}$) and after resampling, $\hat{\boldsymbol{x}}_{t\mid t}$ and $\boldsymbol{P}_{t\mid t}$, (using the weights $\lambda^{(i)}_t$ and particles $\hat{\boldsymbol{x}}^{(i)}_{t\mid t}$).

28.4 Applying the Particle Filter to the Stochastic Volatility Model with Leverage and Jumps

Pitt, Malik, and Doucet apply the particle filter to the *stochastic volatility with leverage and jumps (SVLJ)* model [89–91, 109]. The model has the general form of Taylor's [126] with two modifications. For $t \in \mathbb{N}^0$, let y_t denote the log-return on an asset and x_t denote the log-variance of that return. Then

$$y_t = \epsilon_t e^{x_t/2} + J_t \varpi_t, \tag{28.2}$$

$$x_{t+1} = \mu(1-\phi) + \phi x_t + \sigma_v \eta_t, \tag{28.3}$$

where μ is the mean log-variance, ϕ is the persistence parameter, σ_v is the volatility of log-variance. The first change to Taylor's original model is the introduction of correlation between ϵ_t and η_t:

$$\begin{pmatrix}\epsilon_t\\ \eta_t\end{pmatrix} \sim \mathcal{N}(0, \Sigma), \quad \Sigma = \begin{pmatrix}1 & \rho\\ \rho & 1\end{pmatrix}.$$

The correlation ρ is the leverage parameter, and it is a well-known stylised fact that, in general, $\rho < 0$. The second change is the introduction of jumps in (28.2): $J_t \in \{0, 1\}$ is a Bernoulli counter with intensity p (thus p is the jump intensity parameter), $\varpi_t \sim \mathcal{N}(0, \sigma_J^2)$ determines the jump size (thus σ_J is the jump volatility parameter).

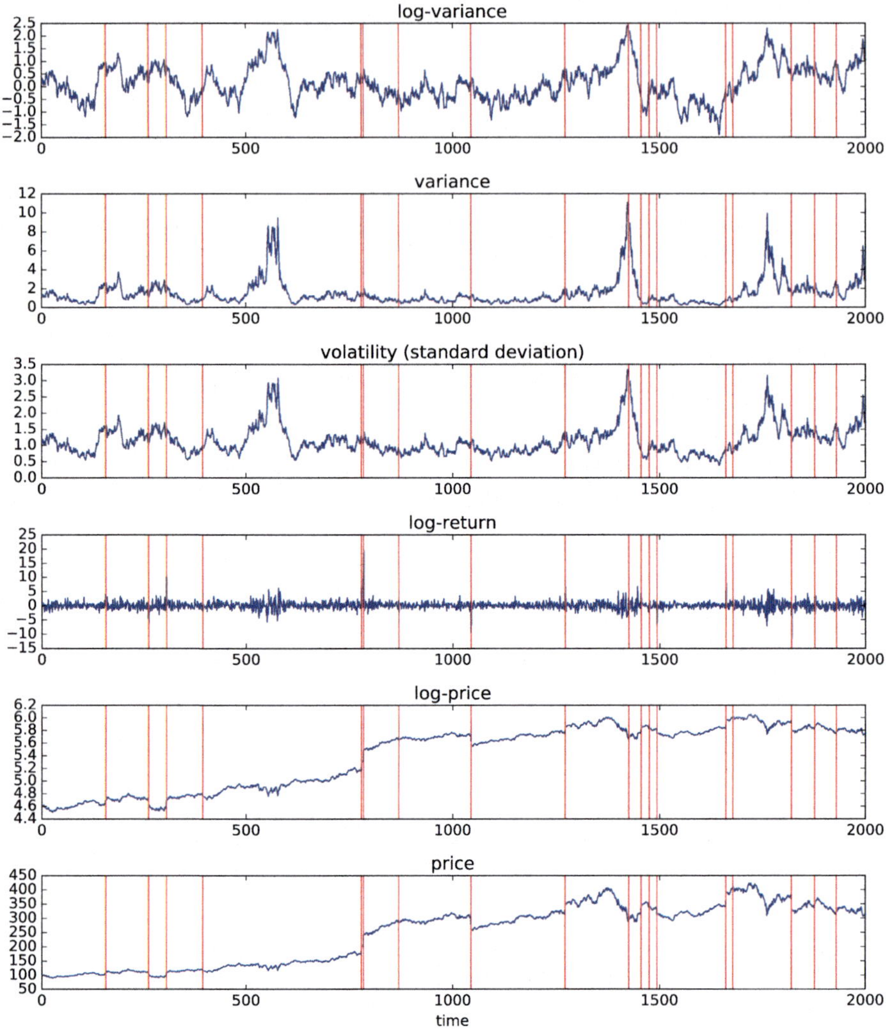

Fig. 28.1 An example of the SVLJ data generated using StateSpaceModelDataGenerator of the BayesTSA library. Here we have used the same parameters as those in Fig. 1 in [109], namely $\mu = 0.25$, $\phi = 0.975$, $\sigma_v^2 = 0.025$, $\rho = -0.8$, $p = 0.01$, $\sigma_J^2 = 10$. The *vertical red lines* indicate the times at which the simulated jumps occur. We assume that the initial price is 100

We obtain a *stochastic volatility with leverage (SVL)*, but no jumps, if we delete the $J_t \varpi_t$ term or, equivalently, set p to zero. Taylor's original model is a special case of SVLJ with $p = 0$, $\rho = 0$. In Fig. 28.1 we show an example of the SVLJ data generated using the StateSpaceModelDataGenerator of BayesTSA.

Here we use the convention adopted in [52, 72, 105, 109] and designate the disturbance that propagates the log-variance between the times t and $t+1$ by η_t. In much of the literature [53, 113, 135] this disturbance is designate η_{t+1}, so (28.3) becomes

$$x_{t+1} = \mu(1-\phi) + \phi x_t + \sigma_v \eta_{t+1}. \tag{28.4}$$

In discrete-time econometric models the choice between (28.3) and (28.4) is a matter of convention. The important point to realise is that, whichever convention that we use, in this model, it is the disturbance on the log-return y_t and the disturbance that propagates the log-variance from x_t to x_{t+1} that are correlated. If the disturbance on the log-return y_t were instead correlated with the disturbance that propagated the log-variance from x_{t-1} to x_t, we would have a different correlation structure and therefore a different model. Throughout the literature, y_t is the log-return that is related, through (28.2), to x_t. In [135], although the convention (28.4) is used, (28.3) is described as "clearer and more consistent". We shall stick with (28.3), so in our case $\mathsf{Cor}\left[\epsilon_t, \eta_t\right] = \rho$, the rest of the disturbances being independent. The differences between the two conventions used in the literature to designate the noises and the two different correlation structures are summarised in Fig. 28.2.

From the filtering perspective, the latent, unobserved log-variance is the state and the log-return, which is observable on the markets, the observation. Using the fact that the disturbances in (28.2) and (28.3) are conditionally Gaussian, they write

$$\eta_t = \rho\epsilon_t + \sqrt{1-\rho^2}\xi_t \tag{28.5}$$

where $\xi_t \overset{i.i.d.}{\sim} \mathcal{N}(0,1)$, turning (28.3) into

$$x_t = \mu(1-\phi) + \phi x_{t-1} + \sigma_v \rho \epsilon_{t-1} + \sigma_v \sqrt{1-\rho^2}\xi_{t-1}. \tag{28.6}$$

In the absence of jumps, $\epsilon_t = y_t e^{-x_t/2}$, and (28.6) can be further rewritten as

$$x_t = \mu(1-\phi) + \phi x_{t-1} + \sigma_v \rho y_{t-1} e^{-x_{t-1}/2} + \sigma_v \sqrt{1-\rho^2}\xi_{t-1}.$$

From this we see that, even in the absence of jumps, $p(x_t \mid x_{t-1}, y_{t-1})$ is highly nonlinear, so particle methods are prime candidates for devising a filtering scheme.

Notice that

$$\begin{aligned} &p(\epsilon_t \mid x_t, y_t) \\ &= \delta(y_t e^{-x_t/2})\mathbb{P}\left[J_t = 0 \mid x_t, y_t\right] + \varphi\left(\epsilon_t; \mu_{\epsilon_t \mid J_t=1}, \sigma^2_{\epsilon \mid J_t=1}\right)\mathbb{P}\left[J_t = 1 \mid x_t, y_t\right]. \end{aligned}$$

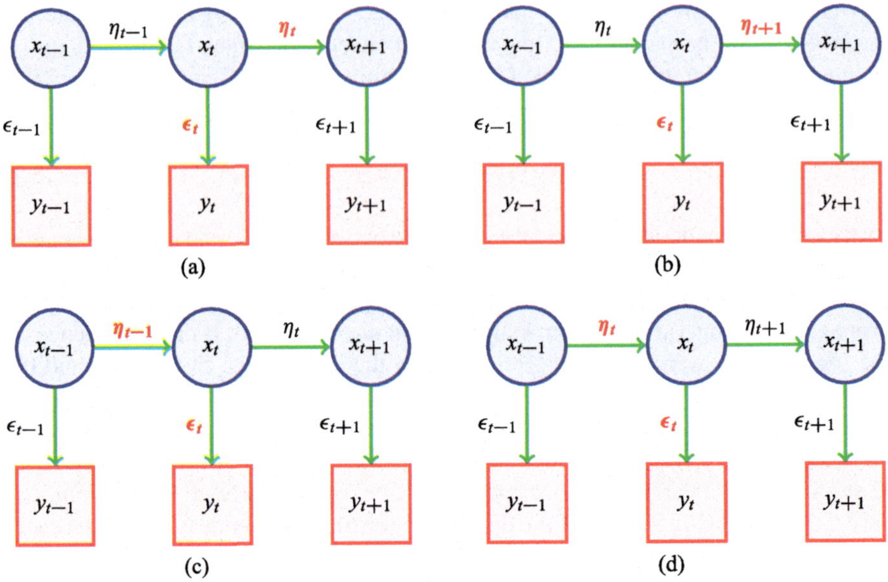

Fig. 28.2 The two conventions used in designating disturbances (noises) in econometric models and the two different correlation structures between the disturbances. Within each subfigure we show the two disturbances that are assumed to be correlated in *bold red*; all the other disturbances are assumed to be uncorrelated. Subfigures (**a**) and (**c**) show the first convention, where the disturbance that propagates x_t to x_{t+1} is ascribed to time t and so referred to as η_t. Subfigures (**b**) and (**d**) show the second convention, where the disturbance that propagates x_t to x_{t+1} is ascribed to time $t+1$ and so referred to as η_{t+1} Subfigures (**a**) and (**b**) show the first correlation structure (CS1), in which the disturbance ϵ_t, involved in y_t (y_t being a function of x_t and ϵ_t), is correlated with the disturbance that propagates x_t to x_{t+1}. Subfigures (**c**) and (**d**) show the second correlation structure (CS2), in which the disturbance ϵ_t, involved in y_t (y_t being a function of x_t and ϵ_t), is correlated with the disturbance that propagates x_{t-1} to x_t. SVL(J) has the first correlation structure (CS1), the contemporaneous dependence in the language of [135], sometimes referred to as *correlation at the same time* in stochastic filtering literature [88]. CS2 is the inter-temporal dependence in the language of [135]. In this stochastic filtering literature this is referred to as *correlation one time step apart* [88]

Pitt et al. design a procedure for simulating ϵ_t from the mixture density $p(\epsilon_t \mid x_t, y_t)$, which is then used in (28.6) to propagate the state. It is described in detail in Appendix B of [89]; we have implemented it in BayesTSA. In the same Appendix they derive the values of the conditional moments $\mu_{\epsilon_t \mid J_t=1}$ and $\sigma^2_{\epsilon \mid J_t=1}$.

The initial distribution of x_0 is taken to be $\mathcal{N}\left(0, \sigma_v^2/(1-\phi^2)\right)$. This, then, leads to the following modification of Algorithm 9 for this special case with nonadditive, correlated noises:

Algorithm 11 An adaptation of the particle filter (Algorithm 9) developed by Pitt et al. for the SVLJ model

1. *Initialisation step*: At time $t = 0$, draw M i.i.d. particles from the initial distribution $\mathcal{N}\left(0, \sigma_v^2/(1-\phi^2)\right)$. Also, initialise M normalised (to 1) weights to an identical value of $\frac{1}{M}$. For $i = 1, 2, \ldots, M$, the samples will be denoted $\hat{x}^{(i)}_{0\,|\,0}$ and the normalised weights $\lambda^{(i)}_0$.
2. *Recursive step*: At time $t \in \mathbb{N}^*$, let $(\hat{x}^{(i)}_{t-1\,|\,t-1})_{i=1,\ldots,M}$ be the particles generated at time $t-1$.

   a. *Importance sampling*:

      i. A. For $i = 1, \ldots, M$, sample $\hat{\epsilon}^{(i)}_{t-1}$ from $p(\epsilon_{t-1} \mid x_{t-1} = \hat{x}^{(i)}_{t-1\,|\,t-1}, y_{t-1})$. (If no y_{t-1} is available, as at $t = 1$, sample from $p(\epsilon_{t-1} \mid x_{t-1} = \hat{x}^{(i)}_{t-1\,|\,t-1})$).

         B. For $i = 1, \ldots, M$, sample $\hat{x}^{(i)}_{t\,|\,t-1}$ from $p(x_t \mid x_{t-1} = \hat{x}^{(i)}_{t-1\,|\,t-1}, y_{t-1}, \hat{\epsilon}^{(i)}_{t-1})$.

      ii. For $i = 1, \ldots, M$, compute the non-normalised weights

$$\omega^{(i)}_t := \lambda^{(i)}_{t-1} \cdot p_{y_t}(y_t \mid \hat{x}^{(i)}_{t\,|\,t-1}), \tag{28.7}$$

   using the observation density

$$p(y_t \mid \hat{x}^{(i)}_{t\,|\,t-1}, p, \sigma_J^2) = (1-p)\left[\left(2\pi e^{\hat{x}^{(i)}_{t\,|\,t-1}}\right)^{-1/2} \exp\left(-y_t^2/(2e^{\hat{x}^{(i)}_{t\,|\,t-1}})\right)\right] +$$
$$p\left[\left(2\pi e^{\hat{x}^{(i)}_{t\,|\,t-1}} + \sigma_J^2\right)^{-1/2} \exp\left(-y_t^2/(2e^{\hat{x}^{(i)}_{t\,|\,t-1}} + \sigma_J^2)\right)\right],$$

   and the values of the normalised weights before resampling ('br')

$${}^{\text{br}}\lambda^{(i)}_t := \frac{\omega^{(i)}_t}{\sum_{k=1}^{M} \omega^{(k)}_t}.$$

   b. *Resampling* (or *selection*): For $i = 1, \ldots, M$, use an appropriate resampling algorithm (such as Algorithm 10) sample $\hat{x}^{(i)}_{t\,|\,t}$ from the mixture

$$\sum_{k=1}^{M} {}^{\text{br}}\lambda^{(k)}_t \delta(x_t - \hat{x}^{(k)}_{t\,|\,t-1}),$$

   where $\delta(\cdot)$ denotes the Dirac delta generalised function, and set the normalised weights after resampling, $\lambda^{(i)}_t$, according to the resampling algorithm.

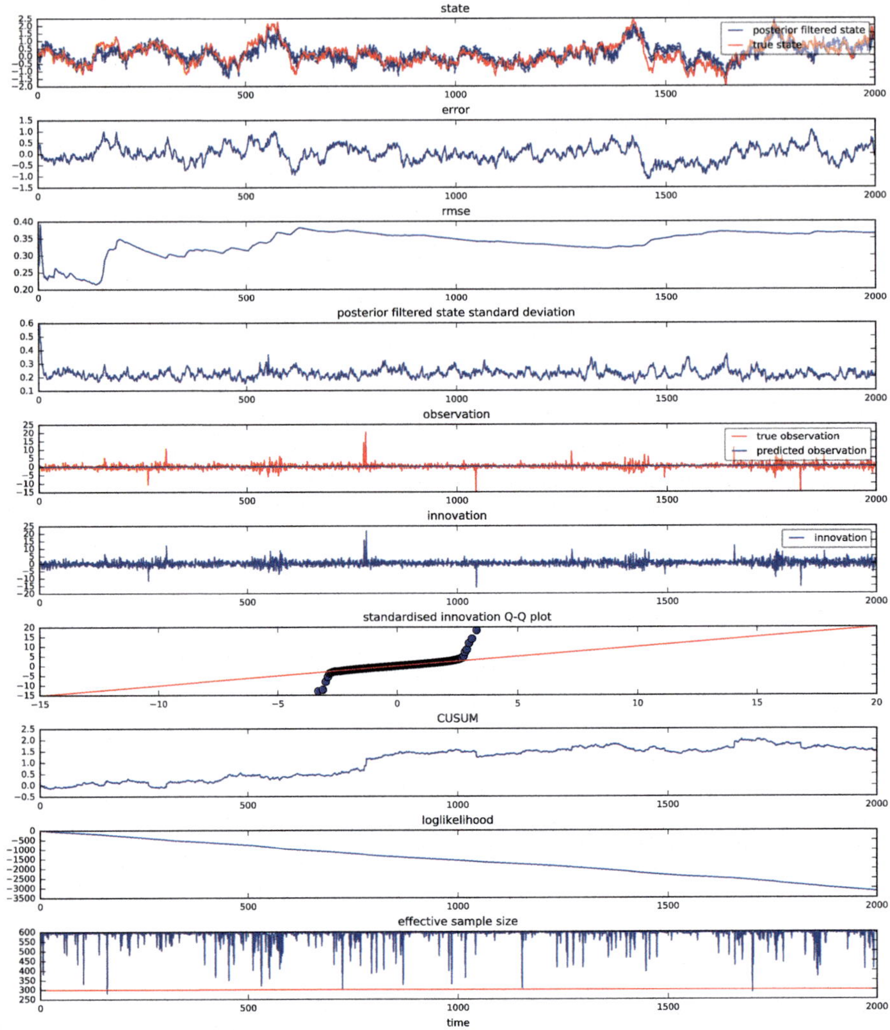

Fig. 28.3 The result of applying Algorithm 11 to the SVLJ data generated using StateSpaceModelDataGenerator of the BayesTSA library. Here we have used the same parameters as those in Fig. 1 in [109], namely $\mu = 0.25$, $\phi = 0.975$, $\sigma_v^2 = 0.025$, $\rho = -0.8$, $p = 0.01$, $\sigma_J^2 = 10$. We assume that the parameters are known exactly, i.e. we use the same parameters in the particle filter as the ones that we used to generate the data. The figure includes the diagnostic plots produced by BayesTSA. The tails on the Q-Q plot deviate from normal significantly due to the substantial jumps ($\sigma_J^2 = 10$)

The result of applying this algorithm to simulated data is shown in Fig. 28.3.

28.5 The Kálmán Filter

The model considered in Sect. 28.4 is nonlinear with correlated Gaussian state process noise and observation noise. Let us now consider a simpler and particularly important specialisation of the general state-space model of Sect. 28.2—the *linear-Gaussian state-space model*. For $t \in \mathbb{N}^*$, suppose that our d_X-dimensional state vector is driven by d_W-dimensional noise and evolves according to the process model that is specified by the single equation—the *process (evolution, transition) equation*

$$\boldsymbol{X}_t = \boldsymbol{F}_t \boldsymbol{X}_{t-1} + \boldsymbol{a}_t + \boldsymbol{W}_t \boldsymbol{w}_t, \tag{28.8}$$

and our d_Y-dimensional observation vector, incorporating d_V-dimensional observation noise, is related to the state by the observation model (or, in this case, the *observation equation*)

$$\boldsymbol{Y}_t = \boldsymbol{H}_t \boldsymbol{X}_t + \boldsymbol{b}_t + \boldsymbol{V}_t \boldsymbol{v}_t, \tag{28.9}$$

where $\boldsymbol{F}_t \in \mathbb{R}^{d_X \times d_X}$, $\boldsymbol{W}_t \in \mathbb{R}^{d_X \times d_W}$, $\boldsymbol{a}_t \in \mathbb{R}^{d_X}$; whereas $\boldsymbol{w}_t \sim \mathcal{N}(\boldsymbol{0}, \boldsymbol{Q}_t)$ are $\mathbb{R}^{d_W}$-valued random variables; $\boldsymbol{H}_t \in \mathbb{R}^{d_Y \times d_X}$, $\boldsymbol{V}_t \in \mathbb{R}^{d_Y \times d_V}$, $\boldsymbol{b}_t \in \mathbb{R}^{d_Y}$; whereas $\boldsymbol{v}_t \sim \mathcal{N}(\boldsymbol{0}, \boldsymbol{R}_t)$ are $\mathbb{R}^{d_V}$-valued random variables. We assume that $\boldsymbol{X}_0 \sim \mathcal{N}\left(\hat{\boldsymbol{x}}_{0\,|\,0}, \boldsymbol{P}_{0\,|\,0}\right)$ and the random variables $\{\boldsymbol{X}_0, \boldsymbol{w}_1, \ldots, \boldsymbol{w}_t, \boldsymbol{v}_1, \ldots, \boldsymbol{v}_t\}$ are mutually independent.

The matrices $\boldsymbol{F}_t$, $\boldsymbol{a}_t$, $\boldsymbol{W}_t$, $\boldsymbol{Q}_t$, $\boldsymbol{H}_t$, $\boldsymbol{b}_t$, $\boldsymbol{V}_t$, $\boldsymbol{R}_t$ are sometimes referred to as *system matrices*. When they don't depend on t, i.e., for all t, $\boldsymbol{F}_t = \boldsymbol{F}$ fixed, etc., the system is called *time invariant*.

We could, equivalently, write (28.8) and (28.9), respectively, as

$$\boldsymbol{X}_t \mid \boldsymbol{X}_{t-1} \sim \mathcal{N}\left(\boldsymbol{F}_t \boldsymbol{X}_{t-1} + \boldsymbol{a}_t, \boldsymbol{W}_t \boldsymbol{Q}_t \boldsymbol{W}_t^{\top}\right),$$
$$\boldsymbol{Y}_t \mid \boldsymbol{X}_t \sim \mathcal{N}\left(\boldsymbol{H}_t \boldsymbol{X}_t + \boldsymbol{b}_t, \boldsymbol{V}_t \boldsymbol{R}_t \boldsymbol{V}_t^{\top}\right),$$

or say that

$$p_{\chi_t} = \varphi\left(\boldsymbol{F}_t \boldsymbol{X}_{t-1} + \boldsymbol{a}_t, \boldsymbol{W}_t \boldsymbol{Q}_t \boldsymbol{W}_t^{\top}\right), \quad p_{\gamma_t} = \varphi\left(\boldsymbol{H}_t \boldsymbol{X}_t + \boldsymbol{b}_t, \boldsymbol{V}_t \boldsymbol{R}_t \boldsymbol{V}_t^{\top}\right).$$

The *Kálmán filter* (Algorithm 12) [67, 70] is an analytic solution to the filtering problem in this setting.

Algorithm 12 Kálmán filter

For each $t \in \mathbb{N}^*$, repeat the following two steps:

1. *Prediction step*:
   a. Predicted (prior) state estimate: $\hat{\boldsymbol{x}}_{t \mid t-1} = \boldsymbol{F}_t \hat{\boldsymbol{x}}_{t-1 \mid t-1} + \boldsymbol{a}_t$.
   b. Predicted (prior) error covariance: $\boldsymbol{P}_{t \mid t-1} = \boldsymbol{F}_t \boldsymbol{P}_{t-1 \mid t-1} \boldsymbol{F}_t^{\mathsf{T}} + \boldsymbol{W}_t \boldsymbol{Q}_t \boldsymbol{W}_t^{\mathsf{T}}$.
2. *Update* (or *correction*) *step*:
   a. Predicted observation: $\hat{\boldsymbol{y}}_{t \mid t-1} = \boldsymbol{H}_t \hat{\boldsymbol{x}}_{t \mid t-1} + \boldsymbol{b}_t$.
   b. Innovation (or observation residual): $\tilde{\boldsymbol{y}}_t = \boldsymbol{y}_t - \hat{\boldsymbol{y}}_{t \mid t-1}$.
   c. Innovation covariance: $\boldsymbol{S}_t = \boldsymbol{H}_t \boldsymbol{P}_{t \mid t-1} \boldsymbol{H}_t^{\mathsf{T}} + \boldsymbol{V}_t \boldsymbol{R}_t \boldsymbol{V}_t^{\mathsf{T}}$.
   d. (Optimal) Kálmán gain: $\boldsymbol{K}_t = \boldsymbol{P}_{t \mid t-1} \boldsymbol{H}_t^{\mathsf{T}} \boldsymbol{S}_t^{-1}$.
   e. Updated (posterior) state estimate: $\hat{\boldsymbol{x}}_{t \mid t} = \hat{\boldsymbol{x}}_{t \mid t-1} + \boldsymbol{K}_t \tilde{\boldsymbol{y}}_t$.
   f. Updated (posterior) error covariance: $\boldsymbol{P}_{t \mid t} = (\boldsymbol{I} - \boldsymbol{K}_t \boldsymbol{H}_t) \boldsymbol{P}_{t \mid t-1}$.

Proposition 28.1 (Some Properties of the Kálmán Filter)

1. $\boldsymbol{P}_{t \mid t-1}$ *is a faithful representation of the prior error covariance:* $\boldsymbol{P}_{t \mid t-1} = \mathsf{Cov}\left[\hat{\boldsymbol{x}}_{t \mid t-1} - \boldsymbol{X}_t\right]$;
2. $\boldsymbol{P}_{t \mid t}$ *is a faithful representation of the posterior error covariance:* $\boldsymbol{P}_{t \mid t} = \mathsf{Cov}\left[\hat{\boldsymbol{x}}_{t \mid t} - \boldsymbol{X}_t\right]$;
3. $\boldsymbol{S}_t$ *is a faithful representation of the innovation covariance:* $\boldsymbol{S}_t = \mathsf{Cov}\left[\tilde{\boldsymbol{y}}_t\right]$;
4. *the prior state estimate is unbiased:* $\mathbb{E}\left[\hat{\boldsymbol{x}}_{t \mid t-1} - \boldsymbol{X}_t\right] = 0$;
5. *the posterior state estimate is unbiased:* $\mathbb{E}\left[\hat{\boldsymbol{x}}_{t \mid t} - \boldsymbol{X}_t\right] = 0$;
6. *The solution to the filtering problem at time t,* $\hat{\boldsymbol{x}}_{t \mid t}$, *is:*
   a. causal *(or* based on the observations*): the estimate is obtained using only the observations* $\boldsymbol{y}_s$ *for* $s \leq t$;
   b. optimal*:* $\hat{\boldsymbol{x}}_{t \mid t}$ *minimises the* mean square error (MSE), $\mathbb{E}\left[\|\boldsymbol{X}_t - \hat{\boldsymbol{x}}_{t \mid t}\|_2^2\right]$, *so it is a* minimum mean square error estimate (MMSE)*;*
   c. online*: for an arbitrary time* $t \in \mathbb{T}$, *the estimate is available (prior, if we haven't seen the observation yet, otherwise posterior).*

The iterative equation for the predicted (prior) error covariance,

$$\boldsymbol{P}_{t \mid t-1} = \boldsymbol{F}_t (\boldsymbol{P}_{t-1 \mid t-2} - \boldsymbol{K}_t \boldsymbol{H}_t \boldsymbol{P}_{t-1 \mid t-2}) \boldsymbol{F}_t^{\mathsf{T}} + \boldsymbol{W}_t \boldsymbol{Q}_t \boldsymbol{W}_t^{\mathsf{T}} \tag{28.10}$$

is a discrete-time, time-varying *matrix Riccati equation*. From this equation we notice[7] that the covariance calculations are independent of the state estimate calculations occurring elsewhere in the filter. Therefore it is possible to perform the covariance calculations separately, offline. There is a rich mathematical theory of matrix Riccati equations, which underly much of filtering and control [1]. In particular, it is known that (28.10) converges to a steady state covariance,

[7]This won't hold, e.g. for the extended Kálmán filter, which we consider next.

$\boldsymbol{P}_{\infty}$, provided that the system is *observable*, a term coined by Kálmán [68, 69], which means, informally, that the state of the system can be determined from the observations.

From (28.10) we notice that the covariance calculations are independent of the state calculations, so can be performed offline, i.e. before the filter is applied to observations arriving in real time.

It is worth noting here that Kálmán's original thinking [67] wasn't formally Bayesian. It was Ho and Lee [57] that would later reinterpret this filter from the Bayesian perspective. There are several Bayesian ways to derive the Kálmán filter [6]. For an overview of the filtering theory from the point of view of a Bayesian, see [18].

28.6 Some Examples of Linear-Gaussian State Space Models

We shall now examine some situations where we get Kálmán-style, i.e. linear-Gaussian, state-space models of the kind considered in Sect. 28.5. Before we turn to financial applications of Kálmán filtering, let us consider a more basic and paradigmatic example from mechanics—the classical Newtonian system [119, Example 5.1].

28.6.1 A Non-financial Example: The Newtonian System

For simplicity, we shall disregard the process noise. Let r be the position of a particle, v its velocity, and a its (constant) acceleration. Then

$$\begin{pmatrix} dr/dt \\ dv/dt \\ da/dt \end{pmatrix} = \begin{pmatrix} 0 & 1 & 0 \\ 0 & 0 & 1 \\ 0 & 0 & 0 \end{pmatrix} \begin{pmatrix} r \\ v \\ a \end{pmatrix}.$$

The state of our system is given by the vector $\boldsymbol{X}_t = (r, v, a)^{\mathsf{T}}$. Denote the matrix of ones and zeros above by $\boldsymbol{A}$. Then the evolution of the state is described by the matrix differential equation $\frac{d\boldsymbol{X}_t}{dt} = \boldsymbol{A}\boldsymbol{X}_t$. By analogy with the scalar ordinary differential equation, its solution is $\boldsymbol{X}_t = e^{\boldsymbol{A}t}\boldsymbol{x}_0$, where $\boldsymbol{X}_t = (r_t, v_t, a_t)^{\mathsf{T}}$ is the state of the system at time t and the *matrix exponential* is defined by the (matrix) Taylor series expansion

$$e^{\boldsymbol{A}t} := \sum_{j=0}^{\infty} \frac{(\boldsymbol{A}t)^j}{j!}.$$

Note that, for $s \leq t \in \mathbb{T}$, $X_t = e^{A(t-s)}X_s$. Let us now discretise this equation using $h_k := t - s$ as the time interval between the time ticks $k-1$ and k (for $k \in \mathbb{N}^*$). We get $\boldsymbol{X}_k = \boldsymbol{F}_k\boldsymbol{X}_{k-1}$, where

$$\boldsymbol{F}_k := e^{\boldsymbol{A}h_k} = \sum_{j=0}^{\infty} \frac{(\boldsymbol{A}h_k)^j}{j!} = \begin{pmatrix} 1 & 0 & 0 \\ 0 & 1 & 0 \\ 0 & 0 & 1 \end{pmatrix} + \begin{pmatrix} 0 & 1 & 0 \\ 0 & 0 & 1 \\ 0 & 0 & 0 \end{pmatrix} h_k + \frac{1}{2} \begin{pmatrix} 0 & 0 & 1 \\ 0 & 0 & 0 \\ 0 & 0 & 0 \end{pmatrix} h_k^2$$

$$= \begin{pmatrix} 1 & h_k & h_k^2/2 \\ 0 & 1 & h_k \\ 0 & 0 & 1 \end{pmatrix},$$

since all the powers of $\boldsymbol{A}$ greater than 2 are zero matrices. We observe that this process model fits the Kálmán filter paradigm, a hint as to why this paradigm is so widely applicable in practice.

28.6.2 Autoregressive Moving Average Models

Consider the ubiquitous *autoregressive moving average* ARMA(p, q) model, which is often fitted to financial time series:

$$y_t = \phi_1 y_{t-1} + \ldots + \phi_p y_{t-p} + \eta_t + \theta_1 \eta_{t-1} + \ldots + \theta_q \eta_{t-q},$$

where $\eta_t \sim \mathcal{N}(0, \sigma^2)$. This model includes as special cases all AR(p) and MA(q) models. We account for the possibility that the model is nonstationary (an autoregressive *integrated* moving average, ARIMA). There are several ways of representing this model as a linear-Gaussian state-space model with a view to applying the Kálmán filter. The approach that we give here is by far the most common [11, 12, 47, 50]. The earlier approach by Pearlman [107], while more efficient, would require one to deal with correlated process and observation noises. Set $m := \max(p, q + 1)$, $\phi_i := 0$ for $i > p$, $\theta_i := 0$ for $i > q$. Then we get (28.8) and (28.9) with $d_X = m$, $d_Y = 1$, $d_W = 1$, and no observation noise, and

$$\boldsymbol{X}_t = \begin{pmatrix} y_t \\ \phi_2 y_{t-1} + \ldots + \phi_p y_{t-m+1} + \theta_1 \eta_t + \ldots + \theta_{m-1} \eta_{t-m+2} \\ \phi_3 y_{t-1} + \ldots + \phi_p y_{t-m+2} + \theta_2 \eta_t + \ldots + \theta_{m-1} \eta_{t-m+3} \\ \vdots \\ \phi_m y_{t-1} + \theta_{m-1} \eta_t \end{pmatrix} \in \mathbb{R}^{m \times 1},$$

$$F = \begin{pmatrix} \phi_1 & 1 & 0 & \cdots & 0 \\ \phi_2 & 0 & 1 & & 0 \\ \vdots & \vdots & \ddots & \ddots & \vdots \\ \phi_{m-1} & 0 & 0 & & 1 \\ \phi_m & 0 & 0 & \cdots & 0 \end{pmatrix} \in \mathbb{R}^{m \times m},$$

$$W = \begin{pmatrix} 1 & \theta_1 & \cdots & \theta_{m-1} \end{pmatrix}^{\mathsf{T}} \in \mathbb{R}^{m \times 1},$$

$$w_t = \eta_t, \quad Q_t = \sigma^2, \quad H = \begin{pmatrix} 1 & 0 & \ldots & 0 \end{pmatrix} \in 1 \times m, \quad b_t = 0, \quad V_t = 0.$$

If y_t is stationary, then $X_t \sim \mathcal{N}(\mathbf{0}, P)$ with P given by the equation $P = FPF^{\mathsf{T}} + \sigma^2 WW^{\mathsf{T}}$, so we can set the initial state and error covariance to $\mathbf{0}$ and P, respectively. For a detailed discussion of applying the Kálmán filter in this particular case, see [25, 138] (Fig. 28.4).

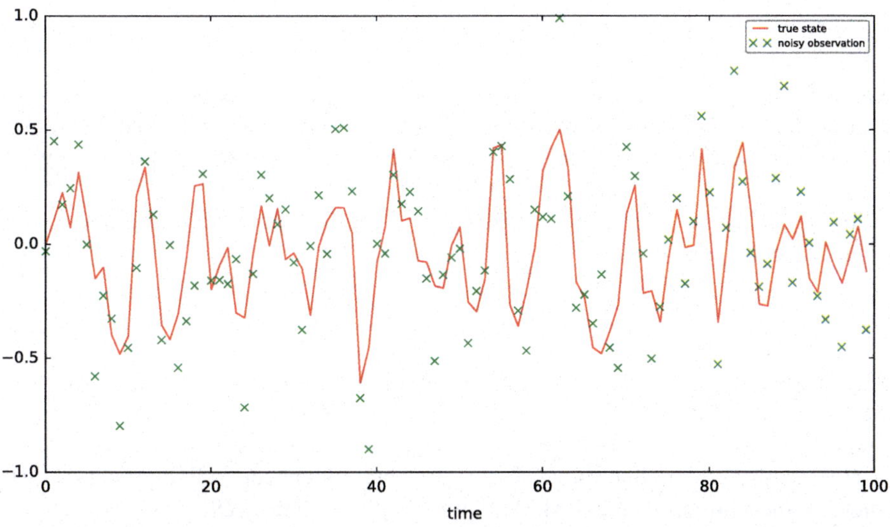

Fig. 28.4 An ARMA(2, 1) time series generated using StateSpaceModelDataGenerator of the BayesTSA library. Its parameters are $\phi_1 = 0.75$, $\phi_2 = -0.4$, $\theta_1 = 0.7$, and $\sigma^2 = 0.1$. We have also generated noisy observations of this series by adding to it white Gaussian noise with variance $\sigma^2/2$

28.6.3 Continuous-Time Stochastic Processes: The Wiener Process, Geometric Brownian Motion, and the Ornstein-Uhlenbeck Process

Let us now consider how we can construct state-space models by discretising some continuous-time stochastic processes that are of particular importance in finance [110, Chap. 4], [118, Chap. 3]. We begin with the one-dimensional variance-scaled Wiener process with drift μ and volatility σ, given by the SDE

$$dX_t = \mu\, dt + \sigma\, dW_t,$$

Let us now discretise the solution of this SDE, for $s \leq t \in \mathbb{T}$,

$$X_t = X_s + \mu(t - s) + \sigma(W_t - W_s).$$

by setting $h_k := t-s$ as the time interval between the time ticks $k-1$ and k ($k \in \mathbb{N}^*$) to obtain a special case of the linear-Gaussian state-space model of Sect. 28.5,

$$X_k = F_k X_{k-1} + a_k + w_k, \tag{28.11}$$

where $F_k = 1$, $a_k = \mu h_k$, $w_k \sim \mathcal{N}\left(0, \sigma^2 h_k\right)$. It is not necessary to make the continuous time intervals, corresponding to each discrete time tick, equal. We must, however, be careful, to scale the drift and process noise variance appropriately, as they are functions of the time step h_k (Fig. 28.5).

To handle the case of geometric Brownian motion (GBM) with percentage drift μ' and percentage volatility σ',

$$dS_t = \mu' S_t\, dt + \sigma' S_t\, dW_t,$$

we transform the process by taking the natural logarithm, $X_t := \ln S_t$, obtaining the Wiener process with the initial value $X_0 = \ln S_0$, drift $\mu = \mu' - (\sigma')^2/2$, and volatility σ'. We can then apply the Kálmán filter to the transformed process.

Now consider the one-dimensional Ornstein-Uhlenbeck (OU) process [110, Chap. 4], the stationary Gauss-Markov process given by the SDE

$$dX_t = \theta(\mu - X_t)\, dt + \sigma\, dW_t,$$

where $X_t \in \mathbb{R}$, $X_0 = x_0$, and $\theta > 0$, μ and $\sigma > 0$ are constants. The solution to this SDE is well-known:

$$X_t = x_0 e^{-\theta t} + \mu(1 - e^{-\theta t}) + \int_0^t \sigma e^{-\theta(t-u)}\, dW_u.$$

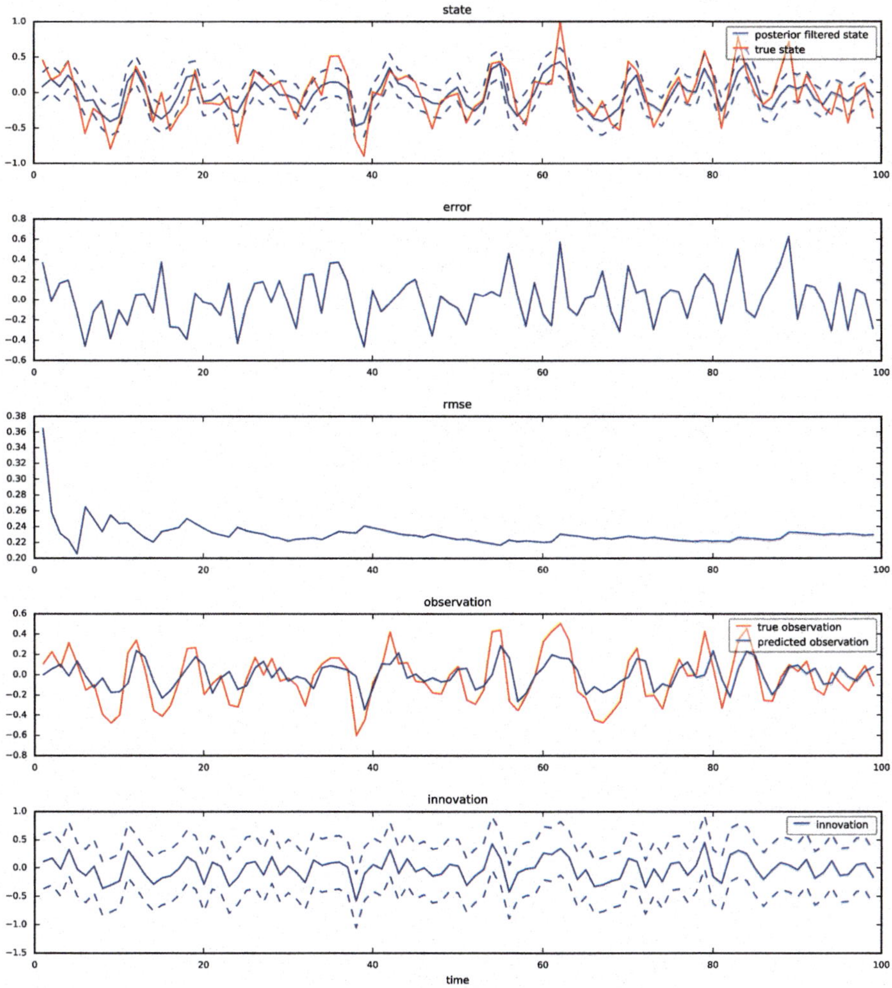

Fig. 28.5 The result of applying the Kálmán filter (Algorithm 12) to the noisy observation in Fig. 28.4 using the state-space model of Sect. 28.6.2 plus additive observation noise with true parameters taken as known

An Itō integral, $\int_s^t f(u)\,dW_u$, of a deterministic integrand, $f(u)$, is a Gaussian random variable with mean 0 and variance $\int_0^t f^2(u)\,du$. In our case, $f(u) = \sigma e^{-\theta(t-u)}$, and $\int_0^t f^2(u)\,du = \frac{\sigma^2}{2\theta}\left(1 - e^{-2\theta t}\right)$. Since this Markov process is homogeneous, its transition density depends only upon the time difference. Setting, for $s \leq t \in \mathbb{T}$, $h_k := t - s$ as the time interval between the time ticks $k-1$ and k (for $k \in \mathbb{N}^*$), we again get the discretised process model equation (28.11), this time with $F_k = e^{-\theta h_k}$, $a_k = \mu(1 - e^{-\theta h_k})$, $w_k \sim \mathcal{N}\left(0, \frac{\sigma^2}{2\theta}\left(1 - e^{-2\theta h_k}\right)\right)$. Again, we observe that this process model fits the Kálmán filter paradigm.

These results readily extend to higher dimensions. The d-dimensional correlated Wiener process with drift follows the SDE

$$d\boldsymbol{X}_t = \boldsymbol{\mu}\,dt + \boldsymbol{L}\,d\boldsymbol{W}_t,$$

where the drift (or infinitesimal mean) $\boldsymbol{\mu}$ is a d-dimensional real column vector, L is a $(d \times k)$-dimensional real matrix $\boldsymbol{L}\boldsymbol{L}^{\intercal} = \boldsymbol{\Sigma} \in \mathbb{R}^{d\times d}$ is positive-definite, and $\boldsymbol{W}$ is a k-dimensional standard Wiener process. The matrix $\boldsymbol{\Sigma}$ is referred to as the infinitesimal covariance matrix. Discretising with the timestep $h_k := t - s$, as above, we get the linear-Gaussian state-space model representation

$$\boldsymbol{X}_k = \boldsymbol{F}_k \boldsymbol{X}_{k-1} + \boldsymbol{a}_k + \boldsymbol{w}_k, \tag{28.12}$$

where $\boldsymbol{F}_k = \boldsymbol{I}_{d\times d}$, $\boldsymbol{a}_k = h_k\boldsymbol{\mu}$, $\boldsymbol{w}_k \sim \mathcal{N}(\boldsymbol{0}_{d\times d}, h_k\boldsymbol{\Sigma})$.

The SDE for the d-dimensional version of the OU process [98],

$$d\boldsymbol{X}_t = -\boldsymbol{\Theta}(\boldsymbol{X}_t - \boldsymbol{\mu})\,dt + \boldsymbol{L}\,d\boldsymbol{W}_t,$$

where $\boldsymbol{\Theta} \in \mathbb{R}^{d\times d}$, has the solution

$$\boldsymbol{X}_t = e^{-\boldsymbol{\Theta}t}\boldsymbol{x}_0 + (\boldsymbol{I} - e^{-\boldsymbol{\Theta}t})\boldsymbol{\mu} + \int_0^t e^{\boldsymbol{\Theta}(s-t)}\boldsymbol{L}\,d\boldsymbol{W}_s.$$

Hence, for $s \leq t \in \mathbb{T}$,

$$\mathbb{E}[\boldsymbol{X}_t] = \mathbb{E}\left[e^{-\boldsymbol{\Theta}t}\boldsymbol{x}_0 + (\boldsymbol{I} - e^{-\boldsymbol{\Theta}t})\boldsymbol{\mu} + \int_0^t e^{\boldsymbol{\Theta}(s-t)}\boldsymbol{L}\,d\boldsymbol{W}_s\right] = e^{-\boldsymbol{\Theta}t}\boldsymbol{x}_0 + (\boldsymbol{I} - e^{-\boldsymbol{\Theta}t})\boldsymbol{\mu},$$

$$\mathsf{Cov}(\boldsymbol{X}_s, \boldsymbol{X}_t) = \mathbb{E}\left[\left(\int_0^s e^{\boldsymbol{\Theta}(u-s)}\boldsymbol{L}\,d\boldsymbol{W}_u\right)\left(\int_0^t e^{\boldsymbol{\Theta}(v-t)}\boldsymbol{L}\,d\boldsymbol{W}_v\right)^T\right],$$

$$\begin{aligned}\mathsf{Var}(\mathbf{X}_t) &= \mathbb{E}\left[\left(\int_0^s e^{\boldsymbol{\Theta}(u-t)}\boldsymbol{L}\,d\boldsymbol{W}_u\right)\left(\int_0^s e^{\boldsymbol{\Theta}(u-t)}\boldsymbol{L}\,d\boldsymbol{W}_u\right)^T\right] \\ &\overset{\text{Itô isometry}}{=} \int_0^t e^{\boldsymbol{\Theta}(u-t)}\boldsymbol{\Sigma}\,e^{\boldsymbol{\Theta}^T(u-t)}\,du.\end{aligned}$$

Meucci [98], citing his correspondence with Van der Werf, shows how to compute this integral using tensor calculus, obtaining

$$\mathrm{vec}(\mathsf{Var}(\boldsymbol{X}_t)) = (\boldsymbol{\Theta} \oplus \boldsymbol{\Theta})^{-1}\left(\boldsymbol{I} - e^{-(\boldsymbol{\Theta}\oplus\boldsymbol{\Theta})t}\right)\mathrm{vec}(\boldsymbol{\Sigma}),$$

whence $\mathsf{Var}(\boldsymbol{X}_t)$ is obtained simply by 'unstacking' the columns of $\text{vec}(\mathsf{Var}(\boldsymbol{X}_t))$. Hence we again obtain the discretisation (28.12) using, as in the one-dimensional case, the fact that this Markov process is homogeneous.

The OU process is ubiquitous in finance and has applications in statistical arbitrage and cointegration [98]. In practice, care is needed when implementing numerics dealing with the matrix exponentials that result [100].

28.7 The Extended Kálmán Filter

We have now seen the two main algorithms of stochastic filtering: the particle filter of Sect. 28.3 and the Kálmán filter of Sect. 28.5. Since the invention of these two algorithms, researchers have come up with numerous new algorithms. Many of these algorithms fall into two categories: extensions of the particle filter and extensions of the Kálmán filter. The scope of this work does not permit us to consider many of these algorithms. In the sequel we shall consider only two important extensions of the Kálmán filter—the extended Kálmán filter and the Gaussian assumed density filter.

For $t \in \mathbb{N}^*$, suppose that, instead of being given by Eqs. (28.8) and (28.9), our process and observation models are, respectively,

$$\boldsymbol{X}_t = \boldsymbol{f}_t(\boldsymbol{X}_{t-1}, \boldsymbol{w}_t), \quad \boldsymbol{Y}_t = \boldsymbol{h}_t(\boldsymbol{X}_t, \boldsymbol{v}_t),$$

where

$$\boldsymbol{f}_t : \mathbb{R}^{dx} \times \mathbb{R}^{dw} \to \mathbb{R}^{dx}, \quad \boldsymbol{f}_t : \boldsymbol{X}, \boldsymbol{w} \mapsto \boldsymbol{f}_t(\boldsymbol{X}, \boldsymbol{w}),$$
$$\boldsymbol{h}_t : \mathbb{R}^{dx} \times \mathbb{R}^{dv} \to \mathbb{R}^{dY}, \quad \boldsymbol{h}_t : \boldsymbol{X}, \boldsymbol{v} \mapsto \boldsymbol{h}_t(\boldsymbol{X}, \boldsymbol{v})$$

are differentiable functions. As before, $\boldsymbol{w}_t \sim \mathcal{N}(\boldsymbol{0}, \boldsymbol{Q}_t)$ are $\mathbb{R}^{dw}$-valued random variables, $\boldsymbol{v}_t \sim \mathcal{N}(\boldsymbol{0}, \boldsymbol{R}_t)$ are $\mathbb{R}^{dv}$-valued random variables. Also as before, we assume that $\boldsymbol{X}_0 \sim \mathcal{N}(\hat{\boldsymbol{x}}_{0\,|\,0}, \boldsymbol{P}_{0\,|\,0})$ and the random variables $\{\boldsymbol{X}_0, \boldsymbol{w}_1, \ldots, \boldsymbol{w}_t, \boldsymbol{v}_1, \ldots, \boldsymbol{v}_t\}$ are mutually independent.

We can linearise these models by means of the truncated Taylor expansions: of $\boldsymbol{f}_t$ around $(\hat{\boldsymbol{x}}_{t-1\,|\,t-1}, \boldsymbol{0})$,

$$\boldsymbol{X}_t = \boldsymbol{f}_t(\hat{\boldsymbol{x}}_{t-1\,|\,t-1}, \boldsymbol{0}) + \left.\frac{\partial \boldsymbol{f}_t}{\partial \boldsymbol{x}}\right|_{(\hat{\boldsymbol{x}}_{t-1\,|\,t-1}, \boldsymbol{0})} (\hat{\boldsymbol{x}}_{t-1\,|\,t-1} - \boldsymbol{X}_t) + \left.\frac{\partial \boldsymbol{f}_t}{\partial \boldsymbol{w}}\right|_{(\hat{\boldsymbol{x}}_{t-1\,|\,t-1}, \boldsymbol{0})} \boldsymbol{w}_t + \ldots,$$

and of $\boldsymbol{h}_t$ around $(\hat{\boldsymbol{x}}_{t\,|\,t-1}, \boldsymbol{0})$,

$$\boldsymbol{Y}_t = \boldsymbol{h}_t(\hat{\boldsymbol{x}}_{t\,|\,t-1}, \boldsymbol{0}) + \left.\frac{\partial \boldsymbol{h}_t}{\partial \boldsymbol{x}}\right|_{(\hat{\boldsymbol{x}}_{t\,|\,t-1}, \boldsymbol{0})} (\hat{\boldsymbol{x}}_{t\,|\,t-1} - \boldsymbol{X}_t) + \left.\frac{\partial \boldsymbol{h}_t}{\partial \boldsymbol{v}}\right|_{(\hat{\boldsymbol{x}}_{t\,|\,t-1}, \boldsymbol{0})} \boldsymbol{v}_t + \ldots.$$

Algorithm 13 Extended Kálmán filter

For each $t \in \mathbb{N}^*$, define the following Jacobian matrices:

$$\boldsymbol{F}_t := \left.\frac{\partial \boldsymbol{f}_t}{\partial \boldsymbol{x}}\right|_{(\hat{\boldsymbol{x}}_{t-1\,|\,t-1},\boldsymbol{0})}, \quad \boldsymbol{W}_t := \left.\frac{\partial \boldsymbol{f}_t}{\partial \boldsymbol{w}}\right|_{(\hat{\boldsymbol{x}}_{t-1\,|\,t-1},\boldsymbol{0})},$$

$$\boldsymbol{H}_t := \left.\frac{\partial \boldsymbol{h}_t}{\partial \boldsymbol{x}}\right|_{(\hat{\boldsymbol{x}}_{t\,|\,t-1},\boldsymbol{0})}, \quad \boldsymbol{V}_t := \left.\frac{\partial \boldsymbol{h}_t}{\partial \boldsymbol{v}}\right|_{(\hat{\boldsymbol{x}}_{t\,|\,t-1},\boldsymbol{0})};$$

repeat the following two steps:

1. *Prediction step*:
   a. Predicted (prior) state estimate: $\hat{\boldsymbol{x}}_{t\,|\,t-1} = \boldsymbol{f}_t(\hat{\boldsymbol{x}}_{t-1\,|\,t-1}, \boldsymbol{0})$.
   b. Predicted (prior) error covariance: $\boldsymbol{P}_{t\,|\,t-1} = \boldsymbol{F}_t\boldsymbol{P}_{t-1\,|\,t-1}\boldsymbol{F}_t^{\mathsf{T}} + \boldsymbol{W}_t\boldsymbol{Q}_t\boldsymbol{W}_t^{\mathsf{T}}$.
2. *Update* (or *correction*) *step*:
   a. Predicted observation: $\hat{\boldsymbol{y}}_{t\,|\,t-1} = \boldsymbol{h}_t(\hat{\boldsymbol{x}}_{t\,|\,t-1}, \boldsymbol{0})$.
   b. Innovation (or observation residual): $\tilde{\boldsymbol{y}}_t = \boldsymbol{y}_t - \hat{\boldsymbol{y}}_{t\,|\,t-1}$.
   c. Innovation covariance: $\boldsymbol{S}_t = \boldsymbol{H}_t\boldsymbol{P}_{t\,|\,t-1}\boldsymbol{H}_t^{\mathsf{T}} + \boldsymbol{V}_t\boldsymbol{R}_t\boldsymbol{V}_t^{\mathsf{T}}$.
   d. Kálmán gain: $\boldsymbol{K}_t = \boldsymbol{P}_{t\,|\,t-1}\boldsymbol{H}_t^{\mathsf{T}}\boldsymbol{S}_t^{-1}$.
   e. Updated (posterior) state estimate: $\hat{\boldsymbol{x}}_{t\,|\,t} = \hat{\boldsymbol{x}}_{t\,|\,t-1} + \boldsymbol{K}_t\tilde{\boldsymbol{y}}_t$.
   f. Updated (posterior) error covariance: $\boldsymbol{P}_{t\,|\,t} = (\boldsymbol{I} - \boldsymbol{K}_t\boldsymbol{H}_t)\boldsymbol{P}_{t\,|\,t-1}$.

This approximation gives rise to the *extended Kálmán filter* (Algorithm 13) [55, 93, 96, 119, 121].

28.8 An Example Application of the Extended Kálmán Filter: Modelling Credit Spread

One way to quantify the credit risk of a risky bond is in terms of the *zero volatility spread*, or *Z-spread* for short. It is obtained by solving for z_{t_0} in

$$P_{t_0} = \sum_{i=1}^{n} \frac{\mathrm{cf}_{t_i}}{\left(1 + \left(s^i_{t_0} + z_{t_0}\right)\delta\right)^i},$$

where P_{t_0} is the dirty market price of the bond at time t_0, cf_{t_i} is the cash flow generated by the bond at time t_i, $s^i_{t_0}$ is the zero-coupon swap rate of appropriate maturity for this cashflow, δ is the frequency of the cashflows expressed as a fraction of the year. For a given issuer, we are interested in modelling the term structure of the Z-spreads across the universe of that issuer's bonds. The Z-spread is then viewed as a function of $\tau \geq t_0$: $z_{t_0}(\tau; \theta_{t_0})$, τ being the time of maturity or, for example, the modified duration, and a d-dimensional vector of parameters. Its dependence on t_0

indicates that the Z-spread curve (as a function of τ) evolves as time progresses. One of our tasks, then, is to keep estimating θ_{t_0} as t_0 moves on.

Suppose that, for a particular issuer, we have a universe of K bonds with Z-spreads $z_{t_0}^{(1)}, \dots, z_{t_0}^{(K)}$ and maturities (or modified durations, etc.) $\tau^{(1)}, \dots, \tau^{(K)}$. Each of these Z-spreads may not lie exactly on the Z-spread curve $z_{t_0}(\tau)$ due to the idiosyncracies of that particular bond, so we allow there to be an *idiosyncratic spread*, $\lambda_{t_0}^{(k)}$, $k \in \{1, \dots, K\}$:

$$z_{t_0}^{(k)} = z_{t_0}(\tau^{(k)}; \theta_{t_0}) + \lambda_{t_0}^{(k)}.$$

As indicated by their dependence on t_0, the idiosyncratic spreads also evolve over time. We will have to keep computing their updated estimates as well.

Thus our latent state at t_0 is the vector

$$x_{t_0} = \left((\theta_{t_0})_1 \cdots (\theta_{t_0})_d \; \lambda_{t_0}^{(1)} \cdots \lambda_{t_0}^{(K)} \right)^{\mathsf{T}} \in \mathbb{R}^{d_X},$$

$d_X = d + K$. We are observing individual bond prices, thus $d_Y = 1$ and our observation at time t_0 is $y_{t_0} = P_{t_0}^{(k)}$, the dirty[8] market price of bond k for some $k \in \{1, \dots, K\}$.

The function h that maps our state to the corresponding observation is given by

$$h^{(k)}(x_{t_0}) = P_{t_0}^{(k)}(z_{t_0}^{(k)}) = \sum_{i=1}^{n^{(k)}} \frac{\mathrm{cf}_{t_i}^{(k)}}{\left(1 + \left(s_{t_0}^i + z_{t_0}(\tau^{(k)}; \theta_{t_0}) + \lambda_{t_0}^{(k)}\right)\delta^{(k)}\right)^i}.$$

Thus, given our state vector, we can evaluate $h^{(k)}(x_{t_0})$. Note that the equation above also depends on the appropriate zero-coupon swap rates $s_{t_0}^i$. These are fast-moving and can be provided exogenously.

Let the scalar parameter α_{t_0} be a particular element of our parameter vector θ_{t_0}, so it is $(\theta_{t_0})^j$ for some $j \in \{1, \dots, d\}$. Then

$$\begin{aligned}
\left.\frac{\partial h^{(k)}}{\partial \alpha_{t_0}}\right|_{x_{t_0}} &= \left.\frac{\partial h^{(k)}}{\partial z_{t_0}}\right|_{x_{t_0}} \cdot \left.\frac{\partial z_{t_0}}{\partial \alpha_{t_0}}\right|_{x_{t_0}} \\
&= \left(-\delta^{(k)} \sum_{i=1}^{n^{(k)}} \frac{i \cdot \mathrm{cf}_{t_i}^{(k)}}{\left(1 + \left(s_{t_0}^i + z_{t_0}(\tau^{(k)}) + \lambda_{t_0}^{(k)}\right)\delta^{(k)}\right)^i} \right) \left.\frac{\partial z_{t_0}}{\partial \alpha_{t_0}}\right|_{\tau^{(k)}},
\end{aligned}$$

[8]Chances are that our market feeds are providing us with clean prices. We should therefore remember to convert these prices to dirty prices before we proceed.

where $\frac{\partial z_{t_0}}{\partial \alpha_{t_0}}$ can be computed analytically for many simple curve models, such as Nelson–Siegel [102]. Similarly, for $j \in \{1, \ldots, K\}$,

$$\frac{\partial h^{(k)}}{\partial \lambda_{t_0}^{(j)}}\bigg|_{x_{t_0}} = \begin{cases} \frac{\partial h^{(k)}}{\partial z_{t_0}}\Big|_{x_{t_0}} \cdot \frac{\partial z_{t_0}}{\partial \lambda_{t_0}^{(j)}}\Big|_{x_{t_0}} = -\delta^{(k)} \sum_{i=1}^{n^{(k)}} \frac{i \cdot \mathrm{cf}_{t_i}^{(k)}}{\left(1+\left(s_{t_0}^{i}+z_{t_0}(\tau^{(k)})+\lambda_{t_0}^{(k)}\right)\delta^{(k)}\right)^{i}}, & \mathrm{j} = \mathrm{k}; \\ 0, & \text{otherwise.} \end{cases}$$

We need to assume suitable dynamics for the process model. For example, one may start with diffusive curve parameters and mean-reverting idiosyncratics. In practice, prices observed in financial markets include costs, which may or may not be known. Observation noise can be used to model this uncertainty.

28.9 Outlier Detection in (Extended) Kálmán Filtering

Note that the predicted observation is distributed as

$$\mathcal{N}\left(\boldsymbol{H}_t \hat{\boldsymbol{x}}_{t \mid t-1} + \boldsymbol{b}_t, \boldsymbol{H}_t \boldsymbol{P}_{t \mid t-1} \boldsymbol{H}_t^{\intercal}\right).$$

Just as we can assign a z-score to $\boldsymbol{y}_t$ if $\boldsymbol{y}$ is one-dimensional, we can assign a Mahalanobis distance to it (which does correspond to the z-score in the one-dimensional case). In general, the *Mahalanobis norm* of a vector $\boldsymbol{y}$ with respect to $\mathcal{N}(\boldsymbol{\mu}, \boldsymbol{\Sigma})$ is given by

$$\|\boldsymbol{y}\|_{\mathcal{N}(\boldsymbol{\mu}, \boldsymbol{\Sigma})} = \sqrt{(\boldsymbol{y} - \boldsymbol{\mu})^{\intercal} \boldsymbol{\Sigma}^{-1} (\boldsymbol{y} - \boldsymbol{\mu})}.$$

It measures the distance of $\boldsymbol{y} \in \mathbb{R}^{d_Y}$ from the centroid (multidimensional mean) of the distribution. $\|\boldsymbol{y}\|^2_{\mathcal{N}(\boldsymbol{\mu}, \boldsymbol{\Sigma})}$ follows the χ^2-distribution with m degrees of freedom. Thus we can set a cut-off for $\boldsymbol{y}_t$, e.g. on the basis of the 0.975th quantile of the χ^2-distribution.

28.10 Gaussian Assumed Density Filtering

In the extended Kálmán filter algorithm, the possibly nonlinear and non-Gaussian process and observation models are approximated by means of linearisation. One could use moment matching as an alternative. By matching the moments of the normal distribution one obtains the *Gaussian assumed density filter*, also known as the *Gaussian filter* [60, 81, 94, 115, 133].

When the noises are additive, the Gaussian moment matching approximation of an additive transform (Algorithm 14) is used to derive the Gaussian filter with additive noise (Algorithm 16). Otherwise, for non-additive noises, the Gaussian moment matching approximation of a non-additive transform (Algorithm 15) is used to derive the Gaussian filter with non-additive noise (Algorithm 17).

Algorithm 16 applies when the process and observation models are, respectively,

$$\boldsymbol{X}_t = \boldsymbol{f}_t(\boldsymbol{X}_{t-1}) + \boldsymbol{w}_t, \quad \boldsymbol{Y}_t = \boldsymbol{h}_t(\boldsymbol{X}_t) + \boldsymbol{v}_t,$$

with $\boldsymbol{f}_t : \mathbb{R}^{d_X} \to \mathbb{R}^{d_X}$, $\boldsymbol{h}_t : \mathbb{R}^{d_X} \to \mathbb{R}^{d_Y}$. Algorithm 17 applies in the more general case when

$$\boldsymbol{X}_t = \boldsymbol{f}_t(\boldsymbol{X}_{t-1}, \boldsymbol{w}_t), \quad \boldsymbol{Y}_t = \boldsymbol{h}_t(\boldsymbol{X}_t, \boldsymbol{v}_t),$$

with $\boldsymbol{f}_t : \mathbb{R}^{d_X} \times \mathbb{R}^{d_X} \to \mathbb{R}^{d_X}$, $\boldsymbol{h}_t : \mathbb{R}^{d_X} \times \mathbb{R}^{d_Y} \to \mathbb{R}^{d_Y}$. As before, $\boldsymbol{w}_t \sim \mathcal{N}(\boldsymbol{0}, \boldsymbol{Q}_t)$ are $\mathbb{R}^{d_X}$-valued random variables, $\boldsymbol{v}_t \sim \mathcal{N}(\boldsymbol{0}, \boldsymbol{R}_t)$ are $\mathbb{R}^{d_Y}$-valued random variables. Also as before, we assume that $\boldsymbol{X}_0 \sim \mathcal{N}\left(\hat{\boldsymbol{x}}_{0\,|\,0}, \boldsymbol{P}_{0\,|\,0}\right)$ and the random variables $\{\boldsymbol{X}_0, \boldsymbol{w}_1, \ldots, \boldsymbol{w}_t, \boldsymbol{v}_1, \ldots, \boldsymbol{v}_t\}$ are mutually independent.

Algorithm 14 Gaussian moment matching approximation of an additive transform [115, Algorithm 6.1]

The moment matching-based Gaussian approximation to the joint distribution of the $\mathbb{R}^{d_X}$-valued random variable $\boldsymbol{X}$ and the transformed $\mathbb{R}^{d_Y}$-valued random variable $\boldsymbol{Y} = \boldsymbol{g}(\boldsymbol{X}) + \boldsymbol{Z}$, where $\boldsymbol{X} \sim \mathcal{N}(\boldsymbol{m}, \boldsymbol{P})$ and $\boldsymbol{Z} \sim \mathcal{N}(\boldsymbol{0}, \boldsymbol{\Sigma})$ is given by

$$\begin{pmatrix} \boldsymbol{X} \\ \boldsymbol{Y} \end{pmatrix} \overset{\text{approx.}}{\sim} \mathcal{N}\left(\begin{pmatrix} \boldsymbol{m} \\ \boldsymbol{\mu} \end{pmatrix}, \begin{pmatrix} \boldsymbol{P} & \boldsymbol{\Xi} \\ \boldsymbol{\Xi}^\top & \boldsymbol{S} \end{pmatrix}\right),$$

where

$$\boldsymbol{\mu} = \int \boldsymbol{g}(\boldsymbol{x})\varphi(\boldsymbol{x}; \boldsymbol{m}, \boldsymbol{P})\, d\boldsymbol{x},$$

$$\boldsymbol{S} = \int (\boldsymbol{g}(\boldsymbol{x}) - \boldsymbol{\mu})(\boldsymbol{g}(\boldsymbol{x}) - \boldsymbol{\mu})^\top \varphi(\boldsymbol{x}; \boldsymbol{m}, \boldsymbol{P})\, d\boldsymbol{x} + \boldsymbol{\Sigma},$$

$$\boldsymbol{\Xi} = \int (\boldsymbol{x} - \boldsymbol{m})(\boldsymbol{g}(\boldsymbol{x}) - \boldsymbol{\mu})^\top \varphi(\boldsymbol{x}; \boldsymbol{m}, \boldsymbol{P})\, d\boldsymbol{x}.$$

Algorithm 15 Gaussian moment matching approximation of a possibly non-additive transform [115, Algorithm 6.2]

The moment matching-based Gaussian approximation to the joint distribution of the $\mathbb{R}^{d_X}$-valued random variable $\boldsymbol{X}$ and the transformed $\mathbb{R}^{d_Y}$-valued random variable $\boldsymbol{Y} = \boldsymbol{g}(\boldsymbol{X}, \boldsymbol{Z})$, where $\boldsymbol{X} \sim \mathcal{N}(\boldsymbol{m}, \boldsymbol{P})$ and $\mathbf{Z} \sim \mathcal{N}(\mathbf{0}, \boldsymbol{\Sigma})$ is given by

$$\begin{pmatrix} \boldsymbol{X} \\ \boldsymbol{Y} \end{pmatrix} \overset{\text{approx.}}{\sim} \mathcal{N}\left(\begin{pmatrix} \boldsymbol{m} \\ \boldsymbol{\mu} \end{pmatrix}, \begin{pmatrix} \boldsymbol{P} & \boldsymbol{\Xi} \\ \boldsymbol{\Xi}^{\intercal} & \boldsymbol{S} \end{pmatrix} \right),$$

where

$$\begin{aligned}
\boldsymbol{\mu} &= \int \boldsymbol{g}(\boldsymbol{x}, z) \varphi(\boldsymbol{x}; \boldsymbol{m}, \boldsymbol{P})\, \varphi(z; \mathbf{0}, \boldsymbol{\Sigma})\; d\boldsymbol{x}\, dz, \\
\boldsymbol{S} &= \int (\boldsymbol{g}(\boldsymbol{x}, z) - \boldsymbol{\mu})(\boldsymbol{g}(\boldsymbol{x}, z) - \boldsymbol{\mu})^{\intercal} \varphi(\boldsymbol{x}; \boldsymbol{m}, \boldsymbol{P})\, \varphi(z; \mathbf{0}, \boldsymbol{\Sigma})\; d\boldsymbol{x}\, dz, \\
\boldsymbol{\Xi} &= \int (\boldsymbol{x} - \boldsymbol{m})(\boldsymbol{g}(\boldsymbol{x}, z) - \boldsymbol{\mu})^{\intercal} \varphi(\boldsymbol{x}; \boldsymbol{m}, \boldsymbol{P})\, \varphi(z; \mathbf{0}, \boldsymbol{\Sigma})\; d\boldsymbol{x}\, dz.
\end{aligned}$$

Algorithm 16 Gaussian filter with additive noise [115, Algorithm 6.3]

The prediction and update steps of the additive noise Gaussian assumed density (Kálmán) filter are:

1. *Prediction step*:
   a. Predicted (prior) state estimate: $\hat{\boldsymbol{x}}_{t\,|\,t-1} = \int \boldsymbol{f}_t(\boldsymbol{x}_{t-1}) \varphi\left(\boldsymbol{x}_{t-1}; \hat{\boldsymbol{x}}_{t-1\,|\,t-1}, \boldsymbol{P}_{t-1\,|\,t-1}\right) d\boldsymbol{x}_{t-1}$.
   b. Predicted (prior) error covariance: $\boldsymbol{P}_{t\,|\,t-1} = \int (\boldsymbol{f}_t(\boldsymbol{x}_{t-1}) - \hat{\boldsymbol{x}}_{t\,|\,t-1})(\boldsymbol{f}_t(\boldsymbol{x}_{t-1}) - \hat{\boldsymbol{x}}_{t\,|\,t-1})^{\intercal} \varphi\left(\boldsymbol{x}_{t-1}; \hat{\boldsymbol{x}}_{t-1\,|\,t-1}, \boldsymbol{P}_{t-1\,|\,t-1}\right) d\boldsymbol{x}_{t-1} + \boldsymbol{Q}_t$.
2. *Update* (or *correction*) *step*:
   a. Predicted observation: $\hat{\boldsymbol{y}}_{t\,|\,t-1} = \int \boldsymbol{h}_t(\boldsymbol{x}_t) \varphi\left(\boldsymbol{x}_t; \hat{\boldsymbol{x}}_{t\,|\,t-1}, \boldsymbol{P}_{t\,|\,t-1}\right) d\boldsymbol{x}_t$.
   b. Innovation (or observation residual): $\tilde{\boldsymbol{y}}_t = \boldsymbol{y}_t - \hat{\boldsymbol{y}}_{t\,|\,t-1}$.
   c. Innovation covariance: $\boldsymbol{S}_t = \int (\boldsymbol{h}_t(\boldsymbol{x}_t) - \hat{\boldsymbol{y}}_{t\,|\,t-1})(\boldsymbol{h}_t(\boldsymbol{x}_t) - \hat{\boldsymbol{y}}_{t\,|\,t-1})^{\intercal} \varphi\left(\boldsymbol{x}_t; \hat{\boldsymbol{x}}_{t\,|\,t-1}, \boldsymbol{P}_{t\,|\,t-1}\right) d\boldsymbol{x}_t + \boldsymbol{R}_t$.
   d. Crosscovariance: $\boldsymbol{\Xi}_t = \int (\boldsymbol{x}_t - \hat{\boldsymbol{x}}_{t\,|\,t-1})(\boldsymbol{h}_t(\boldsymbol{x}_t) - \hat{\boldsymbol{y}}_{t\,|\,t-1})^{\intercal} \varphi\left(\boldsymbol{x}_t; \hat{\boldsymbol{x}}_{t\,|\,t-1}, \boldsymbol{P}_{t\,|\,t-1}\right) d\boldsymbol{x}_t$.
   e. (Optimal) Kálmán gain: $\boldsymbol{K}_t = \boldsymbol{\Xi}_t \boldsymbol{S}_t^{-1}$.
   f. Updated (posterior) state estimate: $\hat{\boldsymbol{x}}_{t\,|\,t} = \hat{\boldsymbol{x}}_{t\,|\,t-1} + \boldsymbol{K}_t \tilde{\boldsymbol{y}}_t$.
   g. Updated (posterior) error covariance: $\boldsymbol{P}_{t\,|\,t} = \boldsymbol{P}_{t\,|\,t-1} - \boldsymbol{K}_t \boldsymbol{\Xi}_t \boldsymbol{K}_t^{\intercal}$.

Algorithm 17 Gaussian filter with possibly non-additive noise [115, Algorithm 6.4]

The prediction and update steps of the possibly non-additive noise Gaussian assumed density (Kálmán) filter are:

1. *Prediction step*:
   a. Predicted (prior) state estimate: $\hat{\boldsymbol{x}}_{t\,|\,t-1} = \int \boldsymbol{f}_t(\boldsymbol{x}_{t-1}, \boldsymbol{q}_t)\varphi\left(\boldsymbol{x}_{t-1}; \hat{\boldsymbol{x}}_{t-1\,|\,t-1}, \boldsymbol{P}_{t-1\,|\,t-1}\right) \varphi\left(\boldsymbol{q}_t; \boldsymbol{0}, \boldsymbol{Q}_t\right)\, d\boldsymbol{x}_{t-1}\, d\boldsymbol{q}_t.$
   b. Predicted (prior) error covariance: $\boldsymbol{P}_{t\,|\,t-1} = \int (\boldsymbol{f}_t(\boldsymbol{x}_{t-1}, \boldsymbol{q}_t) - \hat{\boldsymbol{x}}_{t\,|\,t-1})(\boldsymbol{f}_t(\boldsymbol{x}_{t-1}, \boldsymbol{q}_t) - \hat{\boldsymbol{x}}_{t\,|\,t-1})^{\mathsf{T}} \varphi\left(\boldsymbol{x}_{t-1}; \hat{\boldsymbol{x}}_{t-1\,|\,t-1}, \boldsymbol{P}_{t-1\,|\,t-1}\right) \varphi\left(\boldsymbol{q}_t; \boldsymbol{0}, \boldsymbol{Q}_t\right)\, d\boldsymbol{x}_{t-1}\, d\boldsymbol{q}_t.$

2. *Update* (or *correction*) *step*:
   a. Predicted observation: $\hat{\boldsymbol{y}}_{t\,|\,t-1} = \int \boldsymbol{h}_t(\boldsymbol{x}_t, \boldsymbol{r}_t)\varphi\left(\boldsymbol{x}_t; \hat{\boldsymbol{x}}_{t\,|\,t-1}, \boldsymbol{P}_{t\,|\,t-1}\right) \varphi\left(\boldsymbol{r}_t; \boldsymbol{0}, \boldsymbol{R}_t\right)\, d\boldsymbol{x}_t\, d\boldsymbol{r}_t.$
   b. Innovation (or observation residual): $\tilde{\boldsymbol{y}}_t = \boldsymbol{y}_t - \hat{\boldsymbol{y}}_{t\,|\,t-1}.$
   c. Innovation covariance: $\boldsymbol{S}_t = \int (\boldsymbol{h}_t(\boldsymbol{x}_t, \boldsymbol{r}_t) - \hat{\boldsymbol{y}}_{t\,|\,t-1})(\boldsymbol{h}_t(\boldsymbol{x}_t, \boldsymbol{r}_t) - \hat{\boldsymbol{y}}_{t\,|\,t-1})^{\mathsf{T}} \varphi\left(\boldsymbol{x}_t; \hat{\boldsymbol{x}}_{t\,|\,t-1}, \boldsymbol{P}_{t\,|\,t-1}\right) \varphi\left(\boldsymbol{r}_t; \boldsymbol{0}, \boldsymbol{R}_t\right)\, d\boldsymbol{x}_t\, d\boldsymbol{r}_t.$
   d. Crosscovariance: $\boldsymbol{\Xi}_t = \int (\boldsymbol{x}_t - \hat{\boldsymbol{x}}_{t\,|\,t-1})(\boldsymbol{h}_t(\boldsymbol{x}_t, \boldsymbol{r}_t) - \hat{\boldsymbol{y}}_{t\,|\,t-1})^{\mathsf{T}} \varphi\left(\boldsymbol{x}_t; \hat{\boldsymbol{x}}_{t\,|\,t-1}, \boldsymbol{P}_{t\,|\,t-1}\right) \varphi\left(\boldsymbol{r}_t; \boldsymbol{0}, \boldsymbol{R}_t\right)\, d\boldsymbol{x}_t\, d\boldsymbol{r}_t.$
   e. (Optimal) Kálmán gain: $\boldsymbol{K}_t = \boldsymbol{\Xi}_t \boldsymbol{S}_t^{-1}.$
   f. Updated (posterior) state estimate: $\hat{\boldsymbol{x}}_{t\,|\,t} = \hat{\boldsymbol{x}}_{t\,|\,t-1} + \boldsymbol{K}_t \tilde{\boldsymbol{y}}_t.$
   g. Updated (posterior) error covariance: $\boldsymbol{P}_{t\,|\,t} = \boldsymbol{P}_{t\,|\,t-1} - \boldsymbol{K}_t \boldsymbol{\Xi}_t \boldsymbol{K}_t^{\mathsf{T}}.$

Ito and Xiong [60] were the first to point out the considerable generality of this approach. They showed that the unscented Kalman filter, then recently discovered by Julier and Uhlmann, [62, 63, 128], remedying the deficiencies of the extended Kálmán filtering approach, could be regarded as a special case of the Gaussian filter. The integrals appearing in the Gaussian filter can be computed by means of various powerful quadrature and cubature methods [3, 133] when they cannot be computed analytically. The Gauss–Hermite Kálmán filter (GHKF) and cubature Kálmán filter (CKF) can also be seen as approximations of the Gaussian filter. For details, see [115, Chap. 6].

28.11 Parameter Estimation

Let us now suppose that our state-space model is parameterised by some parameter vector $\boldsymbol{\theta} \in \mathbb{R}^{d_\theta}$, $d_\theta \in \mathbb{N}^*$. For example, in the case of the linear-Gaussian state-space model of the Kálmán filter—equations (28.8) and (28.9)—this would amount to saying that (at least some of) $\boldsymbol{F}_t, \boldsymbol{a}_t, \boldsymbol{W}_t, \boldsymbol{Q}_t, \boldsymbol{H}_t, \boldsymbol{b}_t, \boldsymbol{V}_t, \boldsymbol{R}_t$ are actually functions of $\boldsymbol{\theta}$: $\boldsymbol{F}_t(\boldsymbol{\theta}), \boldsymbol{a}_t(\boldsymbol{\theta}), \boldsymbol{W}_t(\boldsymbol{\theta}), \boldsymbol{Q}_t(\boldsymbol{\theta}), \boldsymbol{H}_t(\boldsymbol{\theta}), \boldsymbol{b}_t(\boldsymbol{\theta}), \boldsymbol{V}_t(\boldsymbol{\theta}), \boldsymbol{R}_t(\boldsymbol{\theta})$. Similarly for all the other special cases of the general state-space model. We may not know the true value of this parameter. How do we estimate it? In other words, how do we *calibrate* the model?

Suppose that we have a time series of observations (historical, generated, etc.), $\boldsymbol{y}_1, \boldsymbol{y}_2, \ldots, \boldsymbol{y}_T$, $T \in \mathbb{T} = \mathbb{N}^*$, and we would like to use this data to calibrate the state-space model. The frequentist approach to parameter estimation relies on the (joint) probability density function of the observations, which depends on the parameters, $p(\boldsymbol{y}_1, \boldsymbol{y}_2, \ldots, \boldsymbol{y}_T; \boldsymbol{\theta})$. We can regard this as a function of $\boldsymbol{\theta}$ with $\boldsymbol{y}_1, \boldsymbol{y}_2, \ldots, \boldsymbol{y}_T$ fixed, $p(\boldsymbol{y}_1, \boldsymbol{y}_2, \ldots, \boldsymbol{y}_T; \boldsymbol{\theta}) := \mathscr{L}(\boldsymbol{\theta})$—the *likelihood function*. In the context of filtering this likelihood function is referred to as *marginal* likelihood, since the hidden states, $\boldsymbol{x}_1, \boldsymbol{x}_2, \ldots, \boldsymbol{x}_T$, are margined out. Our goal, then, is to find the *maximum likelihood estimator (MLE)*, $\hat{\boldsymbol{\theta}}_{ML}$, which is that value of $\boldsymbol{\theta}$ that maximises the likelihood function. In most practical applications one needs to resort to numerical methods, perhaps quasi-Newton methods, such as Broyden-Fletcher-Goldfarb-Shanno (BFGS) [42, Sect. 4.5.2], to find $\hat{\boldsymbol{\theta}}_{ML}$. Each evaluation of the objective function, $\mathscr{L}(\boldsymbol{\theta})$, requires us to run the stochastic filter over the observations $\boldsymbol{y}_1, \boldsymbol{y}_2, \ldots, \boldsymbol{y}_T$, which, depending on the state-space model, can be quite costly computationally. By the chain rule (i), and since we are dealing with a Markov chain (ii),

$$p(\boldsymbol{y}_1, \ldots, \boldsymbol{y}_T) \overset{\text{(i)}}{=} \prod_{t=1}^{T} p(\boldsymbol{y}_t \mid \boldsymbol{y}_0, \ldots, \boldsymbol{y}_{t-1}) \overset{\text{(ii)}}{=} \prod_{t=1}^{T} \int p(\boldsymbol{y}_t \mid \boldsymbol{x}_t) p(\boldsymbol{x}_t \mid \boldsymbol{y}_0, \ldots, \boldsymbol{y}_{t-1})\, d\boldsymbol{x}_t.$$

Here we have omitted the dependence of all the probability densities on $\boldsymbol{\theta}$, e.g. we should have really written $p(\boldsymbol{y}_1, \ldots, \boldsymbol{y}_T; \boldsymbol{\theta})$. For the Kálmán filter, this becomes [50, 51]

$$\begin{aligned} p(\boldsymbol{y}_1, \ldots, \boldsymbol{y}_T) &= \prod_{t=1}^{T} \int \varphi\left(\boldsymbol{y}_t; \boldsymbol{H}_t \boldsymbol{x}_t + \boldsymbol{b}_t, \boldsymbol{V}_t \boldsymbol{R}_t \boldsymbol{V}_t^{\intercal}\right) \varphi\left(\boldsymbol{x}_t; \hat{\boldsymbol{x}}_{t \mid t-1}, \boldsymbol{P}_{t \mid t-1}\right) d\boldsymbol{x}_t \\ &= \prod_{t=1}^{T} \varphi\left(\boldsymbol{y}_t; \boldsymbol{H}_t \hat{\boldsymbol{x}}_{t \mid t-1} + \boldsymbol{b}_t, \boldsymbol{V}_t \boldsymbol{R}_t \boldsymbol{V}_t^{\intercal} + \boldsymbol{H}_t \boldsymbol{P}_{t \mid t-1} \boldsymbol{H}_t^{\intercal}\right) \\ &= \prod_{t=1}^{T} \varphi\left(\boldsymbol{y}_t; \hat{\boldsymbol{y}}_{t \mid t-1}, \boldsymbol{S}_t\right). \end{aligned}$$

In fact, the final equation holds for the other related filters, including the extended Kálmán and Gaussian assumed density filter (up to the approximation error). Since the natural logarithm function is increasing, maximising the *log-likelihood function*, $\ln(\mathscr{L}(\boldsymbol{\theta}))$, is equivalent to maximising $\mathscr{L}(\boldsymbol{\theta})$. The log-likelihood, then, is given by

$$\begin{aligned} \ln(\mathscr{L}(\boldsymbol{\theta})) &= \ln\left(\prod_{t=1}^{T} \varphi\left(\boldsymbol{y}_t; \hat{\boldsymbol{y}}_{t \mid t-1}, \boldsymbol{S}_t\right)\right) = \sum_{t=1}^{T} \ln\left(\varphi\left(\boldsymbol{y}_t; \hat{\boldsymbol{y}}_{t \mid t-1}, \boldsymbol{S}_t\right)\right) \\ &= -\frac{1}{2}\left(T \cdot d_Y \cdot \ln(2\pi) + \sum_{t=1}^{T} \ln(\det(\boldsymbol{S}_t)) + \sum_{t=1}^{T} \tilde{\boldsymbol{y}}_t^T \boldsymbol{S}_t^{-1} \tilde{\boldsymbol{y}}_t\right). \end{aligned}$$

(Recall that d_Y is the number of elements in each observation vector $\boldsymbol{y}_t$.) Notice that, after receiving each observation $\boldsymbol{y}_t$, we can update the likelihood by adding to it the term

$$-\frac{1}{2}\left(d_Y \cdot \ln(2\pi) + \ln(\det(\boldsymbol{S}_t)) + \tilde{\boldsymbol{y}}_t^T \boldsymbol{S}_t^{-1} \tilde{\boldsymbol{y}}_t\right),$$

having initialised it to zero before receiving any observations, as $\tilde{\boldsymbol{y}}_t = \tilde{\boldsymbol{y}}_t(\boldsymbol{\theta})$ and $\boldsymbol{S}_t = \boldsymbol{S}_t(\boldsymbol{\theta})$ are byproducts of running the update (correction) step of the filter. This log-likelihood is referred to as the *prediction error decomposition form* in [50, p. 126]. Further details on its use for statistical inference about unknown parameters can be found in [46].

For the particle filter, we can estimate the log-likelihood function from the non-normalised weights:

$$p(\boldsymbol{y}_1, \ldots, \boldsymbol{y}_T) = \prod_{t=1}^{T} \int p(\boldsymbol{y}_t \mid \boldsymbol{x}_t) p(\boldsymbol{x}_t \mid \boldsymbol{y}_0, \ldots, \boldsymbol{y}_{t-1})\, d\boldsymbol{x}_t \approx \prod_{t=1}^{T} \left(\frac{1}{M} \sum_{k=1}^{M} \omega_t^{(k)}\right),$$

whence

$$\ln(\mathcal{L}(\boldsymbol{\theta})) = \ln\left\{\prod_{t=1}^{T}\left(\frac{1}{M}\sum_{k=1}^{M}\omega_t^{(k)}\right)\right\} = \sum_{t=1}^{T} \ln\left(\frac{1}{M}\sum_{k=1}^{M}\omega_t^{(k)}\right). \tag{28.13}$$

This was first proposed by Kitagawa [73, 74] for the purposes of approximating $\hat{\boldsymbol{\theta}}_{ML}$. Malik et al. [109] point out the practical difficulties which result when using (28.13) as an objective function in an optimiser. In the resampling (or selection) step of the particle filter, we are sampling from a discontinuous empirical distribution function. Therefore $\ln(\mathcal{L}(\boldsymbol{\theta}))$ will not be continuous as a function of $\boldsymbol{\theta}$. To remedy this, they rely on an alternative, continuous, resampling procedure described in [91]. A quasi-Newton method is then used to find $\hat{\boldsymbol{\theta}}_{ML}$ for the parameters $\boldsymbol{\theta} = (\mu, \phi, \sigma_v^2, \rho, p, \sigma_J^2)^{\mathsf{T}}$ of the SVLJ model discussed in Sect. 28.4.

28.12 Relationship with Markov Chain Monte Carlo Methods

Let us briefly discuss how filtering methods relate to *Markov chain Monte Carlo methods (MCMC)*[9]—a vast subject in its own right, therefore our discussion will be cursory at best. The technique takes its origin from the work of Nicholas

[9]"Markov chain" in MCMC refers to the Markov chains that are produced in algorithms such as the Gibbs sampler (see below) rather than the Markov chains that we talked about in Sect. 28.2.

Metropolis (1915–1999), Marshall N. Rosenbluth (1927–2003) and his first wife, Arianna W. Rosenbluth, Edward Teller[10] (1908–2003) and his wife, Augusta H. Teller (1909–2000), at Los Alamos on simulating a liquid in equilibrium with its gas phase [97]. The discovery came when its authors realised that, instead of simulating the exact dynamics, they could simulate a certain Markov chain with the same equilibrium distribution.

Proceeding along the lines of [72], Meyer and Yu [99] demonstrate how MCMC techniques can be used to estimate the parameters of the SVL model. They calibrate the parameters to the time series of observations of daily mean-adjusted log-returns, $y_1, \ldots, y_T$. They obtain the joint prior density

$$p(\boldsymbol{\theta}, x_0, \ldots, x_T) = p(\boldsymbol{\theta}) p(x_0 \mid \boldsymbol{\theta}) \prod_{t=1}^{T} p(x_t \mid x_{t-1}, \boldsymbol{\theta})$$

by successive conditioning. Here $\boldsymbol{\theta} = (\mu, \phi, \sigma_v^2, \rho)^\intercal$ is, as before, the vector of the model's parameters. They assume prior independence of the parameters and choose the same priors as in [72] for μ, ϕ, and σ_v^2, and a uniform prior for ρ. The observation model (28.2) and the conditional independence assumption give the likelihood

$$p(y_1, \ldots, y_T \mid \boldsymbol{\theta}, x_0, \ldots, x_T) = \prod_{t=1}^{T} p(y_t \mid x_t),$$

and the joint posterior distribution of the *unobservables* (the parameters $\boldsymbol{\theta}$ and the hidden states $x_0, \ldots, x_T$; in the Bayesian perspective these are treated identically and estimated in a similar manner) follows from Bayes's theorem; for the SVL model, this posterior satisfies

$$\begin{aligned} p(\boldsymbol{\theta}, x_0, \ldots, x_T \mid y_1, \ldots, y_T) &\propto p(\mu) p(\phi) p(\sigma_v^2) p(\rho) \\ &\prod_{t=1}^{T} p(x_{t+1} \mid x_t, \mu, \phi, \sigma_v^2) \prod_{t=1}^{T} p(y_t \mid x_{t+1}, x_t, \mu, \phi, \sigma_v^2, \rho), \end{aligned}$$

where $p(\mu), p(\phi), p(\sigma_v^2), p(\rho)$ are the appropriately chosen priors,

$$x_{t+1} \mid x_t, \mu, \phi, \sigma_v^2 \sim \mathcal{N}\left(\mu(1-\phi) + \phi x_t, \sigma_v^2\right),$$

$$y_t \mid x_{t+1}, x_t, \mu, \phi, \sigma_v^2, \rho \sim \mathcal{N}\left(\frac{\rho}{\sigma_v} e^{x_t/2} \left(x_{t+1} - \mu(1-\phi) - \phi x_t\right), e^{x_t}(1-\rho^2)\right).$$

[10] Edward Teller is known in the West as "the father of the hydrogen bomb".

Meyer and Yu use the software package BUGS[11] [87, 122] to represent the resulting Bayesian model as a *directed acyclic graph (DAG)*, where the nodes are either constants (denoted by rectangles), stochastic nodes (variables that are given a distribution, denoted by ellipses), or deterministic nodes (logical functions of other nodes); the arrows either indicate stochastic dependence (solid arrows) or logical functions (hollow arrows). This graph helps visualise the conditional (in)dependence assumptions and is used by BUGS to construct full univariate conditional posterior distributions for all unobservables. It then uses Markov chain Monte Carlo algorithms to sample from these distributions.

The algorithm based on the original work [97] is now known as the *Metropolis algorithm*. It has been generalised by Hastings (1930–2016) to obtain the *Metropolis-Hastings algorithm* [54] and further by Green to obtain what is known as the *Metropolis-Hastings-Green algorithm* [45]. A popular algorithm based on a special case of the Metropolis-Hastings algorithm, known as the *Gibbs sampler*, was developed by the brothers Geman [39] and, independently, Tanner and Wong [125].[12] It was further popularised by Gelfand and Smith [38]. Gibbs sampling and related algorithms [41, 111] are used by BUGS to sample from the univariate conditional posterior distributions for all unobservables.

As a result we perform Bayesian estimation—obtain estimates of the *distributions* of the parameters μ, ϕ, σ_v^2, ρ—rather than frequentist estimation, where a single value of the parameters vector, which maximises the likelihood, $\hat{\boldsymbol{\theta}}_{ML}$, is produced, as we saw in Sect. 28.11. Stochastic filtering, sometimes in combination with MCMC, can be used for both frequentist and Bayesian parameter estimation [18]. Filtering methods that update estimates of the parameters online, while processing observations in real-time, are referred to as *adaptive filtering* (see [20, 101, 116, 127] and references therein).

28.13 Prediction

The Kálmán filter equations provide a natural way of predicting the observation $\boldsymbol{y}_{T+1}$ given all the observations up to time T, $\boldsymbol{y}_1, \boldsymbol{y}_2, \ldots, \boldsymbol{y}_T$, $T \in \mathbb{T} = \mathbb{N}^*$. Indeed, the predicted state estimate equation yields

$$\hat{\boldsymbol{x}}_{T+1 \mid T} = \boldsymbol{F}_{T+1}\hat{\boldsymbol{x}}_{T \mid T} + \boldsymbol{a}_{T+1},$$

while the predicted observation equation gives us the *one-step-ahead prediction*

$$\hat{\boldsymbol{y}}_{T+1 \mid T} = \boldsymbol{H}_{T+1}\hat{\boldsymbol{x}}_{T+1 \mid T} + \boldsymbol{b}_{T+1}.$$

[11] An acronym for Bayesian inference Using Gibbs Sampling.

[12] Sometimes the Gibbs sampler is referred to as *data augmentation* following this paper.

The innovation is precisely the difference between the actual observation and the corresponding one-step-ahead prediction.

The *l-step-ahead prediction*, $l \in \mathbb{N}^*$, is obtained by applying the predicted state estimate equation l times, then taking expectations, to obtain

$$\boldsymbol{x}_{T+l \mid T} = \left(\prod_{j=1}^{l} \boldsymbol{F}_{T+j} \right) \boldsymbol{x}_{T \mid T} + \sum_{j=1}^{l-1} \left(\prod_{i=j+1}^{l} \boldsymbol{F}_{T+i} \right) \boldsymbol{a}_{T+j} + \boldsymbol{a}_{T+l},$$

then applying the predicted observation equation,

$$\boldsymbol{y}_{T+l \mid T} = \boldsymbol{H}_{T+l} \hat{\boldsymbol{x}}_{T+l \mid T} + \boldsymbol{b}_{T+l}.$$

In practice, including software implementations, the easiest way to obtain the l-step ahead prediction (and associated uncertainties) is by applying the prediction step of the filter l times, then running the prediction observation, innovation, and innovation covariance substeps of the update step.

To obtain one-step-ahead predictions in particle filters [56, p. 199], [108], we begin by estimating the one-step ahead density

$$p(\boldsymbol{y}_{T+1} \mid \boldsymbol{y}_1, \ldots, \boldsymbol{y}_T) \approx \sum_{i=1}^{M} \left\{ \frac{1}{K} \sum_{k=1}^{K} p_{Y_{T+1}}(\boldsymbol{y}_{T+1} \mid \hat{\boldsymbol{x}}^{(i,k)}_{T+1 \mid T}) \right\} \lambda^{(i)}_T .$$

Here $\hat{\boldsymbol{x}}^{(i,k)}_{T+1 \mid T}$ are drawn from $\tau_{T+1}(\cdot \mid \hat{\boldsymbol{x}}^{(i)}_{T \mid T})$. In [108] the authors recommend setting K to a suitable number greater than one. l-step-ahead predictions can be obtained by propagating the particles through the Markov transition kernels $\tau_{T+1}, \ldots, \tau_{T+l}$ and then using the result to approximate $p(\boldsymbol{y}_{T+l} \mid \boldsymbol{y}_1, \ldots, \boldsymbol{y}_T)$. It is then possible to sample from the resulting mixture distribution.

28.14 Diagnostics

When the true state is known, e.g. when applying the filters to simulated data, the *root mean square error (RMSE)* can be computed for each time T as

$$\sqrt{\sum_{t=1}^{T} \|\boldsymbol{x}^{\text{true}}_t - \hat{\boldsymbol{x}}_{t \mid t}\|_2^2}$$

and is a very useful diagnostic. Another useful value to look at is the log-likelihood of Sect. 28.11. The RMSE and log-likelihood are useful for comparing different filtering algorithms or different parameterisations of the same filtering algorithm.

To assess the health of the Kálmán (extended Kálmán, Gaussian assumed density) filter state-space model, we can look at the *standardised residuals*,

$$\tilde{\boldsymbol{y}}_t^{\text{st}} := \boldsymbol{S}_t^{-1/2}\tilde{\boldsymbol{y}}_t.$$

In many cases, $d_Y = 1$, so we simply have the scalar equation $\tilde{y}_t^{\text{st}} = \tilde{y}_t/\sqrt{S_t}$. Otherwise, the innovation covariance matrix, $\boldsymbol{S}_t$, should be symmetric and positive definite, so that[13]

$$\boldsymbol{S}_t^{-1/2} = \boldsymbol{U}_t\boldsymbol{\Lambda}_t^{-1/2}\boldsymbol{U}_t^{\mathsf{T}},$$

where $\boldsymbol{\Lambda}_t$ is the diagonal matrix of all the eigenvalues of $\boldsymbol{S}_t$, and $\boldsymbol{U}_t$ is the orthogonal matrix whose columns are the normalised eigenvectors associated with the eigenvalues in the corresponding diagonal entries of $\boldsymbol{\Lambda}_t$. The standardised residuals $\tilde{\boldsymbol{y}}_t^{\text{st}}$ should be uncorrelated and follow the standard (possibly multivariate, if $d_Y > 1$) normal distribution. In other words, the sequence $\tilde{\boldsymbol{y}}_t^{\text{st}}$ should constitute *white noise*.[14] We can use, for example, the Ljung-Box test [86] to confirm that the residuals do indeed have zero autocorrelation. We can use the standard normality tests, such as the Q-Q plot or the Jarque-Bera test [7], to detect deviations from normality, which indicates that the model is misspecified. The most basic (but still useful) test consists in visually inspecting the plot of $\tilde{\boldsymbol{y}}_t^{\text{st}}$ over time, the so-called *standardised plot*.

For a comprehensive treatment of Kálmán filter diagnostics and model selection we refer the reader to [50, Chap. 5].

The degeneracy (see Sect. 28.3) of particle filters can be measured by an estimate of the *effective sample size* [92, Sect. 3.5.2], [106]:

$${}^{\text{eff}}M_t = \frac{1}{\sum_{i=1}^{M}({}^{\text{br}}\lambda_t^{(i)})^2}.$$

A smaller ${}^{\text{eff}}M_t$ indicates a larger variance of the weights, i.e. more degeneracy, at time t. In the extreme cases, when, for all $i = 1, \ldots, M$, ${}^{\text{br}}\lambda_t^{(i)} = \frac{1}{M}$, ${}^{\text{eff}}M_t = M$; when, for all but one particle the weights are zero, ${}^{\text{eff}}M_t = 1$. Thus we could monitor ${}^{\text{eff}}M_t$ and spot the degeneracy when it drops below a certain threshold; [92] suggests $M/2$.

---

[13]By the result that the mathematicians calls the *spectral theorem* and mechanical engineers the *principal axis theorem*. As is well known, all the eigenvalues of a positive definite matrix are positive.

[14]The innovations, $\tilde{\boldsymbol{y}}_t$, by construction, form the innovation process of Wold's decomposition. For this reason, the Kálmán and related filters are known as *whitening* filters: the possibly correlated sequence of observations $\boldsymbol{y}_t$ is transformed into a white noise process.

28.15 Further Reading

The subjects of state-space models and stochastic filtering have been studied from different angles by stochastic analysts, mechanical and electrical engineers, statisticians and econometricians, both frequentist and Bayesian. In our brief overview of the subject we have included references to a very small subset of papers on the subject. In this section we shall provide some further bibliographical references—to textbooks and monographs. Our list will of necessity be very incomplete, so we apologise in advance.

Among the first books on state-space models written from an engineering standpoint are [2, 61, 114]. A statistician's (econometrician's) early expositions of the subject can be found in [50, 134]. From the Bayesian standpoint, the subject is examined in [130] and later, in more detail, in [35]. Other statistical texts examining state-space modelling include [29, 30, 47]. Applications of state-space models in economics and finance are discussed in depth in the monograph [137]. A look at finance-specific applications of stochastic filtering oriented towards the practitioner is given in [10]. Practical applications to financial multi-factor models are discussed in [23].

Classic texts on the deeper mathematical aspects of stochastic filtering include [24, 64]. A more recent exposition of the technicalities of the subject from the standpoint of stochastic analysis is [5]. One of the de facto standard introductory textbooks on stochastic analysis [104] dedicates an individual chapter (Chap. 6) to the subject. Bucy, co-inventor of linear-Gaussian filtering, has published his lectures on the mathematics of filtering and Riccati equations [14]. [115] is a recent graduate-level introduction to the mathematics of both linear and nonlinear filtering, whereas [32] is a primer on the mathematics specifically of the Kálmán filter; [19] is an update on an early (1987) text on Kálmán filtering with applications. A straightforward introductory tutorial on Kálmán filtering is available online [129].

The applied and computational aspects are discussed in, inter alia, [55, 119]. Classic texts on the subject include an electrical engineer's [93]. These texts look at Kálmán filtering in much greater depth than is possible here. Particle filtering methods are reviewed in [27, 28]. See [26] for a comparison of resampling methods. A visual explanation of how they work can be found in [33, p. 407].

MCMC methods were only considered in passing, so we cannot provide a detailed bibliography here. A good, fairly succinct introduction to MCMC is [40]. For details we refer the reader to [13, 22, 37, 137]. For an introduction to applications of MCMC in econometrics, see [80]. Many practical details of Bayesian analysis of stochastic volatility models are mentioned in [99]. The free BUGS, WinBUGS, OpenBUGS and JAGS remain in active use in many fields outside econometrics.[15] We refer the reader to WinBUGS documentation [87], a fairly detailed tutorial [9], and a text [103] intended for practitioners.

[15] According to [71, p. 44], these software products are "the most widely used MCMC engine[s] currently available".

Acknowledgements Paul Bilokon and James Gwinnutt would like to acknowledge Martin Zinkin; Paul Bilokon would also like to acknowledge Dan Crisan, Rob Smith (Citigroup) and William Osborn. These former colleagues have taught us many things about the practicalities of stochastic filtering in algorithmic trading.

References

1. Abou-Kandil, H., Freiling, G., Ionescu, V., Jank, G.: Matrix Riccati Equations in Control and Systems Theory. Springer Basel AG, Basel (2003)
2. Anderson, B.D.O., Moore, J.B.: Optimal Filtering. Prentice-Hall, Englewood Cliffs (1979)
3. Arasaratnam, I., Haykin, S.: Cubature Kalman filters. IEEE Trans. Autom. Control **54**(6), 1254–1269 (2009)
4. Bachelier, L.J.-B.A.: Théorie de la spéculation. Ann. Sci. de l'École Norm. Supér. **3**(17), 21–86 (1900)
5. Bain, A., Crisan, D.: Fundamentals of Stochastic Filtering. Stochastic Modelling and Applied Probability, vol. 60. Springer, Berlin (2009)
6. Barker, A.L., Brown, D.E., Martin, W.N.: Bayesian estimation and the Kalman filter. Comput. Math. Appl. **30**(10), 55–77 (1995)
7. Bera, A.K., Jarque, C.M.: Model specification tests: a simultaneous approach. J. Econ. **20**(1), 59–82 (1982)
8. Berzuini, C., Best, N.G., Gilks, W.R., Larizza, C.: Dynamic conditional independence models and Markov Chain Monte Carlo methods. J. Am. Stat. Assoc. **92**(440), 1403–1412 (1997)
9. Best, N., Mason, A., Li, P.: Introduction to Bayesian analysis using WinBUGS February (2011). http://www.bias-project.org.uk/WB2011Man/bayes-intro-2011-slides.pdf
10. Bhar, R.: Stochastic Filtering with Applications in Finance. World Scientific, Singapore (2010)
11. Box, G.E.P., Jenkins, G.M., Reinsel, G.C.: Time Series Analysis. Prentice-Hall, Upper Saddle River (1994)
12. Brockwell, P.J., Davis, R.A.: Time Series: Theory and Methods. Springer, Berlin (1987)
13. Brooks, S., Gelman, A., Jones, G.L., Meng, X.-L. (eds.): Handbook of Markov Chain Monte Carlo. Handbooks of Modern Statistical Methods. Chapman and Hall, Boca Raton (2011)
14. Bucy, R.S.: Lectures on Discrete Time Filtering. Springer, Berlin (1994)
15. Cameron, R.H., Martin, W.T.: Transformation of Wiener integrals under translations. Ann. Math. **45**, 386–396 (1944)
16. Cappé, O., Moulines, E., Rydén, T.: Inference in Hidden Markov Models. Springer Series in Statistics. Springer, Berlin (2005)
17. Carpenter, J., Clifford, P., Fearnhead, P.: An improved particle filter for non-linear problems. IEE Proc. Radar Sonar Navig. **146**(1), 2–7 (1999)
18. Chen, Z.: Bayesian filtering: From Kalman filters to particle filters, and beyond. Statistics **182**(1), 1–69 (2003)
19. Chui, C.K., Chen, G.: Kalman Filtering with Real-Time Applications, 4th edn. Springer, Berlin (2009)
20. Crisan, D., Míguez, J.: Nested particle filters for online parameter estimation in discrete-time state-space Markov models. ArXiv:1308.1883 (2013)
21. Crisan, D., Míguez, J.: Particle-kernel estimation of the filter density in state-space models. Bernoulli **20**(4), 1879–1929 (2014)
22. Dagpunar, J.S.: Simulation and Monte Carlo with Applications in Finance and MCMC. Wiley, New York (2007)
23. Darolles, S., Duvaut, P., Jay, E.: Multi-factor Models and Signal Processing Techniques: Application to Quantitative Finance. Digital Signal and Image Processing Series. Wiley, New York (2013)

24. Davis, M.H.A.: Lectures on Stochastic Control and Nonlinear Filtering. Springer, Berlin (1984)
25. de Jong, P., Penzer, J.: The ARIMA model in state space form. Research Report 40, Department of Statistics, London School of Economics, Houghton Street, London, August (2000). http://www.lse.ac.uk/statistics/documents/researchreport40.pdf
26. Douc, R., Cappé, O., Moulines, E.: Comparison of resampling schemes for particle filtering. In: Proceedings of the 4th International Symposium on Image and Signal Processing and Analysis, pp. 64–69 (2005)
27. Doucet, A., Godsill, S., Andrieu, C.: On sequential Monte Carlo sampling methods for Bayesian filtering. Stat. Comput. **10**, 197–208 (2000)
28. Doucet, A., de Freitas, N., Gordon, N.: An introduction to sequential Monte Carlo methods. In: Sequential Monte Carlo Methods in Practice, pp. 3–14. Springer, Berlin (2001)
29. Durbin, J., Koopman, S.J.: Time Series Analysis by State Space Methods, 1st edn. Oxford University Press, Oxford (2001)
30. Durbin, J., Koopman, S.J.: Time Series Analysis by State Space Methods, 2nd edn. Oxford Statistical Science Series, vol. 38. Oxford University Press, Oxford (2012)
31. Efron, B., Tibshirani, R.J.: An Introduction to the Bootstrap. Monographs on Statistics and Applied Probability, vol. 57. Champan & Hall/CRC, Boca Raton (1994)
32. Eubank, R.L.: A Kalman Filter Primer. Statistics: A Series of Textbooks and Monographs, vol. 186. Chapman & Hall/CRC, Boca Raton (2006)
33. Fernández-Madrigal, J.-A., Claraco, J.L.B.: Simultaneous Localization and Mapping for Mobile Robots: Introduction and Methods. IGI Global (2013)
34. Feynman, R.P., Leighton, R.B., Sands, M.: The Feynman Lectures on Physics, Mainly Mechanics, Radiation, and Heat, vol. 1. Addison-Wesley Publishing, Boston (1963)
35. Frühwirth-Schnatter, S.: Finite Mixture and Markov Switching Models. Springer Series in Statistics. Springer, Berlin (2006)
36. Fujisaki, M., Kallianpur, G., Kunita, H.: Stochastic differential equations for the non linear filtering problem. Osaka J. Math. **9**, 19–40 (1972)
37. Gamerman, D., Lopes, H.F.: Markov Chain Monte Carlo: Stochastic Simulation for Bayesian Inference. Texts in Statistical Science, 2nd edn. Chapman and Hall, Boca Raton (2006)
38. Gelfand, A.E., Smith, A.F.M.: Sampling-based approaches to calculating marginal densities. J. Am. Stat. Assoc. **85**(410), 398–409 (1990)
39. Geman, S.J., Geman, D.: Stochastic relaxation, Gibbs distributions, and the Bayesian restoration of images. IEEE Trans. Pattern Anal. Mach. Intell. **6**, 721–741 (1984)
40. Geyer, C.J.: Introduction to Markov Chain Mone Carlo. In: Handbook of Markov Chain Monte Carlo, pp. 3–49. Handbooks of Modern Statistical Methods. CRC Press, Boca Raton (2011)
41. Gilks, W.R., Wild, P.: Adaptive rejection sampling for Gibbs sampling. Appl. Stat. **41**, 337–348 (1992)
42. Gill, P.E., Murray, W., Wright, M.H.: Practical Optimization. Emerald Group Publishing Limited, West Yorkshire (1982)
43. Girsanov, I.V.: On transforming a certain class of stochastic processes by absolutely continuous substitution of measures. Theory Probab. Appl. **5**(3), 285–301 (1960)
44. Gordon, N.J., Salmond, D.J., Smith, A.F.M.: Novel approach to nonlinear/non-Gaussian Bayesian state estimation. In: IEE Proceedings F (Radar and Signal Processing) (1993)
45. Green, P.J.: Reversible jump Markov chain Monte Carlo computation and Bayesian model determination. Biometrika **82**(4), 711–732 (1995)
46. Hamilton, J.D.: State-space models. In: Handbook of Econometrics, vol. IV, pp. 3039–3080. Elsevier Science, Amsterdam (1994)
47. Hamilton, J.D.: Time Series Analysis. Princeton University Press, Princeton (1994)
48. Hammersley, J.M., Morton, K.W.: Poor man's Monte Carlo. J. R. Stat. Soc. Ser. B (Methodol.) **16**(1), 23–38 (1954)
49. Handschin, J.E., Mayne, D.Q.: Monte Carlo techniques to estimate the conditional expectation in multi-stage non-linear filtering. Int. J. Control **9**(5), 547–559 (1969)

50. Harvey, A.C.: Forecasting, Structural Time Series Models and the Kalman Filter. Cambridge University Press, Cambridge (1989)
51. Harvey, A.C., Phillips, G.D.A.: Maximum likelihood estimation of regression models with autoregressive-moving average disturbances. Biometrika **66**(1), 49–58 (1979)
52. Harvey, A.C., Shephard, N.: Estimation of an asymmetric stochastic volatility model for asset returns. J. Bus. Econ. Stat. **14**(4), 429–434 (1996)
53. Harvey, A.C., Ruiz, E., Shephard, N.: Multivariate stochastic variance models. Rev. Econ. Stud. **61**, 247–264 (1994)
54. Hastings, W.K.: Monte Carlo sampling methods using Markov chains and their applications. Biometrika **57**(1), 97–109 (1970)
55. Haykin, S. (ed.): Kalman Filtering and Neural Networks. Wiley, New York (2001)
56. Herbst, E.P., Schorfheide, F.: Bayesian Estimation of DSGE Models. Princeton University Press, Princeton (2016)
57. Ho, Y.C., Lee, R.C.K.: A Bayesian approach to problems in stochastic estimation and control. IEEE Trans. Autom. Control **9**(4), 333–339 (1964)
58. Hürzeler, M., Künsch, H.R.: Monte Carlo approximations for general state-space models. J. Comput. Graph. Stat. **7**(2), 175–193 (1998)
59. Isard, M., Blake, A.: Condensation – conditional density propagation for visual tracking. Int. J. Comput. Vis. **29**(1), 5–28 (1998)
60. Ito, K., Xiong, K.: Gaussian filters for nonlinear filtering problems. IEEE Trans. Autom. Control **45**(5), 910–927 (2000)
61. Jazwinski, A.H.: Stochastic Processes and Filtering Theory, vol. 64. Elsevier, Amsterdam (1970)
62. Julier, S.J., Uhlmann, J.K.: A general method for approximating nonlinear transformations of probability distributions. Technical report, Robotics Research Group, Department of Engineering Science, University of Oxford, Oxford (1996)
63. Julier, S.J., Uhlmann, J.K.: A new extension of the Kalman filter to nonlinear systems. In: Proceedings of AeroSense: The 11th International Symposium on Aerospace/Defense Sensing, Simulation and Controls, Multi Sensor Fusion, Tracking and Resource Management, pp. 182–193 (1997)
64. Kallianpur, G.: Stochastic Filtering Theory. Applications of Mathematics, vol. 13. Springer Science+Business Media, LLC, New York (1980)
65. Kallianpur, G., Striebel, C.: Stochastic differential equations in stochastic estimation problems. In: Krishnaiah, P.R. (ed.) Proceedings of the Second International Symposium on Multivariate Analysis, Dayton, OH, 1968. Academic, New York (1969)
66. Kallianpur, G., Striebel C.: Stochastic differential equations occurring in the estimation of continuous parameter stochastic processes. Theory Probab. Appl. **14**(4), 567–594 (1969)
67. Kálmán, R.E.: A new approach to linear filtering and prediction problems. Trans. ASME — J. Basic Eng. Ser. D **82**, 35–45 (1960)
68. Kálmán, R.E.: On the general theory of control systems. In: Proceedings of the First International Congress of IFAC, Moscow (1960)
69. Kálmán, R.E.: Mathematical description of linear dynamical systems. J. Soc. Ind. Appl. Math. Ser. A Control **1**(2), 152–192 (1963)
70. Kálmán, R.E., Bucy, R.S.: New results in linear filtering and prediction theory. J. Basic Eng. **83**, 95–108 (1961)
71. Kéry, M., Schaub, M.: Bayesian Population Analysis Using WinBUGS: A Hierarchical Perspective. Academic, New York (2012)
72. Kim, S., Shephard, N., Chib, S.: Stochastic volatility: likelihood inference and comparison with ARCH models. Rev. Econ. Stud. **65**(3), 361–393 (1998)
73. Kitagawa, G.: A Monte Carlo filtering and smoothing method for non-Gaussian nonlinear state space models. In: Proceedings of the 2nd U.S.-Japan Joint Seminar on Statistical Time Series Analysis, pp. 110–131 (1993)
74. Kitagawa, G.: Monte Carlo filter and smoother for non-Gaussian nonlinear state space models. J. Comput. Graph. Stat. **5**(1), 1–25 (1996)

75. Kolmogorov, A.N.: Grundbegriffe der Wahrscheinlichkeitrechnung. Ergebnisse der Mathematik und ihrer Grenzgebiete **2**(3), 1–62 (1933)
76. Kolmogorov, A.N.: Interpolation and extrapolation. Bull. de l'académie des sciences de U.S.S.R., Ser. Math. **5**, 3–14 (1941)
77. Kolmogorov, A.N.: Foundations of the Theory of Probability. Chelsea Publishing Company, New York (1956)
78. Kushner, H.: Nonlinear filtering: the exact dynamical equations satisfied by the conditional mode. IEEE Trans. Autom. Control **12**(3), 262–267 (1967)
79. Lai, T.L., Bukkapatanam, V.: State-space models: applications in economics and finance. In: Adaptive Filtering, Nonlinear State-Space Models, and Applications in Finance and Econometrics. Statistics and Econometrics for Finance, pp. 3–22. Springer, Berlin (2013)
80. Lancaster, T.: An Introduction to Modern Bayesian Econometrics. Blackwell Publishing, Malden (2004)
81. Lewis, F.L.: Optimal Estimation with an Introduction to Stochastic Control Theory. Wiley, New York (1986)
82. Linzhou, X., Xin-Hua, Z., Shao-Qing, Y., Wen-Tao, F.: An efficient particle filter with variable number of particles for bearings-only tracking. In: 2010 IEEE 10th International Conference on Signal Processing (ICSP) (2010)
83. Liptser, R.S., Shiryaev, A.N.: Non linear filtering of Markov diffusion processes. Trudy Matematicheskogo Instituta Stekolva **104**, 135–180 (1968)
84. Liptser, R.S., Shiryaev, A.N.: Statistics of Random Processes. Springer, Berlin (1992)
85. Liu, J.S., Chen, R.: Sequential Monte Carlo methods for dynamic systems. J. Am. Stat. Assoc. **93**(443), 1032–1044 (1998)
86. Ljung, G.M., Box, G.E.P.: On a measure of a lack of fit in time series models. J. Am. Stat. Assoc. **65**, 1509–1526 (1970)
87. Lunn, D.J., Thomas, A., Best, N., Spiegelhalter, D.: WinBUGS – a Bayesian modelling framework: concepts, structure and extensibility. Stat. Comput. **10**, 325–337 (2000)
88. Ma, L., Wang, H., Chen, J.: Analysis of Kalman filter with correlated noises under different dependence. J. Inf. Comput. Sci. **7**(5), 1147–1154 (2010)
89. Malik, S., Pitt, M.K.: Modelling stochastic volatility with leverage and jumps: A simulated maximum likelihood approach via particle filtering. Warwick Economic Research Papers 897, The University of Warwick, Department of Economics, Coventry CV4 7AL, April (2009)
90. Malik, S., Pitt, M.K.: Modelling stochastic volatility with leverage and jumps: A simulated maximum likelihood approach via particle filtering. document de travail 318, Banque de France Eurosystème, February (2011)
91. Malik, S., Pitt, M.K.: Particle filters for continuous likelihood evaluation and maximisation. J. Econ. **165**, 190–209 (2011)
92. Maskell, S.: An introduction to particle filters. In: State Space and Unobserved Component Models: Theory and Applications, pp. 40–72. Cambridge University Press, Cambridge (2004)
93. Maybeck, P.S.: Stochastic Models, Estimation and Control, Volume 1. Academic, New York (1979)
94. Maybeck, P.S.: Stochastic Models, Estimation and Control, Volume 2. Academic, New York (1982)
95. Mayne, D.Q.: A solution of the smoothing problem for linear dynamic systems. Automatica **4**(2), 73–92 (1966)
96. McElhoe, B.A.: An assessment of the navigation and course corrections for a manned flyby of mars or venus. IEEE Trans. Aerosp. Electron. Syst. **AES-2**(4), 613–623 (1966)
97. Metropolis, N., Rosenbluth, A.W., Rosenbluth, M.N., Teller, A.H., Teller, E.: Equation of state calculations by fast computing machines. J. Chem. Phys. **21**, 1087–1092 (1953)
98. Meucci, A.: Review of statistical arbitrage, cointegration, and multivariate Ornstein-Uhlenbeck. Working paper, symmys.com, January (2010)
99. Meyer, R., Yu, J.: BUGS for a Bayesian analysis of stochastic volatility models. Econ. J. **3**, 198–215 (2000)

100. Moler, C., van Loan, C.: Nineteen dubious ways to compute the exponential of a matrix, twenty-five years later. SIAM Review, 45(1), 2003.
101. Naesseth, C.A., Lindsten, F., Schön, T.B.: Nested sequential Monte Carlo methods. In: Proceedings of the 32nd International Conference on Machine Learning (2015)
102. Nelson, C.R., Siegel, A.F.: Parsimonious modeling of yield curves. J. Bus. **60**(4), 473–489 (1987)
103. Ntzoufras, I.: Bayesian Modeling Using WinBUGS. Wiley Series in Computational Statistics. Wiley, New York (2009)
104. Øksendal, B.: Stochastic Differential Equations: An Introduction with Applications, 6th edn. Universitext. Springer, Berlin (2010)
105. Omori, Y., Chib, S., Shephard, N., Nakajima, J.: Stochastic volatility with leverage: fast likelihood inference. August (2004)
106. Orhan, E.: Particle filtering. August (2012). http://www.cns.nyu.edu/~eorhan/notes/particle-filtering.pdf,
107. Pearlman, J.G.: An algorithm for the exact likelihood of a high-order autoregressive-moving average process. Biometrika **67**(1), 232–233 (1980)
108. Pitt, M.K., Shephard, N.: Filtering via simulation: auxiliary particle filters. J. Am. Stat. Assoc. **94**(446), 590–599 (1999)
109. Pitt, M.K., Malik, S., Doucet, A.: Simulated likelihood inference for stochastic volatility models using continuous particle filtering. Ann. Inst. Stat. Math. **66**, 527–552 (2014)
110. Platen, E., Heath, D.: A Benchmark Approach to Quantitative Finance. Springer Finance. Springer, Berlin (2006)
111. Ritter, C., Tanner, M.A.: Facilitating the Gibbs sampler: the Gibbs stopper and the Griddy-Gibbs sampler. J. Am. Stat. Assoc. **87**(419), 861–868 (1992)
112. Rubin, D.B.: Comment: a noniterative sampling/importance resampling alternative to the data augmentation algorithm for creating a few imputations when fractions of missing information are modest: the SIR algorithm. J. Am. Stat. Assoc. **82**(398), 542–543 (1987)
113. Ruiz, E.: Quasi-maximum likelihood estimation of stochastic volatility models. J. Econ. **63**(1), 289–306 (1994)
114. Sage, A.P., Melsa, J.L.: System Identification. Mathematics in Science and Engineering, vol. 80. Elsevier, Amsterdam (1971)
115. Särkkä, S.: Bayesian Filtering and Smoothing. Cambridge University Press, Cambridge (2013)
116. Sayed, A.H.: Adaptive Filters. Wiley-Interscience, New York (2008)
117. Shiryaev, A.N.: Stochastic equations of non linear filtering of Markovian jump processes. Problemy Peredaci Informacii **3**, 3–32 (1966)
118. Shreve, S.E.: Stochastic Calculus for Finance. Volume II: Continuous-Time Models. Springer, Berlin (2004)
119. Simon, D.: Optimal State Estimation: Kalman, H_{∞}, and Nonlinear Approaches. Wiley, New York (2006)
120. Smith, A.F.M., Gelfand, A.E.: Bayesian statistics with tears: a sampling-resampling perspective. Am. Stat. **46**(2), 84–88 (1992)
121. Smith, G.L., Schmidt, S.F., McGee, L.A.: Application of statistical filter theory to the optimal estimation of position and velocity on board a circumlunar vehicle. NASA technical report R-135, National Aeronautics and Space Administration (NASA) (1962)
122. Spiegelhalter, D., Thomas, A., Best, N., Gilks, W.: BUGS 0.5: Bayesian inference Using Gibbs Sampling Manual (version ii). MRC Biostatistics Unit, Institute of Public Health, Robinson Way, Cambridge CB2 2SR, August (1996)
123. Stratonovich, R.L.: Optimum nonlinear systems which bring about a separation of a signal with constant parameters from noise. Radiofizika **2**(6), 892–901 (1959)
124. Stratonovich, R.L.: Application of the Markov processes theory to optimal filtering. Radiofizika **5**(11), 1–19 (1960)
125. Tanner, M.A., Wong, W.H.: The calculation of posterior distributions by data augmentation. J. Am. Stat. Assoc. **82**(398), 528–540 (1987)

126. Taylor, S.J.: Financial returns modelled by the product of two stochastic processes, a study of daily sugar prices. In: Time Series Analysis: Theory and Practice, pp. 203–226. North-Holland, Amsterdam (1982)
127. Vega, L.R., Rey, H.: A Rapid Introduction to Adaptive Filtering. Springer Briefs in Electrical and Computer Engineering. Springer, Berlin (2013)
128. Wan, E.A., van der Merwe, R.: The unscented Kalman filter for nonlinear estimation. In: The IEEE Adaptive Systems for Signal Processing, Communications and Control Symposium (2000)
129. Welch, G., Bishop, G.: An introduction to the Kalman filter. Technical Report TR 95-041, Department of Computer Science, University of North Carolina at Chapel Hill, Chapel Hill, NC 27599-3175, July (2006). http://www.cs.unc.edu/~welch/media/pdf/kalman_intro.pdf
130. West, M., Harrison, J.: Bayesian Forecasting and Dynamic Models, 2nd edn. Springer, Berlin (1997)
131. Wiener, N.: The interpolation, extrapolation and smoothing of stationary time series. Report of the services 19, research project dic-6037, MIT, February (1942)
132. Wiener, N.: Extrapolation, Interpolation, and Smoothing of Stationary Time Series. MIT Press, Cambridge (1949)
133. Wu, Y., Hu, D., Wu, M., Hu, X.: A numerical-integration perspective on Gaussian filters. IEEE Trans. Signal Process. **54**(8), 2910–2921 (2006)
134. Young, P.C.: Recursive Estimation and Time-Series Analysis: An Introduction. Springer, Berlin (1984)
135. Yu, J.: On leverage in a stochastic volatility model. J. Econ. **127**, 165–178 (2005)
136. Zakai, M.: On the optimal filtering of diffusion processes. Zeitschrift für Wahrscheinlichkeitstheorie und Verwandte Gebiete **11**(3), 230–243 (1969)
137. Zeng, Y., Wu, S. (eds.): State-Space Models: Applications in Economics and Finance. Statistics and Econometrics for Finance. Springer, Berlin (2013)
138. Zivot, E., Wang, J.: Modeling Financial Time Series with S-PLUS. Springer, Berlin (2006)

Chapter 29
Using Python to Analyse Financial Markets

Saeed Amen

Abstract In this chapter we discuss the benefits of using Python to analyse financial markets. We discuss the parallels between the stages involved in solving a generalised data science problem, and the specific case of developing trading strategies. We outline the general stages of developing a trading strategy. We briefly describe how open source Python libraries finmarketpy, findatapy and chartpy aim to tackle each of these specific stages. In particular, we discuss how abstraction can be used to help generate clean code for developing trading strategies, without the low level details of data collection and data visualisation. Later, we give Python code examples to show how we can download market data, analyse it, and how to present the results using visualisations. We also give an example of how to implement a backtest for a simple trend following trading strategy in Python using finmarketpy.

29.1 Why Analyse Markets in Python?

When analysing financial markets, traders are often faced with conflicting objectives. Furthermore, these objectives can vary significantly between traders. For high frequency traders, it is crucial that they can quickly digest market data and execute trades based on that analysis as soon as possible. This necessitates the use of languages such as Java or C++. Furthermore, in some cases they might turn to more specialised programming techniques such as using GPU or FPGA. For traders active at lower frequencies, the speed of execution is less important. However, what is common amongst all traders is that they all need to develop trading strategies before they put them into production. The panacea for traders is a language which allows them to quickly develop a trading strategy and put it into production. C++ could be one candidate? It compiles down to machine code, however, it can be time consuming for implementation. R is popular amongst statisticians and boasts a very wide array of statistical libraries. However, it is slow and also not ideal for the implementation of large scale systems.

S. Amen (✉)
Cuemacro Limited, Level39, One Canada Square, London E14 5AB, UK
e-mail: saeed@cuemacro.com

M. Ehrhardt et al. (eds.), *Novel Methods in Computational Finance*,
Mathematics in Industry 25, DOI 10.1007/978-3-319-61282-9_29

In this context, Python can be seen as a language of compromises. First, it is a general purpose language, like C++ or Java (unlike R). It is obviously not as quick as C++ or Java, hence is not likely to be a candidate for a high frequency trading production system. However, it is possible to engineer large libraries in Python, given it has lots of features that implement the object oriented paradigm of coding. Python can call C++ libraries, if there are specific bottlenecks in your code that need to be very fast. There's also the Cython[5] library, which enables you to write Python-like code which compiles down into machine code. The cloud has also brought down the cost of computation, so if we want to speed up our Python code and we can parallelise it, it is relatively inexpensive to throw more cores and memory at it. Speed is not purely a matter of execution but also the time it takes to develop an idea. From the perspective of development time, Python is a "quick" language.

29.2 Python Data Analysis Ecosystem

Another benefit of using Python for doing market data analysis is that there is also a rich ecosystem of Python libraries that deal with data analysis. Here we describe some of the most well known libraries used in Python based data science, beginning with the SciPy stack.

29.2.1 Core Libraries (SciPy Stack)

- IPython [9]—Interactive console for Python (part of Jupyter—which also supports many other languages)
- NumPy [11]—Adds support for high level efficient operations on matrices, which are much quicker than using for loops in Python
- pandas [12]—Time series library, which grew out of work at AQR, which provides common operations for time series, such as aligning, joining etc.
- SciPy library [17]—Operations for numerical analysis and optimisation
- Sympy [18]—For symbolic mathematical operations

29.2.2 Visualisation

- bokeh [3]—JavaScript based library for graphics, similar to plotly.
- matplotlib [10]—Most well known Python based visualisation library. Whilst the interface can seem complex, it is a very flexible library.

- plotly [13]—JavaScript based visualisation library, which also allows sharing of charts to the web or in a private environment. Also accessible from many other languages, including R and Matlab.

29.2.3 Machine Learning

- PyMC3 [14]—Probabilistic programming in Python
- scikit-learn [16]—Most well known library for Python based machine learning
- TensorFlow [19]—Library for machine learning, which has been open sourced by Google

29.2.4 Front-Ends

- Flask [8]—Micro web framework to provide a relatively straightforward front end for Python computations
- xlwings [20]—Uses Excel as a front end for Python based computation

29.2.5 Databases and Market Data

- arctic [1]—Man-AHL's open sourced Python library provides a front end for efficient storing of pandas dataframes in MongoDB
- BLPAPI [2]—Bloomberg's Open Source API for accessing Bloomberg market data
- Quandl [15]—Open source market data provider

29.3 Parallels Between Solving Data Science Problems and Developing Trading Strategies

If we think of any data science problem, it tends to involve several steps. Below, we outline these steps albeit from the perspective of developing a trading strategy.

- Step 1—The first step is in a sense the most important. This involves formulating our hypothesis. In the case of developing a trading strategy, this involves brainstorming to think of a trading rationale. Is our trading strategy based on some sort of traditional factor, such as trend following (and if so, what is the rationale for this factor)? Are we trying to model a certain trader behaviour, such as specific flows? We can think of the rationale as effectively pruning the search space of possible trading strategies. A well thought out rationale can also help to reduce the chances that our final result is the result of excessive data mining (where "help" is the key word!). In practice, coming up with a hypothesis, is something where real world knowledge of markets and a deep experience of what price action can mean is most important.

- Step 2—Once we have our hypothesis, we need data. Typically, we need price data at the very least to compute the returns of the market asset we wish to trade. In many cases, we need other data to generate a trading signal. This can range from fundamental data on the economy to more unusual data sources such as news data and social media.
- Step 3—With both a hypothesis and data, we can now create our trading signal and also backtest our trading rule if we are developing a systematic trading rule. This involves looking at historical data to see how well our trading signal performed. We can split up our analysis between in-sample and out-of-sample sections Alternatively, we might be doing market analysis to inform our decision a discretionary trading decision.
- Step 4—Lastly, once we have completed our backtest, we need to present our results, in written form, along with some data visualisation. A well thought out visualisation can tell us more about any trading rule than a massive table of numbers. Furthermore, the importance of visualisation is not purely confined to the research stage. Understanding how our trading model is performing live is key in helping us to risk manage it, and make adjustments if necessarily over time.

29.4 Building a Python Markets Ecosystem

Over a year ago, I started a Python based library, pythalesians to facilitate my own analysis of markets, which centres around building systematic trading strategies. I eventually open sourced that library, and subsequently split it into several more specialised rewritten libraries listed below. In general, smaller more specialised libraries are more successful as open source projects. It is easier for potential users to understand what a library does when it is smaller and it is easier to identify how each library helps with solving each stage of the data science problem. When a library is very large and encompasses too much functionality, it can become difficult to understand precisely what it does. It is also easier to find contributors to an open source project when its purpose is better defined. The libraries are built upon many other open source Python libraries in particular from the SciPy stack.

- finmarketpy (for backtesting)[7]
- findatapy (for collecting market data)[6]
- chartpy (for data visualisation) [4]

29.5 Designing a Financial Market Research Platform: Abstraction Is the Key

Whilst I've extolled the virtues of Python, for having a rich data science ecosystem, in many cases we can have many choices for which underlying library to use. For example, if we wish to visualise data, we could for example use Plotly, Bokeh or Matplotlib (and this is a curtailed list). Each of these libraries has a different API. If we want to switch between these libraries, we would need to rewrite all our code. The same is true if we are downloading data from multiple sources, each data provider has a different API. If we end up using multiple APIs for visualisation and data collection, our code could become very messy indeed.

Furthermore, it will make it difficult to concentrate on what is the most important part of our problem, creating a trading signal. In finmarketpy, findatapy and chartpy, I've used abstraction to hide away the low level APIs for visualisation and data collection. In their place, I've create a common higher level API. Hence, to the end user, downloading data from Bloomberg or Quandl should look very similar, save for the change of a single keyword. Hence, it allows the trader to concentrate on the more pressing issue of developing a trading algorithm, rather than fiddling around with lower level details. This approach also makes it easier to maintain, in particular when we want to add new data sources or ways of visualising data.

29.6 Event Driven Code for Backtesting or Not?

Furthermore, for common tasks such as backtesting, I've also created templates that make it quicker to generate new ideas, simply define the signal and the assets which are traded. When creating a backtesting environment, the key question is whether you want to make it event driven or not. By event driven, we mean that every new tick of market data triggers a computation which decides whether or not to execute a trade. From a production perspective, this is preferable, because we can use the same code for production or backtesting. If we are simply using a system to research a trading strategy (or trading at low frequencies), it is possible to adopt a simpler approach which involves collecting all our signal data at the outset and all our return data, and multiplying this data together to generate a historical backtest. This is the approach I've adopted for finmarketpy. In languages like Python, there will also be significant speed increase in doing this, given we can vectorise our code more easily in this approach, exploiting libraries like NumPy. Event driven code, by contrast would be much slower in a language such as Python.

29.7 Collecting Market Data and Visualising

We now demonstrate how we can use Python to collect market data using findatapy, using several external sources including Bloomberg.

We use pandas dataframes as our structures to hold the time series data. In our first example, we collect daily data from Quandl for S&P500, which is a free data source. The Market class acts as an interface to lower level market data APIs. We construct a MarketDataRequest object to describe the nature of the market data we want to download.

```
from findatapy.market import MarketDataRequest,
  Market, MarketDataGenerator
market = Market(market_data_generator
                    =MarketDataGenerator())

md_request = MarketDataRequest(
  start_date="01␣Jan␣2005",
  data_source='quandl',
  tickers=['S&P500'],
  fields=['close'],
  vendor_tickers=['YAHOO/INDEX_GSPC'],
  vendor_fields=['Close'])

df_sp = market.fetch_market(md_request)
```

We now repeat the exercise for Bloomberg for S&P500, which requires a valid Bloomberg data license.

```
md_request = MarketDataRequest(
  start_date="01␣Jan␣2005",
  data_source='bloomberg',
  tickers=['S&P500'],
  fields=['close'],
  vendor_tickers=['SPX␣Index'],
  vendor_fields=['PX_LAST'])

df_sp_bbg = market.fetch_market(md_request)
```

As we can see that the code is virtually identical in both cases, the only difference are the vendor specific tickers the data source keyword. We can also download tick data, this time from a retail FX broker, using similar code.

```
md_request = MarketDataRequest(
  start_date='14␣Jun␣2016',
  finish_date='15␣Jun␣2016',
  category='fx',
```

```
  fields=['bid', 'ask'],
  freq='tick',
  data_source='dukascopy',
  tickers=['EURUSD'])

df_retail = market.fetch_market(md_request)
```

Now that we have seen how we can download market data, we discuss ways in which we can analyse and plot this data. Here we show how we can use Python to visual time series. In this example, we collect intraday USD/JPY data from Bloomberg and also the event times for the US employment report for the past few months from Bloomberg.

```
import pandas, datetime
from datetime import timedelta
start_date = datetime.date.today()-timedelta(days=180)

md_request = MarketDataRequest(
  start_date=start_date,
  category='fx',
  freq='intraday',
  data_source='bloomberg',
  tickers=['EURUSD'],
  fields=['close'],
  cache_algo='internet_load_return')

df_fx_bbg = market.fetch_market(md_request)

from findatapy.timeseries import Calculations
calc = Calculations()
df_fx_bbg = calc.calculate_returns(df_fx_bbg)

# fetch NFP times from Bloomberg
md_request = MarketDataRequest(
  start_date=start_date,
  finish_date=finish_date,
  category="events",
  freq='daily',
  data_source='bloomberg',
  tickers=['NFP'],
  fields=['release-date-time-full'],
  vendor_tickers=['NFP␣TCH␣Index'],
  cache_algo='internet_load_return')

df_event_times = market.fetch_market(md_request)
```

```
df_event_times =
  pandas.DataFrame(index=df_event_times
        ['NFP.release-date-time-full'])
```

We now have the raw data available, the next step is to do an event study, where we analyse the moves in USD/JPY during the 3 h following the release of the US employment report. We also do the same for intraday volatility

```
from finmarketpy.economics import EventStudy

df_event = EventStudy().
  get_intraday_moves_over_custom_event(df_fx_bbg,
                        df_event_times)
df_event['Avg'] = df_event.mean(axis=1)
```

Finally, we can plot (see Fig. 29.1) the various event study time series, using the popular matplotlib library. Alternatively, if we were displaying on a webpage, we might prefer the plotly library. We can see that the bulk of the volatility occurs over the actual data release and quickly dissipates.

```
from chartpy import Chart, Style

style = Style()
style.scale_factor = 3
style.file_output = 'eurusd-nfp.png'
style.title = 'EURUSD␣spot␣moves␣over␣recent␣NFP'

style.color = 'Blues'; style.color_2 = []
style.y_axis_2_series = []
style.display_legend = False

style.color_2_series = [df_event.columns[-2],
                        df_event.columns[-1]]
style.color_2 = ['red', 'orange']
style.linewidth_2 = 2
style.linewidth_2_series = style.color_2_series

chart = Chart(engine='matplotlib')
chart.plot(df_event * 100, style=style)
```

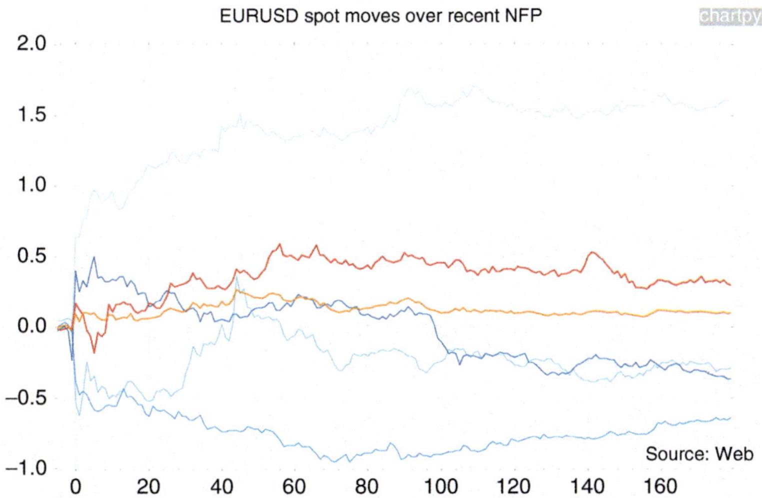

Fig. 29.1 EUR/USD moves during 3 h following recent the US employment reports

Another popular analysis, involves understanding the seasonality of an asset. We can use S&P500, which we downloaded earlier, together with the Seasonality class which is part of finmarketpy (Fig. 29.2).

```
from finmarketpy.economics import Seasonality
df_ret = calc.calculate_returns(df_sp)
day_of_month_seasonality =
  Seasonality().bus_day_of_month_seasonality(df_ret,
                                   partition_by_month=False)
day_of_month_seasonality =
  calc.convert_month_day_to_date_time(
                                day_of_month_seasonality)

style = Style()
style.date_formatter = '%b'
style.title = 'S&P500␣seasonality'
style.scale_factor = 3
style.file_output = "sp-seasonality.png"

chart = Chart()
chart.plot(day_of_month_seasonality, style=style)
```

Fig. 29.2 S&P500 plotted on a seasonal basis

29.8 Visualising a Volatility Surface

Sometimes, more complicated plots might be relevant. One example of this can be seen with FX volatility. Implied volatility is quoted for a range of both strike and tenor combinations. An efficient way to plot this is using a surface. Below, we show how to download FX volatility surface data from Bloomberg and how to plot it (see Fig. 29.3) in chartpy using a plotly backend.

```
md_request = MarketDataRequest(
  start_date=datetime.datetime.now()-timedelta(days=1),
  data_source='bloomberg',
  cut='NYC',
  category='fx-vol-market',
  tickers=['EURUSD'])

df_vol = market.fetch_market(md_request)

from findatapy.market import FXVolFactory

df_vs = FXVolFactory().extract_vol_surface_for_date(
                                df_vol, 'EURUSD', -1)
```

```
from chartpy import Chart, Style

style = Style(title="EURUSD␣vol␣surface",
                    source="chartpy", color='Blues',
  file_output = 'eurusd-volsurface.png')

chart = Chart(df=df_vs, chart_type='surface',
                                        style=style)
chart.plot(engine='plotly')
```

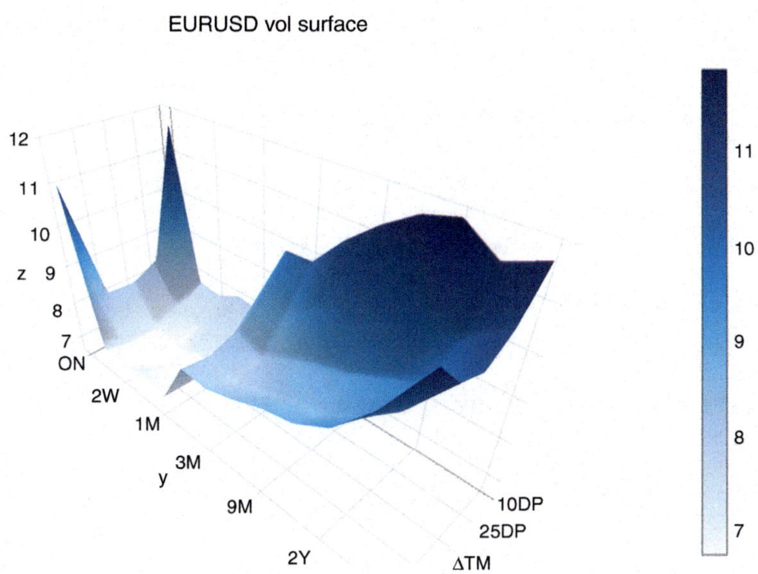

Fig. 29.3 EUR/USD implied volatility surface in mid-October 2016

29.9 Backtesting a Trading Strategy

In this section we show how to create a backtest for a basic trend following style strategy for FX. We extend the abstract class TradingModel to create our backtest, creating the TradingModelFXTrend_Example class. First, we do all the appropriate imports of other Python modules that we shall use later. In the init function, we define some parameters for where we want to output our results, the name of our strategy and also the underlying plotting engine we shall use for displaying the results.

```
import datetime

from findatapy.market import Market,
            MarketDataGenerator, MarketDataRequest
from finmarketpy.backtest import TradingModel,
                                 BacktestRequest
from finmarketpy.economics import TechIndicator

class TradingModelFXTrend_Example(TradingModel):
  def __init__(self):
    super(TradingModel, self).__init__()

    self.market = Market(market_data_generator
                         =MarketDataGenerator())
    self.DUMP_PATH = ''
    self.FINAL_STRATEGY = 'FX␣trend'
    self.SCALE_FACTOR = 1
    self.DEFAULT_PLOT_ENGINE = 'matplotlib'
    self.br = self.load_parameters()

    return
```

Now we define the parameters for the backtest including the start and end date. We also define leverage parameters here on the signal and portfolio level. For each asset we define a volatility target of 10%. The idea behind this is that we equalise the risk in each asset we are trading. If we do not do this, we are essentially allocating more risk in the higher volatility assets. We also apply a final volatility target for the whole portfolio too. In practice, given we don't know what future realised volatility will be, we are unlikely to hit our target exactly.

```
  def load_parameters(self):

    br = BacktestRequest()

    br.start_date = "04␣Jan␣1989"
    br.finish_date = datetime.datetime.utcnow()
```

```
    br.spot_tc_bp = 0.5
    br.ann_factor = 252
    br.plot_start = "01␣Apr␣2015"

    br.calc_stats = True
    br.write_csv = False
    br.plot_interim = True
    br.include_benchmark = True

    # have vol target for each signal
    br.signal_vol_adjust = True
    br.signal_vol_target = 0.1
    br.signal_vol_max_leverage = 5
    br.signal_vol_periods = 20
    br.signal_vol_obs_in_year = 252
    br.signal_vol_rebalance_freq = 'BM'
    br.signal_vol_resample_freq = None

    # have vol target for portfolio
    br.portfolio_vol_adjust = True
    br.portfolio_vol_target = 0.1
    br.portfolio_vol_max_leverage = 5
    br.portfolio_vol_periods = 20
    br.portfolio_vol_obs_in_year = 252
    br.portfolio_vol_rebalance_freq = 'BM'
    br.portfolio_vol_resample_freq = None

    # tech params
    br.tech_params.sma_period = 200

    return br
```

We load up all the market data next. For simplicity we shall be using spot data from Quandl. We do note however, that in practice, using spot data for calculation of FX returns is only an approximation given it doesn't include carry.

```
  def load_assets(self):
    full_bkt = ['EURUSD', 'USDJPY', 'GBPUSD', 'AUDUSD',
      'USDCAD', 'NZDUSD', 'USDCHF', 'USDNOK', 'USDSEK']

    basket_dict = {}

    for i in range(0, len(full_bkt)):
      basket_dict[full_bkt[i]] = [full_bkt[i]]
    basket_dict['FX␣trend'] = full_bkt
```

```
    br = self.load_parameters()
    self.logger.info("Loading␣asset␣data...")
    vendor_tickers = ['FRED/DEXUSEU', 'FRED/DEXJPUS',
      'FRED/DEXUSUK', 'FRED/DEXUSAL', 'FRED/DEXCAUS',
      'FRED/DEXUSNZ', 'FRED/DEXSZUS', 'FRED/DEXNOUS',
      'FRED/DEXSDUS']
    md_request = MarketDataRequest(
      start_date = br.start_date,
      finish_date = br.finish_date,
      freq = 'daily',
      data_source = 'quandl',
      tickers = full_bkt,
      fields = ['close'],
      vendor_tickers = vendor_tickers,
      vendor_fields = ['close'],
      cache_algo = 'internet_load_return')

    asset_df = self.market.fetch_market(md_request)
    spot_df = asset_df
    spot_df2 = None

    return asset_df, spot_df, spot_df2, basket_dict
```

Next, we calculate the signal. This involves calculating a 200D simple moving average. If spot is above the moving average it triggers a buy signal and if it is below, we define a sell signal. This is predefined signal, so it can be defined with very few lines of code.

```
  def construct_signal(self, spot_df, spot_df2,
                                   tech_params, br):

    tech_ind = TechIndicator()
    tech_ind.create_tech_ind(spot_df, 'SMA',
                                      tech_params)
    signal_df = tech_ind.get_signal()

    return signal_df
```

We also define a benchmark for our trading strategy which is simply being long EUR/USD in this case (mainly for simplicity).

```
  def construct_strategy_benchmark(self):
    tsr_indices = MarketDataRequest(
    start_date = self.br.start_date,
      finish_date = self.br.finish_date,
```

```
        freq = 'daily',
        data_source = 'quandl',
        tickers = ["EURUSD"],
        vendor_tickers=['FRED/DEXUSEU'],
        fields = ['close'],
        vendor_fields = ['close'],
        cache_algo = 'internet_load_return')

    df = self.market.fetch_market(tsr_indices)
    df.columns = [x.split(".")[0] for x in df.columns]

    return df
```

We can now kick off the actual calculation, by instantiating the TradingModelFXTrend_Example class and running a few commands to do the number crunching. We also plot some of the results.

```
if __name__ == '__main__':
  model = TradingModelFXTrend_Example()
  model.construct_strategy()
  model.plot_strategy_pnl()
```

We display some out the plot output in (see Fig. 29.4) from our backtest.

As a next step we do some sensitivity analysis using the TradeAnalysis class from finmarketpy. We examine how much transaction costs impact returns. For higher frequency strategies, transaction costs can make up a larger proportion of returns, given they trade more rapidly.

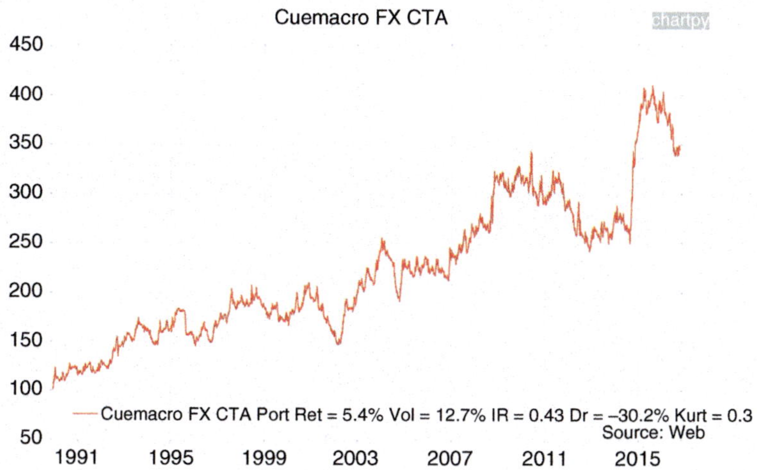

Fig. 29.4 Cumulative returns of FX trend following strategy

```
from finmarketpy.backtest import TradeAnalysis
ta = TradeAnalysis()
tc = [0, 0.25, 0.5, 0.75, 1, 1.25, 1.5, 1.75, 2.0]
ta.run_tc_shock(strategy, tc = tc)
```

We also apply a sensitivity analysis, to understand how removing volatility targeting impacts returns. We see in Fig. 29.5 that using volatility targeting significantly improves risk returns for this strategy.

```
parameter_list = [{'portfolio_vol_adjust': True,
  'signal_vol_adjust' : True},
  {'portfolio_vol_adjust': False, 'signal_vol_adjust'
                                            : False}]
pretty_portfolio_names = ['Vol␣target',
                                'No␣vol␣target']
parameter_type = 'vol␣target'

ta.run_arbitrary_sensitivity(strategy,
  parameter_list=parameter_list,
  pretty_portfolio_names=pretty_portfolio_names,
  parameter_type=parameter_type)
```

Whenever, you create a trading strategy, it can be useful to apply sensitivity analysis to our parameters, to try to understand how robust the strategy is. If a strategy only "works" for a very specific set of parameters, it could suggest that

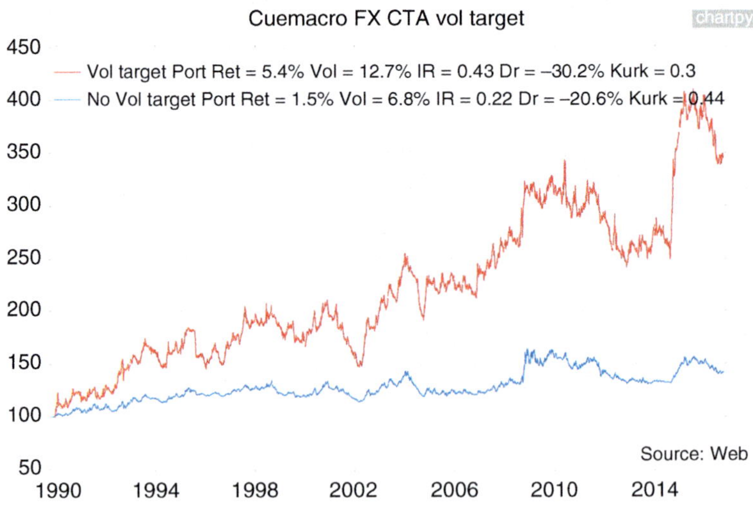

Fig. 29.5 Sensitivity analysis of FX trend depending on vol targeting

we have gone a bit overboard on data mining (unless we can find a very good explanation to why other parameters should not work).

29.10 Conclusions

We have discussed the various benefits of using Python to analyse financial markets. Whilst its speed of execution is slower than languages such as Java or C++, it has a rich ecosystem for data science. It is also quicker to prototype ideas in Python. More broadly we discussed the general steps we adopt when creating a trading strategy, which can be viewed as a data science problem. Later, we briefly discussed our approach to building an Python based ecosystem for developing trading strategies and analysing market data. Lastly, we gave some practical Python based examples showing how to download and analyse data. We also went through an example demonstrating how to create a simple trend following trading strategy in FX.

References

1. arctic: High performance datastore for time series and tick data. http://github.com/manahl/arctic
2. BLPAPI: Bloomberg market data library. https://www.bloomberglabs.com/api/libraries/
3. bokeh: Python interactive visualization library. http://bokeh.pydata.org/en/latest/
4. chartpy: Python data visualisation library. http://www.github.com/cuemacro/chartpy
5. Cython C: Extensions for python. http://www.cython.org
6. findatapy: Python finanical data library. http://www.github.com/cuemacro/findatapy
7. finmarketpy: Python financial trading library. http://www.github.com/cuemacro/finmarketpy
8. flask: Micro web framework. http://flask.pocoo.org
9. IPython: Enhanced interactive console. http://www.ipython.org
10. matplotlib: Python plotting library. http://www.matplotlib.org
11. NumPy: Package for scientific computing with Python. http://www.numpy.org
12. pandas: Python data analysis library. http://pandas.pydata.org
13. ploty: Collaboration platform for modern data science. http://plot.ly
14. PyMC3: Probabilistic programming in python. http://github.com/pymc-devs/pymc3
15. Quandl: Market data API. http://www.quandl.com/
16. scikit-learn: Machine learning in python. http://scikit-learn.org/stable/
17. SciPy library: Fundamental library for scientific computing. https://www.scipy
18. Sympy: Symbolic mathematics. http://www.sympy.org/en/index.html
19. TensorFlow: Open source library for machine intelligence. https://www.tensorflow.org
20. xlwings: Python for excel. http://www.xlwings.org/

Chapter 30
The STRIKE Computational Finance Toolbox

Christof Heuer, Pedro Pólvora, José Silva, Matthias Ehrhardt, Michael Günther, and E. Jan W. ter Maten

Abstract The STRIKE Computational Finance Toolbox (CFT) is one of the output results of the "ITN STRIKE—Novel Methods in Computational Finance" and is concerned with combining the research output of the network. For this purpose the implemented models (MATLAB and PYTHON) of the PhD students and postdocs are collected and user interfaces are developed for a convenient use of those programs. Two different interfaces have been implemented. The first interface combines those submissions, which are dealing with one-dimensional spatial models, e.g. a problem settings with one stock price. With the second user interface two-dimensional programs can be used, which consist of problem settings with two spatial variables, e.g. one stock price and it's volatility, or two stock prices.

30.1 Introduction

The structure of this chapter is as follows. In the first section we will discuss the user interface for the models with one spatial dimension. The spatial dimension in these cases is the price of a stock price S. The options in these cases include European Call and Put options as well as American options. The second section is concerning itself with the user interface for models with two spatial dimensions, which means that not necessary all spatial dimensions are stock prices. In each of those two sections we describe the user interface in detail. After this we introduce each in the Computational Finance Toolbox included model and the implemented numerical solvers for these models Finally, we will conclude and give an outlook.

C. Heuer (✉) • J. Silva • M. Ehrhardt • M. Günther • E.J.W. ter Maten
Lehrstuhl für Angewandte Mathematik/Numerische Analysis, Bergische Universität Wuppertal, Wuppertal, Germany
e-mail: heuer.chr@googlemail.com; jose.pedro.moreira.da.silva@gmail.com; ehrhardt@math.uni-wuppertal.de; guenther@math.uni-wuppertal.de; termaten@math.uni-wuppertal.de

P. Pólvora
Department of Applied Mathematics and Statistics, Comenius University Bratislava, Bratislava, Slovakia
e-mail: pedro.polvora@gmail.com

M. Ehrhardt et al. (eds.), *Novel Methods in Computational Finance*, Mathematics in Industry 25, DOI 10.1007/978-3-319-61282-9_30

30.2 Models with One Spatial Variable

In this section we want to show and explain the structure of the user interface, which combines the research output of models containing one spatial variable. The included models are the linear Black–Scholes (BS) model, non-linear Black–Scholes models as well as a model with coupled liquid an illiquid states of a market. The included option types are European as well as American options. Figure 30.1 shows the results of the calculation of an American Call in the BS model, when using the front-fixing method, as an example. The figure also shows the complete user interface for methods with one spatial dimension. The following sections are organised as follows. First, we describe the Computational Finance Toolbox user interface for models with one spatial dimension in detail. For this, we describe each part of the user interface individually. Afterwards, we discuss the numerical methods, separated by model. For this, we show the differential equations of the models and briefly explain the numerical methods the user can choose in each setting. We additionally mention manuscripts for further reading regarding the numerical schemes.

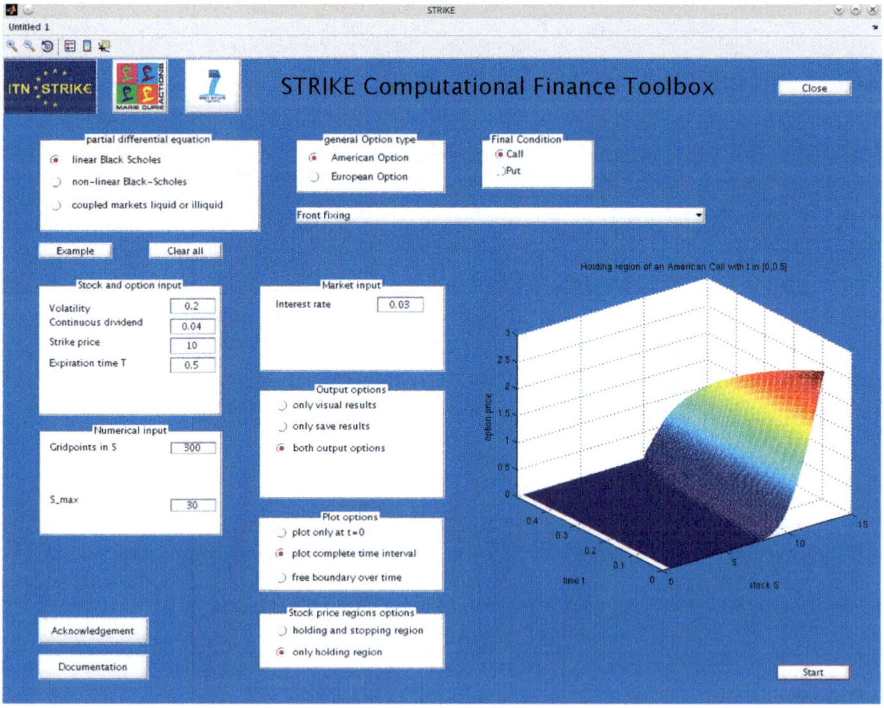

Fig. 30.1 STRIKE toolbox for one spatial variables

30.2.1 General Description of the User Interface

In this section we describe the settings of the Computational Finance Toolbox (CFT) for models with one-dimensional spatial domain. Figure 30.1 shows an example setting, where a front fixing technique is used to approximate the value of an American option. We have a closer look at each selection field as well as the input fields of the user interface. Figure 30.1 shows the complete CFT after an American Call is calculated. In the following, we describe the separate parts of the CFT for problem settings with one spatial domain individually.

At the top left of the user interface we can see the basic configuration of the CFT for problem settings with one spatial variable. In Fig. 30.2 the user can choose from three different general model settings, namely the linear BS model, a non-linear BS model as well as a model, where liquid and illiquid markets are coupled. On the right hand side of this field, the user can choose when the option can be exercised, see Fig. 30.3. If the linear BS model is chosen, it is possible to choose between American and European options. In the case of non-linear BS models and the model distinguishing between a liquid and illiquid market only European options are possible.

Next to the selection of the expiration rights of the option it is possible to choose whether the option should be a Call or a Put, see Fig. 30.4. On the right side of the choice of the final condition it is possible to choose, whether the market is in a liquid or illiquid state, see Fig. 30.5. This choice is only needed when a coupled liquid and illiquid market is chosen. It is thus invisible in all other cases.

Underneath the choice of the final condition or the liquidity of the market the numerical schemes are chosen, see Fig. 30.6. The number of choices is depending

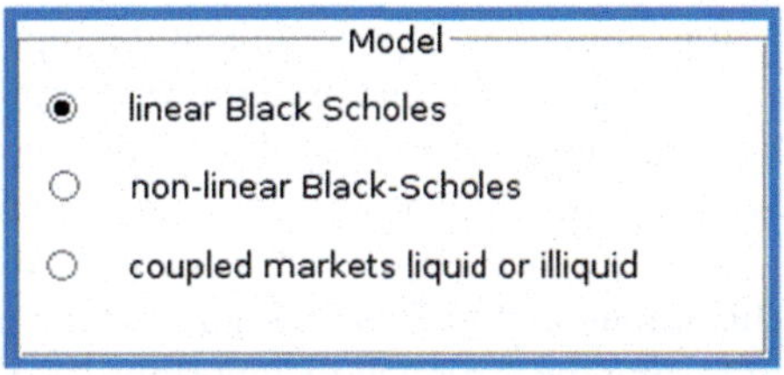

Fig. 30.2 Model selection

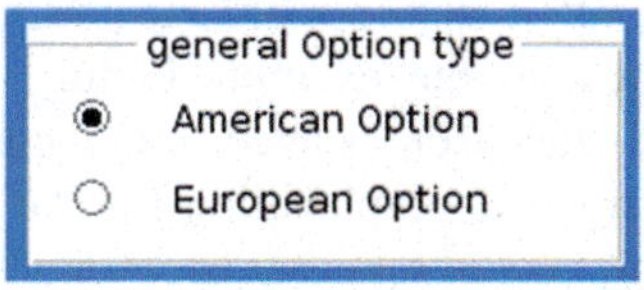

Fig. 30.3 Exercise rights

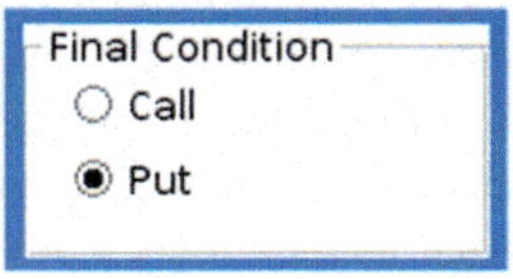

Fig. 30.4 Final condition

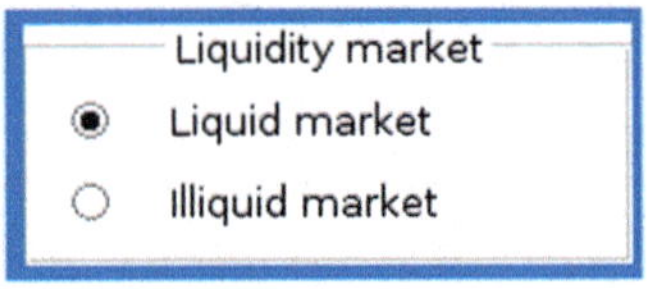

Fig. 30.5 Liquidity of market

Fig. 30.6 Popup menu for the numerical method

Fig. 30.7 Example and clear all button

on the choice of the discussed model and the exercise rights of the option. For an American option in the linear BS setting the only possibility is to use front fixing, whereas for the non-linear BS models the user can choose from different numerical schemes. The possible selections are discussed in Sects. 30.2.2–30.2.4.

Since this user interface is not only meant for people who already know every presented single financial model or numerical scheme, it is important to present the user the opportunity to use pre-defined test variables in the calculation. This can be done using the Example button, see Fig. 30.7, which is located underneath the field for choosing the model. This way it is possible for the user to get familiar with the models as well as numerical schemes by seeing directly possible input variables and their impact.

It is important to mention that the input variables are checked for each problem setting and the program only shows the error message of the firstly determined error. This error message then explains which bounds the parameter has to fulfill. If a user would not have the possibility to use and then change pre-defined input parameters, it might be that the user chooses several parameters out of their bounds at once. This might lead to frustration as a chain of error messages might show, each after correcting the error and then pushing the start again.

It is also possible to clear all input variables with the according button right next to the example button, see Fig. 30.7. Since we want to introduce models to the user as well, we included a Documentation button. Pushing this button shows an explanation for the chosen combination of the model, option type and numerical scheme. The Acknowledgement button is located underneath the field for the numerical input and above the Documentation button. Both of these buttons are shown in Fig. 30.8.

After choosing the general settings, so the model, the option type as well as the numerical scheme, one typically has to specify the input variables. This could be done, as mentioned above, with a default example. But if the user is already familiar

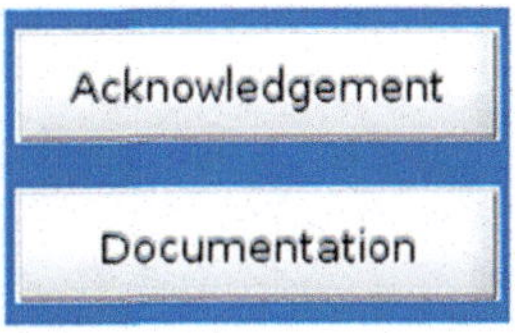

Fig. 30.8 Acknowledgement and documentation button

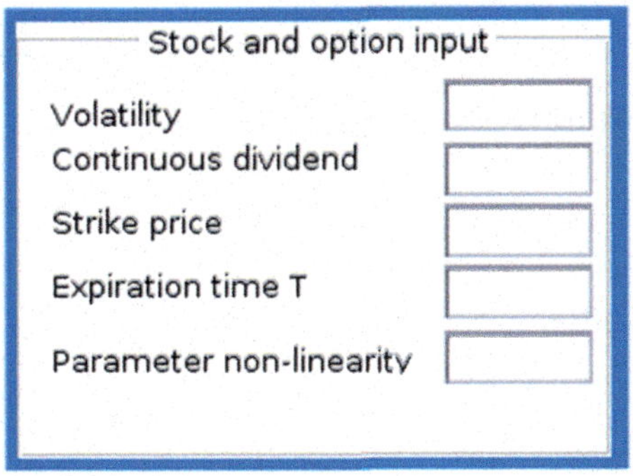

Fig. 30.9 Stock and option input

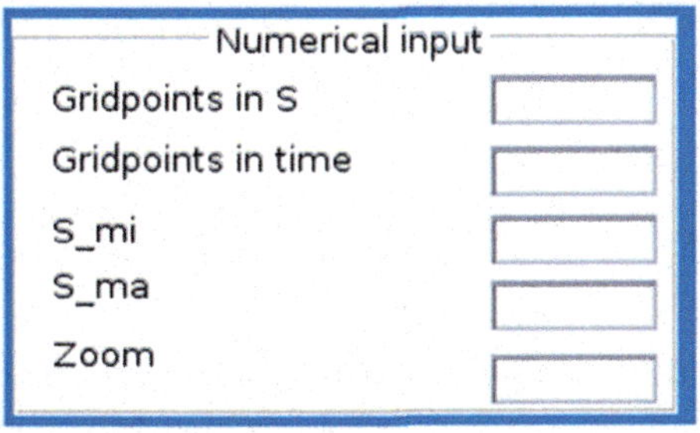

Fig. 30.10 Numerical input

with the chosen combination of model, option type and numerical solver, the user can directly choose suited input variables. The input field for the stock as well as the option parameters, see Fig. 30.9, is located underneath the example button. These input variables can be the volatility/continuous dividend of the stock price or the strike price/expiration date of the chosen option type. Below the stock and option input the user has to define the numerical input, such as the number of grid points or minimal/maximal values of the stock price, see Fig. 30.10. Finally, also values of the market, such as the risk-free interest rate or the parameter for the change of the market from liquid to illiquid and backwards have to be defined on the right hand side of the stock and option input. The field for the market input can be seen in Fig. 30.11.

Finally, it is necessary to determine which output is wanted. It is possible to only save the results, only show visual results, or both. If a non-equidistant grid is used, it is also the opportunity to only show the transformation grid and thus focus on the structure of the grid-points. With this output option the user can see how much the grid concentrates at the strike price. The output options are located below the market input and can be seen in Fig. 30.12. If the user chooses a visualisation of the option

Fig. 30.11 Market input

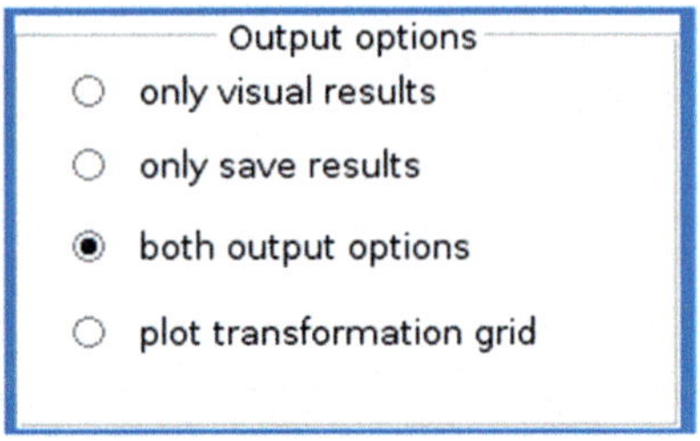

Fig. 30.12 Output options

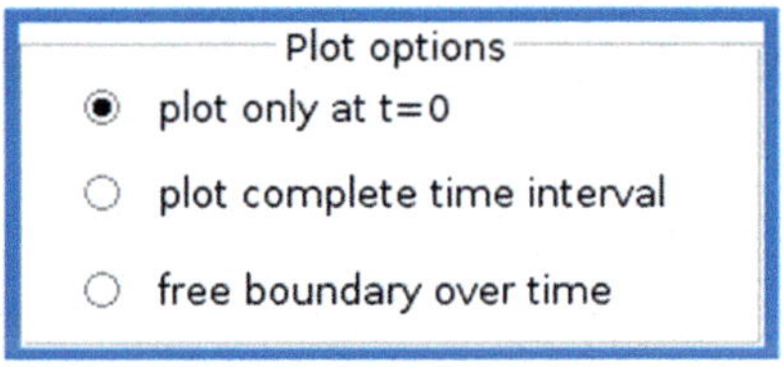

Fig. 30.13 Plot options

price as output, it is also possible to determine whether the results should be shown only for $t = 0$ or for the complete time domain. These options are placed below the output option field, see Fig. 30.13.

30.2.2 Numerical Methods for the Linear Black–Scholes Model

In this section we want to introduce the first possible choice for the model, namely the linear BS model. We first define the partial differential equation of the model. After that we discuss which numerical schemes are possible for which option type. In the linear BS model the partial differential equation (PDE) is given by

$$\frac{\partial V}{\partial t} + \frac{\sigma^2 S^2}{2} \frac{\partial^2 V}{\partial S^2} + rS \frac{\partial V}{\partial S} - rV = 0,$$

see e.g. [3, 12]. The volatility of the stock price $S \geq 0$ is given by $\sigma \geq 0$. The risk-free interest rate is determined by $r \geq 0$. The Final condition of this equation is given by

$$V(S,T) = \begin{cases} \max(S-K,0) & \text{for a Call} \\ \max(K-S,0) & \text{for a Put} \end{cases}$$

with strike price $K > 0$. It is possible to approximate European options as well as American options through the Computational Finance Toolbox.

30.2.2.1 European Options Using Alternating Direction Explicit Method

European Calls can be calculated via Alternating Direction Explicit (ADE) method for an equidistant or non-equidistant grid. It is also possible to show the structure of the non-equidistant grid when choosing the respective method. The following description is taken from [1], only the format has been changed. The *Fichera Theory* focuses on the question of appropriate *boundary conditions* (BCs) for parabolic PDEs degenerating at the boundary. According to the sign of the *Fichera function* one can separate the outflow or inflow part of the solution at the boundary, i.e. it indicates whether one has to supply a BC at the degenerating boundary. It turned out to be very useful for establishing the well-posedness of initial boundary value problems for parabolic PDEs degenerating at the boundary.

We have considered an ADE method, that strongly uses boundary data in the solution algorithm and hence is very sensible to incorrect treatment of boundary conditions. We have implemented this ADE scheme for solving linear and nonlinear BS equations by treating the nonlinearity explicitly. The ADE scheme consists of two explicit sweeps. The 'sweeping procedure' is done from one boundary to another and vice versa. The final solution is defined as an average of these two sweeps after each time step.

ADE schemes are superior to other numerical methods due to their explicitness, stability and the property allowing for an easy parallelisation. Numerical analysis in sense of studying stability and consistency of this method is provided in [2].

For linear models we obtain

- an explicit and unconditionally stable method.
- stability + consistency of order $O\left(k^2 + h^2\right)$
  $\Rightarrow$ convergence of order $O\left(k^2 + h^2\right)$.

The computational effort using ADE schemes instead of implicit schemes for (non)linear models is highly reduced and stability properties are preserved. For results, see [2] (Figs. 30.14 and 30.15).

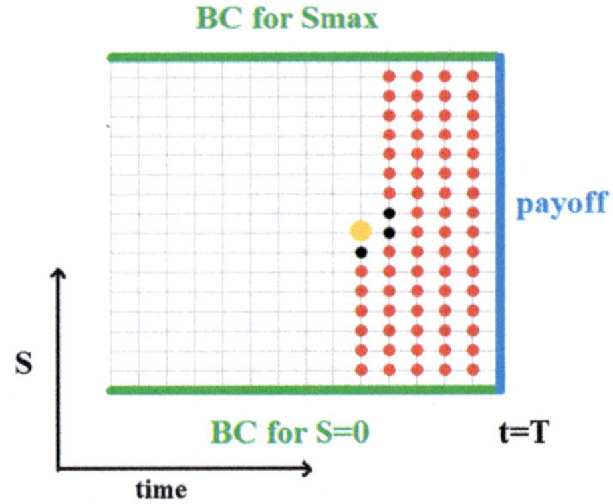

Fig. 30.14 Upward swing

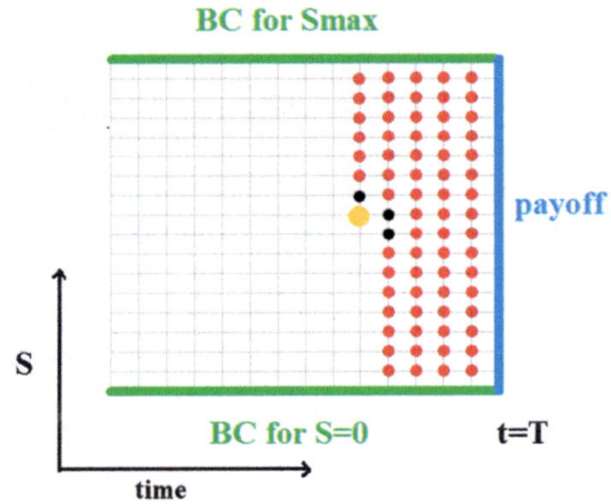

Fig. 30.15 Downward swing

30.2.3 Numerical Methods for Non-linear Black–Scholes Models

In this category two different models are discussed. The first model is the Frey and Patie model, the second model consists of the BS PDE with non-linear volatility. For both of these models we will discuss the included discretisation methods. For the Frey and Patie model, the ADE method is used once with an equidistant or non-equidistant grid. For the non-linear BS model with non-linear volatility we will discuss two different approaches, both using the Newton method.

30.2.3.1 Using the ADE Method for the Frey and Patie Model

The first non-linear Black–Scholes model we discuss is given by the Frey and Patie model,

$$\frac{\partial V}{\partial t} + \frac{1}{2}\frac{\sigma^2 S^2}{[1-\lambda(S)S]^2}\frac{\partial^2 V}{\partial S^2} = 0.$$

We consider a European Call option, so

$$V(S,T) = \max(S-K,0)$$

with strike price $K > 0$. Again, the description is taken from [1], only the format has been changed.

The *Fichera Theory* focuses on the question of appropriate *boundary conditions* (BCs) for parabolic partial differential equations (PDEs) degenerating at the boundary. According to the sign of the *Fichera function* one can separate the outflow or inflow part of the solution at the boundary, i.e. it indicates whether one has to supply a BC at the degenerating boundary. It turned out to be very useful for establishing the well-posedness of initial boundary value problems for parabolic PDEs degenerating at the boundary.

We have considered an ADE method, that strongly uses boundary data in the solution algorithm and hence is very sensible to incorrect treatment of boundary conditions. We have implemented this ADE scheme for solving linear and nonlinear BS equations by treating the nonlinearity explicitly. The ADE scheme consists of two explicit sweeps. The 'sweeping procedure' is done from one boundary to another and vice versa. The final solution is defined as an average of these two sweeps after each time step.

ADE schemes are superior to other numerical methods due to their explicitness, stability and the property allowing for an easy parallelisation. Numerical analysis in sense of studying stability and consistency of this method is provided.

For nonlinear models (Frey and Patie, Barles and Soner model)

- only conditional stability, but in each time step we only need to solve a *scalar* nonlinear equation.

The computational effort using ADE schemes instead of implicit schemes for (non)linear models is highly reduced and stability properties are preserved. For results see [2] (Figs. 30.16 and 30.17).

Fig. 30.16 Upward swing

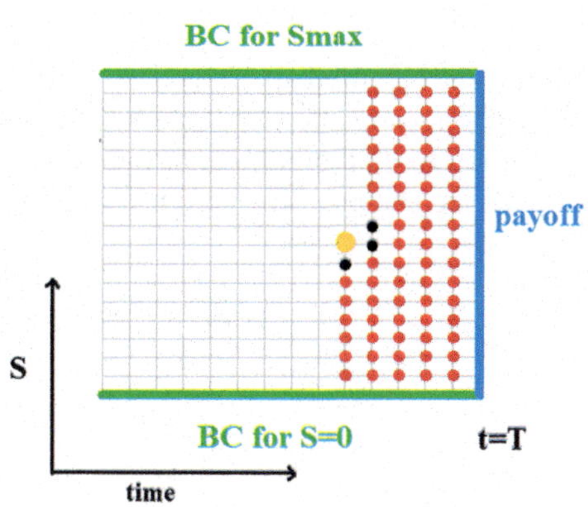

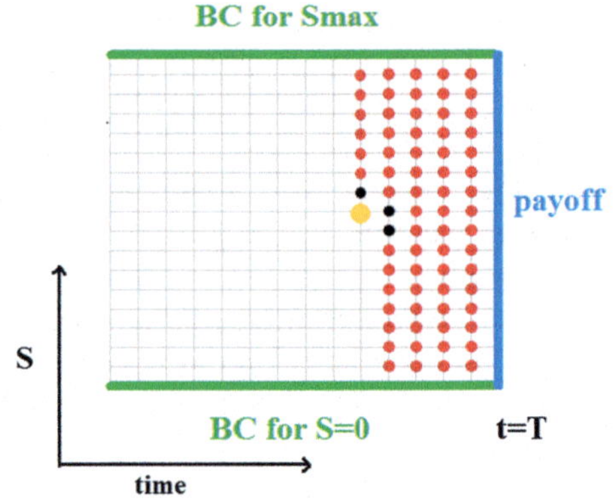

Fig. 30.17 Downward swing

30.2.3.2 Using Newton-Methods for the Non-linear Black–Scholes Model with Non-linear Volatility

The second choice in the partial differential equation field is given by the non-linear Black–Scholes model. For

$$\frac{\partial V}{\partial t} + \frac{\sigma(V)^2 S^2}{2}\frac{\partial^2 V}{\partial S^2} + rS\frac{\partial V}{\partial S} - rV = 0, \tag{30.1}$$

where the non-linearity is given by $\sigma(V) = \tilde{\sigma}(1 - \sin(V))$, one numerical scheme uses the central difference operator and implicit time discretisation and then Newton iteration to approximate the solution, whereas the other program first uses linearisation. Both numerical schemes can be used to approximate European Calls and Puts, so

$$V(S,T) = \begin{cases} \max(S-K,0) & \text{for a Call} \\ \max(K-S,0) & \text{for a Put.} \end{cases}$$

The following description is taken from [14], only the format has been changed.

- Newton's method approach 1:

  The idea of the first approach to Newton's method is to use finite difference method to do temporal and spatial discretizations on Eq. (30.1) with an implicit scheme. The problem will turn to solve the solution of a function $G(V)$ defined as

$$G(V^{n+1}) = V^n - H(V^{n+1})V^{n+1} = 0. \tag{30.2}$$

  To solve Eq. (30.2), one needs to calculate the Jacobian matrix of G for the updating of the approximate.

Table 30.1 The error between NM1 and NM2

$\Delta S = 0.1$	2×10^{-6}
$\Delta S = 0.05$	2×10^{-6}
$\Delta S = 0.01$	1×10^{-6}

- Newton's method approach 2:

  Another approach for Newton's method is to perform linearization first, and then do discretization. The idea is to consider the equation

$$F(V_t, V_S, V_{SS}, V) = V_t + \frac{1}{2}\sigma^2(V)\, S^2\, V_{SS} + rS\, V_S - rV = 0. \tag{30.3}$$

  Supposing that $(V_t^*, V_S^*, V_{SS}^*, V^*)$ is an approximate solution, the linearization of (30.3) leads to

$$\begin{aligned} 0 &= F(V_t^* + e_t, V_S^* + e_S, V_{SS}^* + e_{SS}, V + e) \\ &\approx F(V_t^*, V_S^*, V_{SS}^*, V^*) + \\ &\quad (\frac{\partial F}{\partial V_t^*}e_t + \frac{\partial F}{\partial V_S^*}e_S + \frac{\partial F}{\partial V_{SS}^*}e_{SS} + \frac{\partial F}{\partial V^*}e). \end{aligned} \tag{30.4}$$

  A discretization of the correction term e with an implicit scheme may be used to solve the value e and to do updating.

An example for calculating European Put options with transaction costs (Case I above) is provided here. We fix the time discretization $\Delta t = 0.001$ with given parameters

$$a = 0.001, r = 0.03, \tilde{\sigma} = 0.2, S_{max} = 10, K = 5.5, T = 1, tol = 10^{-6}.$$

We compare the results obtained by NM1, which is from approach 1, and NM2, which is from approach 2. The error is the maximum norm between the option values from both solvers at $t = 0$ (Table 30.1).

In the end, the aim is to find a robust strategy to speed up the whole computation. From the complexity analysis, we can understand that the most time-consuming part from NM1 is to calculate the Jacobian matrix, and similarly in NM2 we have to do calculations to obtain the matrix for updating (Fig. 30.18).

30.2.4 Numerical Methods for Coupled Liquid and Illiquid Market

The description of the numerical scheme as well as the used model are given in [13], where only the format has been changed.

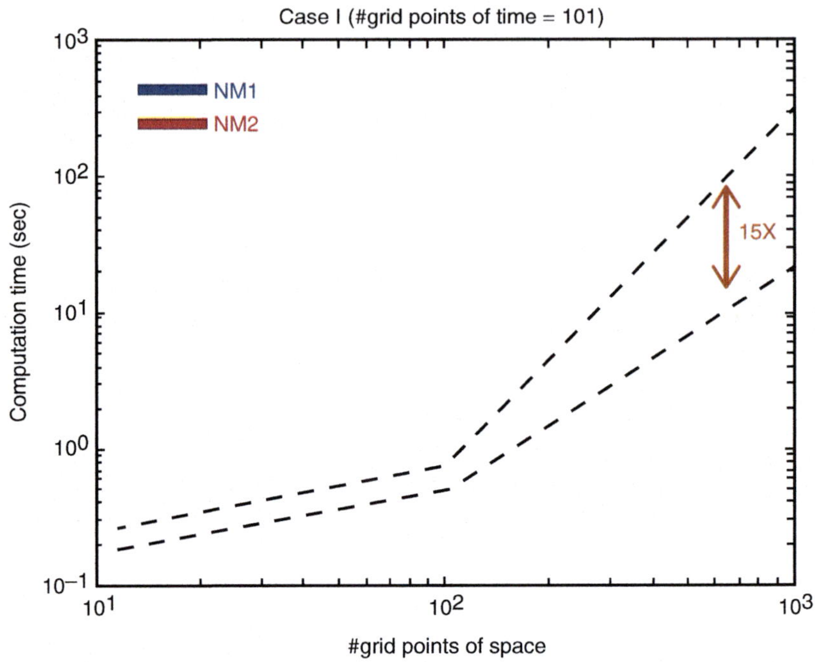

Fig. 30.18 *Blue line* is computation time for NM1. *Red line* is computation time for NM2

In most modern option pricing, models are characterised by discontinuous payoff functions and volatility of underlying may be very low. Risk preferences in optimal investment models may be expressed by variables which vanish in the limit. Against this background, numeric solutions are sought that are adapted to resolve the resulting degenerate equations.

We consider a system of ODEs which arises in option pricing for markets that switch between liquid (state 0) and illiquid (state 1):

$$\begin{cases} \gamma(R^0_\tau - \frac{1}{2}\sigma^2 S^2 R^0_{SS}) = -\nu_{01} e^{-\gamma R^1} e^{\gamma R^0} + d_0 + \nu_{01}, \\ \gamma R^1_\tau = -\nu_{10} e^{-\gamma R^0} e^{\gamma R^1} + \nu_{10}, \end{cases}$$

with the terminal conditions $R^i(T, S) = h(S), \quad i = 0, 1.$

We use an implicit approximation for the two equations and linearise the exponential term using Taylor expansion.

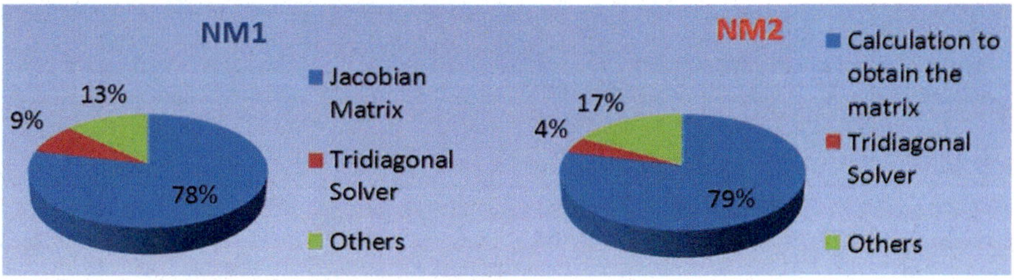

Fig. 30.19 Complexity analysis for both solvers

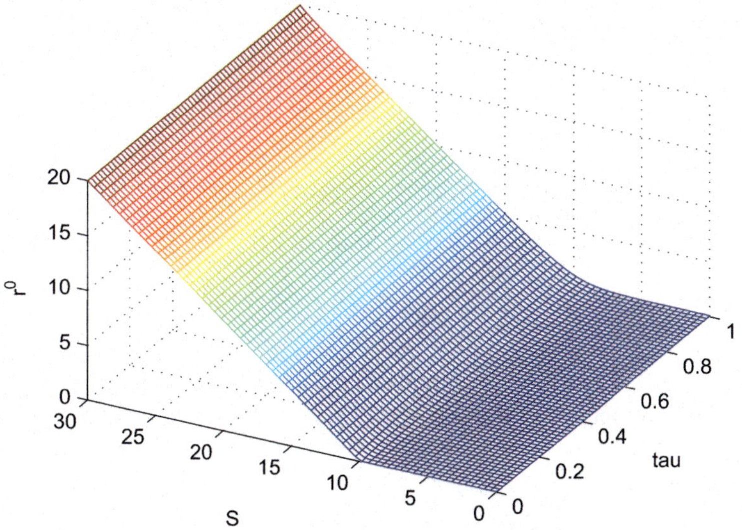

Fig. 30.20 Solution surface of r^0

Another way of looking at this system is recasting it as a Partial Integro-Differential Equation (PIDE):

$$\psi_\tau - \frac{1}{2}\sigma^2 S^2 \psi_{SS} = -a e^{\psi(S,\tau)} \left(c \int_0^\tau e^{-\psi(S,s)} ds + e^{-\gamma h(S)} \right)$$

$$+b - c,$$

$$\psi\,(S,0) = \gamma h(S).$$

The PIDE can be approximated using implicit in time together with a quadrature method for the integral term. A difficulty of this PIDE is that the upper limit of the integral is nonconstant. We use a trapezoidal rule for our quadrature rule together with a Taylor expansion linearisation of the nonlinear term (Fig. 30.19).

Figure 30.20 shows the approximate solution to R^0 which are the dynamics of the price when the market is liquid. As is expected, the surface takes the form of

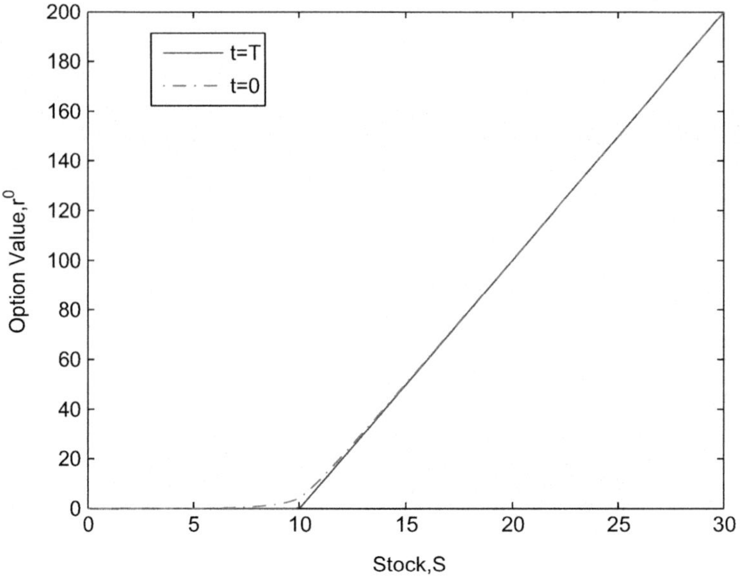

Fig. 30.21 Comparing European option values at issue and maturity in the liquid state

the Black–Scholes surface where the market is assumed to be always liquid. In Fig. 30.21 we compare the value of the option in the liquid state at maturity and issue of the option. As expected, the option value increases with time to maturity.

30.3 Models with Two Spatial Variables

In this section we want to describe the user interface, which combines the implementations of the financial models with two spatial variables. Figure 30.22 shows the example calculation of the value of a butterfly option in the linear two-dimensional model by Black and Scholes.

This section is organised as follows. First, we describe the user interface and give an overview on the functionality of the Computational Finance Toolbox for models with two spatial dimensions. After this, we describe the numerical methods for each included model. The first numerical schemes we describe are implemented for the linear two-dimensional Black–Scholes model with Finite Difference schemes, from upwind-discretisation to high-order compact schemes using a combination of the implicit BDF4 and Crank-Nicolson in the time-discretisation. Thereafter we describe an ADI approach for a non-linear two-dimensional Black–Scholes model, where the non-linearity is given by the correlation.

Then, we describe Alternating Direction Implicit (ADI) schemes as well as finite difference approaches, up to high-order compact schemes, for the Heston model.

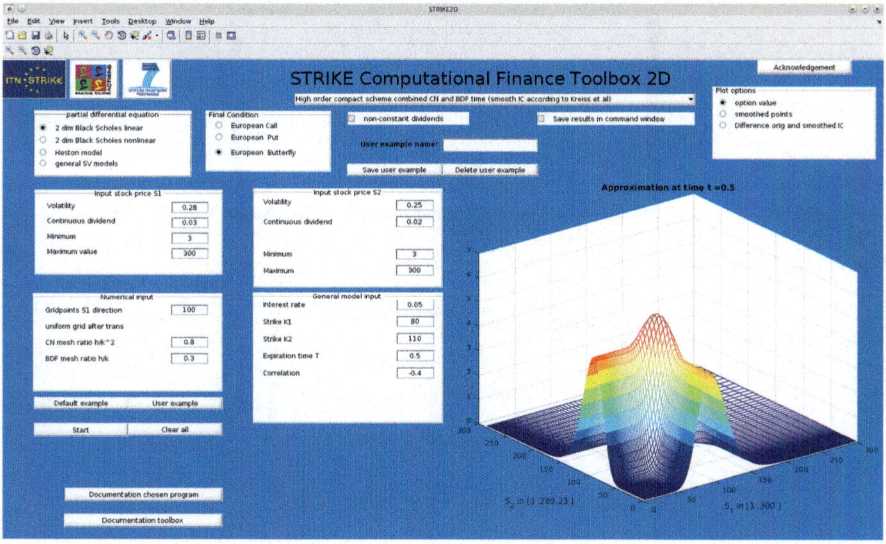

Fig. 30.22 STRIKE toolbox for two spatial variables

Finally, we discuss finite difference schemes for a rather general class of Stochastic Volatility models, starting from the standard second-order central discretisation and up to high-order compact schemes using a combination of the implicit BDF4 and Crank-Nicolson time-discretisation.

30.3.1 General Description of the User Interface

In this section we give an overview on the user interface of the Computational Finance Toolbox for models with two-dimensional spatial domain. In Fig. 30.22 we can see the user interface after calculating a European Butterfly option in the linear two-dimensional Black–Scholes model. In the following we have a closer look on all the separate parts of the user interface. We start with the general model settings, then explain the parameter input fields and finally discuss the output options. A description of the possible numerical schemes in each model setting is given in the Sects. 30.3.2–30.3.5.

In the top-left corner we can see again the logos of the STRIKE network, Marie Curie Actions as well as the seventh framework program. These buttons lead to the respective web pages. Underneath them we can see the field, where the user can choose the overall model. These models include the linear or a non-linear two-dimensional Black–Scholes model, the Heston model or a family of stochastic volatility models, see Fig. 30.23. The next part the user has to define is the final

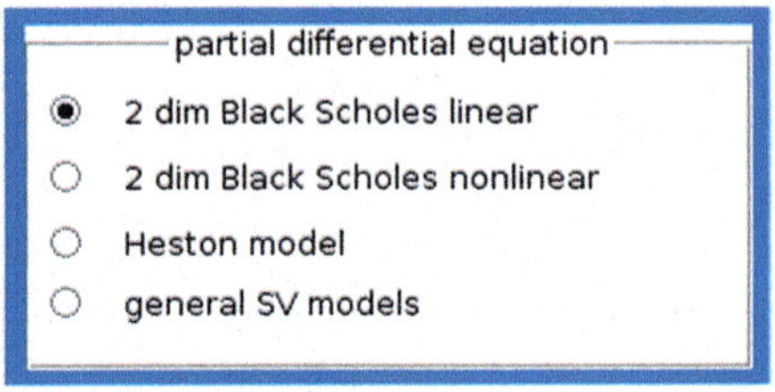

Fig. 30.23 Model selection

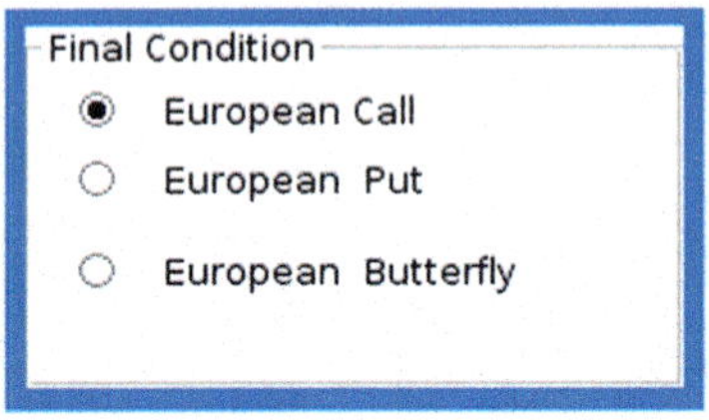

Fig. 30.24 Final condition

Fig. 30.25 Popup menu for the numerical method

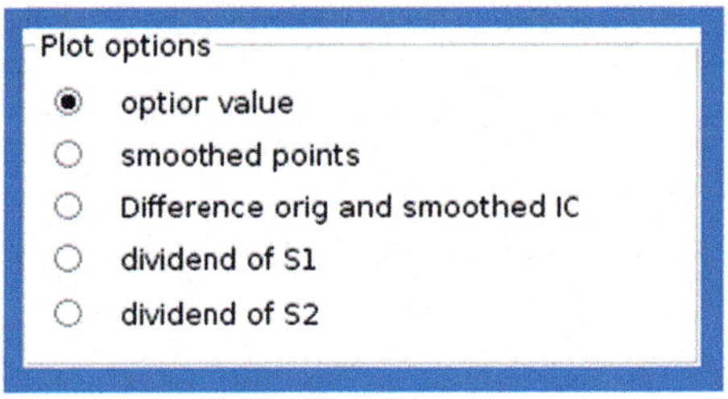

Fig. 30.26 Output options

condition of the European option. Depending on the given model, the European options include Call options, Put options as well as Butterfly options, see Fig. 30.24.

Naturally, the next selection is the type of the numerical method. A popup menu on the right side of the choice for the final conditions gives the user the opportunity to select the wanted method. Figure 30.25 shows an example menu, when a European Call in the Heston model is chosen. The first choice in this setting is the Douglas ADI method.

In the far right of the user interface the plot options can be chosen, see Fig. 30.26. Depending on the numerical method these plot options may contain the option price, the non-constant dividend yield, gridpoints where smoothing of the initial condition is applied or the influence of this smoothing on the initial condition.

Fig. 30.27 Checkbox non-constant dividend

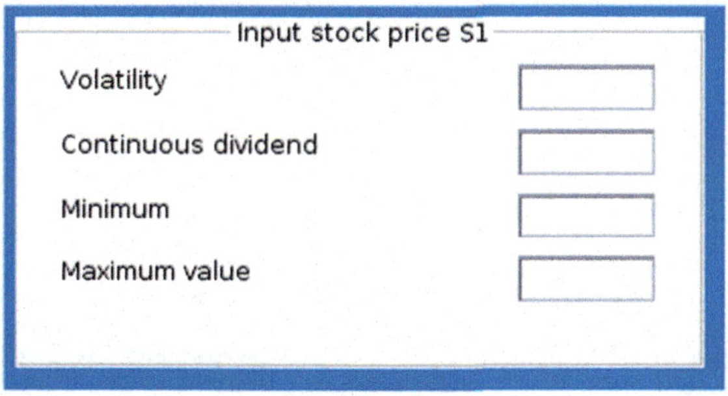

Fig. 30.28 S_1 in 2D BLS model

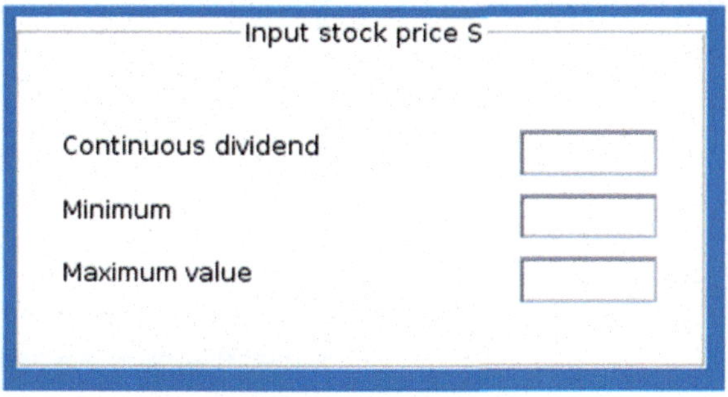

Fig. 30.29 Stock in SV model

The non-constant dividend yield can be activated through a check box right below the choice of the numerical method, see Fig. 30.27, when a high-order compact or the standard second order discretisation is chosen in combination with the linear Black–Scholes model. The dividend is constant, if this check box is not selected or a different combination of model and numerical method is used. The check box is invisible, if it is not possible to choose a non-constant dividend.

In Figs. 30.28 and 30.29 we can see the field where the user can define the parameter values regarding the first spatial variable. Figure 30.28 shows the input field for the first spatial domain, when the underlying model is the linear or non-linear Black–Scholes model. The minimum and maximum values for the stock price have to be defined. as well as the constant values for the volatility and the continuous dividend of the first stock price. The field for the continuous dividend vanishes in the case where the continuous dividend is chosen to be space dependent. When a stochastic volatility model is used the volatility of the stock price is not constant any more, but a random variable itself, which means that the value of the volatility does not have to be defined and is thus invisible in these cases, see Fig. 30.29.

In Fig. 30.30 we can see the input field for the second spatial variable when a linear or non-linear Black–Scholes model is used. This means that the same parameters have to be defined as for the first stock price. In the case of a stochastic

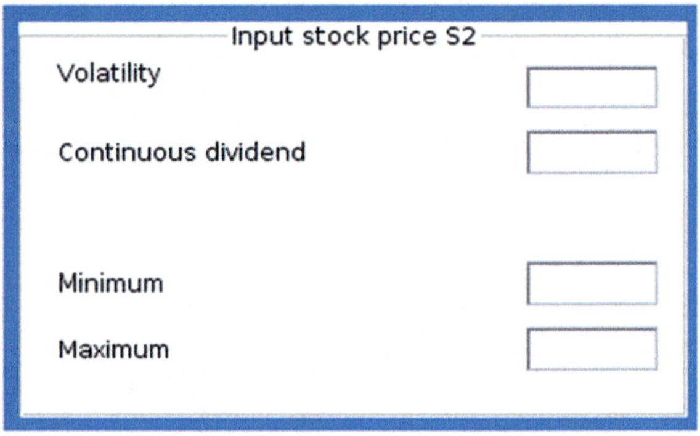

Fig. 30.30 S_2 in 2D BLS model

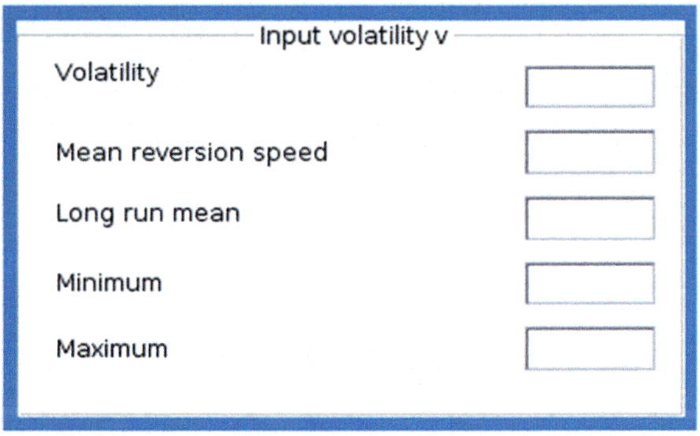

Fig. 30.31 Volatility in SV model

volatility model the field changes, see Fig. 30.31. Then the volatility, the mean reversion speed as well as the long run mean of the volatility have to be specified. Additionally, the user has to chose a minimum and maximum value for the volatility, see Fig. 30.31.

Figure 30.32 shows the field where the user can define the additional model parameters, such as the risk-free interest rate, the expiration date of the option, the strike price of the option or the correlation between the two spatial variables.

In Fig. 30.33 we can see the field for the input of the numerical parameters, such as the number of grid points in the spatial and time directions. In both fields the input variables vary due to the given numerical algorithm and underlying model. Again, only necessary input variables are visible.

Once all the necessary input variables have been defined, the user might want to save these values in order to be able to re-use them. This can be done using the buttons and input field, which can be seen in Fig. 30.34. These are located underneath the popup menu for the selection of the numerical methods. The user simply writes a name for the parameter setting in the field next to "User example name:" and can then save this setting with the corresponding button. A combination of the name defined by the user and the current combination of model, final condition and numerical method is used to save these example values. Thus it is possible to re-use a specific name when the combination of model, final condition

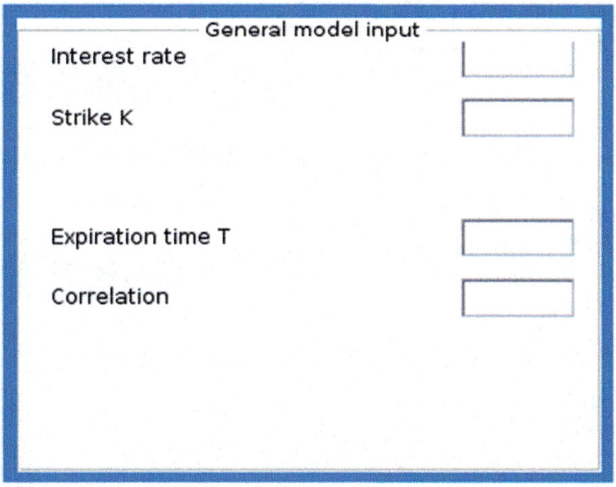

Fig. 30.32 Additional model input

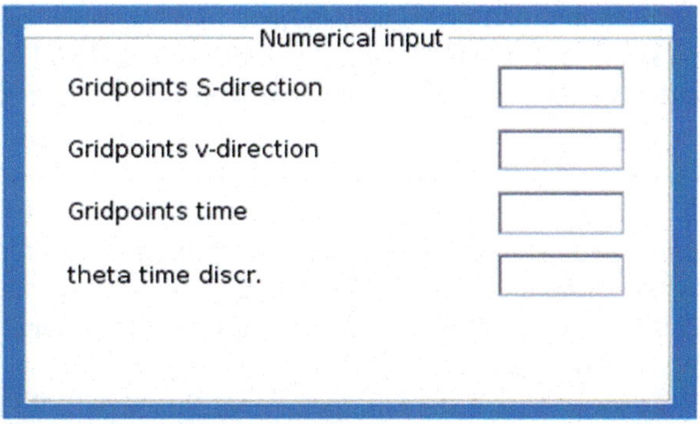

Fig. 30.33 Numerical input

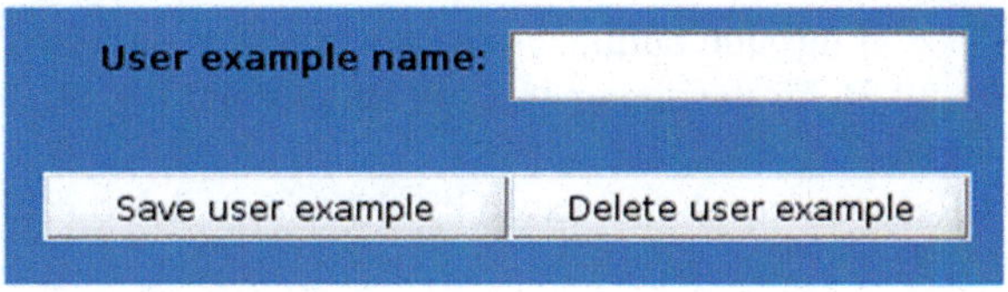

Fig. 30.34 Popup menu for the numerical method

and numerical method is not identical. It is also possible to delete an already saved parameter setting. For one has to select the corresponding combination of model and numerical method and write again the name in the according field. Pushing the Delete user example button then erases the parameter setting.

In Fig. 30.35 we can see a number of buttons located underneath the fields for the numerical input. Since we included several numerical methods and underlying models it is important to provide the user with default examples for the parameter settings. If the corresponding button is used, all necessary variables for the chosen combination of model and numerical method are defined and the program can be started, which is done with the start button. This way one can get an idea on the magnitude of the input variables if the user is not familiar with the model. But

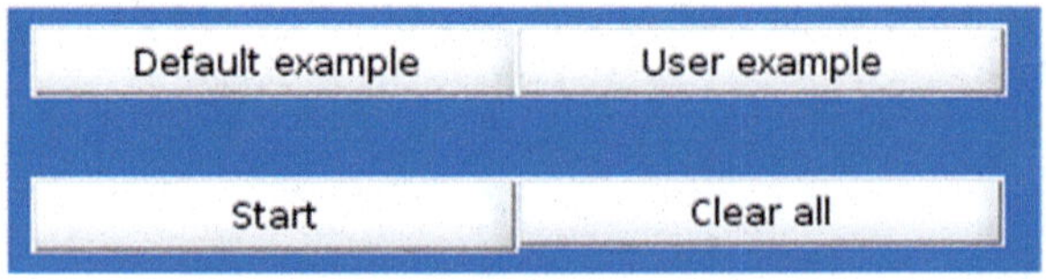

Fig. 30.35 Popup menu for the numerical method

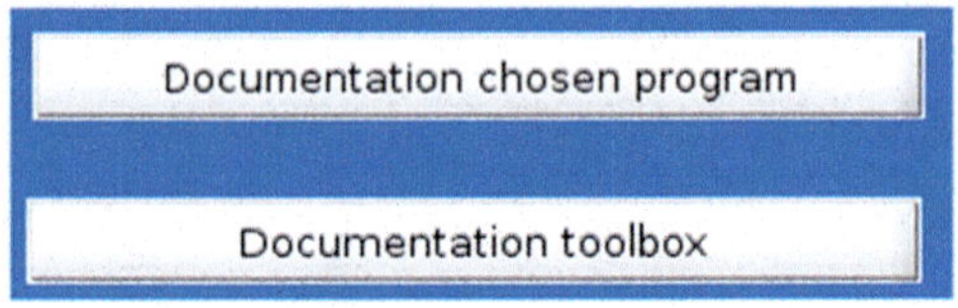

Fig. 30.36 Documentation buttons

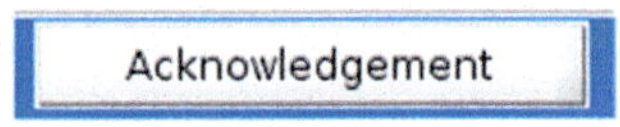

Fig. 30.37 Acknowledgement button

the user can load previously defined user examples as well. Note, the name of the parameter setting for the combination of model and numerical method has to exist and the name has to be written in the appropriate field next to "User example name:", see Fig. 30.34.

Additionally, we included a clear all button and of course a start button. If started, the program checks whether the input parameters are in line with the setting of the model and the numerical method. If a parameter should not fit the restrictions given by either the model or the numerical method, the program shows an error message. This message explains the restrictions on this parameter. Only the first error found by the program is shown.

Descriptions of the Computational Finance Toolbox and the chosen numerical method for the current financial model as well as Acknowledgements can be accessed through buttons in the lower left or upper right corner of the toolbox, see Fig. 30.36 for the documentation buttons and Fig. 30.37 for the Acknowledgements button.

Results may also be saved by checking a tick box above the visual results.

30.3.2 The Linear Two-Dimensional Black–Scholes Model

In this section we want to discuss the first included model, namely the two-dimensional Black–Scholes model, see e.g. [16]. The partial differential equation of the linear two-dimensional Black–Scholes model is given by

$$\begin{aligned} 0 =& \frac{\partial V}{\partial t} - \frac{\sigma_1^2 S_1^2}{2}\frac{\partial^2 V}{\partial S_1^2} - \rho\sigma_1\sigma_2 S_1 S_2 \frac{\partial^2 V}{\partial S_1 \partial S_2} - \frac{\sigma_2^2 S_2^2}{2}\frac{\partial^2 V}{\partial S_2^2} \\ & - \left(r - \delta_1(S_1)\right) S_1 \frac{\partial V}{\partial S_1} - \left(r - \delta_2(S_2)\right) S_2 \frac{\partial V}{\partial S_2} + rV \end{aligned} \tag{30.5}$$

for $(S_1, S_2, t) \in \Omega \times [0, T)$, where $\sigma_i \geq 0$ and $\delta_i(S_i) \geq 0$ are the volatility and continuous dividend of the asset $S_i > 0$ for $i = 1, 2$, respectively, $r > 0$ the risk-free interest rate and $\rho \in [-1, 1]$ is the correlation between the two stock prices. The continuous dividend δ_i is either constant or given by

$$\delta_i(S_i) = \hat{\delta}_i \frac{\tanh\left(\zeta_i(S_i - \hat{S}_i)\right) + \tanh(\zeta_i \hat{S}_i)}{2} \tag{30.6}$$

with $\zeta_i > 0$, $\hat{\delta}_i \geq 0$ and $\hat{S}_i \in [S^{(i)}_{\min}, S^{(i)}_{\max}]$ for $i = 1, 2$, see e.g. [6].

The payoff functions are given by

$$V_T(S_1, S_2) = \begin{cases} (\omega_1 S_1 + \omega_2 S_2 - K)^+ \\ (K - \omega_1 S_1 - \omega_2 S_2)^+ \\ \left(\bar{S} - K_1\right)^+ - 2\left(\bar{S} - K_{12}\right)^+ + \left(\bar{S} - K_2\right)^+ \end{cases}$$

for a Call, Put or Butterfly option with $\bar{S} = \max(S_1, S_2)$ and positive strike prices K, K_1, K_2 and $K_{12} = (K_1 + K_2)/2$.

In the following we describe those numerical methods, which can be used within the Computational Finance Toolbox for models with two-dimensional spatial domain to approximate the solution of the two-dimensional Black–Scholes model.

30.3.2.1 High-Order Compact Schemes and Compact Central Second-Order Scheme

First, we restrict the spatial domain to

$$\Omega = [S^{(1)}_{\min}, S^{(1)}_{\max}] \times [S^{(2)}_{\max}, S^{(2)}_{\max}],$$

where

$$0 < S^{(i)}_{\min} < K \text{ (Call/Put)} \quad \text{or} \quad 0 < S^{(i)}_{\min} < \min(K_1, K_2) \text{ (Butterfly)}$$

and

$$K < S^{(i)}_{\max} < \infty \text{ (Call/Put)} \quad \text{or} \quad \max(K_1, K_2) < S^{(i)}_{\max} < \infty \text{ (Butterfly)}$$

for $i = 1, 2$.

Using $\hat{K} = K$ (Call/Put) or $\hat{K} = K_{12}$ (Butterfly) the transformations

$$x_i = \frac{\min(\sigma_1, \sigma_2)}{\sigma_i} \ln\left(\frac{S_i}{\hat{K}}\right), \qquad \tau = T - t, \qquad u = e^{-r\tau} \frac{V}{K}$$

are applied to (30.5) and the initial conditions. This is the reason why the minimal stock prices have to be positive instead of non-negative.

The program uses a uniform grid G_h with step-size $h > 0$ in each direction, where with the input parameter $N \in \mathbb{N}$ it holds $h = \left(x_{\max}^{(1)} - x_{\min}^{(1)}\right)/(N-1)$. On this grid we apply a fourth-order spatial discretisation in terms of h, which only uses the compact stencil, or the standard second-order discretisation with the central difference operator, see [6]. The semi-discrete scheme reads

$$\sum_{j_1=i_1-1}^{i_1+1} \sum_{j_2=i_2-1}^{i_2+1} \left[M_{j_1 j_2} \partial_\tau U_{j_1 j_2}(\tau) + P_{j_1 j_2} U_{j_1 j_2}(\tau)\right] = g(x, \tau),$$

where $M_{j_1 j_2}$ and $P_{j_1 j_2}$ are constant in time. They are space-dependent when using (30.6), constant otherwise. Crank-Nicolson type time discretisation with step-size $\Delta\tau = ch^2$ and $c > 0$ is then applied to the semi-discrete scheme, see [6, 7]. Alternatively, the Crank-Nicolson type time discretisation is only applied to the semi-discrete scheme in order to calculate the starting values for the BDF4 scheme, which is then applied with $\Delta\tau = c_2 h$ and $c_2 > 0$. This leads to overall $\mathcal{O}(N)$ grid points in time and a consistency order of 4 in terms of h.

The transformed smoothed initial condition is given by

$$\tilde{u}_0(x_1, x_2) = \frac{1}{h^2} \int_{-3h}^{3h} \int_{-3h}^{3h} \Phi_4\left(\frac{x}{h}\right) \Phi_4\left(\frac{y}{h}\right) u_0(x_1 - x, x_2 - y)\, \mathrm{d}x\mathrm{d}y,$$

for any $(x_1, x_2) \in G_h$ and step size $h > 0$, where $\Phi_4(x)$ denotes the Fourier inverse of

$$\hat{\Phi}_4(\omega) = \left(\frac{\sin\left(\frac{\omega}{2}\right)}{\frac{\omega}{2}}\right)^4 \left[1 + \frac{2}{3}\sin^2\left(\frac{\omega}{2}\right)\right]$$

and u_0 the original initial condition, see [11]. Thus the smoothed initial condition $\tilde{u}_0 \xrightarrow{h \to 0} u_0$. A complete derivation of this discretisation as well as some examples for European options is given in [6].

This program has several output options. Firstly, it is possible to show the non-continuous dividend for each spatial direction. Figure 30.38 shows an example with $\zeta = 0.05$, $\hat{\delta} = 2\%$ and $\hat{S} = 130$.

In Fig. 30.39 the grid points where smoothing is applied are shown for a Call/Put. It has to be assured that at least three grid points are above, below, left and right of the function of the non-differentiable points in order to only smooth the initial condition, where it is necessary. This means the number of grid points where u_0 has to be smoothed can be reduced to $\mathcal{O}(N)$, whereas the total number of grid points is in $\mathcal{O}(N^2)$. In the example in Fig. 30.39 this leads to a saving of about 86.67% of the computation for smoothing.

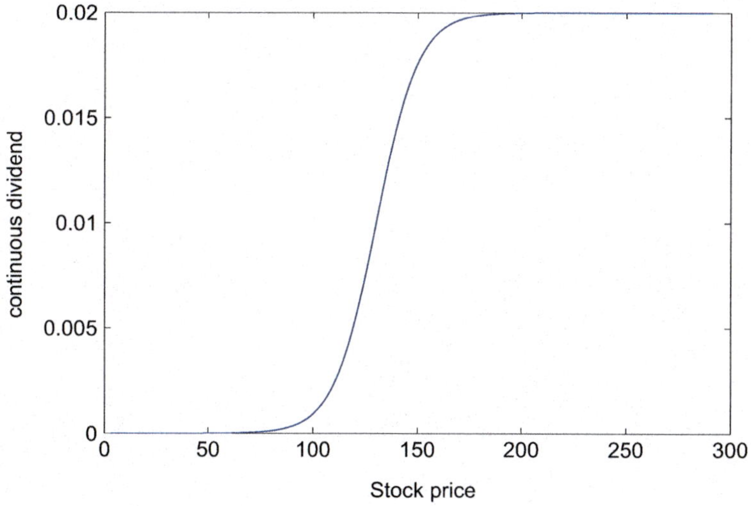

Fig. 30.38 Example non-constant dividend

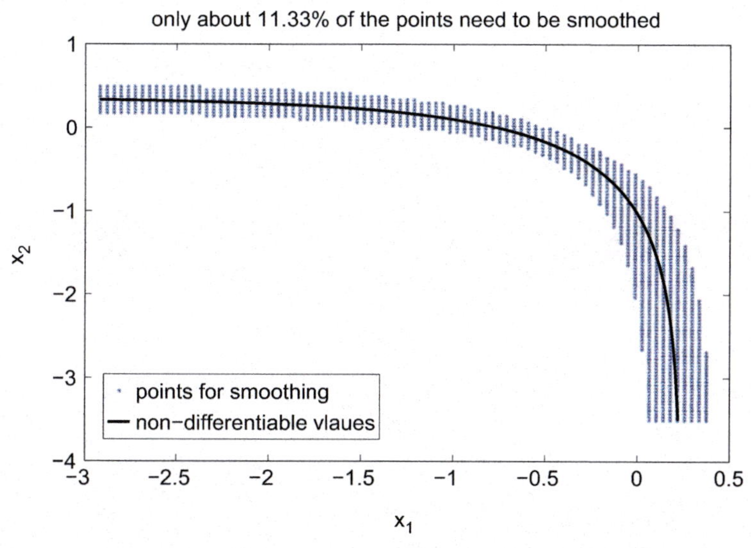

Fig. 30.39 Smoothing points Call/Put

Figure 30.40 shows the points where smoothing is necessary for a Butterfly option. It is important to mention that only for a very small amount of grid points (green points) a smoothing in both spatial directions is necessary. For the green points one has to assure that at least three points are above, below, left and right of the corner points of the non-differentiable points, which means that maximal number of grid points, which need to be smoothed in both direction is not depending

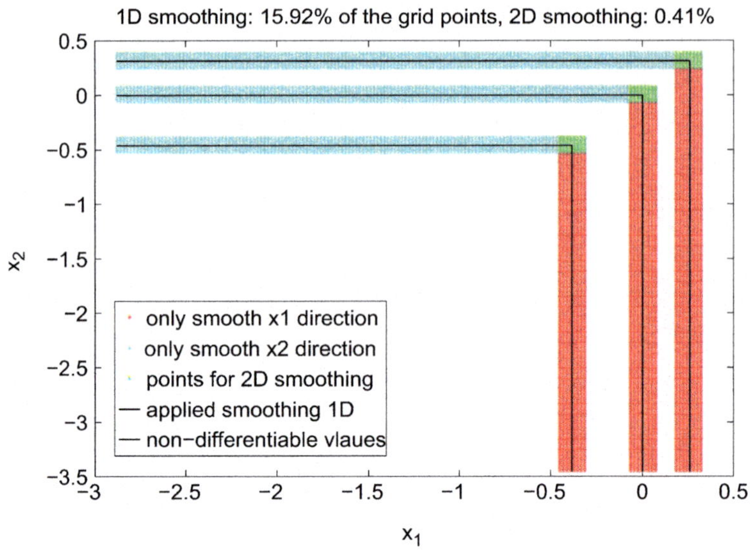

Fig. 30.40 Smoothing points Butterfly

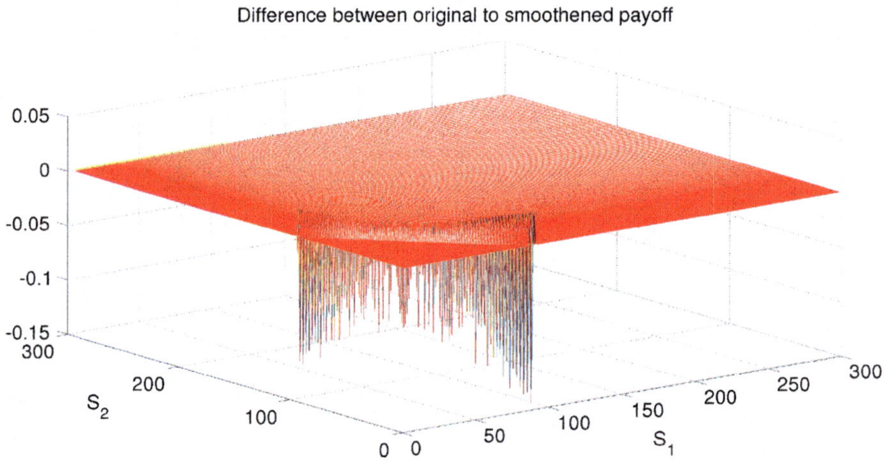

Fig. 30.41 Difference original and smoothed payoff Call option

on h. For the blue and red points it is sufficient to smooth only in one spatial direction. Smoothing the initial condition on a grid point only in one spatial direction is much quicker than smoothing it in both spatial directions.

It is also possible to show the difference between the smoothed initial condition $\tilde{V}_T$, which is $\tilde{u}_0$ re-transformed, and the original payoff V_T, see Fig. 30.41 for a Call, and Fig. 30.42 for a Butterfly.

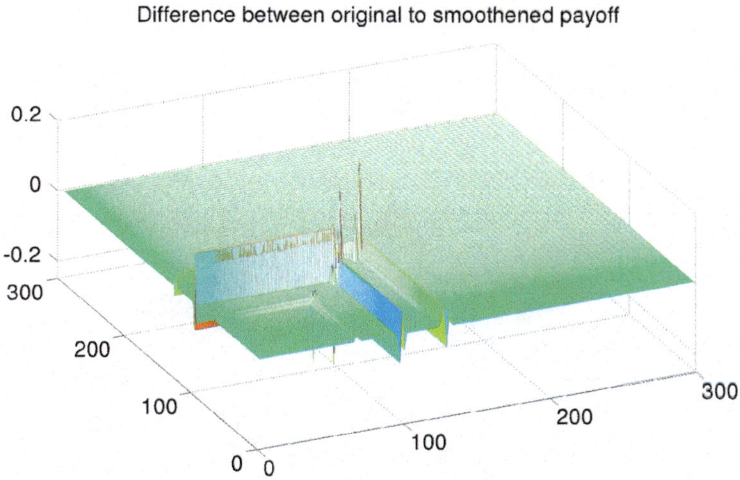

Fig. 30.42 Difference original and smoothed payoff Butterfly option

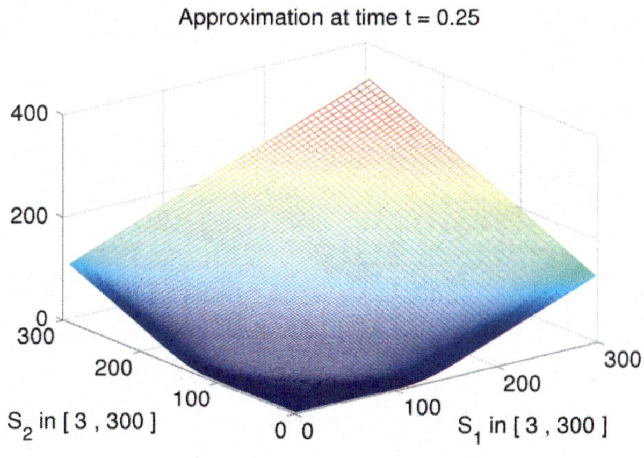

Fig. 30.43 European Call option

Finally, the solution of the two-dimensional Black Scholes partial differential equation is an output option as well, see Fig. 30.43 for a Call using the high-order compact scheme including the BDF4 in the discretisation process,

Figure 30.44 for a Put using the high-order compact scheme which only uses Crank-Nicolson time discretisation and Fig. 30.45 for a Butterfly, where a standard second order discretisation in space was used in combination with Crank-Nicolson discretisation in time.

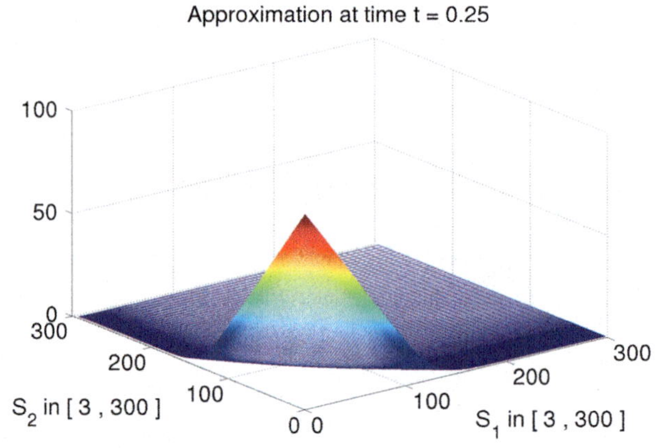

Fig. 30.44 European Put option

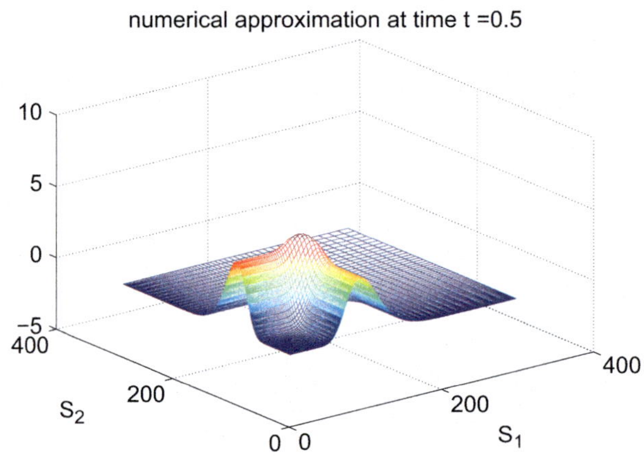

Fig. 30.45 European Butterfly option

30.3.3 The Non-linear Two-Dimensional Black–Scholes Model

The partial differential equation of the non-linear two-dimensional Black–Scholes model is given by

$$\frac{\partial u}{\partial t} = \frac{1}{2}\left(\sigma_1^2 s_1^2 \frac{\partial^2 u}{\partial s_1^2} + 2\rho(\Gamma)\sigma_1\sigma_2 s_1 s_2 \frac{\partial^2 u}{\partial s_1 \partial s_2} + \sigma_2^2 s_2^2 \frac{\partial^2 u}{\partial s_2^2}\right)$$

$$+ r\left(s_1 \frac{\partial u}{\partial s_1} + s_2 \frac{\partial u}{\partial s_2}\right) - ru \text{ for } (s_1, s_2, t) \in \Omega \times (0, T],$$

$$\rho(\Gamma) = \begin{cases} \rho^- & \text{if } \Gamma > 0, \\ \rho^+ & \text{if } \Gamma < 0, \end{cases} \quad \Gamma = \frac{\partial^2 u}{\partial s_1 \partial s_2}, \quad \Omega = (0, S_{\max})^2.$$

We select $u(s_1, s_2, 0) = (s - K_1)^+ - 2(s - K)^+ + (s - K_2)^+$ with $s = \max(s_1, s_2)$ and $K = (K_1 + K_2)/2$ which is a *butterfly payoff*, see e.g. [15]

30.3.3.1 ADI Time-Stepping for the Uncertain Correlation Black–Scholes PDE

The description of this program is taken from [15], only the format has been changed. For the numerical valuation of two-asset options we consider a 2D nonlinear degenerate convection-diffusion-reaction problem with a cross derivative term. The boundary conditions are homogeneous on the far field boundaries and natural on the degenerate boundaries.

To alleviate the nonsmoothness of the payoff at the strikes K_1, K, K_2 we construct a suitable rectangular spatial grid. For each coordinate s_j a uniform mesh is taken in $[\frac{1}{2}K_1, \frac{3}{2}K_2]$ and outside this interval the mesh sizes gradually increase. Next a second-order central finite difference discretization is applied for all derivatives in the PDE w.r.t. s_1 and s_2. This leads to a large stiff nonlinear ODE system,

$$U'(t) = F(U(t)) = A(U(t))U(t), \quad 0 < t \leq T,$$

with given matrix function $A(\cdot)$ and initial vector $U(0) = U_0$. Splitting the function $F = F_0 + F_1 + F_2$ with $F_0(\xi) = A_0(\xi)\xi$, $F_1(\xi) = A_1\xi$, $F_2(\xi) = A_2\xi$, the *Modified Craig-Sneyd (MCS) scheme* [9, 10] then generates approximations U_n to $U(n\Delta t)$ for $n \in \mathbb{N}$. It is a modern time-stepping scheme of ADI type where the (nonlinear) cross derivative part A_0 is conveniently treated in an explicit manner and for $j = 1, 2$ the (essentially one-dimensional) part A_j in the s_j-direction in an implicit manner. The implicit stages stabilize the explicit stages.

We take $K_1 = 35$, $K_2 = 45$, $T = 0.5$, $S_{\max} = 200$, $\sigma_1 = \sigma_2 = 0.5$, $r = 0.05$, $\rho^- = 0.4$, $\rho^+ = 0.6$. Figure 30.46 shows the numerical results for the option value at $t = T$ and Fig. 30.47 for the corresponding cross derivative.

The MCS scheme and similar ADI schemes are highly efficient compared to common un-split schemes. Our present research is devoted to studying their fundamental properties, notably stability, monotonicity and convergence.

30.3.4 The Heston Model

The Heston model is a prominent example of a stochastic volatility model with partial differential equation

$$\begin{aligned} 0 = &\frac{\partial V}{\partial t} + \frac{v S^2}{2}\frac{\partial^2 V}{\partial S^2} + \rho\sigma v S \frac{\partial^2 V}{\partial S \partial v} + \frac{\sigma^2 v}{2}\frac{\partial^2 V}{\partial v^2} \\ &+ (r - \delta)\, S \frac{\partial V}{\partial S} + \kappa\,(\theta - v)\,\frac{\partial V}{\partial v} - rV, \end{aligned} \tag{30.7}$$

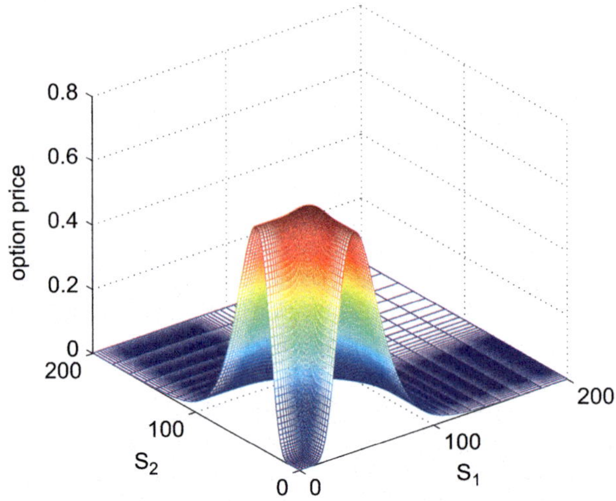

Fig. 30.46 The numerical option value function

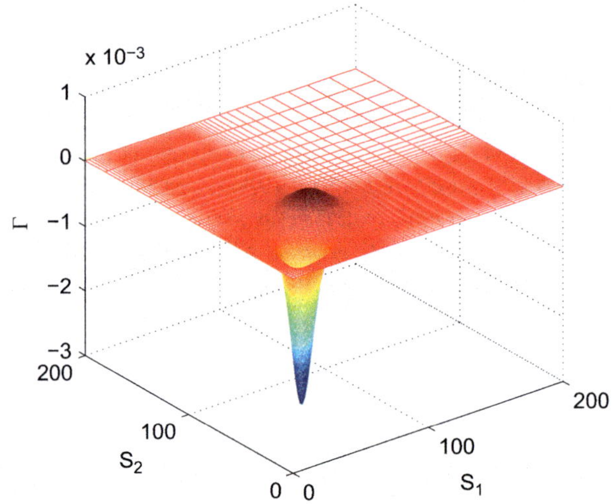

Fig. 30.47 The cross derivative

with $(S, v, t) \in [S_{\min}, S_{\max}] \times [v_{\max}, v_{\max}] \times (0, T]$. We demand $0 < S_{\min} < S_{\max} < \infty$ as well as $0 < v_{\min} < v_{\max} < \infty$. The continuous dividend of the stock price S is given by $\delta \geq 0$. Additionally, $\kappa > 0$ is the mean-reversion speed, $\theta > 0$ is the long-run mean and $\sigma \geq 0$ is the volatility of the volatility v, $r > 0$ the risk-free interest rate, $T > 0$ the expiration time and $\rho \in [-1, 1]$ is the correlation between the stock price and its volatility, see [8]. The payoff functions are given by

$$V_T(S_1,S_2) = \begin{cases} (S-K)^+ \\ (K-S)^+ \\ (S-K_1)^+ - 2\,(S-K_{12})^+ + (S-K_2)^+ \end{cases}$$

for a Call, Put or Butterfly option with positive strike prices K, K_1, K_2 and $K_{12} = (K_1+K_2)/2$ are in the interval $[S_{\min}, S_{\max}]$.

30.3.4.1 High-Order Compact Schemes and Compact Central Second-Order Scheme

Firstly, we use $\hat{K} = K$ (Call/Put) or $\hat{K} = K_{12}$ (Butterfly) and apply the transformations

$$x = \ln\left(\frac{S}{\hat{K}}\right), \qquad y = \frac{v}{\sigma}, \qquad \tau = T - t, \qquad u = e^{-r\tau}\frac{V}{K}$$

to (30.7) and the final conditions. The transformed problem setting is then solved and its solution later re-transformed. The program uses a uniform grid G_h in space with step-size $h > 0$ in each direction, where with the input parameter $N \in \mathbb{N}$ it holds $h = (x_{\max} - x_{\min})/N$. On this grid we apply the central difference operator for the spatial discretisation, which only uses the compact stencil and has second order consistency. The semi-discrete scheme reads

$$\sum_{j_1=i_1-1}^{i_1+1}\sum_{j_2=i_2-1}^{i_2+1}\left[M_{j_1,j_2}\partial_\tau U_{j_1,j_2}(\tau) + P_{j_1,j_2}U_{j_1,j_2}(\tau)\right] = g(x,\tau),$$

where M_{j_1,j_2} and P_{j_1,j_2} are depending on the variable y, but neither on x nor on τ. In the same way as for the two-dimensional linear Black–Scholes model, it is possible to apply BDF4 scheme with $\Delta\tau = c_2 h$ and $c_2 > 0$ to the above mentioned semi-discrete scheme for a high-order compact spatial discretisation. The necessary starting values are generated using Crank-Nicolson type time discretisation with step-size $\Delta\tau = c_1 h^2$ and $c_1 > 0$. This leads to overall $\mathcal{O}(N)$ grid points in time and a consistency order of 4 in terms of h.

Alternatively, it is possible to only apply the Crank-Nicolson (CN) type time discretisation, see [5]. In this case there are overall $\mathcal{O}\left(N^2\right)$ grid points in time. For the standard central second order spatial discretisation it is only possible to use the Crank-Nicolson time discretisation.

The transformed smoothed initial condition is given by

$$\tilde{u}_0(x,y) = \frac{1}{h}\int_{-3h}^{3h} \Phi_4\left(\frac{x_1}{h}\right) u_0(x-x_1,y)\,\mathrm{d}x_1,$$

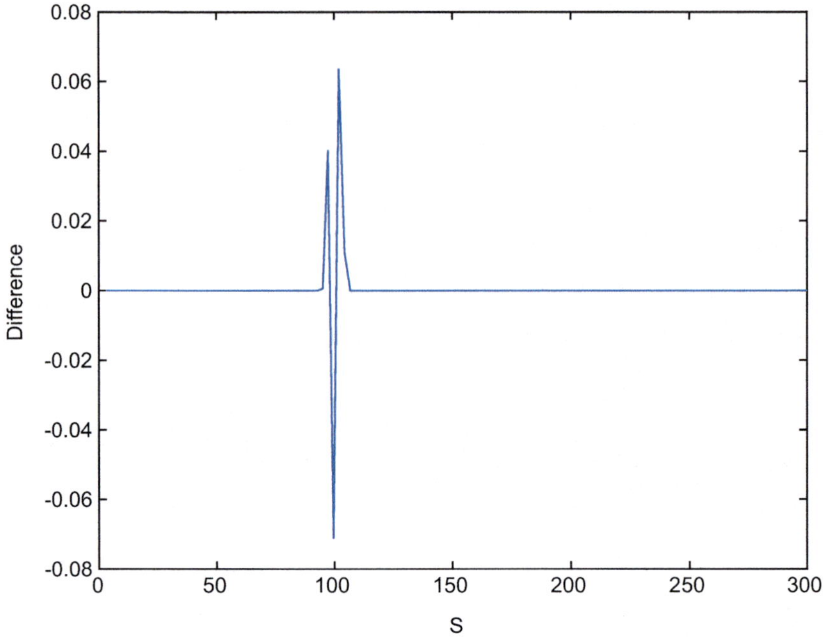

Fig. 30.48 Difference original and smoothed payoff Put option

for any $(x, y) \in G_h$ and step size $h > 0$, where $\Phi_4(x)$ denotes the Fourier inverse of

$$\hat{\Phi}_4(\omega) = \left(\frac{\sin\left(\frac{\omega}{2}\right)}{\frac{\omega}{2}}\right)^4 \left[1 + \frac{2}{3}\sin^2\left(\frac{\omega}{2}\right)\right]$$

and u_0 the original initial condition, see [11]. The smoothing is only applied in x-direction, as u_0 is constant in y-direction. Thus, the smoothed initial condition $\tilde{u}_0 \xrightarrow{h \to 0} u_0$.

For all following examples we chose $\delta = 0.03$, $\kappa = 0.02$, $\theta = 0.25$, $\sigma = 0.25$, $\rho = -0.4$ and $r = 0.05$. We have $S \in [3, 300]$ and $v \in [0.01, 0.4]$ and use $N = 200$ as well as $c_1 = 0.8$ and $c_2 = 0.3$. Since this program uses a smoothed initial condition, one output option is the difference between the original and the smoothed initial condition. Figure 30.48 shows this difference for a European Put. The difference is only non-zero in the area around the strike price, which is the only non-differentiable point.

In Fig. 30.49 we can see the difference between the original and smoothed initial condition for a Butterfly option. We can see a wider support of the difference, as there are three non-differentiable points in the initial condition.

Additionally, one can approximates the values of European Puts or Butterfly options. A European Call could be calculated using the Call-Put parity and the

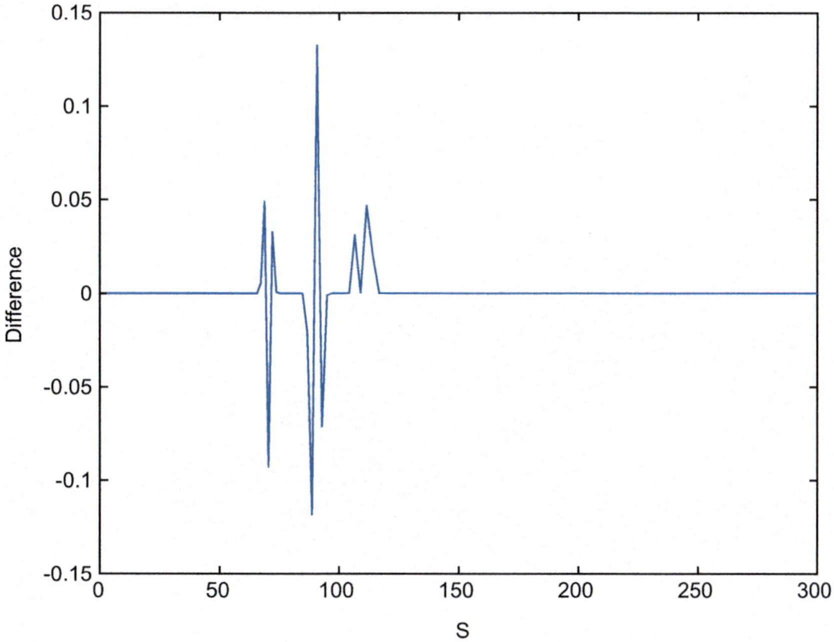

Fig. 30.49 Difference original and smoothed payoff Butterfly option

value of the European Put option. In Fig. 30.50 we approximate a European Put with $K = 100$.

For the approximation of the Butterfly option in Fig. 30.51 we used the strike prices $K_1 = 70$ and $K_2 = 110$.

30.3.4.2 ADI Schemes (Douglas scheme and Modified Craig-Sneyd Scheme)

We applied two ADI-type schemes, Douglas (D) and Modified Craig-Sneyd (MCS), to solve the ODE system resulting from the spatial discretization of (30.7). For the Douglas scheme the user can calculate an European Put option, whereas for the Modified Craig-Sneyd scheme the calculation of both European Put and European Call is possible.

The Douglas scheme consists of an explicit predictor stage followed by two implicit corrector stages

$$\begin{cases} Y_0 = U_{n-1} + \Delta t F\left(t_{n-1}, U_{n-1}\right) \\ Y_j = Y_{j-1} + \theta \Delta t \left[F_j\left(t_n, Y_j\right) - F_j\left(t_{n-1}, U_{n-1}\right)\right], \; j = 1, 2 \\ U_n = Y_2 \end{cases}$$

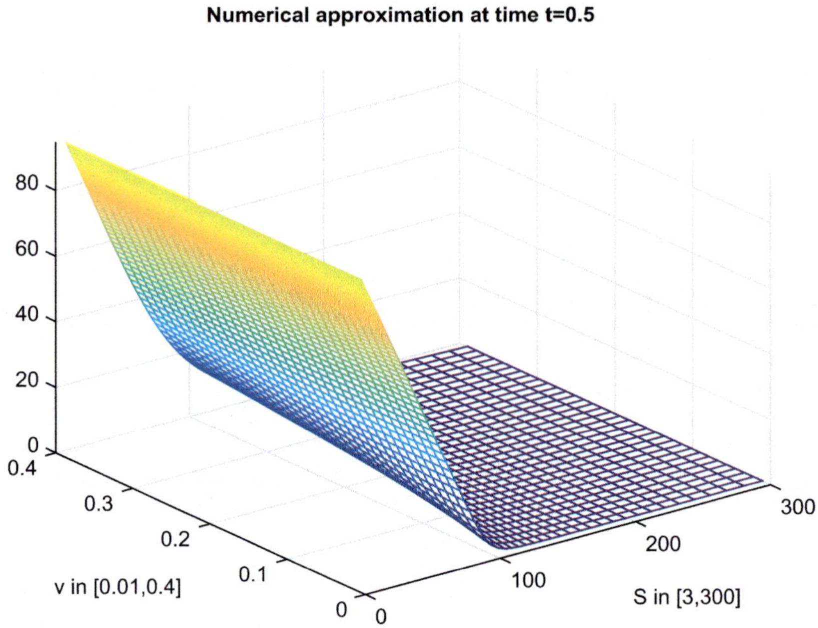

Fig. 30.50 European Put option second order discretisation

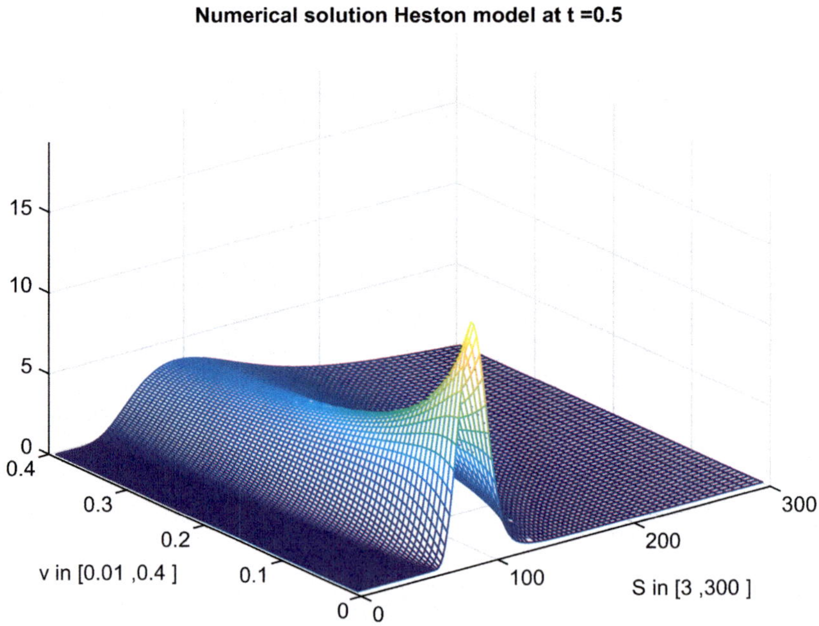

Fig. 30.51 European Butterfly option using HOC schemes, CN and BDF4

Its order of consistency is 1 whenever the mixed-derivatives term is nonzero, otherwise it exhibits consistency of order 2 for $\theta = \frac{1}{2}$.

The Modified Craig-Sneyd is an unconditionally stable scheme of order 2 for $\theta \geq \frac{1}{2}$.

$$\begin{cases} Y_0 &= U_{n-1} + \Delta t F(t_{n-1}, U_{n-1}) \\ Y_j &= Y_{j-1} + \theta \Delta t \left[F_j(t_n, Y_j) - F_j(t_{n-1}, U_{n-1}) \right] \qquad j = 1, 2, \ldots, k \\ \hat{Y}_0 &= Y_0 + \Delta t A_0 \left[F_0(t_n, Y_3) - F_j(t_{n-1}, U_{n-1}) \right] \\ \tilde{Y}_0 &= \hat{Y}_0 + \left(\frac{1}{2} - \theta \right) \Delta t \left[F(t_n, Y_3) - F(t_{n-1}, U_{n-1}) \right] \\ \tilde{Y}_j &= \tilde{Y}_{j-1} + \theta \Delta t \left[F_j(t_n, \tilde{Y}_j) - F_j(t_{n-1}, U_{n-1}) \right] \qquad j = 1, 2, \ldots, k \\ V_n &= \tilde{Y}_3 \end{cases}$$

We used a non-uniform spatial grid based on the hyperbolic sine with focus around $\nu = 0$ and $S = K$.

By default, a spatial domain $\Omega = [0, 15] \times [0, 30K]$ is selected (Figs. 30.52 and 30.53).

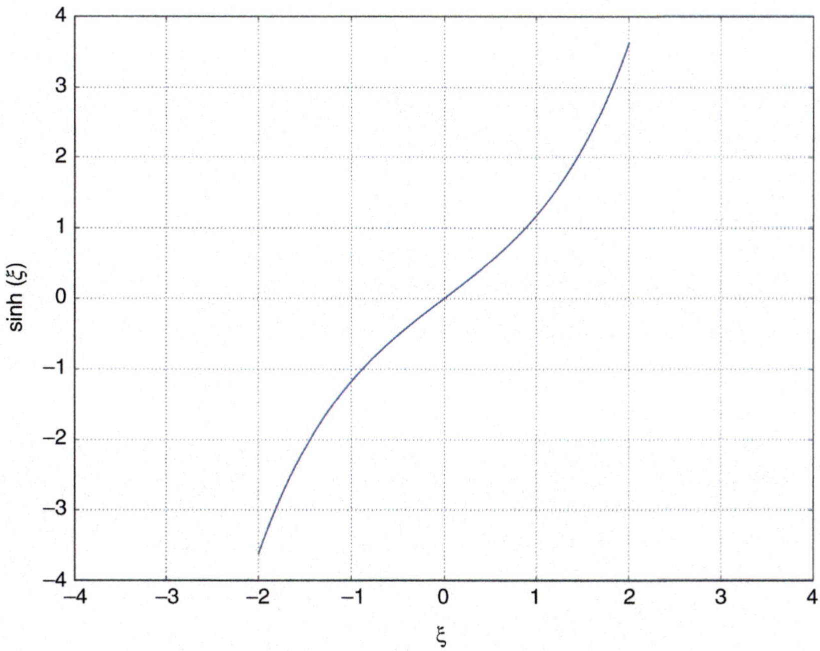

Fig. 30.52 Hyperbolic sine transformation

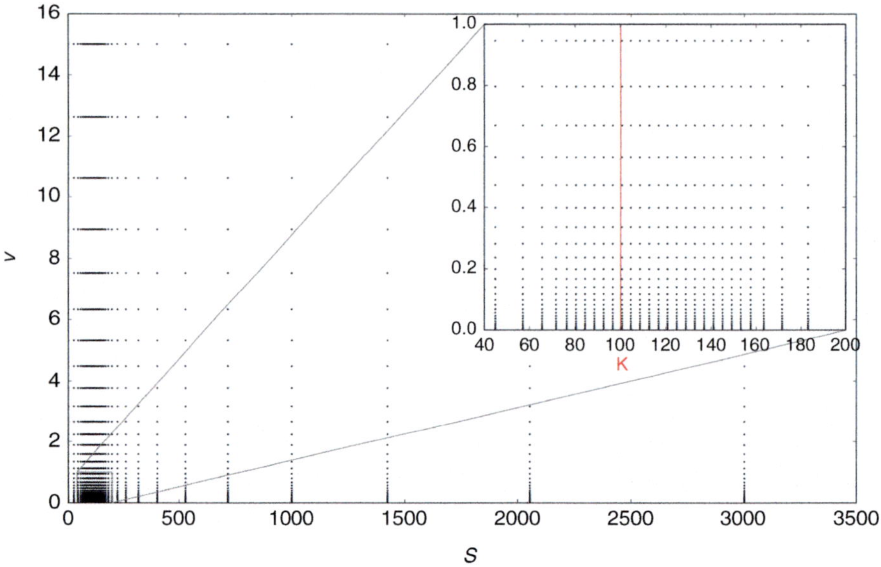

Fig. 30.53 Spatial grid

We have several numerical inputs available:

1. n_S—number of grid points in the S direction.
2. n_v—number of grid points in the v direction.
3. n_t—number of grid points in the t direction.
4. c—concentration of points around K in the S direction.
5. d—concentration of points around 0 in the v direction.

The financial input variables are given by

1. κ—speed of mean-reversion
2. η—long-term mean of variance
3. σ—volatility of variance process
4. ρ—correlation
5. r_d—domestic interest rate
6. r_f—foreign interest rate
7. K—Strike Price
8. T—Maturity (years)

The program has several possible outputs. The most common one is the surface of option values for different S_0 and v_0. Additionally, also the main greeks are available, namely:

1. Delta—Sensitivity of option price to current underlying price
2. Gamma—Second-order sensitivity of option price to underlying price
3. Vega—Sensitivity of option price to volatility

30.3.5 The General Stochastic Volatility Model

The option price in a general stochastic volatility model follows the partial differential equation

$$\begin{aligned} 0 =& \frac{\partial V}{\partial t} + \frac{vS^2}{2}\frac{\partial^2 V}{\partial S^2} + \rho\sigma v^{b+\frac{1}{2}} S \frac{\partial^2 V}{\partial S \partial v} + \frac{\sigma^2 v^{2b}}{2}\frac{\partial^2 V}{\partial v^2} \\ &+ (r-\delta) S \frac{\partial V}{\partial S} + \kappa v^a (\theta - v)\frac{\partial V}{\partial v} - rV, \end{aligned} \tag{30.8}$$

with

$$(S, v, t) \in [S_{\min}, S_{\max}] \times [v_{\max}, v_{\max}] \times (0, T],$$

where

$$a \geq 0, \quad b \in \left(0, \frac{3}{2}\right].$$

We demand $0 < S_{\min} < S_{\max} < \infty$ as well as $0 < v_{\min} < v_{\max} < \infty$. for $(S, v, t) \in \Omega \times (0, T]$, The continuous dividend of the stock price $S > 0$ is denoted by $\delta \geq 0$. Additionally, $\kappa > 0$ is the mean-reversion speed, $\theta > 0$ is the long-run mean and $\sigma \geq 0$ is the volatility of the volatility $v > 0$, $r > 0$ the risk-free interest rate, $T > 0$ the expiration time and $\rho \in [-1, 1]$ is the correlation between the stock price and its volatility, see e.g. [4]. The payoff functions are given by

$$V_T(S_1, S_2) = \begin{cases} (K - S)^+ \\ (S - K_1)^+ - 2\,(S - K_{12})^+ + (S - K_2)^+ \end{cases}$$

for a Put or Butterfly option with positive strike prices K, K_1, K_2 and $K_{12} = (K_1 + K_2)/2$.

30.3.5.1 High-Order Compact Schemes and Compact Central Second-Order Scheme

Firstly, we use $\hat{K} = K$ (Put) or $\hat{K} = K_{12}$ (Butterfly) Eq. (30.8) is transformed using $\tau = T - t$, $u = e^{r\tau} V/K$ as well as

$$x = \ln\left(\frac{S}{\hat{K}}\right), \qquad y = \frac{\ln(v)}{\sigma}$$

in the case $b = 3/2$ and otherwise

$$x = \left(\frac{3}{2} - b\right) \ln\left(\frac{S}{\hat{K}}\right), \qquad y = \frac{v^{\frac{3}{2}-b}}{\sigma}.$$

The program uses a uniform grid G_h in space with step-size $h > 0$ in each direction, where with the input parameter $N \in \mathbb{N}$ it holds $h = (x_{\max} - x_{\min})/N$. On this grid we apply the central difference operator for the spatial discretisation, which only uses the compact stencil and has second order consistency. Applying the spatial discretisation according to [6] the fourth-order accurate semi-discrete scheme reads

$$\sum_{j_1=i_1-1}^{i_1+1} \sum_{j_2=i_2-1}^{i_2+1} \left[M_{j_1,j_2} \partial_\tau U_{j_1,j_2}(\tau) + P_{j_1,j_2} U_{j_1,j_2}(\tau)\right] = g(x, \tau),$$

where M_{j_1,j_2} and P_{j_1,j_2} are depending on the variable y, but neither on x nor on τ.

A Crank-Nicolson type time discretisation with step-size $\Delta\tau = c_1 h^2$ and $c_1 > 0$ is applied to the semi-discrete scheme in order to calculate the starting values for the BDF4 scheme, which is then applied with $\Delta\tau = c_2 h$ and $c_2 > 0$. This leads to overall $\mathcal{O}(N)$ grid points in time and a consistency order of 4 in terms of h.

The transformed smoothed initial condition is given by

$$\tilde{u}_0(x, y) = \frac{1}{h} \int_{-3h}^{3h} \Phi_4\left(\frac{x_1}{h}\right) u_0(x - x_1, y)\, \mathrm{d}x_1,$$

for any $(x, y) \in G_h$ and step size $h > 0$, where $\Phi_4(x)$ denotes the Fourier inverse of

$$\hat{\Phi}_4(\omega) = \left(\frac{\sin\left(\frac{\omega}{2}\right)}{\frac{\omega}{2}}\right)^4 \left[1 + \frac{2}{3}\sin^2\left(\frac{\omega}{2}\right)\right]$$

and u_0 the original initial condition, see [11]. The smoothing is only applied in x-direction, as u_0 is constant in y-direction. Thus, the smoothed initial condition $\tilde{u}_0 \xrightarrow{h\to 0} u_0$.

For all following examples we chose $a = 1$, $b = 1.25$, $\delta = 0.03$, $\kappa = 0.02$, $\theta = 0.25$, $\sigma = 0.25$, $\rho = -0.4$ and $r = 0.05$. We also use $S \in [3, 300]$, $v \in [0.01, 0.4]$, $T = 0.25$, $N = 100$ as well as $c_1 = 0.8$ and $c_2 = 0.3$. Since this program uses a smoothed initial condition (IC), one output option is the difference between the original and the smoothed initial condition. Figure 30.54 shows this difference (re-transformed) for a European Put, where the initial condition does not depend on y. The difference is only non-zero in the area around the strike price, which is the only non-differentiable point.

In Fig. 30.55 we can see the difference (re-transformed) between the original and smoothed initial condition for a Butterfly option. Again, the initial condition does

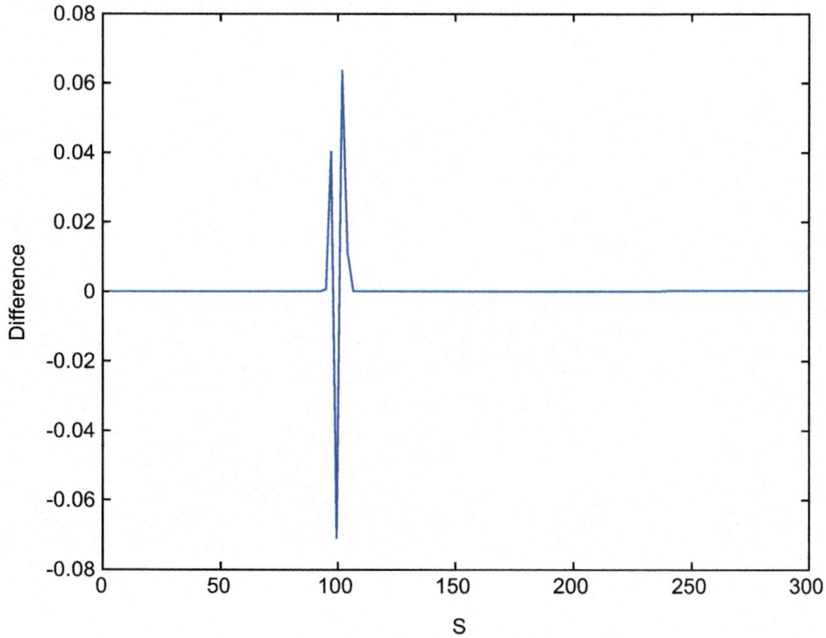

Fig. 30.54 Difference original and smoothed IC European Put option

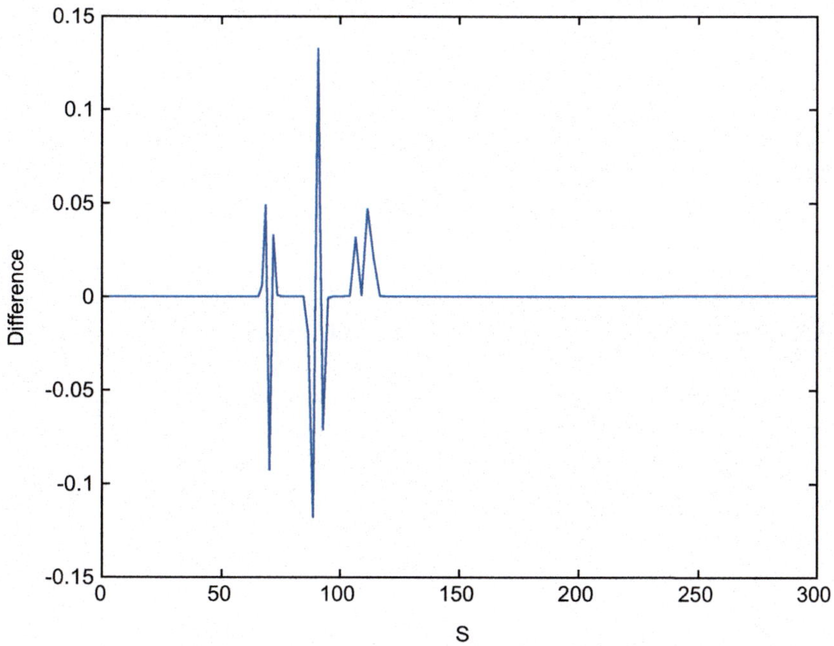

Fig. 30.55 Difference original and smoothed IC European Butterfly option

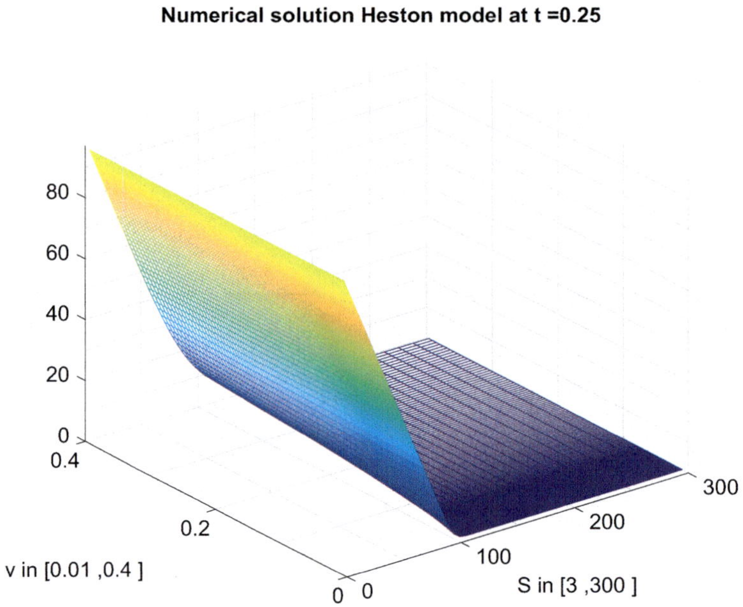

Fig. 30.56 European Put option

not depend on the variable y and thus neither the smoothed initial condition. We can see a wider support of the difference, as there are three non-differentiable points in the initial condition. If the spatial step-size h is small enough, we can see that the support of the difference is given by three disjunct intervals.

Besides the difference between the smoothed and the original initial condition it is of course possible to calculate option prices, namely the values of European Puts or Butterfly options. In Fig. 30.56 we approximate a European Put with $K = 100$.

For the approximation of the Butterfly option in Fig. 30.57 we use $K_1 = 70$ and $K_2 = 110$.

30.3.5.2 ADI Schemes (Modified Craig-Sneyd Scheme)

We applied two ADI-type schemes, Douglas (D) and Modified Craig-Sneyd (MCS), to solve the ODE system resulting from the spatial discretization of (30.8). It is possible to approximate European Put and European Call options.

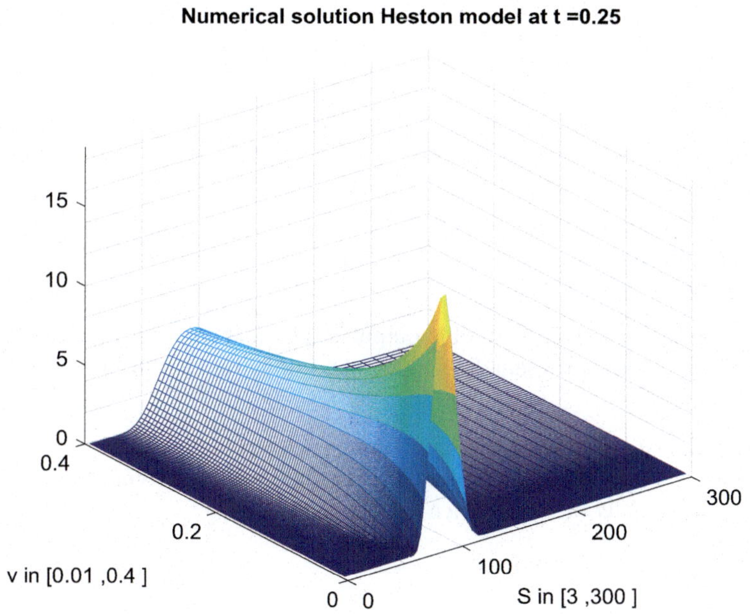

Fig. 30.57 European Butterfly option

The Modified Craig-Sneyd scheme is an unconditionally stable scheme of order 2 for $\theta \geq \frac{1}{2}$.

$$\begin{cases} Y_0 &= U_{n-1} + \Delta t F\left(t_{n-1}, U_{n-1}\right) \\ Y_j &= Y_{j-1} + \theta \Delta t \left[F_j\left(t_n, Y_j\right) - F_j\left(t_{n-1}, U_{n-1}\right)\right] \qquad j = 1, 2, \ldots, k \\ \hat{Y}_0 &= Y_0 + \Delta t A_0 \left[F_0\left(t_n, Y_3\right) - F_j\left(t_{n-1}, U_{n-1}\right)\right] \\ \widetilde{Y}_0 &= \hat{Y}_0 + \left(\frac{1}{2} - \theta\right) \Delta t \left[F\left(t_n, Y_3\right) - F\left(t_{n-1}, U_{n-1}\right)\right] \\ \widetilde{Y}_j &= \widetilde{Y}_{j-1} + \theta \Delta t \left[F_j\left(t_n, \widetilde{Y}_j\right) - F_j\left(t_{n-1}, U_{n-1}\right)\right] \qquad j = 1, 2, \ldots, k \\ V_n &= \widetilde{Y}_3 \end{cases}$$

We used a non-uniform spatial grid based on the hyperbolic sine with focus around $\nu = 0$ and $S = K$.

By default, a spatial domain $\Omega = [0, 15] \times [0, 30K]$ is selected (Figs. 30.52 and 30.53).

We have several numerical inputs available:

1. n_S—number of grid points in the S direction.
2. n_ν—number of grid points in the ν direction.
3. n_t—number of grid points in the t direction.
4. c—concentration of points around K in the S direction.
5. d—concentration of points around 0 in the ν direction.

The financial input variables are given by

1. κ—speed of mean-reversion
2. η—long-term mean of variance
3. σ—volatility of variance process
4. ρ—correlation
5. r_d—domestic interest rate
6. r_f—foreign interest rate
7. K—Strike Price
8. T—Maturity (years)

The program has several possible outputs. The most common one is the surface of option values for different S_0 and v_0. Additionally, also the main greeks are available, namely:

1. Delta—Sensitivity of option price to current underlying price
2. Gamma—Second-order sensitivity of option price to underlying price
3. Vega—Sensitivity of option price to volatility

30.4 Conclusions

The presented user interface focuses on one- and two-dimensional option pricing models. The (already existing or future) implementations of numerical schemes for further one- or two-dimensional or even higher-dimensional models for option pricing as well as other implementations of numerical schemes for further financial models still have to be implemented.

Acknowledgements The authors were partially supported by the European Union in the FP7-PEOPLE-2012-ITN Program under Grant Agreement Number 304617 (FP7 Marie Curie Action, Project Multi-ITN STRIKE—Novel Methods in Computational Finance)

References

1. Bučková, Z., Ehrhardt, M.: Splitting methods on special meshes. ECMI Newsletter 2014, pp. 71–72 (2014)
2. Bučková, Z., Ehrhardt, M., Günther, M.: Alternating direction explicit methods for convection diffusion equations. Acta Math. Univ. Comenian. **84**(2), 309–325 (2015)
3. Black, F., Scholes, M.: The pricing of options and corporate liabilities. J. Polit. Econ. **81**, 637–659 (1973)
4. Christoffersen, P., Jacobs, K., Mimouni, K.: Models for S&P500 dynamics: Evidence from realized volatility, daily returns, and option prices. Rev. Financ. Stud. **23**, 3141–3189 (2010)
5. Düring, B., Fournié, M.: High-order compact finite difference scheme for option pricing in stochastic volatility models. J. Comput. Appl. Math. **236**(17), 4462–4473 (2012)
6. Düring, B., Heuer, C.: High-order compact schemes for parabolic problems with mixed derivatives in multiple space dimensions. SIAM J. Numer. Anal. **53**(5), 2113–2134 (2015)

7. Düring, B., Heuer, C.: High-order compact schemes for Black–Scholes basket options. In: Russo, G., Capasso, V., Nicosia, G., Romano, V. (Eds.) Progress in Industrial Mathematics at ECMI 2014, Springer (2016).
8. Heston, S.L.: A closed-form solution for options with stochastic volatility with applications to bond and currency options. Rev. Financ. Stud. **6**(2), 327–343 (1993)
9. in 't Hout, K., Valkov, R.: ADI time-stepping for the uncertain correlation Black–Scholes PDE. In: Simos, T.E., et. al. (eds.) AIP Conference Proceedings (2014)
10. in 't Hout, K., Welfert, B.D.: Unconditional stability of second-order ADI schemes applied to multi-dimensional diffusion equations with mixed derivative terms. Appl. Numer. Math. **59**, 677–692 (2009)
11. Kreiss, H.-O., Thomee, V., Widlund, O.: Smoothing of initial data and rates of convergence for parabolic difference equations. Commun. Pure Appl. Math. **23**, 241–259 (1970)
12. Merton, R.C.: Theory of rational option pricing. Bell J. Econ. Manag. Sci. **4**, 141–183 (1973)
13. Mudzimbabwe, W., Vulkov, L.: Fitted operator methods and special meshes in computational finance. ECMI Newsletter 2014, pp. 77–78 (2014)
14. Tan, S.-H., Lai, C.-H.: Newton-like method for nonlinear option pricing. ECMI Newsletter 2014, pp. 85–86 (2014)
15. Valkov, R., in 't Hout, K.: ADI time-stepping for the uncertain correlation Black–Scholes PDE. ECMI Newsletter 2014, pp. 88–89 (2014)
16. Wilmott, P.: Derivatives. The Theory and Practice of Financial Engineering. Wiley, Chichester, UK (1998)

Index

M. Ehrhardt et al. (eds.), *Novel Methods in Computational Finance*,
Mathematics in Industry 25, DOI 10.1007/978-3-319-61282-9

Printed by Printforce, the Netherlands